T0327601

Essentials of Advanced Circuit Analysis: A Systems Approach

Essentials of Advanced Circuit Analysis

A Systems Approach

Djafar K. Mynbaev
New York City College of Technology
of the City University of New York

This edition first published 2024

Registered Office
John Wiley & Sons, Inc., 111 River Street, Hoboken, NJ 07030, USA

For details of our global editorial offices, customer services, and more information about Wiley products visit us at www.wiley.com.

Wiley also publishes its books in a variety of electronic formats and by print-on-demand. Some content that appears in standard print versions of this book may not be available in other formats.

Library of Congress Cataloging-in-Publication Data
Names: Mynbaev, Djafar K., author.
Title: Essentials of advanced circuit analysis : a systems approach / Djafar K. Mynbaev, New York City College of Technology of the City University of New York.
Description: Hoboken, NJ, USA : John Wiley & Sons, Inc., 2024. | Includes bibliographical references and index.
Identifiers: LCCN 2023024442 (print) | LCCN 2023024443 (ebook) | ISBN 9781119847229 (hardback) | ISBN 9781119847236 (ePDF) | ISBN 9781119847243 (epub)
Subjects: LCSH: Electric circuit analysis. | Electronic circuits. | Electrical engineering--Mathematics.
Classification: LCC TK454 .M96 2024 (print) | LCC TK454 (ebook) | DDC 621.3815--dc23/eng/20230718
LC record available at https://lccn.loc.gov/2023024442
LC ebook record available at https://lccn.loc.gov/2023024443

Cover Image: © John W Banagan/Getty Images
Cover Design: Wiley

Set in 9.5/12.5pt STIXTwoText by Integra Software Services Pvt. Ltd, Pondicherry, India

To Bronia

Brief Contents

Contents

Preface

Rationale (Advanced Circuit Analysis)

A circuit analysis, a discipline of electrical engineering, finds the voltage across and current through all the circuit components provided that a circuit operates in a steady-state regime and the input signals are either dc or sinusoidal. An advanced circuit analysis does the same work but includes into consideration the transient process and any input signal, which complicates obtaining the answers drastically. A system's approach, while performing the advanced circuit analysis, focuses on finding the input–output relationship in a circuit. Therefore, this book searches the complete (transient plus steady-state) output in response to the circuit's excitation by an arbitrary input.[1]

There are two strategic objectives of this textbook: first, to focus on the fundamentals of advanced circuit analysis with the balance between a systems theoretical approach and the practical concerns in the current technological issues; second, to create a textbook that is not merely a source of information but is an instrument that will teach the readers why real-life engineering problems appear and what are the strategies and techniques for finding their solutions. Let's discuss the issues outlined above.

Modern electrical engineering technology changes rapidly, whereas its theoretical foundation evolves slower. Thus, finding the right balance between theory and practice is the primary objective of academic courses. Our book offers such a balance by concentrating on the subject's fundamentals and relating these fundamentals to professional responsibilities through text discussions and—mainly—examples. Such an approach is hardly innovative since many textbooks do this. What makes our book unique is that the examples are not merely the exercises in plugging given numbers into equations but the problems those students will meet in their professional careers. Many examples are woven into the text fabric; those offered in a formal problem–solution format are accompanied by thorough discussions pinpointing the solution's advantages, drawbacks, limitations, and implications. Thus, our examples serve as essential teaching tools.

Most textbooks traditionally serve as sources of information by introducing physical laws, deriving equations, and explaining how devices and systems work. This function is still valuable, but in the Internet era, when all information is just a click away, its importance is diminishing. Our book, still providing necessary information, *teaches* the reader why real-life engineering problems surface and how they are solved. We explain the need for a specific task, show possible approaches to

1 See Section 4.1 for an in-depth discussion of this statement.

meet the challenge, discuss methods to pursue, and consider their possible implementation. In other words, we do not merely present ready-to-implement solutions but encourage our readers to participate in finding and applying them. The objective is to teach our readers the approaches to finding solutions, the skills that remain with professionals throughout their careers regardless of changes in technology.

An engineer's head is not a warehouse whose shelves an educator must fill with laws, formulas, and instructions on how to do it. Today, all the needed factual information can be easily obtained online, so such an approach to education is a method from the past. What engineers can't find online is the ability to analyze the situation, formulate the problem, find the optimal solution, and verify the solution's validity. And they must do all these steps by considering not a single task at hand but the operation of an entire system whose part this task is.

Pedagogical Issues

Discussion of most topics includes three levels of difficulty: basic, introductory, and advanced. This method allows instructors to choose the proper tier for an individual student (personalization) and gives the student a chance to switch among the levels depending on their progress (adaptive learning). Such an approach also gives the instructors the latitude to cover all the material in a single semester if they work with advanced students or present it more leisurely over an entire year.

More often than not, students study new material without having a solid background, and therefore they need to compensate for this deficiency by memorizing facts, equations, and laws. Since we introduce each topic's basics in the book's Part 1, students can readily refresh their memory without additional sources.

The textbook shows how a problem arises from a real-life need, what approaches can be taken to tackle the problem, what reasoning and logical steps scientists and engineers take to solve it, and what they do to implement it. Historical notes and short biographies of the greatest scientists and engineers also show the students that the problems discussed in this text stemmed from real-life situations and required tremendous efforts by people who created the technology we enjoy today.

Industry surveys consistently show that one of the significant shortcomings of new college graduates is their inability to see a problem and ask questions. The newly minted professionals frequently take everything for granted and often believe that their responsibilities are to plug given numbers into given equations or follow the instructions when operating with actual circuitry. But, in solving a problem or designing a device or circuit, engineers must ask: What is the solution to the problem? Is this the best solution? To what limits does this approach (equation) work? What could be wrong? How does my solution affect other parts of a system to which my circuit or devices belong? Asking these questions, the engineers would consider all possible situations their device and circuit can meet. This ability is critical for an engineer or technologist, so nurturing this ability is one of our objectives. We typically start every new topic by asking questions about the need for its discussion. Furthermore, questions are included at the end of each subject. In short, we encourage the readers to learn what real-life problems are posed and how to solve them best.

Organization of Our Textbook and Instructional Concerns

The book consists of three parts that are subdivided into ten chapters. Each part covers a specific area—background, time-domain, and frequency-domain—of the advanced circuit analyses. The part comprises several chapters that deepen and broaden the topic, as the table of contents attests.

This structure allows the students to start at the primary level and strive for higher and higher levels.

Since almost every chapter is self-sufficient, an instructor can choose the branches to suit an individual class's needs. The chapters covering the basics of various topics can be merged to create a primary course. This course can be used at a sophomore level at any college, including a community college. Consequently, junior and senior courses can be devised from the introductory and advanced chapters.

The book utilizes several *pedagogical devices*, such as extensive discussions for all the examples, questions in the text that encourage students to think outside the box, chapter-opening notes, section and chapter summaries, and historical references. Sidebars and Appendixes present auxiliary but essential information. A reader can skip them without breaking the flow of the text mainstream; however, they enhance the book contents and deepen understanding of the material.

Homework problems are based on real-life questions that students will encounter in the workplace. These problems will also require students to comprehend an entire concept, not just solve an equation or understand how a specific circuit works. Such knowledge will help students develop a professional approach to solving practical problems. The assignments also include questions that require essay answers, which will help readers learn how to present their results in writing, a long-lost skill that is in high demand in the industry these days. Notably, the problems and questions not only test the student's ability to plug numbers into memorized formulas but also gauge their knowledge of theory through its applications to real-world practice. Of course, we include design-oriented problems because we think the design-oriented approach must be the engineer's hallmark. In addition, many problems in the Questions and Problems sections are, in essence, the assignments for mini-projects.

This text extensively employs MATLAB and Multisim. MATLAB is used to automate the calculations, solve the algebraic, matrix, and differential equations, and plot the graphs. But we have constantly reminded our readers that MATLAB is only a mathematical processing machine, and its outputs (results) are as good as its inputs (our manually derived formulas or calculations). In other words, MATLAB results cannot verify our answers. An independent tool must be exploited to validate the results, and Multisim is the such tool. Hence, Multisim simulations are widely utilized throughout the book. Not only do they verify the results of derivations and calculations, but—even more importantly—they lead to developing laboratory exercises.

The MATLAB and Multisim applications for the laboratory exercises have accompanied my Advanced Circuit Analysis course that I taught for many years, and I transferred my experience into the text, including its examples. As a result, each example can serve as a basis for a laboratory exercise because it contains all the necessary information for the lab. For instance, the examples provide instructions on performing transient circuit analysis with Multisim, and they contain detailed demonstrations of how to make measurements with MATLAB and Multisim graphs. Another example is the demonstration of how to configure the circuit's initial conditions with Multisim in Sidebar 5S.1. We think that the ability to anticipate the expected results of an experiment is a vital engineering skill; appropriately, many examples contain prediction parts. All in all, to develop a lab manual, an instructor simply needs to use the example description. What's more, even assignments to many problems can serve as laboratory exercises too.

We focus on the analysis of the passive circuits. They are still an invaluable part of contemporary electronics. Thanks to advances in research, development, and manufacturing, the passive components reduce their sizes to the micro and nano levels, significantly improve the accuracy and stability of their parameters, and increase their long-term steadiness and resistance to harsh ambient conditions. Thus, deep literacy in the operation of passive circuits and their components is a must feature of modern electrical and electronics engineers. Consider the following statement from the publication discussing the future in the transistor design: This approach "... also increases the transistor's capacitance, thereby sapping some of its switching speed.[2]" A mere understanding of this statement, let alone the capability to provide its quantitative analysis, requires fundamental knowledge of the passive RC circuit operation.

I trust the book will be appealing to professionals who want to refresh their memory on the subject matter and take a new look at their everyday work.

By default, the book relies on extensive application of modern technology in a classroom, for the laboratory experiments, and for the personal use.

New Jersey
April 2023

Djafar K. Mynbaev

2 Marko Radosavljevic and Jack Kavalieros, "Taking Moore's Law to New Heights," *IEEE Spectrum*, December 2022, pp. 32–37.

Acknowledgments

This book could not exist without the help and support of many people. I am forever grateful for all the assistance I have received—this is as much your success as mine.

The first group that must be mentioned is that of my professional colleagues. The list starts with the President of New City College of Technology, Dr. Russel K. Hotzler, and his administration. Their gracious support extended well beyond the preparation of this book. Professional discussions and friendly conversations with my colleagues in the Department of Electrical and Telecommunications Engineering Technology inspired me to delve deeper into our vocation's engineering and academic areas.

In addition to many people who directly or indirectly contributed to this book's completion, several colleagues gave me a hand when I was working on specific topics or applications. Mr. Chi Jau Yuan, my long-term associate at our department, helped me with several issues in Multisim circuit simulations. Mr. Alex Ovrutsky, a friend and a computer guru for many years, resolved numerous problems that are inevitably encountered when working with computer applications. Dr. Jacob Sloujitel generously shared his knowledge in teaching matrices in his mathematics courses. Dr. Muhammad Ali Ummy taught the other *Advanced Circuit Analysis* class offered by our department for many years. Naturally, we discussed the teaching material, pedagogical problems, and their solutions. These stimulating conversations certainly affected this book, especially Chapters 7 and 8.

There are two more people whose help in my work cannot be overstated. Both were my students and grew to become high-level professionals capable of helping me with MATLAB applications. Dr. Vitaly Sukharenko, after graduation, became my research associate in the application of plasmonics in telecommunications. We published several scientific papers that made a helpful contribution to his doctoral thesis. Along the way, Vitaly helped me with MATLAB applications when I was writing my preceding and current books. Ms. Ina Tsikhanava, while studying at our department, participated in my research too. Her academic credentials won her a NASA internship. After graduation, she built a prosperous professional career. She wrote numerous MATLAB scripts for both my recently published and present book. I am happy to publicly express my deep appreciation to them both for many years of beneficial collaboration.

The above appreciative listing includes, of course, my students whose curiosity about the subject, desire for more in-depth learning, and general reaction to my teaching inspired me to write this book.

Mr. Brett Kurzman, Commissioning Editor at Wiley, has kept faith in me during many years of our collaboration, which included a long journey with my previous book, *Essentials of Modern Communications*, published in 2020, and slower than desired preparation of the current book. Others at Wiley also provided friendly support. To put it succinctly, my experience working with Wiley was nothing but pleasant.

Finally, this book is dedicated to my wife, Bronia (Bronislava). During the time spent writing this book (and the years spent writing the others) my family patiently endured not having my full attention and involvement in many events and activities. But no one experienced this more than Bronia, who had to deal with this on a day-to-day basis. Thus, this dedication is my way of asking forgiveness from Bronia and, through her, from the whole family. I'm all yours now, but you might miss the days when I had my head in my manuscript.

About the Companion Website

This book is accompanied by a companion website:

www.wiley.com/go/Mynbaev/AdvancedCircuitAnalysis

This website includes the answers to the questions and solutions to the exercises given in the text. It also contains the solutions to some of the most challenging problems. In addition, the qualified instructors can find here the solution manual.

Part 1

Background – Steady-State Analysis of Electrical Circuits

1

Components, Topologies, and Basic Laws of Electrical Circuits

1.1 Introduction

The textbook is written for those who are familiar with primary circuit analysis. Nonetheless, in Part 1, we provide a brief review of the basics of this topic to release the reader from the necessity to seek any additional sources for understanding the central part of the book. Since the transient processes in the electrical circuit are the main focus of this manuscript—this is what an *advanced circuit analysis* is all about—this review is done at the angle of the circuit transitions from one state to the other.

To facilitate reading the text that follows, Table 1.1 reminds the powers of 10 and their designations.

1.2 Main Electrical Parameters: Current and Voltage; Power and Energy

Definitions and Explanations

1.2.1 Electrical Charge and Electrical Current

The fundamental entity of electricity is an *electrical charge*. It is a natural property of matter responsible for all electric phenomena. Everyone is familiar with such phenomena produced by the electrical charges as lightning or sparks we encounter when wearing wool clothing. The smallest known unit charge is carried by an *electron*, an elementary particle. The *unit* of electrical charge is the charge of one electron, which is equal to $q = -1.6 \cdot 10^{-19} C$, where C stands for *coulomb*,[1] the *International System of Units, SI*, electrical charge unit. We agree to consider this electrical charge negative. There are also positive electrical charges. *The like charges repel one another; the charges of the opposite polarities attract each other*. The *Coulomb law* defines the force at which the electrical charges interact as

$$F_C(N) = K_e \frac{Q_1 \cdot Q_2}{d^2}, \tag{1.1}$$

1 **Charles-Augustin de Coulomb** (1736–1806), a French physicist who in 1785–1789 experimentally proved the law named after him. Coulomb's law states that two electric charges of the opposite signs are attracted, and two charges of the same signs are repulsed, and the force between these two charges is proportional to the product of the charges and inversely proportional to the square of the distance between them. The unit of electric charge is named in his honor.

Essentials of Advanced Circuit Analysis: A Systems Approach, First Edition. Djafar K. Mynbaev.

Companion Website: www.wiley.com/go/Mynbaev/AdvancedCircuitAnalysis

Table 1.1 Powers of 10 and their designations.

Name	Math notation	Number	SI symbol	SI prefix
Quintillion	10^{18}	1,000,000,000,000,000,000	E	Exa
Quadrillion	10^{15}	1,000,000,000,000,000	P	Peta
Trillion	10^{12}	1,000,000,000,000	T	Tera
Billion	10^{9}	1,000,000,000	G	Giga
Million	10^{6}	1,000,000	M	Mega
Thousand	10^{3}	1,000	k	Kilo
one	10^{0}	1		
Thousandth	10^{-3}	0.001	m	Milli
Millionth	10^{-6}	0.000 001	μ	Micro
Billionth	10^{-9}	0.000 000 001	n	Nano
Trillionth	10^{-12}	0.000 000 000 001	p	Pico
Quadrillionth	10^{-15}	0.000 000 000 000 001	f	Femto
Quintillionth	10^{-18}	0.000 000 000 000 000 001	a	Atto

where $K_e \approx 8.988\times10^9\,\mathrm{N}\cdot\mathrm{m}^2\cdot\mathrm{C}^{-2}$ is the constant, Q_1 and Q_2 are the signed magnitudes of the charges, and d is the distance between charges. A negative force sign means that the charges are attracted, and the positive sign indicates the repulsive force. The total electrical charge in a given system (and generally, in the universe) is conserved. This statement is known as the *law of conservation of electrical charge.*

Electrical charge lies in the foundations of electrical current, electrical voltage, and electromagnetic field. We start with electrical current.

Exercise Two electric charges have $Q_1 = 4\cdot10^{-9}C$ and $Q_2 = -4\cdot10^{-9}C$. They are placed at the distance of 2.54 cm. At what force do they attract each other? Answer: $F_C = -22.3\cdot10^{-5}(N)$. (Is this large or small force?)

Mini sidebar—Coulomb barrier

We know from our school physics courses that charged particles of the same sign repulse one another, and this phenomenon is attributed to Coulomb. However, as almost every law of nature, this law has its limitations. If these similarly charged particles collide, having extremely high energy levels, they can overcome the repulsion law and come close enough to start a nuclear reaction. This phenomenon is fundamental for developing nuclear fusion, a potential source of clean energy. After all, it is a fusion reaction that powers the sun. Nonetheless, the dream of creating a controlled nuclear fusion has remained elusive despite the tremendous efforts of the world's most advanced research and development institutions. However, most recently American scientists achieved the controlled fusion reaction, in which generated energy exceeds the energy spent on the reaction ignition! It was the major breakthrough in solving this fundamental problem, though the reaction lasted a fraction of a nanosecond (a billionth of a second).

Electrical current is a stream of unit charges. In electrical circuits, which are the subject of this book, these charges are carried by the negatively charged *electrons*. As any stream, the current "flows" through an enclosure. In this case, the enclosure is a conductor in various forms. Everyone is familiar with an electrical wire, the most popular solid-state current conductor. In this book, we will imply this type of conductor when considering the current flow.

How can we measure the strength of a current? Since it is a stream of electrons, each carrying a unit negative charge q, it's reasonable to think that the greater the number of the charges, the stronger the current. Using I, a standard designation for a current, we can write that the current (its strength, in fact) is proportional to the number of charges, N i.e.,

$$I \sim N \cdot q. \tag{1.2}$$

Thus, the total electrical charge, Q, collectively carried by all N electrons involved, constitutes the total amount of electricity that exists in a given conductor's locality, i.e.,

$$Q = N \cdot q(C). \tag{1.3}$$

But an electrical current is a stream; therefore, its strength must be measured by the amount of electricity passing a specific point in one second, which means

$$I(A) = \frac{\Delta Q(C)}{\Delta t(s)}, \tag{1.4a}$$

where A stands for *ampere*.[2] Equation 1.4a shows the average current measured over the interval $\Delta t(s)$. The instantaneous current value can be calculated as

$$i(A) = \frac{dQ(t)}{dt}\left(\frac{C}{s}\right). \tag{1.4b}$$

Therefore,

> the strength of the electric current is measured in *ampere, A, which is a flow of one coulomb of electricity per one second* through the cross-section of a wire.

This definition is based on the value of the elementary charge q given above and included in the International System of Units (SI) since 2019.

Note that the above consideration means that the term *electrical current* implies a *physical phenomenon* (the flow of charges) and a *measure of this phenomenon* (the strength or time rate of this flow).

Electrical current and charge are mutually dependent, which can be formulated as follows: *One coulomb is the amount of electrical charge that flows per one second through a conductor's cross-section when the current's strength is one ampere.* (Alternatively, one *ampere* is the electrical current created by one coulomb streaming through a conductor's cross-section per one second.) One

---

2 **André-Marie Ampère** (1775–1836), a French physicist whose theoretical and experimental research led to the founding in 1820s the new science, electrodynamics, which today is called electromagnetism. He formulated Ampere's law, which states that the mutual action of two lengthy current-carrying wires is proportional to their lengths and the intensities of their currents. Ampere's contribution to the new science was such that James Clerk Maxwell, who brought electromagnetism to its modern form, called him "Newton of electricity." No wonder his name was attached to one of the most important electrical units—the strength of an electrical current.

coulomb consists of 6.24×10^{18} elementary charges q; this number is inverse to the absolute value of the electron charge. As we can see, a coulomb is a tremendous amount of electricity.

Is a 1-ampere current large or small amount? It depends on applications: Commercial power lines deliver about 100 A to a typical house at 220-volts *ac* (*alternating current*), one-room individual air conditioner requires 15 A ac, a cell phone charger can deliver 1.7 A at 9-volts *dc* (*direct current*), a 60-watt equivalent LED lightbulb needs 0.2 A, whereas its incandescing counterpart typically required 0.5A at 110-volts ac. The integrated circuits (IC) that constitute majority of modern electronics operate at milliamperes, *mA*, or microamperes, μA, and individual transistor in these ICs needs a few nanoamperes, *nA*, to act. Later, we will introduce the unique machinery, *Large Hadron Collider (LHC)*, that consumes thousands of amperes.

Exercise How many elementary (unit) electrical charges flow through a conductor's cross-section area if the measured current is 1(*A*)? Answer: 6.24×10^{18} or one coulomb.

We've assumed that an *electrical current* is a stream of electrons, carriers of negative charges. But about 200 years ago, when scientists started investigating this new phenomenon, electricity, they hypothesized that the electrical particles are positive. Since then, we show that current flows in an electrical circuit from a positive terminal of a source to the negative one. This designation is called *conventional flow* because it reflects not a reality but convention among electrical engineers. The actual direction of the flow of the negative charge carriers is called *electron flow*. Figure 1.1 illustrates the concept of two flows in an electrical circuit. It is critical to know this distinction to avoid any confusion; it's imperative for the circuits that include semiconductors, where the current is delivered by both negative (electrons) and positive (*holes*) charge carriers.

Considering electrical current as a stream of electrons is a convenient and easy-to-visualize presentation, but reality is far from this rudimentary model. First, it's necessary to highlight again that electrical current is the process of passing an electrical charge from one place to the other within a conductor. Secondly, we need to know that there are two mechanisms for this passing.

One mechanism is a *drift electrical current*, I_d, which indeed is a flow of electrons inside a metallic wire. This current can be calculated as

$$I_d(A) = \frac{Q}{T} = q\frac{nAx}{T} = qnAv_d, \tag{1.5}$$

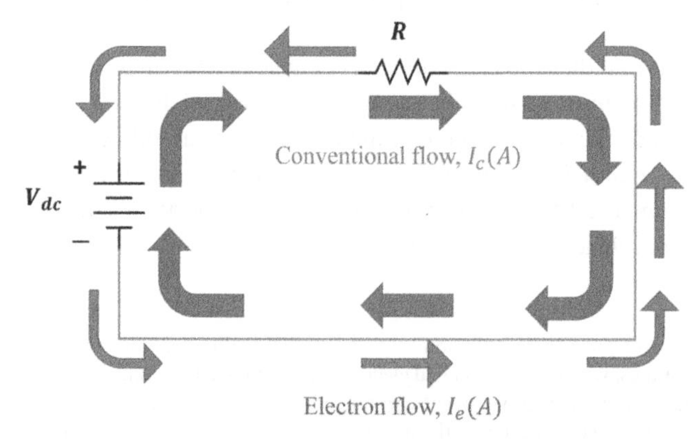

Figure 1.1 Conventional (clockwise) and electron (counterclockwise) flows of electrical current in an electrical circuit. Designations: V_{dc} is a battery (dc source) and R is a resistor.

where $q(C)$ is a unit charge, $n\left(\frac{C}{m^3}\right)$ is charge carriers volume density, $A(m^2)$ is conductor's cross-section, $x(m)$ is the wire length, and the drift velocity is given by $v_d\left(\frac{m}{s}\right) = \frac{x}{T}$. Thus, in (1.5) $n \times A$ is the number of charge carriers in the given cross-section of a wire, qnA is the total electrical charge that passes the cross-section A at a given instant, and $v_d\left(\frac{m}{s}\right) = \frac{x}{T}$ is the average drift velocity of electrons. Under typical electrical laboratory conditions, v_d is about $10^{-4}\frac{m}{s}$. (How many centimeters per hour?) This means that students can complete their laboratory exercise for three hours, but the electrons pushed into a switch at the beginning of this experiment would still sit in the switch. These numbers tell us that the drift current is not a significant factor in the whole operation of a regular electrical circuit. (However, a drift current might play an essential role in various electrical applications that are outside of our interest.)

How then, you might correctly ask, a character on a computer screen can immediately appear after clicking the keyboard button? How can light shine as soon as we turn the switch on? How can a telephone line support our live conversation without noticeable delays? We know that all these processes are provided by electrical circuits. The answer lies in existence of the other, main mechanism of delivering electrical current (signal in this case) called *electrical charge passing*. Consider an electrical wire completely filled with free charges. When we push into this wire an additional charge, this charge will be transferred from one electron to the other almost immediately, at the velocity close to the speed of light. The new charge at the "entrance" will replace one charge at the "exit," thus saving the total amount of charges within the conductor. (Recall the law of conservation of charge.) Figure 1.2, which visualizes these explanations, shows the kind of material particles within a tube. Still, in reality, there are only the electrical charges inside a conductor, and no mechanical motion is involved in this process. In truth, the process consists of transferring a charge from one locality to the next. And even more precise, *the passing-charge process is the change in the electrical field*. This is how the electrical current propagates within a metallic wire. However, in our model, an electrical current is a flow of charges, which Figure 1.2 demonstrates.

André-Marie Ampère discovered that electrical current flowing through a wire creates a *magnetic field* surrounding this wire. Given that both—electrical and magnetic—components of the electromagnetic field are produced by electrical charge, Ampere's discovery is the other confirmation of the above statement.

Which mechanism—drift current or passing charge—plays the main role in electrical current? We hope you have a clear answer to this question now. Yet, despite all its limitations, the model of electrical current as a stream of electrons helps explain many details describing the electrical current behavior in electrical circuits. Bearing in mind all the constraints of this model, we will continue to use it in future discussions.

Equation 1.5 describing the drift current can be partly applied to the passing charge model because it highlights the role of a wire's cross-section in transferring a current. Though it's intuitively clear

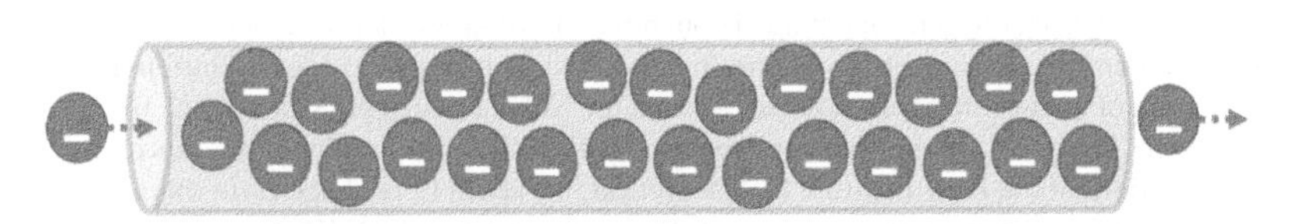

Figure 1.2 Charge passing in an electrical wire.

that the greater the wire cross-section, the more electricity can flow through this wire per second; (1.5) puts this statement on a solid mathematical footing. The wire sizes (cross-sections) are measured by their diameters, and the instruments used for their measurements are called *gauges*. Obviously, for a power line delivering 100 A of the current, the wire diameter must be much greater than that of a laptop charger carrying less than 2 A. Today, *wire gauge* also denotes the standard wire diameters collected in the *American Wire Gauge (AWG) system*. Similar system is also maintained by the *International Electrotechnical Commission (IEC)*, and both systems are interrelated.

To continue using the electron-flow model for an electrical current, we must explain from where these free electrons appear. Metallic wires (solid-state conductors), which we consider in this book, consist of atoms whose nuclei are orbiting by the electrons. We refer to the well-known *Bohr's*[3] *atom planetary model* of an atom shown in Figure 1.3a. An atom's nucleus is composed of tightly packed *protons* and *neutrons*. The protons and neutrons have approximately the same mass, 1.67×10^{-27} kilograms, which is extremely small. However, the electron's mass is much smaller and can be considered negligible compared to this value. Hence, the mass of the whole atom is determined by the mass of its nucleus. In contrast, the atom's size is defined by the electron orbits because the typical nucleus's size is about $100{,}000$ times smaller than the orbit of the outmost electron. Protons carry a unit positive charge equal to that of an electron, $q^{+}=1.6\cdot10^{-19}C$. Since the neutrons have no charge, and the number of protons equals the number of electrons, the atom is electrically neutral. Coulomb forces attract electrons to the nucleus, thus forming the atom as a unit of matter. In metals (conductors), the farthest electrons can easily leave an atom; this is where free electrons in conductors come from. The modern view on the atom is more sophisticated and will be discussed shortly.

Question Why can only the farthest but not all electrons easily leave an atom?

In metals, the nuclei are in fixed positions, making the metallic lattice, but the electrons can freely move due to conductor's natural properties. Applying an external electric force to the wire, we can make these electrons flow in one direction, thus creating an electrical current. (In insulators, the electrons are strongly bounded, and we cannot make them flow.) This elemental model explains why only metals can conduct electricity. The best conductors are silver, copper, and gold. The specific applications in contacts and other circuit connections that require long-time stability and work well in harsh environments use gold. Interestingly, the organizers of the 2020 Olympic games in Japan used gold extracted from recycled smartphones and laptops to produce the Olympic medals. This fact highlights the scale of using gold in modern electronics.

The stream of free electrons—electrical current, that is—experiences *resistance* of the metal through which it flows. The resistance value depends on two factors. First, it is the natural property of the material. Silver, copper, and gold have the smallest *resistivity* measured in $\Omega.m$, though aluminum's resistivity is just slightly greater than that of gold. Secondly, the resistivity of the same material increases with temperature rise. Consider a simplistic model of electrical current as of the flow of

3 **Niels Bohr** (1885–1962) was born in Copenhagen, Denmark. Though he is most known for developing the atom model, where he first applied the quantum physics concept, Bohr, in fact, was a figure equal to Albert Einstein by his contribution to the development of 20th-century physics. He received the Nobel Prize in physics in 1922. In addition to working on his research, Bohr was a mentor and collaborator of almost all outstanding physicists of his time; they worked for various durations at the Bohr Institute for Theoretical Physics in Copenhagen. He was a champion for the international scientific collaboration, and his contribution to creating CERN (European Organization for Nuclear Research), the home of the Large Hadron Collider (LHC), could not be overestimated.

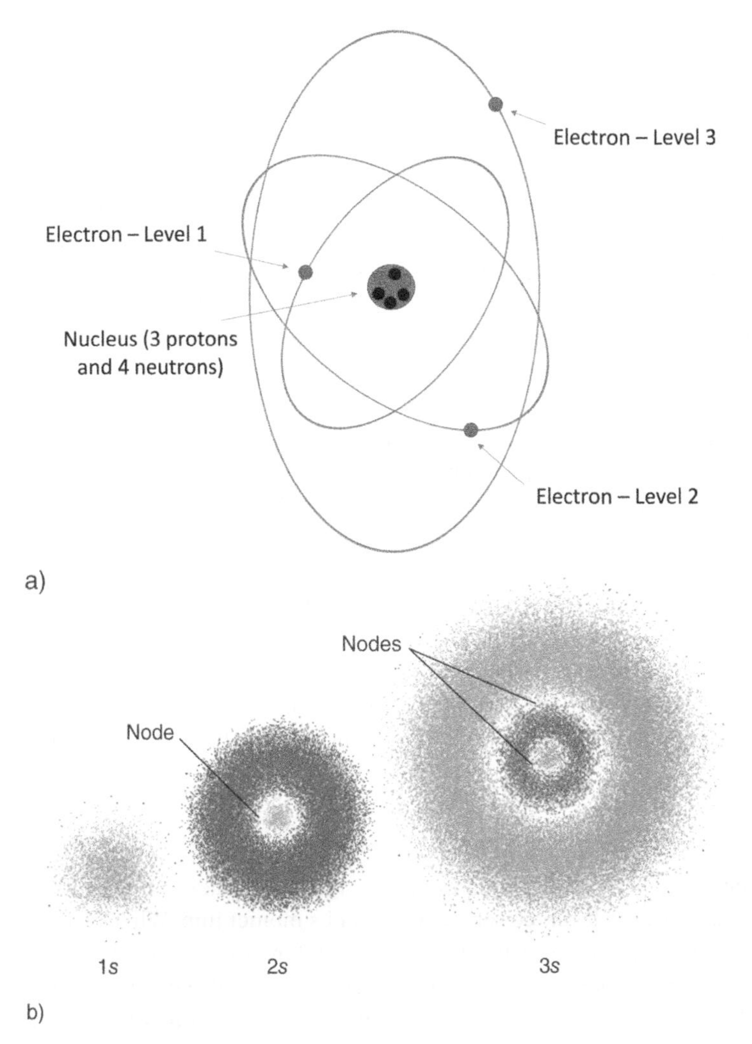

Figure 1.3 Atom models: a) Bohr planetary model; b) quantum-mechanical model of an atom. Not to scale. (Open source: *UCDavis Chemwiki, CC BY-NC-SA 3.0 US.)*

electrons again: The increase in a wire temperature causes the nuclei to vibrate around their fixed positions at the greater amplitudes, leaving less space for the flow of the free electrons. In other words, thermal nuclei vibrations impede the streaming of electrical current. The resistivity caused by the metal property can be improved only by artificial manipulations of this property, which means creating new materials. Though this goal is hypothetically within the reach of modern technology, the scale of the electricity use makes these manipulations economically unjustified so far. Decreasing the thermal resistivity seems can be "easily" obtained by lowering conductor's temperature. However, there are two obstacles to this solution. First, the stream of electrons pushes atoms in general and their nuclei in particular to move; these additional motions are the increased heat of a wire. (Now you know why a charger of a smartphone is warm, or even hot, when it charges your gadget.) In other words, the sheer presence of electrical current increases resistivity. Secondly, simply freezing a conductor implies placing each conductor in this world into a special cooling camera, which is practically impossible. However, special applications make such a solution a necessity.

Fortunately, overcoming electrical resistance is possible thanks to the phenomenon called *superconductivity*. More than a hundred years ago, it was experimentally shown that some metals jump to a superconductive, no-resistance state after being cooled below $20°K(-253°C)$. When in this state, the metal can conduct an electrical current indefinitely thanks to its zero *resistance*. Today, this technology has matured enough to be employed since 2008 at the *Large Hadron Collider (LHC)*, the world's most giant machine and the most sophisticated scientific instrument. The LHC is a particle accelerator; its function is to accelerate the beam of elementary particles to the speed close to the light speed and collide two high-energy beams to investigate the tiniest particles of the matter. It is a 27-km ring of gigantic tubes buried underground at a 100-meter depth. A tremendously strong magnetic field must be created to accelerate these particles to the required velocity. Making such a magnetic field requires a huge electrical current. (Remember about the Ampere's discovery related to electrical current and magnetic field? See his biography early in this section.) In fact, currents up to 12,000 amperes are used in LHC today; this value is planned to rise to 27,000 amperes in the near future. Creating, delivering, and employing currents of this scale would be impossible without using superconductivity. To support the LHC operation, all magnet coils are kept cooled up to $1.9°K$ or $-271.1°C$. We can't provide a better example of the importance of the thermal issues in using the electrical current.[4]

Let's return to the model of an atom. The *Bohr's atom planetary model* is convenient for explaining the fundamental processes associated with electrical current, but it has several limitations. For one, it considers an *electron* as a particle orbiting a nucleus, whereas the electron is anything but a particle. When we say "particle," we imply the extremely tiny solid speck, like a dust particle, whose coordinates, that is, the particle's *position*—$x(m), y(m), z(m)$—and the *momentum*—$mv\left(n \cdot s = \frac{kg.m}{s}\right)$—can be readily determined. However, today, an atom and its components can be fully described only by quantum mechanics whose laws are based, in particular, on the *Heisenberg's uncertainty principle*. This principle states that the more accurately we determine the position of an elementary particle, the less precisely we can find its momentum. What's more, both of the above quantities are measured in probabilistic terms. Therefore, in quantum mechanics, we cannot determine the electron position; we can only ascertain the *probability* of finding the electron in a specific location. For example, instead of saying that an electron is located at $x = 2 \pm 0.01 nm$, we must say that the electron is located at $x = 2 \pm 0.01 nm$ with a *probability* of 86%. The same approach is valid for the momentum.

Figure 1.3b visualizes this concept. It shows that each electron doesn't move around its nucleus along a single planar path, the *orbit*; instead, it locates somewhere in the three-dimensional region (space) surrounding the nucleus. This region is called the *orbital*. Figure 1.3b shows three orbitals—1s, 2s, and 3s—instead of three orbits presented in Figure 1.3a. The color of each orbital differs in intensity; the greater the color intensity, the higher the probability of finding an electron within this orbital's location. Figure 1.3b also shows the uncolored regions, *nodes*, where the probability of finding an electron is zero.

These remarks are intended to show the fundamental limitations of the Bohr planetary model of an atom. Nevertheless, this model is still widely applied thanks to its usefulness for explaining the primary relationships in the electrical circuits, specifically, an electrical current.

---

4 Visit https://home.cern/science/accelerators/large-hadron-collider to read more on the LHC and the crucial role of superconductivity in its operation.

Important: We must comprehend that we live in the observable world, where we can visualize all processes and subjects; in contrast, the processes and subjects of quantum mechanics are not observable, and we can only present them in the kind of visual analogies.

Finally, the last question in the discussion of electrical current: What physical phenomenon forces electrons to move along a wire, thus creating the stream of charges? In one word—*voltage*.

Question Niels Bohr's atom model demonstrated in Figure 1.3a requires that an electron changes its energy level only discretely, visually depicted as electron location at an individual orbit. The modern view of this model shown in Figure 1.3b highlights that an electron is not located precisely at a fixed orbit but spread in a kind of cloud called an orbital. How is the Bohr central concept—discreetness of energy change—observed in the modern atom model?

Electrical current—a summary:

- For this book, we assume that electrical current is a stream of electrons.
- Quantitatively, current reflects the rate at which electrons flow through a certain point of the circuit. Specifically, one *ampere* is the electrical current created by one coulomb of electrical charge, streaming through a conductor's cross-section per one second. Mathematically, the instantaneous value of electrical current is defined as

$$i(A) = \frac{dQ(C)}{dt(s)}. \tag{1.4bR}$$

- We understand that an electron-flow model has the following limitations:
  - We accept that the current flows from positive to negative terminals (*conventional flow*), whereas in reality, electrons flow from negative to the positive terminal.
  - Though the electron-flow model of an electrical current is considered as the main one, we remember that the primary mechanism of transferring charge along a conductor is *electrical charge passing*.
  - The Bohr planetary atom model enables us to explain the source of free electrons in conductors (metals); however, today, based on the advances in quantum mechanics, an atom is described by a more accurate model shown in Figure 1.3b.

1.2.2 Electrical Voltage

As we comprehend from the preceding section, it's necessary to move electrical charge carriers in one direction through a wire to create an electrical current. The forces must be applied to those charge carriers to make this motion. Where do these forces come from? Every electrical circuit includes a dc or ac *source*. Consider, for example, an *electrical battery* shown in Figure 1.1. The battery generates an excess of electrons on its negative terminal (this is why it is denoted as negative), whereas its positive terminal has a deficiency of electrons, which means that it is charged positively. To establish charge equilibrium in the circuit, the negative charge carries electrons flow to the positive terminal. This *electron flow* makes up an electrical current, as shown in Figure 1.1.

> Thus, *the difference in the number of electrons between the negative and positive terminals of a battery forces the electrons to move through a circuit, creating an electrical current.*

As you are trying to complete this cycle mentally, you will immediately ask what happens when the electrons arrive at the battery's positive terminal. Well, the difference in the number of electrons between terminals should decrease; however, the *battery itself continues producing electrons and collecting them at the negative terminal.* By doing so, the battery supports the difference in charges between its terminals at the required level.

Therefore, the difference in electrical charges between the two battery terminals creates the forces pushing the electrons to move, thus creating an electrical current.

The difference in electrical charge between two points of a wire is called **voltage.**

More accurately,

voltage *is the quantitative measure of the difference in electrical charge, or electrical potential, between two points of an electrical circuit. The unit of voltage is* ***volt, V.***

It is named after Alessandro Volta, an Italian scientist who at the end of the 18th century invented an electrical battery and made several significant contributions to developing and understanding electricity.

Historically, *voltage* is called ***electromotive force, emf,*** because the difference in electrical charge between two points in an electrical circuit creates the force that moves the electrons from one circuit's point to the other. (So, voltage is not the force, but it creates a force.) But when a force, $F(N)$, moves an object over a distance, $d(m)$, this force does *work*, $W(N.m)$; i.e.,

$$W(N \cdot m) = F(N) \cdot d(m). \tag{1.6}$$

We need to recollect that work has the same unit, *joule*, J, as *energy*, $E(J)$ because $N \cdot m = J$. The same unit of work and energy means that they have common ground; indeed, *energy is the capacity to do the work.* Quantitatively, *work done to move an object is equal to the energy spent on this work.* Work and energy commonly share the same notation, $W(J)$; however, more often than not, electrical energy is designated as $E(J)$.

Therefore, the *work done, or energy spent,* by an electromotive force to move a unit charge between two points in an electrical circuit is related to *voltage* as

$$Voltage(V) = \frac{Work(J)}{Charge(C)} = \frac{Electrical\,Energy(J)}{Charge(C)}, \tag{1.7}$$

which means that $V = \frac{J}{C}$. We can conclude that moving the charge carries over a distance (that is, creating an electrical current) requires spending some electrical energy. This energy comes from a source (battery, for example) that generates voltage.

Equation 1.7 enables us to present the other view at voltage: *Voltage is energy need to be spent (or work done) to move an electrical charge from one point of a circuit to the other.*

$$Voltage(V) \cdot Charge(C) = Work(J) = Electrical\,Energy(J) \tag{1.8}$$

To reflect the dynamic nature of voltage, its instantaneous value can be presented as

$$v(V) = \frac{dW(J)}{dq(C)} = \frac{dE(J)}{dq(C)}. \tag{1.9}$$

The higher the voltage of the source, the greater the electrical current produced by the source at the same circuit. The famous Ohm's law expresses this relationship as

$$V(volt) = R(ohm) \cdot I(amper), \tag{1.10}$$

where $R(\Omega)$ is the constant resistance. See Figures 1.1 and 1.4. We will discuss Ohm's law in Section 1.5.

Whether 1 volt is a big or small amount? The ac voltage in power transmission lines delivering electricity from power plants to the local distribution centers ranges from 100 kV to 1000 kV and higher. Our homes are supplied by 110 V or 220 V ac. The voltage of an automotive battery is traditionally 12 V dc. A smartphone's charger typically provides 9 V dc, and a flashlight battery commonly delivers 1.5 V dc. The average voltage in a human cell is 70 mV = 0.07 V dc.

Voltage can be measured between two points of a circuit, for example, between terminals A and B of a resistor, as shown in Figure 1.4. If V_A is greater than V_B, $V_A > V_B$, then V_{AB} *is the voltage drop across the resistor*. Strictly speaking,

$$V_{AB} = V_A - V_B, \tag{1.11}$$

where V_A and V_B are voltages of points A and B with respect to ground, V_0. However, $(V_A - V_0) - (V_B - V_0) = V_{AB}$. Clearly, voltmeters show the voltage drop V_{AB} directly between two points. This measure is the most used in circuit analysis. Be aware that in the parlance of the electrical engineering industry the term *voltage* commonly means *voltage drop*. In Figure 1.4, the voltage drop across the resistor is obviously equal to the voltage, V_{dc}, produced by the battery. What's more, in this circuit, V_{dc} is the electromotive force, $emf(V)$, that makes the current flows. To highlight the role of voltage as an electromotive force in contrast to a voltage drop, Ohm's law given in (1.10) must be rewritten as

$$emf(V) = R \cdot I. \tag{1.12}$$

Equation 1.12 states that the greater the voltage (electromotive force), the greater the current in a simple $emf - R$ circuit, provided that the resistance $R(\Omega)$ is constant.

Voltage is a sophisticated subject, and it usually takes time and practice to comprehend this entity fully. An analogy should help; one of the most popular is comparison voltage and potential energy. When we elevate a weight above the ground, we provide the weight with potential energy,

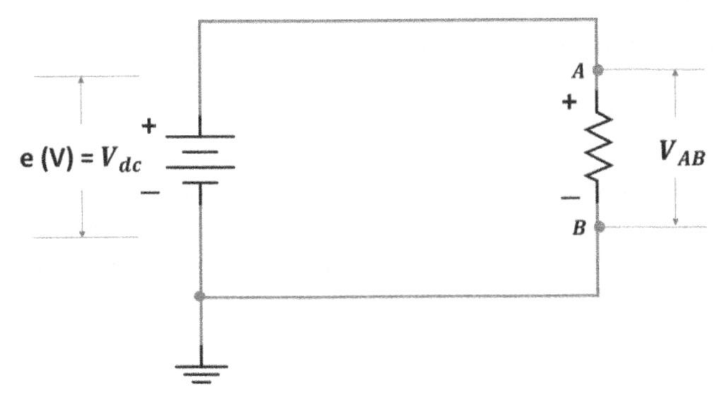

Figure 1.4 Electrical circuit containing a battery and a resistor. (Designation: $e(V) \equiv emf(V)$.)

which is the capacity to do the work. If the weight is released, it moves down doing the work, as (1.6) describes. Similarly, when we charge a battery, we create its capacity to do the work. When we connect the battery to a circuit, an electrical current starts to flow, which results in spending the energy or doing the work.

Question Figure 1.4 shows voltage V_{dc} next to the battery. Does it mean that the battery produces (generates) voltage?

Exercise Consider the circuit shown in Figure 1.4. What current the battery creates in the circuit if $R = 4\ \Omega$ and $V_{dc} = 12V$? Show the conventional direction of the current low.

***Voltage*—a summary:**

- Voltage is an entity that makes electrical charges move, thus, creating an electrical current.
- Voltage is the work done, or energy spent, to move a unit electrical charge between two points of a circuit. It is measured in volts, $V = \frac{J}{C}$.
- Mathematically, the voltage instantaneous value is expressed through the work $W(J)$ done or energy $E(J)$ spent over the electrical charge $q(C)$ as

$$v(V) = \frac{dW(J)}{dq(C)} = \frac{dE(J)}{dq(C)}. \tag{1.9R}$$

- Voltage drop across a resistor's terminal A and B is $V_{AB} = V_A - V_B$.
- Ohm's law establishes the relationship between voltage, V, and current, I, in a single resistor, R, dc circuit as

$$V(volt) = R(ohm) \cdot I(ampere) \tag{1.8R}$$

A word of caution: A capital letter V denotes electrical voltage, the voltage drop across a circuit element, battery value, a signal, and voltage's SI unit. To avoid confusion, we'll use subscripts to denote the voltage of a specific function, for example, V_R for the voltage drop across a resistor, V_{dc} or V_{ac} for battery value, and V_{in} for an input signal. Capital V will be reserved for the unit designation.

1.2.3 Electrical Energy and Electrical Power

We know that electrical energy is expressed through voltage and charge as

$$Voltage(V) \cdot Charge(C) = Work(J) = Electrical\,Energy(J). \tag{1.8R}$$

This formula, however, doesn't tell us how fast energy is spent. We know from physics that, in general, the time rate of energy spending is *power, p (W)*. Power is measured in *watt, W*. (This unit is named after Scottish engineer *James Watt* (1736–1819), who invented a steam engine, thus enabling the Industrial Revolution.) Applying this knowledge to the electrical world, we can write

$$p(W) = \frac{dE(J)}{dt(s)}. \tag{1.13}$$

Using the chain rule and (1.4b) and (1.9) enables us to rewrite (1.13) as

$$p(W) = \frac{dE(J)}{dq(s)} \cdot \frac{dq(J)}{dt(s)} = v(V) \cdot i(A). \tag{1.14a}$$

Thus, electrical power is expressed through voltage and current. If a dc voltage drop across resistor $R(\Omega)$ is equal to $V_{AB} = 5(V)$ and the current flowing through the resistor is $I = 0.4(A)$, then electrical power delivered to this resistor is

$$P = V_{AB} \cdot I = 2(W).$$

This power will be dissipated by the resistor.

Equation 1.14a works only for dc circuits, where $v(V)$ and $i(A)$ are constant. We usually denote the dc values by capital letters; hence (1.14a) should be written as

$$P(W) = V \cdot I. \tag{1.14b}$$

For ac resistive circuits, power is given by

$$p_R(t)[W] = v_R(t) \cdot i_R(t) = \frac{v_R^2}{R} = R \cdot i_R^2, \tag{1.14c}$$

where Ohm's law is applied. We will see how electrical power and energy are calculated for capacitive and inductive ac circuits in Section 1.3.

Exercise Consider the electric circuit shown in Figure 1.4. What power is dissipated by resistor $R = 10(k\Omega)$ if $V_{dc} = 12(V)$? Answer: $P = 14.4(mW)$.

To conclude Section 1.1, we want to explain that we deliberately avoid considering examples when discussing electrical charge, current, voltage, energy, and power. A curious reader, carefully reading this section, will undoubtedly notice that possible examples would be the simple exercises in plugging numbers into the given formulas. Our goal was to save time on such elementary operations. Still, we encourage our readers to do these calculations to get better feeling what level of values are involved in the operations of each of these subjects. Provided exercises and exemplifying computations weaving into the text should help do these calculations.

1.3 Passive Components: Resistors, Capacitors, and Inductors

This section is devoted to the resistors, capacitors, and inductors called the passive circuit components (elements). The term *passive* means that these elements operate by the virtue of signals only, and they don't need an external power to work. For example, a resistor causes the voltage drop just because current enters and flows through it. In contrast, a transistor cannot amplify an entry signal without an external power supply V_{CC}.

1.3.1 Resistance and Resistor

Resistance, as the term suggests, is the material's ability to resist, oppose, or withstand the current flow. A resistor is a passive circuit component that makes resistance.

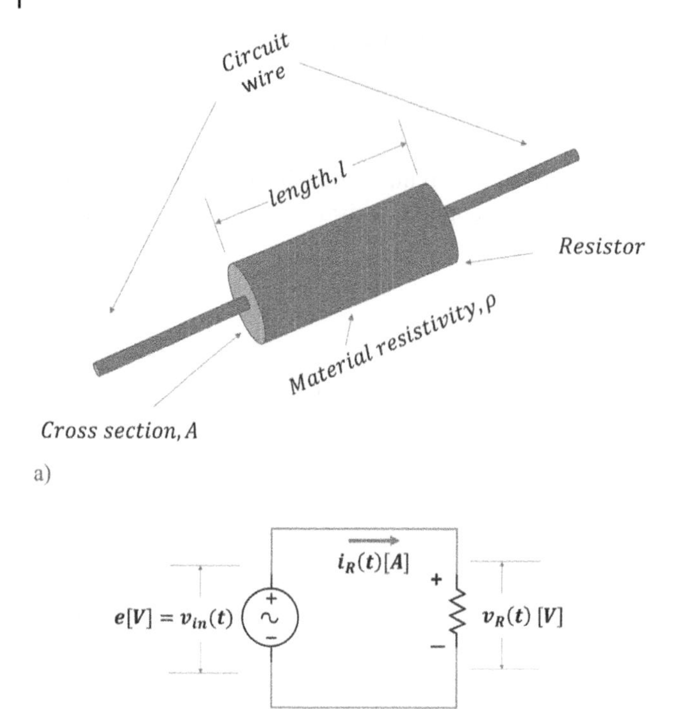

a)

$i_R(t)[A]$

$e[V] = v_{in}(t)$

$v_R(t)\ [V]$

b)

Figure 1.5 Typical lumped resistor: a) Construction; b) resistor in circuit.

A lumped resistor is typically a cylindrical piece of a special material that partly conducts and partly opposes the flow of electric charges. A visual presentation of a resistor is shown in Figure 1.5a. A resistor electrical circuit is presented in Figure 1.5b.

Each slice of a resistor opposes the charge flow, which implies that the longer the resistor, the greater the total resistance. On the other hand, the greater the resistor's cross section, the more space for charge flow, the less total resistance. Mainly, the total resistance depends on the material property called *resistivity*. Conductors allow a charge flow with little resistance, whereas insulators impede the flow almost completely. The above reasoning brings us to the following formula of a resistance, $R(\Omega)$:

$$R(\Omega) = \rho\frac{l}{A}, \tag{1.15}$$

where $\rho(\Omega \cdot \text{m})$ is the material resistivity, $l(m)$ is the resistor's length, and $A(m^2)$ is the resistor's cross-section area. The SI resistance unit is ohm[5]; it is denoted by a Greek letter Ω (omega).

5 **Georg Simon Ohm** (1789–1854) was a German scientist who, in 1827, introduced his famous law stating that the current flowing through a wire is proportional to voltage and inversely proportional to resistance. This law, which became fundamental in the electricity theory, was coldly received by his scientific contemporaries, partly because the electricity of that time was in its infancy. Over the years, the significance of Ohm's law became evident and received full recognition. In 1841, Ohm was awarded by Copley medal by the Royal Society of London, and the year later, he was made a foreign member of this society. He worked as a school teacher when he made his discovery but finished his professional career as a professor at a prestigious Munich university. The SI unit of resistance is named in his honor.

Equation 1.15 suggests that the resistance can be easily controlled by manipulating the resistor length and cross-section area; the resistivity, however, remains the natural property of a material. Table 1.2 clarifies this statement and hints at how to make a resistor: If you make the small cylinder shown in Figure 1.5 from carbon instead of silver, its resistivity increases by thousand times. We must bear in mind that modern technology can create new, artificial materials with the desired properties, including their resistivity. Therefore, Table 1.2 is just a guideline rather than an all-inclusive reference list. (Revisit Section 1.2, where resistance was discussed concerning electrical current flow.)

Table 1.2 puts the material classification on the numerical footing: Now, we can readily distinguish between conductors and semiconductors by saying that the conductors exhibit resistivity in the order of 10^{-8} $(\Omega \cdot sm)$, the semiconductor's resistivity ranges from 10^{-5} to 10^{2} $(\Omega \cdot \text{m})$, whereas the insulators have $\rho(\Omega \cdot \text{m})$ from 10^{10} and above.

What is the resistance of a silver wire of 1-meter long and 1-centimeter radius of a cross-section? Apply (1.15) to compute

$$R = \rho \frac{l}{A} = 1.6 \cdot 10^{-8} (\Omega \cdot m) \frac{1(m)}{3.14.(0.1)^2} = 0.51 \cdot 10^{-6} (\Omega).$$

The value of 0.5 $\mu\Omega$ is infinitesimally small resistance. For the vast majority of electrical and electronic circuits, we can consider such resistance as zero. What current will flow in a circuit where the terminals of a 12-volt dc battery are connected with such silver wires? Ohm's law enables us to calculate

$$I = \frac{V_{dc}(V)}{R(\Omega)} = \frac{12(V)}{0.51 \cdot 10^{-6}(\Omega)} = 23.5 \cdot 10^{6} (A).$$

The value of 23.5 million amperes is an astronomically colossal current that cannot practically exist because it dissipates such an enormous amount of power even on this infinitesimally small resistance that the wire evaporates instantaneously. Indeed, applying (1.14c) gives $P = R \cdot I^2 = 281.6 \cdot 10^6 (W)$, which is power of a medium-sized power plant! Recall from Section 1.2 that even the Large Hadron Collider, the machine built at the cutting edge of modern technology, plans to use "only" $27 \cdot 10^3$ amperes in its operations. It would be able to do so exclusively due to using superconductive circuits with zero resistance.

Table 1.2 Resistivity of various materials.

Material	Resistivity, $\rho(\Omega \cdot \text{m})$	Application
Silver	$1.6 \cdot 10^{-8}$	Conductor
Copper	$1.7 \cdot 10^{-8}$	Conductor
Gold	$2.4 \cdot 10^{-8}$	Conductor
Aluminum	$2.8 \cdot 10^{-8}$	Conductor
Carbon	$3.6 \cdot 10^{-5}$	Semiconductor
Silicon	$47 \cdot 10^{2}$	Semiconductor
Glass	$1.0 \cdot 10^{12}$	Insulator

The conclusion we draw from our example is simple: The circuit becomes practically shortened if the battery terminals are connected with a conductive wire. An example of a *short circuit* is shown in Figure 1.6a.

What is the resistance of a glass "wire" of the same dimensions as in the preceding example? What current does flow through the circuit shown in Figure 1.6b if we connect terminals A and B with this glass wire? The calculations with $\rho = 1.0 \cdot 10^{12}(\Omega.\mathrm{m})$ give $R = 31.8 \cdot 10^{12}(\Omega)$, which in reality means the infinite resistance, and $I = 0.38 \cdot 10^{-12}(A)$, which is an infinitesimally small value. Practically, no current will flow through this circuit. We consider it as an *open circuit*; its schematic is shown in Figure 1.6b.

To summarize, a short circuit *and an* open circuit *are the models of real circuits where the resistance of the wire goes either to zero or to infinity, respectively.*

The resistance $R(\Omega)$ of a resistor depends on its *temperature*. As we discussed in Section 1.2, when temperature increases, the amplitude of atom oscillations increases too, hindering the charges stream. For conductors, this dependence can be presented as

$$R_T(\Omega) = R_0(\Omega)\left[1 + \alpha(T - T_0)\right]. \tag{1.16a}$$

Here, $R_T(\Omega)$ and $R_0(\Omega)$ are the resistances at the current temperature T and at the initial reference temperature T_0, respectively, and $\alpha\left(\frac{1}{degree}\right)$ is the temperature coefficient. This coefficient for copper, for example, is equal to $\alpha = 0.004\left(\frac{1}{degree}\right)$. For computations, we assume that l *and* A in (1.15) remain constant, and all temperature effects are expressed by α. Then, (1.16a) takes the form

$$\rho_T(\Omega) = \rho_0(\Omega)\left[1 + \alpha(T - T_0)\right], \tag{1.16b}$$

where ρ_0 is the reference resistivity given in Table 1.2. Thus, if the copper wire temperature increases from $20°C$ to $50°C$, its resistance becomes

$$\rho_T(\Omega) = \rho_0(\Omega)\left[1 + \alpha(T - T_0)\right] = 1.7 \cdot 10^{-8}\left[1 + 0.004(30)\right] = 1.9 \cdot 10^{-8}(\Omega \cdot \mathrm{m}).$$

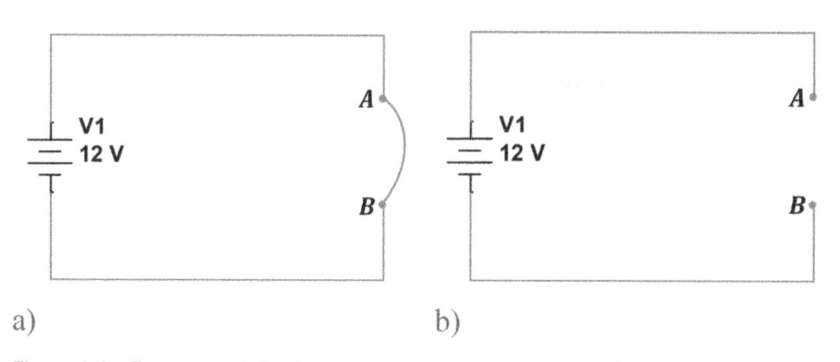

Figure 1.6 Two types of circuits: (a) Short; (b) open.

The resistivity change from $1.7.10^{-8}(\Omega)$ to $1.9.10^{-8}(\Omega)$ is indeed microscopic; more accurately, it is "nanoscopic" because it equals 2 $n\Omega$.m ! However, in power lines, temperature variations cause significant changes in the cable resistances. Fortunately, Equation 1.16a enables us to compute the resistance changes and take the adequate measures to mitigate their effects. It's important to know that (1.16a) and (1.16b) are valid only for *conductors*; resistances of semiconductors and other materials used in the electrical and electronic industry depend on temperature nonlinearly.

Resistance also depends on the current *frequency*. Low-frequency and dc currents flow through the entire volume of a wire; this is why (1.15) includes the wire cross-section as a vital factor. However, as the frequency increases, the current flow concentrates along the outer wire surface; this phenomenon is called the *skin effect*. The higher the frequency, the smaller the depth of the "skin," where current flows. In other words, the higher the frequency, the smaller the utilized cross-section area, the greater the resistance. This effect becomes significant at the megahertz and gigahertz range of frequencies. In addition, conductive wires show capacitive and inductive properties at the higher frequencies, making them frequency-dependent elements, as we'll learn soon. There are some other phenomena that make the resistance frequency-dependent, but they are outside of the mainstream of this book. The main point we need to take is that *a resistor does not always exhibit the resistance value shown on its marking*.

This discovery implies that we must bear in mind all the above factors when using Ohm's law; otherwise, the obtained results might be inaccurate.

Question How temperature variations in circuit wire can affect the accuracy of Ohm's law?

Despite the resistance's dependence on temperature and frequency (and other factors, such as radiation and mechanical stress) in reality, we can presume that the resistor possesses linear

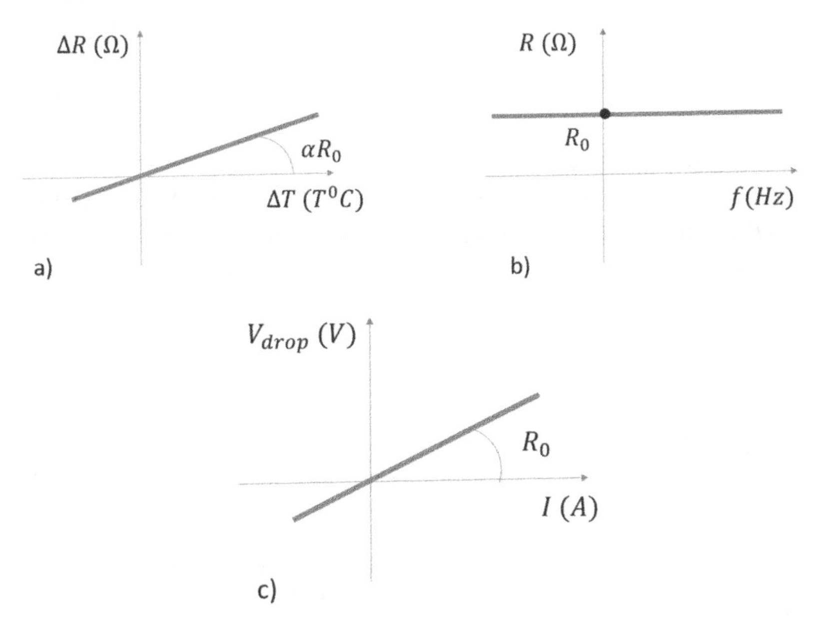

Figure 1.7 Characteristics of a linear resistor: a) Temperature dependence; b) frequency dependence; c) Ohm's law.

properties when the variations of all these disturbing factors are negligible. Thus, unless it specifies otherwise, we will work with *linear resistor, the model* whose main characteristics are shown in Figure 1.7.

The obvious transformation of (1.16a) gives

$$R_T(\Omega) - R_0(\Omega) = \alpha R_0(\Omega)(T - T_0)]$$

or

$$\Delta R(\Omega) = \alpha R_0(\Omega)\left[\Delta T(^0C)\right]. \tag{1.17}$$

Figure 1.7a visualizes (1.17) by showing that the resistance change, $\Delta R(\Omega)$, linearly depends on the change in temperature, $\Delta T(^0C)$, and the slope of this line is $\alpha R_0\left(\frac{\Omega}{^0C}\right)$. However, (1.17) is derived for conductors, whereas resistors are made from composition of metal and carbon. Therefore, temperature dependence of a resistor is more sophisticated than (1.17) and Figure 1.7a show. Nevertheless, for most applications we can accept the linear model of a resistor and use (1.17) as its mathematical description.

At the low ac frequencies, we can neglect the skin effect and other frequency-dependent phenomena in resistors and accept the linear $R-f$ resistor's model shown in Figure 1.7b.

Finally, temperature fluctuations and frequency variations cause the changes in $R(\Omega)$; therefore, Ohm's law should be written as

$$V_{drop}(V) = R(t)[\Omega] \cdot I(A). \tag{1.18}$$

If we can neglect those variations in the resistance, then Ohm's law must be shown as Figure 1.7c depicts it, namely, by a straight line in $V-I$ coordinate system.

A resistance describes the *opposition* to the current flow, but it's also worth to measure the *ease* at which current flows. This characteristic is inverse to resistance and called *conductance*, G(S). It is given by

$$G(S) = \frac{1}{R(\Omega)}. \tag{1.19}$$

Here S stands for siemens,[6] which is the SI unit of conductance.

If you touch a lumped resistor in a working laboratory circuit, most likely it will be warm. This is because the electrons collide with atoms when moving through the vibrating lattice of a wire and lose their energy. This energy eventually turns to heat. To simplify calculations, Equation 1.13 can be rewritten in the discrete members as

$$\Delta P(W) = \frac{\Delta E(J)}{\Delta t(s)}. \tag{1.20}$$

This equation states that energy $\Delta E(J)$ spent by the electrical current flowing through a resistor over time interval $\Delta t(s)$ is equal to power $\Delta P(W)$ dissipated by the resistor. On the other hand, if

6 **Ernst Werner von Siemens** (1816–1892), a German engineer, inventor, and industrialist. His numerous inventions include the pointer telegraph and the concept of a loudspeaker. He developed electrical generators and motors and used them to build the first electrical railway and elevator. He founded a company called (you guessed it) Siemens; it is still one of the major players in today's electrical and electronic industry. The SI unit of conductivity, siemens, is named after him.

the voltage drop across the resistor is V volts and the current flowing through this resistor is I amperes, then Equation 1.14b expresses the power dissipated by this resistor as

$$P(W) = V(V) \cdot I(A). \tag{1.14bR}$$

For ac resistive circuits, we can retrieve (1.14c) as

$$p_R(t)[W] = v_R(t) \cdot i_R(t) = \frac{v_R^2}{R} = R \cdot i_R^2. \tag{1.14cR}$$

Exercise What is the resistance of an incandescent light bulb in the USA 120-volt residential line if its power is 60 (W)? Answer: $R = 240(\Omega)$.

Combining (1.20) and (1.14b) gives

$$\Delta E(J) = \Delta P(W) \cdot \Delta t(s) = V(V) \cdot I(A) \cdot \Delta t(s). \tag{1.21}$$

To interpret this equation, let's consider the circuit shown in Figure 1.8. First, this figure demonstrates the *passive sign convention for power*: If the current enters the component's positive terminal, then this component absorbs and dissipates the power. If the current enters the component's negative terminal, this terminal supplies the power. Indeed, the current $I(A)$ enters the positive terminals of resistors $R_1(\Omega)$ and $R_2(\Omega)$, therefore these passive components dissipate power. In contrast, the current $I(A)$ enters the battery's negative terminal, and the battery supplies the power to the circuit.

Secondly, Figure 1.8 demonstrates the meaning of the term *voltage drop*. The battery supplies $E = V_{dc}(V)$ to the entire circuit. This voltage is distributed between the circuit components $R_1(\Omega)$ and $R_2(\Omega)$, so that

$$V_{dc}(V) = V_{R1} + V_{R2}.$$

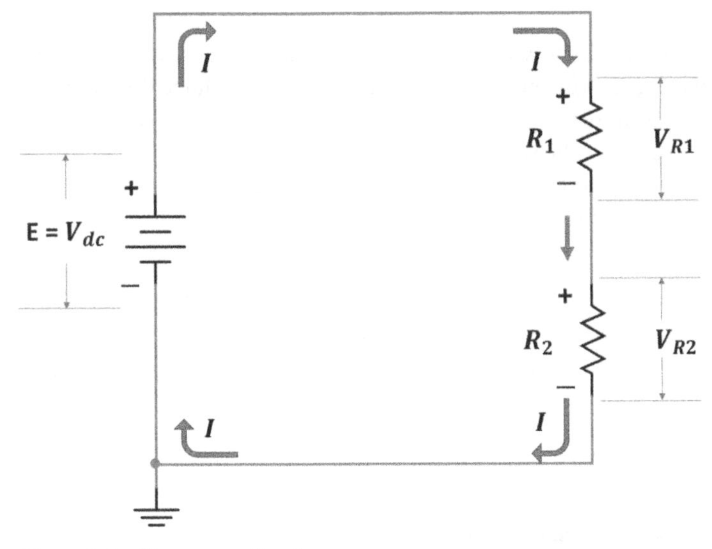

Figure 1.8 Sign convention for power.

The current $I(A)$ is the same for both resistors; therefore, according to Ohm's law, the voltage drops across these resistors are

$$V_{R1}(V) = I(A) \cdot R_1(\Omega) \text{ and } V_{R2}(V) = I(A) \cdot R_2(\Omega).$$

These two formulas show that the voltage supplied to the circuit by the battery drops across each resistor proportionally to their resistances. This voltage drops are compensated by the *voltage rise* occurs at the battery.

Note that Figure 1.8 and its discussion are given for a dc current. This is why the capital letters were used to designate the electrical quantities. Nevertheless, all the conclusions are valid for the resistive ac circuit too.

To sum up the discussion of resistance and resistor, it's worth noting that the lumped resistors are at the extinction phase along with the electrical circuits built with discreet components. Those circuits are replaced by integrated circuits at any possible area of applications. Nevertheless, as a material property, resistance continues to play a major role in designing and manufacturing all equipment based on electrical phenomena. As long as electrical current flows through a circuit, it experiences resistances either at the lumped components or at the wires, connectors, and other circuit elements.

Resistance and resistor—a summary

- Resistance is the material's property to resist, oppose, or withstand the current flow; it is also a phenomenon presenting this property.
- A resistor is a passive lumped component of an electrical circuit that makes resistance.
- The main characteristic of resistor's material is resistivity, $\rho(\Omega \cdot \text{m})$ whose value ranges from 10^{-8} for conductors, to 10^{0} for semiconductors, to 10^{12} for insulators. The total material resistance is proportional to resistivity.
- A short circuit is a model where two terminals are connected by the wire whose resistance tends to zero; an open circuit is a model whose two terminals are connected by the material whose resistance goes to infinity.
- Resistance depends on temperature, frequency, and other external factors. However, when these factors vary slightly, we can assume that the resistance is constant, which leads to a linear model of a resistor.
- When current flows through a resistor, the voltage across this resistor drops, reflecting the amount of energy spent for pushing the current through the resistor. This voltage drop is compensated by the voltage rise at the source (e.g., battery). The energy for this rise is consumed from the internal or external processes in the source.
- Electrical energy spent on a resistor per unit time describes power dissipated by the resistor. This power is governed by (1.14b), $P(W) = \dfrac{dE(J)}{dt(s)} = V \cdot I.$
- If current enters the positive terminal of a circuit component, this component absorbs and dissipates power, as a resistor does; otherwise, the component produces power, as a source (e.g., battery) does.
- Conductance shows the ease with which the component transfers the current; conductance is reversed to the resistance.

1.3.2 Capacitance and Capacitor

A capacitor is a circuit passive reactive component that can store electrical energy. It consists of two parallel conducting plates separated by a *dielectric*, a special type of insulator. Figure 1.9a visualizes this definition, and Figure 1.9b shows a capacitor in an electrical circuit.

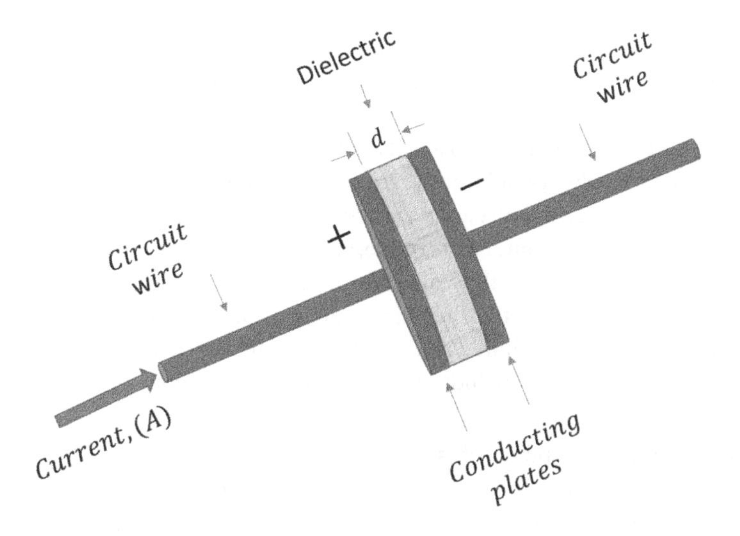

a)

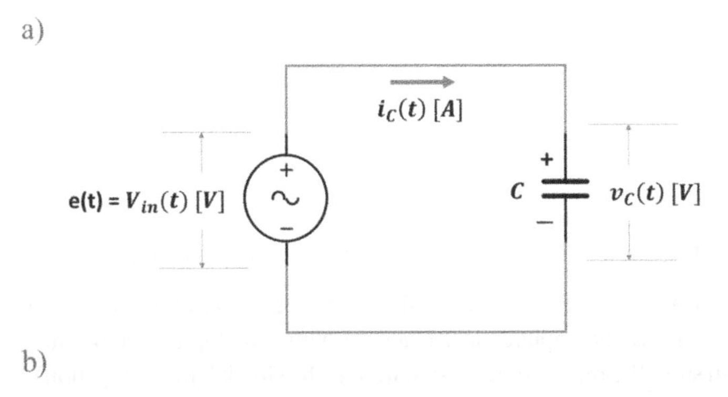

b)

Figure 1.9 Capacitor: a) Construction; b) connection in a circuit.

When a capacitor is connected to a battery, the positive charges accumulate at one plate, and the negative charges accumulate at the opposite conducting plate. However, due to the dielectric separating the plates, charges cannot flow through a capacitor. Thus, the capacitor stores the supplied energy. In the circuit shown in Figure 1.9b, the voltage across the capacitor is equal to the voltage supplied by the battery if the capacitor has been connected to the battery for a long time. The process of accumulating the supplied electrical energy by a capacitor is called *charging*. A fully charged capacitor looks like a battery; it can give energy to the other circuit components for a limited time while it exhausts all stored energy. The release of accumulated energy by the capacitor is called *discharging*. If a charged capacitor is connected to the ground for long time, it will fully discharge.

The amount of electrical charge accumulated at the capacitor's plates, $Q(C)$, is proportional to the applied voltage, $v(V)$, as

$$Q(C) = C(F) \cdot v(V), \tag{1.22}$$

where $C(F)$ is the *capacitance*, and F stands for *farad*.[7] (Unfortunately, the electrical engineering industry uses capital letter C to denote different things: *coulomb* for the unit of charge, *Celsius* for the centigrade temperature scale, and *capacitance* for the capacitor ability to accumulate electrical charge. We must be careful with these notations to avoid confusion.)

Considering a capacitor operation, it's always helpful to remember the *law of conservation of electrical charge*. It states that the total charge in any closed electrical system (the number of positive charges minus all negative charges) is constant. For example, suppose the number of negative charges within the capacitor's plates decreases. In this case, the current (the stream of negative charges) flows from the capacitor into a circuit. Still, the current arriving at the other capacitor plate brings negative charges that keep the total charge constant between the capacitor plates. At a greater scale, the total charge of the circuit that includes this capacitor remains constant.

Exercise The stray capacitance of a joint point of an electronic circuit is equal to $12pF$. If the applied voltage is 3.3V, how many electrons are collected at this joint? Answer: $245.5.10^6$ *electrons*; *i.e.*, 246 *millions os electrons*. (Hint: Use (1.22) and the $6.2.10^{18}$ number of electrons in one coulomb, as discussed in Section 1.2.)

A capacitance is described by the following formula:

$$C(F) = \frac{\varepsilon\left(\frac{F}{m}\right).A(m^2)}{d(m)}, \tag{1.23}$$

where $\varepsilon\left(\frac{F}{m}\right)$ is the permittivity of a dielectric, $A(m^2)$ is the area of a conducting plate, and $d(m)$ is the width of a dielectric, as shown in Figure 1.9b. *Permittivity* describes the ability of a material to accumulate electrical charge in response to the applied electrical field. Thus, the higher the permittivity, the more electrical energy can store the material. Its designation is the Greek letter ε (epsilon). The vacuum permittivity is $\varepsilon_0 = 8.854 \cdot 10^{-12}\left(\frac{F}{m}\right)$. Typically, the permittivity of the materials, ε, is measured by the relative value, $\varepsilon_r = \frac{\varepsilon}{\varepsilon_0}$. Thus, the relative permittivity of air is 1, paper 1.5, carbon 2.5, mica between 3 and 6, and glass 5. The listed permittivity values are approximate because they depend on temperature, EM field frequency, and other factors. Dielectrics used in capacitors have a high permittivity, which increases the capacitors' ability to store electrical energy.

---

7 **Michael Faraday** (1791–1867), one of the most famous British chemists and physicists. The man of very humble origin, who suffered from starvation in his childhood, he learned the basics of reading and writing in a Sunday church school. This was his only formal education, but he eventually became one of the most knowledge men of the age by self-studying. He started working as a technician at the Royal Institution of London in 1813 and worked here for 45 years. Initially, he worked as an assistant to Sir Humphry Davy, the prominent British chemist. Under Davy's advisory and later independently, Faraday made many discoveries and developments in chemistry. In 1820s, his scientific interest shifted to electricity. After making many inventions and discoveries in this field, he developed theory of electromagnetic induction and confirmed his development experimentally. This result is known today as Faraday's law of induction; Maxwell later generalized it and included it in his four equations that constitute theory of the electromagnetic field. Faraday's law of induction is the theoretical foundation of such applications as electrical motors, electrical generators, electrical transformers, induction stoves, and many more. Thus, just discovering the law of induction would make Faraday's name immortal. Michael Faraday was one of the greatest experimentalists who ever lived because he designed his experiments based on his deep understanding the nature of the subject.

Dielectric is an insulating material that doesn't allow conduction current to flow through because it doesn't have enough free electrons. It does, however, experience *polarization* when placed in an electric field. In this context, polarization means that the positive electrical charges shift slightly in the field direction, whereas negative ones shift in the opposite direction. Thus, polarization creates an internal electric field inside the dielectric. Thanks to polarization, a dielectric material stores more electrical charges than vacuum, which means that a capacitor can store more electrical energy. This is the reason for placing dielectrics in capacitors. Examples of dielectric include dry air, nitrogen, ceramic, glass, and plastic. In contrast to the dielectric, an *insulator*, though also blocking conduction current, can't experience polarization and therefore doesn't react anyhow when placed into the electric field. Examples of insulators include rubber and wood.

Let's return to discussion of (1.23). Clearly, the greater the area of a conducting plate, the more charges it can accommodate. Also, the closer the capacitor plates to one another, the more electrical energy can be stored between them.

What a capacitor's area would be needed to obtain $C = 1(F)$ if the plates are separated by a piece of paper whose $d = 100\mu m$? The paper permittivity is $\varepsilon\left(\frac{F}{m}\right) = \varepsilon_r \cdot \varepsilon_0 = 1.5.8.854 \cdot 10^{-12} = 13.281 \cdot 10^{-12}$. Thus, from (1.23) we get

$$A(m^2) = \frac{C(F)\cdot d(m)}{\varepsilon\left(\frac{F}{m}\right)} = \frac{10.10^{-4}}{13.281.10^{-12}} = 0.753.10^8.$$

Converting the result in square miles (1 square meter is $3.861.10^{-7}$ square miles) results in

$$A(sqmi) = 29.073.$$

Such an area is typically occupied by an American city with 150,000 citizens. Fortunately, the area of a capacitor's plates should not be as flat as an area of a town. Engineers use many "tricks" to design a capacitor with a vast surface area and a small overall size. What's more, electronic circuits exploit capacitors with micro-, nano-, and pico-capacitances, which reduces the required areas in a million, billion, and trillion times.

We must realize that the above considerations are valid only for *dc* circuits. This is why we use the upper case to denote current and voltage. Our discussion shows that a *capacitor works as an open circuit in dc applications.* This statement explains why one of the most ubiquitous capacitor applications blocks the unwanted dc signals in countless communication devices. All these statements are valid only after *charging* and *discharging* processes (to be discussed shortly) are fully completed.

How does a capacitor behave when the applied voltage is *ac*? Let's rewrite (1.22) as $Q(t) = Cv(t)$ to stress that the charge stream created by the variable applied voltage is also variable. Differentiating both sides of the new version of (1.22) gives

$$\frac{dQ(t)}{dt}\left[\frac{C}{s}\right] = C\frac{dv(t)}{dt}.$$

But $\frac{dQ(t)}{dt}\left[\frac{C}{s}\right] = i(t)[A]$ by definition given in (1.14b). Therefore, $i - v$ relationship for a capacitor in the *ac* circuit is given by

$$i_C(t)[A] = C\frac{dv_C(t)}{dt}. \tag{1.24}$$

Exercise Consider a capacitor whose $C = 18(nF)$. What is its current if the applied voltage is given by $v_C(t) = 5\sin(6t)(V)$? Answer: $i_C(t) = 540\cos(6t)(nA)$.

Equation 1.24 is the fundamental formula governing capacitor behavior in an ac circuit. Let's consider its most significant implications:

- Equation 1.24 shows that no current flows through a capacitor if the voltage across the capacitor is dc because $\frac{dV}{dt} = 0$ *if V is constant*. Though we declared this fact before, Equation 1.24 proves it in a mathematical form.
- On the same note, (1.24) states that $i_C(t)[A]$ is not zero when $v_C(t)$ is variable (that is, ac). To understand the physics behind this phenomenon, consider two points: First, recall Ampere's law, which asserts that a conduction current flowing through a wire creates a magnetic field around this wire. Secondly, know that an experiment shows that when the current applied to a capacitor's plates is ac, the variable magnetic field is created within the capacitor. This result means that the new source of the variable magnetic field is created by the changing charges between the capacitor's plates. According to Ampere's law, this new source cannot be anything else but a current. It is caused by the shift (polarization) positive and negative charges in a dielectric under the applied electric field. Maxwell[8] called this phenomenon a *displacement current.* He added this term to Ampere's law and included this generalized formula into his equations. A *displacement* is a mathematical quantity that reflects the effect of external and internal (polarization, remember?) electrical fields on a dielectric. A displacement current is proportional to a rate of displacement change in time, very much as conduction current is proportional to the rate of the charge change, which justifies the use of the term "current" for the former member. It can be shown mathematically that a displacement current carries the dimension "ampere" and is equal to the applied conduction current. However, this current has no electrons drift and no charge flow, and in this regard term "current" is confusing.
- Thus, when an ac conduction current is applied at a capacitor, changing charges inside the capacitor create a displacement "current," which transfers all attributes of the applied current through the capacitor. In fact, the sum of the conduction and displacement current is continuous despite a capacitor's interruption of the conduction current. The distinction between these two types of current for the circuit operation is unnoticeable.
- It follows from (1.24) that *voltage across a capacitor, $v_C(t)$, cannot change abruptly*. For example, if $v_C(t)$ would be able to jump from 0 (V) to 1 (V) for $t = 0$, its change rate (that is, a capacitor current) should be infinity, which is impossible. Therefore, (1.24) is the foundation of the *capacitor voltage continuity*. In other words, the value of a capacitor voltage existed at the moment before the input voltage change, $t = 0^-$, is the same as it will be at the instant after the change, $t = 0^+$. The common way to describe this statement in formula is

$$v_C\,|_{t=0^-} = v_C\,|_{t=0^+}\,. \tag{1.25}$$

8 **James Clerk Maxwell** (1831–1879) was a British (Scottish) physicist whose place in modern science is ranked with Isaac Newton and Albert Einstein thanks to the fundamental nature of his contributions. Maxwell's equations that form the contemporary understanding of the electromagnetic field are familiar today to every electrical engineer in the world. Any current wireless connection from the radio to Wi-Fi is based on his theory. Besides this fundamental result, Maxwell significantly contributed to many areas of modern science, such as thermodynamics, gas theory, astronomy, and color photography. He published his first scientific paper aged 14 and continued to work through the last day of his life.

The importance of this point becomes evident in the analysis of a capacitor's behavior under the switching conditions in the chapters that follow. The physics behind (1.25) is that all charges that make $v_C(t)$ cannot accumulate at the capacitor's plates or disappear from the plates instantaneously by their very nature.

- Even though the voltage across a capacitor, $v_C(t)$, cannot change instantaneously, the current through a capacitor can. In formula, this statement means

$$i_C\,|_{t=0^-} \neq i_C\,|_{t=0^+}. \tag{1.26}$$

Therefore, when a switch turns on, the capacitor current value can change from zero to the given amount abruptly.

- Equation 1.24 leads to the formula for calculating the voltage across the capacitor when its current is known:

$$v_C(t)[V] = \frac{1}{C}\int_{-\infty}^{t} i_C(t)dt. \tag{1.27}$$

It shows the voltage accumulated across the capacitor when its current runs for the interval between t_0 and t. To calculate this voltage value, Integral (1.27) can be split into two parts: from $-\infty$ to t_0 and from t_0 to t. Thus,

$$v_C(t)\,[V] = \frac{1}{C}\int_{-\infty}^{t_0} i_C(t)dt + \frac{1}{C}\int_{t_0}^{t} i_C(t)dt = v_C(t_0) + \frac{1}{C}\int_{t_0}^{t} i_C(t)dt. \tag{1.28}$$

The member $v_C(t_0)$ in (1.28) represents the voltage accumulated for the interval from $-\infty$ to t_0. This means that a capacitor can save the information; there is a *capacitor memory*.

(Students performing exercises on circuit analysis in an academic laboratory often see a capacitor as a bulky circuit element. However, a capacitor plays a significant role as a memory element in the micro- and nanosized integrating circuits forming the state-of-the-art processors used today in quantum sensors and the artificial intelligence computer units.[9])

The above discussion helps to understand capacitor charging and capacitor discharging processes in a dc circuit: When a switch turns on, and the input dc voltage is applied to a capacitor, the charges start accumulating at the capacitor's plates, and the voltage across the capacitor, $v_C(t)$ $[V]$, begins increases. If the external voltage is applied to a capacitor long enough, the capacitor will be fully charged, and $v_C(t)$ becomes equal to the applied voltage. In contrast, the current through a capacitor, $i_C(t)$ $[A]$, abruptly reaches its maximum because the change of $v_C(t)$ will be maximal at the moment when the switch activates. As $v_C(t)$ continues to grow, the current continues to decrease because the change rate of $v_C(t)$ will decrease. When the voltage across the capacitor reaches its maximum value and becomes constant, the current becomes zero.

Exercise The voltage applied to capacitor whose $C = 2(\mu F)$ is given by $v_C(t) = 12(1 - e^{-3t})(V)$. What current flows through capacitor? Answer: $i_C(t) = 72e^{-3t}(\mu A)$. (Suggestion: Prove that this exercise describes a discharging process of a capacitor. Hint: Consider $i_C(t)$ values at $t = 0(s)$ and let $t \to \infty(s)$.)

9 IEEE Spectrum, June 2022, Samuel K. More, "3 Paths to 3D Processors," pp. 25–29.

When the voltage and current are applied to a capacitor, the power delivered to it is defined by (1.14a) as

$$p(t)[W] = v(t) \cdot i(t). \tag{1.14aR}$$

Using (1.24), we can rewrite (1.14a) as

$$p_C(t)[W] = v_C(t) \cdot i_C(t) = Cv\frac{dv}{dt}. \tag{1.29}$$

An *ideal capacitor* doesn't dissipate its power; it receives the power from a circuit and returns it to the circuit.

Recall Equation 1.13,

$$p(t)[W] = \frac{dE(t)[J]}{dt(s)}. \tag{1.13R}$$

This equation enables us to calculate electric energy stored in a capacitor based on (1.29) as

$$E_C(t) = \int_{-\infty}^{t} p_C(t)dt = C\int_{-\infty}^{t} v_C \frac{dv_C}{dt}dt = C\int_{v(-\infty)}^{v(t)} v_C dv_C = \frac{1}{2}Cv_C^2(t)\Big\{_{v(-\infty)}^{v(t)} = \frac{1}{2}Cv_C^2(t). \tag{1.30a}$$

The lower limit of (1.30a) is zero, $v(-\infty) = 0$, because a capacitor is not charged at $t \Rightarrow -\infty$. Therefore,

$$E_C(t) = \frac{1}{2}Cv_C^2(t). \tag{1.30b}$$

You are recalled that each bit of information being transferred from one place to the other must be carried by an electrical or optical signal. Since a capacitor can store electrical energy, it can store information because electrical energy is an information career. In other words, a capacitor can be used as a *memory* element. This statement is an additional support to our conclusion made at the discussion of (1.28): Memory element is, indeed, one of the most ubiquitous capacitor's applications.

For the circuit analysis, it's necessary to have three capacitor characteristics—the $i-v$ relationship, dependence $C(F)$ on *temperature*, and the change with *frequency*. (Refer to Figure 1.7, where such characteristics are shown for a resistor.) The $i-v$ relationship is given by (1.24); it depends on the nature of $v(t)$, as discussed above. If the capacitance is constant, $i - \frac{dv}{dt}$ graph is a straight line, and a capacitor is *linear*. However, in reality, $C(F)$ depends on voltage, so the capacitance in Equation 1.24 might be a variable. The capacitance also depends on temperature; this dependence is strongly affected by a dielectric material used in a capacitor. We need to consult the specifications sheet for every particular capacitor to get the correct result of the capacitance calculations. This book assumes a *linear capacitor* model.

Does a *capacitance* depend on frequency? Equation 1.23 defining capacitance doesn't contain this dependence. In general, we can consider that capacitance doesn't change with the frequency for a wide range of frequencies. However, we know that a *capacitor's operation* strongly depends on the frequency: When $f(Hz) = 0$, the capacitor blocks current completely; when $f(Hz) > 0$, the capacitor "conducts" the current. What's more, (1.24) states that the higher the rate of voltage change, the greater the capacitor current. Obviously, the rate of voltage change increases with the increase of its frequency. Therefore, the greater the frequency of the applied capacitor voltage, the less the capacitor opposition to the flowing current. To describe this capacitor property, a new characteristic called *capacitive reactance*, X_C, also called the capacitor reactive impedance, is introduced as

$$X_C(\Omega) = \frac{1}{2\pi fC} = \frac{1}{\omega C}. \tag{1.31}$$

Reactance's dimension is ohm because it is a measure of opposition to the current flow. It is inversely proportional to frequency. Dependence of $X_C(\Omega)$ on frequency is visualized in Figure 1.10. This capacitor's characteristic is crucially important for circuit analysis, as seen throughout this book. Due to this property, a capacitor is called a *reactive component*.

Question Can you suggest any practical device that can utilize the frequency-discriminating capacitor's property?

Compare Figures 1.7b and 1.10 to recall that the resistor and its resistance do not depend on frequency at all. This frequency-dependence difference between resistors and capacitors plays a vital role in circuit analysis.

Equations 1.24 and 1.31 and Figure 1.10 assert that in the dc case a capacitor blocks current completely; in this case its model is an *open circuit*. When a high-frequency signal is applied to the capacitor, its "resistance" (capacitive reactance) tends to zero, and its model becomes a short circuit. Refer to Figure 1.9b and sketch these two models.

Exercise When the current flowing through a capacitor has frequency 100 Hz, the capacitive reactance is $X_C = 1(\Omega)$. What is the capacitance? Answer: $C = 1.6(mF)$.

So far, we've considered a capacitor as a lumped circuit element. But even if a circuit doesn't contain a capacitor, it can—and in most cases does—experiences capacitor's effect through

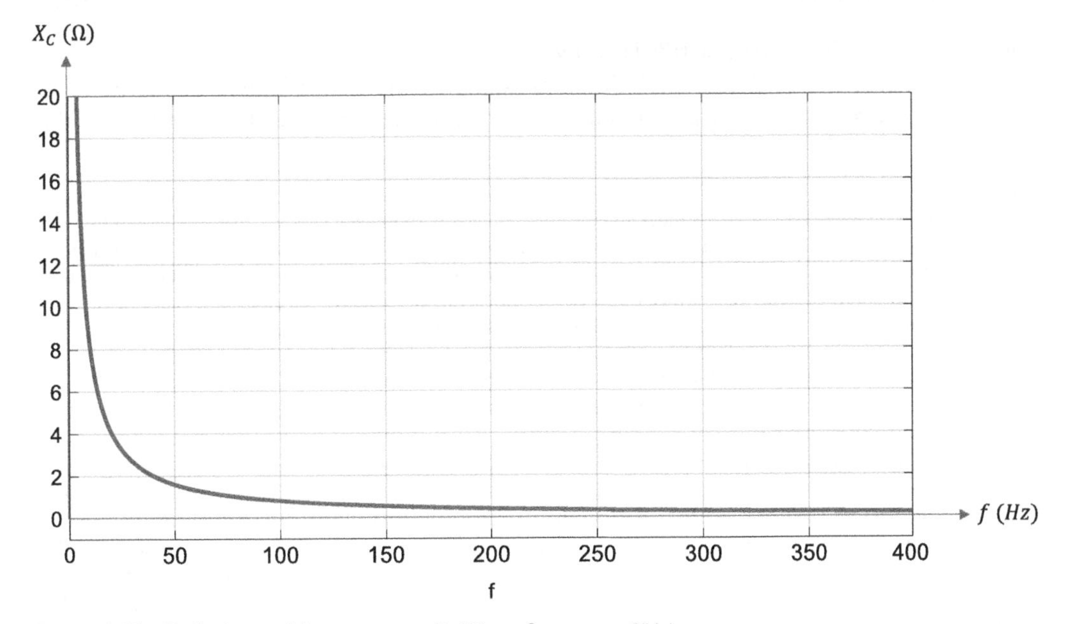

Figure 1.10 Typical capacitive reactance, $X_C(\Omega)$, vs. frequency, f(Hz).

distributed or *stray capacitances* that appears when two conductors stand close to one another. These stray capacitances might affect the circuit operation significantly. Bear this point in mind.

1.3.3 Inductance and Inductor

An inductor is a circuit passive, reactive component that can store energy in its magnetic field. It's typically made from a coil implemented by a conducting wire that wound around a cylindrical core. This structure is called a *solenoid.* Figure 1.11a shows a typical inductor's construction, whereas Figure 1.11b demonstrates an inductor as a circuit element.

When an inductor is connected to an ac source, current flows through the coil and creates a magnetic field that concentrates inside the coil. This phenomenon is governed by *Faraday's law.*

Let's be known that $B(T)$ is the strength of the magnetic field, where T stands for *tesla,*[10] the unit of this strength. This magnetic field creates a *magnetic flux* $\Psi(Wb)$ propagating in all directions.

The unit of magnetic flux, *weber,*[11] is defined as $Wb = T \cdot m^2$. Having these remarks, we can refer to Faraday's law that relates an *electromotive force,* $\varepsilon(V)$, to magnetic flux, $\Psi(Wb)$, as

$$\varepsilon(V) = -\frac{d\Psi(Wb)}{dt}. \tag{1.32}$$

The law states that time-changing magnetic flux when penetrates through a coil creates an electromotive force, emf, that opposes the change in Ψ. (Equation 1.32 justifies the following standard definition of weber: *One weber is the magnetic flux,* Ψ, *which, linking a coil of one turn, would produce in it an electromotive force of 1 volt if its time rate would be* $\frac{d\Psi}{dt} = 1\left(\frac{Wb}{s}\right)$.)

Consider the inductor shown in Figure 1.11a. Its total magnetic flux, Ψ, is created by the current flowing through the N turns of the coil. It's reasonable to assume that Ψ is proportional to the current $i_L(t)$. Thus,

$$\Psi(Wb) = kNi_L(t), \tag{1.33}$$

where k is the coefficient to be clarified shortly.

However, Equation 1.33 does not tell the whole story because only effective part of the entire magnetic flux creates the emf inside an inductor. To take this fact into account, we introduce the

---

10 **Nicola Tesla** (1856–1943) was an engineer and inventor. He was born in what is now Croatia, educated in Europe, where he started his professional career, and moved to the United States in 1884. Since that time, he has lived and worked in America. Tesla's inventions are too numerous to list here; it's sufficed to say that he invented and—with the support of George Westinghouse's Electric Company—implemented the alternating-current approach to deliver electric power for industrial and customer needs. In doing so, he won the battle with no less than Thomas Edison, who promoted a direct current for the same purpose. Encyclopedia Britannica states that three Nobel Prize recipients at Tesla's funeral addressed their tribute to "one of the outstanding intellects of the world who paved the way for many of the technological developments of modern times." SI unit *tesla* is named in tribute to this engineering genius.

11 **Wilhelm Eduard Weber** (1804–1891) was a German physicist. He studied at the University of Halle. Aged 27, Weber became a professor at the University of Göttingen—the most prestigious German university of the time—by recommendation of Carl Gauss. In cooperation with Gauss, he developed and built an electromagnetic telegraph that successfully operated at the university campus. Weber made many significant contributions to the theory and practice of electromagnetism. It's worth mentioning that he developed the electrical and magnetic unit system together with Gauss. He also discovered the role of the speed of light in electromagnetics. The unit of magnetic flux is named in his honor.

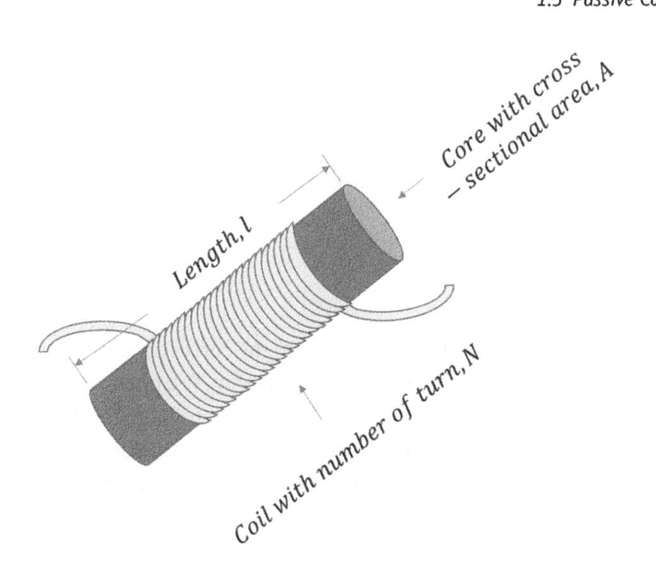

a)

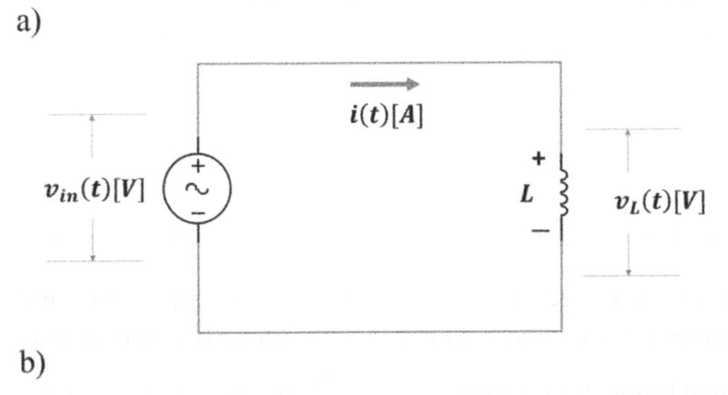

b)

Figure 1.11 Inductor: a) Typical construction; b) inductor in a circuit.

flux linkage, $\wedge$ (Wb), the effective part of Ψ that links an inductor's solenoid and creates the emf. Surely, $\wedge$ is proportional to Ψ and the number of the solenoid turns N; that is

$$\wedge(Wb) = N\Psi. \tag{1.34}$$

Equation 1.34 explains why $\wedge$ and Ψ carry the same unit: They relate one to the other by the dimensionless coefficient N.

Now, the effective part of magnetic field making the useful work in an inductor, the *flux linkage*, can be formulated as

$$\wedge(Wb) = kN^2 i_L(t). \tag{1.35}$$

Denoting the proportionality coefficient between current $i_L(t)$ and magnetic flux $\wedge$ as L and calling it *inductance*, we state

$$\wedge(Wb) = Li_L(t). \tag{1.36}$$

From (1.32) and (1.36) we get

$$\varepsilon(V) = -L\frac{di_L(t)}{dt}. \tag{1.37a}$$

Now, we need to recall that emf is simply a voltage; in our consideration, it is the voltage, $v_L(t)$, across the inductor. Thus, we obtain the essential formula, governing $v - i$ relationship of an inductor:

$$v_L(t) = L\frac{di_L(t)}{dt}. \tag{1.37b}$$

But why does (1.37a) contains a negative sign and (1.37b) doesn't, you might correctly ask. The answer is that the experiment shows that the voltage induced across an inductor, $v_L(t)$, *opposes* the change (increase or decrease) in the current that induces this voltage. This fact is reflected in the polarity rule shown in Figure 1.11b. (Note that this polarity rule is also valid for a resistor and a capacitor. See Figures 1.4 and 1.9b.) Therefore, we can state that an *inductor is a circuit element that resists the change in the current flowing through it.*

The *inductance*, $L(H)$, introduced in (1.36) is not just a coefficient; it is the inductor property to oppose the change of an inductor's current, $i_L(t)$. Its value depends on an inductor's structure, core material, and inductor size. For a solenoid shown in Figure 1.11a, the inductance formula is as follows:

$$L(H) = \frac{\mu N^2 A}{l}. \tag{1.38}$$

Here, $H = \frac{volt \cdot second}{ampere}$ is the inductance unit called *henry*,[12] $\mu\left(\frac{H}{m}\right)$ is *permeability*, which is an ability of a material to react on an applied magnetic field positively, N is the number of wire turns in a solenoid, $A(m^2)$ is the solenoid's cross-section area, and $l(m)$ is its length. Similarly to *permittivity*, permeabilities of materials are measured by relative permeability, $\mu_r = \frac{\mu}{\mu_0}$, where μ_0 is the permeability of free space (vacuum), $\mu_0 = 4\pi \times 10^{-7}$ (H/m).

Equation 1.38 explains that coefficient k introduced in (1.33) is given by $k = \frac{\mu A}{l}$ for a solenoid. This equation also asserts that a coil inductance is proportional to permeability, which suggests how to increase inductance: If the coil core is air (free space, that is), its relative permeability is 1, but if the core is made from iron, then its permeability increase in hundreds of times, and inductance increases proportionally. The increase in N boosts the inductance exponentially, and the ratio $\frac{A}{l}$ multiplies the inductance too.

---

12 **Joseph Henry** (1797–1878) was an American physicist and engineer who made a significant contribution to the theory and practice of electromagnetics. In fact, he discovered the law of electromagnetic induction and applied this law to building practical devices even before Michael Faraday. Unfortunately, he did not publish his discovery timely. At the peak of his career, he became a professor at Princeton University. Henry's academic reputation was such that he was appointed the first secretary of the Smithsonian Institution, the first American scientific organization created by the Congress of the United States. The SI unit of inductance, *henry*, was named in honor of him.

Equation 1.37b enables us to derive $i-v$ relationship in an inductor as

$$i_L(t)[A]=\frac{1}{L}\int_{-\infty}^{t} v_L(t)dt=\frac{1}{L}\int_{-\infty}^{t_0} v_L(t)dt+\frac{1}{L}\int_{t_0}^{t} v_L(t)dt=i_L(t_0)+\frac{1}{L}\int_{t_0}^{t} v_L(t)dt. \quad (1.39)$$

The member $i_L(t_0)$ in (1.39) states that an inductor has memory similarly to a capacitor. See (1.28). In the early days of analog computers in the 1950s, the small inductors served as magnetic memory cells.

Power delivered to the inductor is given by

$$p_L(t)[W]=v(t).\ i(t)=L\frac{di}{dt}i. \quad (1.40)$$

Compare (1.40) with (1.29). Like a capacitor, an *ideal inductor is lossless*, which means *it doesn't dissipate power but returns it to the circuit.*

Energy stored by an inductor in its magnetic field can be derived from (1.40) as

$$E_L(t)[J]=\int_{-\infty}^{t} p_L(t)dt=L\int_{-\infty}^{t} i\frac{di}{dt}dt=L\int_{i(-\infty)}^{i(t)} idi=\frac{1}{2}Li^2(t). \quad (1.41)$$

You will recall that (1.30a) shows that the electrical energy is stored by a capacitor. Compare that equation with (1.41).

The set of equations from (1.37a) through (1.41) describes the main inductor properties. Below, we discuss them from the advanced circuit analysis standpoint:

- It follows from (1.37b) that for *dc current*, an inductor works as a piece of wire, that is as a *short circuit*. But the same piece of wire starts working as an inductor as soon as *ac current* starts flowing through it because this current creates magnetic field. Don't forget that the wire resistance, even to dc current, is minimal but still finite. It is called *ohmic resistance*, or *winding resistance*. Typically, it is in the order of units or even fractions of ohm; still, it might be a factor in circuit analysis at the low frequencies. If zero ohmic resistance is assumed, the inductor is considered an *ideal*.
- If an inductor current, $i_L(t)$, changes abruptly, then its voltage, $v_L(t)$, should go to infinity, as Equation 1.37b shows. Therefore, the *inductor current cannot change instantaneously*; that is,

$$i_L|_{t=0^-}=i_L|_{t=0^+}. \quad (1.42)$$

This inductor property is analogous to the capacitor property in the continuity of its voltage. See (1.25) and its discussion for comparison with (1.42). Can you explain what physics is behind this inductor property? However, the voltage across an inductor, $v_L(t)$, can change instantaneously. For example, if a dc circuit in Figure 1.11b would contain a switch and this switch turns on, at this moment, $v_L(t)$ changes from zero to V_{in}. Therefore,

$$v_L|_{t=0^-}\neq v_L|_{t=0^+}\ . \quad (1.43)$$

- *Inductance* of a fabricated inductor remains constant, according to (1.38). Then, (1.37b) states that the relationship between $v_L(t)$ and $\frac{di_L(t)}{dt}$ is linear. Nevertheless, in real-life situations, the inductance can change due to variations in temperature, the value of an applied voltage, and the frequency of an input signal. We, however, neglect these secondary effects and will consider a *linear inductor* throughout this book.

- We learned that an inductor opposes to changes in the induced current, which implies that it resists passing an ac signal. How can we describe this "resistance?" Such an inductor's property is called *inductive reactance*, $X_L(\Omega)$, also known as inductive reactive impedance, is given by

$$X_L(\Omega) = 2\pi fL. \tag{1.44}$$

The inductive reactance, $X_L(\Omega)$, is proportional to the signal frequency, $f(Hz)$. Thus, the graph X_L-vs.-f is a straight line, in contrast to the capacitive reactance shown in Figure 1.10. It follows from (1.44) that at high frequencies, the inductor can block the input signal almost completely; in other words, the inductor can serve as an open circuit. (Take, for example, $L = 1(mH)$ and compute $X_L(\Omega)$ at $f = 1(MHz)$ and at $1(GHz)$.)

- We mentioned previously that such phenomenon as capacitance can be found either in a lumped circuit element, capacitor, or as a stray effect in a fabricated circuit. As opposed to a capacitance, an inductance belongs exclusively to a lumped inductor; a stray inductance is very seldom phenomenon.
- Let's mention the inductors applications: From a practical engineering standpoint, it's important to know that capacitances and resistances are mostly fabricated as interwoven parts of integrated circuits, ICs. Unhappily, inductors can't be built in the IC forms; they are still standing-alone, discrete components. Fortunately, modern technology enables us to build inductors in miniature designs that sized in millimeters; such inductors can be used in IC-based devices. On the opposite end of inductors' applications, they are giant coils consuming thousands-amperes current, as in the case of the *Large Hadron Collider (LHC)* mentioned in Section 1.2. You can easily imagine many other inductor applications ranging in scale from micro-sized components to massive constructions.
- What do inductors and capacitance have in common? They *react* to the changes, not to the inducing forces' values. Specifically, the current will flow through a capacitor only if the applied voltage changes, as (1.24) states: $i_C(t)[A] = C\frac{dv_C(t)}{dt}$. Analogously, the voltage will appear across an inductor only if the current flowing through it varies, which is described by (1.37b) as $v_L(t)[V] = L\frac{di_L(t)}{dt}$. In other words, if $v_C(t) = constant$, then $i_C(t) = 0$; similarly, if $i_L(t) = constant$, then $v_L(t) = 0$. However, if $i_R(t) = constant$, then still $v_R(t) = R \cdot i_R(t)$. Therefore, capacitors and inductors *react* to the change in their inducing forces in contrast to resistors whose values depend only on the inducing force values. Now you understand why capacitors and inductors are called the *reactive components*, but resistors are not.

1.3.4 Comparison Characteristics of Three Passive Components—Resistor, Capacitor, and Inductor

The above discussion of three main passive circuit components is summarized in Table 1.3.

Table 1.3 Three passive components—a summary.

	Resistor	Capacitor	Inductor
The circuit component	Resistor is a passive lumped component of an electrical circuit that makes resistance.	A capacitor is a passive, reactive lumped component that stores energy in its electrical field and resists sudden changes in voltage.	An inductor is a passive, reactive lumped component that stores energy in its magnetic field and opposes abrupt changes in current.
Circuit with a component: sign rules	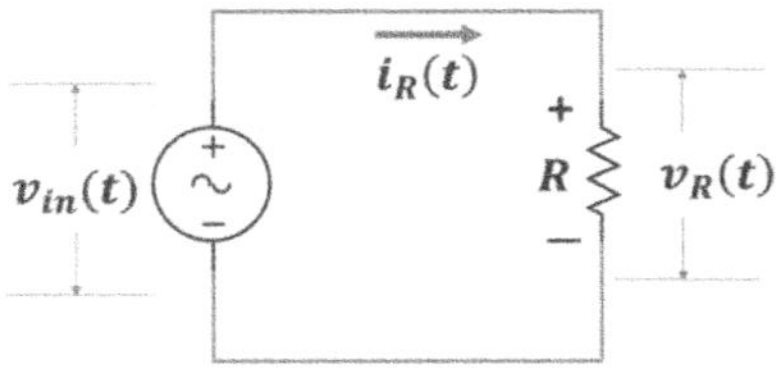 $v_{in}(t)$, $i_R(t)$, R, $v_R(t)$ (+ −) Figure 1.5b	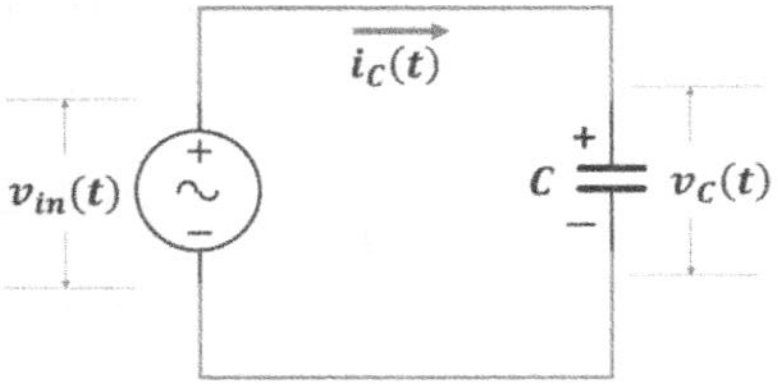 $v_{in}(t)$, $i_C(t)$, C, $v_C(t)$ (+ −) Figure 1.9b	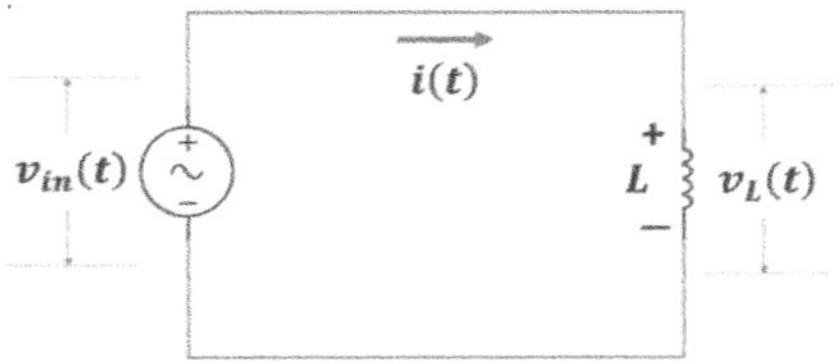 $v_{in}(t)$, $i(t)$, L, $v_L(t)$ (+ −) Figure 1.11b
Main phenomenon	Resistance, $R(\Omega)$, is the material's property to resist, oppose, or withstand the current flow.	Capacitance, $C(F)$, is the capacitor's ability to accumulate electrical charges, as given by (1.22), $q(coloumb) = C(F)v(V)$.	Inductance, L(H), is the inductor's property to oppose the change of inductor's current, $i_L(t)$, as given by (1.36), $\varepsilon(V) = -L\frac{di_L(t)}{dt}$.
Main characteristic	Resistance (1.5) $R(\Omega) = \rho(\Omega\cdot\text{m})\frac{l(m)}{A(m^2)}$	Capacitance (1.23) $C(F) = \frac{\varepsilon\left(\frac{F}{m}\right).A(m^2)}{d(m)}$	Inductance (1.38) $L(H) = \frac{\mu N^2 A}{l}$
Reactances—dependence on frequency	None	$X_C(\Omega) = \frac{1}{2\pi fC} = \frac{1}{\omega C}$ (1.31)	$X_L(\Omega) = 2\pi fL = \omega L$ (1.44)
i − *v relationship*	$i_R(t)[A] = \frac{v_R(t)}{R}$ Ohm's law (1.8)	$i_C(t)[A] = C\frac{dv_C(t)}{dt}$ (1.24)	$v_L(t)[V] = L\frac{di_L(t)}{dt}$ (1.37)

(Continued)

Table 1.3 (Continued)

	Resistor	Capacitor	Inductor
$v-i$ relationship	$v_R(t)[V] = R \cdot i_R(t)$ Ohm's law (1.8)	$v_C(t)[V] = v_C(t_0) + \frac{1}{C}\int_{t_0}^{t} i_C(t)dt$ (1.28)	$i_L(t)[A] = i_L(t_0) + \frac{1}{L}\int_{t_0}^{t} v_L(t)dt$ (1.39)
Linear component (model)	Yes	Yes	Yes
Continuity conditions	Both voltage $v_R(t)$ and current $i_R(t)$ can change abruptly.	Capacitor's voltage obeys continuity rule, $v_C(t=0^-) = v_C(t=0^+)$, (1.25). Current can change instantaneously, $i_C(t=0^-) \neq i_C(t=0^+)$, (1.26).	Inductor's current obeys continuity rule, $i_L(t=0^-) = i_L(t=0^+)$, (1.42). Voltage can change abruptly, $v_L(t=0^-) \neq v_L(t=0^+)$, (1.43).
Short circuit and open circuit	Short circuit if $R \to 0$. Open circuit if $R \to \infty$.	Short circuit if $f \to \infty$ and thus $X_C(\Omega) \to 0$. Open circuit if $f \to 0$ and therefore $X_C(\Omega) \to \infty$.	Short circuit if $f \to 0$ and thus $X_L(\Omega) \to 0$. Open circuit if $f \to \infty$ and therefore $X_L(\Omega) \to \infty$.
Power delivered to (dissipated by) the component	$p_R(t)[W]$ $= v_R(t) \cdot i_R(t)$ $= \frac{v_R^2}{R} = R.i_R^2$. (1.14c)	$p_C(t)[W]$ $= v_C(t) \cdot i_C(t) = Cv\frac{dv}{dt}$ (1.29) No power dissipation by ideal capacitor.	$p_L(t)[W]$ $= v(t) \cdot i(t) = Li\frac{di}{dt}$. (1.40) No power dissipation by ideal inductor.
Energy	$\Delta E_R(J) = P_R.\Delta t$ $= V_R.I_R.\Delta t(s)$. (1.21)	$E_C(t) = \frac{1}{2}Cv_C^2(t)$ (130b)	$E_L(t) = \frac{1}{2}Li^2(t)$ (1.41)
Memory	None	Yes, it stores energy in electrical field, as member $v_C(t_0)$ in (1.28) shows.	Yes, it stores energy in magnetic field, as member $i_L(t_0)$ in (1.39) shows.

1.4 Series and Parallel Circuits; Voltage and Current Divider Rules

1.4.1 Electrical Circuits

An electrical circuit is an assembly of electrical components. All these components, or elements, are interconnected to provide closed paths for currents. An electrical circuit is also called a *network* because it is a set of nodes (elements) connected by links (conductors). Alternatively, a circuit can be considered a *system* or interconnection of components performing a specific task.

Every circuit combines its elements in different *circuit topologies*, the combinations of series and parallel connections. This section introduces these connections and the primary rules in working with such circuits. But first, it's necessary to briefly review the basics of electrical circuits. Figure 1.12 shows an electrical circuit and its components (elements).

A circuit starts with a *ground*, the element symbolizing *zero voltage potential*. Voltage of any circuit component is measured with respect to the ground, as Section 1.2 describes. The second necessary element is a *source*, also called *input*, *excitation*, or *drive*. From a system standpoint, it is an input; Figure 1.12a shows specifically the voltage input denoted as $v_{in}(t)$. A circuit is built to connect many components (resistors, capacitors, inductors, op-amps, and others). Figure 1.12 demonstrates the circuit with four components. When components are connected in sequence (in series) by a single two-terminal line, this assembly is called *circuit branch*. Figure 1.12 contains the following four branches: *ab*, *be*, *cd*, and *af*. Connection *bc* is not a branch because it doesn't contain circuit component, and points *b* and *c* have the same voltage potential because they are connected by an ideal wire. This is also true for connections *de* and *ef*; their mutual potential is zero. The point where two or more branches are connected is called a *node*. Points *a*, *b*, *e*, and *f* in Figure 1.12 are nodes; points *c* and *d* are not because *c* electrically coincides with point *b* and *d* with *e*. Finally, the closed path in a circuit enabling current to flow is called a *circuit loop*. The circuit in Figure 1.12 contains two loops: One goes through nodes *a-b-e-f*, and the other connects nodes *b-c-d-e*.

Question The circuit shown in Figure 1.12 has two loops. However, branch *be* belongs to both of them. How then can we consider these loops independently?

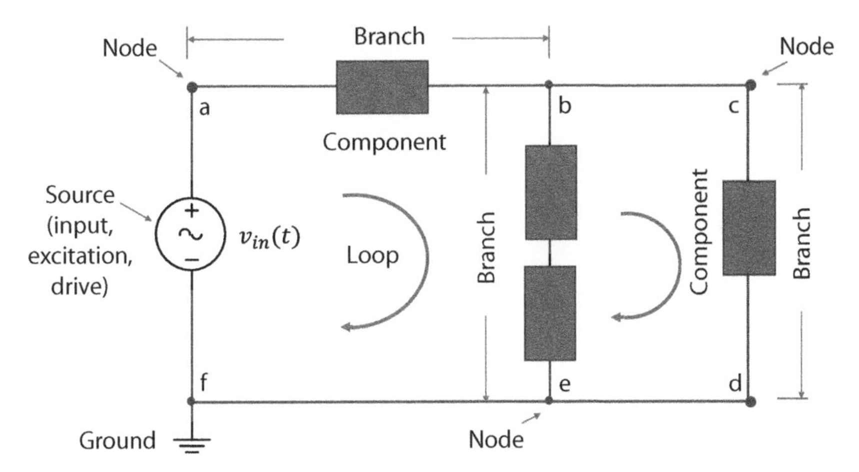

Figure 1.12 Introduction to electrical circuit: Circuit and its main elements.

One important point of discussion, an *equivalent circuit*, can be explained as follows: Refer to Figure 1.6a that shows terminals A and B connected by a wire with zero resistance. Engineers say that this circuit is shortened. Obviously, this shortened circuit cannot exist in reality; it models the capacitor when the source's frequency goes to infinity (remember $X_C(\Omega) = \frac{1}{2\pi fC}$?) or an inductor if circuit's input is V_{dc} because $X_L(\Omega) = 2\pi fL$. Thus, this short circuit is an equivalent circuit for the situations mentioned above. And what does the circuit in Figure 1.6b with open A-B terminals model? You guessed it: The dc circuit with a capacitor connected to A and B or an inductor at these terminals when the input's frequency goes to infinity. Therefore, Figure 1.6b is the equivalent circuit for the above-described circuits.

This brief introduction refers to current and voltage discussed in Section 1.1 and to passive circuit components resistors, capacitors, and inductors considered in Section 1.2. Series and parallel connections are the subjects of this section, sources are presented in Section 1.4, and primary circuit laws will be introduced in Section 1.5.

1.4.2 Series and Parallel Resistors and Voltage and Current Divider Rules

What if we need to build a circuit with two or even more components? How can we connect them? There are two basic ways to connect any circuit components: *series* and *parallel*. We start studying this subject with the *resistors' series connections*, which are shown in Figures 1.13a through 1.13d.

Regardless of their orientations and locations in the circuits, all these resistors are connected in series. Why? Here is the definition:

Circuit components are connected in series if they carry the same current.

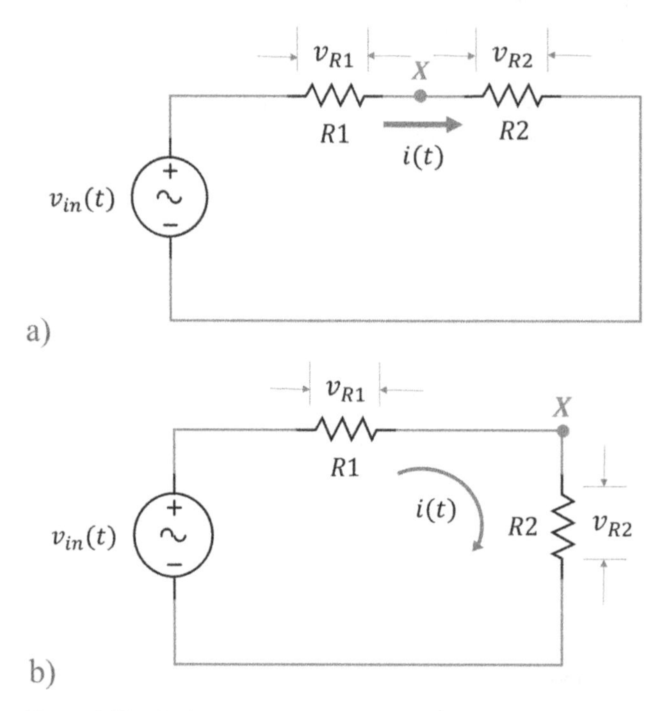

Figure 1.13 Resistors connected in series: a) Two resistors in one line; b) two resistors in the adjacent lines; c) two resistors in the opposite lines; d) three resistors in three circuit lines.

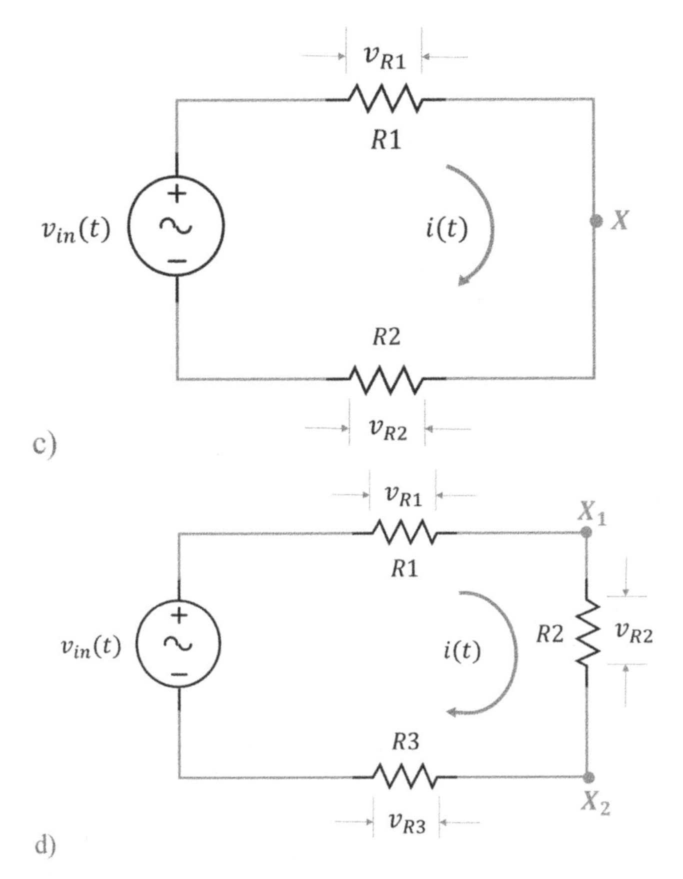

Figure 1.13 (Cont'd)

Examine all figures shown under the umbrella of Figure 1.13 and confirm that all resistors have the same current flowing through them. In order to carry the same current, two adjacent resistors must share one node to which no other connections are attached. Such a connection point is called a *simple node*. Thus, saying that two components are connected through a simple node means giving the other definition of a series connection. Simple nodes are denoted in all Figures 1.13 by X mark.

Pay particular attention to Figure 1.13d: Visually, it might make an impression that three resistors are connected in a sophisticated topology. However, the close examining of the figure reveals that R_1 and R_2 are connected by a simple node X_1, and R_2 and R_3 are connected by a simple node X_2. Therefore, all three resistors carry the same current, which means they are connected in series.

So, the current is one for all components connected in series, but what about voltage? Let's repeat that Ohm's law states

$$v_k(V) = R_k \cdot i_k, \text{where} \, k = 1,2,3,\ldots n. \tag{1.45}$$

Thus, the voltages across R_1 and R_2 are $v_{R1}(V) = R_1 \cdot i$ and $v_{R2}(V) = R_2 \cdot i$, as Figure 1.13a shows.

Therefore, the first conclusion we can make from the discussion of a *series circuit* is the following:

> *When several resistors are connected in series, they carry the same current, but the voltage drop across each one depends on the value of a specific resistor.*

In general, this rule holds for the series connections of all passive components, namely resistors, capacitors, and inductors.

There is an important implication of the above consideration. Examine Figure 1.14, where many resistors are connected in series. The input voltage, v_{in}, is distributed among all the resistors' voltages so that

$$v_{in} = v_{R1} + v_{R2} + v_{R3} + \ldots + v_{Rn}. \tag{1.46}$$

(Equation 1.46 is, in fact, one of the versions of Kirchhoff's voltage law (KVL), which will be discussed in the next section.) After applying Ohm's law, (1.46) becomes

$$v_{in} = v_{R1} + v_{R2} + v_{R3} + \ldots + v_{Rn} = R_1 \cdot i + R_2 \cdot i + R_3 \cdot i + \ldots R_n \cdot i = \sum\nolimits_1^n R_k \cdot i. \tag{1.47}$$

Denoting

$$R_1 + R_2 + R_3 + \ldots R_n = \sum_1^n R_k = R_{total}, \tag{1.48}$$

enables us to state that a circuit with n series resistors can be replaced by a circuit with one equivalent resistor R_{total} defined by (1.48). This statement is visualized in Figure 1.14.
One must realize that (1.48) is a rule for summation of resistors connected in series.

Equations 1.47 and 1.48 enable us to derive a voltage divider rule, another vital tool in circuit analysis. This rule helps find the value of a voltage drop across an individual component among others connected in series. Since the current is one for all components in a series circuit, it can be calculated by using (1.45) and (1.48) as

$$i = \frac{v_{in}}{R_1 + R_2 + R_3 + \ldots R_n} = \frac{v_{in}}{R_{total}}. \tag{1.49}$$

Thus,

$$v_k(V) = R_k \cdot i = \frac{R_k}{R_1 + R_2 + R_3 + \ldots R_n} v_{in} = \frac{R_k}{R_{total}} v_{in}. \tag{1.50}$$

Equation 1.50 is a ***voltage divider rule***; it allows finding individual voltage by knowing only an input voltage and all circuit's resistors.

Exercise Consider the circuit shown in Figure 1.13. Compute V_{R3} if $V_{in} = 24(V)$ and $R_1 = 3(\Omega)$, $R_2 = 4(\Omega)$, and $R_3 = 5(\Omega)$. Answer: $V_{R3} = 10(V)$.

The other way to connect components in electrical circuits is called the *parallel connection*. *Parallel resistors* are pictured in a set of figures under the umbrella of Figure 1.15.

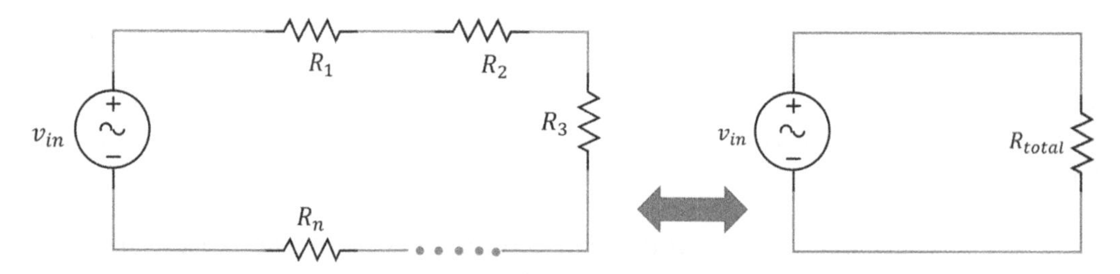

Figure 1.14 A set of series-connected resistors can be replaced by one equivalent resistor, R_{total}.

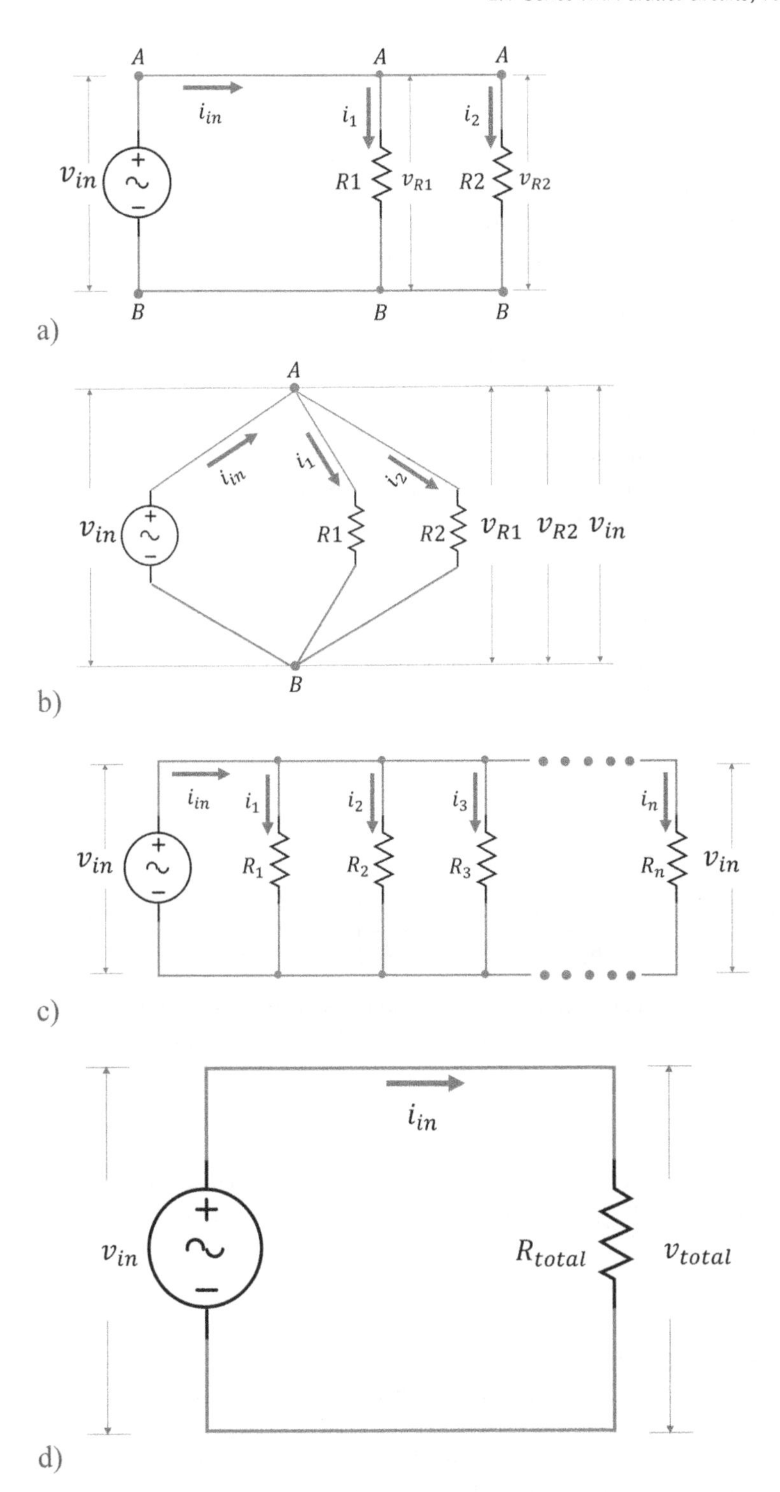

Figure 1.15 Parallel resistors: a) Standard circuit view; b) factual circuit presentation; c) summation of parallel resistors; d) equivalent circuit of resistors connected in parallel.

The main feature of parallel connection is that all the connected components share the same two nodes. Figure 1.15a explicitly demonstrates this point: The source and resistors R_1 and R_2 partake the same nodes A and B. Figure 1.15a shows the standard presentation of a parallel circuit. You will recall that we assume that the *ideal wires* with zero resistances wire our circuits. This means that all nodes A have the same voltage with respect to the ground, which is also valid with all nodes B. To highlight this point, Figure 1.15b presents a nontraditional but factual view of the circuit with parallel connections. It shows that, in reality, all circuit components are connected to only two nodes.

From consideration of the topologies shown in Figures 1.15a and 1.15b follows that *all parallel circuit components have the same voltage.* This feature is often viewed as the definition of a parallel connection.

The parallel circuit topology implies that the voltage is the same for all components, but these components must share the current generated by a source. Figures 1.15a and 1.15b demonstrate this point clearly: Current i_{in} generated by the source splits into current i_1 flowing through resistor R_1 and current i_2 flowing through resistor R_2 so that

$$i_{in}(A) = i_1 + i_2. \tag{1.51a}$$

Note the branches with resistors R_1 and R_2 make two different circuit loops. If a circuit contains n parallel resistors, the input current splits into n individual currents, as Figure 1.15c shows. Thus, (1.51a) for the circuit in consideration becomes

$$i_{in}(A) = i_1 + i_2 + i_3 + \ldots + i_n = \sum\nolimits_1^n i_k. \tag{1.51b}$$

Applying Ohm's law, $i_k(A) = \frac{v_{in}}{R_k}$, to (1.51b) results in

$$i_{in}(A) = \frac{v_{in}}{R_1} + \frac{v_{in}}{R_2} + \frac{v_{in}}{R_3} + \ldots + \frac{v_{in}}{R_n} = \left(\frac{1}{R_1} + \frac{1}{R_2} + \frac{1}{R_3} + \ldots + \frac{1}{R_n}\right) v_{in}. \tag{1.52}$$

Note that (1.52) holds only for a parallel circuit, where all components have the same voltage v_{in} .

Equation 1.52 gives the summation rule for the resistors connected in parallel, namely

$$\frac{1}{R_{total}} = \frac{1}{R_1} + \frac{1}{R_2} + \frac{1}{R_3} + \ldots + \frac{1}{R_n}. \tag{1.53}$$

This rule allows for modeling a practical circuit with n individual resistors by an equivalent circuit containing only one equivalent resistor R_{total}. Such a circuit is shown in Figure 1.15d, and its $v - i$ relationship is given by

$$v_{in}(V) = i_{in} \cdot R_{total}. \tag{1.54}$$

Remember that R_{total} is calculated employing (1.53) but not a simple summation, as it does in a series circuit. Don't forget that (1.53) calculates $\frac{1}{R_{total}}$, so you need to inverse the result to obtain R_{total}.

Let's consider the problem: Calculate R_{total} for the circuit shown in Figure 1.15a if $R_1 = 3\Omega$ and $R_2 = 6\Omega$. The problem and its solution are mathematically written as

$$\frac{1}{R_{total}} = \frac{1}{R_1} + \frac{1}{R_2} = \frac{R_2 + R_1}{R_1 \cdot R_2} = \frac{9}{18} = \frac{1}{2}. \tag{1.55}$$

Answer: $R_{total} = 2\Omega$.

To denote two resistors connected in parallel, we use the following standard notation, $\boldsymbol{R_1 \parallel R_2}$. In this problem, we can write our calculations as

$$R_1 \parallel R_2 = R_{total} = \frac{R_1 \cdot R_2}{R_2 + R_1} = 2(\Omega). \tag{1.56}$$

Observe that the notation $R_k \parallel R_m$ is used to calculate the value $R_{total}\,(\Omega)$, as (1.56) shows, whereas the general rule for summation of parallel resistors given in (1.53) enables us to find only $\frac{1}{R_{total}(\Omega)}$.

It's easy to derive the formula for three resistors:

$$\frac{1}{R_{total}(\Omega)} = \frac{1}{R_1} + \frac{1}{R_2} + \frac{1}{R_3} = \frac{R_2 \cdot R_3 + R_1 \cdot R_3 + R_1 \cdot R_2}{R_1 \cdot R_2 \cdot R_3} \tag{1.57a}$$

and

$$R_1 \parallel R_2 \parallel R_3 = R_{total}(\Omega) = \frac{R_1 \cdot R_2 \cdot R_3}{R_2 \cdot R_3 + R_1 \cdot R_3 + R_1 \cdot R_2}. \tag{1.57b}$$

It's possible to extend (1.57b) for any number of resistors, but the formula becomes cumbersome. Calculations of R_{total} by simply adding the fractions $\frac{1}{R_k}$ are more straightforward.

Exercise What is R_{total} for the circuit shown in Figure 1.15c if $R_1 = 3\Omega$, $R_2 = 6\Omega$ and $R_3 = 9\Omega$? Answer: 0.61Ω.

Observe that the total resistance is always smaller than the minimal resistance in a parallel connection. If all n parallel resistances are equal, then the total resistance is

$$R_{total}(\Omega) = \frac{R}{n}. \tag{1.58}$$

Equation 1.19 introduced the *conductance* as

$$G(S) = \frac{1}{R(\Omega)}. \tag{1.19R}$$

It might help to manipulate with parallel resistances using $G(S)$.

Equations 1.52, 1.53, and 1.54 enable us to compute an *individual current* if we know only an *input current* and all the *resistances*. Indeed, an individual current i_k is given by

$$i_k(A) = \frac{v_k}{R_k} = \frac{v_{in}}{R_k} = \frac{i_{in} \cdot R_{total}}{R_k} = \frac{i_{in} \cdot R_{total}}{R_k} = \frac{\frac{1}{R_k}}{\frac{1}{R_{total}}} i_{in} = \frac{\frac{1}{R_k}}{\left(\frac{1}{R_1} + \frac{1}{R_2} + \frac{1}{R_3} + \ldots + \frac{1}{R_n}\right)} i_{in}. \tag{1.59a}$$

For two resistors in parallel, (1.59a) takes the form

$$i_1(A) = \frac{R_2}{(R_1 + R_2)} i_{in}$$

and

$$i_2(A) = \frac{R_1}{(R_1 + R_2)} i_{in}. \tag{1.59b}$$

Equations 1.59a and 1.59b are the mathematical presentations of the ***current divider rule***. Along with the *voltage divider rule*, (1.59a) is a powerful tool simplifying mathematical manipulations in circuit analysis.

Exercise What is i_2 for the circuit shown in Figure 1.15a if $R_1 = 3\Omega$, $R_2 = 6\Omega$, and $i_{in} = 0.9A$? Answer: 0.3A.

1.4.3 Series-Parallel Resistors—Examples

In general, an electrical circuit inevitably contains combinations of series and parallel resistors. The preceding section prepares us to consider examples of analysis of such combinations.

Example 1.1 Finding an equivalent (total) resistance.

Problem: Find the equivalent resistor for the circuit shown in Figure 1.16a.

Solution: The goal is to find the total (equivalent) resistor for the set of resistors in this circuit. It's always advisable to start the search from the end (output) of a circuit. In this case, summing the series resistors R_5 and R_6 gives

$$R_{5+6} = 5 + 6 = 11\ \Omega.$$

The new circuit is shown in Figure 1.16b. Then, combining parallel resistors R_3 and R_4 results in

$$R_{3\|4} = R_3 \| R_4 = 1.7\ \Omega$$

as Figure 1.16c demonstrates. In this newly combined branch, resistors R_2 and $R_{3\|4}$ are connected in series and can be replaced by one resistor $R_{2+3\|4}$ as

$$R_{2+3\|4} = R_2 + R_{3\|4} = 3.7\ \Omega.$$

Figure 1.16d presents the appropriately modified circuit. Now, parallel resistors $R_{2+3\|4}$ and R_{5+6}

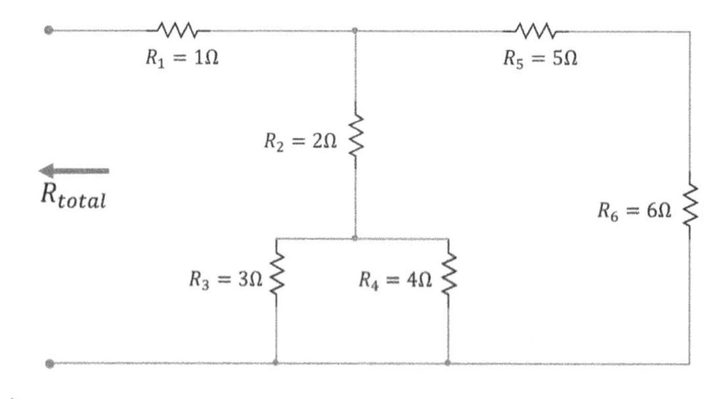

a)

Figure 1.16 Finding the total resistance for Example 1.1: a) The original circuit; b) series resistors R_5 and R_6 summed up making R_{5+6}; c) parallel resistors R_3 and R_4 combined making $R_{3\|4}$; d) series resistors R_2 and $R_{3\|4}$ summed up making $R_{2+3\|4}$; e) parallel resistors $R_{2+3\|4}$ and R_{5+6} combined making $(R_{2+3\|4} \| R_{5+6})$; f) series resistors R_1 and $(R_{2+3\|4} \| R_{5+6})$ summed up making a circuit with one total (equivalent) resistor R_{total}.

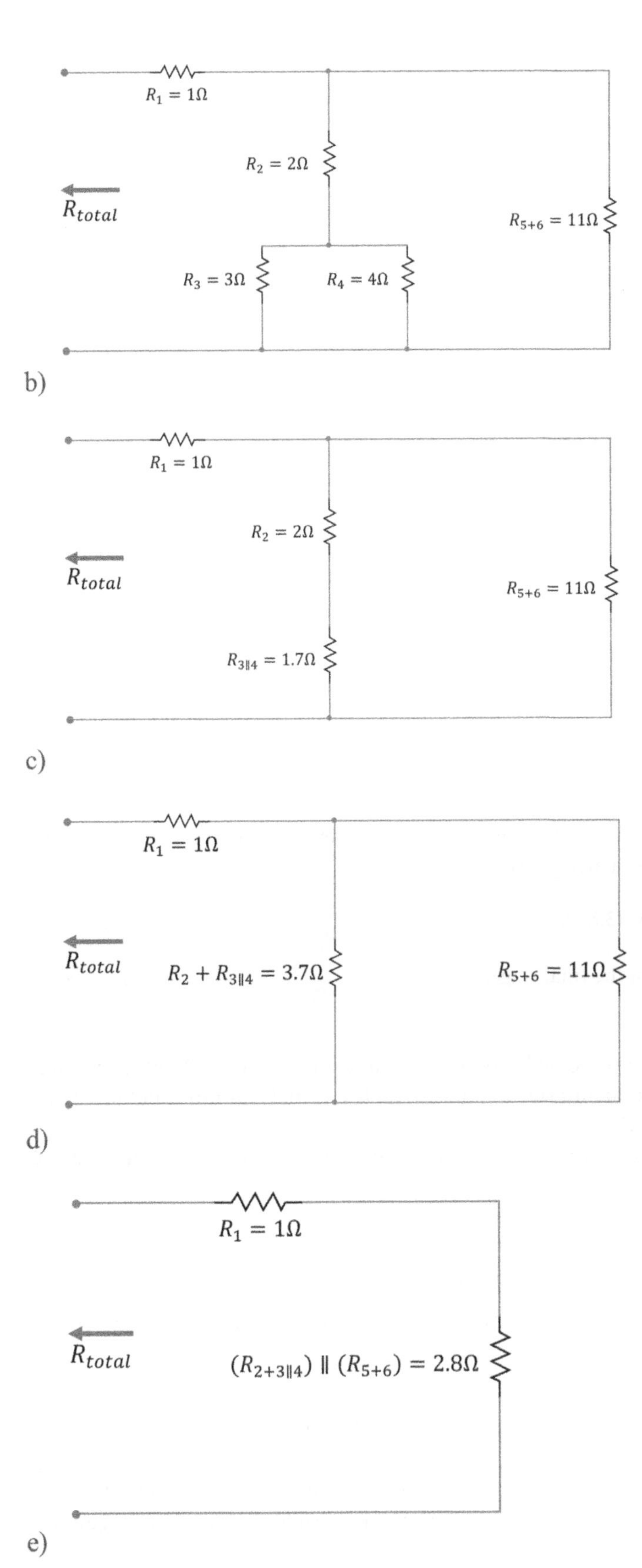

Figure 1.16 (Cont'd)

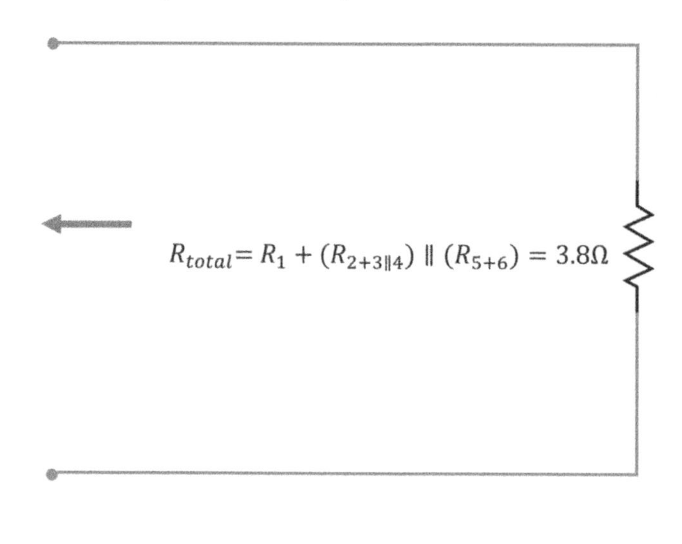

Figure 1.16 (Cont'd)

have to be combined to make one resistor $\left(R_{2+3\|4} \,\|\, R_{5+6}\right)$, i.e.,

$$\left(R_{2+3\|4} \,\|\, R_{5+6}\right) = \frac{R_{2+3\|4} \cdot R_{5+6}}{R_{2+3\|4} + R_{5+6}} = 2.8\ \Omega.$$

See Figure 1.16e. Finally, the total (equivalent) resistor, R_{total}, is found by summing up R_1 and $\left(R_{2+3\|4} \,\|\, R_{5+6}\right)$ connected in series, which gives

$$R_{total} = R_1 + \left(R_{2+3\|4} \,\|\, R_{5+6}\right) = 3.8\ \Omega.$$

Figure 1.16f shows the circuit in the final stage.
The problem is solved.

Discussion: Carefully review the presented process of finding an equivalent (total) resistance, R_{total}. Though all the steps seem straightforward, it takes practice to make us proficient in solving such problems correctly and quickly.

The next example enables us to learn how to find voltages across and currents through the individual circuit components.

Example 1.2 Finding voltages across and currents through the series-parallel resistive circuit.

Problem: Find voltages across and currents through all the individual resistors in the circuit shown in Figure 1.17a.

Solution: Figure 1.17a is an original circuit discussed in Example 1.1 and shown in Figure 1.16a. The only difference is that Figure 1.17a demonstrates 12-volt dc source (input) and all the nodes are marked by the lower-case letters. The analysis of this circuit starts with finding its total (equivalent) voltage and current, which can be done using the circuit obtained in Figure 1.16f. Its reproduction in Figure 1.17b shows that the total voltage, V_{ab}, is equal to V_{in}, and the total current, I_{in}, is given by

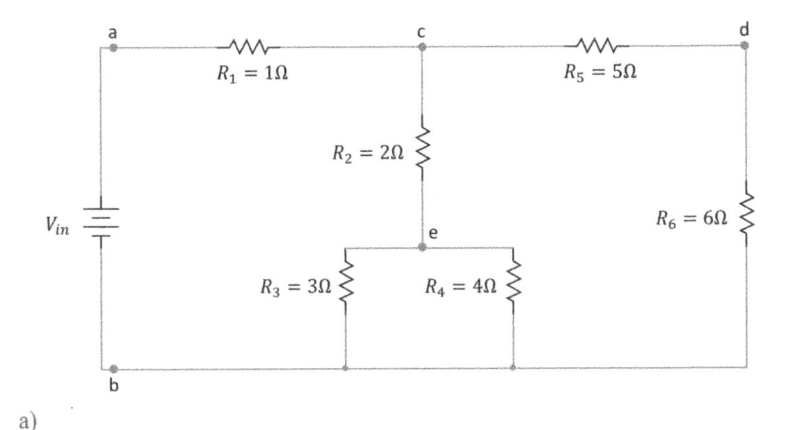

a)

a

$I_{in} = 3.2A$

$V_{ab} = 12.0V$

$V_{in} = 12\ V$

$R_{total} = R_1 + (R_{2+3\|4}) \parallel (R_{5+6}) = 3.8\Omega$

b

b)

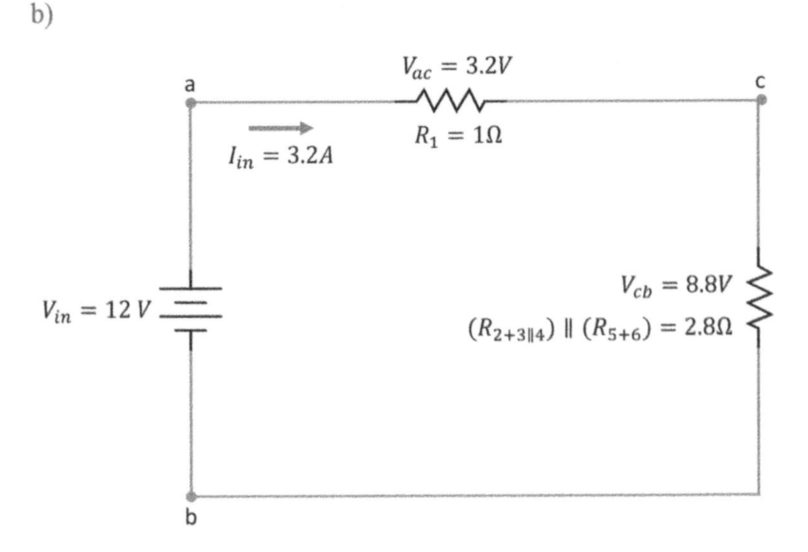

c)

Figure 1.17 Voltages across and currents through the series-parallel resistive circuit for Example 1.2: a) Original circuit with node designations; (b) equivalent circuit with total resistance R_{total}; c) circuit with series resistors R_1 and ($R_{2+3\|4} \| R_{5+6}$); d) circuit with parallel resistors $R_{2+3\|4}$ and R_{5+6}; e) circuit with series resistors R_2 and $R_{3\|4}$ and parallel to them resistor R_{5+6}; f) almost completely restored circuit with only one combined resistor R_{5+6} left; g) fully restored original circuit with all currents and voltages shown.

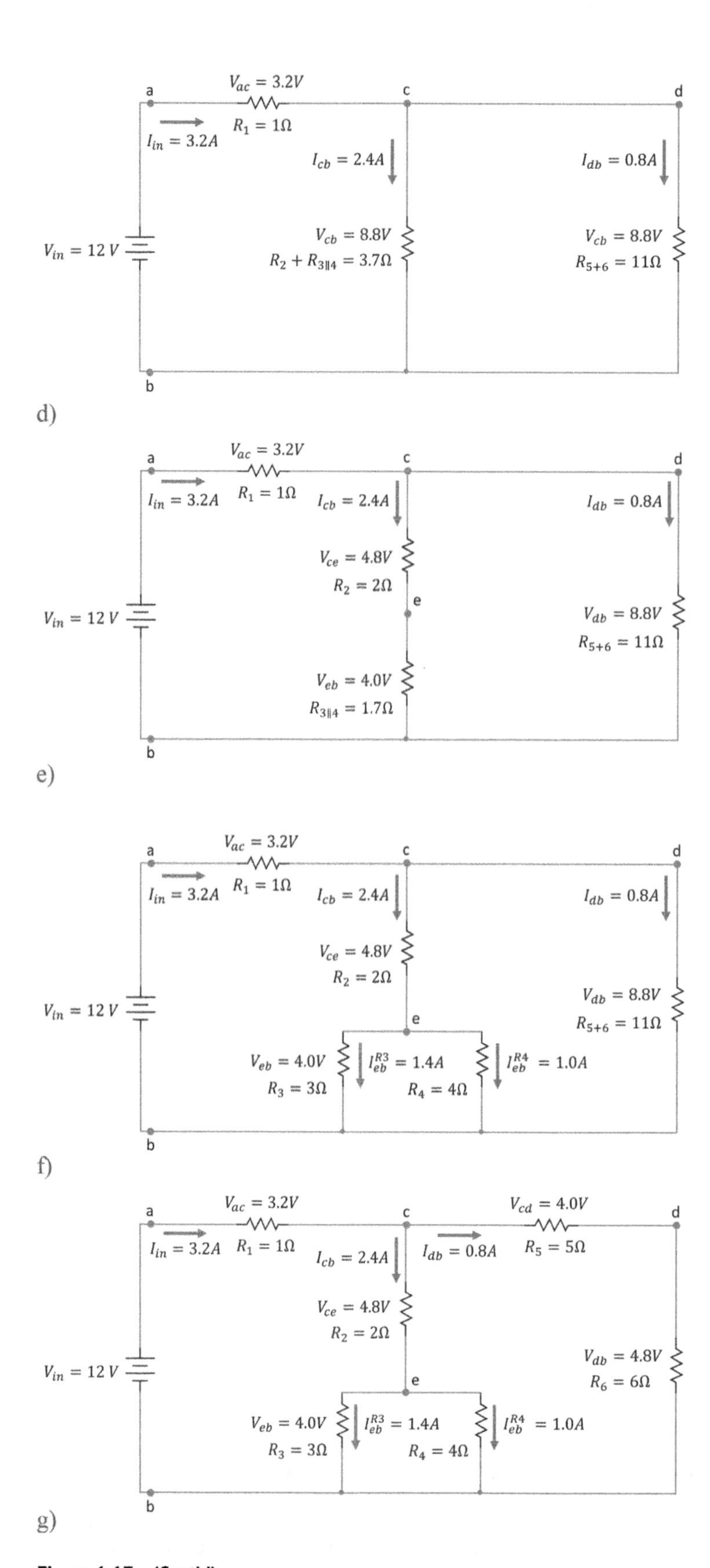

Figure 1.17 (Cont'd)

$$I_{in} = \frac{V_{in}}{R_{total}} = \frac{12}{3.8} = 3.2(A).$$

Then restoring step-by-step the given circuit in the order reversed to that shown in Figures 1.16b through 1.16f, enables us to find consequently all required currents and voltages. Thus, Figure 1.17c shows that replacing R_{total} with series resistors $R_1 = 1\Omega$ and $\left(R_{2+3\|4} \,\|\, R_{5+6}\right) = 2.8\Omega$ allows for finding V_{R1} as

$$V_{R1} = V_{ac} = I_{in} \cdot R_1 = 3.2(V)$$

and

$$V_{cb} = I_{in} \cdot \left(R_{2+3\|4}\right) \| \left(R_{5+6}\right) = 3.2(A) \cdot 2.8(\Omega) = 8.8(V).$$

Our calculations must meet the condition $V_{in} = V_{ac} + V_{cb}$, and they do so: $3.2(V) + 8.8(V) = 12(V)$.

By the next step back, we return to the circuit with a series resistor R_1 and two parallel branches with resistors $R_{2+3\|4}$ and R_{5+6}. This circuit shown in Figure 1.17d allows for calculation of two currents

$$I_{cb} = \frac{V_{cb}}{R_{2+3\|4}} = \frac{8.8(V)}{2.8(\Omega)} = 2.4(A)$$

and

$$I_{db} = \frac{V_{db}}{R_{2+3\|4}} = \frac{8.8(V)}{11(\Omega)} = 0.8(A).$$

Note that $2.4(A) + 0.8(A) = 3.2(A)$ as it must be.

Continuing the circuit restoration, we replace the *cb* branch by two branches *ce* and *eb*, as Figure 1.17e displays. This step enables us to calculate V_{ce} and V_{eb} voltages by applying the *voltage divider rule*:

$$V_{ce} = \frac{R_2}{R_2 + R_{3\|4}} V_{cb} = \frac{2}{3.7} 8.8 = 4.8\,(V)$$

and

$$V_{eb} = \frac{R_{3\|4}}{R_2 + R_{3\|4}} V_{cb} = \frac{1.7}{3.7} 8.8 = 4.0\,(V).$$

Alternatively, V_{ce} and V_{eb} can be calculated using *Ohm's law* as

$$V_{ce} = I_{cb} \cdot R_2 = 2.4(A) \cdot 2.0(\Omega) = 4.8\,(V)$$

and

$$V_{eb} = I_{cb} \cdot R_{3\|4} = 2.4(A) \cdot 1.7(\Omega) = 4.0\,(V).$$

Figures 1.17d and 1.17e show that $V_{ce} + V_{eb}$ must be equal to V_{cb}. Our calculated values meet this requirement as $4.8(V) + 4.0(V) = 8.8(V)$.

The next notch is to restore two parallel branches with $R_3 = 3(\Omega)$ and $R_4 = 4(\Omega)$. Using the *current divider rule*, we compute

$$I_{eb}^{R3} = \frac{1/R_3}{\left(1/R_3 + 1/R_4\right)} I_{cb} = \frac{0.33}{0.58} 2.4 = 1.4(A)$$

and

$$I_{eb}^{R4} = \frac{1/R_4}{\left(1/R_3 + 1/R_4\right)} I_{cb} = \frac{0.25}{0.58} 2.4 = 1.0(A).$$

By summing $I_{eb}^{R3} + I_{eb}^{R4} = I_{cb}$, or recalculating the numbers using Ohm's law, we verify that our result is correct.

Finally, splitting resistor R_{5+6} into two original resistors R_5 and R_6, computing their voltages as $V_{cd} = V_{R5} = 4.0(V)$ and $V_{db} = V_{R6} = 4.8(V)$, and verifying the result, we fully recover the given circuit with all the required currents and voltages obtained. This circuit is shown in Figure 1.17g.

The problem is solved.

Discussion: Here are several significant thoughts we can deduct from this and preceding examples:

- Examples 1.1 and 1.2 are simply two stages of one process—*circuit analysis*. Thus, if the problem is to determine all currents and voltages in a circuit, we must start with Stage 1 (i.e., Example 1.1) to find R_{total}, which opens the door for the total solution.
- Remember to verify each and every calculation, as was done in these examples.
- Don't forget about checking the units. This step is important in confirming the solution's correctness.
- Introductory circuit analysis is not a sophisticated procedure, but it takes a lot of practice to perform it quickly and correctly. So, practice by creating your examples and solving the problems attached to this chapter.

1.4.4 Series and Parallel Capacitors and Inductors

Electrical circuits include not only resistors but also capacitors and inductors. Let's consider the rules governing the series and parallel connections of these circuit components.

Series connections of capacitors are shown in Figure 1.18a, and parallel connections of capacitors are demonstrated in Figure 1.18b.

How can we derive the rules for finding equivalent (total) capacitors in series and parallel connections? Refer to $i - v$ relationships, as we did in the resistor cases. In a series connection, shown in Figure 1.18a, all capacitors have the same current but different voltages. Hence, applying (1.27) gives

$$v_{Ck}(t)[V] = \frac{1}{C_k} \int_{-\infty}^{t} i_{in}(t) dt, \text{ where } k = 1,2,3,\ldots,n. \tag{1.60}$$

Since

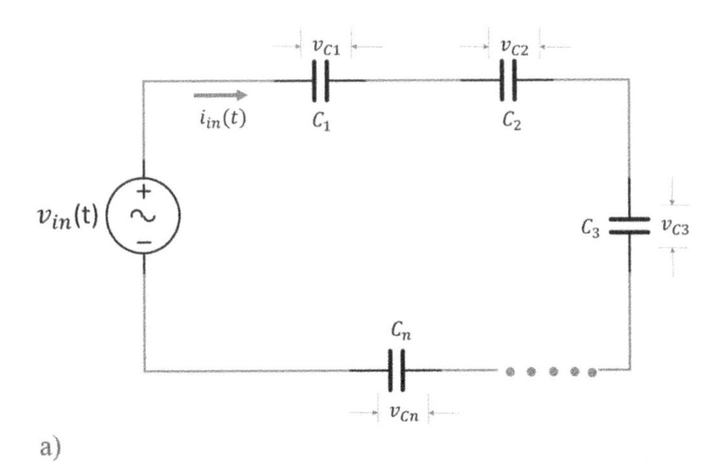

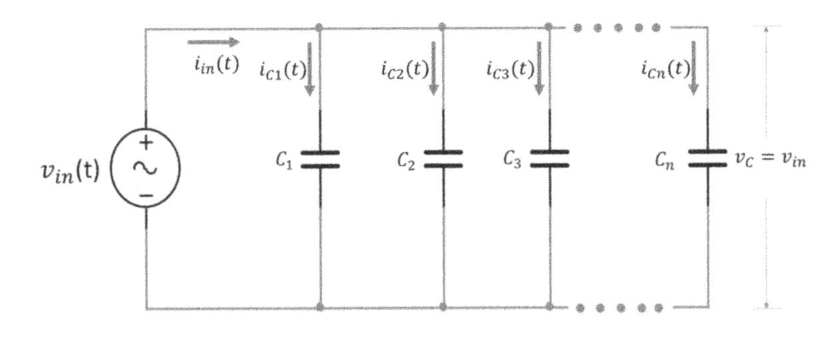

Figure 1.18 Series and parallel capacitors in electrical circuit: a) Series capacitors; b) parallel capacitors.

$$\begin{aligned} v_{in}(V) &= v_{C1} + v_{C2} + v_{C3} + \ldots + v_{Cn} \\ &= \left(\frac{1}{C_1} + \frac{1}{C_2} + \frac{1}{C_3} + \ldots + \frac{1}{C_n}\right)\int_{-\infty}^{t} i_{in}(t)dt = \frac{1}{C_{total}^{series}}\int_{-\infty}^{t} i_{in}(t)dt, \end{aligned} \tag{1.61}$$

where the total (equivalent) capacitance of the circuit is given by

$$\frac{1}{C_{total}^{series}} = \frac{1}{C_1} + \frac{1}{C_2} + \frac{1}{C_3} + \ldots + \frac{1}{C_n}. \tag{1.62}$$

Therefore, Equation 1.62 provides the rule of dealing with capacitors connected in series. A curious reader is encouraged to sketch the schematic of a circuit being equivalent to that shown in Figure 1.18a. In other words, you are invited to visualize (1.61). Refer to Equations from 1.53 through 1.57b to refresh your memory on how to work with the set of the fractions like $\frac{1}{C_1} + \frac{1}{C_2} + \frac{1}{C_3} + \ldots + \frac{1}{C_n}$.

Exercise Three capacitors whose $C_1 = 1(\mu F)$, $C_2 = 2(\mu F)$, *and* $C_3 = 3(\mu F)$ are connected in series. Compute the total capacitance. Answer: $C_{total}^{series} = 1.8(\mu F)$.

Consider capacitors connected in parallel that are shown in Figure 1.18b. It's clear that the input current is distributed among all the circuit branches, i.e.,

$$i_{in}(A) = i_{C1} + i_{C2} + i_{C3} + \ldots + i_{Cn}. \tag{1.63}$$

Refer to (1.24)

$$i_{Ck}(A) = C_k \frac{dv(t)}{dt}, \text{where } k = 1, 2, 3, \ldots, n. \tag{1.24R}$$

Plug (1.24) in (1.63) and find

$$\begin{aligned} i_{in}(A) &= C_1 \frac{dv(t)}{dt} + C_2 \frac{dv(t)}{dt} + C_3 \frac{dv(t)}{dt} + \ldots + C_n \frac{dv(t)}{dt} \\ &= (C_1 + C_2 + C_3 + \ldots + C_n) \frac{dv(t)}{dt}. \end{aligned} \tag{1.64}$$

Now, derive the formula for total (equivalent) capacitor for parallel connections as

$$C_{total}^{parallel}(F) = C_1 + C_2 + C_3 + \ldots + C_n. \tag{1.65}$$

Exercise Three capacitors whose $C_1 = 1(\mu F),\ C_2 = 2(\mu F),\ and\ C_3 = 3(\mu F)$ are connected in parallel. Compute the total capacitance. Answer: $C_{total}^{parallel} = 6(\mu F)$.

Now consider the circuits with *inductors* shown in Figures 1.19a and 1.19b. For a *series connection*, all inductors have the same current, and the sum of all voltages across individual inductors is equal to the input voltage, i.e.,

$$v_{in}(V) = v_{L1} + v_{L2} + v_{L3} + \ldots + v_{Ln}. \tag{1.66}$$

Refer to (1.37) to recall that

$$v_{Lk}(V) = L_k \frac{di(t)}{dt}, where\ k = 1, 2, 3, \ldots, n. \tag{1.37R}$$

Therefore,

$$\begin{aligned} v_{in}(V) &= L_1 \frac{di(t)}{dt} + L_2 \frac{di(t)}{dt} + L_3 \frac{di(t)}{dt} + \ldots + L_4 \frac{di(t)}{dt} = (L_1 + L_2 + L_3 + \ldots + L_n) \frac{di(t)}{dt} \\ &= L_{total}^{series} \frac{di(t)}{dt}. \end{aligned} \tag{1.67}$$

Thus, we obtain the equivalent (total) inductance for a series connection as

$$L_{total}^{series}(H) = (L_1 + L_2 + L_3 + \ldots + L_n). \tag{1.68}$$

For parallel connections, we employ the fact that the voltage across all individual inductors is the same, but the currents are different. Then, using (1.39), we get

$$\begin{aligned} & i_{in}(A) = i_{L1} + i_{L2} + i_{L3} + \ldots + i_{Ln} \\ & = \left(\frac{1}{L_1} + \frac{1}{L_2} + \frac{1}{L_3} + \ldots + \frac{1}{L_n} \right) \int_{-\infty}^{t} v_{in}(t)dt = \frac{1}{L_{total}^{parallel}} \int_{-\infty}^{t} v_{in}(t)dt. \end{aligned} \tag{1.69}$$

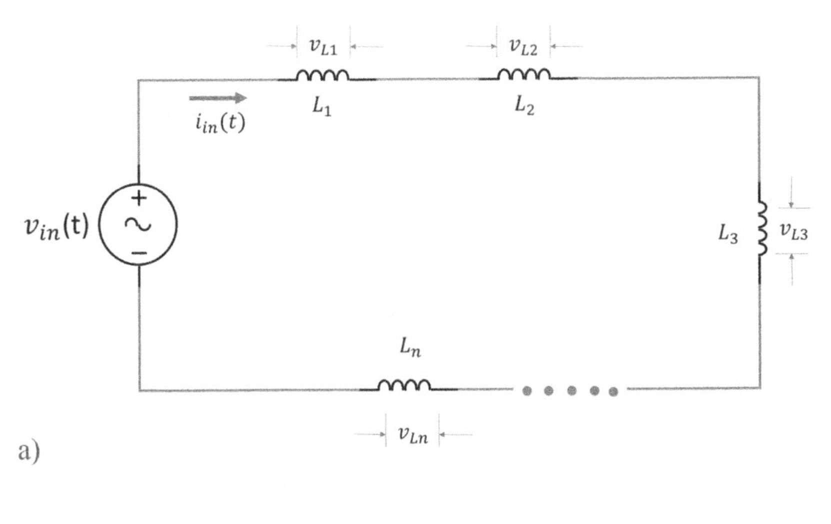

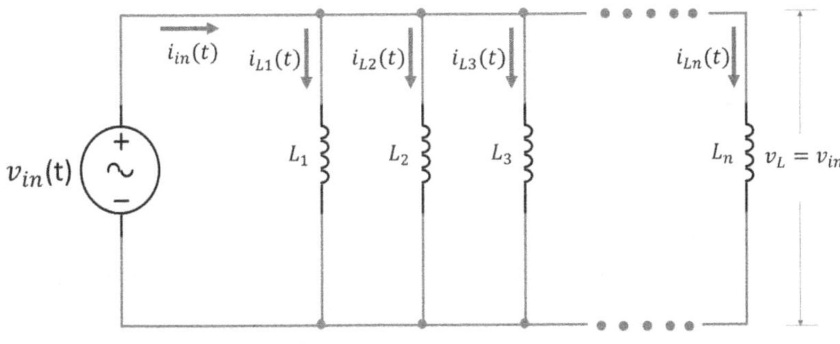

Figure 1.19 Electrical circuit with inductors: a) Series connection; b) parallel connection.

Therefore,

$$\frac{1}{L_{total}^{parallel}} = \frac{1}{L_1} + \frac{1}{L_2} + \frac{1}{L_3} + \ldots + \frac{1}{L_n}. \tag{1.70}$$

Since we've studied how to calculate total series and parallel quantities in the immediately preceding subsections, there is no need to repeat this material.

Exercise The given circuit contains three inductors whose inductances are $L_1 = 0.1H$, $L_2 = 0.2H$, and $L_3 = 0.3H$. What is the total circuit inductance if the inductors are connected in series? In parallel? Answers: $L_{total}^{series} = 0.6H$ *and* $L_{total}^{parallel} = 18.3H$.

In Examples 1.1 and 1.2 we found voltages across and currents through all the circuit elements for *resistive* circuits. Providing such an analysis for the circuits containing *reactive* (capacitors and inductors) components is not a simple task. On the one hand, for dc circuits, capacitors operate as open circuits, and inductors work as short circuits, which makes their analysis trivial. On the other hand, for ac circuits, we must involve the capacitor and inductor reactances, $X_C(\Omega)$ *and* $X_L(\Omega)$, that depend on the signal frequency. Most importantly, these reactances are complex quantities and require higher-level mathematics than simple algebra used for resistive circuits. Therefore, we have to postpone an ac analysis of reactive circuits until Chapter 3.

1.5 Sources (Inputs, Excitations, or Drives)

1.5.1 Sources

Each circuit contains various components in diverse networking, but every circuit has a *source* that enables it to perform its task. From a system standpoint, such a source is called an *input* because it is a signal that the circuit processes to produce an output. The source is also called an *excitation* because it excites the circuit's operation or a *drive* since it activates the circuit operation.

A *source is an element that makes the circuit operational by supplying power or providing an input signal*. Your smartphone's battery is the source of power, whereas the antenna of this gadget delivers an input signal. From a system-analysis standpoint, a source is an input whose relationship with the system's output is the purpose of this analysis.

For advanced circuit analysis, we need to recall that there is a *voltage source* and *current source* providing power and input signals to the circuit. The voltage input (source) generates an assigned voltage, $v_{in}(t)[V]$, across its terminals and supplies it to a load circuit. Family of Figures 1.20 demonstrates these definitions. The dc source is shown in Figure 1.20a, and an ac input is presented in Figure 1.20b. The voltage value supplied by the input doesn't depend on the load current and therefore doesn't depend on the load variations.

The first examples of *dc voltage sources* are the batteries. They range in voltage scale from microvolts to megavolts. Modern technology heavily depends on batteries. It's sufficed to refer to such batteries applications as remotely operational microsensors, electric automobiles, and power batteries accumulating electric energy from the farms of solar panels. Nevertheless, there are plenty of other voltage sources. Our laptops are fed, besides the battery, by an ac adaptor that rectifies the ac voltage from a power line to the dc voltage required by the computer.

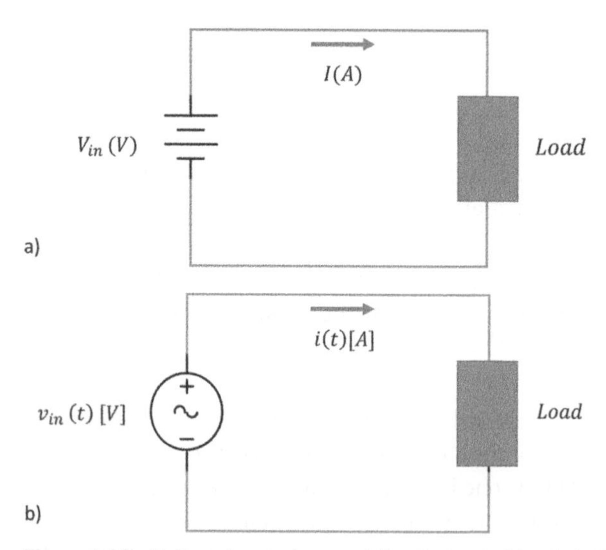

Figure 1.20 Voltage inputs (sources) for the loads (circuits): a) dc source; b) ac source.

The first example of an *ac voltage source* is a *power line* found everywhere, from our homes to offices to industrial settings. In addition, all signals delivering information are also examples of ac voltage inputs.

Performing your laboratory exercises, you definitely use a *signal (function) generator*, a standing alone device that generates various electrical signals (sinusoidal, square wave, etc.) This generator is an example of an *independent voltage source* because it internally generates its signal separately from a load circuit's operation. There are also *dependent (controlled) voltage sources* whose output is controlled by an external signal. This control signal can be either voltage or current by its nature. One of the most familiar examples of such sources is a *voltage-controlled oscillator, VCO*, whose output is controlled by an external voltage. Among its numerous applications, allegedly most popular is a phase-locked loop system, PLL, where the feedback from the load circuit adjusts the output of the VCO signal. (See, for example, Mynbaev and Scheiner 2020, pp.746–750.) The general presentation of dependent and independent voltage sources is given in Figure 1.21.

Question Is a charger of your smartphone dependent or independent voltage source? Explain. Answer: Independent because it generates an assigned voltage value independently on the operation of the smartphone and any other external devices.

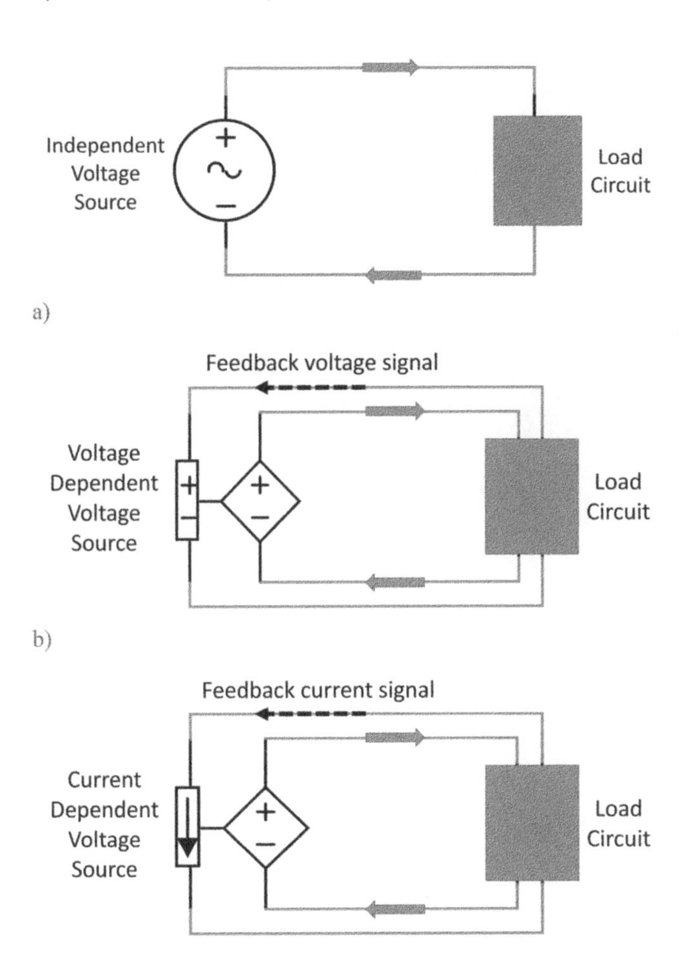

Figure 1.21 Voltage sources: a) Independent source; b) voltage-controlled voltage source; c) current-controlled voltage source. (Sources are shown in Multisim notations.)

Considering the *current sources*, it's worth mentioning that they have much in common with the voltage sources. Figures 1.22a through 1.22c demonstrate the definitions of current sources.

The main features of a current source are as follows:

- The current source generates an assigned current, $i_{in}(t)[A]$, and supplies it to a load circuit.
- The generated current value doesn't depend on the load voltage and the load variations.

Examples of current sources include the photoelectric cells and collector currents of transistors.

It must be pointed out that the above discussion refers to the *ideal sources*.

1.5.2 Series and Parallel Sources, Ideal and Real Sources, and Source Transformation

Three more points are worth consideration.

1) First, the voltage *sources* can be connected in series, and the current sources can be connected in parallel. These configurations enable us to increase the output of the sources. Figures 1.23a

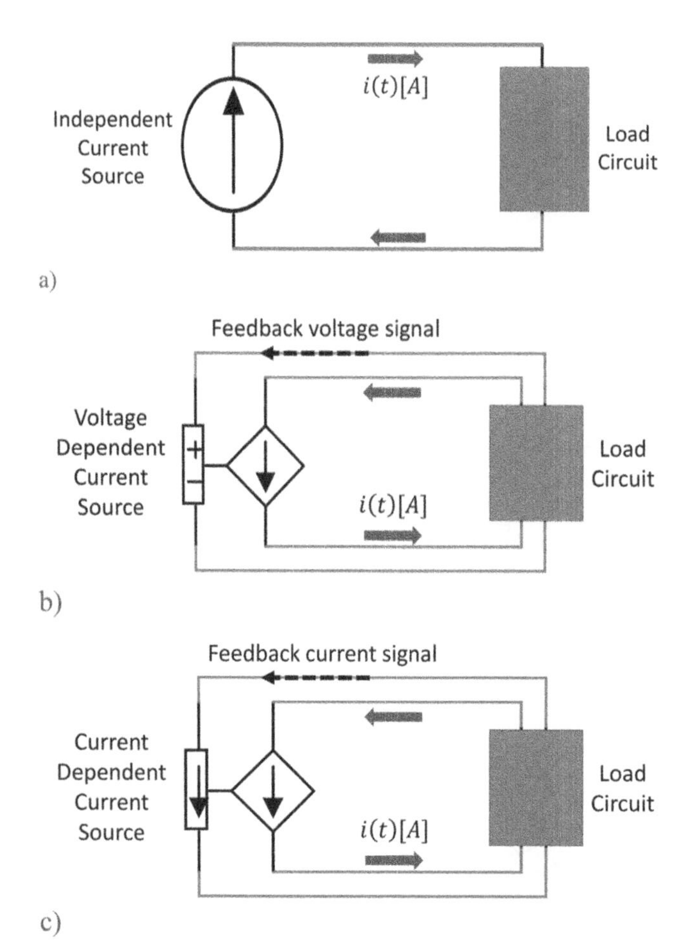

Figure 1.22 Current sources: a) Independent source; b) voltage-controlled current source; c) current-controlled current source. (Sources are shown in Multisim notations.)

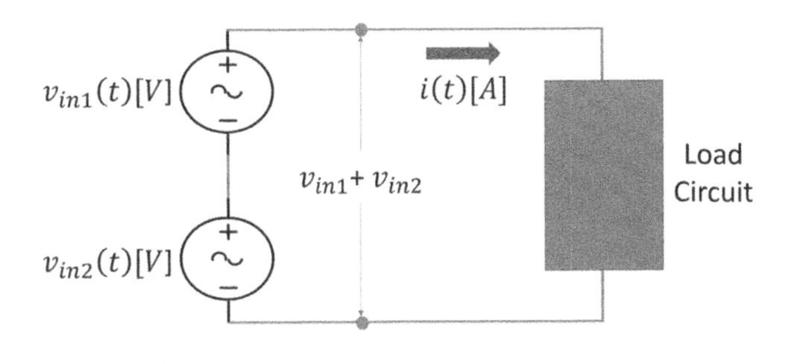

a)

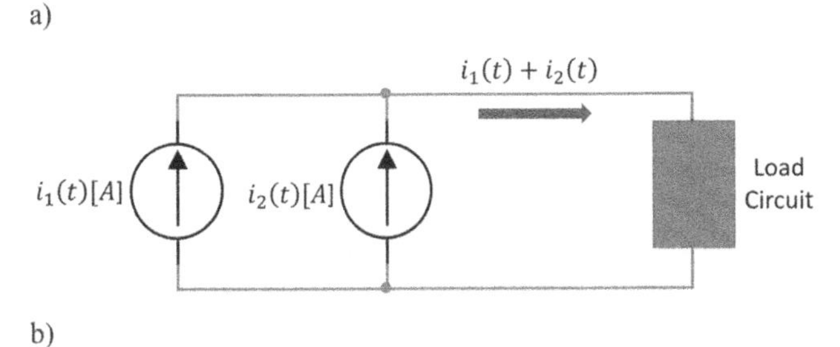

b)

Figure 1.23 Series and parallel connections of the sources: a) The voltage sources connected in series; b) the current sources connected in parallel.

and 1.23b illustrate the concept. Two voltage sources connected in series supply the sum of their voltages to the load circuit, as shown in Figure 1.23a. Two current sources connected in parallel feed the load circuit with the sum of their currents, as illustrated in Figure 1.23b. However, *these summation rules are valid only for sources*; series and parallel connections of the passive components (resistors, capacitors, and inductors) obey different regulations, as discussed in Section 1.3. Another particularity of the source connections is that the voltage sources must be never connected in parallel, and the current sources cannot be wired in series.

Question Can the controlled (dependent) voltage sources be connected in series? Controlled current sources in parallel?

2. The second point is about the difference between an ideal and a real (practical) source. The *ideal sources* are shown in Figures 1.20 through 1.23 and described in the preceding part of this section. In reality, however, every source includes its internal resistance, $R_{int}(\Omega)$. Figure 1.24a shows that an internal resistor, R_{int1}, is connected in series in a *practical (real) voltage source.* Figure 1.24b demonstrates that resistor R_{int2} is wired in parallel in a *practical (real) current source.* The load circuit's viewpoint at the sources is shown in Figure 1.24c.

How important is taking into account the internal resistances of the sources? In other words, whether the difference between ideal and practical sources affects the results of circuit analysis? Let's take ideal and practical voltage sources given in Figures 1.21a and 1.24a, respectively, replace the symbolic load with a loading resistor R_l for simplicity, display them in Figure 1.25, and analyze their operations.

The ideal voltage source delivers the load voltage

$$v_l^{ideal}(V) = v_{in}, \tag{1.71a}$$

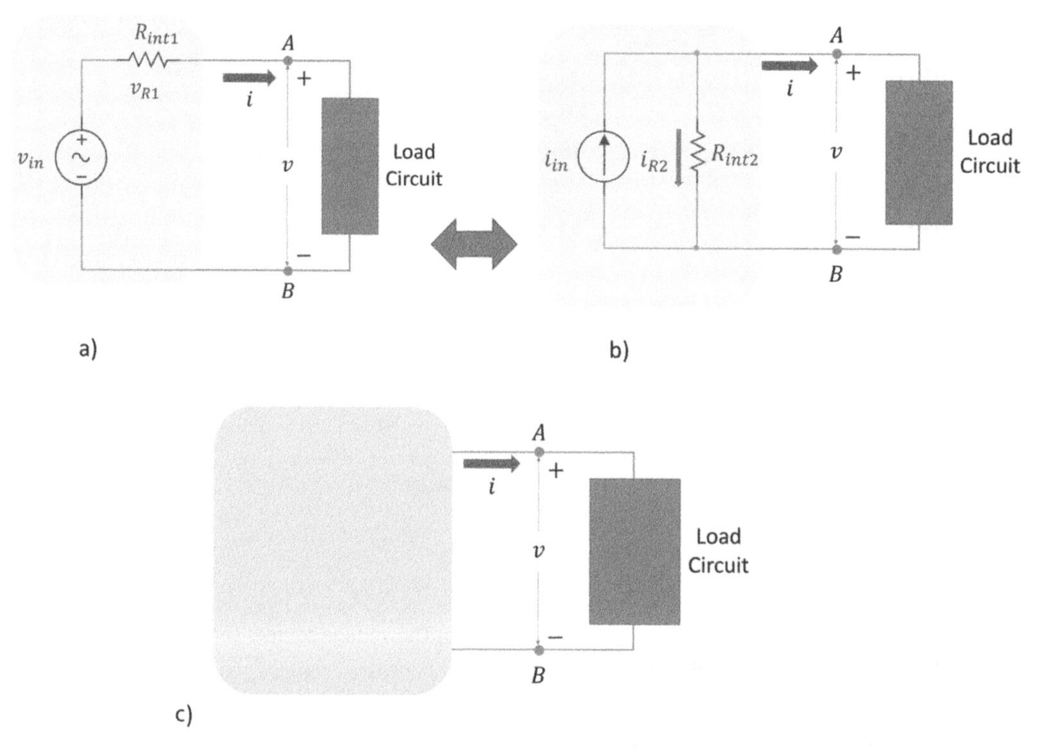

Figure 1.24 Real (practical) voltage and current sources and their transformations: a) Real voltage source (shaded); b) real current source (shaded); c) the load circuit viewpoint at a source. (The source is shaded.)

as shown in Figure 1.25a. The current provided by the ideal source is

$$i_l^{ideal}(A) = \frac{v_{in}}{R_L}. \tag{1.71b}$$

The voltage v_{in} generated by a real (practical) voltage source is distributed between R_{int} and R_l, that is, $v_{in} = v_{Rint} + v_l^{real}$. See Figure 1.25b. Thus,

$$v_l^{real}(V) = v_{in} - v_{Rint}. \tag{1.72}$$

In other words, the load receives only part of the voltage generated by the source. To increase the received load voltage, we need to decrease v_{Rint} and ideally turn it zero. But $v_{Rint} = i_l \cdot R_{int}$; therefore, to make $v_{Rint} = 0$, we need to eliminate R_{int}, that is, build an ideal source. Note that $i_l(A) = \frac{v_{in}}{R_{int} + R_l}$. So, the conclusion is

We need to minimize the voltage source's internal resistance to maximize the voltage delivered to the load.

Exercise The internal resistance of a 9-volt source is $R_{int} = 12(\Omega)$, whereas the load resistance is given as $R_l = 3(\text{k}\Omega)$. What voltage does a load receive really? Answer: $8.96(V)$.

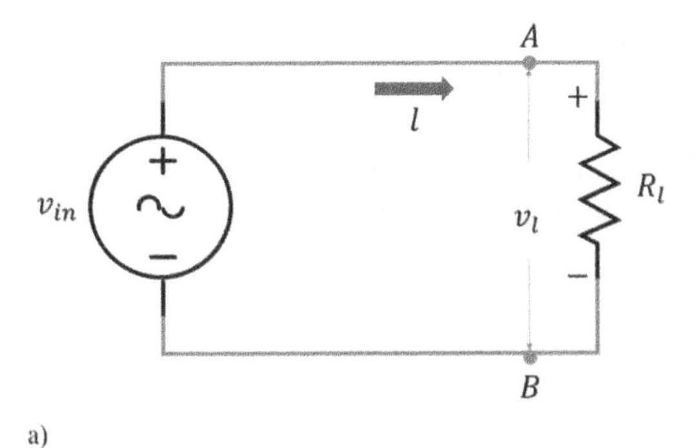

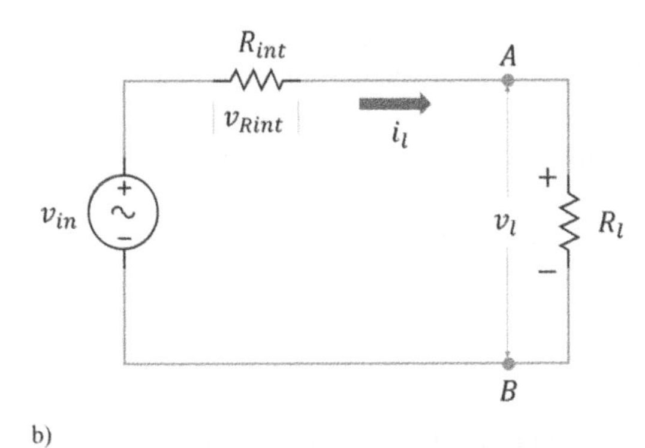

Figure 1.25 Comparison voltage sources: a) Ideal voltage source; b) real (practical) voltage source.

The current flowing through R_l due to the real voltage source is given by

$$i_l^{real}(A) = \frac{v_{in}}{R_{int} + R_l}. \tag{1.73}$$

Equation 1.73 also highlights the need to minimize R_{int} of a voltage source. (Can you explain why?)

Figures 1.26a and 1.26b show an ideal and real (practical) current source, respectively. Can you guess by mere observation of Figure 1.26b what should we do to maximize the performance of a practical current source? Yes, you are right: We need to increase R_{int} as much as possible. Here is why: As Figure 1.26 shows, the input current splits between R_{int} and R_l; i.e.,

$$i_{in}(A) = i_{Rint} + i_l. \tag{1.74}$$

Ideally, it should be

$$i_{in}(A) = i_l, \tag{1.75}$$

as Figure 1.26a demonstrates. Therefore, we need to eliminate i_{Rint} to maximize delivery from a current source, which can be achieved by letting $R_{int} \to \infty$. We leave it for you to calculate the voltages in the circuit shown in Figure 1.26b. So, the conclusion is

> *To maximize the input current delivered to the load, we need to tend the current source's internal resistance to infinity.*

Exercise What current will be delivered to the load in Figure 1.26b if $i_{in} = 9(mA)$ and $R_{int} = 12(\Omega)$ and $R_l = 3(k\Omega)$? Answer: $i_l = 0.036(mA)$. (Hint: Apply (1.59b)). Discussion: If a current-driven circuit shown in Figure 1.26b keeps the same values ($R_{int} = 12(\Omega)$ and $R_l = 3(k\Omega)$) as a voltage-driven circuit presented in Figure 1.25b, then the load in Figure 1.26b receives only smallest portion of the input current because the source's internal resistance draws the most of i_{in}. This is why we need to have $R_{int}(\Omega) \rightarrow \infty$.)

Bear these two critical points in mind when analyzing the practical circuits.

3. The third and final point of this discussion is about source transformation. The function of each source is to supply a load circuit with the required voltage, v, and current, i, as Figures 1.24a and 1.24b show. From this viewpoint, the load circuit doesn't "care" which source—voltage or current—supplies the needed energy; it only "cares" about values of voltage and current it receives. This load circuit viewpoint is presented in Figure 1.24c. But if we need to substitute voltage source for the current one, or vice versa, we must provide their equivalency. The procedure and conditions of this sameness is called *source transformation*.

Under what conditions do a voltage source and a current source deliver equal voltage v and current i to the load circuit? Consider the source circuit (shaded) in Figure 1.24a: The voltage source produces v_{in} that is distributed between v_R and v, that is,

$$v_{in} = v_{Rint1} + v. \tag{1.75}$$

On the other hand, Figure 1.24b shows that current i_{in} splits into i_{R2} and i, which means

$$i_{in} = i_{Rint2} + i. \tag{1.76}$$

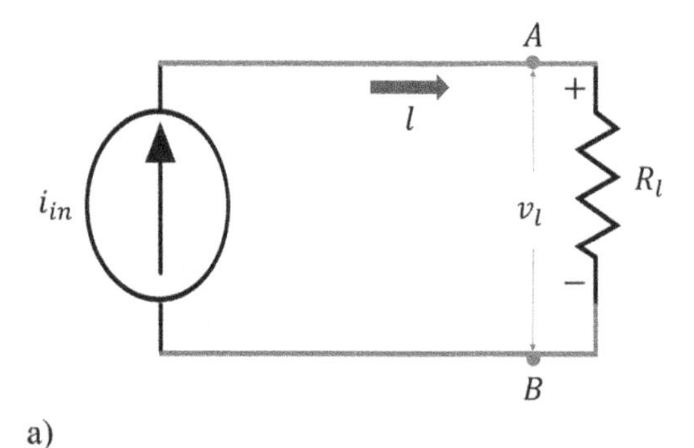

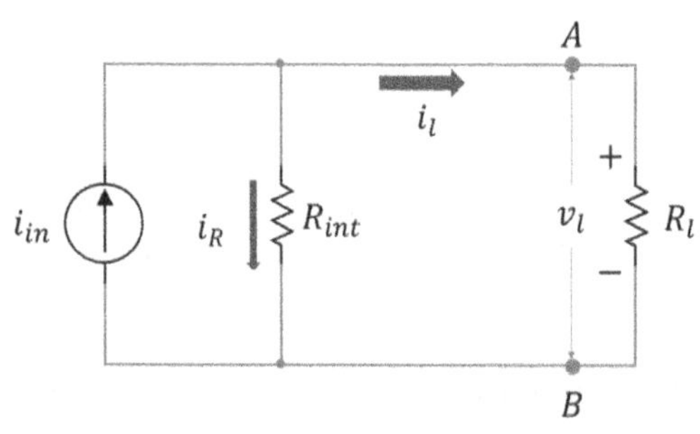

Figure 1.26 (a) Ideal and (b) real (practical) current sources.

To discover the relationship between these two equations, we must determine where current i is hidden in (1.75) and voltage v is reposed in (1.76). The first answer appears after rewriting (1.75) as $v = v_{in} - v_{R1}$, examining Figures 1.24a, and applying Ohm's law to v_{R1} as $v_{R1} = i \cdot R_{in1}$. Thus, we obtain

$$v = v_{in} - i \cdot R_{int1}. \tag{1.77}$$

By observation of Figure 1.24b and application of Ohm's law to i_{R2}, we can rewrite (1.76) as

$$i_{in} = i_{Rint2} + i = \frac{v}{R_{int2}} + i.$$

Now, we can derive the voltage v equation for a current source as

$$v = i_{R2} \cdot R_{int2} - i \cdot R_{int2}. \tag{1.78}$$

Equations 1.77 and 1.78 display the voltage v supplied to the load circuit, which is expressed through two different right-hand sides (RHSs), an impossible entity. Thus, these RHSs must be equal, which can be achieved only under two conditions:

$$(1) \quad R_{int1} = R_{int2} = R$$

and

$$(2) \quad v_{in} = i_{R2} \cdot R_{int2} = i_{R2} \cdot R. \tag{1.79}$$

Therefore, a voltage source can be replaced by a current source if and only if the conditions (1.79) are met.

We need to reiterate that here the voltage and current sources are equivalent only from the load circuit perspective; in other words, they are equivalent only because they supply the same v and i to the load circuit. We will see throughout the book that transformed sources are helpful models for advanced circuit analysis, but we will remember that the sources themselves are not equivalent. Also, remember to follow the polarities shown in Figure 1.24a and 1.24b.

To put this discussion on a practical footing, let's consider an example.

Example 1.3 Source transformation.

Problem: The source circuit containing the voltage and the current sources is shown in Figure 1.27a. Transform this circuit to the equivalent source so that the voltage provided at the $A - B$ open load terminals will be $v_L = 6(V)$.

Solution: Since the voltage source has a series resistor, it's easy to transform it into a current source. Then, it will be effortless to combine two current sources. Then we need to determine the value of the equivalent resistor. To implement this plan, we transform the 12-V source and its 6-Ω series resistor (shaded) into a current source and a 6-Ω parallel resistor by following the explanations illustrated in Figures 1.24a, 1.24b, and 1.24c. This transformed current source is shown in Figure 1.27b. The value of the new $I2$ current is calculated according to Ohm's law as:

$$I_2 = \frac{V1}{R1} = \frac{12}{6} = 2(A).$$

Combining $R1$ and $R2$ parallel resistors gives $R3$. The schematic of the new current source is shown in Figure 1.27c. The voltage at the $A - B$ terminals is the voltage across the resistor $R3$

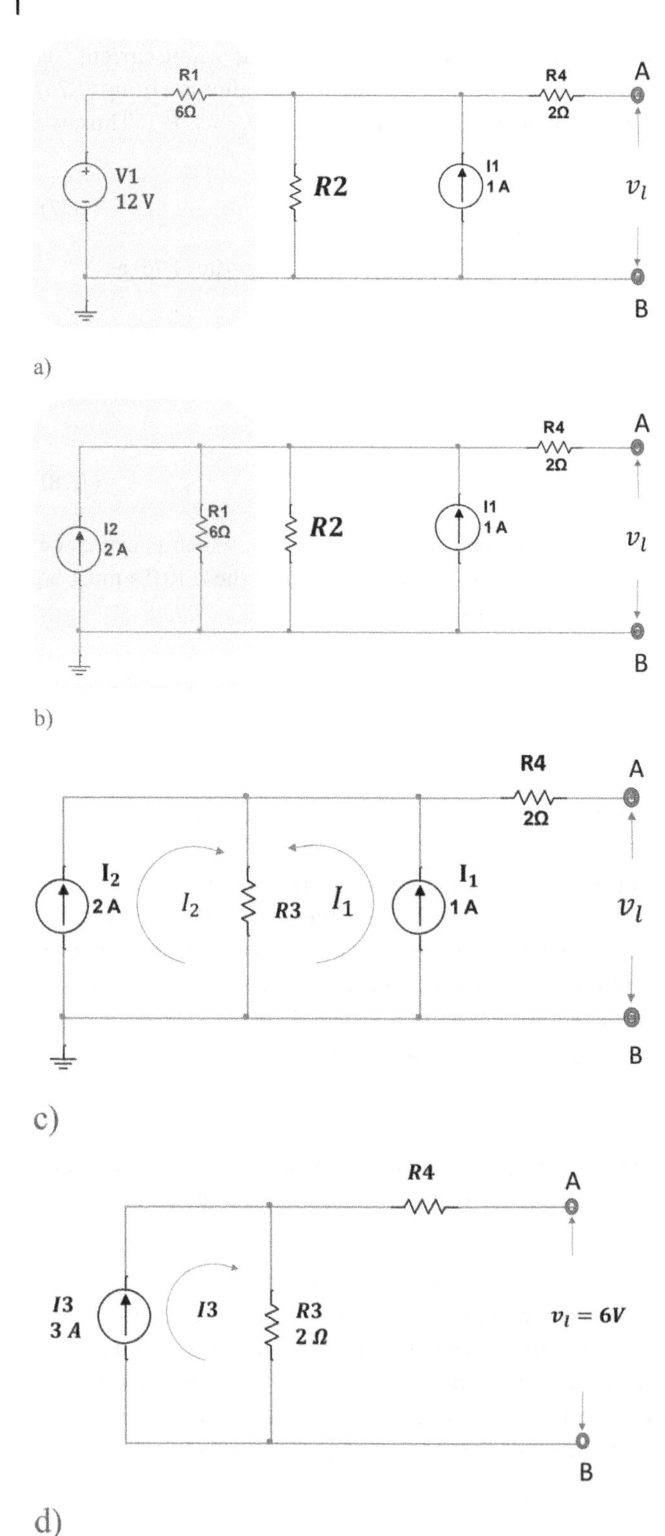

Figure 1.27 Source transformation for Example 1.3: a) Actual voltage source (shaded); b) transformed current source (shaded); c) new circuit with two current sources; d) equivalent current source circuit.

because the circuit is open at these terminals, and no current flows through $R4$. Figure 1.27c shows that the current flowing through $R3$ is the sum of $I1$ *and* $I2$, hence,

$$v_{R3} = v_L = (I1 + I2) \cdot R3 = I3 \cdot R3 = 3 \cdot R3 \textit{ must be equal to } 6V.$$

Thus, $R3$ must be equal to 2Ω, which requires to set $R2 = 3\Omega$ because $R1 \| R2 = 6 \| R2 = 2 \Rightarrow R2 = 3$. The final equivalent circuit is given in Figure 1.27d.

The problem is solved.

1.6 Primary Circuit Laws (Ohm's and Kirchhoff's Laws)

1.6.1 Circuit Analysis

What is circuit analysis? It is a theory whose subject is an electrical circuit. As with any engineering theory, circuit analysis is a set of mathematical methods. These methods aim to find unknown electrical quantities of a given electrical circuit. The quantities in question are voltages across and currents through the circuit components based on the intended circuit's topology and values of the components. Why do we need it, or what is the goal of the circuit analysis? To predict the results of the circuit operation. Without the circuit analysis, we would have to rely on the trial-and-error method, endlessly varying the schematics, types of components, and component values in the attempts to design and build the circuit that operates as required. The circuit analysis enables us to theoretically predict how a given circuit will work and forecast the circuit performance when the circuit's schematic, components, or their values change.

What quantities are the subject of the circuit analysis? First, currents, voltages, and power—basic parameters of the circuit operation, and second, the values of the circuit passive (resistors, capacitors, and inductors) and active (operational amplifier) components.

What is the basis of the circuit analysis? It is a theory founded on the primary circuit analysis laws that, in turn, are based on the fundamental laws of physics.

As with any theory, the circuit analysis is accompanied by the practice whose task is to confirm or disprove the theory predictions. The practice does so by making measurements of the circuit parameters and comparing them with the theoretical predictions.

The sample of the circuit analysis is given in Examples 1.1 and 1.2 of Section 1.4. In Example 1.1, we applied the rules for combining series and parallel resistors to obtain the circuit's total (equivalent) resistance. In Example 1.2, using the total resistance, we find all voltages across and currents through each circuit resistor employing Ohm's law, voltage divider, and current divider rules. It's worth revisiting these examples and observing how all aspects of the circuit analysis listed above are implemented over there. Primary and advanced circuit analysis will be the main subject of the following chapters; hence, we will learn all the aspects of the circuit analysis.

Note: The work with circuits requires two approaches: circuit analysis and circuit design. The circuit analysis works with a given circuit as described in the first part of this section, and the circuit design aims to project a new circuit that will operate as required. Element of design is demonstrated in Example 1.3 of Section 1.5, where the circuit schematic is given, but resistor value must be found to make a source deliver six volts to the load. As Example 1.3 demonstrates, circuit analysis and circuit design are closely related and work in unison to achieve the same goal—to create a circuit performing as required.

1.6.2 Ohm's Law

Ohm's law is introduced in (1.12) of Section 1.3, where the short bio of Georg Ohm is also presented. It has a straightforward meaning: The voltage across the resistor is equal to resistance times current flowing through the resistor; i.e.,

$$V(V) = R(\Omega) \cdot I(A). \tag{1.80}$$

Refer to Section 1.2, where we discuss the voltage nature: Voltage creates a force that moves charge carriers through a circuit, thus creating an electrical current. Therefore, the greater the voltage, the more the current. In other words, a current is proportional to voltage, which is precisely what Ohm's law states. Refer to Figures 1.4 and 1.5b to visualize this consideration.

When in 1827 Ohm established this law by making quantitative measurements, it was not obvious to his peers; even worse, it met a hostile reaction of that scientific community. Fortunately, with further development of electricity, the law was firmly proven and accepted as one of the fundamental ground rules of the field.

Since this law is discovered by experiment, it is called *empirical law*. *Ohm's law* presented in (1.80) implies that $R(\Omega)$ is an ideal, linear resistance. The circuits containing only such resistances are often called *ohmic circuits*. An ideal $R(\Omega)$ does not depend on any external factors such as temperature, signal frequency, or voltage level, as discussed in Section 1.3. Therefore, Ohm's law can be applied to ac circuits too, and it can be written as

$$v(t)[V] = R(\Omega) \cdot i(t)[A]. \tag{1.81}$$

The concept of proportionality between voltage and current, expressed in Ohm's law, is also applied to reactive circuits, containing capacitors and inductors. In this case, the *resistance* must be replaced by the *reactive impedance (reactance)* $X_C(f)[\Omega]$ or $X_L(f)[\Omega]$. As discussed in Section 1.3, capacitors and inductors require differential operations and complex-numbers manipulations to describe their behaviors. This approach is postponed until Chapter 3. A curious reader might be interested to know that the concept of Ohm's law is applied not only to the circuit's parameters but even to the quantities of the electromagnetic field.

Finally, it must be emphasized that Ohm's law is one of the fundamental rules governing a circuit analysis; thus, we cannot overestimate its importance. Examples 1.1, 1.2, and 1.3 fully attest to this point.

1.6.3 Kirchhoff's Voltage Law

Kirchhoff[13] discovered voltage and circuit laws in his student-day study; he continued to elaborate on this concept in his PhD work.

Kirchhoff's voltage law (KVL) states that the algebraic sum of all the voltages along the closed loop of a circuit is equal to zero; i.e.,

---

13 **Gustav Robert Kirchhoff** (1824–1887) was a German physicist known today to every electrical engineer thanks to his voltage and current laws. While a student, he formulated these laws and then made them the subject of his Ph.D. thesis. Besides, Kirchhoff contributed to modern science much more than these electrical laws. He (along with Bunsen) innovated the spectrum analysis and his study of the black body radiation led Max Planck to revolutionize physics by establishing quantum mechanics. Kirchhoff's career included such prestigious positions as professors at Heidelberg and Berlin universities, the best Germany educational and research institutions.

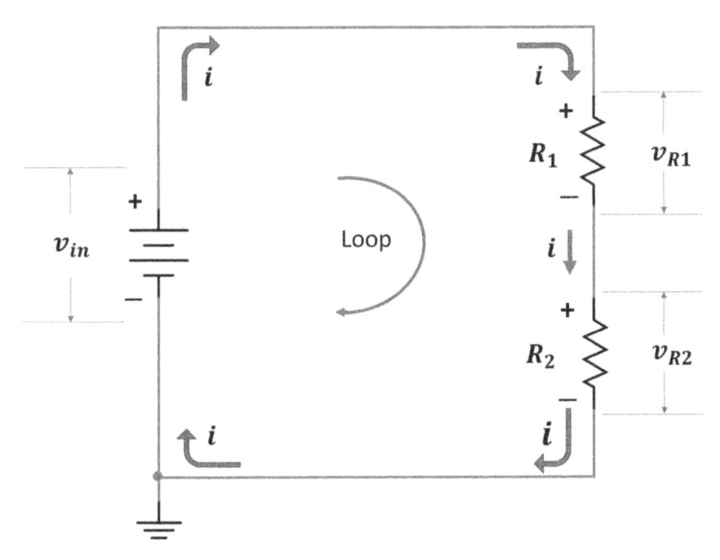

Figure 1.28 Sign convention for voltages.

$$\sum_1^n v_k = 0, \text{ where } k = 1,2,3,\ldots,n. \tag{1.82}$$

Consider, for example, Figure 1.28: Here, the sum is

$$-v_{in} + v_{R1} + v_{R2} = 0. \tag{1.83a}$$

Pay attention to the *voltage sign convention*: (1) A looping direction follows the conventional current flow, and (2) if the current enters a component's positive terminal, the voltage across the component is positive; otherwise, it is negative. This is why v_{in} in (1.83a) is negative, but v_{R1} and v_{R2} are positive. (Figure 1.28 is a modified version of Figure 1.8.)

From the system standpoint, *KVL* should be presented as an *input–output relationship*, which for Figure 1.28 becomes

$$v_{in} = v_{R1} + v_{R2}. \tag{1.83b}$$

In this format, the law states that the input voltage is distributed among all the voltages of the series components. Indeed, voltage entering a closed circuit from a single source cannot disappear or be created within this circuit. Thus, (1.83b) emphasizes that the physics underlining KVL is the *law of conservation of energy*. (Refer to the discussion of the voltage nature in Section 1.2.)

KVL is a powerful tool in circuit analysis. Here is an example of how to employ this tool in practice.

Example 1.4 KVL application to the analysis of a series-parallel circuit.

Problem: Employing only KVL, find voltages V_{ce}, V_{eb}, and V_{db} across the appropriate components of the circuit shown in Figure 1.29.
Solution: Applying KVL to Loop 1 gives

$$-V_{in} + V_{ac} + V_{ce} + V_{be} = 0.$$

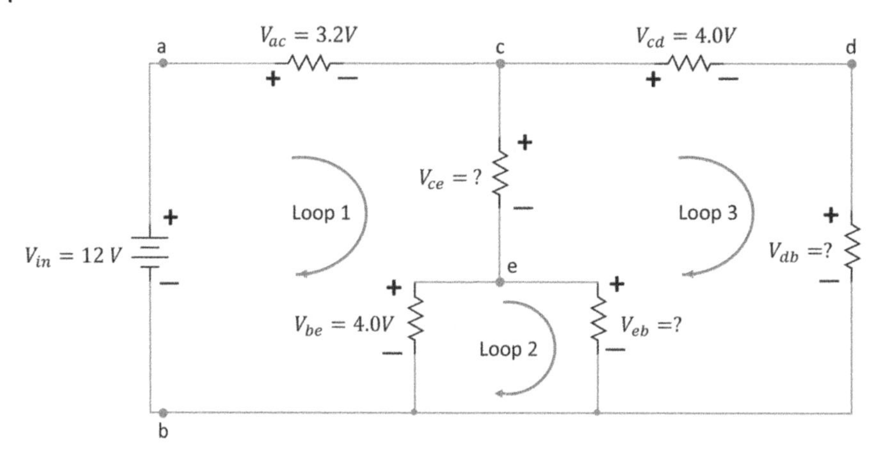

Figure 1.29 Circuit schematic for KVL application in Example 1.4.

Straightforward computations result in

$$V_{ce} = 4.8V$$

Loop 2 produces

$$-V_{be} + V_{eb} = 0 \Rightarrow V_{be} = V_{eb} = 4.0V.$$

This is the well-expected result because the voltage between e and b terminals is the same for two resistors in parallel. In fact, this is a proof of the voltage sameness rule for parallel connections.

Finally, application of KVL to Loop 3 yields

$$-V_{ce} + V_{cd} + V_{db} - V_{eb} = 0 \Rightarrow V_{db} = V_{ce} - V_{cd} + V_{eb} = 4.8V.$$

The problem is solved.

Discussion:

- The first point to take away from this example is how to form the voltage equation of a circuit loop depending on the voltage polarities of a component. The second point deals with the voltage across the component shared by neighboring loops, in this example, voltage V_{ce}: It must be included in the equations composed for both loops. The final and vital point is about polarities: They determine the signs of all voltages participating in a loop equation.
- This problem can be set in an alternative way: Consider Figure 1.17Ra where the same circuit is presented with the values of all resistors. Can we determine all voltages using only KVL and voltage divider rule? Yes, we can. The set of slightly modified Figures 1.17Ra through 1.17Rg along with the comments delivers the solution.

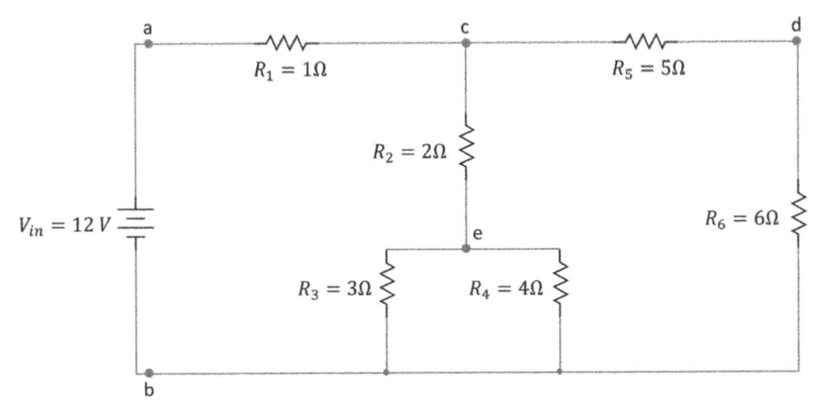

a) This is the given circuit.

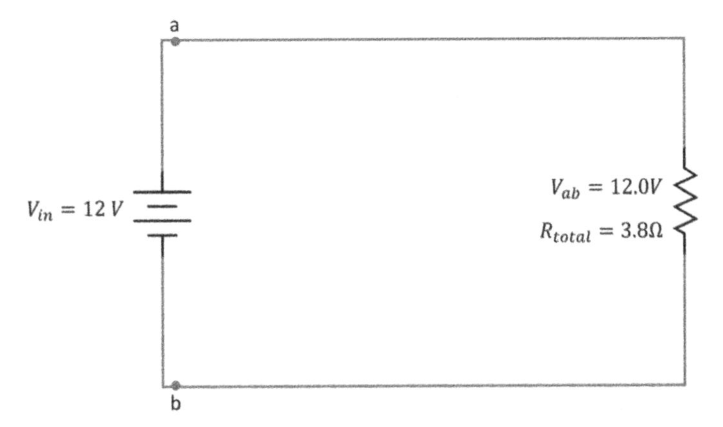

b) This is the equivalent circuit whose total (equivalent) resistance is determined in Example 1.2. KVL gives $-V_{in} + V_{ab} = 0$; thus the voltage across R_{total} is found.

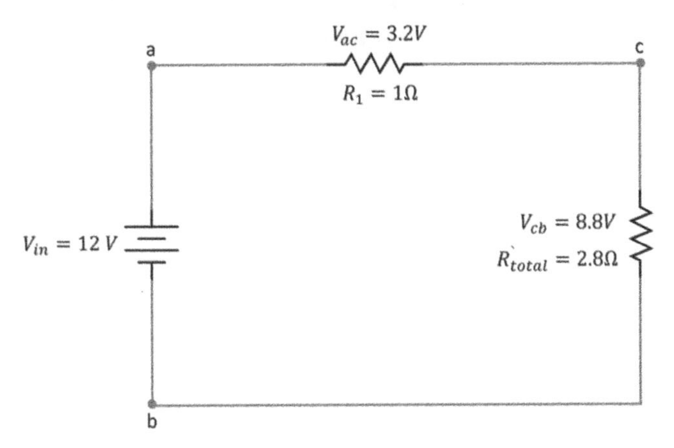

c) Recover $R_1 = 1\ \Omega$ and find the voltage across it using voltage divider rule: $V_{ac} = V_{R1} = \frac{R_1}{R_{total}} V_{in} = 3.2\ V$.

Figure 1.17R Applying KVL and the voltage divider rule to the analysis of series-parallel resistive circuit: a) Original circuit; b) the total resistance of the original circuit; c) finding the voltage across R_1; d) recovering two parallel branches of the original circuit; e) finding V_{ce} and V_{eb}; f) showing $V_{eb} = V_{R3} = V_{R4}$; g) obtaining the two remaining voltages V_{cd} and V_{db} to recover the original circuit.

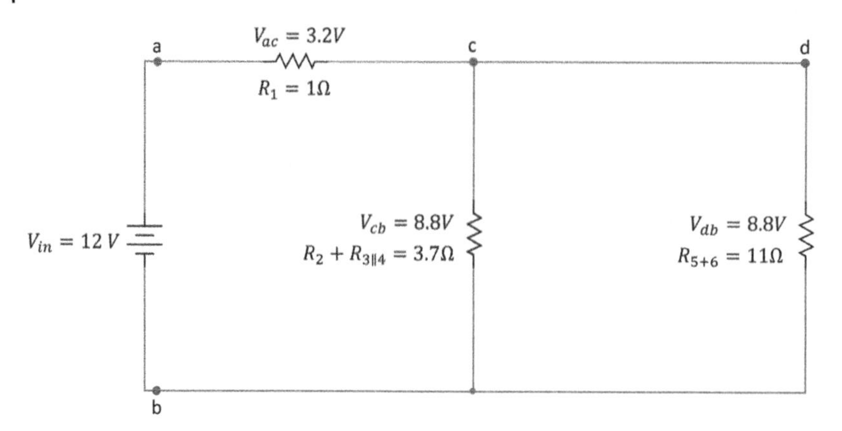

d) Recover two parallel branches V_{cb} *and* V_{db} along with their voltages and resistances.

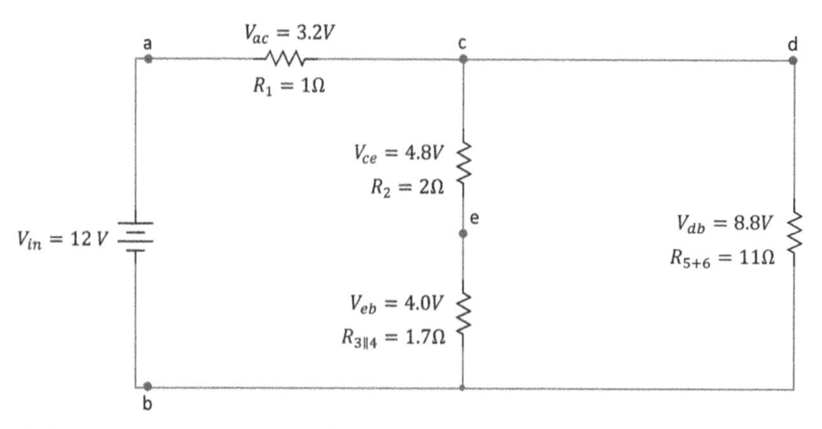

e) Finding the voltages $V_{ce} = V_{R2}$ and $V_{eb} = V_{R_{3\|4}}$ across two series resistors by applying voltage divider rule.

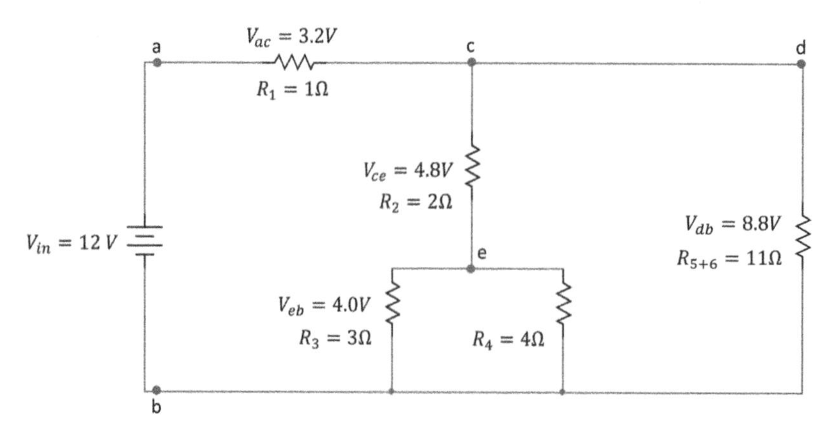

f) Showing $V_{eb} = V_{R3} = V_{R4}$.

Figure 1.17R (Cont'd)

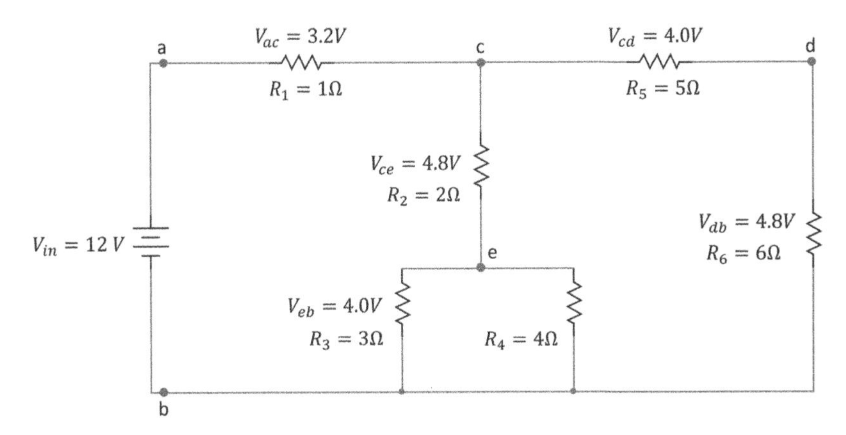

g) Obtaining $V_{cd} = V_{R5}$ and $V_{ab} = V_{R6}$ to recover the original circuit.

Figure 1.17R (Cont'd)

The results of our calculations can be verified by comparing them with the values obtained in Example 1.2.

1.6.4 Kirchhoff's Current Law

Kirchhoff's current law (KCL) states that the algebraic sum of all currents entering and leaving a node is equal to zero, i.e.,

$$\sum_{1}^{n} i_k^a = 0, \tag{1.84}$$

where $k = 1,2,3,\ldots,n$ is the number (designation) of a current entering or leaving the node a. Refer to Figure 1.30a: The input current, i_{in}, enters node b, where it splits into two outgoing currents, i_{R1} and i_{R2}. Hence, for this circuit, *KCL* takes the following form,

$$i_{in} - i_{R1} - i_{R2} = 0. \tag{1.85a}$$

To stress the fact that current in a closed circuit with a single source cannot disappear or be created, (1.85a) should be rewritten as

$$i_{in} = i_{R1} + i_{R2}. \tag{1.85b}$$

This form better visualizes the input–output relationship among the currents. It also emphasizes that *KCL* is based on the *law of conservation of charge.*

Pay attention at the current sign convention: The current that enters a given node is considered positive, whereas the current leaving the node is negative. (**Exercise:** Mark the signs of all currents shown in Figure 1.30b.)

What if a node is a junction point of several branches? KCL answers to this question: The sum of the input currents must be equal to the sum of the output currents. Hence, (1.84) can be presented as

$$\sum_{1}^{n} i_{kin} = \sum_{1}^{n} i_{lout}, \text{ where } k,l = 1,2,3,\ldots,n, \tag{1.86}$$

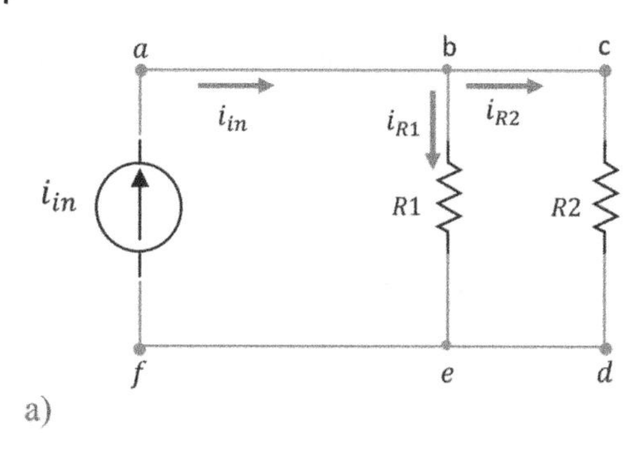

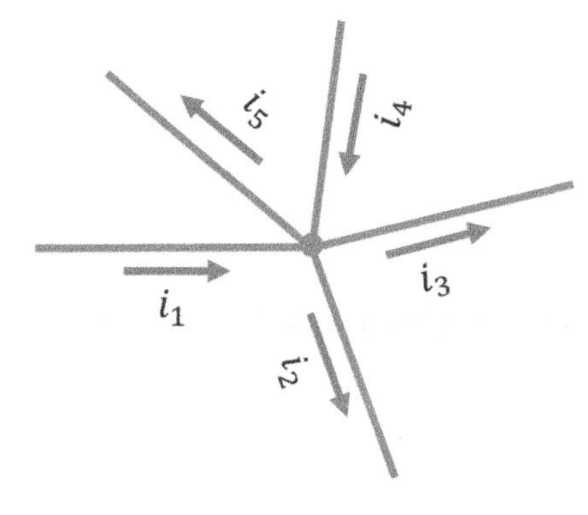

Figure 1.30 Currents distribution in a circuit: a) Two parallel branches; b) a node with several branches.

which means that $\sum_{1}^{n} i_{kin} - \sum_{1}^{n} i_{lout} = 0$, as required by KCL.

Figure 1.30b shows the example of the node with two input currents and three output currents. In this example

$$i_1 + i_4 = i_2 + i_3 + i_5.$$

Example 1.5 below demonstrates how to apply KCL to the analysis of a resistive circuit.

Example 1.5

Application of KCL to the analysis of a series-parallel circuit.

Problem: Find unknown currents in the circuit shown in Figure 1.31.
Solution: The starting point of this analysis is finding I_{R1} because this current is further distributed among all the branches. By the circuit layout,

$$I_{R1} = I_{in} = 3.2A.$$

Applying KCL to node c yields

$$I_{R1} = I_{R2} + I_{R5},$$

from which current I_{R2} is calculated as

$$I_{R2} = I_{R1} - I_{R5} = 2.4A.$$

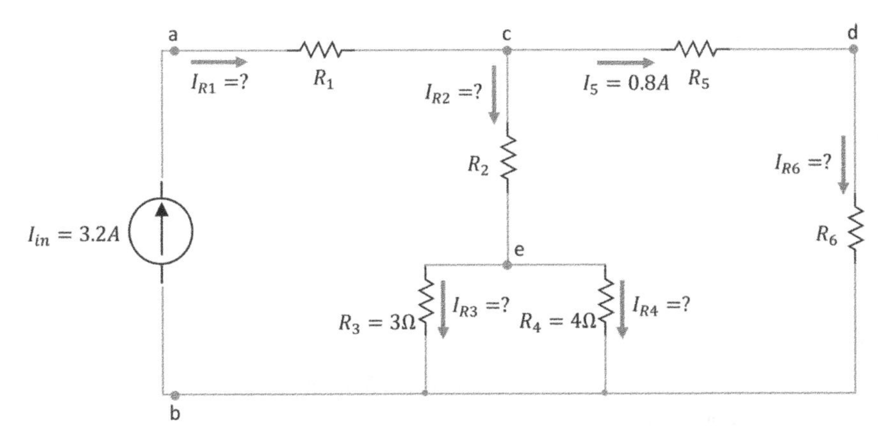

Figure 1.31 Application of KCL to the analysis of a series-parallel resistive circuit for Example 1.5.

Current I_{R3} can be readily found using current divider rule as

$$I_{R3} = \frac{R_4}{R_3 + R_4} I_{R2} = 1.4A.$$

Thus, $I_{R4} = I_{R2} - I_{R3} = 1.0A$, according to KCL applied to node e. Alternatively, $I_{R4} = \frac{R_3}{R_3 + R_4} I_{R2} = 1.0A$, as before.

Finally, $I_{R6} = I_{R5} = 0.8A$ because resistors R_5 and R_6 are connected in series.
The problem is solved.

Discussion: All the calculations are straightforward and don't require further discussion. It takes analysis and practice to become proficient in these manipulations and perform them quickly and accurately. To verify our results, compare them with the values obtained in Example 1.2 of Section 1.4.

Don't underestimate the power of these three fundamental circuit laws—Ohm's law, KVL, and KCL. Though from this introductory discussion, they might seem the things from the past, when all circuits were only dc and understanding of their operations was at the rudimentary level, such an impression would be completely wrong. These laws are well and alive and actively used in cutting-edge electronic technology. One example: The recent publication in the IEEE magazine concerning neural-network processing is entitled[14] *Ohm's law + Kirchhoff's current law = Better AI*. Here, *IEEE* stands for *Institute of Electrical and Electronic Engineers*, the world's largest professional society, and *AI* stands for *Artificial Intelligence.*

Questions and Problems for Chapter 1

*Questions and problems marked by asterisks require knowing the material of the entire Chapter 1, not only a specific section, and they might also require additional online research.

Section 1.1 *Main electrical parameters: Current and voltage; power and energy*

14 IEEE Spectrum, December 2021, pp. 44–50.

1 Write down the following numbers in numerals:
- 12 kHz
- 34 MHz
- 45 GHz
- 56THz
- 7 mF
- 8 μF
- 9 nF
- 11 pF

2 What is an electric charge?

3 Charges $Q_1 = 9mC$ and $Q_2 = -5mC$ are placed at $d = 6cm$. What is the force of their interaction? Will they attract to one another or repel?

4 Consider electrical current:
a) What is it?
b) How can we measure its strength?
c) The electrical current strength, $I(A)$, is proportional to unit charge q. Can we increase this charge to increase $I(A)$? Explain.
d) Equation 1.4a shows that $I(A)$ is inversely proportional to the time interval $\Delta t(s)$. Does it mean we can increase the electrical current by shortening $\Delta t(s)$? If your answer is yes, explain why it is so.

5 What is the relationship between one coulomb and one ampere?

6 What do terms *conventional flow* and *electron flow* refer to?

7 Electrical current is the flow of charge carriers. What are the polarities of these carriers?

8 Electrical current is defined as a flow of charge carriers. However, Gustav Kirchhoff "demonstrated that current flows through a conductor at the speed of light[15]". How is it possible that the charge carriers (particles) travel at the speed of light?

9 What is charge passing?

10 Can we increase the current strength $I(A)$ by increasing a wire gauge? Explain.

11 What is the difference between conductors and isolators from their atomic structure point of view?

12 What is resistance? How can we decrease it?

13 Define electrical voltage.

14 Why is one terminal of a battery marked as positive and the other as negative?

15 Is there any difference between voltage and electromotive force, emf?

16 What is the relationship between voltage and electrical energy?

17 Is there any relationship between voltage and current?

18 In reference to the voltage across a resistor, we always call it "voltage drop," which implies that it should be a voltage rise. Is it true? If your answer is yes, where does this rise occur?

19 Does a battery produce (generate) voltage? Explain.

---

15 www.britanica.com.

Section 1.2 *Passive components: resistors, capacitors, and inductors*

20 What is the difference between resistance and resistor?

21 Consider the resistance of a copper wire:
a) What is its R value if the wire is 1-meter long and 1-centimeter radius of a cross-section?
b) What current will flow in a circuit where 12-volt dc battery terminals are connected with such copper wires? Is this a big or small current?

22 What is the resistance at the terminals of a short circuit? An open circuit?

23 *What current will flow through the circuit whose 12-volt battery terminals are connected by optical fiber?

24 How much does the resistance of copper wire change if the ambient temperature changes from $-50°C$ to $50°C$? Is this a large or small change?

25 What is a skin effect? At what frequency range—GHz or kHz—does it affect a resistance significantly? Why?

26 Does Ohm's law depend on the ambient temperature? Explain.

27 In this book, the model called *linear resistor* is assumed. List all the main features of this model.

28 What is conductance? How does its unit relate to ohm?

29 What power does a 1- $k\Omega$ resistor dissipate if connected to 12-V dc source? What energy does this resistor consume for one second?

30 The text says that lumped resistors are almost disappeared from modern electronics. Why do we still need to study them?

31 Examine Figure 1.8, where polarity signs shown at each circuit element: How do you determine whether a component absorbs or dissipates power?

32 Consider capacitor:
a) What is it?
b) Sketch the construction of a typical capacitor.
c) How does a capacitor operate in a dc circuit?

33 What are the charging and discharging operations of a capacitor?

34 What charge does capacitor accumulate if its capacitance is $C = 12(nF)$ and the applied voltage is $V_{dc} = 12(V)$?

35 Why do we need to use three different types of material—metallic, semiconductor, and dielectric—in the circuit components?

36 What is permittivity? What is relative permittivity?

37 *According to (1.23), capacitance is inversely proportional to the area of a conducting plate. Research online to discover what measures the engineers use to increase the plate area without significantly increasing the capacitor size.

38 Consider mathematical formulas of the $i - v$ relationship for a resistor and a capacitor: Why is this formula for the resistor algebraic but for the capacitor differential?

39 A capacitor works as an open circuit for dc circuit. However, (1.24) shows that the current can flow through capacitor. Why? What physics is behind this phenomenon?

40 The text refers to the *capacitor voltage continuity*. Explain what this phenomenon is and why it is essential.

41 *Consider an RC circuit shown in Figure 1.P41a. As shown, the capacitor was charged up to 12 V, and then the switch flipped. The record of the experimentally performed discharging process is shown in Figure 1.P41b:

a) How does the *capacitor voltage continuity* is implemented here? Explain and mark points $v_C|_{t=0^-}$ *and* $v_C|_{t=0^+}$ in the figure.

b) How does the current through the capacitor change in this case? Sketch the graph $i_c(t)$ and show points $i_C(t=0^-)$ *and* $i_C(t=0^+)$ in the figure.

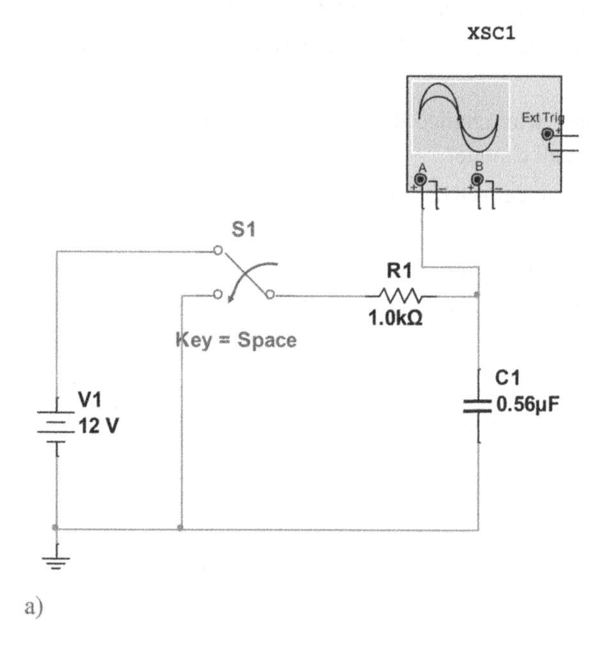

a)

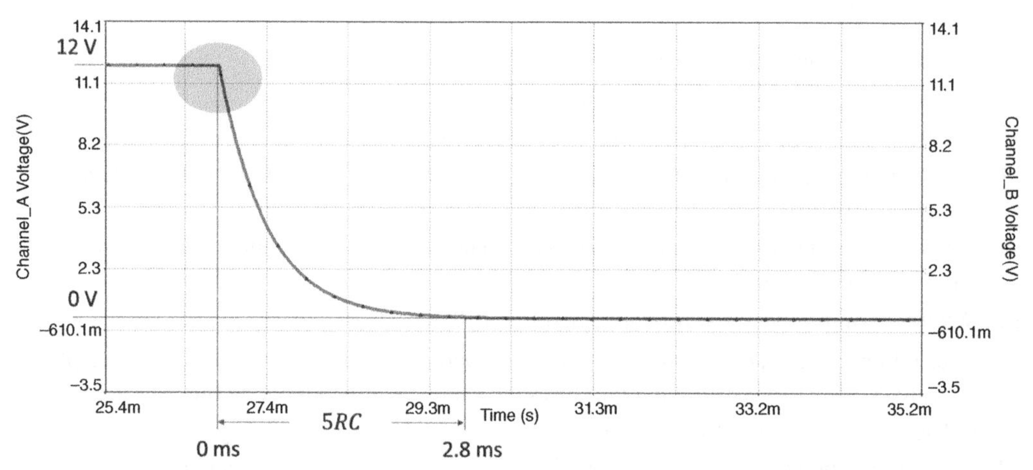

b)

Figure 1.P41 Capacitor voltage continuity for Problem 1.41: a) The schematic of the RC circuit; b) discharging process; c) visual answer to Problem 1.41a explaining the capacitor voltage continuity; d) visual answer to Problem 1.41b concerning capacitor current instantaneous change.

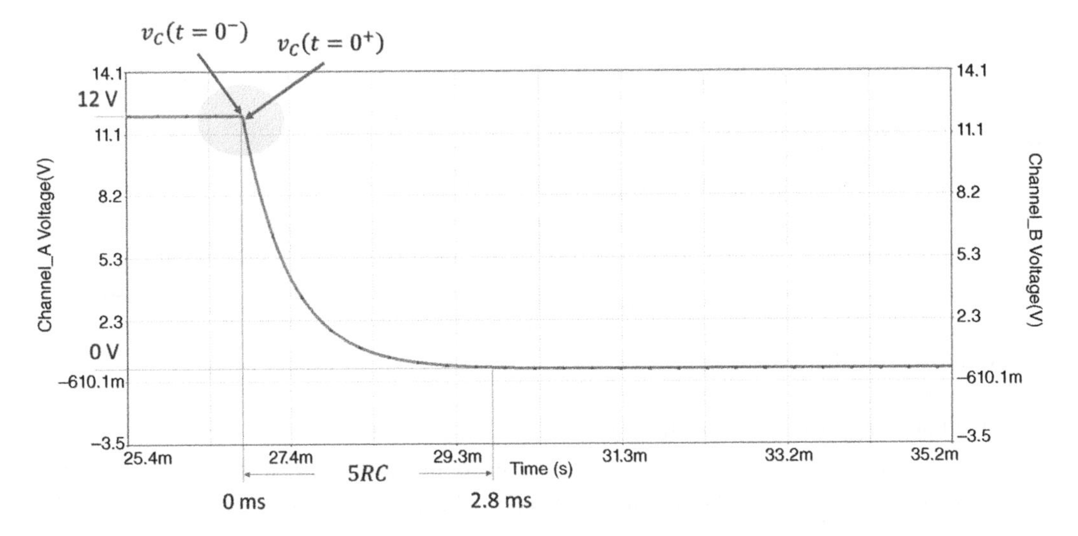

c)

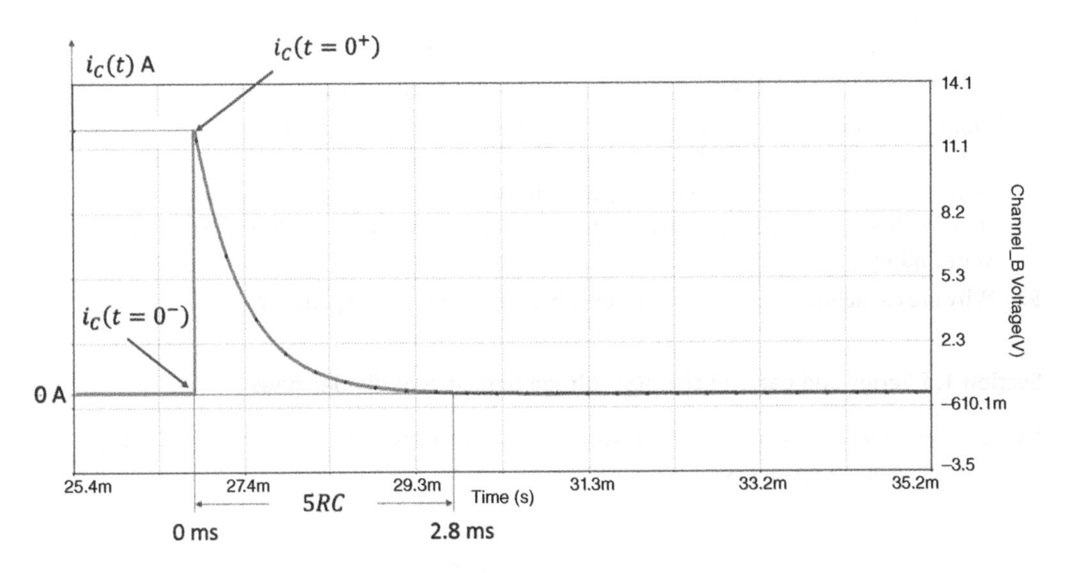

d)

Figure 1.P41 (Cont'd)

42 If the ac current flowing through a capacitor is given, can we find the voltage across the capacitor? If the answer is yes, create a numerical example.

43 A 12-nF capacitor is in the series circuit whose source generates $v_{in(t)} = 5\sin(4t)$. What power is delivered to this capacitor?

44 A 12-mF capacitor is in the circuit whose source generates $v_{in}(t) = 5\sin(4t)$. What energy is delivered to this capacitor?

45 What capacitor is called *linear*?

46 List three main capacitor characteristics and explain their meaning.

47 A circuit includes a source and a 12-μF capacitor. What is its reactance if the signal's frequency is 3 (MHz)?

48 What is inductor?

49 The term *electromotive force, emf or* $\varepsilon(V)$, presents in description both capacitor and inductor. Is there any difference between *emf* for a capacitor and *emf* for inductor?

50 Consider a circuit consisting of a source and an inductor whose $L = 5(mH)$: What is the voltage drop across the inductor if (1) $i_{in} = 12(V)$? (2) $i_{in} = 12\sin(4t)$? and (3) $i_{in} = 12\sin(4000t)$?

51 A 5-mH inductor has the current $i_L = 12\sin(400t)$:
a) What is the delivered power?
b) Does the inductor dissipate this power?
c) What energy does this inductor store?

52 A switch connects a 5-mH inductor to the current source with $i_{in} = 12\sin(400t)$. Initially, the switch was open. At $t = 0$, the switch is turned on. What is the inductor's current immediately after $t = 0$?

53 Explain the difference between ideal inductor and linear inductor.

54 What is an inductive reactance of a 5-mH inductor if the signal's frequency is 3 MHz?

55 *Visit the website of the *Large Hadron Collider* and find information of where and how inductors are used in this colossal engineering instrument. Write an essay based on your findings.

56 *Table 1.3 summarizes the discussion of three passive components, resistors, capacitors, and inductors. Power can be delivered to all three elements, but only the resistor dissipates it. Why? Capacitors and inductors have their reactances that carry unit ohm, the same as the resistors have. Make this question the topic of your independent study and write an essay on your findings.

57 Why are capacitor and inductor collectively called *reactive* components?

Section 1.3 Series and parallel circuits; voltage and current divider rules

58 Consider a schematic shown in Figure 1.P58 and identify source, ground, branches, nodes, loops, and all components.

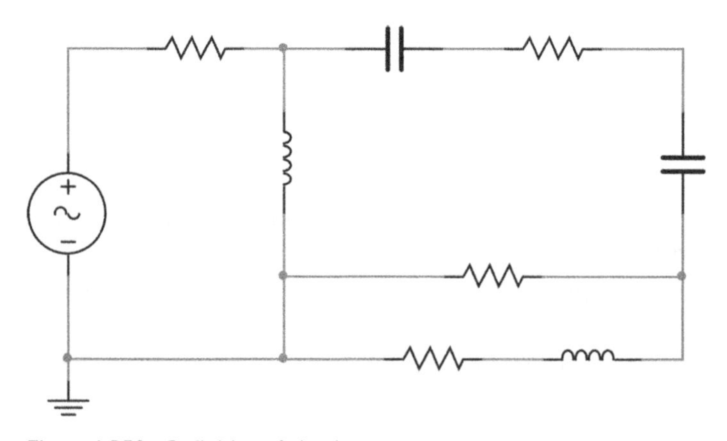

Figure 1.P58 Definition of circuit.

59 Consider the circuit shown in Figure 1.P59:
a) In what topology—series or parallel—the resistors are connected? Prove your answer.
b) Determine the voltage drop across each resistor. Verify your answer.
c) Find the current flowing through each component.

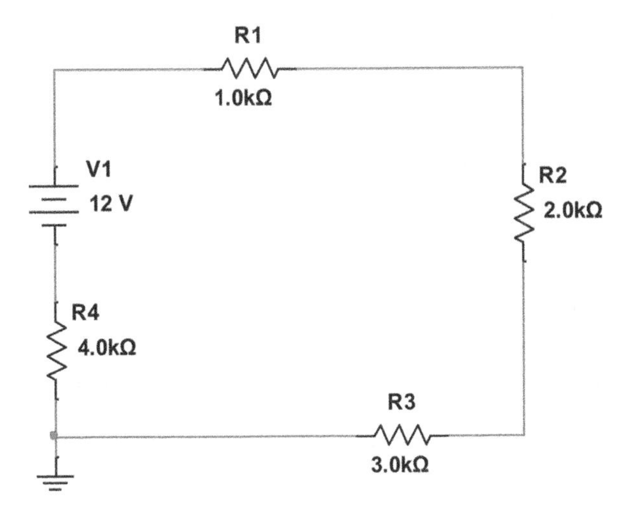

Figure 1.P59 Resistor circuit.

60 A circuit with four resistors is shown in Figure 1.P60. The values are: $V_{in} = 12(V)$, $R_1 = 1(k\Omega)$, $R_2 = 2(k\Omega)$, $R_3 = 3(k\Omega)$, and $R_4 = 4(k\Omega)$.
a) In what topology—series or parallel—the resistors are connected? Prove your answer.
b) Find the current flowing through each component. Verify your answer
c) Determine the voltage drop across each resistor. Verify your answer.

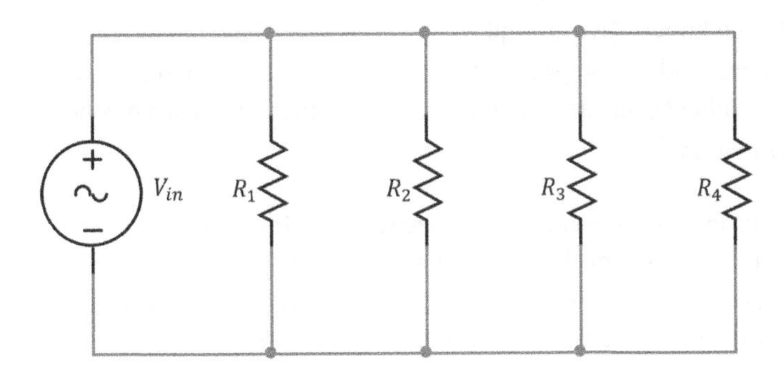

Figure 1.P60 The circuit with four resistors for Problem 1.60.

61 Find the voltage drop and current through each component of the circuit shown in Figure 1.P61.

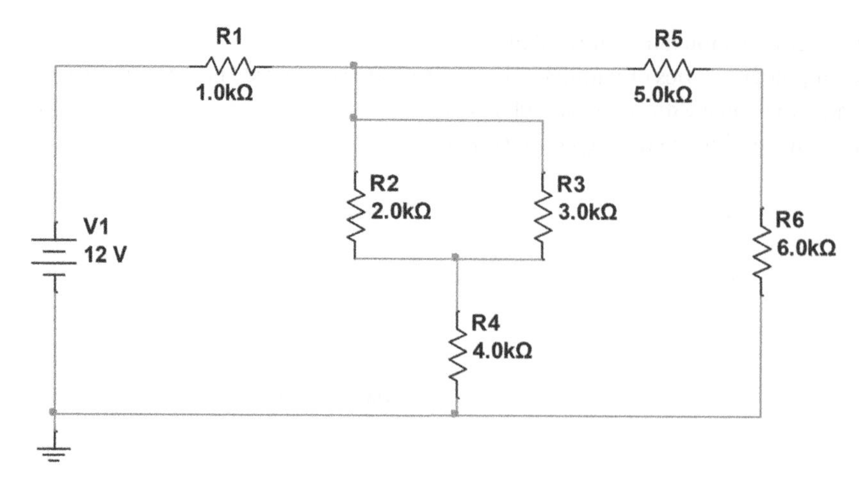

Figure 1.P61 Circuit with series and parallel resistors for Problem 1.61.

62 The circuit contains three capacitors, $C_1 = 10\mu F$, $C_2 = 20\mu F$, and $C_3 = 30\mu F$. What is the total capacitance if the capacitors are connected
a) in series?
b) in parallel?

63 The circuit contains three inductors, $L_1 = 12\mu H$, $L_2 = 24\mu H$, and $L_3 = 36\mu H$. What is the total inductance if the inductors are connected
a) in series?
b) in parallel?

Section 1.4 Sources (inputs, excitations, or drives)

64 What is a source? Give definition and examples.

65 What kind of sources do you know? Give examples.

66 What is the difference between ideal and practical sources in terms their circuitry and ability to be connected with each other? Consider separately voltage sources and current sources.

67 Consider source transformation:
a) What is it? Explain.
b) What are the conditions for transformation of a current source into a voltage one?
c) From what standpoint the transformed sources are equivalent?

68 In Example 1.3, all sources were transformed to the equivalent current source, as Figure 1.27d shows. Transform all sources in Figure 1.27a into equivalent voltage source.

Section 1.5 Primary circuit laws (Ohm's and Kirchhoff's laws)

69 Consider circuit analysis:
a) What is it? Explain.
b) What quantities are the subject of the circuit analysis?
c) What is the basis of the circuit analysis?
d) Give an example of circuit analysis
e) Explain the difference between circuit analysis and circuit design. Give examples.

70 Focus on Ohm's law:

a) If the voltage across a 4-$k\Omega$ resistor is 24 volts, what is the current through this resistor?

b) Why does Ohm's law is called *empirical*? What does it mean?

c) To what circuit—dc or ac—Ohm's law can be applied? Explain.

d) *Michael Faraday mainly started his experiment to verify a concept he already had for the phenomenon he investigated. Did Georg Ohm want to prove the known concept, or did he discover his law by accident measurements?

71 Consider Kirchhoff's voltage law (KVL):

a) What does this law state? Explain using formulas.

b) What is sign convention for voltages and why is it important for KVL?

c) In essence, KVL states that the input voltage is distributed across all the circuit components; that is, the input voltage is conserved. Why then the text refers to the *law of conservation of energy* as a basis of KVL?

72 Refer to Figure 1.P61: Find all voltages using KVL and voltage divider rule. (Hint: See the discussion of Figure 1.17R in Example 1.4.)

73 What does Kirchhoff's current law (KCL) state? What is the physics behind this law?

74 Consider Figure 1.P61: Applying KCL and the current divider rule, find currents flowing through circuit resistors.

What do you take away from studying Chapter 1? Write a one-page essay that highlights the six most significant points of Chapter 1.

2

Methods of DC Circuit Analysis

2.1 Introduction—DC and AC Circuits

In electrical engineering, we consider two types of currents: *direct current, dc*, and *alternating current, ac*. Also, the terms *dc* and *ac* can be applied to voltage, and you won't be surprised when you meet the terms dc and ac circuits. Sometimes, dc and ac are written in capital letters, as we've done in this subtitle; sometimes they are written in lower case, as we show in the text. Unhappily, academia and the industry of the electrical engineering field do not observe any standard designation; thus, in your professional career, you can meet either notation.

With dc, both current and voltage have constant values and directions of propagation (polarities) in each problem. Of course, we can change their directions and values, but then we will have a new problem. A *dc signal requires only one number—its value—to be described*. Polarity is an auxiliary though important parameter. Obviously, circuits whose currents and voltages are dc are called *dc circuits*. In *ac circuits*, current and voltage change their values and polarities in a deterministic or sporadic way. The best known and most widely used is a sinusoidal ac signal whose magnitude varies periodically. *Description of a sinusoidal signal needs three numbers—amplitude, frequency of variations, and a phase shift*. Obviously, the analysis of ac circuits is much more sophisticated than a dc circuit analysis.

In this chapter, we will consider only dc circuits. We remember that a dc circuit containing only a single *capacitor* turns into an *open circuit*, and a similar circuit having only an *inductor* behaves like a *short circuit*. Therefore, the only component that remains operational in a dc circuit is a resistor. No wonder that the circuits considered in this chapter contain only resistors.

This brief introduction explains why we start considering the circuit analysis with dc circuits: It enables us to introduce this analysis's primary laws and methods straightforwardly, minimizing the mathematical complexity.

2.2 Nodal Analysis and Mesh Analysis

Fundamentals of the circuit analysis at an introductory level have been given in Chapter 1. Specifically, that chapter discusses main electrical parameters (current and voltage; power and energy), passive components (resistors, capacitors, and inductors), series and parallel circuits, sources, and primary circuit laws (Ohm's and Kirchhoff's laws). It's worth revisiting that chapter

Essentials of Advanced Circuit Analysis: A Systems Approach, First Edition. Djafar K. Mynbaev.

Companion Website: www.wiley.com/go/Mynbaev/AdvancedCircuitAnalysis

to understand the Chapter 2 material better. Reviewing the *Circuit analysis* subsection in Section 1.5 that discusses the main *what* and *why* is particularly conducive.

To recall, circuit analysis aims to find the voltages across and currents through each circuit component. Ohm's and Kirchhoff's laws enable us to do this, but employing them to analyze sophisticated circuits becomes a cumbersome and time-consuming procedure. This chapter introduces new approaches to and theorems of circuit analysis, of which this section presents two fundamentally important methods—*nodal analysis* and *mesh analysis*.

The concept of both nodal and mesh analysis is to describe an electrical circuit not in terms of $i - v$ equation of each component but using the equations describing the status of every *node* or every *loop* of the circuit. These equations introduce new *circuit variables*. They allow, of course, finding the needed voltages across and currents through all circuit elements. Such an approach simplifies the analysis of circuits, especially the sophisticated ones. No wonder the nodal and mesh methods are one of the most powerful tools in the modern circuit analysis.

Chapter 2 and its sections concentrate on the key ideas and methods of new approaches to circuit analysis, thus continuing the review of the primary circuit analysis technique started in Chapter 1. In other words, we leave the detailed discussion of these methods to the textbooks specialized in basic circuit analysis.

2.2.1 Nodal Analysis

2.2.1.1 Basic Concept

Circuit analysis aims to determine the voltage across and current through each circuit component, and the nodal analysis *is a specific technique for achieving this goal.*

In the nodal analysis, the voltages at the circuit nodes (node voltages) become the circuit variables. To find them, we choose a reference (datum) node and determine all node voltages with respect to this one. Typically, a ground point with zero electrical potential serves as a reference node.

As we recall from Section 1.2, the voltage across a component is the difference in electrical potential between two terminals of this component. Since each component is set between two nodes, its voltage can be readily determined, and its current can be calculated as soon as the voltages of these nodes become known. Thus, obtaining the node voltages is the primary objective of the nodal analysis.

Description of each node voltage requires one independent linear equation. If a circuit contains n nodes, it takes n − 1 independent equations to describe the entire circuit because the voltage value of the reference one is already known. Reducing the number of circuit equations is one of the advantages of nodal analysis.

To explain the procedure of the nodal circuit analysis, let's consider the series-parallel circuit shown in Figure 2.1a. The circuit includes the voltage source V_{in}, the current source I_{in}, and four resistors.

Let's consider the nodal analysis of the circuit shown in Figures 2.1a and 2.1b in a step-by-step format:

1) Chose the ground point as a reference node and marked it 0. Mark two other nodes as Node 1 and Node 2. Designate the node voltages as v_1 and v_2.
2) Remember that v_1 is the potential difference between v_1 and the reference node. Define v_2 similarly. (See the dotted arrows in Figure 2.1a.)
3) *Assume* the flow of the currents as shown in Figure 2.1b. Then apply KCL to Nodes 1 and 2, respectively, and find:

$$\begin{cases} i_1 = i_2 + i_3 \\ i_2 + I_{in} = i_4 \end{cases}. \tag{2.1}$$

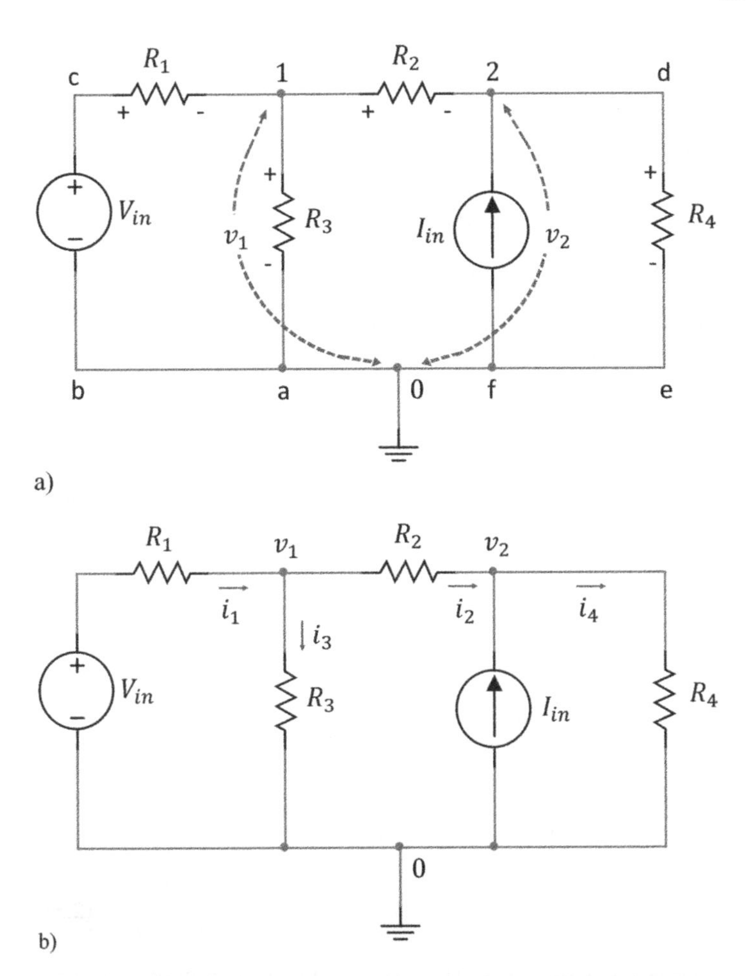

Figure 2.1 Nodal analysis explanations: a) The given circuit and its nodes and components; b) the given circuit with currents.

4) Express the currents through the node and source voltages employing Ohm's law to obtain the following *system of two linear equations*, also known as *simultaneous equations* or *equation system*:

$$\begin{cases} \dfrac{V_{in}-v_1}{R_1} = \dfrac{v_1-v_2}{R_2} + \dfrac{v_1-0}{R_3} \\ \dfrac{v_1-v_2}{R_2} + I_{in} = \dfrac{v_2-0}{R_4} \end{cases}. \tag{2.2}$$

5) Solve this equation system and find v_1 and v_2.
6) Calculate the element voltages using the obtained v_1 and v_2 as

$$v_{R1} = V_{in} - v_1,\ v_{R2} = v_1 - v_2,\ v_{R3} = v_1,\ \text{and}\ v_{R4} = v_2. \tag{2.3}$$

Thus, all component currents and voltages will be determined, and the circuit analysis is complete.

Remember that current directions are assumed, and the actual directions will be obtained after completing the analysis. See the Discussion Section of Example 2.1, where this point is considered and exemplified.

To clarify the application of this theory, let's consider an example.

Example 2.1 Nodal analysis of an electrical circuit with two nodes.

Problem: Perform nodal analysis of the electrical circuit shown in Figure 2.2.

Solution: Figure 2.2 reproduces the circuit presented in Figure 2.1 but with the component values. Performing Steps 1 and 2 of the theory discussed above, we replace (2.2) with its counterpart with numerical values

$$\begin{cases} \frac{12-v_1}{1} = \frac{v_1-v_2}{2} + \frac{v_1}{3} \\ \frac{v_1-v_2}{2} + 0.6 = \frac{v_2}{4} \end{cases}. \tag{2.4}$$

Multiplying the top equation through by six and collecting members yields

$$11v_1 - 3v_2 = 72.$$

Rearranging the bottom equation in a similar manner results in

$$-2v_1 + 3v_2 = 2.4.$$

Hence, instead of (2.4) we have two following simultaneous equations

$$\begin{cases} 11v_1 - 3v_2 = 72 \\ -2v_1 + 3v_2 = 2.4 \end{cases}. \tag{2.5}$$

There are several methods to solve this equation system. Here, we employ the *substitution method* whereas the others are considered in Sidebar S2. Let's multiply the top equation of the system (2.5) by 2 and the bottom one by 11. The result is

$$\begin{cases} 22v_1 - 6v_2 = 144 \\ -22v_1 + 33v_2 = 26.4 \end{cases}. \tag{2.6}$$

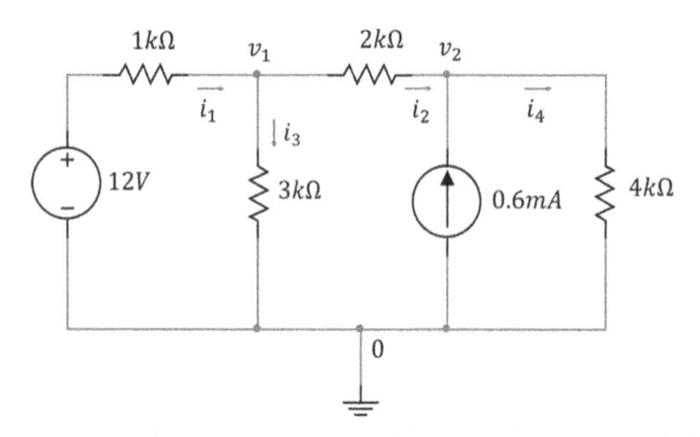

Figure 2.2 Nodal analysis of the electrical circuit for Example 2.1.

Adding these two equations up eliminates v_1 and gives

$$v_2 = 6.31V.$$

Substituting this value of v_2 into one of the equations shown in (2.6) results in $v_1 = 8.27V$. Therefore, the nodal analysis is complete.

Discussion:

- Finding the node voltages enables us to determine the currents and voltages of all the circuit components. As mentioned, this is a great advantage of the nodal analysis because the number of the circuit nodes is always less than the number of elements. In this example, the circuit in Figure 2.2 has two nodes and four resistors. However, getting v_1 and v_2 enables us to calculate all four remaining currents and voltages. Table 2.1 shows all the calculated values along with the employed formulas. Analyze the table carefully; draw the circuit shown in Figure 2.2 and place the values of all elements' voltages and currents.
- *We must remember that without verifying the results, the problem cannot be considered solved correctly*. Checking the results is as important as obtaining them.
- In the case of nodal analysis, the first check is to verify the compliance of the results with KCL. Equation 2.1 displays the KCL conditions for our circuit; plugging in the obtained values verifies the results. Therefore,

$$\begin{cases} i_1 = i_2 + i_3 \\ i_2 + I_{in} = i_4 \end{cases} \Rightarrow \begin{cases} 3.73 = 0.98 + 2.75 \\ 0.98 + 0.6 = 1.58 \end{cases}.$$

Table 2.1 The voltages and currents of all circuit elements in Example 2.1.

Circuit element	Voltage across	Current through	Comment
Voltage source, V_{in}	$V_{in} = 12V$	$i_1 = \frac{12 - v_1}{R_1} = \frac{(12 - 8.27)V}{1k\Omega} = 3.73mA$	The nodal analysis in Example 2.1 yields $v_1 = 8.27V$. $v_2 = 6.31V$.
Current source, I_{in}		$I_{in} = 0.6mA$	
Resistor R_1	$V_{R_1} = 12 - v_1 = 3.73V$	$I_{R_1} = i_1 = \frac{12 - v_1}{R_1} = \frac{(12 - 8.27)V}{1k\Omega} = 3.73mA$	
Resistor R_2	$V_{R_2} = v_1 - v_2 = 8.27 - 6.31$ $= 1.96V$	$I_{R_2} = i_2 = \frac{v_1 - v_2}{R_2} = \frac{(8.27 - 6.31)V}{2k\Omega} = 0.98mA$	
Resistor R_3	$V_{R_3} = v_1 - 0 = 8.27V$	$I_{R_3} = i_3 = \frac{v_1 - 0}{R_3} = \frac{(8.27)V}{3k\Omega} = 2.75mA$	
Resistor R_4	$V_{R_4} = v_2 - 0 = 6.31V$	$I_{R_4} = i_4 = \frac{v_2 - 0}{R_4} = \frac{(6.31)V}{4k\Omega} = 1.58mA$	

As seen, the KCL conditions are met. (Remember, all *currents* in this example are measured in mA). To verify the *voltages*, plug the obtained values from Table 2.1 into (2.4). Bear in mind that such a check is just the repetition of the previous one; in other words, this is not an independent verification of the results.

- How do we know what direction each current has to flow? The *general rule* is *the current flows from the node having higher voltage potential to the node with lower potential*. Refer to Figure 2.2 to see the examples: Current i_1 flows from the source (the highest voltage potential in a circuit) to Node 1, and current i_3 flows from Node 1 to the ground (the lowest voltage potential in the circuit). But what about current i_2? We can reasonably assume that $v_2 < v_1$ because the voltage reaching v_2 experiences two drops starting from the source. However, if we chose a wrong current direction, the solution indicates so by assigning the appropriate sign to the current. For example, if we funnel i_2 in Figure 2.2 in the opposite directions, from v_2 to v_1, then the top equation of (2.4) would take the form

$$\frac{12-v_1}{1}+\frac{v_2-v_1}{2}=\frac{v_1}{3}. \tag{2.4X}$$

Performing the calculations required by (2.4), we discover that LHS of (2.4X) is equal to $-21.6(mA)$, whereas RHS of the same equation equals $21.6(mA)$. Thus, the *supposed direction of i_2 was wrong*.

- The other option to check the results is calculating the balance of supplied and dissipated *power*. The circuit in Figure 2.2 has two sources that release power and four resistors that dissipate power. Current $i_1 = \frac{12-v_1}{1} = 3.73mA$ flows through the voltage source from the negative terminal to the positive one; therefore, the source *produces power*, as shown in Section 1.4. This power is given by

$$p_{Vin} = V_{in} \cdot i_1 = 12(V) \cdot 3.73(mA) = 44.7mW.$$

The current produced by the current source I_{in} should flow, by the general rule, from Node 2 to the ground, but its direction is opposite, which means that this source releasing power. The voltage across the current source is $v_2 = 6.31V$. Therefore,

$$p_{Iin} = v_2 \cdot I_{in} = 6.31(V) \cdot 0.6(mA) = 3.8mW.$$

Therefore, total *delivered power* is equal to

$$p_{Vin} + p_{Iin} = 48.5mW.$$

The dissipated power can be calculated as

$$p_{Rk} = i_{Rk}^2 R_k.$$

Plugging into the above equation the given and obtained numbers yields

$$p_{Rk} = i_1^2 R_1 + i_2^2 R_2 + i_3^2 R_3 + i_4^2 R_4 = 13.91 + 1.92 + 22.69 + 9.99 = 48.5\text{mW}.$$

Review again the circuit in Figure 2.2, analyze closely directions of the currents with respect to the source polarities, and understand our reasoning about releasing or dissipating power. Realize that some sources can *consume power*, as, for example, a charger of your mobile device does.

This is the end of Example 2.1.

It's worth considering the example of the nodal analysis of a more sophisticated electrical circuit.

Example 2.2 Nodal analysis of an electrical circuit with three nodes.

Problem: Find the node voltages v_1, v_2, and v_3 in the circuit shown in Figure 2.3. The values of the source currents and the resistors are: $V_{in} = 6V$ and $I_{in} = 2mA$, $R_1 = 1k\Omega$, $R_2 = 2k\Omega$, $R_3 = 3k\Omega$, $R_4 = 4k\Omega$, and $R_5 = 5k\Omega$.

Solution: We follow the steps presented in the discussion of Figure 2.1. The reference node (ground 0) and designations of three other nodes are shown in Figure 2.3.

The assumed directions of the currents are shown in Figure 2.3. Though we make the most plausible assumptions, this decision is not critical, and it will be resolved by the current signs obtained in the solution. Distinguish between the source current I_{in} and the branch currents i_1, i_2, i_3, and i_4.

Next, apply KCL to all three nodes and find:

At v_1: $i_2 + i_5 = i_1$
At v_2: $i_2 + i_4 = i_3$
At v_2: $I_{in} + i_5 = i_4$

Further, we show how these currents are written in terms of the node voltages:

$$i_1 = \frac{V_{in} - v_1}{R_1},\ i_2 = \frac{v_1 - v_2}{R_2},\ i_3 = \frac{v_2 - 0}{R_3},\ i_4 = \frac{v_3 - v_2}{R_4},\ and\ i_5 = \frac{v_1 - v_3}{R_5}.$$

(An experienced reader can omit this step, but it helps a beginner avoid mistakes.)

Expressing the above KCLs in voltages gives the following system of three equations:

$$\begin{cases} \dfrac{v_1 - v_2}{R_2} + \dfrac{v_1 - v_3}{R_5} = \dfrac{V_{in} - v_1}{R_1} \\ \dfrac{v_1 - v_2}{R_2} + \dfrac{v_3 - v_2}{R_4} = \dfrac{v_2 - 0}{R_3} \\ \dfrac{v_1 - v_3}{R_5} + I_{in} = \dfrac{v_3 - v_2}{R_4} \end{cases}. \tag{2.7}$$

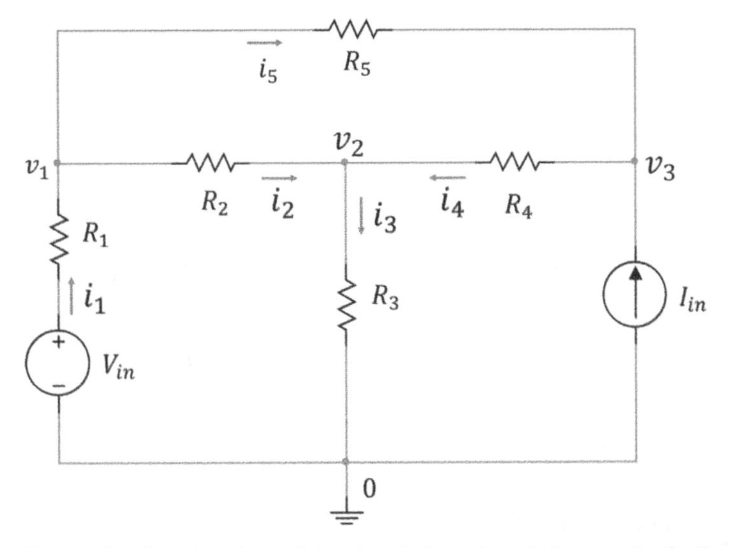

Figure 2.3 Nodal analysis of the electrical circuit with three nodes for Example 2.2.

After collecting the terms, the system of the node equations takes the following form:

$$\begin{cases} v_1\left(\frac{1}{R_1}+\frac{1}{R_2}+\frac{1}{R_5}\right)-\frac{v_2}{R_2}-\frac{v_3}{R_5}=\frac{V_{in}}{R_1} \\ \frac{v_1}{R_2}-v_2\left(\frac{1}{R_2}+\frac{1}{R_3}+\frac{1}{R_4}\right)+\frac{v_3}{R_4}=0 \\ -\frac{v_1}{R_5}-\frac{v_2}{R_4}-v_3\left(\frac{1}{R_4}+\frac{1}{R_5}\right)=I_{in} \end{cases}.$$

Plugging in the component values yields

$$\begin{cases} 17v_1-5v_2-2v_3=10V_{in} \\ 6v_1-13v_2+3v_3=0 \\ -4v_1-5v_2+9v_3=20I_{in} \end{cases}. \tag{2.8}$$

There are several methods to solve such a system of equations besides the substation explored in Example 2.1. All of them are discussed in Sidebar S2.

Here is our equation system whose source values are inserted.

$$\begin{cases} 17v_1-5v_2-2v_3=60 \\ 6v_1-13v_2+3v_3=0 \\ -4v_1-5v_2+9v_3=40 \end{cases}$$

Further, we put it into MATLAB in the shown form to calculate the solution. Below are the code and the result of MATLAB's calculations:

```
>> A = [17 -5 -2;
6 -13 3;
-4 -5 9]; %Define the matrix of coefficients A;
B = [60; 0; 40]; %Define the matrix of constants B;
V=linsolve(A,B) %Command MATLAB to find the solution as matrix
V = [v1;v2;v3];
V =
6.2581
5.2258
10.1290
```

The problem is solved.

Discussion:

- To check our solution, we can verify that the KCL conditions are met at all nodes. Plugging the obtained values of the node voltages and given source values yields:

$$\text{at } v_1\ \frac{v_1-v_2}{R_2}+\frac{v_1-v_3}{R_5}=\frac{V_{in}-v_1}{R_1}\cdot\frac{6.26-5.23}{2}+\frac{6.26-10.13}{5}=\frac{6-6.26}{1}\cdot-0.26=-0.26$$

$$\text{at } v_2\ \frac{v_1-v_2}{R_2}+\frac{v_3-v_2}{R_4}=\frac{v_2-0}{R_3}\cdot\frac{6.26-5.23}{2}+\frac{10.13-5.23}{4}=\frac{5.23}{3}\cdot 1.74=1.74$$

$$\text{at } v_3\ \frac{v_1-v_3}{R_5}+I_{in}=\frac{v_3-v_2}{R_4}\cdot\frac{6.26-10.13}{5}+2=\frac{10.13-5.23}{4}\cdot 1.22=1.22$$

Therefore, our solution passed this check.

- Pay attention that the solution did not require changing the assumed directions of the currents. Thus, these directions are correct.
- You are encouraged to solve this problem by other methods discussed in Sidebar S2, such as Gaussian elimination algorithm and Cramer's rule.

  This is the end of Example 2.2.

Before completing the nodal analysis, we need to consider a special case called a *supernode*.

2.2.1.2 Supernode

Examine circuit in Figure 2.4a: Deriving KCL at the non-reference nodes v_1 and v_2 requires knowledge of current i_v flowing through the voltage source. In circuit analysis, there is no way to know a current flowing through an independent voltage source, which means we have a problem in the analysis of the given circuit. For this case, there is a technique called *supernode*. In a nutshell, *the technique considers the nodes v_1 and v_2 and the voltage source as a single node, a supernode.* This assumption eliminates the need to find i_v and allows for deriving KCL at this imitative node. In Figure 2.4b, the supernode is circumscribed and shaded.

Now, the KCL for the supernode reads:

$$i_1 = i_2 + i_3 + i_4. \tag{2.9}$$

Remember that the goal of the nodal analysis is to find the node voltages. To achieve this goal, we rewrite (2.9) in terms of voltages using Ohm's law as

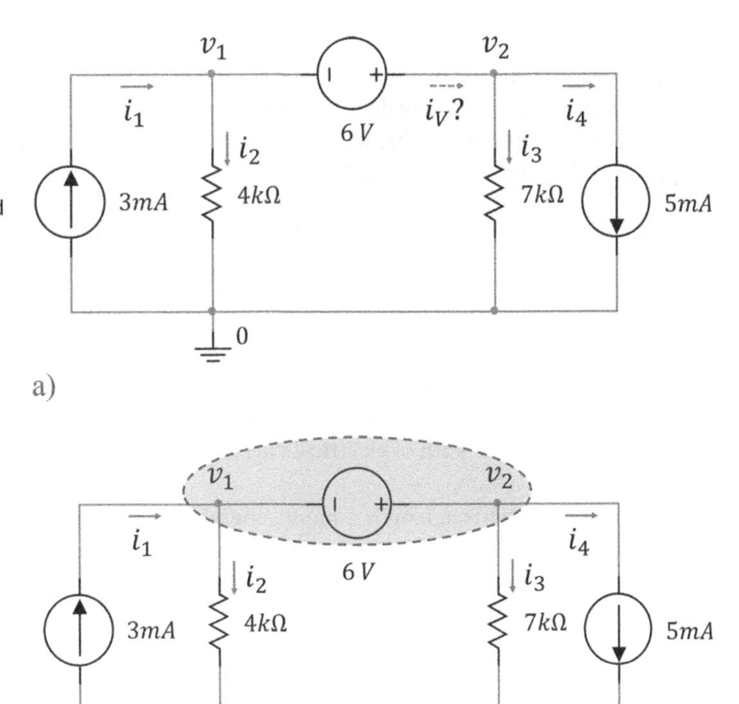

Figure 2.4 Applying the supernode technique: a) The circuit with an independent voltage source connected between two non-reference nodes; b) the same circuit with a supernode (circumscribed and shaded).

$$3mA = \frac{v_1}{4k\Omega} + \frac{v_2}{7k\Omega} + 5mA. \tag{2.10}$$

Equation 2.10 enables us to find

$$v_1 = -\left(\frac{4}{7}v_2 + 8\right)V. \tag{2.11}$$

It seems we can't attain our goal because we get only one equation for two unknown variables, v_1 and v_2. Fortunately, the circuit shows a constraint equation:

$$v_2 - v_1 = 6V. \tag{2.12}$$

Now, we have two simultaneous equations, (2.11) and (2.12). Solving this system yields

$$v_1 = -7.27V \text{ and } v_2 = -1.27V. \tag{2.13}$$

This is how the supernode technique can solve the problem with an independent voltage source connected between two non-reference nodes.

This is the end of a brief review of nodal analysis.

2.2.2 Mesh Analysis

2.2.2.1 Basic Concept

Mesh analysis is another powerful tool in circuit analysis. In many aspects, it mirrors the nodal analysis; however, the *mesh currents* serve as circuit variables in mesh analysis. If we use KCL and Ohm's law to find the nodal voltages in the nodal analysis, then in mesh analysis finding mesh currents calls for using—you guessed it—KVL and Ohm's law.

But what is a *mesh*, and how does it differ from a *loop*? A dictionary explains that "mesh is one of the openings between the threads or cords of a net."[1] However, in electrical circuits, there is a specific meaning of this term, namely,

a mesh is a closed path in a circuit with no other paths inside it.

In contrast, "a *loop* is any closed path through a circuit where no node more than once is encountered."[2] Consider the circuit shown in Figure 2.5a: One path forms Mesh 1, and the other path makes Mesh 2. The path going through the entire circuit constitutes a *loop*. Compare this illustration with the above definitions to firmly comprehend the meaning of these two terms. Also, review the dictionary definition again and see why mesh is just "one of the openings." Section 1.3 introduced a loop as a closed path of electrical current in a circuit. Now, we know that the closed path could be a mesh or a loop.

From these definitions, a loop is a more general entity than a mesh, and the method in consideration should be called *loop analysis* rather than *mesh analysis*. Some authors observe this logic; nevertheless, the electrical engineering community commonly uses the latter.

Here are the main features of a mesh analysis:

- The mesh analysis aims to find mesh currents for all circuit meshes, which enables us to calculate the voltages across and currents through all circuit elements.

---

1 https://www.merriam-webster.com/dictionary/mesh.

2 https://byjus.com/questions/what-is-the-difference-between-loop-and-mesh.

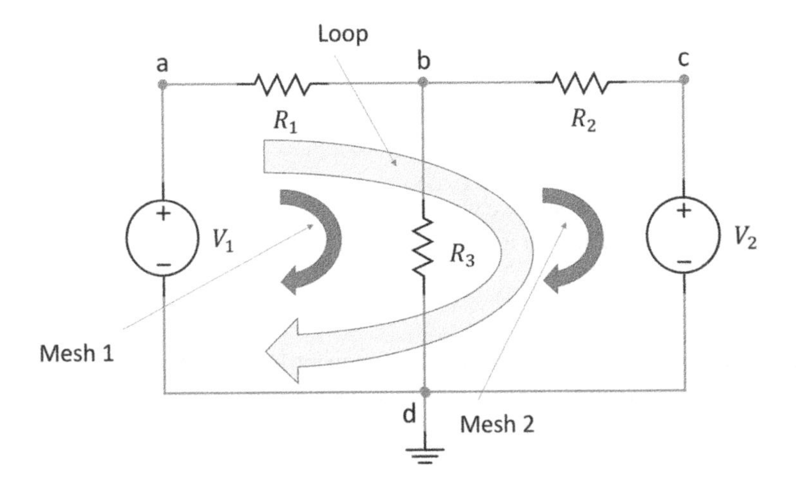

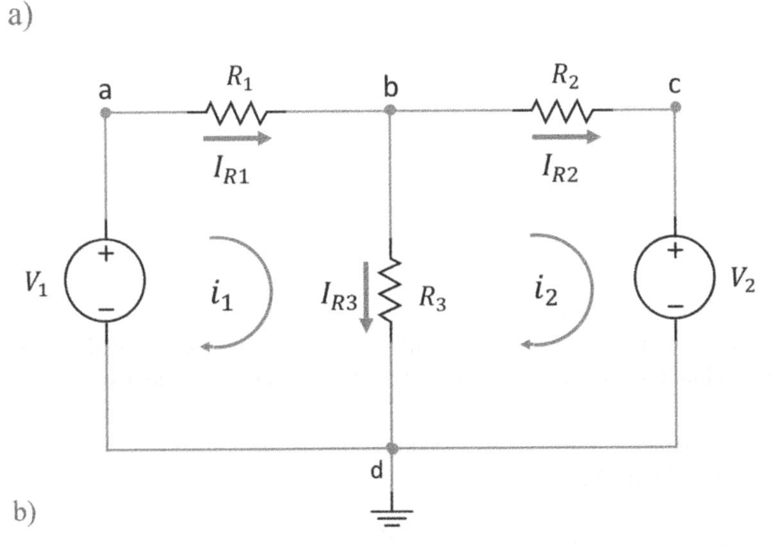

Figure 2.5 Mesh analysis: a) Loop and meshes; b) mesh currents and branch currents.

- There are two types of currents in this analysis, as Figure 2.5b shows: mesh currents i_1 and i_2 and branch currents, I_{R1}, I_{R2}, and I_{R3}. (You will recall that nodal analysis also introduces two types of voltages, node voltages and branch voltages.) As soon as we find the mesh currents, branch currents can be readily determined.
- The mesh analysis starts with assigning mesh currents with arbitrary flow directions. Typically, these directions are clockwise, as in Figure 2.5b.
- Going along each mesh and using KVL, we derive an equation for individual mesh. Repeating this procedure, we derive equations for all meshes of the circuit. These equations are simultaneous; that is, they create the system of equations.
- Applying Ohm's law enables us to rewrite these equations in terms of mesh currents.
- Solving the system of the current equations, we find all mesh currents, thus, achieving the goal of the mesh analysis. Then, calculating the branch voltages and currents becomes a technical matter.

Let's exercise these steps using the circuit given in Figure 2.5b as an example. Applying KVL to Mesh 1 gives

$$V_{R1} + V_{R3} = V_1.$$

Rewriting this equation in terms of the mesh currents yields

$$R_1 \cdot i_1 + R_3 \cdot (i_1 - i_2) = V_1. \tag{2.14}$$

(Pay attention to the current difference: In Mesh 1 we write $i_1 - i_2$, not $i_2 - i_1$!)
Similar approach produces the following equation for Mesh 2

$$-R_3 \cdot i_1 + (R_3 + R_2) \cdot i_2 = -V_2. \tag{2.15}$$

To simplify the calculations, let's assign the values to each circuit component: $V_1 = 12V, V_2 = 6V, R_1 = 1$ $k\Omega$, $R_2 = 2\ k\Omega$, and $R_3 = 3\ k\Omega$. Then (2.14) and (2.15) give the following system of equations:

$$\begin{cases} 4i_1 - 3i_2 = 12V \\ -3i_1 + 5i_2 = -6V \end{cases}. \tag{2.16}$$

Solving this system by either Gaussian elimination method or Cramer's rule produces

$$i_1 = \frac{42}{11} mA$$

and

$$i_2 = \frac{12}{11} mA. \tag{2.17}$$

We did these calculations by both methods, and they produce identical results. Thus, the first check of the solution is performed. An additional result verification is plugging the obtained values of i_1 and i_2 in (2.16) to confirm that these equations turn into identities. A curious reader should do this.

Having attained the circuit variables, i_1 and i_2, we can readily compute the branch values, I_{R1}, I_{R2}, and I_{R3}. Examining the circuit in Figure 2.5b gives

$$I_{R1} = i_1 = \frac{42}{11} mA,\ I_{R2} = i_2 = \frac{12}{11} mA, \text{ and } I_{R3} = i_1 - i_2 = \frac{30}{11} mA.$$

The application of KCL at the node requires

$$I_{R1} = I_{R2} + I_{R3},$$

which is satisfied with our numerical values.

Alternatively to KCL, we can check the results with KVL: Plugging the calculated values in (2.14), we check the KVL satisfaction for Mesh 1 as

$$R_1 \cdot i_1 + R_3 \cdot (i_1 - i_2) = V_1 \cdot \frac{42}{11} + 3 \cdot \left(\frac{42}{11} - \frac{12}{11}\right) = 12V.$$

We will receive $\frac{132}{11} = 12$, which validates our solution. Performing similar calculations for Mesh 2 using (2.15) confirms the correctness of our solution.

This discussion presents an idea of how mesh analysis works in general. To deepen our understanding of this method, let's consider an example. But before this, it's necessary to state that a mesh analysis can be applied only to *planar circuits*. A circuit is called planar if its schematic contains no branches that pass over or under any other branch of this circuit. In this regard, a nodal analysis, which doesn't have such restrictions, is a more powerful tool.

Example 2.3 Application of mesh analysis to a dc circuit.

Problem: Applying mesh analysis to the circuit shown in Figure 2.6, find all branch currents.

Solution: The plan is clear: Apply the mesh analysis and find mesh currents i_1, i_2, and i_3; then calculate the required branch currents and voltages.

Applying KVL to Mesh 1 yields

$$V_{2k\Omega} + V_{4k\Omega} + 8V = 0.$$

Pay attention to the polarities of $8-V$ and $1-V$ batteries: The mesh currents i_1 and i_2 enter their positive terminals, which means these batteries show the voltage drop, not voltage rise. In other words, these batteries consume power, not release it. This is why they are placed in the LHS of the KVL of both meshes.

Plug the mesh currents in Mesh 1 equation and write

$$2\cdot(i_1 - i_3) + 4\cdot(i_1 - i_2) = -8V.$$

Likewise, construct KVL for Mesh 2

$$5\cdot(i_2 - i_3) + 4\cdot(i_2 - i_1) = -1V.$$

In the similar manner, derive the third mesh equations as

$$8\cdot i_3 + 5\cdot(i_3 - i_2) + 2\cdot(i_3 - i_1) = 33V.$$

Pay attention to how we position the mesh currents to calculate the differences in each mesh

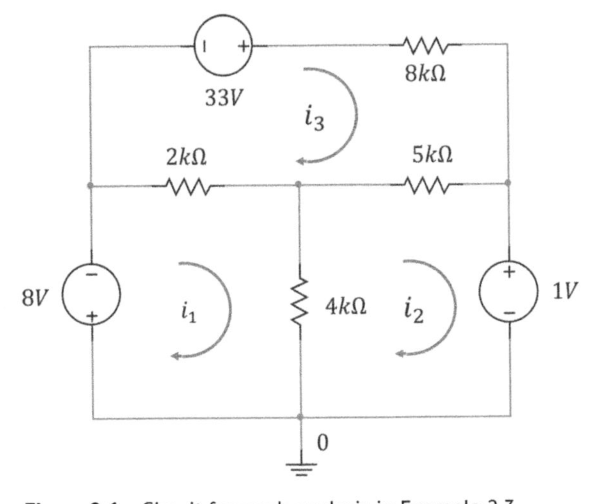

Figure 2.6 Circuit for mesh analysis in Example 2.3.

equation:

- For Mesh 1, current i_1 goes first in all subtractions.
- For Mesh 2, current i_2 is the first member in the difference of currents.
- The similar order is observed for current i_3.

This order for all currents is essential because it determines the current signs in the subsequent equations, as mentioned in the deliberation on (2.11). (See more on this issue in the *Discussion* Section of this example.)

The straightforward manipulations bring us the following three simultaneous mesh equations:

$$\begin{cases} 6i_1 - 4i_2 - 2i_3 = -8 \\ -4i_1 + 9i_2 - 5i_3 = -1 \\ -2i_1 - 5i_2 + 15i_3 = 33 \end{cases}. \tag{2.18}$$

Presenting (2.18) in a matrix format yields (See Sidebar 2S Solving the system of linear equations.)

$$\begin{vmatrix} 6 & -4 & -2 \\ -4 & 9 & -5 \\ -2 & -5 & 15 \end{vmatrix} \begin{vmatrix} i_1 \\ i_2 \\ i_3 \end{vmatrix} = \begin{vmatrix} -8 \\ -1 \\ 33 \end{vmatrix}. \tag{2.19}$$

We need to solve this equation system and find the mesh currents. Let's use Cramer's rule to determine $i_1, i_2, and\, i_3$. Figure 2.7a shows all the details of the main determinant calculations; Figure 2.7b demonstrates the calculations of the minor determinants Δ_1, Δ_2, and Δ_3. (Refer to Sidebar S2.)

You will recall that the solution to (2.18), that is, mesh currents $i_1, i_2, and\ i_3$, by Cramer's rule are determined as

$$i_1 = \frac{\Delta_1}{\Delta} = 1mA,$$

$$i_2 = \frac{\Delta_2}{\Delta} = 2mA,$$

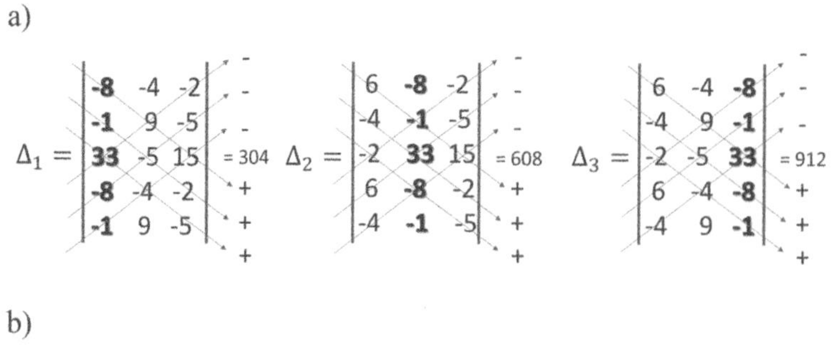

$\Delta =$
6 -4 -2 | -36
-4 9 -5 | -150 } = -426
-2 -5 15 | -240
6 -4 -2 | +810 } = 304
-4 9 -5 | +(-40) } = 730
+(-40)

a)

$\Delta_1 =$ | **-8** -4 -2 | **-1** 9 -5 | **33** -5 15 | **-8** -4 -2 | **-1** 9 -5 | = 304
$\Delta_2 =$ | 6 **-8** -2 | -4 **-1** -5 | -2 **33** 15 | 6 **-8** -2 | -4 **-1** -5 | = 608
$\Delta_3 =$ | 6 -4 **-8** | -4 9 **-1** | -2 -5 **33** | 6 -4 **-8** | -4 9 **-1** | = 912

b)

Figure 2.7 Determinant calculations for Example 2.3: a) Calculation of the main determinant; b) calculations of the particular determinants.

$$i_3 = \frac{\Delta_3}{\Delta} = 3mA.$$

Therefore, the mesh currents are found.

To find all the branches currents and voltages, we examine the circuit in Figure 2.6 and find:

$$I_{2k\Omega} = i_1 - i_3 = -2mA$$
$$I_{4k\Omega} = i_1 - i_2 = -1mA$$
$$I_{5k\Omega} = i_2 - i_3 = -1mA$$
$$I_{8k\Omega} = i_3 = 3mA$$

The current directions in the given circuit and their convention polarity signs are shown in Figure 2.8. To verify the results, we check that KVL is satisfied at each mesh as follows:

Mesh 1: $8V - 2k\Omega \cdot 2mA - 4k\Omega \cdot 1mA = 0$
Mesh 2: $1V + 4k\Omega \cdot 1mA - 5k\Omega \cdot 1mA = 0$
Mesh 3: $-33V + 8k\Omega \cdot 3mA + 5k\Omega \cdot 1mA + 2k\Omega \cdot 2mA = 0.$

The problem is solved.

Discussion:

The key point is to follow the *sign convention rule* discussed in Chapter 1. In this example, we arbitrarily assign clockwise directions for all mesh currents. Calculating the voltage across a component shared by two meshes, we follow the simple rule:

- The minuend is the current of the mesh in review.
- The subtrahend is the current from the shared mesh.
- The difference carries the value and the sign of the current flowing through this component.

For example, for Mesh 1, voltage across $R = 2k\Omega$ shared by Meshes 1 and 3 is calculated as

$$V_{2k\Omega} = R_{2k\Omega} \cdot (i_1 - i_3).$$

Then we calculate the branch current through $R_{2k\Omega}$ as

$$I_{2k\Omega} = i_1 - i_3 = -2mA.$$

The negative sign of this branch current tells us that this current flows against the assigned direction of the mesh current i_1, as Figure 2.8 shows. Equation 2.20 demonstrates this rule:

$$I_{2k\Omega} = i_1 - i_3 = -2\ mA \tag{2.20}$$

Branch current　Subtrahend

Minuend　Difference

Calculating voltage across a component that belongs to only one mesh is straightforward: Its current carries the value and the sign of the appropriate mesh current, and Ohm's law finds its voltage. In our example, such a component is resistor $R = 8k\Omega$ belonging to Mesh 3. Its current is

$$I_{8k\Omega} = i_3 = 3mA,$$

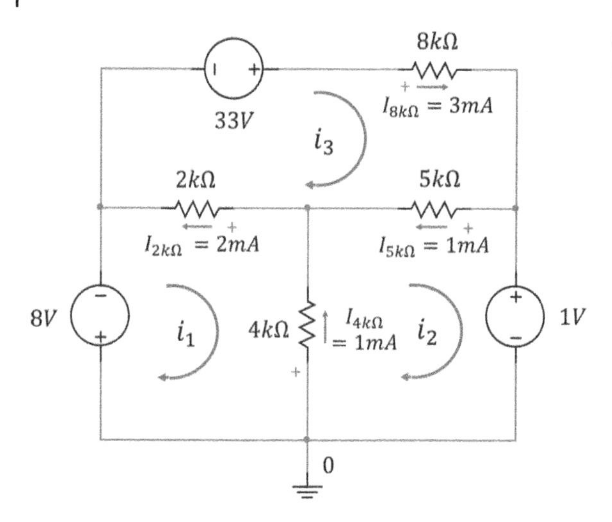

Figure 2.8 The circuit for Example 2.3 with branch currents and their polarities.

and its voltage is

$$V_{8k\Omega} = R_{8k\Omega} \cdot I_{8k\Omega} = 8k\Omega \cdot 3mA = 24V.$$

Understand and follow this procedure; it lets you to avoid mistakes in which an inexperienced person typically is trapped.

This is the end of Example 2.2.

2.2.2.2 Supermesh

Consider the circuit shown in Figure 2.9a. Using a mesh analysis for this circuit requires constructing two KVL equations for Mesh 1 and Mesh 2. But this is impossible because there is no way to know the voltage across $6-mA$ current source. If this difficulty looks familiar to you, you are correct: We encountered a similar problem in the nodal analysis, where we could not determine the current flowing through a voltage source between two non-reference nodes. To circumvent the problem, we created a *supernode* that excluded the tricky part of a circuit. Likewise, in the mesh analysis, we resort to *supermesh* to count out the current source in question.

Consider Figure 2.9b: Now, we loop around the periphery of the entire circuit, ignoring the defending current source. The excluded area is shaded, and the new loop—supermesh—is highlighted. Going along this supermesh, we construct the new KVL. However, we use the assigned mesh current at each resistor to compute its voltage. Thus,

$$\begin{aligned}
&-5V + V_{2k\Omega} + V_{3k\Omega} + V_{4k\Omega} \\
&= -5V + i_1 \cdot R_{2k\Omega} + i_2 \cdot (R_{3k\Omega} + R_{4k\Omega}) \\
&\Rightarrow -5V + 2i_1 + 7i_2 = 0.
\end{aligned}$$

Still, we have two unknown variables, i_1 *and* i_2, and only one equation. To add the second equation, we can use KCL at node A, which gives

$$i_1 + 6mA = i_2.$$

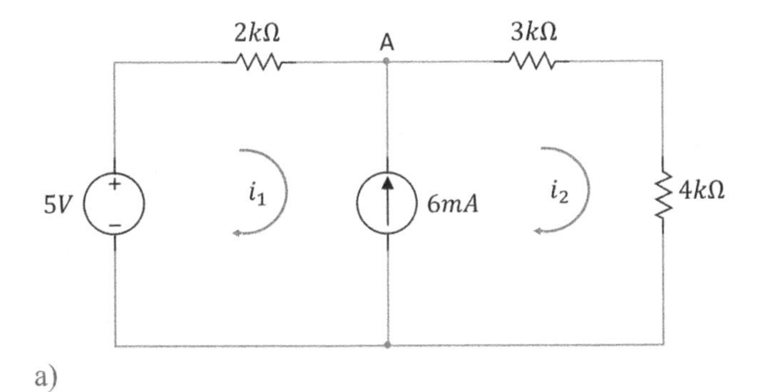

a)

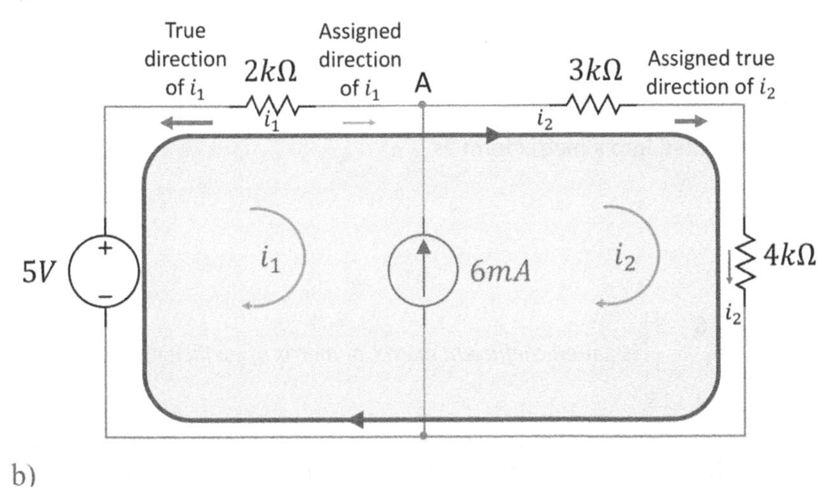

b)

Figure 2.9 Supermesh: a) The need for supermesh; b) supermesh technique.

Now, we have the equation system,

$$\begin{cases} 2i_1 + 7i_2 = 5 \\ i_1 - i_2 = -6 \end{cases}. \tag{2.21}$$

Solving (2.30) yields

$$i_1 = -4.1mA \text{ and } i_2 = 1.9\ mA.$$

A negative sign of i_1 indicates that this mesh current's assigned direction is incorrect. Figure 2.9b shows the assigned and the true directions of i_1. The correct directions of i_1 and i_2 make KCL at node A convincible: 6-mA current delivered by the source to this node splits into 4.1-mA current flowing into Mesh 1 and 1.9-mA current streaming into Mesh 2.

This is how a supermesh technique helps to resolve the issue of a defending current source.

In conclusion, it's worth emphasizing that the choice between nodal and mesh analysis lies in the nature of the problem: If nodal voltages are the subjects of a problem, then, naturally, the nodal analysis is a preferred method; for the search of mesh currents, the mesh analysis is a more favorable tool.

Sidebar 2S Solving the System of Linear Equations

Gaussian[3] elimination algorithm and its modifications

Example 2.1 demonstrated how to apply the *substitution method* to solve the system of two equations. However, this method is challenging to apply to the systems of three and more equations, and it is based on intuitive reasoning rather than on a structured computative approach. The *Gaussian elimination algorithm* enables us to solve the system of many equations and offers an analytical, algorithmic technique.

As an example, suppose the nodal analysis of a circuit produces the following three-equation system:

$$\begin{cases} 2v_1 - 6v_2 + 8v_3 = 40 \\ 4v_1 + 3v_2 + 5v_3 = 67 \\ 7v_1 - 9v_2 + 13v_3 = 80 \end{cases}. \tag{2S.1}$$

To start, (2S.1) must be presented into a matrix form as

$$\begin{vmatrix} 2 & -6 & 8 \\ 4 & 3 & 5 \\ 7 & -9 & 13 \end{vmatrix} \begin{vmatrix} v_1 \\ v_2 \\ v_3 \end{vmatrix} = \begin{vmatrix} 40 \\ 67 \\ 80 \end{vmatrix}. \tag{2S.2}$$

Here, the *square matrix* $\begin{vmatrix} 2 & -6 & 8 \\ 4 & 3 & 5 \\ 7 & -9 & 13 \end{vmatrix}$ is called *coefficient matrix, or matrix of coefficients*. Compare the rows and columns of the system equation (2S.1) with those in matrix (2S.2) to understand how this matrix is formed. Note that its *dimension* is 3*x*3, meaning that the matrix has three rows and three columns and learn that such a *matrix* is called *square*. Matrices of variables $\begin{vmatrix} v_1 \\ v_2 \\ v_3 \end{vmatrix}$ and constants $\begin{vmatrix} 40 \\ 67 \\ 80 \end{vmatrix}$ are three-rows and one-column matrices (vectors, in fact) whose dimensions are 3*x*1. Their origins become clear by comparison (2.7) and (2.8).

The equation system (S2.1) can also be presented in the *augmented matrix form* shown in (2S.3). This form saves time and increases clarity in the matrix manipulations.

$$\left(\begin{array}{ccc|c} 2 & -6 & 8 & 40 \\ 4 & 3 & 5 & 67 \\ 7 & -9 & 13 & 80 \end{array}\right). \tag{2S.3}$$

We imply that the reader is familiar with basics of the matrix operations and turn straight to the examples.

---

3 Carl Friedrich Gauss (1777–1855) was a German mathematician and physicist. He made immense contributions in mathematics, physics, and other sciences, including number theory, analysis, differential geometry, geodesy, magnetism, astronomy, and optics. In mathematics, his outstanding achievements earned him the name "Prince of Mathematics;" in physics, there was a Gaussian system of units before the International System of Units (SI units) was introduced in 1960. We still rely on his fundamental theorem of algebra, the least-squares method, and many other results. In nodal analysis, we apply an algorithm known as Gaussian elimination to solve the system of linear equations.

The Gaussian algorithm calls for reducing a square matrix to the *echelon form*, as shown in (2S.4)

$$\begin{vmatrix} a_{11} & a_{12} & a_{13} \\ a_{21} & a_{22} & a_{23} \\ a_{31} & a_{32} & a_{33} \end{vmatrix} \begin{vmatrix} \nu_1 \\ \nu_2 \\ \nu_3 \end{vmatrix} = \begin{vmatrix} B_1 \\ B_2 \\ B_3 \end{vmatrix} \Rightarrow \begin{vmatrix} a_{11} & a_{12} & a_{13} \\ 0 & a_{22} & a_{23} \\ 0 & 0 & a_{33} \end{vmatrix} \begin{vmatrix} \nu_1 \\ \nu_2 \\ \nu_3 \end{vmatrix} = \begin{vmatrix} B_1 \\ B_2 \\ B_3 \end{vmatrix} \tag{2S.4}$$

Original square matrix — *Echelon form of the* original square matrix

Here a_{11} through a_{33} are the coefficients (numbers). From (2S.4), all three variables can be found immediately because $V_3 = \frac{B_3}{a_{33}}$, V_2 is calculated from Row2, *and* V_1 is obtained from Row1.

The algorithm consists of the following steps demonstrated for the $3x3$ matrix shown in (2S.4):

Step 1. Make a_{21}, the leftmost member of the second row, zero.

Step 2. Make a_{31}, the leftmost member of Row3 zero.

Step 3. Make a_{32}, the central member of Row3 zero.

Step 4. Obtain the echelon form of the given matrix and compute the required values: $\boldsymbol{V}_3$, $\boldsymbol{V}_2$, and $\boldsymbol{V}_1$.

Step 5. Check the results by plugging the obtained variables into the original set of equations.

Bear in mind that the algorithm sets the goal to be obtained at each step, whereas the specific manipulations enabling to achieve this goal are left for you, who will carry out this work.

To reduce the original matrix to the echelon form, we can perform either of three operations or combinations of them:

- To multiply and divide a row by any constant except zero.
- To add (subtract) a row to (from) the other row.
- To interchange two rows.

As an example, let's apply the Gaussian elimination algorithm to (2S.2), aiming to reduce it to (2S.4) and find the variables.

Step 1. To turn a_{21} to zero, let's multiply Row 1, $R1$, by Row2, $R2$, subtract $R2$ from the product, and make this result a new $R2$. (Don't forget about the RHS.) That is, taking the given matrices,

$$\begin{vmatrix} 2 & -6 & 8 \\ 4 & 3 & 5 \\ 7 & -9 & 13 \end{vmatrix} \begin{vmatrix} V_1 \\ V_2 \\ V_3 \end{vmatrix} = \begin{vmatrix} 40 \\ 67 \\ 80 \end{vmatrix}, \tag{2S.2R}$$

and performing the suggested manipulations, we arrive to the required form:

$$(2R1) \Rightarrow \begin{vmatrix} 4 & -12 & 16 \\ 4 & 3 & 5 \\ 7 & -9 & 13 \end{vmatrix} \begin{vmatrix} V_1 \\ V_2 \\ V_3 \end{vmatrix} = \begin{vmatrix} 80 \\ 67 \\ 80 \end{vmatrix},$$

$$\text{and } (2R1) - R2 \rightarrow R2 \Rightarrow \begin{vmatrix} 2 & -6 & 8 \\ 0 & -15 & 11 \\ 7 & -9 & 13 \end{vmatrix} \begin{vmatrix} V_1 \\ V_2 \\ V_3 \end{vmatrix} = \begin{vmatrix} 40 \\ 13 \\ 80 \end{vmatrix}.$$

Step 2. To make a_{31}, the leftmost member of R3, zero, we need to eliminate 7 from R3. To achieve this goal, we multiply $R1$ by 7 and $R3$ by 2, subtract $R3$ from $R1$, and make the result the third row. That is,

$$(7R1)\text{ and }(2R3)\Rightarrow\begin{vmatrix}14 & -42 & 56\\0 & -15 & 11\\14 & -18 & 26\end{vmatrix}\begin{vmatrix}v_1\\v_2\\v_3\end{vmatrix}=\begin{vmatrix}280\\13\\160\end{vmatrix},$$

and

$$(7R1)-(4R3)\rightarrow R3\Rightarrow\begin{vmatrix}2 & -6 & 8\\0 & -15 & 11\\0 & -24 & 30\end{vmatrix}\begin{vmatrix}v_1\\v_2\\v_3\end{vmatrix}=\begin{vmatrix}40\\13\\120\end{vmatrix}.$$

Note that we keep $R1$ in its original form, which is achieved by dividing $R1$ by 7.

Step 3. To get a_{32}, the central member of $R3$, zero, we multiply $R2$ by −24 and $R3$ by 15, which produces

$$R2\cdot(-24)\text{ and }R3\cdot(15)\Rightarrow\begin{vmatrix}2 & -6 & 8\\0 & 360 & -264\\0 & -360 & 450\end{vmatrix}\begin{vmatrix}v_1\\v_2\\v_3\end{vmatrix}=\begin{vmatrix}40\\-312\\1800\end{vmatrix}.$$

Adding $R2$ to $R3$ makes new $R3$. Leaving $R2$ in its preceding form, i.e., dividing it by −24, we arrive at the echelon form of the original matrix.

$$R2+R3\rightarrow R3\text{ and }R2/(-24)\Rightarrow\begin{vmatrix}2 & -6 & 8\\0 & -15 & 11\\0 & 0 & 186\end{vmatrix}\begin{vmatrix}v_1\\v_2\\v_3\end{vmatrix}=\begin{vmatrix}40\\13\\1488\end{vmatrix}. \tag{S.5}$$

Step 4. Having obtained (S.5), we can readily compute the required values. First, $v_3=\frac{B_3}{a_{33}}$ must be computed as

$$186v_3=1488\Rightarrow v_3=\frac{1488}{186}=8.$$

Plugging $v_3=8$ into $R2$ of (S.5) results in

$$-15v_2+11\cdot v_3=13\Rightarrow v_2=\frac{-13+11\cdot 8}{15}=5.$$

Finally, v_1 can be calculated from R1 as

$$2v_1-6v_2+8v_3=40\Rightarrow 2v_1-6\cdot 5+8\cdot 8=40\Rightarrow v_1=3.$$

Therefore, the answer is

$$v_1=3,\ v_2=5,\text{ and }v_3=8.$$

Step 5. Check the results by plugging the obtained variables into the original set of equation (2S.1). If the answers are correct, the equations will turn to the set of identities.

$$\begin{cases} 2v_1 - 6v_2 + 8v_3 = 40 \\ 4v_1 + 3v_2 + 5v_3 = 67 \\ 7v_1 - 9v_2 + 13v_3 = 80 \end{cases} \Rightarrow \begin{cases} 2\cdot3 - 6\cdot5 + 8\cdot8 = 40 \\ 4\cdot3 + 3\cdot5 + 5\cdot8 = 67 \\ 7\cdot3 - 9\cdot5 + 13\cdot8 = 80 \end{cases}$$

Though the description of the Gaussian elimination algorithm might seem intimidating, it is, in fact, a fairly straightforward operation. The ease of its implementation will quickly come with practicing.

Modifications of the Gaussian elimination algorithm

There are modern modifications of the Gaussian elimination algorithm. The first of them calls for a final presentation of a given coefficient matrix in a *row-echelon form* shown in (2S.6)

$$\left(\begin{array}{ccc|c} 1 & a_{12} & a_{13} & B_1 \\ 0 & 1 & a_{23} & B_2 \\ 0 & 0 & 1 & B_3 \end{array}\right) \tag{2S.6}$$

Compare this form with (2S.3). Remember, B_1, B_2, *and* B_3 *are constants of a given system of equations received* ***after*** *all the manipulations required by the algorithm.* With this form, $v_3 = B_3$, $v_2 = B_2 - a_{23} \cdot v_3$, and $v_1 = B_1 - a_{12} \cdot v_2 - a_{13} \cdot v_3$.

The Gaussian algorithm's second and most important version is called *the Gauss-Jordan elimination algorithm.* This version calls for transforming a given matrix into *a reduced row-echelon form* presented in (2S.7).

$$\left(\begin{array}{ccc|c} 1 & 0 & 0 & B_1 \\ 0 & 1 & 0 & B_2 \\ 0 & 0 & 1 & B_3 \end{array}\right) \tag{2S.7}$$

As we learn soon, this matrix type is known as *the identity matrix.* Description of the Gaussian elimination algorithm shows the steps that lead to building the given matrix in an echelon form. Likewise, Figure 2S.1 shows the sequence of steps that should be performed to construct a matrix in a reduced row-echelon form but without the step descriptions. We are confident that our readers will quickly develop the necessary details of these manipulations.

As soon as the coefficient matrix is received in a *reduced row-echelon (identity) form*, the original equation system takes the format similar to (2S.2)

$$\begin{vmatrix} 1 & 0 & 0 \\ 0 & 1 & 0 \\ 0 & 0 & 1 \end{vmatrix} \begin{vmatrix} v_1 \\ v_2 \\ v_3 \end{vmatrix} = \begin{vmatrix} C_1 \\ C_2 \\ C_3 \end{vmatrix}. \tag{2S.8}$$

From (2S.8), all variables can be immediately determined as

$$v_1 = C_1,\ v_2 = C_2,\ \text{and}\ v_3 = C_3. \tag{2S.9}$$

This straightforward path to finding the variables explains the popularity of the Gauss-Jordan algorithm today.

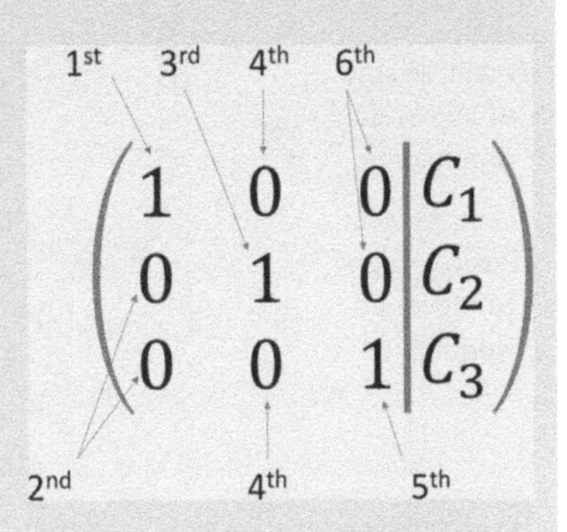

Figure 2S.1 The step sequence to build a matrix in a reduced row-echelon form in the Gauss-Jordan elimination algorithm.

Cramer's rule

This rule is named after Gabriel Cramer (1704–1752), a Genevan mathematician. He published his rule in 1750. The rule enables us to solve the system of linear equations of any order, though the complexity of calculations increases drastically with the order rise. Practically, Cramer's rule works reasonably well for the system of two or three equations.

We will discuss this rule using Equations S.1 and S.2 as examples. These equations are reproduced below for convenience.

$$\begin{cases} 2v_1 - 6v_2 + 8v_3 = 40 \\ 4v_1 + 3v_2 + 5v_3 = 67 \\ 7v_1 - 9v_2 + 13v_3 = 80 \end{cases} \tag{2S.1R}$$

and

$$\begin{vmatrix} 2 & -6 & 8 \\ 4 & 3 & 5 \\ 7 & -9 & 13 \end{vmatrix} \begin{vmatrix} v_1 \\ v_2 \\ v_3 \end{vmatrix} = \begin{vmatrix} 40 \\ 67 \\ 80 \end{vmatrix}. \tag{2S.2R}$$

Here are the steps to find the solution by using the Cramer's rule:

1) The variables $v_1,\ v_2, and\, v_3$ are found as follows:

$$v_1 = \frac{\Delta_1}{\Delta}, v_2 = \frac{\Delta_2}{\Delta}, \text{and} v_3 = \frac{\Delta_3}{\Delta}, \tag{2S.10}$$

where $\Delta, \Delta_1, \Delta_2,$ and Δ_3 are the *determinants*.

2) There are two methods to calculate a determinant:
   - The first one is displayed in Figures 2S.2a and 2S.2b, using determinant Δ as a sample. Here, we choose in sequence every member of the top row of the *coefficient matrix* and multiply them one after another by the determinants of the proper *minor matrix*. A minor 2x2 matrix is formed by crossing out the row and the column containing the multiplying member. Such a matrix is shaded in Figure 2S.2b. A minor determinant is calculated by mul-

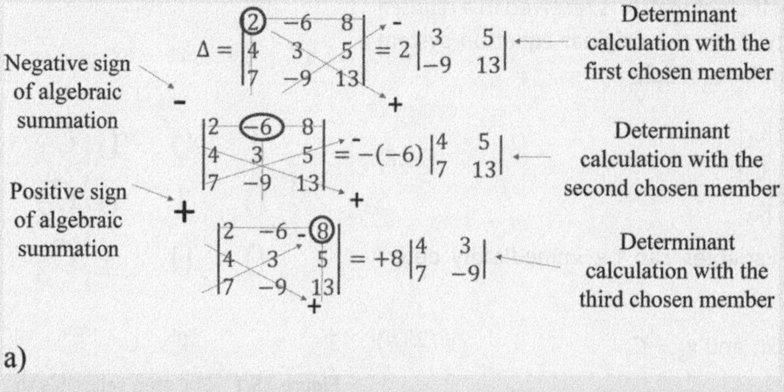

$$\Delta = \begin{vmatrix} 2 & -6 & 8 \\ 4 & 3 & 5 \\ 7 & -9 & 13 \end{vmatrix} = 2\begin{vmatrix} 3 & 5 \\ -9 & 13 \end{vmatrix}$$

$$\begin{vmatrix} 2 & -6 & 8 \\ 4 & 3 & 5 \\ 7 & -9 & 13 \end{vmatrix} = -(-6)\begin{vmatrix} 4 & 5 \\ 7 & 13 \end{vmatrix}$$

$$\begin{vmatrix} 2 & -6 & 8 \\ 4 & 3 & 5 \\ 7 & -9 & 13 \end{vmatrix} = +8\begin{vmatrix} 4 & 3 \\ 7 & -9 \end{vmatrix}$$

Negative sign of algebraic summation

Positive sign of algebraic summation

Determinant calculation with the first chosen member

Determinant calculation with the second chosen member

Determinant calculation with the third chosen member

a)

Figure 2S.2 Calculations of the main determinant of $3x3$ matrix by Laplace expansion: a) The general order of the determinant calculation using minor determinant; b) example of calculations with one chosen member and minor matrix; c) calculations by setting $5x3$ matrix; d) calculations for solving Equation 2S.1 using $5x3$ and $3x5$ matrices.

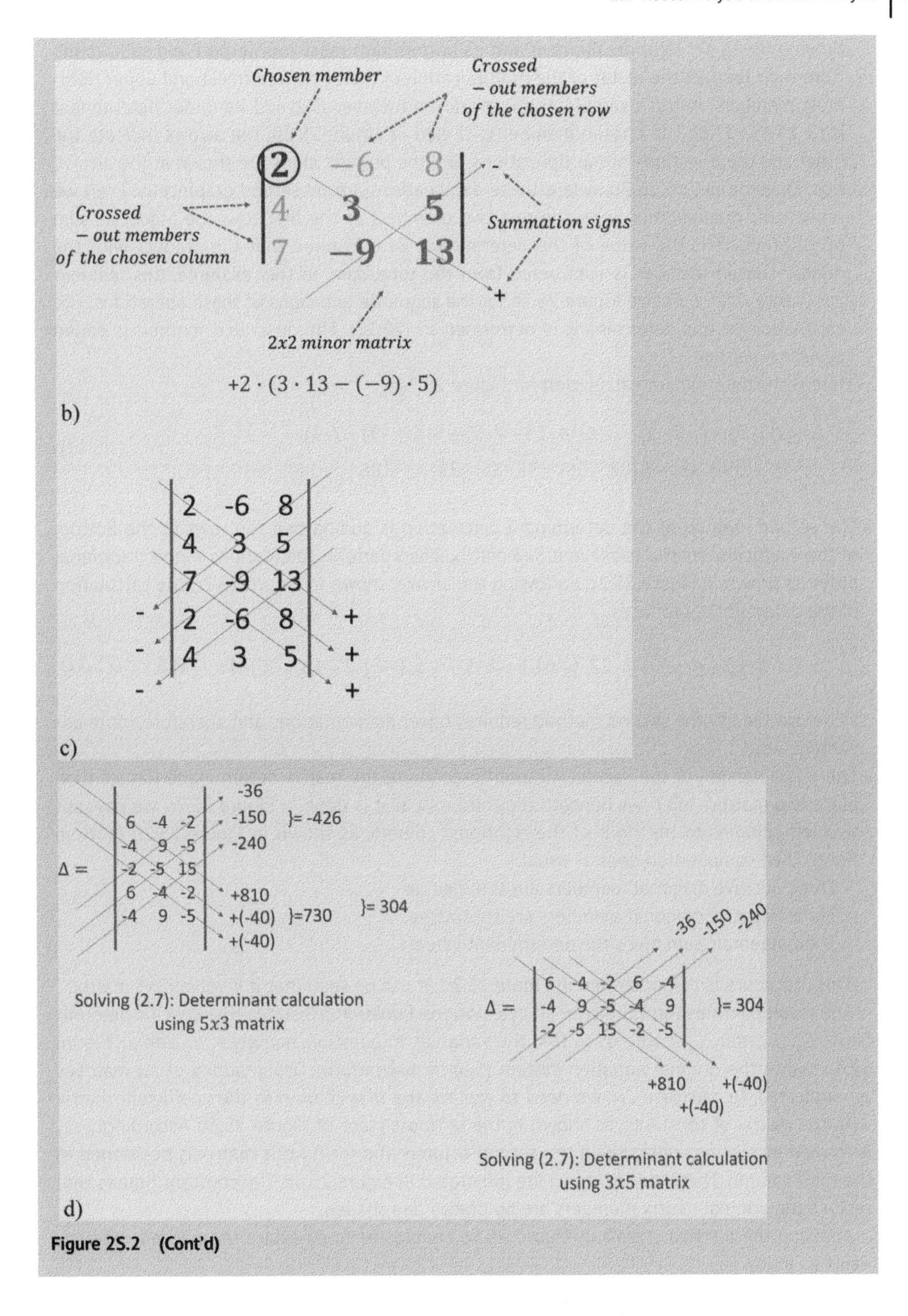

Figure 2S.2 (Cont'd)

tiplying the upper leftmost member and its bottom rightmost counterpart and subtracting from their product the result of the multiplication of the bottom leftmost and upper rightmost members. In this example, the first chosen member is 2, and its minor determinant is $(3\cdot 13-(-9)\cdot 5)$. The chosen member is circled in Figure 2S.2b; the arrows indicate the order and the directions of multiplications, and the product signs are shown at the arrows' tips. Examine Figure 2S.2b, where these explanations are presented graphically. Then we repeat this manipulation with the rest two members of the first row; the algebraic sum of all products is the value of the determinant. Bear in mind that the product with the middle-chosen member is subtracted from the total sum. In this example, this member is $-(-6)(4\cdot 13-7\cdot 5)$. See Figure 2S.2a for the sequence and signs of these operations. The calculation of this determinant is performed in (2S.11). The described method is called *Laplace expansion*.

Here is the determinant calculation for Figure 2S.2a:

$$\begin{aligned}\Delta &= 2\left(3\cdot 13-(-9)\cdot 5\right)-(-6)\left(4\cdot 13-7\cdot 5\right)+8\left(4\cdot(-9)-7\cdot 3\right)=\\ &=2(39+45)+6(52-35)+8(-36-21)=-186.\end{aligned} \tag{S.11}$$

- The second method of the determinant calculation is adding two top rows to the bottom of the coefficient matrix to make it 5*x*3 matrix. Then perform calculations in the traditional order, as shown in Figure 2S.2c. Following the arrows shown in Figure 2S.2c, the calculation of the determinant produces

$$\Delta = 2\cdot 3\cdot 13+4\cdot(-9)\cdot 8+7\cdot(-6)\cdot 5-8\cdot 3\cdot 7-5\cdot(-9)\cdot 2-13\cdot(-6)\cdot 4=-186, \tag{2S.12}$$

as before. Clearly, the second method requires fewer manipulations and, therefore, more advantageous.

- There is a version of the second method to compute the matrix determinant. Rather than place two matrix's top rows beneath the third row, as it is done in Figure 2S.2c, we now put two left columns to the right of the rightmost column, as shown in Figure 2S.2d (bottom right). Then computation goes as usual:
  - Three positive diagonal members are summed up,
  - Three negative diagonal members are subtracted,
  - Their algebraic sum gives the determinant value.

This procedure is demonstrated in Figure 2S.2d; it can be seen that it involves 3*x*5 matrix.

The result confirms the validity of all calculations. Equation S.8 states that specific determinants Δ_1, Δ_2, and Δ_3 enable us to find the variables in question, namely v_1, v_2, and v_3. Figure S2.3a shows the original equation system (S.8) in matrix form. Designations of all matrices are indicated. To calculate Δ_1, we need to replace the first column in the coefficient matrix with the matrix of constants, as shown in the leftmost place of Figure 2S.3b. Accordingly, Δ_2 is formed by placing constants in the second column, and for Δ_3 this matrix is positioned in the third column. These clarifications are illustrated in Figure 2S.3c. The constant figures that replace the original matrix members are boldfaced and shaded.

Applying the method shown in Figure 2S.3c enables us to calculate the needed determinants as follows:

$$\begin{aligned}\Delta_1 &= 40\cdot 3\cdot 13+67\cdot(-9)\cdot 8+80\cdot(-6)\cdot 5\\ &\quad -80\cdot 3\cdot 8-40\cdot(-9)\cdot 5-67\cdot(-6)\cdot 13=-9144+8586=-558.\end{aligned} \tag{2S.13}$$

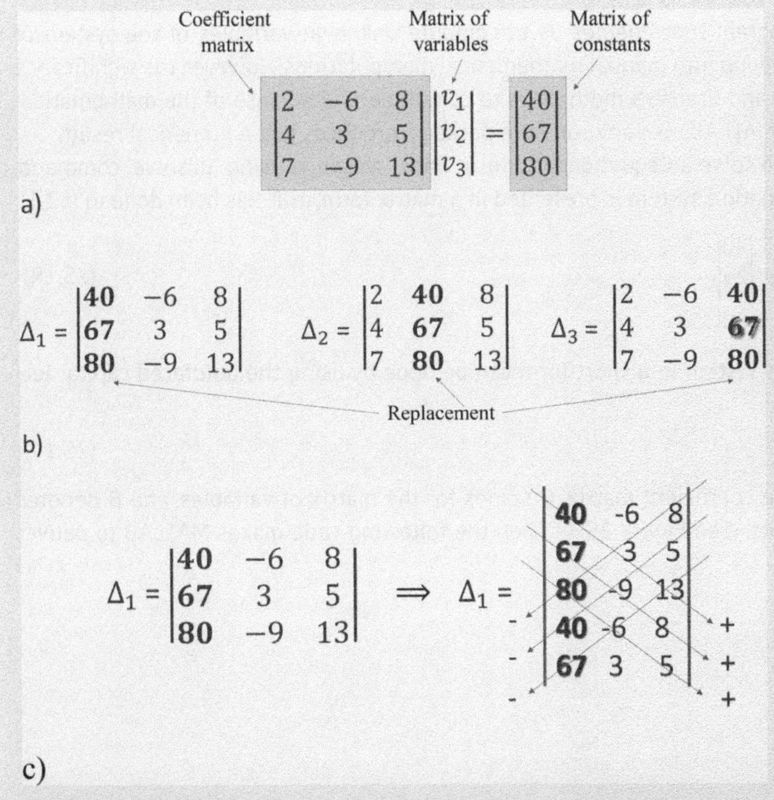

Coefficient matrix — Matrix of variables — Matrix of constants

$$\begin{vmatrix} 2 & -6 & 8 \\ 4 & 3 & 5 \\ 7 & -9 & 13 \end{vmatrix} \begin{vmatrix} v_1 \\ v_2 \\ v_3 \end{vmatrix} = \begin{vmatrix} 40 \\ 67 \\ 80 \end{vmatrix}$$

a)

$$\Delta_1 = \begin{vmatrix} \mathbf{40} & -6 & 8 \\ \mathbf{67} & 3 & 5 \\ \mathbf{80} & -9 & 13 \end{vmatrix} \quad \Delta_2 = \begin{vmatrix} 2 & \mathbf{40} & 8 \\ 4 & \mathbf{67} & 5 \\ 7 & \mathbf{80} & 13 \end{vmatrix} \quad \Delta_3 = \begin{vmatrix} 2 & -6 & \mathbf{40} \\ 4 & 3 & \mathbf{67} \\ 7 & -9 & \mathbf{80} \end{vmatrix}$$

Replacement

b)

$$\Delta_1 = \begin{vmatrix} \mathbf{40} & -6 & 8 \\ \mathbf{67} & 3 & 5 \\ \mathbf{80} & -9 & 13 \end{vmatrix} \Rightarrow \Delta_1 = \begin{vmatrix} \mathbf{40} & -6 & 8 \\ \mathbf{67} & 3 & 5 \\ \mathbf{80} & -9 & 13 \\ \mathbf{40} & -6 & 8 \\ \mathbf{67} & 3 & 5 \end{vmatrix}$$

- - - + + +

c)

Figure 2S.3 Calculation of the minor determinants for Equation 2S.1: a) Matrices designations; b) the column replacements; c) the Δ_1 determinant calculation using 5x3 matrix.

Therefore,

$$v_1 = \frac{\Delta_1}{\Delta} = \frac{-558}{-186} = 3.$$

Similarly,

$$\Delta_2 = 5702 - 6632 = -930 \text{ and } v_2 = \frac{\Delta_2}{\Delta} = \frac{-930}{-186} = 5.$$

Alike,

$$\Delta_3 = -3774 + 2286 = -1488 \text{ and } v_3 = \frac{\Delta_2}{\Delta} = \frac{-1488}{-186} = 8.$$

We can conclude that the Cramer's rule delivers the same results as the Gaussian elimination algorithm, which is expected.

Bear in mind that the volume and the complexity of the calculations with the Gaussian elimination algorithm and its versions increase significantly as the number of equations rises, which is not the case with Cramer's rule. Also, Cramer's rule can be easily automated, which is another advantage of this method.

MATLAB solution

MATLAB offers a program that enables us to compute unknown variables of the system of equations without delving into manual mathematical manipulations. However, the significance of knowing Gaussian and Cramer's methods is to comprehend the sense of the mathematical operations, whereas MATLAB, like any software package, produces just a numerical result.

The simplest way to solve an equation system, or *linear system*, is using "linsolve" command. It implies that the equation system is presented in a matrix form, as it has been done in (S2.1):

$$\begin{vmatrix} 2 & -6 & 8 \\ 4 & 3 & 5 \\ 7 & -9 & 13 \end{vmatrix} \begin{vmatrix} V_1 \\ V_2 \\ V_3 \end{vmatrix} = \begin{vmatrix} 40 \\ 67 \\ 80 \end{vmatrix}. \tag{2S.1R}$$

Presentation of this system in a short form can be done by using the boldfaced capital letters as

$$\boldsymbol{A} \cdot \boldsymbol{V} = \boldsymbol{B}. \tag{2S.14}$$

Here $\boldsymbol{A}$ represents the coefficient matrix, $\boldsymbol{V}$ stands for the matrix of variables, and $\boldsymbol{B}$ denotes the matrix of constants. (See Figure 2S.3.) Then the following code makes MATLAB to deliver the solution:

```
>> A = [2 -6 8;
4 3 5;
7 -9 13];
B = [40;67;80];
V=linsolve(A,B)
V =
3.0000
5.0000
8.0000
```

Thus, MATLAB provides the same answer that we received by manual calculations.

Inverse matrix and the other method of the MATLAB solution

Inverse matrix

In general, a matrix

$$\begin{vmatrix} a_{11} & a_{12} & a_{13} \\ a_{21} & a_{22} & a_{23} \\ a_{31} & a_{32} & a_{33} \end{vmatrix} \begin{vmatrix} x_1 \\ x_2 \\ x_3 \end{vmatrix} = \begin{vmatrix} b_1 \\ b_2 \\ b_3 \end{vmatrix} \tag{2S.15}$$

can be presented in the abbreviated form as

$$\boldsymbol{A} \cdot \boldsymbol{X} = \boldsymbol{B}. \tag{2S.16}$$

Solving this matrix equation means finding the vector $\boldsymbol{X}$, that is x_1, x_2, *and* $\boldsymbol{x}_3$. Equation 2S.16 seems to give a hint: To find X, we should "divide" $\boldsymbol{B}$ by $\boldsymbol{A}$, that is

$$\boldsymbol{X} \neq \frac{\boldsymbol{B}}{\boldsymbol{A}}$$

We deliberately use *not equal to* sign, $\neq$, because there is NO such a mathematical operation as matrix division. Instead, to find $\boldsymbol{X}$, we must multiply $\boldsymbol{B}$ by the *inverse matrix* $\boldsymbol{A}^{-1}$, as (2S.17) shows:

$$X = A^{-1} \cdot B. \tag{2S.17}$$

A matrix A^{-1} is called *inverse* to a *square matrix* A if

$$A \cdot A^{-1} = I,$$

where I is identity matrix given by

$$I = \begin{vmatrix} 1 & 0 & 0 \\ 0 & 1 & 0 \\ 0 & 0 & 1 \end{vmatrix}.$$

Note that an inverse matrix is defined only for square matrices. Review an algebra course to refresh your memory on the calculation of an inverse matrix and the matrix multiplication. (It is easy to apply the Cramer's rule in general and the inverse matrix in particular by using the graphing calculators.)

Surely, you can use MATLAB, which computes for you a required inverse matrix and delivers the answer to your specific problem in nodal and mesh analyses. Finding the inverse matrix enables us to verify our solution to a given problem. It is an unambiguous way to check our result. A curious reader is invited to verify all the results obtained in this section by calculating the inverse matrices.

Here how our example (S.1) works with MATLAB when we use the inverse matrix: We instruct MATLAB to compute the inverse matrix A^{-1} (denoted as AI) and multiply the inverse matrix A^{-1} by matrix of constants B to find the variables matrix of variables B, that is v_1, v_2, and v_3:

```
>>A = sym([2, -6, 8;4, 3, 5;7, -9, 13]); %Presenting matrix A;
observe the syntaxis.
AI = inv(A)
B = [40;67;80]; %Separating the numbers by semicolons, we make B a
3x1 matrix A.
V = AI*B % Command MATLAB to find the variables.
AI =
[-14/31, -1/31, 9/31]
[17/186, 5/31, -11/93]
[ 19/62, 4/31, -5/31] %MATLAB calculates the inverse matrix.
V =
3
5
8 %Here are the answers.
```

Below is the other version of the *MATLAB* code[4] for finding the solution to (S.1) by using the *inverse matrix*:

syms v1 v2 v3% Define the system of equations eqn as eqn1, eqn2, eqn3

```
eqn1 = 2*v1 - 6*v2 + 8*v3 == 40;
eqn2 = 4*v1 + 3*v2 + 5*v3 == 67;
eqn3 = 7*v1 - 9*v2 + 13*v3 == 80;
eqn=[eqn1; eqn2; eqn3]; % This line allows for printing eqn as the
system of equations eqn1,
```

4 Prepared by Ina Tsikhanava.

```
%eqn2, eqn3 defined above.
[eqn,B] = equationsToMatrix([eqn1, eqn2, eqn3], [v1, v2, v3]) %
This function converts
%equations eqn1, eqn2, eqn3 into the matrix form as eqn and B. The
second input v1, v2, v3
%define the independent variables to be found. The LHS of equation
[eqn,B] command to print
%eqn and B in matrix form.
inverse_eqn=inv(eqn) %This line requests MATLAB to invert matrix
eqn (i.e., eqn^(-1))
v=inverse_eqn*B %This line instructs MATLAB to find variables v
(v1,v2,v3) by multiplying
%inverse_eqn and matrix B.
eqn =
[2, -6, 8]
[4, 3, 5]
[7, -9, 13]
B =
40
67
80
inverse_eqn =
[-14/31, -1/31, 9/31]
[17/186, 5/31, -11/93]
[19/62, 4/31, -5/31]
v =
3
5
8
```

Therefore, we have one more confirmation that our solution is correct. This approach is more sophisticated than that using "linsolve" command, but the reward is obtaining the inverse matrix, which might be essential in certain applications.

2.3 Thevenin's Theorem and Norton's Theorem

2.3.1 Thevenin's Theorem

This theorem states that

> *a two-terminal linear circuit, regardless of the number of its components, can be replaced by a circuit consisting of a voltage source, V_{Th}, and a series impedance, Z_{Th}. Here, V_{Th} and Z_{Th} are Thevenin's voltage and impedance to be defined shortly.*

Notes:

1) Here and in our further considerations, we often refer to *linear circuits*. The term *linear* means that all mathematical operations describing the circuit obey two following rules:

$$f(x)+f(y)=f(x+y) \tag{2.22a}$$

and

$$f(kx)=kf(x), \tag{2.22b}$$

or, combining both equations,

$$f(kx)+f(ky)=kf(x+y) \tag{2.22c}$$

where k is a constant. Equation 2.22a demonstrates the *additivity* rule, (2.22b) shows *homogeneity* property, and $f(x)$ that satisfies (2.22c) is called *homogeneous linear function*. For example, the sum of $v_1 = R \cdot i_1$ and $v_2 = R \cdot i_2$ is given by $v_1 + v_2 = R(i_1 + i_2)$ because $v_n = R \cdot i_n$ is a *homogeneous linear function*. However, $y = \sin(3x)$ is **NOT** equal to $3\sin(x)$; also, the sum of $y_1 = \sin(3x)$ and $y_2 = \sin(5x)$ is equal to $y_1 + y_2 = \sin(3x) + \sin(5x)$, but **NOT** to $\sin(3x + 5x)$ because the *sine* is a nonlinear function. *Ohm's law, $i = Rv$*, is a linear proposition because it states that current through a resistor is proportional to the applied voltage, but it does NOT describe $i - v$ relationship in a *diode* whose current is NOT proportional to the voltage.

2) *Terminal* is the end point in a circuit. A *two-terminal circuit* is a standard circuit whose input or output consists of a pair of terminals. Either voltage is applied across these terminals or current flows through these terminals into or out of the circuit. Terminals a and b in Figure 2.10 are the input terminals for the load circuit and they are the output terminals for the source RLC circuit (shaded).

The theorem in discussion is named after Leon Charles Thevenin[5] who introduced it in 1883. Figure 2.10 illustrates this theorem. The actual RLC circuit (shaded) is shown in Figure 2.10a. It is connected with the load at $a - b$ terminals; in essence, this circuit serves as a source for the load. In Figure 2.10b, the Thevenin equivalent circuit, consisting of only a Thevenin's source, V_{Th}, and Thevenin's impedance, Z_{Th}, is shown. Figure 2.10c demonstrates that the load sees the voltage v and current i supplied by the equivalent circuit. In a dc circuit, the v voltage is given by

$$v = V_{Th} - iR_{Th}. \tag{2.23}$$

The circuit simplification achieved by Thevenin's theorem enables us to calculate the voltage across and current through the load without analyzing the entire actual circuit. Indeed, if both circuits shown in Figures 2.10a and 2.10b have the same voltages and currents at their $a - b$ terminals, then the voltage and current provided by the Thevenin equivalent circuit to the load are equal to those supplied by the actual circuit. The application of the Thevenin equivalent circuit becomes especially beneficial at the circuit's design and when a load is variable.

But how can we find those V_{Th} and R_{Th}? There are two approaches to obtain these quantities. First, and the easiest way, is to *measure* them. Review Figure 2.10a again: The actual circuit will remain open at the $a - b$ terminals if we disconnect the load. Thus, the voltage across these terminals is Thevenin's voltage, V_{Th}. Use a digital multimeter, DMM, and measure it. What about the current? The current flowing from terminal a to terminal b *is* the current generated by the actual circuit. Thus, shortening the actual circuit at the terminals a and b and measuring the short-circuit

5 Leon Charles Thevenin (1857–1927) was a French telegraph engineer. After graduation from the Ecole Polytechnique—the premier French engineering school—he worked in telegraphy all his professional life. Based on Ohm's and Kirchhoff's laws, which he studied closely, Thevenin developed his theorem whose acceptance by the world electrical industry much exceeded his expectations. He was a modest and honest man with a variety of interests; he can serve as a role model of a true engineer.

$$v = V_{Th} - iR_{Th}.$$

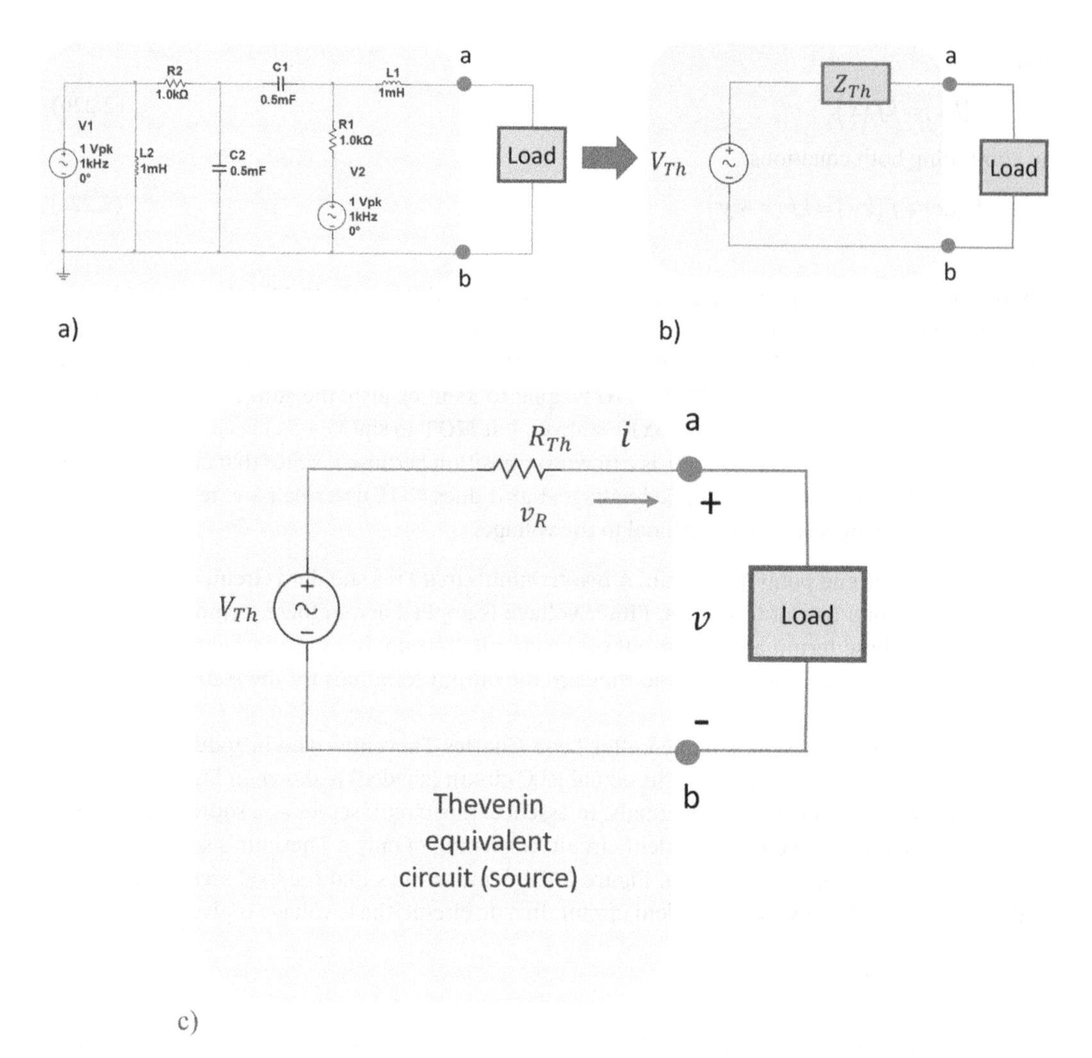

Figure 2.10 Thevenin's theorem: a) Actual circuit (shaded) with a load; b) the Thevenin equivalent circuit; c) Thevenin's voltage and current as seen by the load.

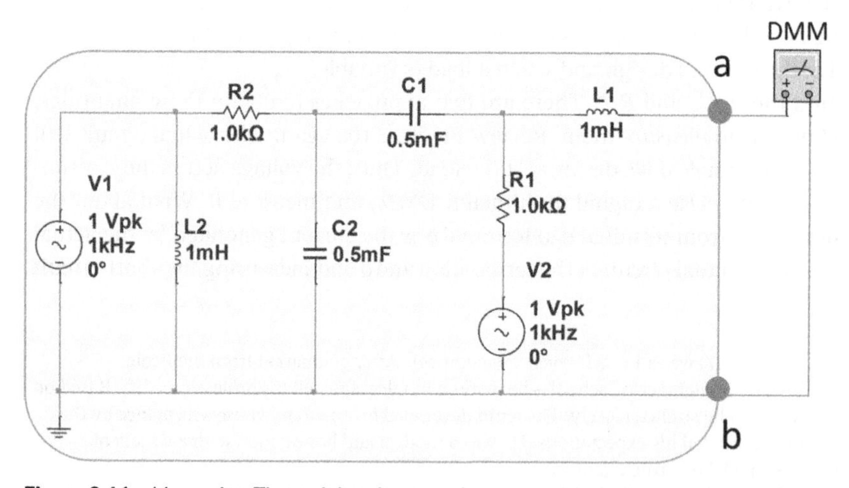

Figure 2.11 Measuring Thevenin's voltage and current with digital multimeter, DMM.

current we obtain I_{Th}. (This procedure means physically attaching the ammeter's leads to a and b terminals.) See Figure 2.11. Thus,

> *to find V_{Th} and I_{Th}, measure voltage at the open-circuit terminals and current at the short-circuit terminals.*

This is where this well-known rule of measuring Thevenin's parameters came from. (Students are rightly afraid of operating with a short circuit because, theoretically, the short-circuit's current tends to infinity. In reality, however, an ammeter has minimal but finite resistance whose value varies from milliohms to several ohms.) Knowing V_{Th} and I_{Th}, we compute

$$R_{Th} = \frac{V_{Th}}{I_{Th}},$$

by Ohm's law. This is how all Thevenin's parameters can be obtained by measurements. In practice, however, to avoid measuring a short-circuit current, it's better to use a *potentiometer*. We need to connect a potentiometer instead of a load and vary its resistance until the voltage across its fixed and variable contacts will be equal to $\frac{V_{Th}}{2}$. The resistance between these two contacts is R_{Th}. Indeed, if $R_{Th} = R_L$ in Figure 2.10b, then V_{Th} is evenly distributed between these two resistances. (Sketch the schematic of the experimental setup for this measurement to clarify this idea.)

Notes:

- Don't confuse the Thevenin voltage V_{Th} measured at the *opened $a-b$ terminals* with voltage v *applied to the load*. This difference is clearly shown in Figure 2.10c and demonstrated by Equation 2.14. A similar statement is valid for the currents: Current i would feed the load, whereas the current I_{Th} is produced by the Thevenin equivalent circuit (source) and measured at the *shortened $a-b$ terminals*. This difference becomes more apparent soon when we'll discuss Norton's theorem.
- Thevenin equivalent circuit provides the characteristics of the original source circuit at the *circuit-load interface* only. It doesn't reproduce any internal characteristics of an actual source circuit. This statement is also valid for the Norton equivalent circuit to be discussed shortly.

Thevenin's theorem is the same for DC and AC circuits; clearly, mathematical manipulations for AC circuits are more sophisticated due to the use of complex numbers. Specifically, we need to use the *phasors* instead of numerical values for currents, voltages and resistances. Also, this procedure works well for *independent sources* only. (To remind, an independent source produces its signal without external control, whereas a *dependent source* is controlled by external voltage or current. In this book, we mainly consider independent sources.)

How can we find the parameters of the Thevenin equivalent circuit by calculations? Figure 2.10b shows that we need to determine Z_{Th} and V_{Th}. Well, Z_{Th}, by its very nature, is the impedance of the source circuit without sources themselves and a load. Thus, *we need to remove the load, leave*

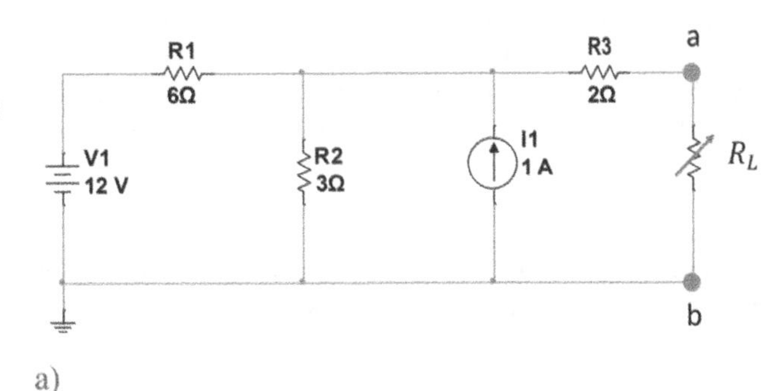

Figure 2.12 Building the Thevenin equivalent circuit for Example 2.4: a) Actual circuit; b) circuit for calculation of R_{Th}; c) circuit for V_{Th} calculation; d) the Thevenin equivalent circuit.

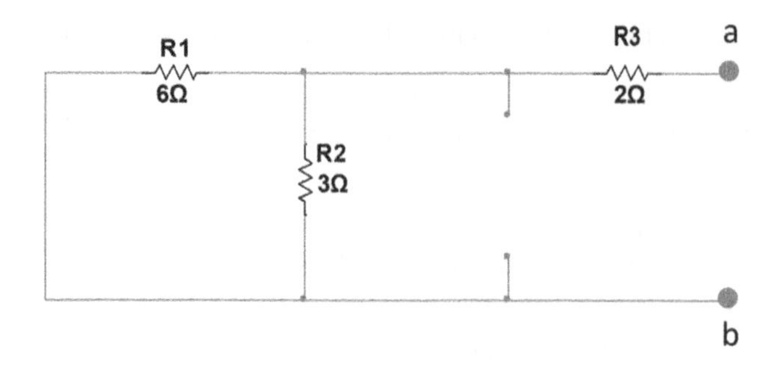

b)

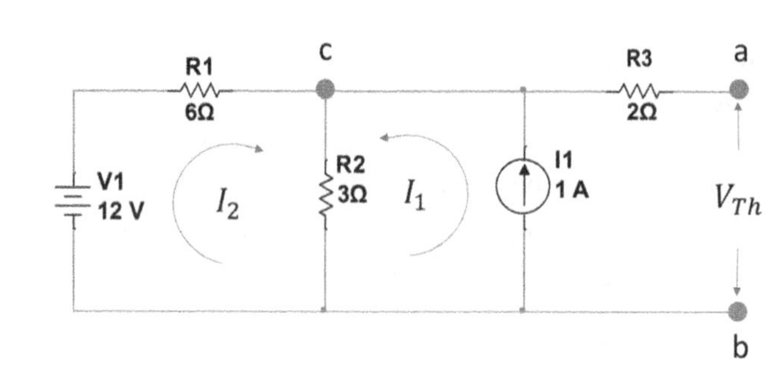

c)

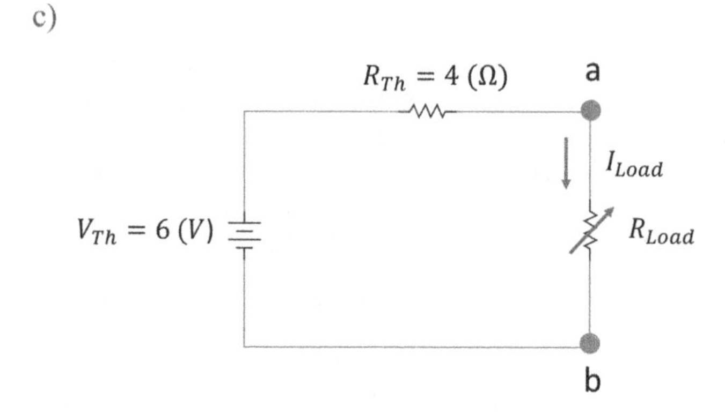

d)

Figure 2.12 **(Cont'd)**

terminals a and b open, and replace all voltage sources by the short circuits and all current sources by the open circuits. Then, we need to calculate Z_{Th} of the remaining passive circuit. See Figure 2.12a. To find V_{Th}, all sources must be placed back, and the calculation of V_{Th} at the open $a-b$ terminals must be performed. See Figure 2.12b. The procedure sounds intimidating, but the following example will mitigate its understanding and implementation.

Example 2.4 Building Thevenin equivalent circuit.

Problem: The actual circuit is shown in Figure 2.12a. Build its Thevenin equivalent, and determine the load current *iA* if R_L changes from 8Ω to 16Ω to 56Ω.

Solution: To build a Thevenin equivalent circuit, we need to find R_{Th} and V_{Th}. The Thevenin resistor, R_{Th}, is a resistor of the given actual circuit without sources and load. Thus, we replace the 12-V dc source with a short circuit and 1-A current source by an open circuit and leave terminals *a* and *b* open, as shown in Figure 2.12b. Then,

$$R_{Th} = 6 \| 3 + 2 = 4\,(\Omega).$$

Thevenin's voltage V_{Th} *is the voltage produced by the actual circuit at its open terminals a* and *b*, as Figure 2.12c shows. The nodal analysis at node *c* of this circuit yields

$$\frac{12 - V_{Th}}{6} + 1 = \frac{V_{Th}}{3}.$$

Note that there is no current through R_3 because $a-b$ terminals are open; this is why $V_C = V_{Th}$. Solving for V_{Th} yields

$$V_{Th} = 6\,(V).$$

Therefore, the required Thevenin equivalent circuit is found and shown in Figure 2.12d.

Current I_{Load} is calculated as

$$I_{Load} = \frac{V_{Th}}{R_{Th} + R_L}.$$

Hence, $I^{8}_{Load} = 0.5A$, $I^{16}_{Load} = 0.3A$, and $I^{56}_{Load} = 0.1A$.

The problem is solved.

Discussion:

- Notice how easily the Thevenin equivalent circuit enables us to calculate currents for a variable load. Without the Thevenin method, we'd have to repeat the entire circuit analysis for each load's value.
- In calculations of the Thevenin's parameters we can use any method of classical circuit analysis. For example, instead of the *nodal analysis* employed in this example, we can use *mesh analysis* as follows:

$$-12 + 6I_2 + 3(I_1 + I_2) = 0 \text{ and } I_1 = 1.$$

Solve for I_2 and get

$$I_2 = 1(A).$$

Now,

$$V_{Th} = 3(I_1 + I_2) = 6\,(V),$$

as before.

The other option is using the *source transformation*. (See Section 1.4.) After transforming $V_1 = 12(V)$ and $R_1 = 6(\Omega)$ into a current source, the actual circuit in Figure 2.12a takes the form shown in Figure 2.13a. Combining two parallel resistors R_1 and R_2, we obtain the circuit shown in Figure 2.13b, from which we can readily find V_{Th} as

$$V_{Th} = (I_1 + I_2)R_3 = 6\,(V).$$

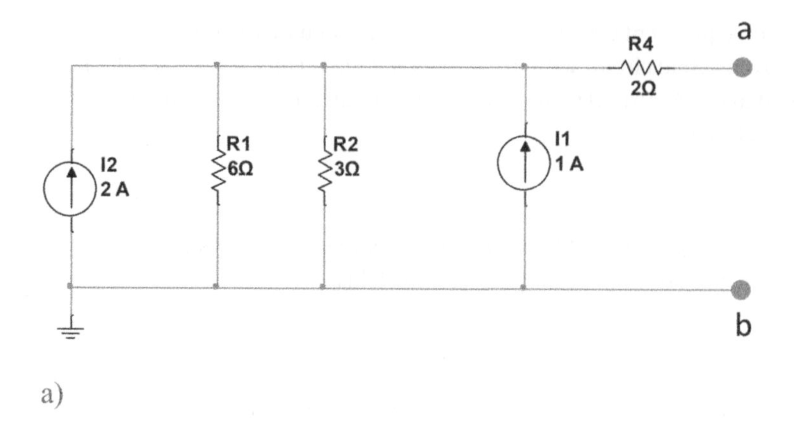

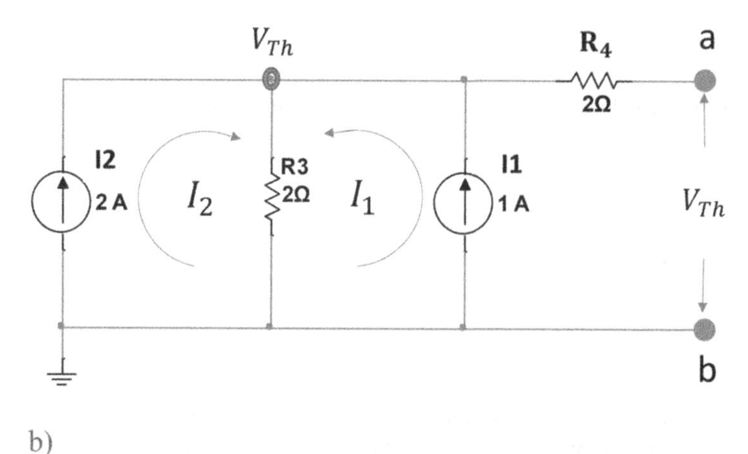

Figure 2.13 Finding V_{Th} by source transformation for Example 2.4: a) Transformation of the 12-V source into 2-A source in the actual circuit; b) the equivalent circuit for the V_{Th} calculation.

- What if we need to find the load resistance that delivers 5 V to the load? Thevenin equivalent circuit makes the solution to this design problem an easy matter. Indeed, Figure 2.12d shows that V_{Load} can be readily found by the voltage-divider rule as

$$V_{Load} = \frac{R_{Load}}{R_{Th} + R_{Load}} V_{Th} = \frac{R_{Load}}{4 + R_{Load}} 6 = 5V.$$

Solving this equation for R_{Load}, we calculate $R_{Load} = 20\ \Omega$. Hence, the application of Thevenin's theorem significantly alleviates designing an electrical circuit.

This is the end of Example 2.4.

2.3.2 Norton's Theorem

This theorem states that

> *a two-terminal linear circuit, regardless of the number of its components, can be replaced by a circuit consisting of a current source, I_N, and a parallel impedance, Z_N.*

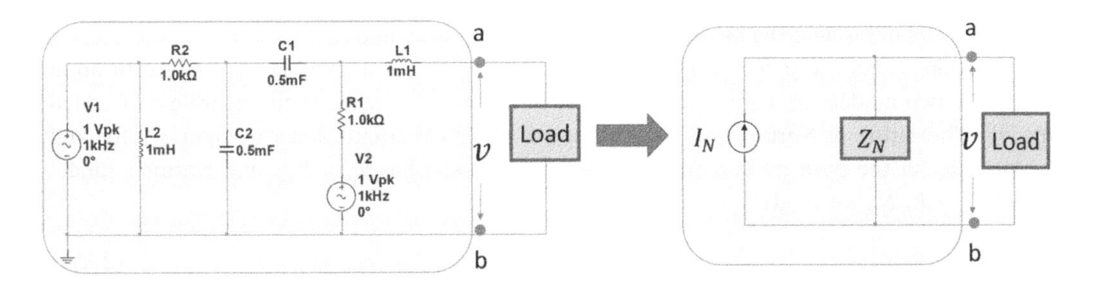

a)

b)

Figure 2.14 Norton's theorem: a) The concept; b) comparison of Thevenin's and Norton's theorems.

Figure 2.14 visualizes Norton's theorem. If we compare this figure with Figure 2.10, where Thevenin's theorem is pictured, we observe that the concept of both theorem is the same: *Replace a complex two-terminal circuit with the circuit consisting of a source and a resistor.* Recall the source transformation discussed in Section 1.4 and realize that *Norton's theorem is a source-transformed Thevenin's theorem.* This statement becomes more evident if we review Figure 2.14b, where these two sources are shown. The resistance $R = R_{Th} = R_N$ is the same for both models because this is the resistance that a load "sees" at the open-circuit $a-b$ terminals of both Thevenin's and Norton's circuits, and these circuits are equivalent.

We should not be surprised by such a similarity between these two theorems; after all, Edward L. Norton,[6] in developing his proposition in 1926, was inspired by Thevenin's theorem.

Reviewing the Thevenin model at the LHS of Figure 2.14b, we see that the voltage v, which this source supplies to a load, is given by

$$v = V_{Th} - v_R = V_{Th} - iR_{Th}, \tag{2.24a}$$

as in (2.23). In addition, from a Norton model, we find

$$i = I_N - i_{RN} = I_N - \frac{v}{R_N}. \tag{2.24b}$$

6 Edward L. Norton (1898–1983) worked for the famous Bell Laboratories most of his professional life. His areas of interest included acoustics and electrical engineering. Graduated from MIT with BS and Columbia University with MS, he was well known for applying theory masterly and his intuition to design a new electrical system for telephony and data transmission. In his work, he extensively used Thevenin's theorem. The idea known today as Norton's theorem was born when he had to design the current-driven circuit. His published legacy includes 19 patents, 3 papers, and numerous technical memorandums, one of which contains his theorem.

These two equations describe the linkage among v and V_{Th} and i and I_N.

To find the relationship between Thevenin's and Norton's parameters, we must bear in mind that these two models are equivalent because a load "sees" the same input regardless of which model—Thevenin's or Norton's—is used to replace an actual circuit. Refer to Figures 2.10a and 2.14a. Thus, at the *open a – b terminals*, Thevenin's model has $V_{oc} = V_{Th}$, and Norton's model shows $V_{oc} = R_N I_N$, which gives

$$V_{Th} = R_N I_N. \tag{2.25a}$$

At the short-circuit terminals, the current for Thevenin's model is $I_{sc} = \frac{V_{Th}}{R_{Th}}$, whereas Norton's model has $I_{sc} = I_N$. Therefore,

$$I_N = \frac{V_{Th}}{R_{Th}}. \tag{2.25b}$$

Equations 2.25a and 2.25b show the relationship between the parameters of Thevenin's and Norton's models. The identity

$$R_{Th} = R_N = R \tag{2.25c}$$

can be readily proved if we consider (2.25a) and (2.25b) as a system of equations.

Let's turn to the examples.

Example 2.5 Building the Norton equivalent circuit

Problem: Find the Norton's equivalent to the circuit shown in Figure 2.15a.

Solution: First, let's find R_N. We repeat the procedure used in working with a Thevenin model in Example 2.3: Leave the open circuit instead of a current source and short circuit in place of a voltage source. The circuit obtained after these transformations is shown in Figure 2.15b. Straightforward calculations produce

$$R_N = (R_2 + R_3) \| R_4 = (2 + 4) \| 3 = 2\ \Omega.$$

This is the Norton resistance as seen by the load.

To find I_N, we shorten the circuit at the $a - b$ terminals, which excludes R_4 from consideration.

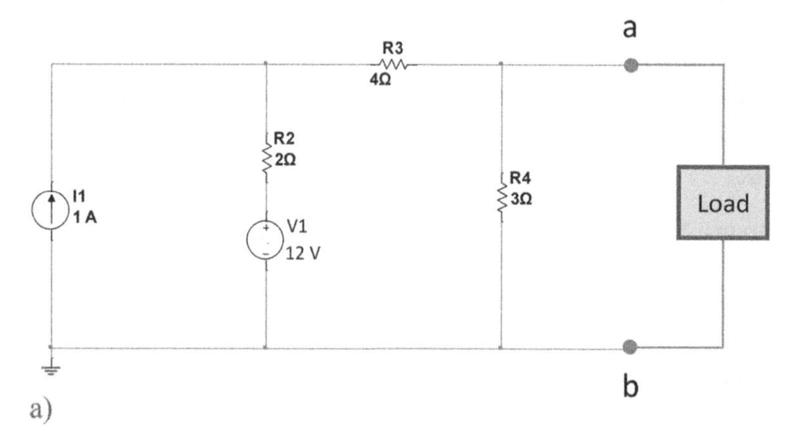

Figure 2.15 Finding the Norton equivalent circuit for Example 2.5: a) Actual circuit; b) finding R_N; c) finding I_N; d) measuring I_N and V_{Th}.

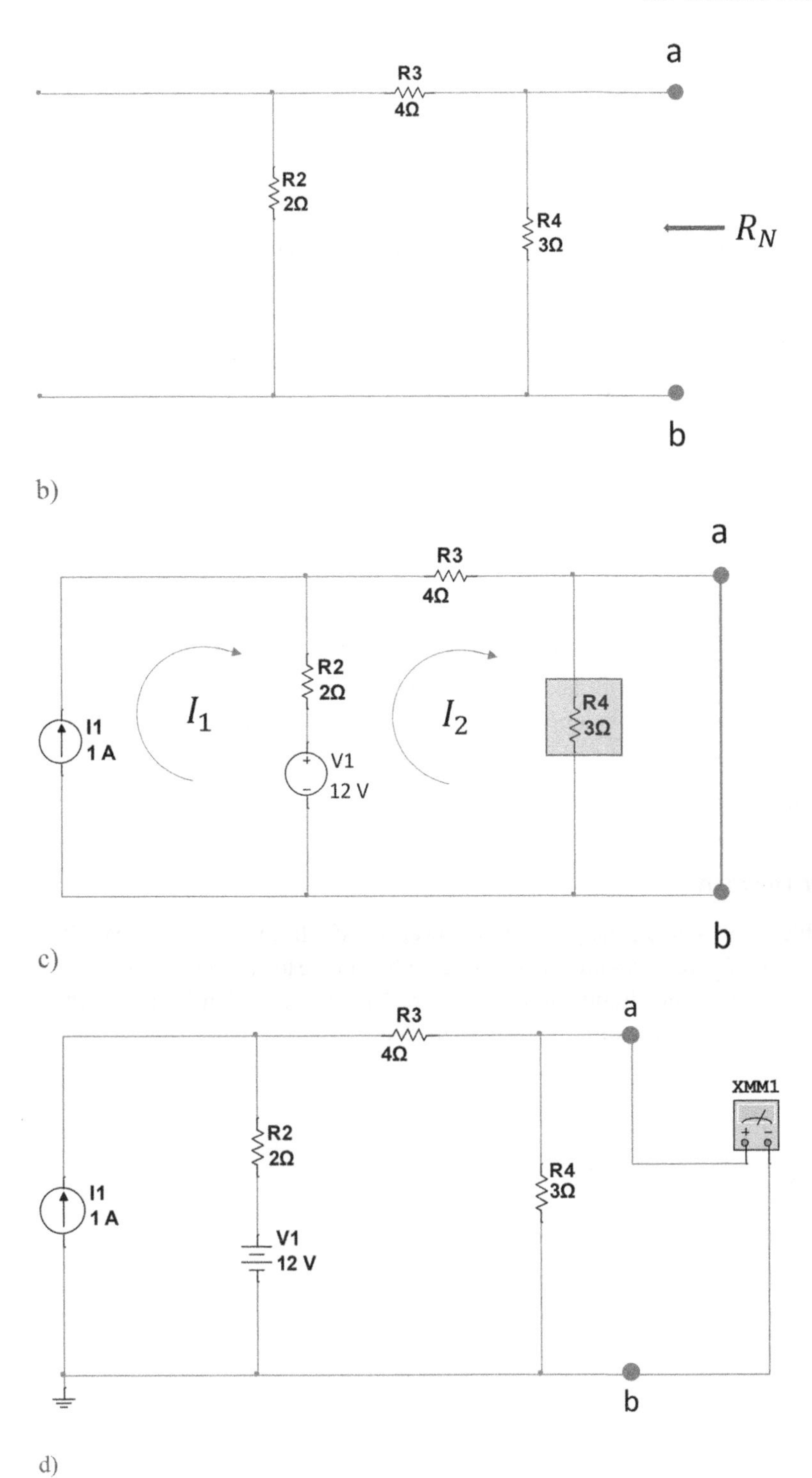

Figure 2.15 (Cont'd)

Then, using the mesh analysis, we compose two following simultaneous equations and calculate the Norton current:

$$\begin{cases} I_1 = 1A \\ I_2(R_2 + R_3) - 12 - I_1 R_2 = 0 \end{cases} \Rightarrow I_2 = I_N = 2.33A.$$

Therefore, we find the Norton equivalent circuit whose schematic is shown previously in Figure 2.14b and whose parameters are

$$R_N = 2\ \Omega \text{ and } I_N = 2.33A.$$

We, of course, can find I_N by other methods, for example as

$$I_N = \frac{v_{oc}}{R_N} = \frac{V_{Th}}{R_{Th}}.$$

The problem is solved.

Discussion:

- Remember that V_{Th} is the voltage measured at the open-circuit terminals a and b and $R_{Th} = R_N$. We leave it for you as an exercise to find that in this example $V_{Th} = 4.66V$.
- If we attach the DMM to the $a-b$ terminals, and measure in sequence current and voltage, we experimentally obtain the Norton parameters calculated above. The experimental setup is shown in Figure 2.15d.

This is the end of Example 2.5.

2.3.3 Power Transfer Theorem

Thevenin theorem enables us to analyze the problem of power transfer from a source to load. If a Thevenin model with V_{Th} and R_{Th} replaces the real source circuit, and the load is represented by its ohmic resistance R_{Load}, we obtain the circuit shown in Figure 2.16. The question is, under what condition the source will transfer the maximum power to the load.

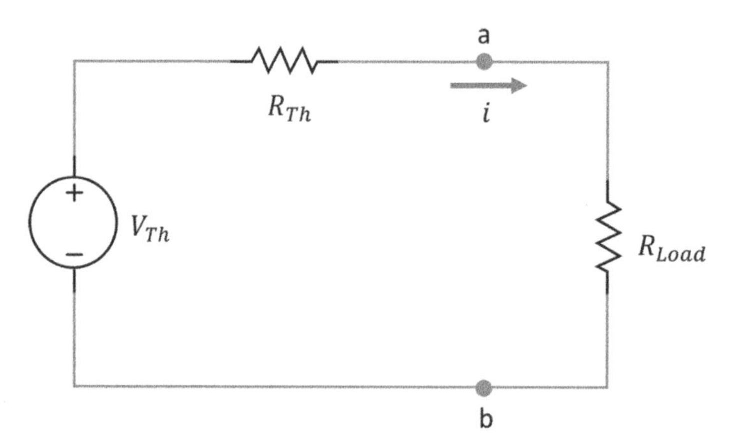

Figure 2.16 Power transfer theorem.

The power delivered to the load is given by

$$p(W) = v \cdot i, \tag{2.26}$$

where v and i are voltage and current at the $a-b$ interface. According to the voltage divider rule, the voltage v is determined as

$$v = \frac{R_{Load}}{R_{Th} + R_{Load}} V_{Th}. \tag{2.27}$$

The current i is calculated by Ohm's law as

$$i = \frac{V_{Th}}{R_{Th} + R_{Load}}. \tag{2.28}$$

Therefore,

$$p(W) = v \cdot i = \frac{R_{Load} \cdot V_{Th}^2}{\left(R_{Th} + R_{Load}\right)^2}. \tag{2.29}$$

Since the source parameters, V_{Th} *and* R_{Th}, are given, the only variable in (2.29) is R_{Load}. Thus, the power delivered to the load will reach its maximum value when

$$\frac{dp}{dR_{Load}} = 0. \tag{2.30}$$

Performing the differentiation yields

$$\frac{dp}{dR_L} = \frac{d}{dR_L}\left(\frac{R_{Load} \cdot V_{Th}^2}{\left(R_{Th} + R_{Load}\right)^2}\right) = \left(\frac{\left(R_{Th} + R_{Load}\right)^2 - 2R_{Load}\left(R_{Th} + R_{Load}\right)}{\left(R_{Th} + R_{Load}\right)^4}\right) V_{Th}^2$$

$$= \left(\frac{R_{Th} - R_{Load}}{\left(R_{Th} + R_{Load}\right)^3}\right) V_{Th}^2 = 0. \tag{2.31}$$

Therefore, *the condition for transferring the maximum power from the source to the load is*

$$R_{Th} = R_{Load}. \tag{2.32}$$

If we plug Condition 2.31 into (2.32), then the maximum power can be computed as

$$p_{max}(W) = v \cdot i = \frac{V_{Th}^2}{4R_{Th}}. \tag{2.33}$$

The following example helps to elaborate on this topic.

Example 2.6 Maximum power transfer from a source to load.

Problem: Consider the circuit shown in Figure 2.16, where $V_{Th} = 60V$, $R_{Th} = 3k\Omega$, and $R_{Load} = 2k\Omega$. Find the actual and the maximum power delivered to the load.

Solution: The actual power delivered to the load is determined by (2.29); its value is

$$p = v \cdot i = \frac{R_{Load} \cdot V_{Th}^2}{\left(R_{Th} + R_{Load}\right)^2} = 0.288W.$$

Application of (2.33) yields the following value of the maximum power:

$$p_{max} = \frac{v_{Th}^2}{4R_{Th}} = 0.3W.$$

The problem is solved.

Discussion

- The numerical result shows that the delivered power doesn't change drastically when the maximum transfer condition is violated. To clarify this point, it's necessary to consider how the delivered power depend on the relationship between R_{Load} and R_{Th}. To find the normalized ratio of p/p_{max} as a function of R_{Load}/R_{Th}, let's divide (2.29) by (2.33). The result is

$$\frac{p}{p_{max}} = \frac{4\frac{R_{Load}}{R_{Th}}}{\left(1 + \frac{R_{Load}}{R_{Th}}\right)^2}. \tag{2.34}$$

Figure 2.17 shows the graph of $\frac{p}{p_{max}}$ vs. $\frac{R_{Load}}{R_{Th}}$. Note that when $\frac{R_{Load}}{R_{Th}} = 1$, then $\frac{p}{p_{max}} = 1$, as expected. This is when transmitted power reaches its peak. However, when R_{Load} tends either to zero or to infinity, the power delivered to the load approaches zero in both cases. Why?

Here is MATLAB script for Figure 2.17 based on (2.34.)

```
Rload_div_Rth = [0.001 0.01 0.1 1 10 100 1000 10000];
• p_div_pmax = (4.*Rload_div_Rth)./((1+Rload_div_Rth).^2);
```

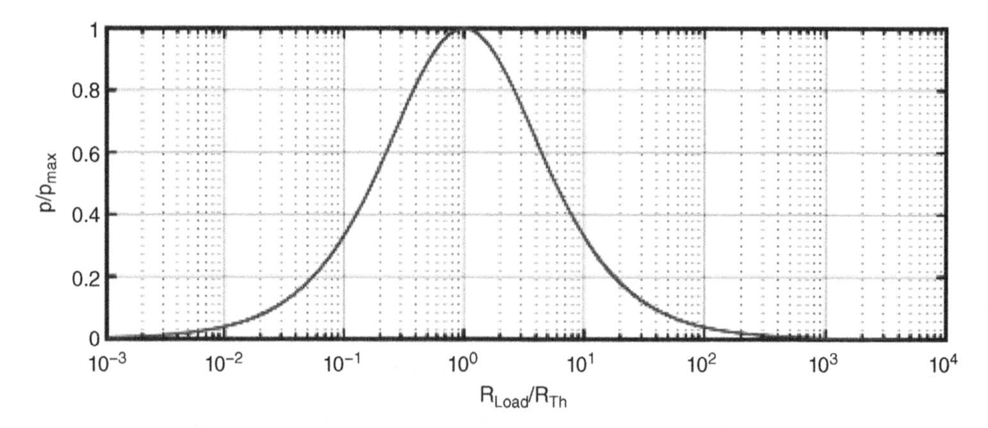

Figure 2.17 Power transfer as a function of $\frac{R_{Load}}{R_{Th}}$.

- % Opens up the window where graphs are plotted
- semilogx(Rload_div_Rth,p_div_pmax) % Plot in semilogarithmic scale
- grid; xlabel('R_L_o_a_d / R_T_h'); ylabel('p/p_m_a_x'); % Labels

(MATLAB code and the graph were prepared by Ina Tsikhanava.)

- The answer is that the power is the product of voltage and current, $p(W) = v \cdot i$. When $R_{Load} \cdot 0$, the $a-b$ interface becomes a short circuit and the current reaches its maximum. The voltage, however, becomes zero, thus turning the delivered power to zero too. Indeed, using (2.27) and (2.28) yields

$$i_{Load}\ R_{Load} \Rightarrow 0 = \frac{V_{Th}}{R_{Th} + R_{Load}} \Rightarrow \frac{V_{Th}}{R_{Th}} = i_{max}(A)$$

and

$$v_{Load}\ R_{Load} \Rightarrow 0 = \frac{R_{Load}}{R_{Th} + R_{Load}} v_{Th} = \frac{\frac{R_{Load}}{R_{Th}}}{1 + \frac{R_{Load}}{R_{Th}}} \frac{V_{Th}}{R_{Th}} \Rightarrow i_{Load} \cdot \frac{R_{Load}}{R_{Th}} \Rightarrow 0(V). \qquad (2.35)$$

On the other hand, when $R_{Load} \Rightarrow \infty$, the $a-b$ interface becomes an open circuit, the voltage gets its maximum value, but the current tends to zero. That is,

$$i_{Load}\ R_{Load} \Rightarrow \infty = \frac{V_{Th}}{R_{Th} + R_{Load}} \Rightarrow 0(A)$$

and

$$v_{Load}\ R_{Load} \Rightarrow \infty = \frac{R_{Load}}{R_{Th} + R_{Load}} V_{Th} \Rightarrow V_{Th}(V) = v_{max}(V). \qquad (2.36)$$

The reader is encouraged to build the graphs of $i_{Load}\left(\frac{R_{Load}}{R_{Th}}\right)$ and $v_{Load}\left(\frac{R_{Load}}{R_{Th}}\right)$ to visualize the trends in the developments of the load current and voltage.

- It's important to realize that the load current and voltage, $i_{Load}(A)$ and $v_{Load}(V)$, are the parameters of a signal delivered from a source to the load. Thus, for our example, the actual values of the load current and voltage are

$$i_{Load}^{actual} = \frac{V_{Th}}{R_{Th} + R_{Load}} = 12(mA)$$

and

$$v_{Load}^{actual} = \frac{R_{Load}}{R_{Th} + R_{Load}} V_{Th} = 24(V).$$

The maximum values of the signal are computed as

$$i_{Load}^{max} = \frac{V_{Th}}{R_{Th}} = 20\,mA$$

and

$$v_{Load}^{max} = V_{Th} = 60(V).$$

- This example and its discussion aim to enhance your understanding of the power transfer theorem.

Questions and Problems for Chapter 2

Introduction–DC and AC Circuits

1 *What is analysis? What is circuit analysis?

2 What do DC and AC stand for?

3 What is the difference between dc and ac circuits?

4 How many numbers do we need to fully describe dc and ac signals? What are these numbers?

5 Why do dc circuits contain only resistors, but not capacitors and inductors? Explain.

Nodal Analysis and Mesh Analysis

6 Chapter 1 introduces the circuit analysis based on Ohm's and Kirchhoff's laws. Why then do we need nodal and mesh analysis?

7 Define a node. Identify nodes and designate their voltages in the circuit shown in Figure 2.P7.

8 What are node voltages? What is a reference node? Show them in Figure 2.P7.

9 Why are the node voltages called *circuit variables*? How can you find the voltage across and current through each circuit element using these circuit variables?

10 Consider Figure 2.P7 where $V1 = 12V, I1 = 2A, R1 = 3\Omega$, $R2 = 4\Omega$, and $R3 = 5\Omega$. Find voltage drop across and current through each resistor.

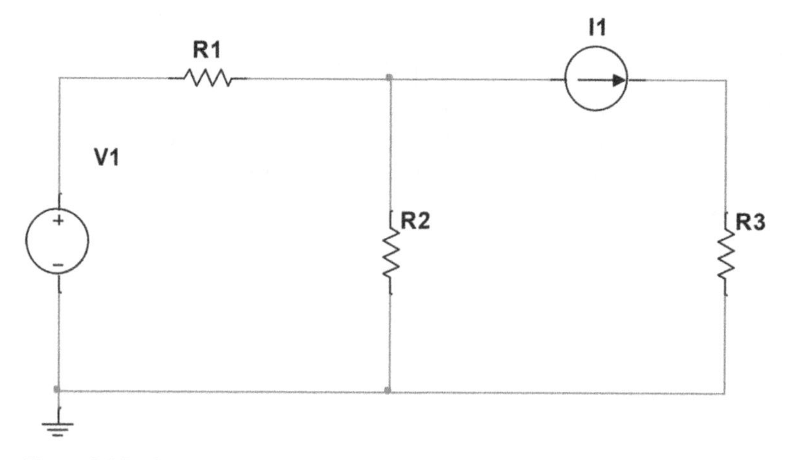

Figure 2.P7 Basic resistive dc circuit.

11 *In Figure 2.P11 find voltage drop across and current through each resistor if $V1 = 12V, I1 = 2A, R1 = 4\Omega$, $R2 = 5\Omega$, $R3 = 6\Omega$, and $R4 = 7\Omega$. Show your solution in a step-by-step format. Verify your answers.

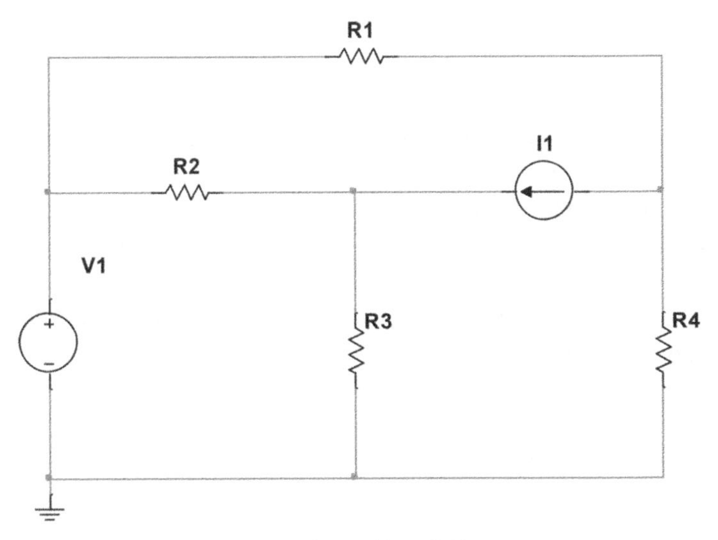

Figure 2.P11 Resistive circuit for Problem 2.11.

12 For Problem 11 build the table with your results similar to Table 2.1. Discuss the table.

13 Using Figures 2.P7 and 2.P11 demonstrate how do you choose the directions of all currents at the start of a nodal analysis? Do you use any reason or your choice is arbitrarily?

14 For Problem 11 demonstrate that your answers are correct comparing delivered and dissipated power.

15 Applying the supernode technique, find voltages and currents for all elements of the circuit shown in Figure 2.P15.

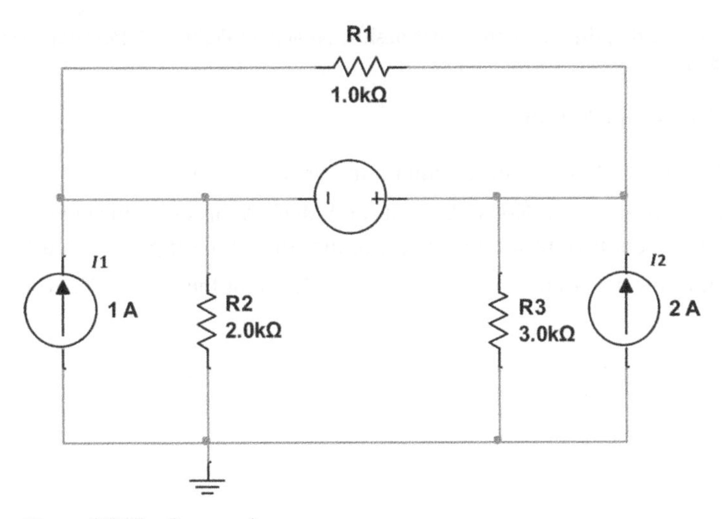

Figure 2.P15 Supernode.

16 Define mesh and loop. Show them on the circuit presented in Figure 2.P15.

17 What is the principal difference between nodal and mesh analyses?

18 Identify all mesh currents and branch currents in Figure 2.P15.

19 How to choose the directions of the mesh currents?

20 Using the mesh analysis, find voltages and currents for all components of the circuit shown in Figure 2.P7 if $V1 = 12V, I1 = 2A, R1 = 3\Omega$, $R2 = 4\Omega$, and $R3 = 5\Omega$. Compare your results with those obtained in Problem 2.10.

21 Consider Figure 2.P11: Applying the mesh analysis, find all voltages and currents of this circuit if $V1 = 12V, I1 = 2A, R1 = 4\Omega$, $R2 = 5\Omega$, $R3 = 6\Omega$, and $R4 = 7\Omega$. Verify your results by comparing them with calculations in Problem 2.11.

22 What is the sign convention and why is it important for the mesh analysis? Illustrate your explanations by using Figure 2.P15.

23 *Consider the circuit shown in Figure 2.P23: Using the supermesh method, find all voltages and currents in this circuit. Verify the results.

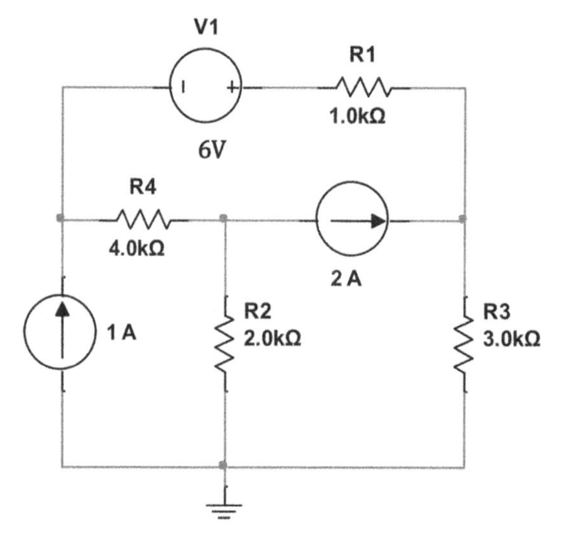

Figure 2.P23 The circuit requiring the use of the supermesh approach for Problem 2.23. (After Irwin and Nelms 2018, pp. 128–129.)

Thevenin's Theorem and Norton's Theorem

24 Whether Ohm's and Kirchhoff's laws are linear equations? Prove your answers.

25 Review all figures presented in section Nodal Analysis and Mesh Analysis of this Questions-and-Problems section and identify only two-terminal circuits. Show their inputs and outputs.

26 Identify input and output on the circuit shown in Figure 2.P26. Plot the general Thevenin's equivalent circuit.

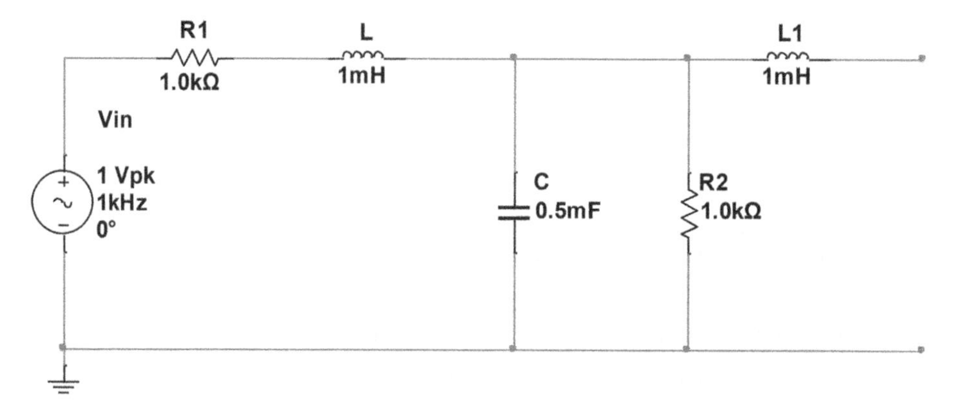

Figure 2.P26 Application of Thevenin's theorem.

27 How can we measure Thevenin's voltage and current? Sketch a circuit to illustrate your explanations.

28 Find V_{Th} and I_{Th} for the circuit given in Figure 2.P28. Show all your manipulations and draw the equivalent circuit for every step of calculations. Sketch Thevenin's equivalent circuit.

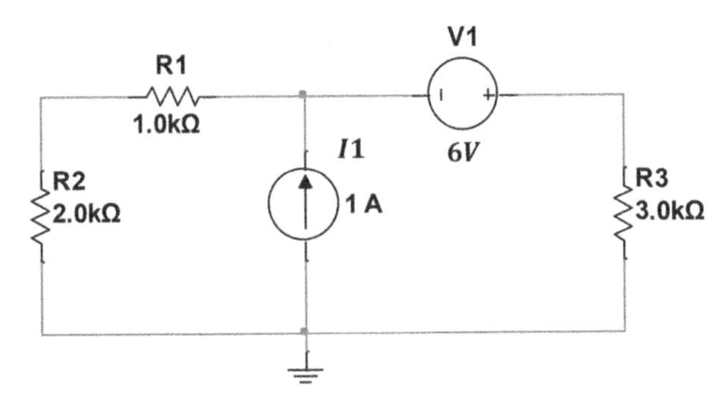

Figure 2.P28 Thevenin's theorem–problem.

29 Whether V_{Th} is the voltage seen by the load? Clarify your answer by sketching the appropriate equivalent circuit.

30 What advantages do offer the use of Thevenin's theorem?

31 What methods of the circuit analysis can we use for calculations of Thevenin's parameters?

32 Solve Problem 2.28 by using mesh analysis, nodal analysis, and source transformation.

33 Consider circuit given in Figure 2.P28: What load resistance do we need to deliver 12 V to the load?

34 Formulate Norton's theorem. Sketch Norton's equivalent circuit for the circuit shown in Figure 2.P26.

35 Compare Thevenin's and Norton's theorems: What do they have in common and how they differ?

36 What is the relationship among the parameters of Thevenin's and Norton's models?

37 Find Norton's parameters and sketch Norton's equivalent to the circuit given in Figure 2.P28.

38 How can we measure Norton's parameters? Sketch a circuit and explain.

39 What does the power transfer theorem state? How can this statement be proved?

40 Use Thevenin's equivalent circuit found in Problem 2.28. What power will be delivered to the load if $R_{Load} = 6k\Omega$? What must be the value of R_{Load} to deliver the maximum power?

3

Basics of AC Steady-State Circuit Analysis

3.1 Steady-State AC Circuit Analysis—Basic Tools

3.1.1 Introduction

What is the difference between dc and ac circuit analysis? As mentioned in Section 2.1, the difference is in the nature of the signals the circuits are operating with. A *dc signal* is constant for any given problem; hence, the only thing we need to know about it is its *value*. (The direction of flowing (polarity) is its second parameter.) The consequence of this feature is that capacitors and inductors aren't used in dc circuits because they work respectively as open or short circuits for dc signals.

In contrast, *ac signals* vary in time; consequently, they need (1) more parameters for their description, and (2) all three passive components (resistors, capacitors, and inductors) work in ac circuits. Therefore, the ac circuit analysis is more complicated than the dc one. On the positive side, most approaches, methods, and rules (theorems) considered for dc circuits in Chapters 1 and 2 conceptually work for ac circuits with appropriate modifications.

Though we refer to ac circuit analysis as the work with a time-varying signal, in this chapter, we will consider only one type of such a signal, a *sinusoid*. As will be seen shortly, a sinusoidal signal predictably varies in time and repeats itself in a specific interval called a *period*. The other constraint in this chapter's consideration is that neither the circuit's nor the sinusoidal signal's parameters do not change for a given problem. In other words, the state of the circuit being analyzed remains steady. This is why this chapter is called a steady-state ac circuit analysis.

With this brief introduction in mind, let's start considering the subject.

3.1.2 Sinusoidal Signals

Every textbook in electrical circuit analysis presents the discussion of a *sinusoidal signal*. A reader can find many thoughtful considerations of this topic in the books listed in our bibliography. To keep you in the same presentation style, we refer you to our textbook (Mynbaev and Scheiner 2020, pp. 110–138), where a sinusoidal signal and its applications are considered in depth.

A *sinusoidal function* is well known and thoroughly investigated in mathematics, physics, and engineering; it is also widely used in many other fields of science and technology. In engineering, this function takes a specific form called a *sinusoidal signal* due to its dependence on time. We assume that the reader is familiar with sinusoidal signals and basic manipulations with them; Sidebar 3S contains a brief reminder on the topic.

Essentials of Advanced Circuit Analysis: A Systems Approach, First Edition. Djafar K. Mynbaev.

Companion Website: www.wiley.com/go/Mynbaev/AdvancedCircuitAnalysis

All the following explanations are illustrated in the family of Figures 3.1. We strongly recommend to constantly compare written and graphic descriptions because visualization of a text helps in a deeper understanding of a topic.

There are two types of sinusoidal signal—cosine and sine. The general formula of this signal for cosine is given by

$$v(t) = Acos(\omega t + \theta). \tag{3.1a}$$

Similarly, for sine function, this signal is

$$v(t) = Asin(\omega t + \theta), \tag{3.1b}$$

and

$$sin\left(\omega t - 90^0\right) = cos(\omega t). \tag{3.1c}$$

Here $v(t)$ is a time-depended electrical signal. We use $v(t)$ to designate voltage, $i(t)$ to denote current, and $p(t)$ to denominate power. Letter A in (3.1a) and (3.1b) denotes a signal *amplitude*, which is the maximum value of the signal magnitude. The amplitude unit depends on the nature of a signal: For $v(t)$ it is volt, V, for $i(t)$, it is ampere, A, and for $p(t)$ it is watt, W.

The Greek letter ω (omega) denotes radian frequency measured in radians per second, $\frac{rad}{s}$. This frequency shows how many radians per second the signal makes. Thus, the product $\omega t(rad)$ shows what angle in radians a signal makes for the observation time $t(s)$. The radian frequency, also called *angular*, relates to well-known *cyclic*, or simply *frequency*, $f\left(\frac{1}{s} = Hz\right)$, as

$$\omega\left(\frac{rad}{s}\right) = 2\pi(rad)f\left(\frac{1}{s}\right). \tag{3.2}$$

Frequency $f(Hz)$ shows *the number of cycles per second* that a signal makes. A number is dimensionless, hence the f unit is $\frac{1}{s}$. This unit is named hertz, Hz, to honor Heinrich Hertz.[1] If a sinusoidal signal has, say, $f = 5(Hz)$, this signal makes 5 cycles per one second. It means that the signal passes one cycle for $\frac{1}{f} = \frac{1}{5} = 0.2$ second. This is the other vital sinusoidal signal parameter called a period. *The period, $T(s)$, of a sinusoidal signal is the duration of one cycle of this signal*: it is equal to

$$T(s) = \frac{1}{f(Hz)}. \tag{3.3}$$

In our example, where $f = 5(Hz)$, the period, $T = \frac{1}{f(Hz)} = 0.2(s)$. A sinusoidal signal completes its cycle for one period, but this cycle corresponds to changing its phase by 2π angle. (Derive the

1 Heinrich Rudolf Hertz (1857–1894) was a German physicist who experimentally proved the existence of electromagnetic waves. These waves were considered as a theoretical exercise in his days, and nobody, including Hertz himself, did see their practical use. As we know today, electromagnetic waves of all frequencies from low-radio spectrum to light play the vital role in modern technology, being the primary means of delivering signals in telecommunications. Besides his major discovery, he contributed in many other fields, and his work with cathode rays and photoelectric effect significantly impacted technology and science.

mathematical proof of this statement from (3.2) and (3.3).) Therefore, the time change of $\frac{T}{2}(s)$ corresponds to changing the signal angle by $\pi = 180^0$, and time alteration in $\frac{T}{4}(s)$ is akin to moving by $\frac{\pi}{2} = 90^0$.

The third parameter of a sinusoidal signal, angle $\theta(rad)$, is called the *initial phase* or *phase shift.* It shows what angle the signal has at $t = 0$. In physics and mathematics, the whole sum $(\omega t + \theta)(rad)$ is called *phase*; in engineering, more often than not the term *phase* means $\theta(rad)$.

There is one more significant parameter of a sinusoidal signal, a *magnitude.* It is the value of the signal at a specific instant. For example, consider a cosine signal whose $A = 2V, f = 5Hz, and\ \theta = 0(rad)$. What is its magnitude at $t = 0.72(s)$? Plugging all numbers into (3.1a) yields

$$v_{t=0.76} = 2\cos(2\pi \cdot 5 \cdot 0.72 + 0) = -1.62(V).$$

Figure 3.1a depicts the cosine signal, $v(t) = v(t) = Acos(\omega t) = 2cos(10\pi t)$. Note that the figure is drawn in $v(t)[V] - t[s]$ axes; therefore, the signal's frequency and its initial phase cannot be shown in this plot. But these values can be computed from the graph's values using (3.1a), namely, $f(Hz) = \frac{1}{T(s)}$ and $\theta = cos^{-1}\left(\frac{v(0)}{A}\right)$, where $v(0)$ is the signal's magnitude at $t = 0(s)$.

Figure 3.1a demonstrates the following signal's parameters:

- amplitude, $A = 2(V)$,
- period, $T = \frac{1}{f} = 0.2(s)$, where $f(Hz) = \frac{\omega\left(\frac{rad}{s}\right)}{2\pi(rad)} = \frac{10\pi}{2\pi} = 5(s)$,
- magnitude at $t = 0.72(s)$, $v_{t=0.76} = 2\cos(2\pi \cdot 5 \cdot 0.72) = -1.62(V)$.

Recall that the figure is drawn in $v(t)[V] - t[s]$ axes. If you want to build this figure in $v(t)[V] - \omega t[rad]$ axes, the horizontal axis will carry *radians* dimension. Do it by modifying the enclosed MATLAB code.

MATLAB script for building Figure 3.1a:

```
>> omega = 2*pi*5;
t = (0:0.01:(10*pi))/omega;
v = 2*cos(omega*t)
figure
h = plot(t,v)
grid
xlabel('time (s)')
ylabel('v(t) = 2cos(31.4t)')
```

Figure 3.1a shows the cosine signal $v(t) = 2 \cdot \cos(2\pi \cdot 5 \cdot t)$. (What is its period?) Figures 3.1b through 3.1d depict sinusoidal signals and illustrate the following definitions: Figure 3.1b visualizes the role of a signal's amplitude, Figure 3.1c shows how a signal's frequency affects its graph, Figure 3.1d demonstrates two sinusoidal signals that have the mutual phase shift of $-\frac{\pi}{2} = -90^0$, which corresponds to the time shift of $-\frac{T}{4}(s)$. Since by definition $\cos(\omega t - 90^0) = \sin(\omega t)$, 3.1d presents, in fact, cosine and sine signals. Why negative 90^0? If we go along the horizontal ωt axis to

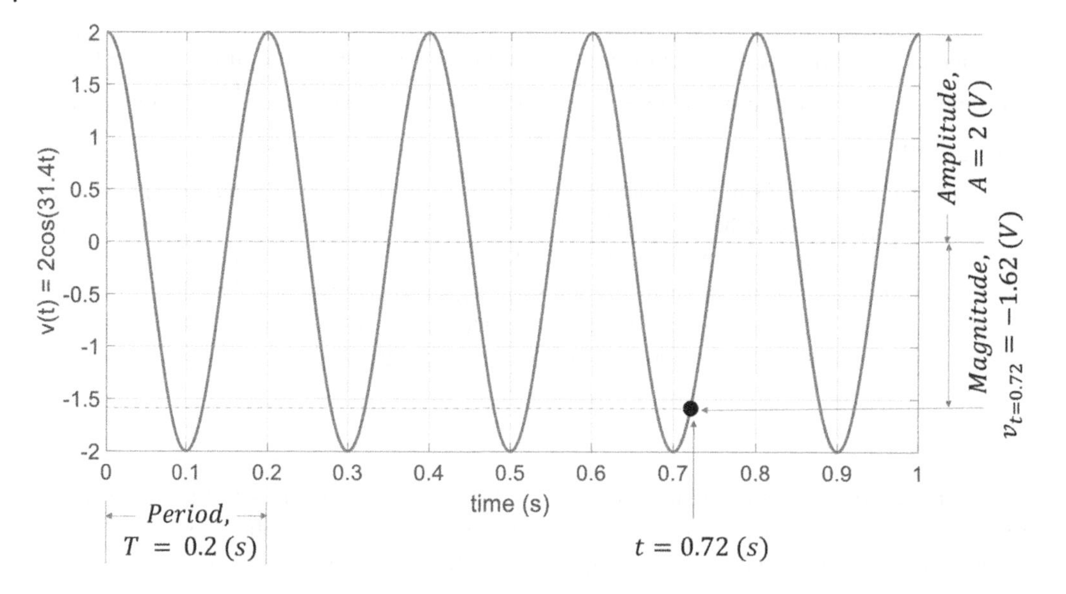

a)

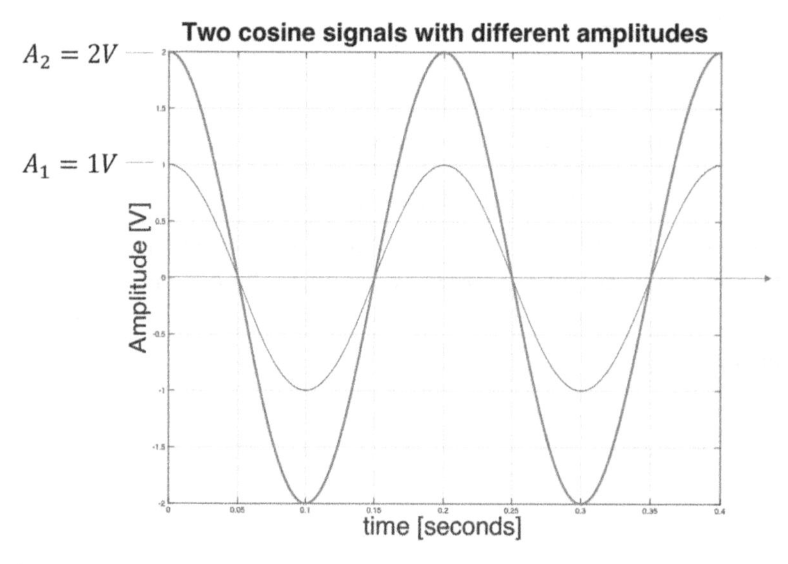

b)

Figure 3.1 A sinusoidal signal and its parameters: a) A cosine signal $v(t) = 2 \cdot \cos(2\pi \cdot 5 \cdot t)$; b) two cosine signals with different amplitudes; c) two cosine signals with different frequencies; d) cosine and sine signals; e) shifting a cosine signal by $-\frac{T}{4}(s) \Rightarrow -90°$. (Graphs 3.1b, 3.1c, and 3.1d are prepared by Ina Tsikhanava.)

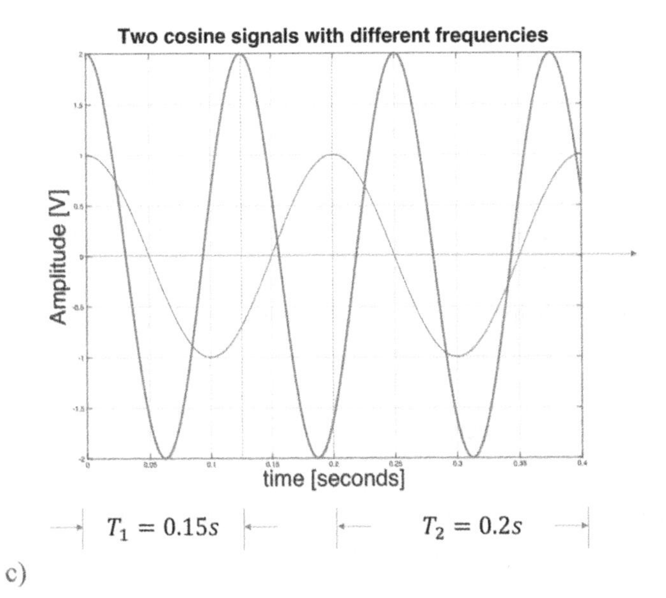

c)

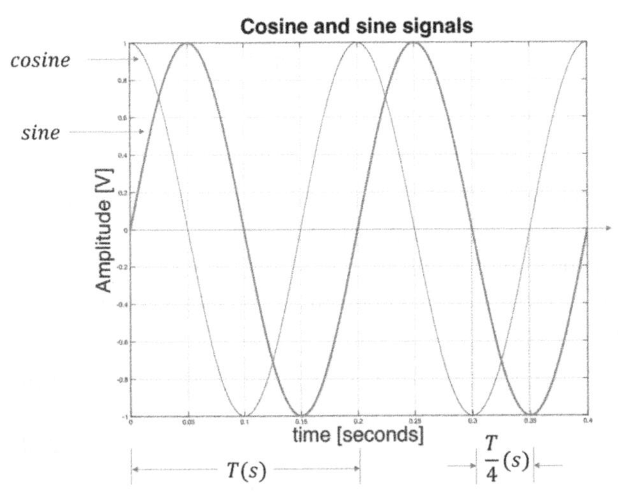

d)

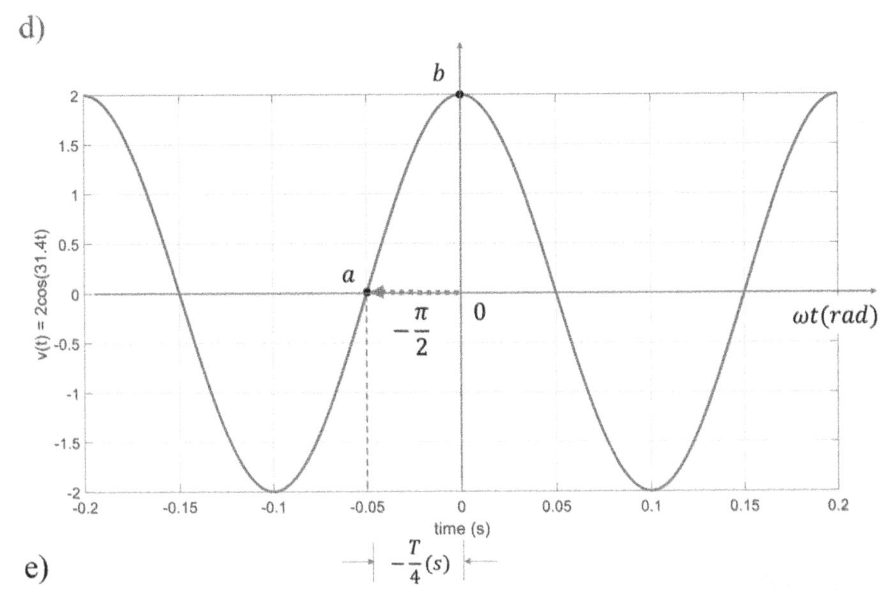

e)

Figure 3.1 (Cont'd)

the left, then the first point where the cosine graph intersects this axis will be $-\frac{\pi}{2} = -90^0$. It is designated a in Figure 3.1e. The signal that starts at point a and goes through point b and further is a sine signal. Compare it with the sine signal showed in Figure 3.1d. This is the graphical proof that shifting cosine at $-\frac{T}{4}(s) \Rightarrow -90^0$ turns it into a sine signal.

This brief review of a sinusoidal signal will help us to smoothly navigate through all mathematical manipulations that involve trigonometric identities.

3.1.3 Sinusoidal Signals and *R, L, and C* Components

Let's analyze a series RLC circuit shown in Figure 3.2. The goal is to find the voltage across and current through each circuit element. In a series circuit, all components share the same current; hence, we start by searching for different voltages. Applying KVL yields

$$v_{R(t)} + v_L(t) + v_C(t) = v_{in}(t). \tag{3.4}$$

This book explores a *systems approach* aiming to relate output and input of a circuit. The output of this circuit is the voltage across the capacitor,

$$v_C(t) \equiv v_{out}(t). \tag{3.5}$$

(You will recall that the sign $\equiv$ means "identical to.") Thus, the task boils down to expressing all members of LHS of (3.4) through $v_C(t)$, therefore, pertaining $v_{out}(t)$ to $v_{in}(t)$.

Since all three circuit components have the same current; i.e., $i_R = i_C = i_L = i$, we need to express their voltages through this current. As we've learned in Chapter 1, the $v - i$ relationships of a resistor, a capacitor, and an inductor are given by

$$v_R = R \cdot i_R \tag{1.18R}$$

$$i_C = C\frac{dv_C}{dt} \tag{1.24R}$$

$$v_L = L\frac{di_L}{dt}. \tag{1.37R}$$

Plugging $i = C\frac{dv_C}{dt}$ into v_R and v_L produces

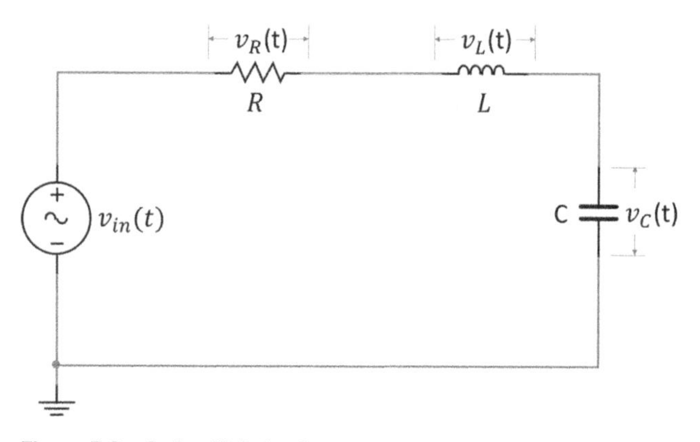

Figure 3.2 Series RLC circuit.

$$v_R = R \cdot i = RC\frac{dv_C}{dt}$$

and

$$v_L = L\frac{di}{dt} = LC\frac{d}{dt}\left(\frac{dv_C}{dt}\right) = LC\frac{d^2v_C}{dt^2}. \tag{3.6}$$

Ultimately, we insert (3.6) into (3.4) and obtain

$$v_{R(t)} + v_L(t) + v_C(t) = v_{in}(t) \Rightarrow LC\frac{d^2v_C}{dt^2} + RC\frac{dv_C}{dt} + v_C = v_{in}(t). \tag{3.7}$$

Rewriting (3.7) in the input–output terms brings

$$LC\frac{d^2v_{out}(t)}{dt^2} + RC\frac{dv_{out}(t)}{dt} + v_{out}(t) = v_{in}(t). \tag{3.8}$$

Equation 3.8 is the *second-order differential equation* describing the RLC circuit in its entirety. Chapters 4, 5, and 6 will discuss the differential equations and their applications to the circuit analysis in depth. Here, we want to stress that characterization of a time-varying electrical circuit (means, an ac analysis) requires the use of differential equations as soon as the reactive components L and C are involved. See (1.24) and (1.37). However, in future chapters, we will learn that dealing with differential equations is not an easy subject, and therefore, working with differential equations complicates an ac circuit analysis drastically.

It seems that in the case of sinusoidal signals, working with the differential equations would not be that difficult because differentiation and integration of the sinusoidal signals are the easy operations—the derivative and integral of sinusoidal signals are the other sinusoidal signals. However, a close look reveals that even this case has its complexity. Let's take this careful glance.

Permit the current in the RLC circuit in Figure 3.2 be a cosine signal,

$$i = A\cos(\omega t).$$

To determine $v_C(t) \equiv v_{out}(t)$, we refer to (1.28),

$$v_C(t)[V] = v_C(t_0) + \frac{1}{C}\int_{t_0}^{t} i(t)dt, \tag{1.28R}$$

and, assuming $v_C(t_0) = 0$ and $t_0 = 0$, find the output voltage as

$$v_{out}(t) = \frac{1}{C}\int_0^t i(t)dt = \frac{1}{C}\int_0^t A\cos(\omega t)dt = \frac{A}{C}\sin(\omega t)\Big|_0^t = \frac{A}{C}\sin(\omega t). \tag{3.9}$$

Plugging (3.9) in (3.8) yields

$$LC\frac{d^2v_{out}(t)}{dt^2} + RC\frac{dv_{out}(t)}{dt} + v_{out}(t) = v_{in}(t) \Rightarrow -LA\omega^2\sin(\omega t) + RC\omega\cos(\omega t) + \frac{A}{C}\sin(\omega t) = v_{in}(t). \tag{3.10}$$

Equation 3.10, which fully describes an RLC circuit, requires adding and subtracting sine and cosine signals. This is not an easy task. For (3.10), the only direct method of finding the algebraic sum of the LHS sinusoids is to add them graphically. It can be done by placing all three sinusoids on the same set of axes and adding or subtracting their magnitudes at every point of the horizontal (here, time) axis. We strongly recommend you perform this operation to comprehend this technique fully. Also, this try should convince you that this is not a practically acceptable method.

To sum up, an ac analysis of an RLC circuit reveals two new vitally essential features that we did not encounter in analyzing dc circuits, namely:

1) We must use the differential equations to describe an ac circuit in its entirety;
2) We must add or subtract sinusoidal signals to perform the required manipulations with the circuit currents and voltages.

Both features make an ac analysis a challenging undertaking. It took the genius of Charles Steinmetz[2] to resolve these issues by introducing the phasors and the use of complex numbers in ac analysis.

3.1.4 Phasors

Equation 3.10 shows that using differential equations in ac circuit analysis becomes a manageable issue when ac signals are the sinusoidal functions. But how to resolve the problem of manipulations with the sinusoids? To answer this question, refer to how we must do this manually: Add the magnitudes of all addends. Therefore, all what we need to know are the magnitudes of the sinusoids to be algebraically added. Is it possible to discern these magnitudes without investigating the actual sinusoidal waveforms? The answer is yes, provided using the *phasors*.

Consider the left-hand side of Figure 3.3a: A radius vector of length A rotates around a fixed center 0 at the constant angular velocity $\omega\left(\frac{rad}{s}\right)$. This vector covers angle $\omega t(rad)$ for the time interval $t(s)$. If the vector initially shifted at angle θ at $t = 0(s)$, then its whole angle at $t(s)$ will be $\alpha = (\omega t + \theta)(rad)$, as shown in Figure 3.3b. Let's turn to the right-hand side of Figure 3.3a: The projection of vector A at the vertical axis at every angle α is placed on the $A\cos\alpha - \alpha$ set of axes. When $\alpha = 0(degree)$, the projection is A_0; when $\alpha = 30^0$, the projection is A_{30}, and so on. Thus, the projections A_0, A_{30}, A_{60},…,A_{360} placed at $\alpha = 0^0, \alpha = 30^0, \alpha = 60^0, \ldots, \alpha = 360^0$ in RHS of Figure 3.3a are the magnitudes of a cosine signal at the given angles. In other words, Figure 3.3a shows how a rotating vector A relates to a cosine signal. See, for example, the following link[3] for a dynamic illustration of this relationship.

We call vector $\boldsymbol{A}$ a *rotating phasor*. Why do we need to introduce this phasor? You will recall that our task is to add or subtract sinusoidal signals of the same frequency. Hence, if a phasor fully represents a sinusoid, then adding or subtracting vectors will be simpler than doing the same with sinusoids.

2 Charles Proteus Steinmetz (1865–1923) was an American engineer and mathematician. He was born and educated in Germany and emigrated to the USA aged 24. For most of his professional life in the US he worked for General Electric Company. He was deformed from the birth and stayed only four feet tall. However, he is remembered for his intellectual results, forever transforming the field of electrical engineering. Concerning the subject of this book, Steinmetz replaced sophisticated, calculus-based mathematics used for the circuit analysis in that time by simple algebraic equations relied on complex numbers. He also made seminal contribution in a transient theory of electrical circuits. The above several sentences cannot describe the contribution in electrical engineering made by the man who amassed 197 patents and authorized more than ten classical books. This is why we want to attract attention of our readers to the biography and achievements of this great engineer and scientist.

3 https://lpsa.swarthmore.edu/BackGround/phasor/phasor.html

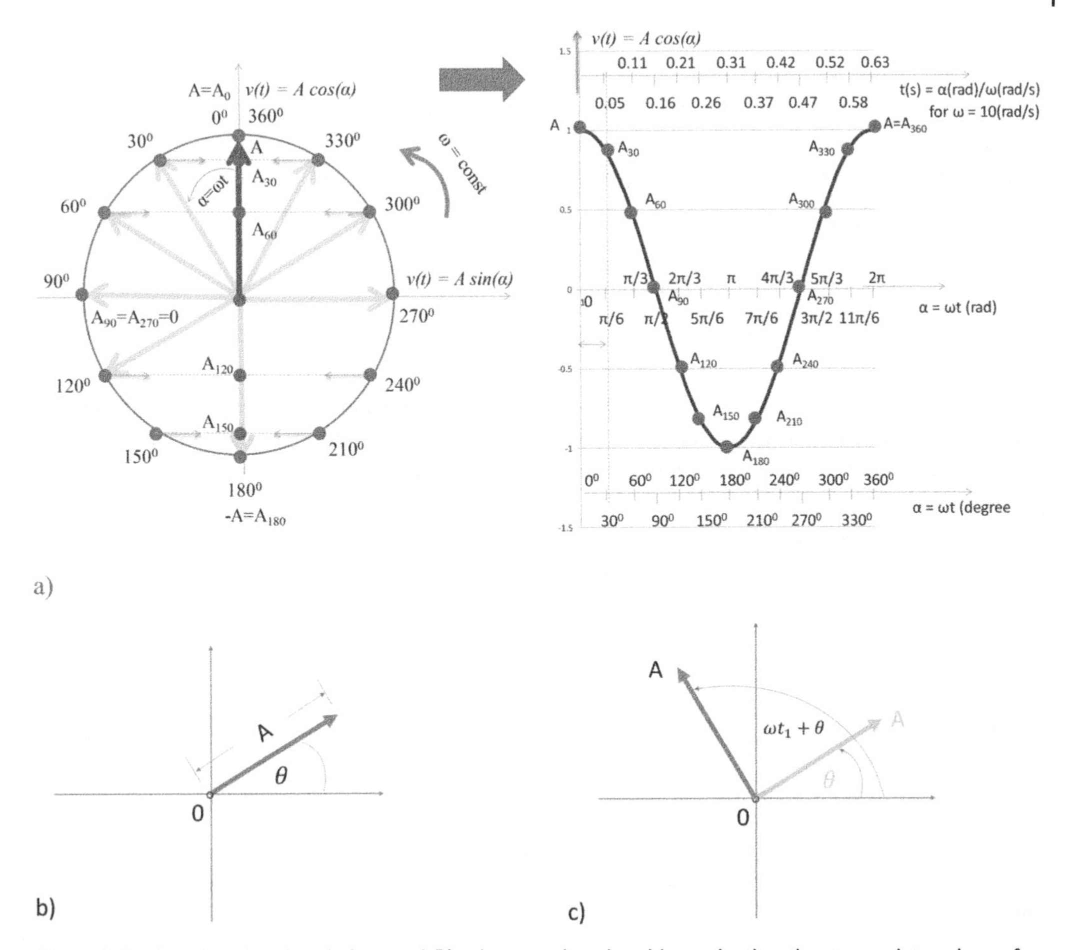

Figure 3.3 A cosine signal and phasor: a) Plotting a cosine signal by projecting the appropriate values of a rotating vector; b) a phasor at the initial position; c) the rotating phasor at the instant t_1. (Adapted from Mynbaev and Scheiner, Wiley, 2020, p. 141.)

The shorthand way to present a sinusoid as a phasor is using a complex number; that is,

$$v(t) = A\cos(\omega t + \theta) \Leftrightarrow \Re \boldsymbol{V} = Ae^{j(\omega t + \theta)}. \tag{3.11}$$

The exponent property permits presenting a rotating phasor $\Re \boldsymbol{V}$ as

$$\Re \boldsymbol{V} = Ae^{j(\omega t + \theta)} = e^{j\omega t}Ae^{j\theta} = e^{j\omega t}\boldsymbol{V}, \tag{3.12}$$

where the *phasor* $\boldsymbol{V}$ is given by

$$\boldsymbol{V} = Ae^{j\theta}. \tag{3.13}$$

This phasor is shown in Figure 3.3b. (The boldfaced letters denote complex numbers, as usual.) Remember that we manipulate with the sinusoids of the same frequency; thus, the factor $e^{j\omega t}$ is common for all of them. Therefore, the phasor $\boldsymbol{V} = Ae^{j\theta}$ fully represents an individual sinusoid with amplitude A and phase shift θ in the *phasor domain*. For example, if $v(t) = 3\cos(5t + 25^0)$ in the *time domain*, then $\boldsymbol{V} = 3e^{j25}$ represents this cosine in the *phasor domain*. To show a sine signal in the *phasor domain*, we turn the sine into cosine as trigonometric identities in Sidebar 3S demonstrates, namely, $cos(\varphi \pm 90^0) = \mp sin\varphi$. Thus, $3\sin(5t+25^0) = 3\cos(5t+25^0-90^0) = 3\cos(5t-65^0)$, which results in $\boldsymbol{V} = 3e^{-j65}$. However, the negative sine signal $-3\sin(5t+25^0)$ gets its cosine as $-3\sin(5t+25^0) = 3\cos(5t+25^0+90^0) = 3\cos(5t+115^0)$, which transforms in $\boldsymbol{V} = 3e^{j115}$.

A phasor $\boldsymbol{V} = Ae^{j\theta}$ can also be written in the *angular (polar) form* as

$$\boldsymbol{V} = Ae^{j\theta} \Leftrightarrow A\angle\theta. \tag{3.14}$$

Formula (3.15) sums up the phasor presentations of a sinusoidal signal:

$$\Re \boldsymbol{V} = Ae^{j(\omega t+\theta)} = e^{j\omega t}\boldsymbol{V} \Rightarrow \boldsymbol{V} = Ae^{j\theta} = A\angle\theta. \tag{3.15}$$

As mentioned, all four basic algebraic operations, such as addition, subtraction, multiplication, and division can be readily performed with phasors. However, calculations needed for these operations call for the presentation of a *phasor* as a *complex number*, and Figure 3.4 visualizes such a presentation.

The phasor (vector) $\boldsymbol{V}$ whose amplitude is A and phase θ is presented as a sum of its real part, $Acos\theta$ and its imaginary part as $Asin\theta$; that is

$$\boldsymbol{V} = Ae^{j\theta} = Acos\theta + jAsin\theta. \tag{3.16}$$

This presentation of a phasor is based on the *Euler's identity*

$$e^{j\theta} = cos\theta + jsin\theta, \tag{3.17}$$

from which the reverse formulas follow as

$$cos\theta = Re\left\{e^{j\theta}\right\} = Re\left\{cos\theta + jsin\theta\right\}$$

and

$$sin\theta = Im\left\{e^{j\theta}\right\} = Im\left\{cos\theta + jsin\theta\right\}. \tag{3.18}$$

We imply that the reader is familiar with the basic operations of the complex numbers; nevertheless, to save your time, Sidebar 3S presents a brief review of this topic.

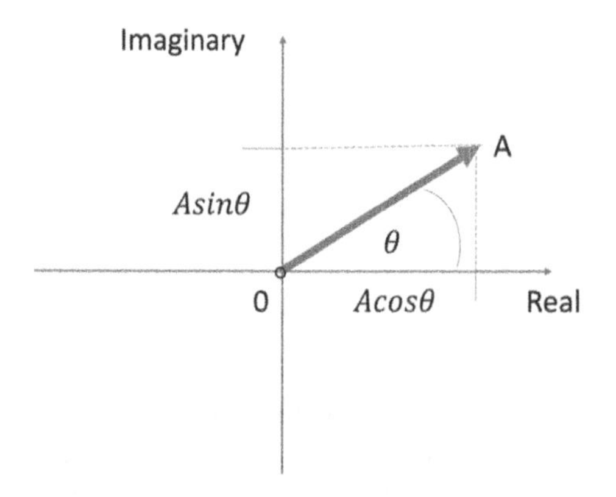

Figure 3.4 Phasor as a complex number.

The summary of this section is as follows:

- An ac circuit analysis involves differential equations because $v-i$ relationships for the reactive components L and C are described by derivatives.
- A steady-state ac circuit analysis is restricted by applications of sinusoidal signals whose derivatives are new sinusoidal signals. Thus, the steady-state circuit analysis permits skipping the differential equations but creates the problem of manipulations with sinusoidal signals.
- The latter problem is resolved by using phasors since they are intimately related to sinusoidal signals, as Figure 3.3a shows. On the other hand, phasors can be presented as complex numbers whose mathematical manipulations are simple and well-known.

Example 3.1 helps to put this theory on a practical footing.

Example 3.1 Manipulations with sinusoidal signals using phasors.
Problem: Two sinusoidal signals, $v_1(t) = 3cos(8t + 30^0)$ and $v_2(t) = 7sin(8t + 50^0)$ are presented to an electrical circuit. The analysis of this circuit requires to determine the following combinations of these signals:

a) sum,
b) difference,
c) product,
d) quotient.

Find these combinations.

Solution: All the necessary manipulations are described by Equations 3.11 through 3.15. Also, refer to Sidebar 3S below, whose formulas carry format 3S.X.

Let's start with presenting the given signals in the phasor and polar (angular) forms:

$$v_1(t) = 3cos\left(8t + 30^0\right) \Rightarrow \Re\boldsymbol{V}_1 = 3e^{j\left(8t+30^0\right)} \Rightarrow \boldsymbol{V}_1 = 3e^{j\left(30^0\right)} \Rightarrow \boldsymbol{V}_1 = 3\angle 30^0$$

and

$$v_2(t) = 7sin\left(8t + 50^0\right) = 7cos\left(8t + 50^0 - 90^0\right) \Rightarrow \Re\boldsymbol{V}_2 = 7e^{j\left(8t+50^0-90^0\right)}$$

$$\Rightarrow \boldsymbol{V}_2 = 7e^{-j\left(40^0\right)} \Rightarrow \boldsymbol{V}_2 = -7\angle 40^0.$$

Having got these given signals in the phasor domain, we can find all the required combinations.

a) The *sum* of two sinusoids in both domains is

$$v_1(t) + v_2(t) \Rightarrow \boldsymbol{V}_1 + \boldsymbol{V}_2.$$

Addition and subtraction are simpler performed by employing the complex numbers in the rectangular forms. Thus, using (3.17), $e^{j\theta} = cos\theta + jsin\theta$, enables us to transit from the exponential forms to the rectangular ones as

$$\boldsymbol{V}_2 = 3\angle 30^0 = 3e^{j\left(30^0\right)} = 3\cos\left(30^0\right) + j3\,\sin\left(30^0\right) = 3\cdot 0.866 + j3\cdot 0.5 = 2.598 + j1.5$$

and

$$\boldsymbol{V}_2 = 7\angle -40^0 = 7e^{-j\left(40^0\right)} = 7\cos\left(40^0\right) - j7\sin\left(40^0\right) = 7\cdot 0.766 - j7\cdot 0.643 = 5.362 - j4.501.$$

Don't forget that $sin\theta = \cos(\theta - 90^0)$, and therefore, if $cos\theta \Rightarrow A\angle\theta$, then $sin\theta \Rightarrow A\angle(\theta - 90^0)$.
The *sum of two phasors* according to (3S.11) is

$$(a_1 + jb_1) + (a_2 + jb_2) = (a_1 + a_2) + j(b_1 + b_2). \quad \text{(3S.11R)}$$

Applying this rule produces

$$\boldsymbol{V}_1 + \boldsymbol{V}_2 = 2.598 + j1.5 + 5.362 - j4.501 = 7.96 - j3.00.$$

This sum can be obtained in a polar form using (3S.7) as

$$\boldsymbol{V}_{1+2} = \boldsymbol{V}_1 + \boldsymbol{V}_2 = \sqrt{\left((7.96)^2 + (3^2)\right)}\angle tan^{-1}\left(\frac{-3}{7.96}\right) = 8.51\angle -20.6^0 = 8.51e^{-j20.6^0}.$$

Obviously, the rotating phasor of $\boldsymbol{V}_{1+2}$ is given by

$$\Re\boldsymbol{V}_{1+2} = 8.51e^{j(8t - 20.6^0)}.$$

Finally, applying (3S.18a), the *time-domain sum of two given sinusoids* is calculated as

$$cos\varphi = Re\left\{e^{j\varphi}\right\}$$

$$\Rightarrow v_{1+2}(t) = 8.51\cos(8t - 20.6^0).$$

The process of summation seems to be long and tedious, but this impression is due to our willingness to show each and every detail of the manipulation. After some practice, you will be able to perform it readily and quickly as demonstrated in the following operation—finding the difference of two sinusoids.

b) The *difference* of the given signal is

$$v_1(t) - v_2(t) \Rightarrow \boldsymbol{V}_1 - \boldsymbol{V}_2.$$

Applying (3S.12),

$$(a_1 + jb_1) - (a_2 + jb_2) = (a_1 - a_2) + j(b_1 - b_2), \quad \text{(3S.12R)}$$

yields

$$\boldsymbol{V}_1 - \boldsymbol{V}_2 = 2.598 + j1.5 - (5.362 - j4.501) = -2.764 + j6.001.$$

Using (3.S.7) results in

$$\boldsymbol{V}_{1-2} = \boldsymbol{V}_1 - \boldsymbol{V}_2 = 6.606\angle -65.27^0 = 6.606e^{-j65.27^0}.$$

Based on a polar form of V_{1-2}, the rotating phasor and the *required time-domain difference* are given by

$$\Re\boldsymbol{V}_{1-2} = 6.606e^{j(8t - 65.27^0)}$$

$$\Rightarrow v_{1-2}(t) = 6.61\cos(8t - 65.27^0).$$

c) *Multiplication* of two sinusoids is easier performed using their phasors in polar (angular) forms; thus, following to (3S.15),

$$\boldsymbol{V}_1 \cdot \boldsymbol{V}_2 = (V_1\angle\theta_1)\cdot(V_2\angle\theta_2) = V_1 \cdot V_2\angle(\theta_1 + \theta_2), \quad \text{(3S15R)}$$

we find

$$\Rightarrow \boldsymbol{V}_1 \cdot \boldsymbol{V}_2 = \left(3\angle 30^0\right)\cdot\left(7\angle -40^0\right) = 21\angle -10^0.$$

The *time-domain product of two given* sinusoids is

$$v_1(t)\cdot v_2(t) = \left(3\cos\left(8t+30^0\right)\right)\cdot\left(7sin\left(8t+50^0\right)\right) = 21\cos\left(8t-10^0\right).$$

d) *Division* performed according to (3S.16),

$$\frac{A\angle\Phi}{B\angle\Theta} = \frac{A}{B}\angle(\Phi-\Theta), \tag{3S.16R}$$

results in

$$\frac{v_1(t)}{v_2(t)} \Rightarrow \frac{\boldsymbol{V}_1}{\boldsymbol{V}_2} = \frac{3\angle 30^0}{7\angle -40^0} = 0.43\angle 70^0.$$

The time-domain quotient of two given sinusoids is

$$\frac{v_1(t)}{v_2(t)} = \frac{3cos\left(8t+30^0\right)}{7sin\left(8t+50^0\right)} = 0.43\cos\left(8t+70^0\right).$$

Thus, the problem is solved.

Discussion:

- Pay special attention to the angle value because, as Figure 3S.1 states, the value and the sign of a calculated phase shift depends on the quadrant in which this phase shift will be located.
- We can illustrate the summation of two sinusoids applying their rectangular forms to the phasor presentation in a complex $Re-Im$ plane. Consider Figure 3.5, where $\boldsymbol{V}_1 = 2.6 + j1.5$ and $\boldsymbol{V}_2 = 5.4 - j4.5$ along with their projections on the real and imaginary axes are shown. We add their approximate values, 2.6 and 5.4 on *Re* axis and 1.5 and –4.5 on the *Im* axis, following Rule 3S.11. Having their sums, $Re = 7.96$ and $Im = -j3.0$, enables us to build the new phasor (vector) $\boldsymbol{V}_{1+2}$ whose amplitude, $V_{1+2} = 8.51$, and phase shift, $\theta = -20.6^0$, are equal to these quantities found by manual calculations in the example. Though this graphical illustration doesn't bring

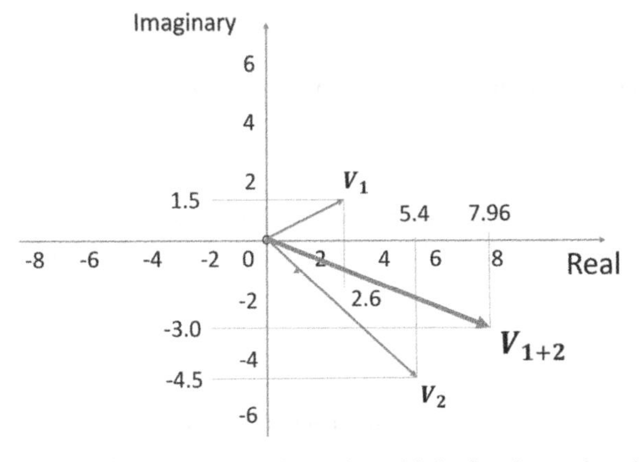

Figure 3.5 Summation of two sinusoids in the phasor domain.

new knowledge about a phasor domain, it delivers deeper understanding to the sense of the mathematical manipulations performed in the phasor domain.

As an exercise, we recommend you to find graphically vector $\boldsymbol{V}_{1-2}$ in the $Re - Im$ plane.

This is the end of Example 3.1.

The other significant feature of a phasor is its *derivative property*. Refer to the presentation of a sinusoidal signal in the phasor domain given in (3.11) and (3.12) as

$$v(t) = A\cos(\omega t + \theta) \Leftrightarrow \Re\boldsymbol{V} = Ae^{j(\omega t+\theta)} = e^{j\omega t}Ae^{j\theta} = e^{j\omega t}\boldsymbol{V}. \tag{3.12R}$$

On the other hand, Equation 3.18 states that

$$A\cos(\omega t + \theta) = Re\left\{Ae^{j(\omega t+\theta)}\right\} = Re\left\{e^{j\omega t}\boldsymbol{V}\right\}. \tag{3.19}$$

Differentiation of this signal produces

$$\frac{d}{dt}\left(A\cos(\omega t + \theta)\right) = \frac{d}{dt}Re\left\{Ae^{j(\omega t+\theta)}\right\} = Re\left\{\frac{d\left(Ae^{j(\omega t+\theta)}\right)}{dt}\right\}$$

$$= Re\left\{\frac{d\left(e^{j\omega t}\boldsymbol{V}\right)}{dt}\right\} = Re\left\{j\omega e^{j\omega t}\boldsymbol{V}\right\} = Re\left\{j\omega e^{j\omega t}Ae^{j(\omega t+\theta)}\right\}. \tag{3.20a}$$

Referring to (3.19) and equating the first and the last members of (3.20a) yields

$$\frac{d}{dt}\left(A\cos(\omega t + \theta)\right) = Re\left\{j\omega e^{j\omega t}Ae^{j(\omega t+\theta)}\right\} = Re\left\{j\omega\Re\boldsymbol{V}\right\}. \tag{3.20b}$$

Therefore, differentiation of a sinusoidal signal is equivalent to multiplying the rotating phasor of this signal by the factor $j\omega$.

This statement describes the phasor derivative property.

Question If the differentiation of a sinusoidal signal is equivalent to multiplying its rotating phasor by a factor $j\omega$, what is a shorthand for presenting the integration of the sinusoidal signal? [Hint: Follow the steps shown in (3.20a) but replace differentiation by the integration. Use an indefinite integral and set the integration constant to zero.]

Sidebar 3S Basic manipulations with sinusoidal functions and complex numbers

3S.1 Sinusoidal Functions

To successfully tackle the incoming problems, we need to recall that the sum of sine and cosine functions having the same frequency can be presented as a sinusoidal signal with a new amplitude and additional phase shift; that is,

$$a\sin\omega t + b\cos\omega t = A\sin(\omega t + \theta). \tag{3S.1}$$

Here, $A = \sqrt{a^2 + b^2}$ and $\tan\theta = \frac{b}{a}$. The proof of (3S.1) is a straightforward: The sum $a\sin\omega t + b\cos\omega t$ can be presented as

$$asin\,\omega t + bcos\,\omega t = A\left(\frac{a}{A}sin\omega t + \frac{b}{A}cos\omega t\right). \tag{3S.2}$$

Let's denote $\frac{a}{A} = cos\theta$ and $\frac{b}{A} = sin\theta$. Then, (3S.2) takes the form

$$asin\omega t + bcos\omega t = A\left(cos\theta\,sin\omega t + sin\theta\,cos\omega t\right) = Asin\left(\omega t + \theta\right). \tag{3S.3}$$

In all the above manipulations, pay special attention to the form of (3S.1), $asin\omega t + bcos\omega t = Asin\left(\omega t + \theta\right)$. Any change in the member sequence or sign will result in different expressions. In general,

$$Asin\left(\omega t \pm \theta\right) = asin\,\omega t \pm bcos\,\omega t$$

and

$$Acos\left(\omega t \pm \theta\right) = acos\,\omega t \mp bsin\,\omega t, \tag{3S.4}$$

where A and θ are defined by the above formulas. (Question: Can you prove this statement?)

In our mathematical manipulations, we often use the following *trigonometric identities*, and it's helpful to have them handy:

$$cos\left(\varphi \pm 90^0\right) = \mp sin\varphi$$

$$sin\left(\varphi \pm 90^0\right) = \pm cos\varphi$$

$$cos\left(\varphi \pm \vartheta\right) = cos\varphi\,cos\vartheta \mp sin\varphi\,sin\vartheta$$

$$sin\left(\varphi \pm \vartheta\right) = sin\varphi\,cos\vartheta \pm cos\varphi\,sin\vartheta$$

$$cos\varphi\,cos\vartheta = \frac{1}{2}\left[cos\left(\varphi + \vartheta\right) + cos\left(\varphi - \vartheta\right)\right]$$

$$sin\varphi\,sin\vartheta = \frac{1}{2}\left[cos\left(\varphi - \vartheta\right) - cos\left(\varphi + \vartheta\right)\right]$$

$$sin\varphi\,cos\vartheta = \frac{1}{2}\left[sin\left(\varphi + \vartheta\right) + sin\left(\varphi - \vartheta\right)\right]$$

$$cos\varphi + cos\vartheta = 2cos\left(\frac{\varphi + \vartheta}{2}\right)cos\left(\frac{\varphi - \vartheta}{2}\right)$$

$$cos\varphi - cos\vartheta = -2sin\left(\frac{\varphi + \vartheta}{2}\right)sin\left(\frac{\varphi - \vartheta}{2}\right)$$

$$sin\varphi + sin\vartheta = 2sin\left(\frac{\varphi + \vartheta}{2}\right)cos\left(\frac{\varphi - \vartheta}{2}\right)$$

$$sin\varphi - sin\vartheta = 2cos\left(\frac{\varphi + \vartheta}{2}\right)sin\left(\frac{\varphi - \vartheta}{2}\right).$$

3S.2 Basics of Complex Numbers

To be prepared for operations with *complex numbers*, it's worthwhile to review their main forms and mathematical manipulations.

An *imaginary unit j* is defined as

$$j^2 = -1, \tag{3S.5a}$$

and therefore,

$$j = \sqrt{(-1)}, \tag{3S.5b}$$

and

$$\frac{1}{j} = \frac{j}{j^2} = -j. \tag{3S.5c}$$

A complex number **Z** can be presented in three following forms:

$$\begin{cases} a\ rectangular\ \ form,\ \mathbf{Z} = a + jb \\ a\ polar(angular)\ form,\ \mathbf{Z} = A\angle\Theta. \\ an\ exponential\ \ form,\ \mathbf{Z} = Ae^{j\theta} \end{cases} \tag{3S.6}$$

Here the *amplitude* (*magnitude*, *modulus*), A, and the phase angle, Θ, are given by

$$\begin{cases} A = \sqrt{a^2 + b^2} \\ \theta = tan^{-1}\left(\frac{b}{a}\right). \end{cases} \tag{3S.7}$$

Note that complex numbers are denoted by boldfaced capital letters, such as **Z**, **A**, or **B**, whereas their amplitudes (moduli) are given in regular fonts, such as Z, A, or B. You can often meet the amplitudes are designated as the absolute values of the complex numbers; that is, $|Z|$ is the amplitude of **Z**, $|A|$ is the modulus od **A**, and so on.

The relationship between rectangular and polar forms is visualized by the diagram shown in Figure 3S.1. Note how the phase angle, Θ, is calculated: For 1st and 2nd quadrants, we count the angle in the counterclockwise (CCW) direction, which makes this angle positive; for the 3rd and 4th quadrants, we might continue counting it in the same direction to keep it positive, or we can count it in the clockwise (CW) direction, which would make it negative.

You will recall that there are *phasor* ***V*** and *rotating phasor* $\Re\boldsymbol{V}$, where a phasor is a vector with the same amplitude A and the phase shift θ, whereas its counterpart is also $A-\theta$ vector but rotating with angular velocity $\omega\left(\frac{rad}{s}\right)$. Thus, the phasor's exponential form is $\boldsymbol{V} = Ae^{j\theta}$ and the $\Re\boldsymbol{V}$ exponential form is $\Re\boldsymbol{V} = Ae^{j(\omega t+\theta)}$. The relationship between a phasor and the complex number in a rectangular form, $a + jb$, is demonstrated in Figure 3S.1. (For more on this topic, see, for example, (Mynbaev and Scheiner 2020, pp. 178–181.)

Also, we need to recall the basic operations with complex numbers, remembering that j•j = −1:

Two complex numbers $\boldsymbol{Z} = a + jb$ and $\boldsymbol{Z}^* = a - jb$ are called *complex conjugates*. The sum of the complex conjugates is a real number,

$$\boldsymbol{Z} + \boldsymbol{Z}^* = (a + jb) + (a - jb) = 2a, \tag{3S.8}$$

and their difference is an imaginary number,

$$\boldsymbol{Z} - \boldsymbol{Z}^* = (a + jb) - (a - jb) = j2b. \tag{3S.9}$$

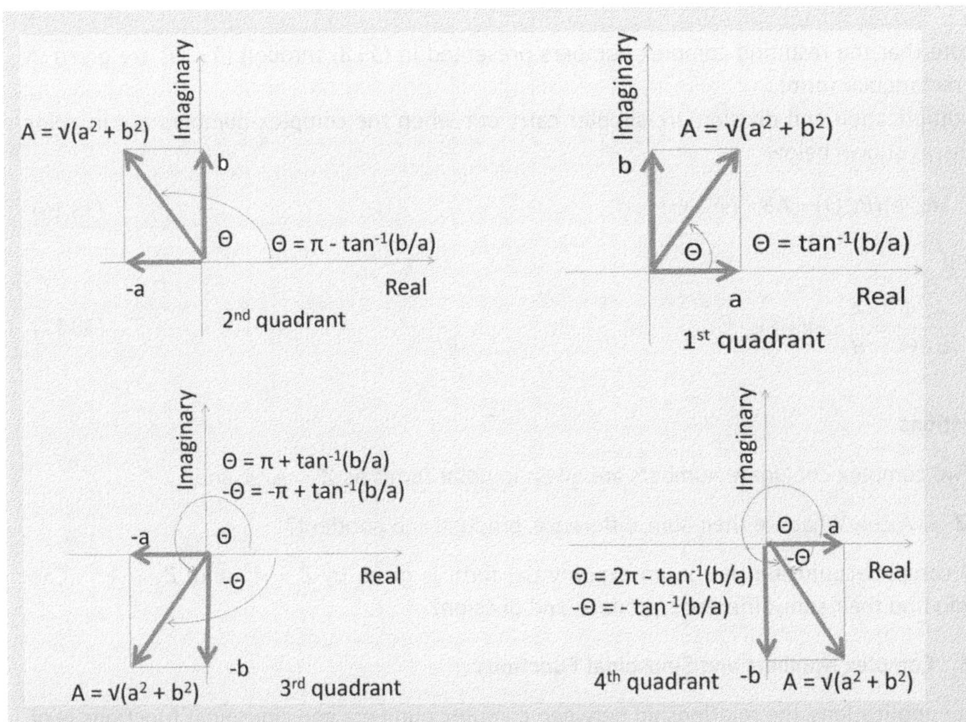

Figure 3S.1 Phasor and a complex number in the rectangular form.

The product of the complex-conjugate pair is a real number,

$$(a+jb)(a-jb)=a^2+b^2, \tag{3S.10}$$

whereas their division results in a complex number,

$$\frac{a+jb}{a-jb}=\frac{(a+jb)(a+jb)}{(a-jb)(a+jb)}=\frac{\left(a_2^2-b_2^2\right)}{\left(a_2^2+b_2^2\right)}+j\frac{2ab}{\left(a_2^2+b_2^2\right)}. \tag{3S.11}$$

In general, the sum of two complex numbers is given by

$$(a_1+jb_1)+(a_2+jb_2)=(a_1+a_2)+j(b_1+b_2), \tag{3S.12}$$

and the difference is

$$(a_1+jb_1)-(a_2+jb_2)=(a_1-a_2)+j(b_1-b_2). \tag{3S.13}$$

The product of two complex numbers is a complex number,

$$(a_1+jb_1)\,(a_2+jb_2)=\,(a_1a_2-b_1b_2)+j(a_1b_2+a_2b_1), \tag{3S.14}$$

and their division produces a complex number too,

$$\frac{a_1+jb_1}{a_2+jb_2}=\frac{(a_1+jb_1)(a_2-jb_2)}{(a_2+jb_2)(a_2-jb_2)}=\frac{(a_1a_2+b_1b_2)}{\left(a_2^2+b_2^2\right)}+j\frac{(a_2b_1-a_1b_2)}{\left(a_2^2+b_2^2\right)}. \tag{3S.15}$$

Note that the resulting complex numbers presented in (3S.8) through (3S.15) are given in the rectangular forms.

Multiplication and division are simpler carry out when the complex numbers are in polar forms, as shown below:

$$(A\angle\Phi)(B\angle\Theta) = AB\angle(\Phi+\Theta) \tag{3S.16}$$

and

$$\frac{A\angle\Phi}{B\angle\Theta} = \frac{A}{B}\angle(\Phi-\Theta). \tag{3S.17}$$

Questions

1) Two complex-conjugate numbers are given in polar forms as $\boldsymbol{Z} = A\angle\Phi$ and $\boldsymbol{Z}^* = A\angle\text{-}\Phi$. What are their sum, difference, product, and quotient?
2) A complex-conjugate pair in an exponential form is given by $\boldsymbol{Z} = A^{j\theta}$ and $\boldsymbol{Z}^* = A^{-j\theta}$. Can you find their sum, difference, product, and division?

3S.3 Complex Numbers and Sinusoidal Functions

In our applications, the relationship between complex numbers and sinusoidal functions is of vital importance. The presentation of a sinusoidal signal as an exponential function is based on the *Euler's identity*

$$e^{j\theta} = cos\theta + jsin\theta, \tag{3S.18}$$

from which the reverse formulas follow as

$$cos\theta = Re\{e^{j\theta}\} = Re\{cos\theta + jsin\theta\} \tag{3S.19a}$$

and

$$sin\theta = Im\{e^{j\theta}\} = Im\{cos\theta + jsin\theta\}. \tag{3S.19b}$$

Taking into account (3S.6) and (3S.18), we can write

$$\boldsymbol{Z} = Ae^{j\theta} = Acos\theta + jAsin\theta, \tag{3S.20a}$$

from which we can derive

$$Re\{\boldsymbol{Z}\} = Acos\theta \tag{3S.20b}$$

and

$$Im\{\boldsymbol{Z}\} = Asin\theta. \tag{3S.20c}$$

All the given equations will considerably simplify our mathematical manipulations in this chapter and in the advanced analysis of electrical circuits involving the complex poles of the s-domain functions.

This section has provided us with the descriptions of sinusoidal signal and their phasor presentation—the primary tools for performing an ac steady-state circuit analysis.

3.2 AC Circuit Analysis of RC, RL, and RLC Circuits with Sinusoids and Phasors

3.2.1 Voltage-Current Relationships of R, L, and C Components for Sinusoidal Signals in the Time and Phasor Domains

We must firmly keep in mind that the material in Sections 3.1 and 3.2 work only for *sinusoidal signals*. This is why we continue to stress that only *steady-state circuit analysis* (means operations with sinusoidal signals) is considered in these sections.

We carefully analyzed the $v-i$ relationships of $R, L,$ and C in Section 1.2. Table 1.3 summarized our analysis; it's a worthwhile effort to revisit this table. In this subsection these general relationships are considered for one specific case—a sinusoidal signal. Thus, we insert a sinusoidal signal in the equations governing these connections. Table 3.1 summarizes the results of such an insertion not only in the time domain but also in the phasor domain.

Here are the explanations of mathematical manipulations presented in Table 3.1:

- To check (3.22), we follow this logic: If $v_C(t)=\frac{A}{C\omega}\sin(\omega t+\theta)$, which we found in (3.23), is correct, then taking its derivative and multiplying it by C, we must receive given $i_C(t)$, according to $i_C(t)[A]=C\frac{dv_C(t)}{dt}$ (1.24). We performed this test, and it went as expected.
- Equation 3.23 is obtained as follows:
  - First, we construct the *rotating phasor* of the resistor current, $\Re \boldsymbol{I}_R$, based on the given input current, $i_R(t)=A_R cos(\omega t+\theta_R)$, by following the phasor definition; i.e., $\Re \boldsymbol{I}_R=A_R e^{j(\omega t+\theta_R)}=e^{j\omega t}A_R e^{j\theta_R}=e^{j\omega t}\boldsymbol{I}_R$, where the *current phasor* is given by $\boldsymbol{I}_R=A_R e^{j\theta_R}$.
- Secondly, we find that the rotating phasor $\Re \boldsymbol{V}_R$ of the resistor voltage is obtained from the rotating current phasor employing definition (3.20), $v_R(t)[V]=ARcos(\omega t+\theta)$, in the phasor domain. Thus,

$$\Re \boldsymbol{V}_R=RA_R e^{j(\omega t+\theta_R)}=Re^{j\omega t}A_R e^{j\theta_R}=Re^{j\omega t}\boldsymbol{I}_R.$$

- Equation 3.24 is derived in a similar manner:
  - The *rotating phasor* of the inductor current, $\Re \boldsymbol{I}_L$, is given by

$$\Re \boldsymbol{I}_L=A_L e^{j(\omega t+\theta_L)}=e^{j\omega t}A_L e^{j\theta_L}=e^{j\omega t}\boldsymbol{I}_L,$$

  where $\boldsymbol{I}_L=A_L e^{j\theta_L}$.
- Before deriving the rotating phasor $\Re \boldsymbol{V}_L$, we replace the member $-sin(\omega t+\theta)$ by $cos(\omega t+\theta+90^0)$ based on the *trigonometric identities* listed in Sidebar 3S. From this point, we receive $\Re \boldsymbol{V}_L$ by following the routine used for derivation of $\Re \boldsymbol{V}_R$, namely,

$$\Re \boldsymbol{V}_L=j\omega LA_L e^{j(\omega t+\theta_L+90^0)}=j\omega Le^{j\omega t}e^{j90^0}A_L e^{j\theta_L}=j\omega Le^{j\omega t}\boldsymbol{I}_L.$$

- To find the capacitor's voltage phasor, shown in (3.25), we turn this time to a different approach based on (1.24):
  - First, find the capacitor current rotating phasor as

$$\Re \boldsymbol{I}_C=A_C e^{j(\omega t+\theta_C)}=e^{j\omega t}A_C e^{j\theta_C}=e^{j\omega t}\boldsymbol{I}_C,$$

  where $\boldsymbol{I}_C=A_C e^{j\theta_C}$.

Table 3.1 Voltage-current relationship of R, L, and C for sinusoidal signals.

Element	General $v-i$ equation		The $v-i$ equations for $i(t) = A\cos(\omega t+\theta)$ in the time domain		The $v-i$ relationship for a sinusoidal signal in the phasor domain	
R	$v_R(t)[V] = Ri_R(t)$	(1.8R)	$v_R(t)[V] = AR\cos(\omega t+\theta)$	(3.20)	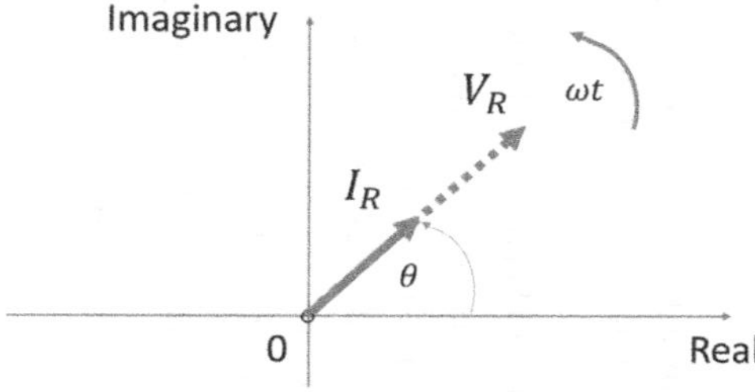 Imaginary, V_R, ωt, I_R, θ, 0, Real $\Re\boldsymbol{I}_R = Ae^{j(\omega t+\theta)} = e^{j\omega t}Ae^{j\theta} = e^{j\omega t}\boldsymbol{I_R}$ $\Re\boldsymbol{V_R} = Re^{j\omega t}\boldsymbol{I_R}$	(3.23)
L	$v_L(t)[V] = L\dfrac{di_L(t)}{dt}$	(1.37R)	$v_L(t)[V]$ $= -AL\omega sin(\omega t+\theta)$ $= AL\omega cos(\omega t+\theta+90^0)$	(3.21)	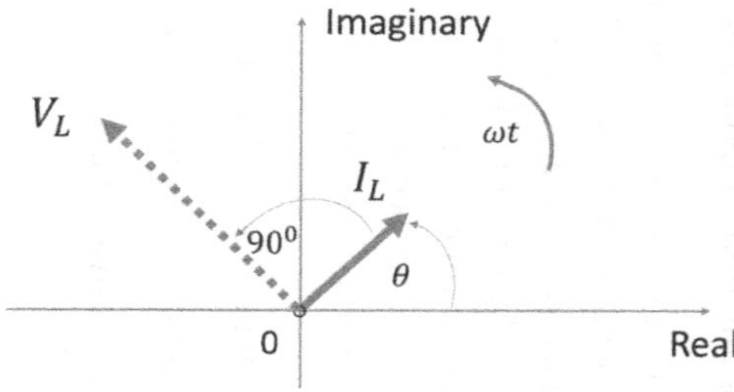 Imaginary, V_L, ωt, I_L, 90^0, θ, 0, Real $\Re\boldsymbol{I_L} = Ae^{j(\omega t+\theta)} = e^{j\omega t}Ae^{j\theta} = e^{j\omega t}\boldsymbol{I_L}$ $\Re\boldsymbol{V_L} = jwLe^{jwt}\boldsymbol{I_L}$	(3.24)

C	$v_C(t)[V] = \frac{1}{C}\int_t^0 i_C(t)dt$ for $t_0 = 0$ and $v_C(t_0) = 0$ (1.28R) or $i_C(t)[A] = C\frac{dv_C(t)}{dt}$ (1.24R)	$v_C(t)[V] = \frac{A}{C\omega}\sin(\omega t + \theta)$ (3.22)	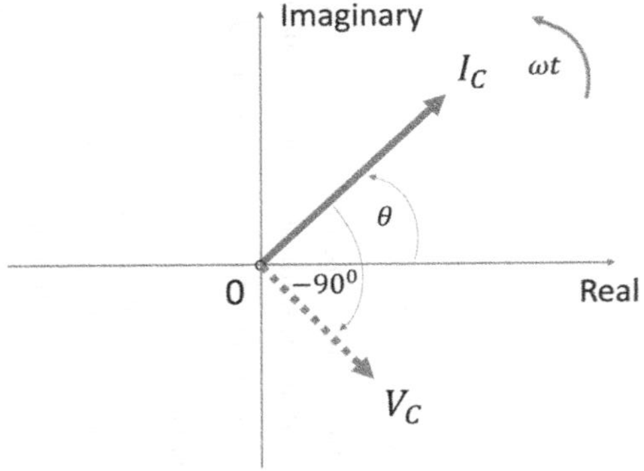 Imaginary, I_C, ωt, θ, -90^0, 0, Real, V_C $\Re \boldsymbol{I}_C = Ae^{j(\omega t+\theta)} = e^{j\omega t}Ae^{j\theta} = e^{j\omega t}\boldsymbol{I}_C$ (3.25) $\Re \boldsymbol{V}_C = -j\frac{1}{wC}e^{jwt}\boldsymbol{I}_C$

- Secondly, recalling that $i_c(t) = Re\{\Re \boldsymbol{I}_c\}$ and $v_C(t) = Re\{\Re \boldsymbol{V}_C\}$, we perform derivation required by (1.24), $i_C(t)[A] = C\frac{dv_C(t)}{dt}$, in the phasor domain as $Re\{\Re \boldsymbol{I}_C\} = Re\{j\omega C \Re \boldsymbol{V}_C\}$.

This expression states that
$\Re \boldsymbol{I}_C = j\omega C \Re \boldsymbol{V}_C$, from which we conclude

$$\Re \boldsymbol{V}_C = \frac{1}{j\omega C}\Re \boldsymbol{I}_C$$

Recalling from (3S.5c) that $\frac{1}{j} = -j$, we arrive at (3.25).

The most important point to take away from examining Table 3.1 is that the *differentiation in the time domain of a sinusoidal signal is equivalent to the multiplication of its phasor by $j\omega$ in the phasor domain, and the integration is equivalent to the division by $\frac{1}{j\omega}$. In other words, an integro-differential equation in the time domain becomes an algebraic equation in the phasor domain.*

This statement is made in Section 3.1, but now Table 3.1 delivers its comparative and geometrical presentation, which helps to illustrate the specific terminology commonly used in *ac circuit analysis*. You will recall that a phasor rotates in a counterclockwise (CCW) direction by convention. The resistor's $\Re \boldsymbol{I}_R$ *and* $\Re \boldsymbol{V}_R$ coincide and rotate together. The rotating voltage phasor of an inductor, $\Re \boldsymbol{V}_L$, is ahead of $\Re \boldsymbol{I}_L$; this is why we say that $\Re \boldsymbol{V}_L$ *leads* $\Re \boldsymbol{I}_L$, or the current phasor of an inductor *lags* its voltage counterpart. Accordingly, the current phasor of a capacitor *leads* its voltage counterpart, meaning that $\Re \boldsymbol{V}_C$ *lags* $\Re \boldsymbol{I}_C$.

To complete discussion of $v - i$ relationships for $R, L,$ and C components with sinusoidal signals, it's necessary to consider the impedances of these components.

3.2.2 Impedances

The $v - i$ relationships presented in Equations 3.22, 3.23, and 3.24 in Table 3.1 are written in terms of rotating phasors because they stem from manipulations with the original sinusoidal signals. However, they can be shown in terms of the phasors by omitting $e^{j\omega t}$ member from those equations as shown in Table 3.2 as Equations 3.26, 3.27, and 3.28.

We remember that $R(\Omega) = \frac{V_R}{I_R}$ is called *resistance*, the measure of how much a resistor opposes to the dc current flow. It is reasonable to assume that

$$\boldsymbol{Z}_R(\Omega) = \frac{\boldsymbol{V}_R}{\boldsymbol{I}_R}$$

does the same for sinusoidal signals, i.e., $\boldsymbol{Z}_R$ *represents the resistance in the phasor domain*. But what quantities $j\omega L$ and $\frac{1}{jC\omega}$ can present if they have an imaginary unit included? As (3.26) shows, for a resistor, $\boldsymbol{Z}_R(\Omega)$ is a proportionality coefficient that relates $\boldsymbol{V}_R$ and $\boldsymbol{I}_R$. Follow this pattern, we accept that $j\omega L$ and $\frac{1}{j\omega C}$ describe the inductor and capacitor opposition to the ac current flow. They are called *impedances* and are denoted as $\boldsymbol{Z}_L$ and $\boldsymbol{Z}_C$, respectively. Their roles in $v - i$ relationships are clear from (3.27) and (3.28), and their formulas are numbered as 3.30, 3.31, and 3.32. See Table 3.2.

Table 3.2 Ohm's laws, impedances, and admittances for R, L, and C.

Component	Ohm's law		Impedance		Admittance	
R	$\mathbf{V}_R = R\mathbf{I}_R = \mathbf{Z}_R\mathbf{I}_R$	(3.26)	$\mathbf{Z}_R(\Omega) = R$	(3.29)	$\mathbf{Y}_R = \frac{1}{\mathbf{Z}_R} = \frac{1}{R}$	(3.32)
L	$\mathbf{V}_L = j\omega L\mathbf{I}_L = \mathbf{Z}_L\mathbf{I}_L$	(3.27)	$\mathbf{Z}_L(\Omega) = j\omega L$	(3.30)	$\mathbf{Y}_L = \frac{1}{\mathbf{Z}_L} = \frac{1}{j\omega L}$	(3.33)
C	$\mathbf{V}_C = \frac{1}{j\omega C}\mathbf{I}_C = \mathbf{Z}_C\mathbf{I}_C$	(3.28)	$\mathbf{Z}_C(\Omega) = \frac{1}{j\omega C} = -j\frac{1}{\omega C}$	(3.31)	$\mathbf{Y}_C = \frac{1}{\mathbf{Z}_C} = j\omega C$	(3.34)

Impedances are boldfaced because they are complex numbers. They can be written in a rectangular form as

$$\mathbf{Z}(\Omega) = G^{-1} + jX, \tag{3.35a}$$

where G^{-1} is called *resistance*, and X *reactance*. For a *resistive impedance*, we have a resistance R,

$$\mathbf{Z}_R(\Omega) = G_R^{-1} + jX_R = R + j0 \Rightarrow G^{-1} = R, \tag{3.35b}$$

an $\mathbf{Z}_L$ has only *inductive reactance* X_L,

$$\mathbf{Z}_L = G_L^{-1} + jX_L = 0 + j\omega L \Rightarrow X_L = \omega L, \tag{3.35c}$$

whereas $\mathbf{Z}_C$ contains only *capacitive reactance* X_C,

$$\mathbf{Z}_C = G_C^{-1} + jX_C = -j\frac{1}{\omega C} \Rightarrow X_C = -\frac{1}{\omega C}. \tag{3.35d}$$

As any complex number, an impedance can be shown in a *polar form* as

$$\mathbf{Z}(\Omega) = Z\angle\theta, \tag{3.36a}$$

where $Z = \sqrt{\left((G^{-1})^2 + X^2\right)}$ and $\theta = tan^{-1}\left(\frac{X}{G}\right)$. Thus,

$$\mathbf{Z}_R(\Omega) = R\angle 0, \tag{3.36b}$$

$$\mathbf{Z}_L(\Omega) = \omega L\angle 90^0, \tag{3.36c}$$

$$\mathbf{Z}_c(\Omega) = \frac{1}{\omega C}\angle -90^0. \tag{3.36d}$$

Comparison of (3.30) with (3.36c) and (3.31) with (3.36d) enables us to write

$$\mathbf{Z}_L = j\omega L = \omega L\angle 90^0, \tag{3.37a}$$

$$\mathbf{Z}_C = -j\frac{1}{\omega C} = \frac{1}{\omega C}\angle -90^0. \tag{3.37b}$$

These equations confirm the general rule: *Multiplying a phasor X by j is equivalent to rotating it by 90^0 (CCW), whereas multiplying X by $-j$ turns the phasor by -90^0 (CW)*. See Table 3.1.

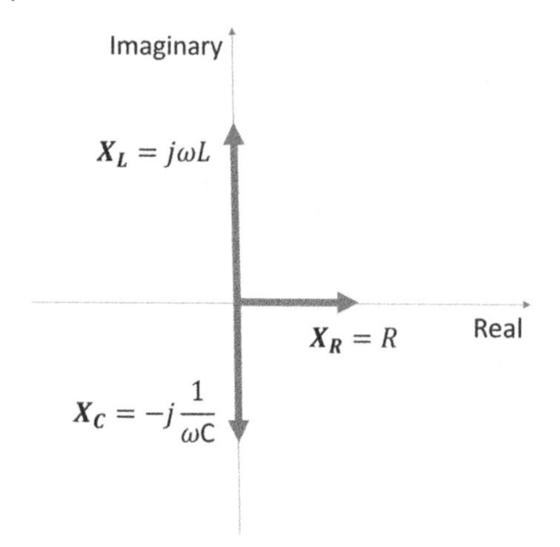

Figure 3.6 Diagram of Z_L, Z_R, *and* Z_C impedances.

Equations 3.36b through 3.37b are visualized in Figure 3.6.

Question We know from (3S.20a) that $Z = Ae^{j\theta} = A\cos\theta + jA\sin\theta$. What are $Re\{Z\}$ and $Im\{Z\}$ in trigonometric terms? Can you express Z_R, Z_L, and Z_C in trigonometric terms?

In contrast to pure resistance, *inductive* and *capacitive reactances* (and therefore their impedances) *depend on frequency*. Equation 3.35c shows that $|X_L|$ is proportional to frequency, and Equation 3.35d along with Figure 1.10 illustrates how $|X_C|$ changes with frequency.

To reiterate, when dealing with resistance, we introduced its reverse quantity, *conductance*, as

$$G(S) = \frac{1}{R(\Omega)}. \tag{1.19R}$$

Likewise, we can introduce the reverse quantity to an impedance; it is called *admittance* and is given by

$$Y = \frac{1}{Z}.$$

The Formulas 3.32, 3.33, and 3.34 for the specific admittances are given in Table 3.2.

From the impedance definition, we can deduct that *the sum, difference, product, and quotient of impedances can be found according to the appropriate rules demonstrated in Sidebar 3S.*

Section 1.3 analyzes the series and parallel dc resistive circuit and shows that such an analysis requires finding an *equivalent resistor* for each circuit in question. It also demonstrates the technique for this finding. Refer specifically to Examples 1.1 and 1.2. A similar method of finding *equivalent impedances* can be applied to the steady-state analysis of ac circuits. The following example demonstrates how it can be done based on the dc circuit analysis technique.

Example 3.2 Finding the equivalent (total) impedance of an RLC circuit.

Problem: Find the equivalent (total) impedance of the circuit shown in Figure 3.7 if the frequency of the input signal is 60 Hz.

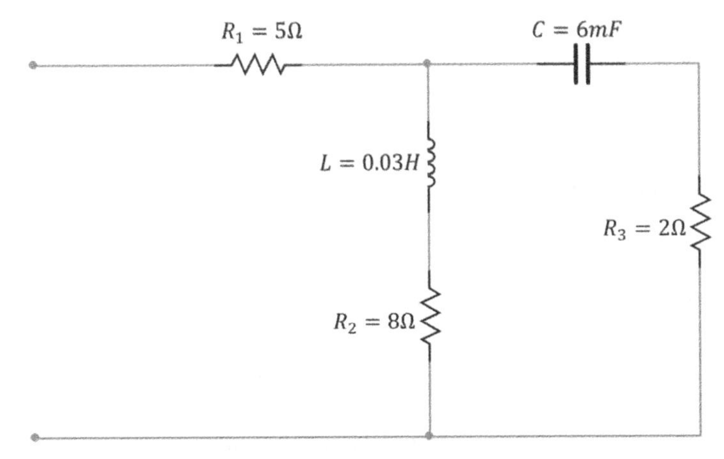

Figure 3.7 Finding the equivalent (total) impedance of the RLC circuit for Example 3.2.

Solution: We rely on the detailed procedure of finding an equivalent resistor demonstrated in Example 1.1 and illustrated in Figure 1.16.

- The impedances of C and R_3 are connected in series, which gives

$$\boldsymbol{Z}_{C+R3} = \boldsymbol{Z}_C + \boldsymbol{Z}_{R3} = -j\frac{1}{\omega C} + R_3 = 2 - j\frac{1}{120\pi \cdot 6 \cdot 10^{-3}} = 2 - j0.44.$$

- The impedances of L and R_2 are also connected in series; hence, their sum is equal to

$$\boldsymbol{Z}_{L+R2} = \boldsymbol{Z}_L + \boldsymbol{Z}_{R2} = j\omega L + R_2 = 8 + j120\pi \cdot 0.03 = 8 + j11.31.$$

- These two summed impedances are in parallel, thus,

$$\boldsymbol{Z}_{C+R3} \,||\, \boldsymbol{Z}_{L+R2} = \frac{10 + j10.87}{20.98 + j19.1} = \frac{(10 + j10.87)(20.98 - j19.1)}{20.98^2 + 19.1^2} = \frac{417.42 + j37.05}{804.97} = 0.52 + j0.05.$$

- Finally, the equivalent (total) impedance is calculated as

$$\boldsymbol{Z}_{eq} = \boldsymbol{Z}_{R1} + \boldsymbol{Z}_{C+R3} \,||\, \boldsymbol{Z}_{L+R2} = 5 + 0.52 + j0.05(\Omega) = 5.52 + j0.05\,(\Omega).$$

  The problem is solved.

You are encouraged to draw a circuit for each step of finding Z_{eq}, similar to what we did in Figure 1.16.

3.2.3 Kirchhoff's Voltage Law in the Phasor Domain

As it is shown in the preceding subsection, Ohm's law and the technique of finding the equivalent impedances work well in the phasor domain if we apply the methods used in dc circuit analysis and replace all the circuit quantities ($V, I, R, L,$ *and* C) by the appropriate phasors ($\boldsymbol{V}, \boldsymbol{I}, \boldsymbol{Z}_R, \boldsymbol{Z}_L,$ **and** $\boldsymbol{Z}_C$). We can expect that the Kirchhoff's laws will work in the phasor domain too after the appropriate transformation.

Let's start with KVL: Extending the law given for dc circuits in (1.82), we can say that the sum of voltages along the closed loop of an ac electrical circuit is equal to zero, i.e.,

$$\sum_1^n v_k(t) = 0, \text{ where } k = 1,2,3,\ldots,n. \tag{3.38a}$$

Don't forget that we restrict our consideration by sinusoidal signals; thus, (3.38a) can be rewritten as

$$\sum_1^n v_k(t) = \sum_1^n A_k \cos(\omega t + \theta_k) = 0, \text{ where } k = 1,2,3,\ldots,n. \tag{3.38b}$$

Equation 3.38b shows the signals of one frequency passing through all elements of the given circuit loop.

Let's transfer this equation into the phasor domain by using (3.11), (3.12), and (3.13):

$$A_k \cos(\omega t + \theta_k) \Rightarrow Re\left\{e^{j\omega t} A_k e^{j\theta_k}\right\} = Re\left\{e^{j\omega t} \mathbf{V}_k\right\}, \tag{3.38c}$$

where phasor $\mathbf{V}_k$ is given by

$$\mathbf{V}_k = A_k e^{j\theta_k}. \tag{3.13R}$$

Recalling (3.11) and plugging (3.38c) in (3.38b) yields

$$\sum_1^n Re\left\{e^{j\omega t} \mathbf{V}_k\right\} = 0, \tag{3.39}$$

which means

$$\sum_1^n Re\{\mathbf{V}_k\} = 0 \Rightarrow \mathbf{V}_1 + \mathbf{V}_2 + \mathbf{V}_3 + \ldots + \mathbf{V}_n = 0 \tag{3.40}$$

because $e^{j\omega t} \neq 0$. Therefore, Equation 3.40 *states that Kirchhoff's voltage law holds in the phasor domain*. Here is the example.

Example 3.3 KVL application in the phasor domain for an RLC circuit.
Problem: A series RLC circuit shown in Figure 3.8 is connected to a preceding stage of a multi-stage circuit. What input voltage, $v_{in}(t)$, will the RLC circuit see? The current flowing through is $i(t) = 1\cos(120\pi t)A$, and the component values are $R = 5.3\Omega$, $L = 14mH$, and $C = 0.5mF$. Apply KVL in the phasor domain. Verify the result by Multisim simulation.

Solution: To start, we need to convert the input current in the phasor form. Using (3.11), we find

$$i(t) = 1\cos(120\pi t) \Rightarrow \Re \mathbf{I} = e^{j120\pi t} 1e^{j0} \Rightarrow I = 1e^{j0} = 1A.$$

Next, we determine the component impedances in the phasor domain as

$$Z_R = 5.3 + j0 = 5.3e^{j0} = 5.3\angle 0^0,$$

$$Z_L = 0 + j120\pi \cdot 0.014 = 0 + j5.278 = 5.278e^{j90^0} = 5.278\angle 90^0,$$

$$Z_C = 0 - \frac{j1}{120\pi \cdot 0.005} = 0 - j5.305 = 5.305e^{-j90^0} = 5.305\angle -90^0.$$

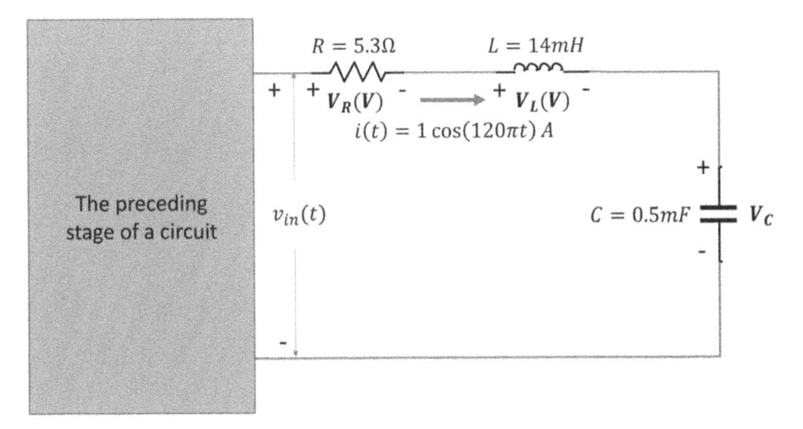

Figure 3.8 An RLC series circuit for considering KVL in the phasor domain for Example 3.2.

Having the component impedances, we can use Ohm's law in the phasor domain to find the voltages across each element as

$$\boldsymbol{V_R} = \boldsymbol{Z_R} \cdot \boldsymbol{I},$$

$$\boldsymbol{V_L} = \boldsymbol{Z_L} \cdot \boldsymbol{I},$$

$$\boldsymbol{V_C} = \boldsymbol{Z_C} \cdot \boldsymbol{I}.$$

Applying Equation 3.40, KVL in the phasor domain, results in

$$\boldsymbol{V_{in}} = \boldsymbol{V_R} + \boldsymbol{V_L} + \boldsymbol{V_C} = \boldsymbol{Z_R} \cdot \boldsymbol{I} + \boldsymbol{Z_L} \cdot \boldsymbol{I} + \boldsymbol{Z_C} \cdot \boldsymbol{I} = (\boldsymbol{Z_R} + \boldsymbol{Z_L} + \boldsymbol{Z_C})\boldsymbol{I}. \quad (3.41)$$

Here $\boldsymbol{I} = A_i e^{j\theta}$. Plugging the computed values in (3.41) yields

$$\boldsymbol{V_{in}} = (5.3 + j0 + 0 + j5.278 + 0 - j5.305)\Omega \cdot 1A$$
$$= (5.3 - j0.027)V \approx 5.3\angle -0.29^0 V = 5.3e^{-j0.29^0} V.$$

Finally,

$$\Re \boldsymbol{V_{in}} = 5.3e^{j(120\pi t - 0.29^0)} \Rightarrow v_{in}(t) = 5.3\cos(120\pi t - 0.29^0)V,$$

and the answer is

$$v_{in}(t) = 5.3\cos(120\pi t - 0.3^0)V.$$

The problem is manually solved.

Discussion:

1) The circuit shown in Figure 3.9 is built to verify the manually obtained result by the Multisim simulation. To run the simulation, employ the following *Multisim commands:*
   - *Stimulate* from the popup menu,
     - *Analyses,*
       - Single frequency analysis,
         - Mark frequency column,
         - Choose the output form: *Magnitude/Phase* or *Real/Imaginary,*
           - Go to Output and choose V(1),
           - Simulate.

Figure 3.9 Multisim simulation of the KVL application in a series RLC circuit or Example 3.3: a) Circuit's schematic; b) magnitude/phase table; c) real/imaginary table.

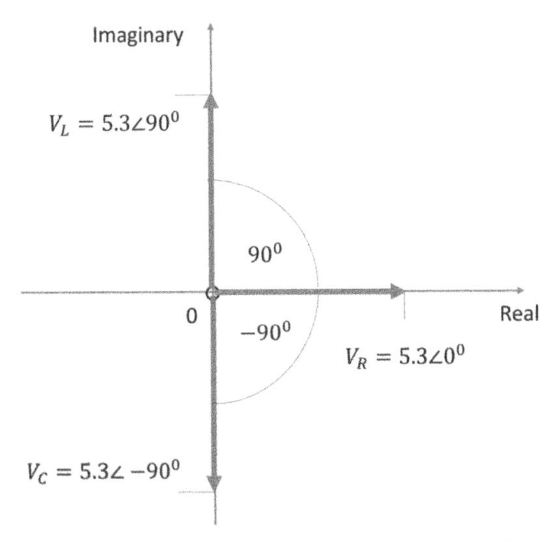

Figure 3.10 Diagram of the phasors for Example 3.2.

For this demonstration, we request Multisim to return both *Magnitude/Phase* and *Real/Imaginary* tables. They confirm our manually obtained results.

2. Why are the imaginary part and phase shift in $\boldsymbol{V_{in}}$ close to zero? Examine Figure 3.10 to see that the phasors $\boldsymbol{V_L}$ and $\boldsymbol{V_C}$ are aligned along the imaginary axis in the opposite directions as they should be. Since we choose their magnitudes to be almost equal at the given frequency, their sum is close to zero. This is a good **exercise**: Choose new values of L and C to make $\boldsymbol{V_L} \neq \boldsymbol{V_C}$, draw the new phasor diagram, and see how the new phasor $\boldsymbol{V_R} + \boldsymbol{V_L} + \boldsymbol{V_C}$ will be oriented. Do all necessary calculations manually. Then use Multisim to verify your result by building an RLC circuit with new component values and simulating its operation in the phasor domain.

Question Can we change the values of X_L *and* X_C without changing L and C?

3. It follows from (3.41) that supplying the input voltage to this circuit, we obtain its total current as

$$\boldsymbol{I} = \frac{\boldsymbol{V_{in}}}{\boldsymbol{Z_R} + \boldsymbol{Z_L} + \boldsymbol{Z_C}} A.$$

Since in our example $\boldsymbol{Z_L} = \boldsymbol{Z_C}$ and $\boldsymbol{V_{in}}$ are constant, we can make $\boldsymbol{I}$ tends to infinity by reducing $\boldsymbol{Z_R} = R$ to zero. Is it possible? Yes, this situation is a well-known phenomenon called *resonance*. The equality of the reactive impedances, $\boldsymbol{Z_L} = \boldsymbol{Z_C}$, is the necessary resonance condition; varying the ohmic resistance changes the amplitude of the resonance reaction.

This is the end of Example 3.3.

3.2.4 Voltage Divider Rule

The circuit shown in Figure 3.11a is almost a replica of that in Figure 3.9a, but the problem is quite different: Figure 3.11a shows that the *voltage source*, rather than the current one, is connected to the circuit. The task is to find the *voltage drops across each circuit's component*.

A similar problem was solved in Chapter 1 by applying the *voltage divider rule* (1.50) for a resistive circuit,

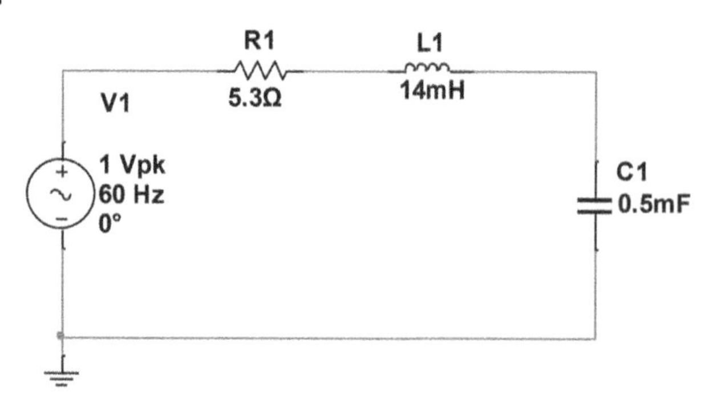

a)

V_R 5.3Ω V_L 14mH V_{in} V_C 0.5mF

b)

Figure 3.11 Voltage divider rule for RLC circuit: a) Circuit's schematic; b) the circuit presentation in the phasor domain.

$$v_k(V) = R_k \cdot i = \frac{R_k}{R_1 + R_2 + R_3 + \ldots R_n} v_{in} = \frac{R_k}{R_{total}} v_{in}. \tag{1.50R}$$

Employing this rule for the reactive RLC circuit can be made provided that all the manipulations are performed in the phasor domain. Thus, in Equation 1.50, we must replace $v_{k(t)}$ by $\boldsymbol{V_k}$, $i(t)$ by $\boldsymbol{I}$, and R_k by $\boldsymbol{Z_k}$, as shown in (3.42)

$$\boldsymbol{V_k}(\boldsymbol{V}) = \boldsymbol{Z_k} \cdot \boldsymbol{I} = \frac{\boldsymbol{Z_k}}{\boldsymbol{Z_1} + \boldsymbol{Z_2} + \boldsymbol{Z_3} + \ldots \boldsymbol{Z_n}} \boldsymbol{V_{in}} = \frac{\boldsymbol{Z_k}}{\boldsymbol{Z_{total}}} \boldsymbol{V_{in}}. \tag{3.42}$$

For the circuit shown in Figure 3.12, this replacement produces

$$\begin{cases} \boldsymbol{V_R}(\boldsymbol{V}) = \dfrac{\boldsymbol{Z_R}}{\boldsymbol{Z_{total}}} \boldsymbol{V_{in}} \\ \boldsymbol{V_L}(\boldsymbol{V}) = \dfrac{\boldsymbol{Z_L}}{\boldsymbol{Z_{total}}} \boldsymbol{V_{in}} \\ \boldsymbol{V_C}(\boldsymbol{V}) = \dfrac{\boldsymbol{Z_C}}{\boldsymbol{Z_{total}}} \boldsymbol{V_{in}} \end{cases}. \tag{3.43}$$

If we borrow from Example 3.3 the component values, the input voltage $v_{in}(t) = 1\cos(120\pi t - 0.3^0)V$, and the calculated impedances,

$$Z_R = 5.3 + j0 = 5.3e^{j0} = 5.3\angle 0^0,$$

$$Z_L = 0 + j5.278 = 5.278e^{j90^0} = 5.278\angle 90^0,$$

$$Z_C = 0 - j5.305 = 5.305e^{-j90^0} = 5.305\angle -90^0,$$

then the input voltage and total impedance in the phasor format become

$$\boldsymbol{V}_{in} = 5.3\angle -0.3^0$$

and

$$\boldsymbol{Z}_{total} = \boldsymbol{Z}_R + \boldsymbol{Z}_L + \boldsymbol{Z}_C = 5.3 - j0.027 = 5.3\angle -0.3^0.$$

All the computed values enable us to calculate the individual voltage drops as

$$\boldsymbol{V}_R = \frac{\boldsymbol{Z}_R}{\boldsymbol{Z}_{total}}\boldsymbol{V}_{in} = \frac{5.3\angle 0^0}{5.3\angle -0.3^0}5.3\angle -0.3^0 = 5.3\angle 0^0 V \Rightarrow v_R(t) = 5.3\cos(120\pi t)V.$$

$$\boldsymbol{V}_L = \frac{\boldsymbol{Z}_L}{\boldsymbol{Z}_{total}}\boldsymbol{V}_{in} = \frac{5.278\angle 90^0}{5.3\angle -0.3^0}5.3\angle -0.3^0 \Rightarrow v_L(t) = 5.278\cos(120\pi t + 90^0)V = -5.278 sin(120\pi t).$$

$$\boldsymbol{V}_C = \frac{\boldsymbol{Z}_C}{\boldsymbol{Z}_{total}}\boldsymbol{V}_{in} = \frac{5.305\angle -90^0}{5.3\angle -0.3^0}5.3\angle -0.3^0 \Rightarrow v_C(t) = 5.305\cos(120\pi t - 90^0)V = 5.305 sin(120\pi t).$$

This example shows how to apply the voltage divider rule in the phasor domain to calculate the individual component voltage drop.

Exercise Draw $v_R(t)$, $v_L(t)$, and $v_C(t)$ on the one set of axes and graphically prove that their sum is indeed equal to $v_{in}(t)$.

3.2.5 Kirchhoff's Current Law and Current Divider Rule in the Phasor Domain

Refer again to Equations 3.11 through 3.13, where the relationships among a sinusoidal voltage signal and its phasor representation are shown. It's clear that these relations hold for current too; so, we can write

$$i(t) = A\cos(\omega t + \theta) \Rightarrow e^{j\omega t}Ae^{j\theta} = e^{j\omega t}\boldsymbol{I},$$

where $\boldsymbol{I} = Ae^{j\theta}$ is the given sinusoidal current in the phasor format. Recall Kirchhoff's current law's discussion in Section 1.5: *Kirchhoff's current law (KCL) states that the algebraic sum of all currents entering and leaving a node is equal to zero*, i.e.,

$$\sum_1^n i_k = 0, \text{ where } k = 1,2,3,\ldots,n. \tag{1.84R}$$

Extending this law to the phasor domain gives

$$\sum_1^n \boldsymbol{I}_k = 0, \text{ where } k = 1,2,3,\ldots,n. \tag{3.44}$$

Equation 3.44 enables us to use KCL in the phasor domain. The following example demonstrates how to apply KCL and voltage divider rule in the phasor domain to find currents through individual circuit elements.

Example 3.4 Application of KCL and voltage divider rule in the phasor domain to find currents through individual circuit elements.

Problem: The parallel RLC circuit, the input voltage, and the values of its components are shown in Figure 3.12. Find the currents flowing through each circuit element.

Solution: On the one hand, we solved a similar problem for a dc resistive circuit in Example 1.2, where we employed KCL, found a total (equivalent) resistance, and used the current divider rule. On the other hand, we've learned in the extant section that all these methods can be applied to an ac reactive circuit provided that all mathematical manipulations are performed in the phasor domain. Hence, we must present all the given circuit parameters in the phasor formats and perform manipulations in the sequence demonstrated in Figure 1.17. Here is how we can do these calculations, bearing in mind that the component values are taken from Example 3.2:

- Find the total (equivalent) impedance:

$$\frac{1}{\boldsymbol{Z}_{total}} = \frac{1}{\boldsymbol{Z}_R} + \frac{1}{\boldsymbol{Z}_L} + \frac{1}{\boldsymbol{Z}_C} = \frac{1}{5.3\angle 0^0} + \frac{1}{5.278\angle 90^0} + \frac{1}{5.305\angle -90^0}$$
$$\Rightarrow \boldsymbol{Z}_{total} \approx 5.3\angle 0^0.$$

- Find the total (equivalent or input) current using Ohm's law:

$$\boldsymbol{I}_{total} = \frac{\boldsymbol{V}_{in}}{\boldsymbol{Z}_{total}} = \frac{1\angle 0^0}{5.3\angle 0^0} = 0.189\angle 0^0 \Rightarrow i_{total}(t) = 0.189\cos(120\pi t)A$$

- Determine the current through each circuit's element:
  We can use the current divider rule derived for a resistive circuit in (1.59a),

$$i_k(A) = \frac{i_{total} \cdot R_{total}}{R_k}, \tag{1.59aR}$$

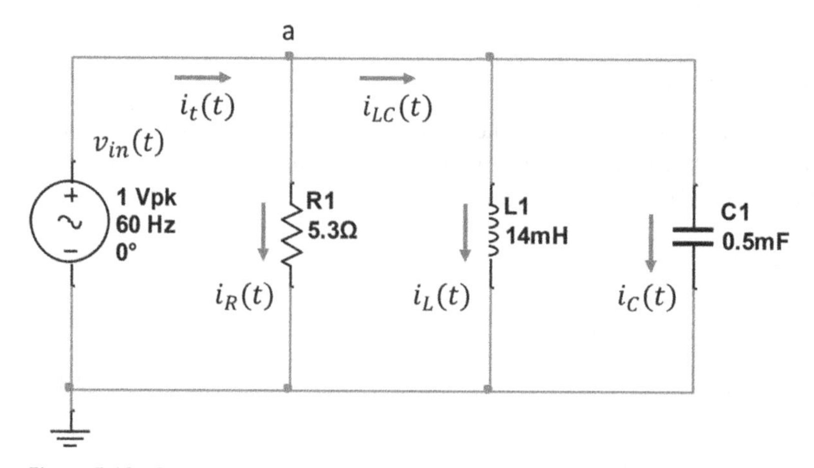

Figure 3.12 Parallel RLC circuit for Example 3.4.

in our problem by rewriting it in the phasor format as

$$\boldsymbol{I}_k = \frac{\boldsymbol{I}_{total} \cdot \boldsymbol{Z}_{total}}{\boldsymbol{Z}_k}(k = 1,2,3,\ldots n). \tag{3.45}$$

Here, (3.45) is the *current divider rule in the phasor domain.* Plugging the given and computed values results in

$$\boldsymbol{I}_R = \frac{0.189\angle 0^0 \cdot 5.3\angle 0^0}{5.3\angle 0^0} = 0.189\angle 0^0 = 0.189 + j0$$
$$\Rightarrow i_R(t) = 0.189\cos(120\pi t)A.$$

$$\boldsymbol{I}_L = \frac{0.189\angle 0^0 \cdot 5.3\angle 0^0}{5.278\angle 90^0} \approx 0.189\angle -90^0 = 0 - j0.189$$
$$\Rightarrow i_L(t) = 0.189\cos\left(120\pi t - 90^0\right)A = 0.189\sin(120\pi t)A.$$

$$\boldsymbol{I}_C = \frac{0.189\angle 0^0 \cdot 5.3\angle 0^0}{5.305\angle -90^0} \approx 0.189\angle 90^0 = 0 + j0.189$$
$$\Rightarrow i_C(t) = 0.189\cos\left(120\pi t + 90^0\right)A = -0.189\sin(120\pi t)A.$$

The problem is solved.

Discussion:

- Check the results:

First, we must confirm that KCL is met at node *a* of the circuit. Starting with the phasor domain, we find that it holds:

$$\boldsymbol{I}_{total} = \boldsymbol{I}_R + \boldsymbol{I}_{LC} = \boldsymbol{I}_R + \boldsymbol{I}_L + \boldsymbol{I}_C$$
$$\Rightarrow 0.189\angle 0^0 = (0.189 + j0) + (0 - j0.189) + (0 + j0.189).$$

Repeating the check in the time domain, we confirm that KCL holds here too:

$$i_{total}(t) = i_R(t) + i_L(t) + i_C(t)$$
$$\Rightarrow 0.189\cos(120\pi t)A = 0.189\cos(120\pi t) + 0.189\sin(120\pi t) - 0.189\sin(120\pi t).$$

- The calculations show that the total input current is equal to the current flowing through the resistor, which means that no current flows through the reactive section of the circuit. Why $\boldsymbol{I}_L + \boldsymbol{I}_C$, or alternatively, $i_L(t) + i_C(t)$, are equal to zero? This is because $\frac{1}{\boldsymbol{Z}_{LC}} = \frac{1}{\boldsymbol{Z}_L} + \frac{1}{\boldsymbol{Z}_C} \to 0$, which means that $\boldsymbol{Z}_{LC} \to \infty$. This is why the whole input current flows through the resistor only. This well-known phenomenon is called *anti-resonance* because at the resonant frequency ($f_r = 60Hz$ in this example), the total impedance reaches its maximum value. How can we prove this statement? If we change the frequency of an input signal, the reactive impedances $X_L = 2\pi fL$ and $X_C = \frac{1}{2\pi fC}$ will change, and no resonance condition will exist. Then, the input current will split among *R*, *L*, and *C*, according to (3.35c).
- We urge you to plot the phasor diagram showing $\boldsymbol{Z}_R$, $\boldsymbol{Z}_L$, *and* $\boldsymbol{Z}_C$ in this example. To do so, consider Figure 3.10 as a pattern.

- **Auxiliary problem**: What will be $i_{total}(t)$ of the circuit shown in Figure 3.11 if the frequency of the input signal changes from 60 Hz to 120 Hz? All other circuit parameters have been left intact.
- **Solution**: Reactive impedances will change as

  $X_L = 2\pi fL = 240\pi \cdot 0.014 = 10.56\Omega$ and $X_C = \dfrac{1}{2\pi fC} = \dfrac{1000}{240\pi \cdot 0{,}5} = 2.65\Omega.$

  Thus,

$$\boldsymbol{Z_{total}} = \boldsymbol{Z_R} + \boldsymbol{Z_L} + \boldsymbol{Z_C} = 5.3 + j0 + 0 + j10.56 + 0 - j2.65 = 5.3 + j7.91 = 9.52\angle 56.2^0\Omega.$$

  Therefore,

$$\boldsymbol{I_{total}} = \frac{\boldsymbol{V_{in}}}{\boldsymbol{Z_{total}}} = \frac{1V}{9.52\angle 56.2^0\Omega} = 0.105\angle -56.2^0\,A. \Rightarrow i_{total}(t) = 0.105\cos\left(240\pi t - 56.2^0\right).$$

Draw the phasor diagram to illustrate this solution and graphically prove that this case has no anti-resonance.

- Calculate $i_R(t),\ i_L(t),\ and\ i_C(t)$ to confirm further that the total (input) current splits among all three circuit elements and find in what proportion it will split.

The circuit analysis in the phasor domain is a powerful and effective method of analyzing ac circuits, as the study of this section should convince you. Unhappily, this technique is restricted to sinusoidal signals only.

3.3 Theorems and Principles of AC Steady-State Circuit Analysis

Chapters 1 and 2 show what *dc circuit analysis* is and how to perform it. The first two sections of Chapter 3 demonstrate how the approach developed for dc circuit analysis can be applied to *ac steady-state circuit analysis* using phasors. Indeed, Sections 3.1 and 3.2 define current, voltage, Ohm's law, KCL, KVL, current divider rule, and voltage divider rule—all entities we need to know to perform a circuit analysis—in the phasor domain. In this section, we want to advance in ac sinusoidal circuit analysis by extending the application of nodal and mesh methods and Thevenin's and Norton's theorems discussed in Chapter 2 to ac circuits. Also, we add new analysis tools—the *superposition* and *duality principles.*

3.3.1 Nodal and Mesh Circuit Analyses

Nodal circuit analysis discussed in Section 2.1 aims to find the voltage across and current through every circuit component by determining the nodal voltages using KCL. The voltages and currents in search are called *branch voltages* and *currents*—review Section 2.1, particularly Example 2.1. We will demonstrate the use of the nodal analysis for the ac steady-state circuit in Example 3.5.

Example 3.5 Nodal analysis in the phasor domain.

Problem: Find the voltage across and current through every element of a series-parallel RLC circuit shown in Figure 3.14a employing the nodal analysis.

Solution: We rely on Example 2.1 with modifications that reflect peculiarities of ac steady-state circuit analysis.

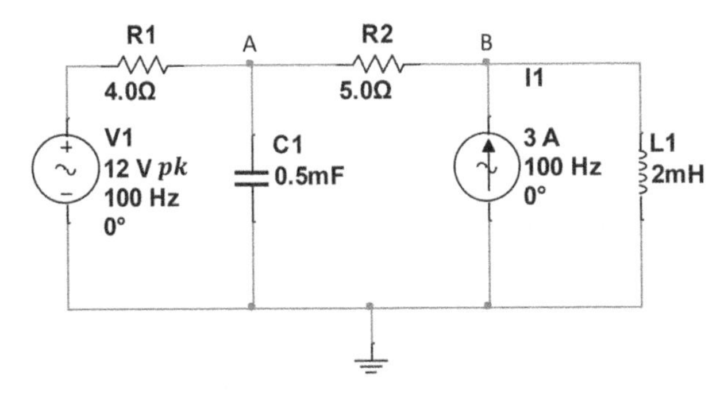

a)

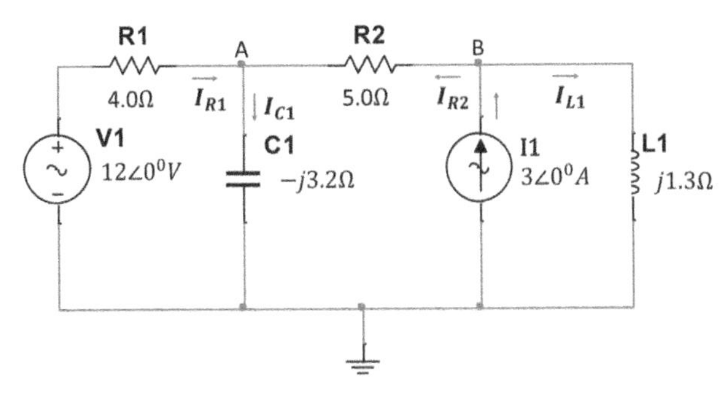

b)

Figure 3.13 Nodal analysis of series-parallel RLC circuit for Example 3.5: a) Circuit schematic; b) the phasor-domain model of the given circuit.

Table 3.3 Time-domain variables and their phasor-domain equivalents in Example 3.5.

Component	Time-domain value	Phasor-domain equivalent
$R_1(\Omega)$	$R_1(\Omega)=4$	$Z_{R1}=R_1(\Omega)+j0(\Omega)=4\Omega$
$R_2(\Omega)$	$R_2(\Omega)=5$	$Z_{R2}=R_1(\Omega)+j0(\Omega)=5\Omega$
$L1(H)$	$X_{L1}=2\pi f L1(\Omega)=1.3\Omega$	$Z_{L1}=0+jX_L(\Omega)=j1.3\Omega$
$C1(F)$	$X_{C1}=\dfrac{1}{2\pi f C1}(\Omega)=3.2\Omega$	$Z_{C1}=0-jX_C(\Omega)=-j3.2\Omega$

- The first step is to convert time-domain variables into their phasor-domain equivalents. Table 3.3 shows these conversions.

  Figure 3.13b presents a given circuit in the phasor domain and the calculated impedances.

- The second step is to apply KCL to nodes A and B as follows:

$$Node\,A \Rightarrow \boldsymbol{I}_{R1}+\boldsymbol{I}_{R2}=\boldsymbol{I}_{C1} \Rightarrow \frac{\boldsymbol{V1}-\boldsymbol{V}_A}{\boldsymbol{Z}_{R1}}+\frac{\boldsymbol{V}_B-\boldsymbol{V}_A}{\boldsymbol{Z}_{R2}}=\frac{\boldsymbol{V}_A-0}{\boldsymbol{Z}_{C1}}$$

$$\Rightarrow \frac{12-\boldsymbol{V}_A}{4}+\frac{\boldsymbol{V}_B-\boldsymbol{V}_A}{5}=\frac{\boldsymbol{V}_A}{-j3.2}.$$

$$Node\,B \Rightarrow \boldsymbol{I}_{R2} + \boldsymbol{I}_{L1} = \boldsymbol{I1} \Rightarrow \frac{\boldsymbol{V}_B - \boldsymbol{V}_A}{\boldsymbol{Z}_{R2}} + \frac{\boldsymbol{V}_B - 0}{\boldsymbol{Z}_{L1}} = \boldsymbol{I1}$$

$$\Rightarrow \frac{\boldsymbol{V}_B - \boldsymbol{V}_A}{5} + \frac{\boldsymbol{V}_B - 0}{\boldsymbol{Z}_{L1}} = 3 \Rightarrow \frac{\boldsymbol{V}_B - \boldsymbol{V}_A}{5} + \frac{\boldsymbol{V}_B}{j1.23} = 3.$$

- Thus, we have the system of two equations to find the two unknown variables—the nodal voltages $\boldsymbol{V}_A$ and $\boldsymbol{V}_B$:

$$\begin{cases} \dfrac{12 - \boldsymbol{V}_A}{4} + \dfrac{\boldsymbol{V}_B - \boldsymbol{V}_A}{5} = \dfrac{\boldsymbol{V}_A}{-j3.2} \\ \dfrac{\boldsymbol{V}_B - \boldsymbol{V}_A}{5} + \dfrac{\boldsymbol{V}_B}{j1.23} = 3 \end{cases}. \tag{3.46}$$

Equation 3.46 is simplified to the following standard form:

$$\begin{cases} (0.45 + j0.31)\boldsymbol{V}_A - 0.2\boldsymbol{V}_B = 3 \\ -0.2\boldsymbol{V}_A + (0.2 - j0.81)\boldsymbol{V}_B = 3 \end{cases}. \tag{3.47}$$

Note that the dimensions of both equations are amperes.

- Presentation of (3.47) in the matrix format results in

$$\begin{bmatrix} (0.45 + j0.31) & -0.2 \\ -0.2 & (0.2 - j0.81) \end{bmatrix} \begin{vmatrix} \boldsymbol{V}_A \\ \boldsymbol{V}_B \end{vmatrix} = \begin{vmatrix} 3 \\ 3 \end{vmatrix}. \tag{3.48}$$

Sidebar S2 in Section 2.1 discusses several methods of solving the system of linear equations in depth. We use MATLAB to obtain the solution. The MATLAB code and the result are shown below:

```
>> A = [0.45 + 0.31i -0.2;
-0.2 0.2 - 0.81i];
B = [3; 3];
V = A\B
V=
6.0186 - 2.0238i
1.6788 + 4.7753i
```

- To verify the calculated result, we conduct the Multisim simulation of the circuit shown in Figure 3.13a. The Multisim commands are those discussed in Example 3.3. Below are the values of the nodal voltages obtained by Multisim.

  MULTISIM: V(2) 6.01017 - j 2.02240 $[\boldsymbol{V}_A]$
  V(3) 1.73018 + j4.84536 $[\boldsymbol{V}_B]$

The MATLAB and Multisim sets of values are close; the difference could be due to approximations and the methods of computations.

- We can compute all the required values using the obtained nodal voltages, as shown in Table 3.4.
- The obtained Multisim-based results satisfy KCL at both Node A and B, as shown below.

$$Node\,A \Rightarrow \boldsymbol{I}_{R1} - \boldsymbol{I}_{R2} = \boldsymbol{I}_{C1} \Rightarrow 1.497 + j0.510 - (0.856 - j1.374) = 0.63 + j1.88$$
$$\Rightarrow 0.64 + j1.88 = 0.63 + j1.88.$$
– Acceptable.

Table 3.4 Formulas and MATLAB-based calculated branch voltages and currents values for Example 3.5.

Component	Voltage across		Current through	
	Formula	Value	Formula	Value
$R1$	$\boldsymbol{V}_{R1}(\boldsymbol{V}) = \boldsymbol{V}1 - \boldsymbol{V}_A$	$\boldsymbol{V}_{R1}(V) = 5.99 + \mathrm{j}2.02$	$\boldsymbol{I}_{R1}(A) = \frac{\boldsymbol{V}_{R1}}{R1} = \frac{5.99 + \mathrm{j}2.02}{4}$	$\boldsymbol{I}_{R1}(A) = 1.497 + j0.510$
$R2$	$\boldsymbol{V}_{R2}(V) = \boldsymbol{V}_B - \boldsymbol{V}_A$	$\boldsymbol{V}_{R2}(V)$ $= -1.73 + \mathrm{j}4.85$ $- 6.01 + \mathrm{j}2.02$ $= -4.28 + \mathrm{j}6.87$	$\boldsymbol{I}_{R2}(A) = \frac{\boldsymbol{V}_{R2}}{\boldsymbol{R}2} = \frac{4.28 - \mathrm{j}6.87}{5}$	$\boldsymbol{I}_{R2}(A) = 0.856 - j1.374$
$L1$	$\boldsymbol{V}_{L1}(V) = \boldsymbol{V}_B$	$\boldsymbol{V}_{L1}(V) = 1.73 + \mathrm{j}4.85$	$\boldsymbol{I}_{L1}(A) = \frac{\boldsymbol{V}_{L1}}{\boldsymbol{Z}_{L1}} = \frac{\boldsymbol{V}_{L1}}{\boldsymbol{j\omega L}} = \frac{1.73 + \mathrm{j}4.85}{j1.3}$	$I_{L1}(A) = 3.73 - j1.33$
$C1$	$\boldsymbol{V}_{C1}(\boldsymbol{V}) = \boldsymbol{V}_A$	$\boldsymbol{V}_{C1}(\boldsymbol{V}) = 6.01 - \mathrm{j}2.02$	$\boldsymbol{I}_{C1}(A) = \frac{\boldsymbol{V}_{C1}}{\boldsymbol{Z}_{C1}} = \frac{\boldsymbol{V}_{C1}}{\boldsymbol{j\omega C}} = \frac{6.01 - \mathrm{j}2.02}{-j3.2}$	$\boldsymbol{I}_{C1}(A) = 0.63 + j1.88$

$$Node\,B \Rightarrow -\boldsymbol{I}_{R2} + \boldsymbol{I}_{L1} = \boldsymbol{I}1 \Rightarrow -0.86 + j1.37 + 3.73 - j1.37 = 2.87. \text{ vs } 3.00. \text{ Acceptable.}$$

This is the end of Example 3.5.

Mesh analysis is also considered in Section 2.1. You are strongly encouraged to thoroughly review that section, mainly Example 2.3. Then, you will recall that the mesh analysis mirrors the nodal analysis in their tasks—finding the mesh currents and calculating the voltages and currents of all circuit elements. No wonder they employ similar approaches. Specifically, the mesh analysis requires using KVL to compose the circuit simultaneous equations, which enables us to determine the mesh currents. We must remember that this approach can be applied to ac circuits provided that the analysis is performed in the phasor domain.

Section 2.1 and Sections 3.1 and 3.2 enable us to consider the ac mesh analysis directly in an example.

Example 3.6 Finding the branch voltages and currents of a series-parallel RLC circuit.
Problem: Find all the branch voltages and currents of the circuit shown in Figure 3.14a using the mesh analysis.

Solution: Figure 3.14a is based on Figure 2.6 shown in Example 2.3; the difference is that the current figure contains only ac sources; the inductor and the capacitor have replaced two resistors, and one resistor is added in Mesh 3. This is why we will follow the methodology used in Example 2.3 to solve Example 3.6.

- First, we build the phasor-domain model of the given circuit, shown in Figure 3.14b.
- Next, we use a mesh analysis to derive the system of equations describing the given circuit.

This time, we apply KVL to Mesh 1 in a slightly different format, enabling us to build the system of equations immediately. Thus, for Mesh 1, we have

$$(3 - j3.2)I_1 + j3.2I_2 - 3I_3 = -8V. \tag{3.49a}$$

Similarly, for Mesh 2, we obtain

$$j3.2I_1 + (5 - j3.2)I_2 - 5I_3 = 1V. \tag{3.49b}$$

Alike, KVL for Mesh 3 takes the following form

$$-3I_1 - 5I_2 + (19 + j1.3)I_3 = 33V. \tag{3.49c}$$

Building the equations in this format, we avoid an extra step that involves creating the current differences used in Example 2.3. Putting Equations 3.49a through 3.49c together, we directly attain the system of three equations describing the given circuit in the phasor domain:

$$\begin{cases} (3 - j3.2)I_1 + j3.2I_2 - 3I_3 = -8 \\ j3.2I_1 + (5 - j3.2)I_2 - 5I_3 = 1 \\ -3I_1 - 5I_2 + (19 + j1.3)I_3 = 33 \end{cases}. \tag{3.50}$$

Equation 3.50 is presented in the matrix format as

$$\begin{vmatrix} (3 - j3.2) & j3.2 & -3 \\ j3.2 & (5 - j3.2) & -5 \\ -3 & -5 & (19 + j1.3) \end{vmatrix} \begin{vmatrix} I_1 \\ I_2 \\ I_3 \end{vmatrix} = \begin{vmatrix} -8 \\ 1 \\ 33 \end{vmatrix}. \tag{3.51}$$

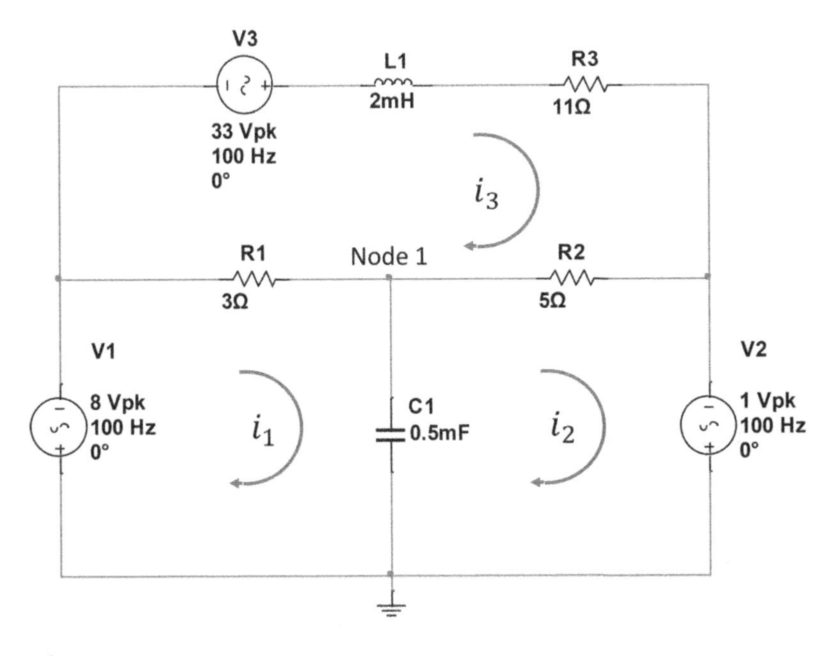

a)

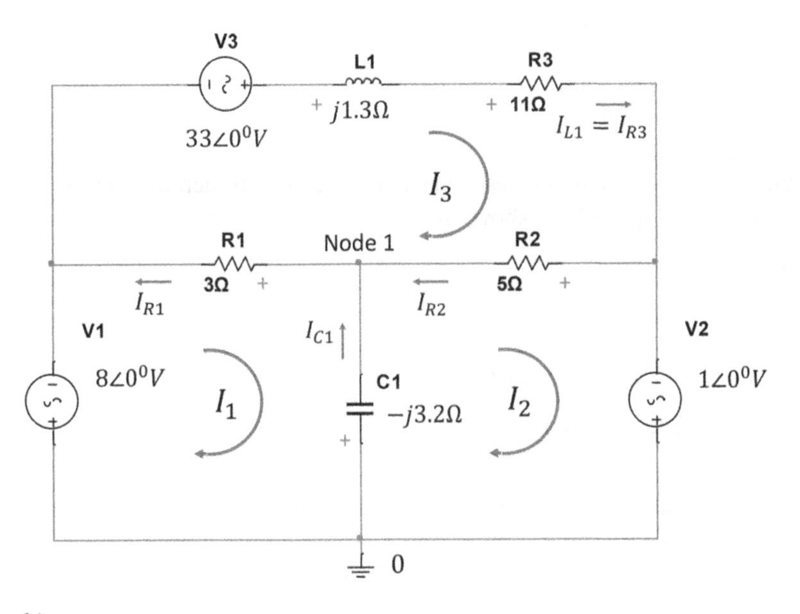

b)

Figure 3.14 Mesh analysis of a series-parallel RLC circuit in Example 3.6: a) The schematic of the given circuit; b) the phasor-domain model of the circuit.

- Sidebar S2 demonstrates several methods of solving such an equation system. To solve (3.51), we choose MATLAB, write the code, and attain the result shown below:

```
>> A = [3-3.2i 3.2i -3;
3.2i 5-3.2i -5;
-3 -5 19+1.3i];
B = [-8; 1; 33];
I = linsolve(A,B)
I =
0.9982 - 1.0570i
1.7308 + 0.1934i
2.3311 - 0.2755i
```

Rewriting the result in the traditional format, we find:

$$I_1(A) = 0.9982 - j1.0570$$
$$I_2(A) = 1.7308 + j0.1934$$
$$I_3(A) = 2.3311 - j0.2755.$$

- Obtained mesh currents enable us to compute the branch voltages and currents. We put the results of these computations in Table 3.5.
- To verify the obtained result, we conduct a Multisim simulation of the circuit shown in Figure 3.14a. The results are presented in Figure 3.15, which shows that the MATLAB calculations of the branch currents closely coincide with that of Multisim simulation. Multisim doesn't show the branch voltages; instead, it displays source and nodal voltages. To compare the MATLAB and Multisim voltages would require additional manual calculations, the work which adds no new information to this example.
- We can add another layer of confidence to our results by checking the satisfaction of the KCL requirement at Node 1, which, as Figure 3.14b shows, is

$$\boldsymbol{I}_{R2} + \boldsymbol{I}_{C1} = \boldsymbol{I}_{R1}(\boldsymbol{A}).$$

Plugging in the obtained values, we find

$$-0.60 + j0.47 + (-0.73 - \text{j}1.25) = -1.33 - j0.78.$$

As we can see from Table 3.5, the KCL at Node 1 is met.

Table 3.5 Computed branch currents and voltages based on the mesh currents calculated by MATLAB.

Component	Current through		Voltage across	
	Formula	**Value**	**Formula**	**Value**
$R1$	$\boldsymbol{I}_{R1}(A) = I_1 - I_3$	$\boldsymbol{I}_{R1}(A) = -1.33 - j0.78$	$\boldsymbol{V}_{R1}(V) = R1 \cdot \boldsymbol{I}_{R1}$	$\boldsymbol{V}_{R1}(V) = -3.99 - j2.34$
$R2$	$\boldsymbol{I}_{R2}(A) = I_2 - I_3$	$\boldsymbol{I}_{R2}(\text{A}) = -0.60 + j0.47$	$\boldsymbol{V}_{R2}(V) = R2 \cdot \boldsymbol{I}_{R2}$	$\boldsymbol{V}_{R2}(V) = 3.00 + j2.35$
$R3$	$\boldsymbol{I}_{R3}(A) = I_3.$	$\boldsymbol{I}_{R3}(A) = 2.33 - j0.27$	$\boldsymbol{V}_{R3}(V) = R3 \cdot \boldsymbol{I}_3$	$\boldsymbol{V}_{R3}(V) = 25.63 - j2.97$
$L1$	$\boldsymbol{I}_{L1}(A) = \boldsymbol{I}_{R3} = \boldsymbol{I}_3$	$\boldsymbol{I}_{L1}(A) = 2.33 - j0.27$	$\boldsymbol{V}_{L1}(V) = j\omega L \cdot \boldsymbol{I}_{L1}$	$\boldsymbol{V}_{L1}(V) = 0.35 + j3.03$
$C1$	$\boldsymbol{I}_{C1}(A) = I_1 - I_2$	$\boldsymbol{I}_{C1}(A) = -0.73 - \text{j}1.25$	$\boldsymbol{V}_{C1}(V) = \boldsymbol{j\omega C} \cdot \boldsymbol{I}_{C1}$	$\boldsymbol{V}_{C1}(V) = -4 + j2.36$

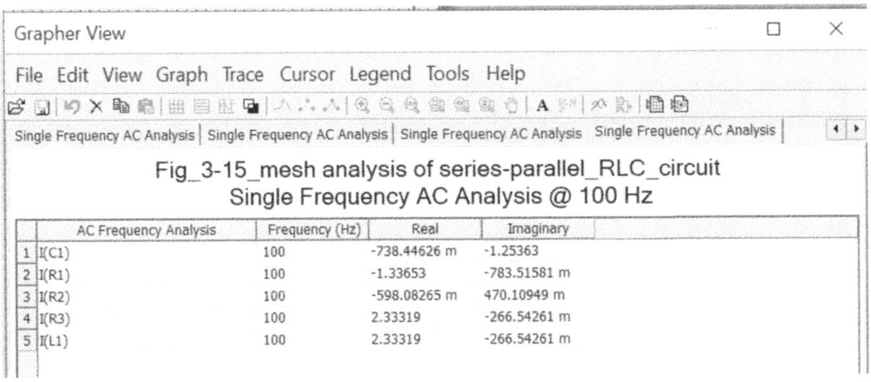

Grapher View

File Edit View Graph Trace Cursor Legend Tools Help

Single Frequency AC Analysis | Single Frequency AC Analysis | Single Frequency AC Analysis | Single Frequency AC Analysis

Fig_3-15_mesh analysis of series-parallel_RLC_circuit
Single Frequency AC Analysis @ 100 Hz

	AC Frequency Analysis	Frequency (Hz)	Real	Imaginary
1	I(C1)	100	-738.44626 m	-1.25363
2	I(R1)	100	-1.33653	-783.51581 m
3	I(R2)	100	-598.08265 m	470.10949 m
4	I(R3)	100	2.33319	-266.54261 m
5	I(L1)	100	2.33319	-266.54261 m

Figure 3.15 The results of Multisim simulation of the RLC circuit in Example 3.6.

- A curious reader is encouraged to perform manually all the remaining calculations, using the hints given in Sidebar S2.
- A comparison of this example and its theory with the material presented in Section 2.2 and Example 2.3 should again convince you that our analysis of the resistive circuits fully prepares us for studying ac steady-state circuits. To persuade yourself, devise and solve an example involving the *supermesh* case.
  This is the end of Example 3.6.

3.3.2 Thevenin's and Norton's Theorems

You will recall that *Thevenin's theorem* is presented in Section 2.2:

> *a two-terminal linear circuit, regardless of the number of its components, can be replaced by a circuit consisting of a voltage source, V_{Th}, and a series impedance, Z_{Th}.*

Note that the theorem doesn't restrict the type of circuit to which it can be applied; thus, it implies that ac steady-state circuits obey this theorem too. This point becomes more apparent when we recall that the mathematical manipulations needed for implementing Thevenin's theorem are based on Ohm's law, KCL, KVL, and nodal and mesh analyses. But this theoretical basis can be applied to the study of ac circuits when used in the phasor format. As we've done with the nodal and mesh analysis, we'll move to the example to demonstrate the applicability of the Section 2.2 material to the ac circuits.

Example 3.7 Thevenin's theorem in ac steady-state analysis.

Problem: Build Thevenin's equivalent of the circuit shown in Figure 3.16a and determine the load current $\boldsymbol{I}_{Load}(A)$ if a) $\boldsymbol{Z}_{Load} = 9\Omega$, b) $\boldsymbol{Z}_{Load} = j9\Omega$, and c) $\boldsymbol{Z}_{Load} = -\boldsymbol{j}9\Omega$.

Solution: Refer to Example 2.4, whose methodology can be directly applied to this example.

- First, in Figure 3.16b, we build the phasor-domain equivalent to the actual circuit. In this figure, the load impedance is removed, and Thevenin's voltage is shown between open terminals $\boldsymbol{a}$ and $\boldsymbol{b}$, according to its definition.
- The equivalent circuit in Figure 3.16b enables us to find *Thevenin's voltage* as

$$\boldsymbol{V}_{Th}(\boldsymbol{V}) = \boldsymbol{V}_a - \boldsymbol{V}_b.$$

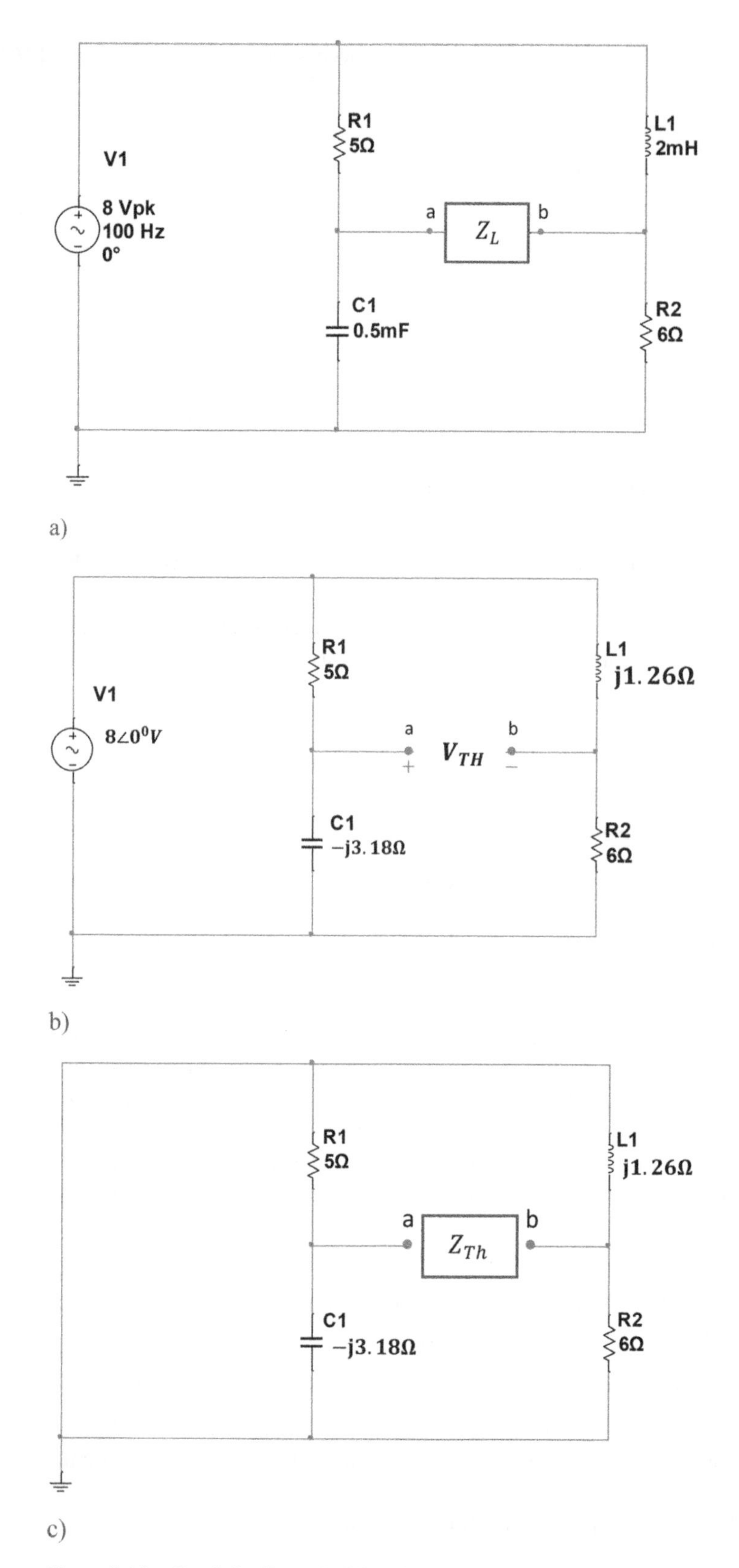

Figure 3.16 Circuit for Example 3.7: a) Schematic of the actual circuit; b) the phasor-domain model of the original circuit with V_{Th}; c) the circuit for calculating Thevenin's impedance, Z_{Th}; d) the required Thevenin equivalent circuit; e) results of Multisim simulation for $R_{Load} = 9(\Omega)$.

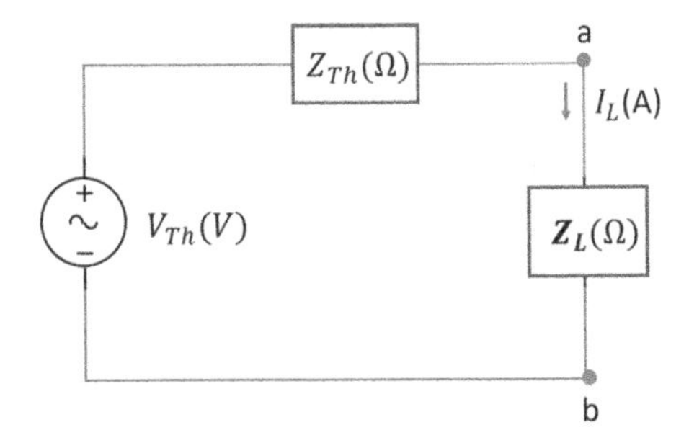

d)

Grapher View

File Edit View Graph Trace Cursor Legend Tools Help

Single Frequency AC Analysis

Fig_3_16e_Thevenin_RLC_bridge_Example 3-7
Single Frequency AC Analysis @ 100 Hz

	AC Frequency Analysis	Frequency (Hz)	Real	Imaginary
1	V(1)	100	8.00000	0.00000
2	V(2)	100	3.53067	-4.36506
3	V(3)	100	7.82777	-2.23943
4	I(C1)	100	1.37132	1.10919
5	I(R1)	100	893.86683 m	873.01105 m
6	I(R2)	100	1.30463	-373.23872 m
7	I(R3)	100	-477.45572 m	-236.18033 m
8	I(L1)	100	1.78208	-137.05839 m
9	I(V1)	100	-2.67595	-735.95266 m

e)

Figure 3.16 (Cont'd)

Observe that $\boldsymbol{V}_a$ is the voltage drop across C1; it can be found by voltage-divider rule as

$$\boldsymbol{V}_a = \frac{\boldsymbol{X}_{C1}}{\boldsymbol{R}1+\boldsymbol{X}_{C1}}\boldsymbol{V}1 = \frac{-j3.2}{5-j3.18}8 = \frac{-j25.6(5+j3.18)}{5^2+3.18^2} = \frac{-j128+81.41}{35.11} = 2.32-j3.65.$$

Similarly,

$$\boldsymbol{V}_b = \frac{\boldsymbol{R}2}{\boldsymbol{R}2+\boldsymbol{X}_{L1}}\boldsymbol{V}1 = \frac{6}{6+j1.26}8 = 7.66-j1.61.$$

Thus,

$$\boldsymbol{V}_{Th}(\boldsymbol{V}) = \boldsymbol{V}_a - \boldsymbol{V}_b = -5.34 - j2.04.$$

- You will recall that finding *Thevenin's impedance*, $\boldsymbol{Z}_{Th}$, requires the replacement of the voltage source by a short circuit Also, the circuit load terminals *a* and *b* must be left open. Figure 3.16c shows the circuit obtained after these operations. (Refer again to Section 2.2 and Example 2.4.) From observation of Figure 3.16c, we conclude that

$$\boldsymbol{Z}_{Th}(\Omega) = \boldsymbol{Z}_1 + \boldsymbol{Z}_2,$$

where $\boldsymbol{Z}_1 = \boldsymbol{R1} \,||\, \boldsymbol{X}_{C1}$ and $\boldsymbol{Z}_2 = \boldsymbol{X}_{L1} \,||\, \boldsymbol{R2}$. (Redraw the circuit in Figure 3.16c to visualize these statements.) Calculating $\boldsymbol{Z}_1$ produces

$$\boldsymbol{Z}_1(\Omega) = \frac{\boldsymbol{R1} \cdot \boldsymbol{X}_{C1}}{\boldsymbol{R1} + \boldsymbol{X}_{C1}} = \frac{5 \cdot (-j3.18)}{5 - j3.18} = \frac{-j15.9(5 + j3.18)}{5^2 + 3.18^2} = \frac{50.56 - j79.5}{35.11} = 1.44 - j2.26.$$

Likewise,

$$\boldsymbol{Z}_2(\Omega) = \frac{\boldsymbol{R2} \cdot \boldsymbol{X}_{L1}}{\boldsymbol{R2} + \boldsymbol{X}_{L1}} = \frac{6 \cdot j1.26}{6 + j1.26} = 0.25 + j1.21.$$

Thus,

$$\boldsymbol{Z}_{Th}(\Omega) = \boldsymbol{Z}_1 + \boldsymbol{Z}_2 = 1.69 - j1.05.$$

- The first part of the problem is solved, and the required Thevenin equivalent circuit is shown in Figure 3.16e. The load current I_{Load} can be computed for various loads by the following formula

$$\boldsymbol{I}_{Load}(\boldsymbol{A}) = \frac{\boldsymbol{V}_{Th}}{\boldsymbol{Z}_{Th} + \boldsymbol{Z}_{Load}}. \tag{3.52}$$

- For $Z_{Load} = 9(\Omega)$, the load current is

$$I_{Load^*9}(A) = \frac{(-5.34 - j2.04)(V)}{(1.69 - j1.05 + 9)(\Omega)} = -0.476 - j0.238(A).$$

The Multisim simulation results presented in Figure 3.16e confirm our calculations: See the framed line for I(R3).

- For the pure inductive load, $Z_{Load} = j9(\Omega)$, the current is

$$I_{Load^*j9}(A) = \frac{(-5.34 - j2.04)(V)}{(1.69 - j1.05 + j9)(\Omega)} = -0.382 + j0.590.$$

- Finally, when the load is the 0.5-mF capacitor, the load current becomes

$$I_{Load^*-j9}(A) = \frac{(-5.34 - j2.04)(V)}{(1.69 - j1.05 - j9)(\Omega)} = 0.1105 - j0.5499.$$

The problem is solved.

Discussion:

- Notice that without the Thevenin method, we'd have to repeat the entire circuit analysis for each load's value. Using Thevenin's theorem makes the computations of the load currents for variable loads an easy matter.
- Verification of the results by Multisim simulation confirms their correctness. A curious reader is encouraged to conduct such a simulation on their own.
- We can find V_{Th} by the *mesh analysis*, as Example 2.4 demonstrates. In our example, the mesh equation can be derived as follows: Figure 3.17 shows the mesh over the top half of the given circuit. This mesh is described by

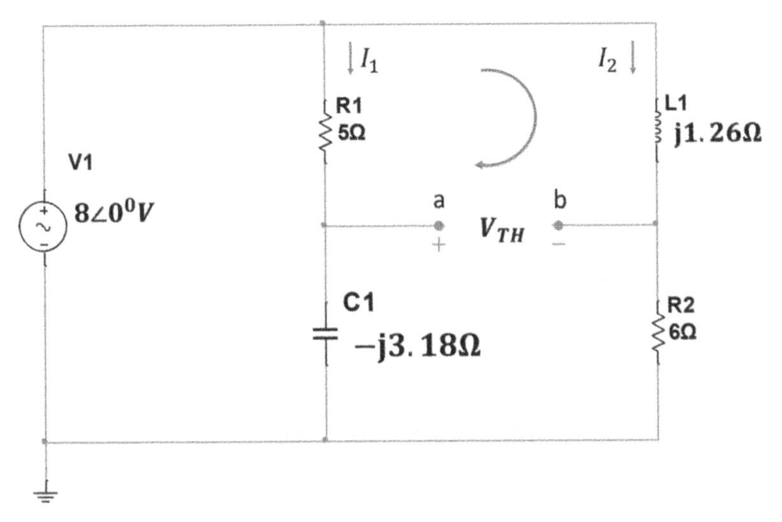

Figure 3.17 Derivation of the mesh equation for Example 3.7.

$$-I_1 \cdot R1 + I_2 \cdot X_{L1} - V_{Th} = 0.$$

Currents I_1 and I_2 are given by

$$I_1(A) = \frac{V1}{R1 + X_{C1}} = 1.139 + j0.725 \text{ and } I_2(A) = \frac{V1}{R2 + X_{L1}} = 1.277 - j0.268.$$

Thus,

$$V_{Th}(V) = I_1 \cdot R1 - I_2 \cdot X_{L1} = -5.357 - j2.016,$$

which is very close to that attained before. The small discrepancies in the V_{Th} values are attributed to the rounding the numbers involved in the calculations.

- In Example 2.3, we discussed how Thevenin's theorem helps design an electrical circuit. Let's revive this point for a steady-state ac circuit presented in Figure 3.16a. The ac problem is what load impedance is needed to deliver 5V to the load? From examining Figure 3.16d, we deduct that

$$V_{Load}(V) = \frac{Z_{Load}}{Z_{Th} + Z_{Load}} V_{Th},$$

according to the voltage-divider rule. Plugging in the known values and solving for Z_{Load} result in

$$Z_{Load}(\Omega) = -2.1125 + j1.3125 = 2.487\angle -31.85^0.$$

The calculated value of the load impedance provides the formal solution to the problem but how this load can be materialized is another matter.

This is the end of Example 3.7.

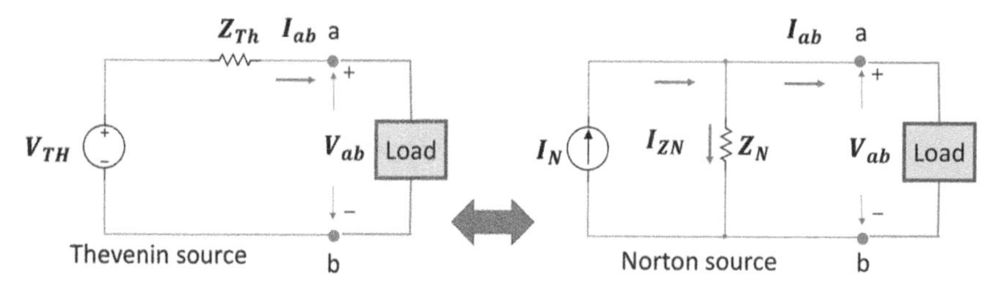

Figure 3.18 Thevenin's and Norton's theorems in the phasor formats.

Section 2.2 quotes *Norton's theorem* as

> *"a two-terminal linear circuit, regardless of the number of its components, can be replaced by a circuit consisting of a current source, I_N, and a parallel impedance, Z_N."*

A comparison of Thevenin's and Norton's theorems shows that the concept of both theorems is the same: *Replace a complex two-terminal circuit with the circuit consisting of a source and an impedance.* Moreover, Norton's theorem is, in essence, a source-transformed Thevenin's theorem. We urge you to revisit Section 2.2 and Example 2.4 to revive your understanding of these statements. Figure 3.18, visualizing these remarks, is the version of Figure 2.15b redrawn in the phasor-domain format.

You are reminded that when terminals $a - b$ are opened, $V_{ab} = \boldsymbol{V_{Th}}$ for Thevenin's model and $V_{ab} = \boldsymbol{Z_N I_N}$ for Norton's model, from which we can conclude that

$$\boldsymbol{V_{Th}}(V) = \boldsymbol{Z_N I_N}. \tag{3.53a}$$

Equation 3.53a *and the following* Equations 3.53b *and* 3.53c, *are the basis of the equivalency of Thevenin's and Norton's models.* When terminals $a - b$ are shorted, the currents flowing through these terminals are $I_{ab} = \frac{\boldsymbol{V_{Th}}}{\boldsymbol{R_{Th}}}$ and $I_{ab} = \boldsymbol{I_N}$ for Thevenin's and Norton's models, respectively. Thus,

$$\boldsymbol{I_N}(A) = \frac{\boldsymbol{V_{Th}}}{\boldsymbol{Z_{Th}}}. \tag{3.53b}$$

Considering (3.53a) and (3.53b) as an equation system, we can readily derive

$$\boldsymbol{Z_{Th}}(\Omega) = \boldsymbol{Z_N} = \boldsymbol{Z}. \tag{3.53c}$$

These three formulas are replicas of (2.34a), (2.34b), and (2.34c) but written in the phasors.

Let's consider an example.

Example 3.8 Building the Norton equivalent circuit in the phasor domain.

Problem: Build the Norton equivalent to the circuit shown in Figure 3.19a. What current will flow through $\boldsymbol{Z_L} = 9\ \Omega$?

Solution: We are adapting the methodology of Example 2.4 to the phasor-domain approach. The phasor-domain model of the actual circuit is shown in Figure 3.19b; terminals a and b provide the connection to a load. Figure 3.19c enables us to calculate $\boldsymbol{Z_N}$ because its sources are replaced, as required by Norton's theorem rule. We find

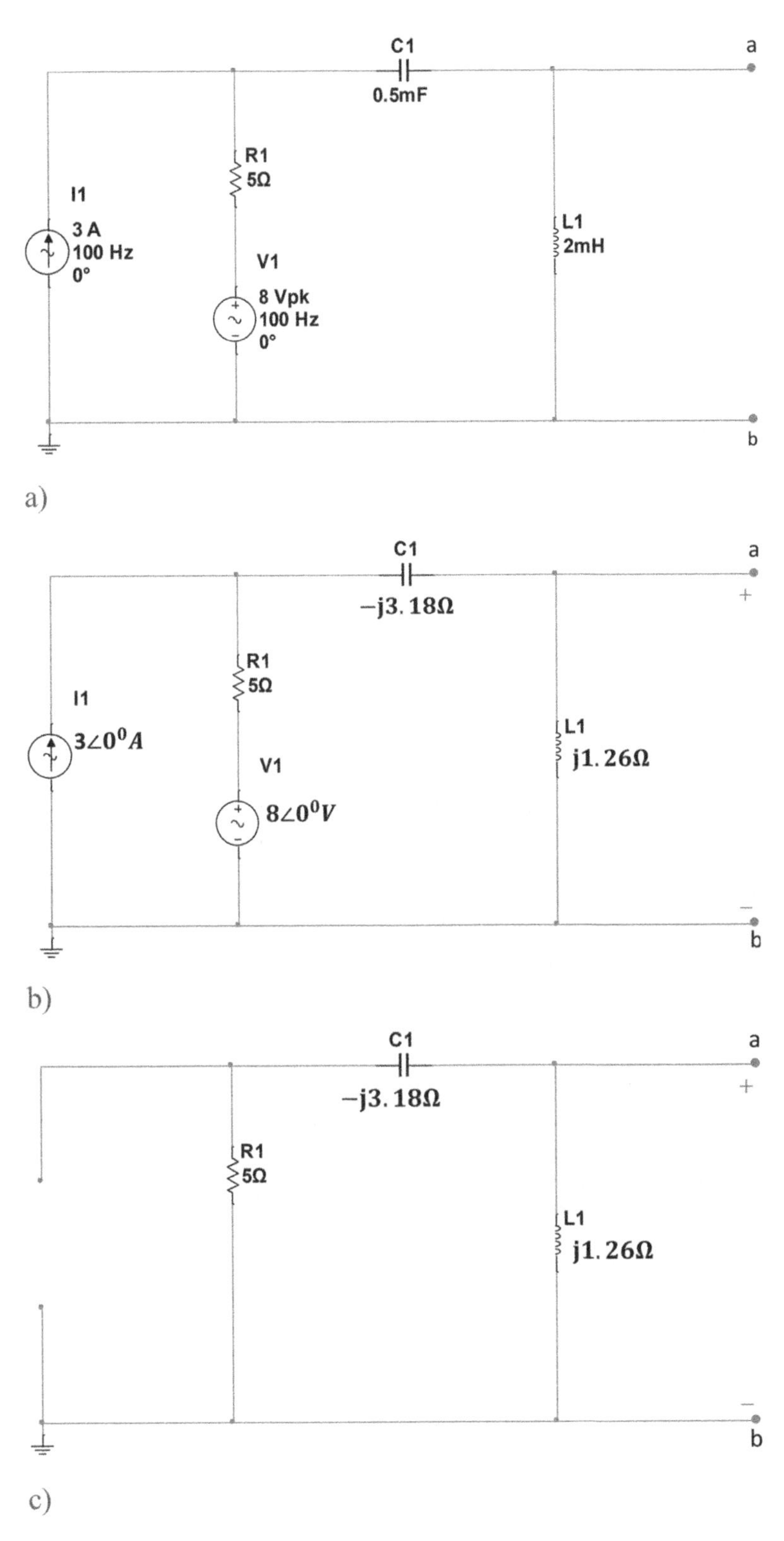

Figure 3.19 Building the Norton equivalent circuit in the phasor domain for Example 3.8: a) Actual time-domain circuit: b) the phasor-domain model; c) finding Z_N; d) finding I_N; e) the Norton equivalent circuit that includes the load; f) the results of Multisim simulation.

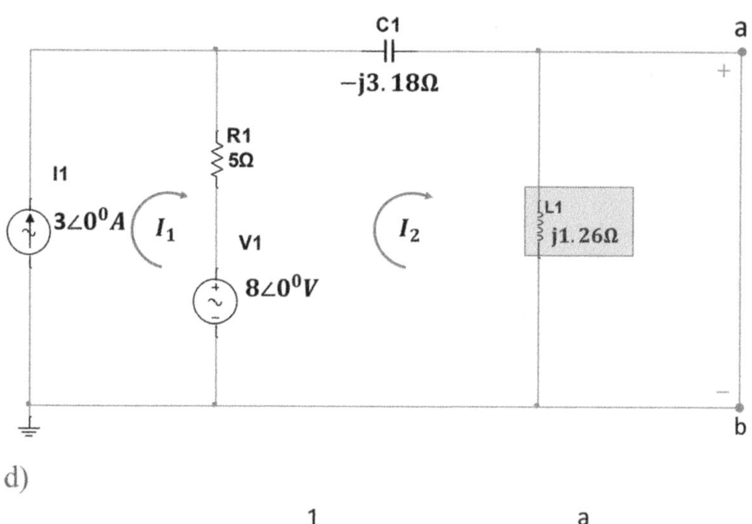

d)

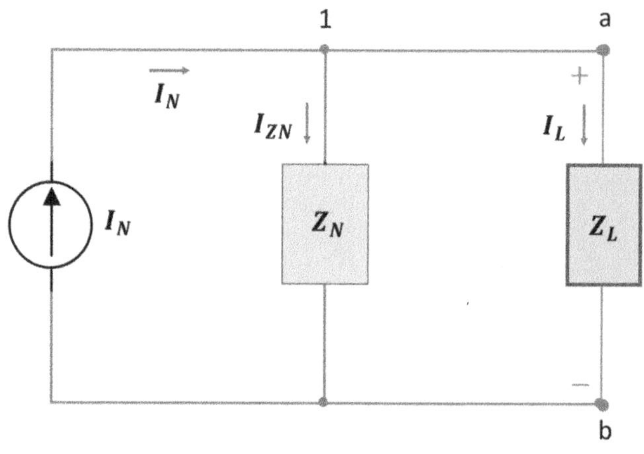

e)

Grapher View

File Edit View Graph Trace Cursor Legend Tools Help

Single Frequency AC Analysis | Single Frequency AC Analysis | Single Frequency AC Analysis

Fig_3_19-f_Norton_RLC_Example 3-8_Z-L-9 ohm
Single Frequency AC Analysis @ 100 Hz

	AC Frequency Analysis	Frequency (Hz)	Real	Imaginary
1	V(1)	100	3.53378	-7.34106
2	V(2)	100	-1.13968	5.05152
3	V(3)	100	8.00000	0.00000
4	I(C1)	100	3.89324	1.46821
5	I(I1)	100	3.00000	0.00000
6	I(R1)	100	-893.24422 m	-1.46821
7	I(R2)	100	-126.63143 m	561.28052 m
8	I(L1)	100	4.01988	906.93078 m
9	I(V1)	100	-893.24422 m	-1.46821

f)

Figure 3.19 (Cont'd)

$$\boldsymbol{Z_N}(\Omega) = (R1 + \boldsymbol{X_{C1}}) \parallel \boldsymbol{X_{L1}} = 0.28 + j1.37.$$

Our next step is calculating the Norton's current $\boldsymbol{I_N}(A)$. As in Example 2.4, we shorten the circuit at $a-b$ terminals to exclude $\boldsymbol{L1}$, and perform the mesh analysis. The result is

$$\begin{cases} I_1 = 3A \\ -V1 - R1 \cdot I_1 + R1 \cdot I_2 + X_{C1} \cdot I_2 = 0 \end{cases} \Rightarrow I_2 = I_N(A) = 3.28 + j2.08.$$

Therefore, the Norton equivalent circuit parameters are found, and the required circuit is shown in Figure 3.19e.

To verify the results, a Multisim simulation with $\boldsymbol{Z_L} = 9\Omega$ was conducted; the answers are presented in Figure 3.19f. To compare the manual calculations with Multisim simulation, we need to find the load current, $\boldsymbol{I_{Load}}$. Examination of Figure 3.19e shows that finding $\boldsymbol{I_{Load}}$ is the exercise in the application of the current-divider rule, namely,

$$\boldsymbol{I_{Load}}(\boldsymbol{A}) = \frac{\boldsymbol{Z_N}}{\boldsymbol{Z_L} + \boldsymbol{Z_N}} \boldsymbol{I_N} = \frac{(0.28 + j1.37)(3.28 + j2.08)}{9.28 + j1.37} = -0.127 + j0.562.$$

Multisim simulation displays this current as $I(R2)$, which is framed in Figure 3.19f. The comparison calculated and simulated values show that they are acceptably close.

The problem is solved.

Discussion:

- Additional confirmation of our results can be attained by verifying KCL at Node 1 in Figure 3.19e. This KCL requires that

  $$\boldsymbol{I_N}(\boldsymbol{A}) = \boldsymbol{I_{ZN}} + \boldsymbol{I_{Load}}.$$

  Calculation of needed $\boldsymbol{I_{ZN}}$ produces

  $$\boldsymbol{I_{ZN}}(\boldsymbol{A}) = \frac{\boldsymbol{Z_L}}{\boldsymbol{Z_L} + \boldsymbol{Z_N}} \boldsymbol{I_N} = 3.4 + j1.51.$$

  Therefore, at Node 1 we have

  $$3.28 + j2.08 = 3.4 + j1.51 + (-0.127 + j0.562),$$

  which is true. Hence, we can rely on our results.

- A curious reader can find $\boldsymbol{I_N}(\boldsymbol{A})$ by leaving the $a-b$ terminals open in Figure 3.19d and calculating $\boldsymbol{V_{Th}}(\boldsymbol{V})$ at these terminals. Then, knowing that $\boldsymbol{Z_N}(\Omega) = \boldsymbol{Z_{Th}}(\Omega)$, as (3.53c) displays, Norton's current is calculated as

  $$\boldsymbol{I_N}(\boldsymbol{A}) = \frac{\boldsymbol{V_{Th}}}{\boldsymbol{Z_{Th}}}.$$

  This exercise would be the other proof of the correctness of our solution.

  This is the end of Example 3.8

3.3.3 Superposition Principle (Theorem)

Consider the circuit shown in Figure 3.20a, which has two inputs, $V1$ and $I1$. Suppose we need to find the response of this circuit.

To solve the problem, we determine from observation that the problem boils down to finding I_1 because I_2 is given, and, knowing the currents, we can compute all voltages.

Using the mesh method, we find for Mesh 1 and Mesh 2:

$$-V1 + R1 \cdot I_1 + R3 \cdot I_1 - R3 \cdot I_2 = 0 \Rightarrow -8 + 3I_1 + 5I_1 - 5I_2 = 0$$
$$I_2 = -3.$$

Thus, the system of two equations is

$$8I_1 - 5I_2 = V1, \tag{3.54a}$$

$$I_2 = -3A, \tag{3.54b}$$

which enables us to calculate I_1 as

$$I_1 = \frac{-7}{8} = -0.875A. \tag{3.54c}$$

All voltages are computed using Ohm's law as $V_{R1} = R1 \cdot I_1 = -2.625V$, $V_{R2} = R2 \cdot I_2 = -12V$, and $V_{R3} = R3 \cdot (I_1 - I_2) = 10.625V$, which completes the analysis of this circuit.

The careful review of the preceding circuit analysis shows that it hints at developing the other approach to find I_1. Let's rewrite (3.54a) and (3.54b) as

$$I_1 = \frac{5}{8}I_2 + \frac{V1}{8}, \tag{3.55a}$$

$$I_2 = -3. \tag{3.55b}$$

Equation 3.55a states that current I_1 is due to the additive actions of two sources, I_2 (I1) and $V1$. Therefore, one part of this current is caused by I_2, and the other is created by $V1$. We can find each part of I_1 independently and then sum them up to compute the whole I_1. This method is called the *principle of superposition*; it is based on the *linearity* of all involved equations. (Refer to Note 1 in Section 2.2 to refresh your memory on linearity; pay particular attention to Equations 2.31a, 2.31b, and 2.31c.)

To implement this principle in our example, let's first exclude the voltage source $V1$ from the given circuit and compute $I_1^{'}$ caused by I_2. Excluding a voltage source means making its voltage zero, which implies a short circuit. Figure 3.20b shows this circuit. Mesh1 produces

$$8I_1^{'} - 5I_2 = 0 \Rightarrow I_1^{'} = \frac{5}{8}I_2 \Rightarrow I_1^{'} = -\frac{15}{8}A.$$

Next, excluding current source—which is making its current zero—requires leaving an open circuit in place of $I1$. See Figure 3.20c, whose Mesh 2 gives

$$8I_1^{''} - V1 = 0 \Rightarrow 8I_1^{''} - 8 = 0 \Rightarrow I_1^{''} = 1A.$$

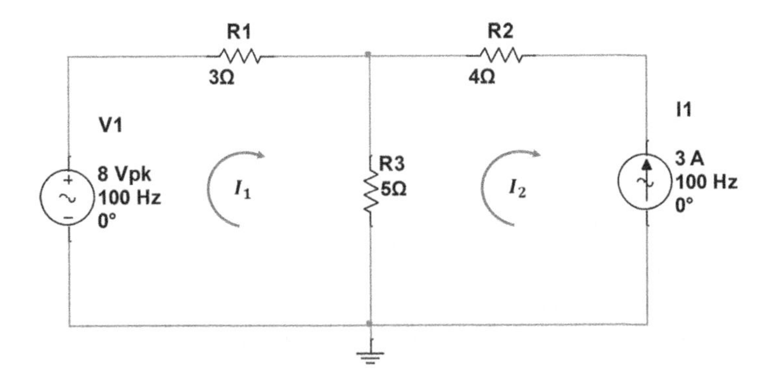

a)

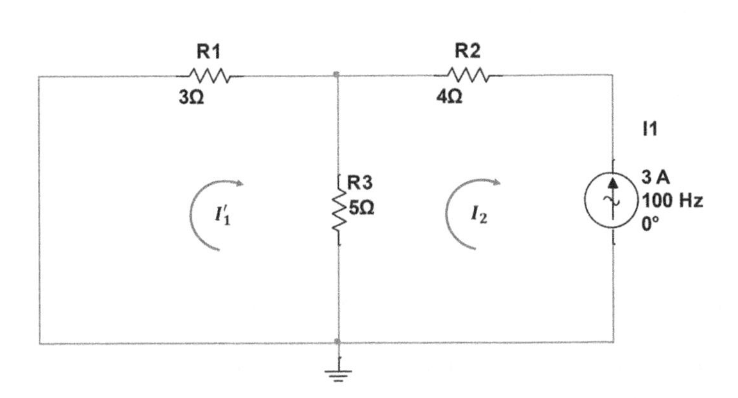

b)

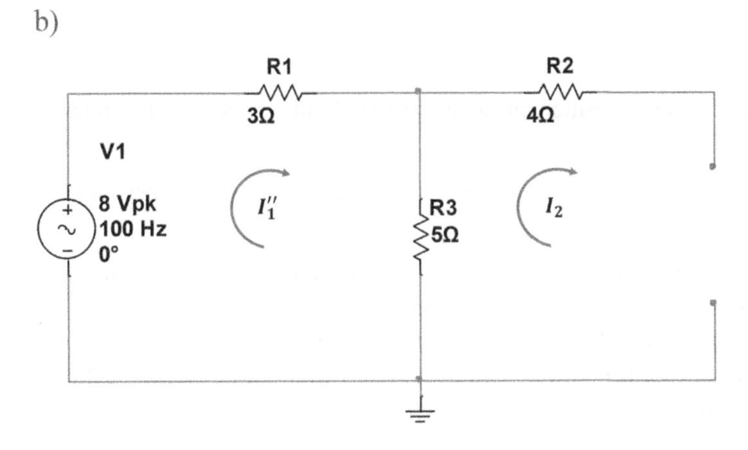

c)

Figure 3.20 Principle of superposition: a) The circuit with two independent sources; b) the circuit with the current source, I_2, only; c) the circuit with the voltage source, $V1$, only.

Finally, by the superposition principle, the total current I_1 is the sum of the obtained parts, that is,

$$I_1 = I_1^{'} + I_1^{''} = -\frac{15}{8} + 1 = -\frac{7}{8} A,$$

as expected.

We can generalize the above discussion as follows:

- The superposition principle (theorem) states that a response of a linear circuit can be obtained by summarizing in sequence the individual responses to all acting alone sources.
- To clarify the preceding general statement, let's denote a circuit parameter (current or voltage) as X_n $(n = 1,2,3,\ldots.)$ and a voltage or current source as Y_m $(m = 1,2,3,\ldots.)$. Permit X_{11}, which is a part of parameter X_1, be independently caused by the source Y_1 only, the other part of the same parameter, X_{12} be independently produced by the source Y_2 only, and so on. Then the whole parameter X_1 can be found as

$$X_1 = X_{11} + X_{12} + X_{13} + \ldots + X_{1m}. \tag{3.56}$$

- To find parameter X_{1m}, all sources except Y_m must be excluded, that is, the short circuits must replace all voltage sources, and the open circuits supplant all current sources.
- All sources must operate at the same frequency, for this is a single-frequency analysis. The value of frequency can be any, including zero.
- All sources to be excluded must be independent; the dependent (controlled) source must be left intact.
- The superposition principle works for both resistive and ac circuits; in the latter case, the analysis is to be performed in the phasor domain.

Let's consider an example.

Example 3.9 Application of the superposition principle to the circuit analysis in the phasor domain.

Problem: Find the response of the circuit shown in Figure 3.21a using the superposition principle.

Solution: From observation, we conclude that the response of this circuit is the current through $L1$ because its terminals would be connected to the following stages of the total hypothetical network. Therefore, we need to find $I_{L1}^{'}$ caused by $V1$ and $I_{L1}^{''}$ created by $I1$, and sum these parts up.

First, we construct the phasor-domain model of this circuit shown in Figure 3.21b. Next, we exclude the current source $I1$. This circuit is presented in Figure 3.21c, whose KVL gives

$$-V1 + \left(R1 + X_{C1} + X_{L1}\right) \cdot I_{L1}^{'} = 0.$$

Plugging the component values in produces:

$$I_{L1}^{'} = \frac{8}{5 - j1.92} = \frac{40 + j15.36}{28.69} = 1.394 + j0.535.$$

Excluding the voltage source yields the circuit given in Figure 3.21d. The current-divider rule helps to calculate $I_{L1}^{''}$ as

$$I''_{L1} = \frac{R1}{R1 + X_{C1} + X_{L1}} I1 = \frac{5}{5 - j1.92} 3 = 2.612 + j1.005.$$

Finally,

$$I_{L1} = I'_{L1} + I'_{L1} = 4.006 + j1.54.$$

Multisim simulation gives $I_{L1}(Multisim) = 4.005 + j1.543$, which is a very close result. Hence, we can consider our solution validated. Write the result in the phasor form.

The problem is solved.

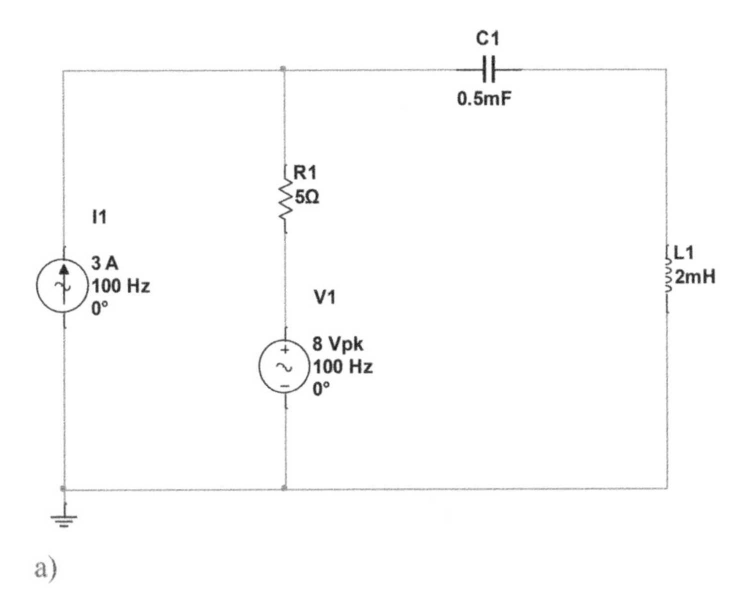

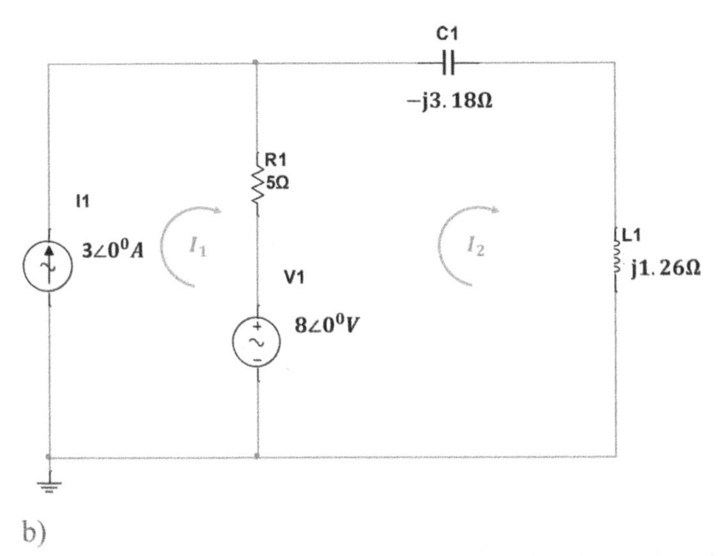

Figure 3.21 Application of the superposition principle to the circuit analysis in Example 3.9: a) Original circuit; b) the phasor-domain model of the original circuit; c) circuit analysis with a voltage source; d) circuit analysis with current source.

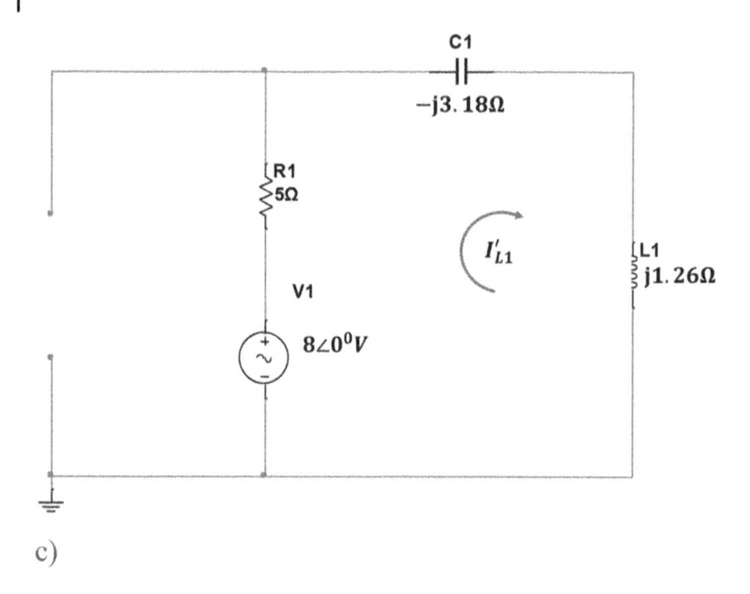

c)

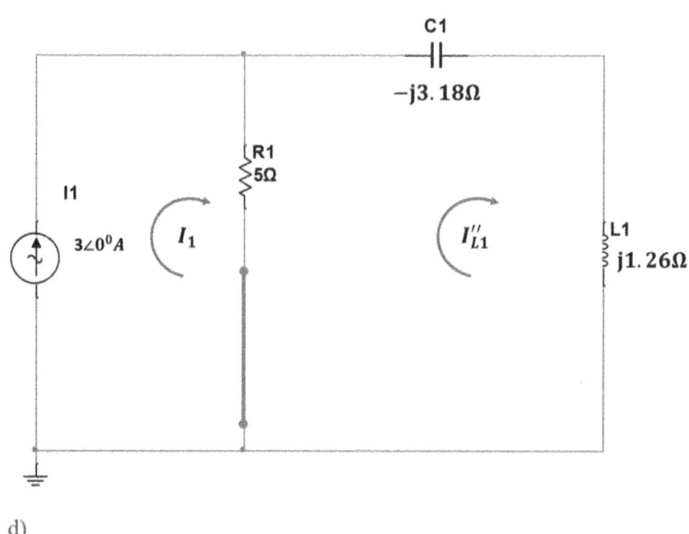

d)

Figure 3.21 (Cont'd)

Discussion:

- The other option to verify our results is to solve the problem by any other method. Let's use, for example, mesh analysis. Figure 3.21b shows of the mesh loops and the directions of their currents. Mesh 1 and Mesh 2 give, respectively,

$$I_1 = 3A.$$

$$-V1 + R1(I_2 - I_1) + X_{C1}I_2 + X_{L1}I_2 = 0 \Rightarrow I_2 = \frac{23}{5 - j1.92} = 4.008 + j1.539.$$

This value is eminently close to our result attained by the superposition method.

- A superposition principle is a potent tool in circuit analysis. However, it has specifics as all other tools we have studied. It is up to us, engineers, to choose the optimal tool for every concrete problem.

Question What is the value of I_{R1} current in the circuit given in Figure 3.21a?

This is the end of Example 3.9.

3.3.4 Duality Principle

In this section, we discuss various methods and techniques that enable us to analyze electrical circuits, and the duality principle belongs to this group too.

Consider, for example, mesh and nodal analysis: Do they rely on similar approaches and mathematical manipulations but with different circuit parameters? Yes, a mesh analysis is based on KVL and operates with voltages, whereas a nodal analysis is based on KCL and performs similar operations but with currents. But what about KCL and KVL? Aren't they found on a similar idea: considering the sum of mesh voltages in KVL and the sum of node currents in KCL?

Have you noticed that in the initial short introduction, we constantly refer to the *duals*, the two electrical entities performing similar tasks? Indeed, current and voltage, KCL and KVL, mesh analysis, and nodal analysis are examples of duals. Table 3.6 presents dual electrical pairs.

Here are several examples of duals and dualities:

- The capacitor and inductor $i - v$ relationships are

$$i_C(A) = C\frac{dv_C}{dt} \Leftrightarrow v_L(V) = L\frac{di_L}{dt},$$

and the formulas for energy stored by each element are

$$W_C(J) = \frac{1}{2}Cv_C^2 \Leftrightarrow W_L(J) = \frac{1}{2}Li_L^2.$$

Table 3.6 Dual electrical pairs.

Charge, Q (C).	Flux linkage, $\wedge(Wb)$.
Current, I (A).	Voltage, V (V).
KCL	KVL
Node	Mesh
Conductance, G (S).	Resistance, R (Ω).
Inductance, L (H).	Capacitance, C (F).
Admittance, Y (S).	Impedance, Z (Ω).
Susceptance, B (S)	Reactance, $X(\Omega)$.
Norton's theorem	Thevenin's theorem
Open circuit	Short circuit
Parallel circuit	Series circuit
Current division	Voltage division

- Series and parallel circuits:
  - Consider series and parallel circuit topologies shown in Figure 3.22a. To describe a series circuit, we apply KVL and obtain

$$V(V)=IZ_1+IZ_2+IZ_3=I(Z_1+Z_2+Z_3).$$

For a parallel circuit, application of KCL produces

$$I(A)=I_1+I_2+I_3=VY_1+VY_1+VY_1=V(Y_1+Y_2+Y_3).$$

Note the duality.

- Consider Figure 3.22. The KVL application produces the following description of the series RLC circuit:

$$v(t)(V)=v_R(t)+v_L(t)+v_C(t)=Ri(t)+L\frac{di(t)}{dt}+\frac{1}{C}\int_0^T i(t)dt.$$

Let's employ the duality principle to obtain the new equation. Using Table 3.6, we exchange voltages and currents, resistance and admittance, capacitance and inductance, and differentiation and integration. The result is

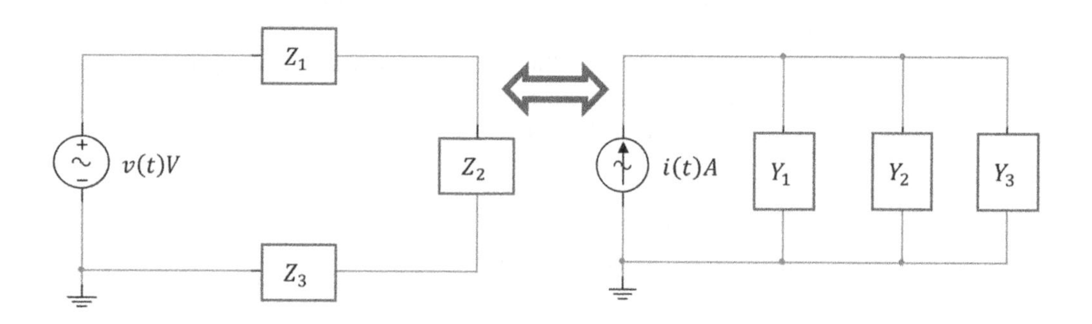

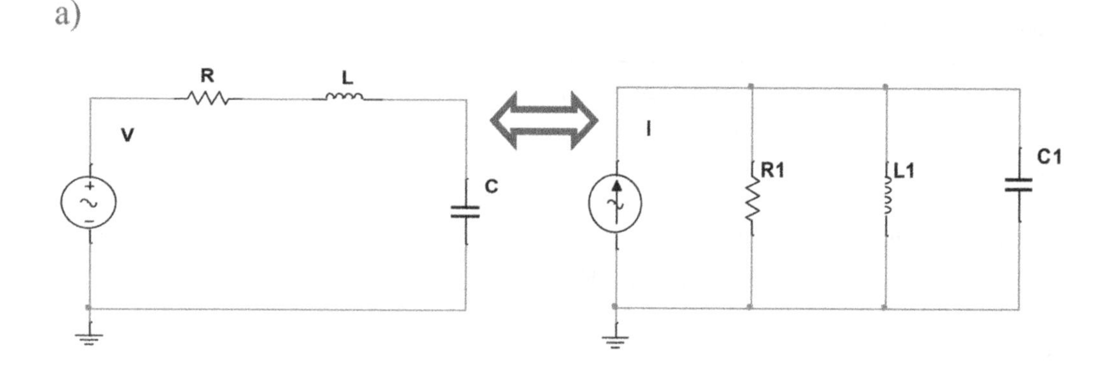

Figure 3.22 Dual circuits: a) Series (left) and parallel (right) general network; b) series and parallel RLC circuits.

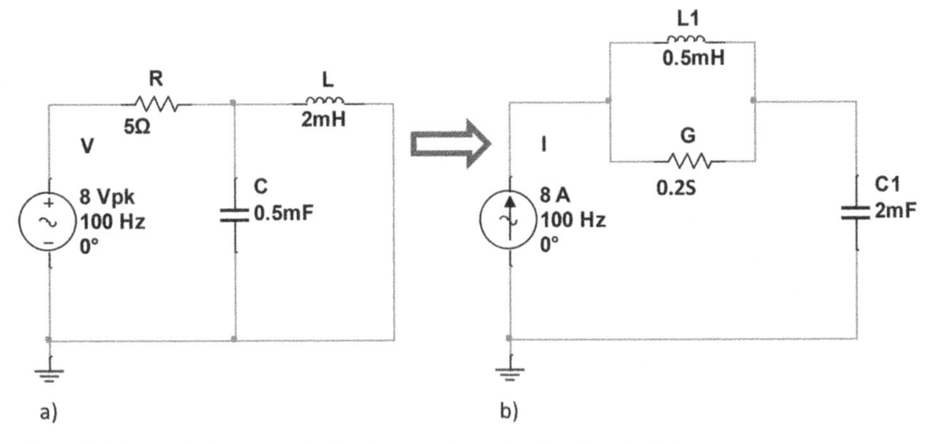

Figure 3.23 Dual electrical circuits for exercise in the duality principle.

$$i(t)(A) = i_R(t) + i_L(t) + i_C(t) = Yv(t) + C\frac{dv(t)}{dt} + \frac{1}{L}\int_T^0 v(t)dt.$$

Not surprisingly, this equation is KCL for the parallel RLC circuit shown in Figure 3.23b. This transfiguration is an example of how the duality principle can save time and effort in the analysis of electrical circuits.

There is a routine that helps to construct a dual circuit when an original one is sophisticated. Some introductory circuit analysis textbooks describe this technique, but our brief review skips this minor point.

Exercise in the duality principle Draw the circuit dual to that shown in Figure 3.23a and show the component values.

Answer: See Figure 3.23b. (Pay attention to all dual transformations:

$V\,source \rightarrow I\,source, R(\Omega) \rightarrow G(S)$, $L(H) \rightarrow C1(F)$, $C(F) \rightarrow L1(H)$, series $R + C \rightarrow$ parallel $G \| L$, parallel $(R + C) \| L \rightarrow$ series $(G \| L1) + C1$. Also, notice the component values.)

Questions and Problems for Chapter 3

3.1 Steady-state AC circuit analysis—basic tools

1 What is the difference between dc and ac circuits? Give examples.
2 How do dc and ac signals differ in their nature? Sketch the examples of a dc and an ac signals.
3 How many parameters do we need to know to fully describe a dc signal? An ac sinusoidal one? What are those parameters?
4 Is a sinusoid a dc or ac signal? Prove your answer by sketching a cosine function (waveform).
5 What is the difference between cosine and sine signal? Use their graphs and formulas to explain.
6 What is frequency of a sinusoidal signal? Why do we use two frequencies—cyclic and angular?
7 How can we learn about a cosine frequency when observe its waveform graph?

8 Can you determine the value of a sine's initial phase y examining its graph? Explain.
9 Sketch the graph of $v(t) = 12\cos(314t + 30^0)$ and show its amplitude and magnitude at $t = 0.3s$.
10 Equation 3.1c states $sin(\omega t - 90^0) = cos(\omega t)$. Why -90^0?
11 Consider Figure 3.2 whose $v_{in}(t) = 12\cos(628t)$, $R = 2k\Omega$, L = 3 mH, and C = 5 mF. Derive the differential equation describing this circuit operation.
12 Using Figure 3.3, explain the relationship between a cosine function and its phasor. Visit Wikipedia's entry Phasor and examine Figure 3.2, where the rotating phasor dynamically relates to a sinusoidal signal. Use this entry and animated figure to enhance your answer. Why a phasor rotates counterclockwise (CCW)?
13 *Consider phasors:
   a) What are they? What are the differences among a rotating phasor, phasor, and a phasor in polar (angular) form?
   b) Why do we need phasors?
   c) It says that the phasors resolve the issue with using the differential equation for the ac circuit analysis. How? Do they help if the input will be a square wave signal? Explain.
   d) Besides helping with differential equations, how else do phasors simplify the ac circuit analysis?
14 Consider the circuit shown in Figure 3.P14 whose $i_{in}(t) = 6\cos(100t)$, $R = 3\Omega$ and L = 20 mH. Find the output voltage using the phasors.

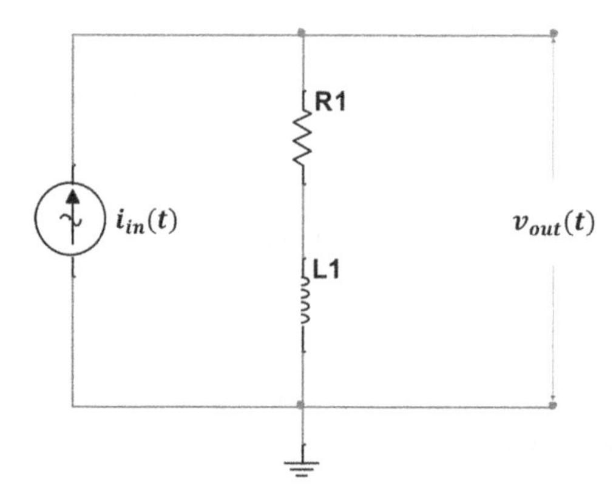

Figure 3.P14 Summing sinusoids with phasors.

15 Using the phasors, find the difference of $v_1(t) = 5cos(20t + 45^0)$ and $v_2(t) = 8sin(20t + 72^0)$. Show the result in the phasor domain and in the time domain. (Hint: See Example 3.1.)
16 Figure 3.2 shows a series RLC circuit whose series current is $i_s(t) = 7\cos(100t)$. Find the sum of voltages $v_R(t)$, $v_L(t)$, *and* $v_C(t)$. (Hint: See (3.10).)

3.2 AC circuit analysis of RC, RL, and RLC circuits with sinusoids and phasors

17 The current flowing through the series RLC circuit given in Figure 3.2 is equal to $i_s(t) = 7\cos(100t + 45^0)$. Depict voltage and current for each circuit element in the phasor domain if $R = 7\Omega$, L = 12 mH, and C = 18 mF. (Hint: See Table 3.1 and accompanying explanations).

18 Consider the series RLC circuit shown in Figure 3.2 and discussed in Problem 17:
a) Find the total impedance of this circuit.
b) Build vector diagram of all impedances.
c) Write down all the impedances in the trigonometric form.

Borrow all values from Problem 17. (Hint: See Table 3.2 and Figure 3.7.)

19 Find the total (equivalent) impedance for the circuit shown in Figure 3.P19. (Hint: See Example 3.2.)

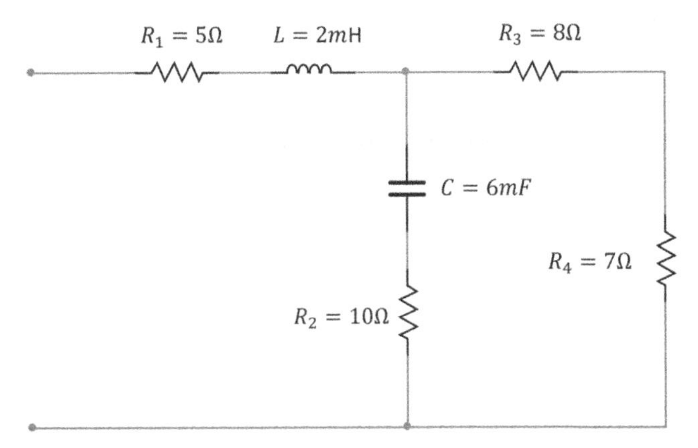

Figure 3.P19 Finding total (equivalent) impedance.

20 For the circuit shown in Figure 3.P20 find the input voltage, $v_{in}(t)$, using the phasor-domain technique. Verify your result by the Multisim simulation of the circuit operation. Sketch the diagram for the phasors you use in this problem. (Hint: See Example 3.3.)

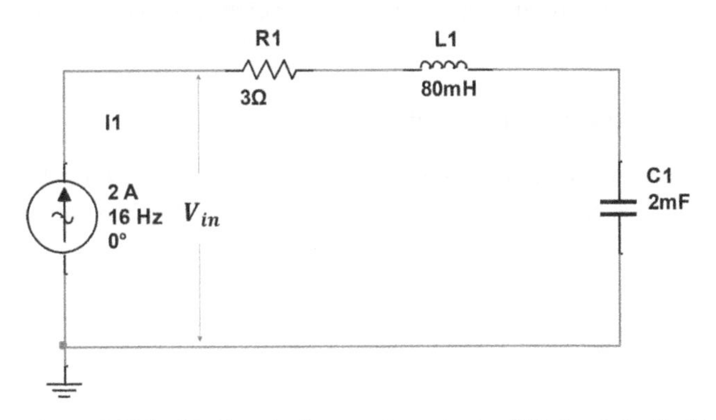

Figure 3.P20 Finding the input voltage in the RLC circuit for Problem 3.20.

21 Use the phasor-domain technique to find the voltage across each individual element of the circuit shown in Figure 3.P21. Sketch the phasor diagram. Check the answer.

22 Determine the individual currents flowing through each component of the RLC parallel circuit shown in Figure 3.P22. Use the phasor method. Draw the phasor diagram. Verify your answer.

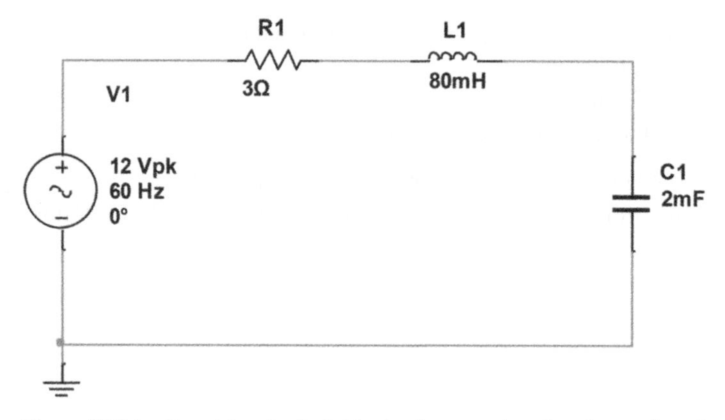

Figure 3.P21 Searching the individual voltages using the phasor-domain technique.

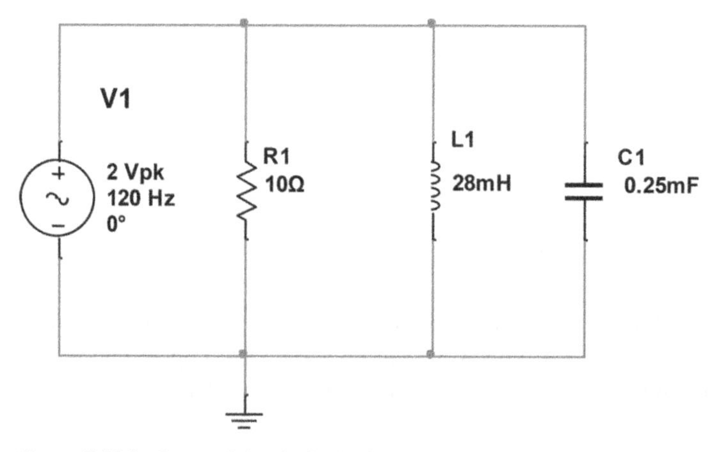

Figure 3.P22 Determining individual currents through each component of the parallel RLC circuit.

3.3 Theorems and Principles of AC steady-state circuit analysis

23 Find voltages and currents for each element of the series-parallel RLC circuit shown in Figure 3.P23. Use nodal method in the phasor domain. Verify the answers with Multisim simulation. (Hint: Refer to Example 3.5.)

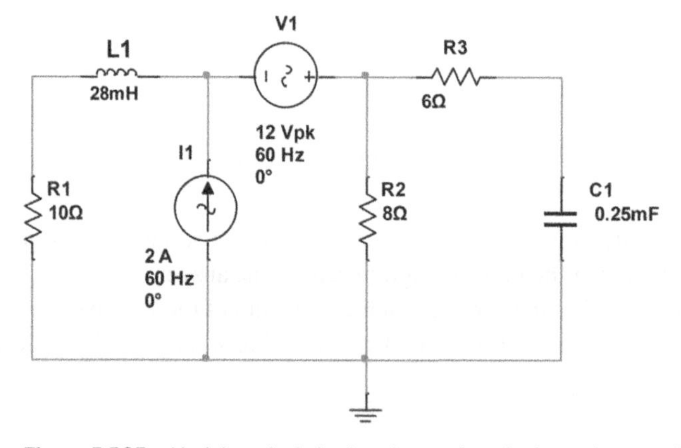

Figure 3.P23 Nodal analysis in the phasor domain for series-parallel RLC circuit.

24 Repeat Problem 3.23 but using the mesh analysis in the phasor domain. Show all meshes. Confirm your answers by simulating the circuit operation with Mutisim. (Hints: (1) See Example 3.6. (2) Apply the supermesh technique to obtain the third equation to find I_{R2} current.)

25 Transform the network shown in Figure 3.P23 into the circuit with one voltage source. Sketch all the intermediate and final circuits and compute the sources and component values. (Hint: Refer to Section 1.4.)

26 In the circuit shown in Figure 3.P26, find the load current $\boldsymbol{I_L}$ if $\boldsymbol{Z_L} = 9(\Omega)$. Use Thevenin's theorem in the phasor domain. Sketch the final circuit. Verify your answer with Multisim simulation. (Hints: (1) Follow Example 3.7. (2) Transform the current source $I1$ into a voltage source and combine voltage sources as Problem 25 demonstrates. This circuit enables you to compute $\boldsymbol{V_{Th}}$. (3) Refer to the original circuit in Figure 3.P26 and compute $\boldsymbol{Z_{Th}}$. (4) Sketch the equivalent circuits at every stage of your analysis. (5) Some numbers can be used from Problems 23 and 24 because those circuits are similar except of $\boldsymbol{R_2}$.

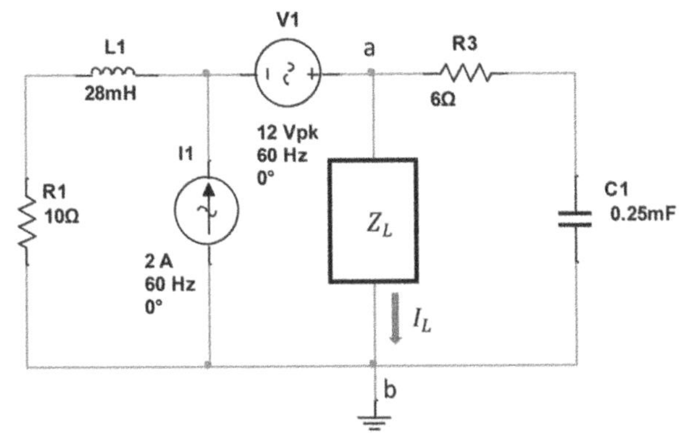

Figure 3.P26 Finding the load current I_L in a series-parallel RLC circuit using Thevenin's theorem in the phasor domain.

27 Use the circuit in Figure 3.P26 to find the load current $\boldsymbol{I_L}$ when $\boldsymbol{Z_L} = 9(\Omega)$ by applying Norton's theorem. Compare your results with those obtained in Problem 26. Alternatively, run the Multisim simulation to verify your answer. (Hint: See Example 3.8.)

28 For the circuit shown in Figure 3.P26, find I_L if $\boldsymbol{Z_L} = 9(\Omega)$. You must use the superposition principle. (Hint: See Example 3.9.)

29 For the circuit given in Figure 3.P29 find its dual counterpart. (Hint: See Figure 3.24.)

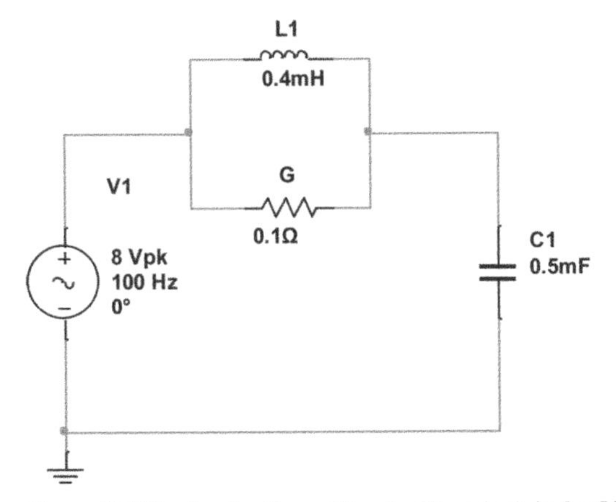

Figure 3P.29 Application of the duality principle for RLC circuit in Problem 3.29.

Part 2

Advanced Circuit Analysis in the Time Domain

4

Advanced Circuit Analysis in Time Domain—I

4.1 Circuit's Response—Statement of Problem

4.1.1 Circuit's Response—What It Is

We start with the introduction of a system. *A dictionary defines the system as* "a group of devices or artificial objects or an organization forming a network especially for distributing something or serving a common purpose." This includes an electrical circuit, a telephone system, a heating system, a highway system, a computer system, a communication system, etc. A technical system may include hardware, software, and firmware parts and their combinations. In this book, we consider electrical circuits as systems; this is why the book carries the subtitle *A Systems Approach*. What distinguishes this approach from a traditional circuit analysis? In short, the conventional circuit analysis concentrates on finding electrical characteristics (current through and voltage across) of each circuit's component. This book focuses on the circuit's input–output relationship.

To represent any system mathematically, we need to rely on the physical laws governing the system's operation and compose the formulas stemming from these laws. In electrical circuits, we need to derive the equations that relate the output signal to the input by describing the circuit's structure and parameters.

The input–output relationship is one of the major system's characteristics; every engineer looks first for the input–output relationship when examining an unknown system. This statement is true for an electrical circuit too. Figure 4.1a illustrates this formulation in general, and Figure 4.1b uses an RC circuit as an example. In general, an input excites a system and causes system's output. In an RC circuit, we want to determine the *output signal* (the voltage across the capacitor in this example), which appears in response to an applied input. Thus, a *circuit's response is the voltage or current representing the circuit's reaction (output) to a given excitation (input) signal.*

The objective of this book is to study how to find theoretically a circuit's response. If we learn that, we would be able *to predict* the response of a circuit, which, in turn, enables us *to design a circuit with the desired response.*

Consider the other series RC circuit shown in Figure 4.2a. Its input is the voltage, $v_{in}(t)$, applied right after the switch, and the output (response), $v_{out}(t)$, is the voltage drop across the capacitor. Input signals and the circuit's responses for two other examples in Figures 4.2b and 4.2c are self-explanatory.

The *traditional circuit analysis's approach* to finding the circuit's output is limited by two (though critical) inputs: *dc* and *sinusoidal inputs*. Both inputs cause a *steady-state* circuit's operation, provided the circuit's components are constant. In reality, the circuit always operates in a

Essentials of Advanced Circuit Analysis: A Systems Approach, First Edition. Djafar K. Mynbaev.

Companion Website: www.wiley.com/go/Mynbaev/AdvancedCircuitAnalysis

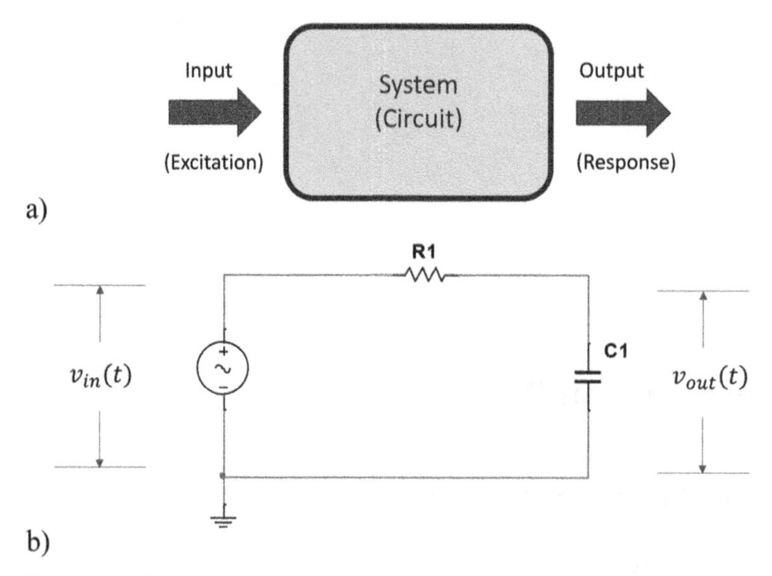

Figure 4.1 System (circuit) response: a) General view; b) the circuit view.

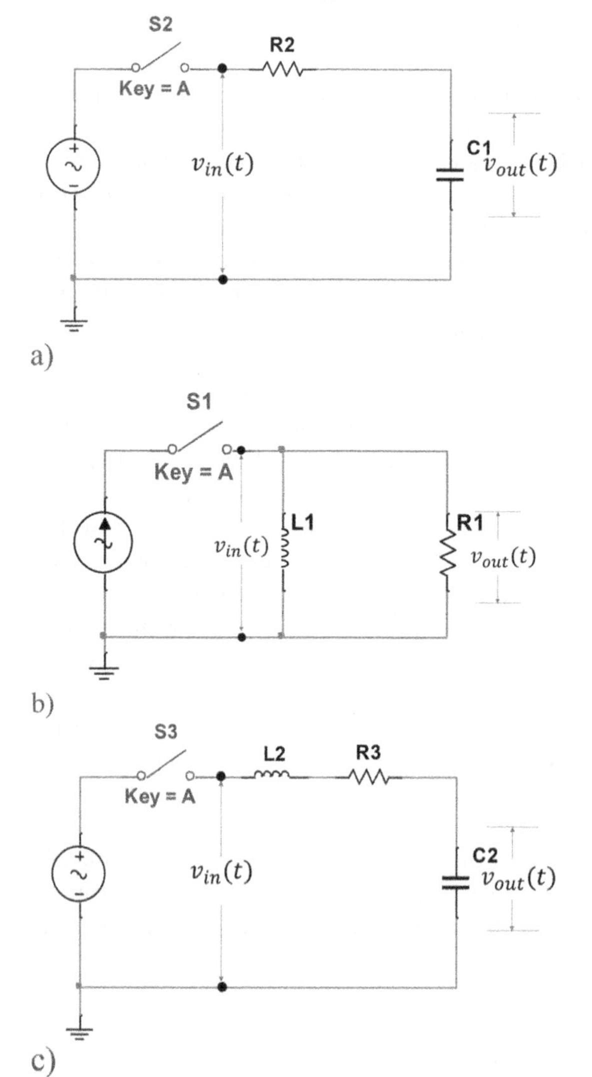

Figure 4.2 Responses of electrical circuits: a) Response of a series RC circuit; b) response of a parallel RL circuit; c) response of a series RLC circuit.

non-steady state, and its inputs can be signals of any type. To describe a circuit's operation in this unrestricted case, a more general method to obtain the circuit's response is developed. This method is called *advanced circuit analysis, and it considers any input signals and both the* steady-state and transient responses of a network to these signals.

4.1.2 Transient Response–What and Why

Consider the RC series circuit shown in Figure 4.1b. The input is a sinusoidal signal,

$$v_{in}(t) = V_{in}\cos(\omega t),$$

whose amplitude is V_{in} and radian frequency is $\omega\left(\frac{rad}{s}\right) = 2\pi f$. What is its output (response)? We know that a *linear time-invariant system*, which an RC circuit is, preserves the waveform of an input signal but changes its parameters. Thus, the RC circuit's output must be a cosine signal with a smaller amplitude and an additional phase shift; i.e.,

$$v_{out}(t) = V_{out}\cos(\omega t + \theta). \tag{4.1}$$

Hence, the problem boils down to finding V_{out} and θ using the given V_{in}, ω, R, and C. An ac circuit analysis considered in Section 3.2 hints that the proper tool for tackling this problem is using phasors. Let $\boldsymbol{V_{in}}$ be an input phasor and $\boldsymbol{V_{out}} \equiv \boldsymbol{V_C}$ be the output one. (We use the boldface to denote a complex number, as usual.) The voltage-divider rule gives

$$\boldsymbol{V_{out}} = \boldsymbol{V_{in}}\frac{\boldsymbol{Z_C}}{\boldsymbol{Z_T}} = \boldsymbol{V_{in}}\frac{-jX_C}{R - jX_C}. \tag{4.2}$$

You will recall that $\boldsymbol{Z_C} = -jX_C = -\frac{j}{\omega C}$ is the *capacitive impedance* and $\boldsymbol{Z_T} = \boldsymbol{Z_R} + \boldsymbol{Z_C} = R - j\omega C$ is the *total circuit's impedance*. Separating the real and imaginary parts of the RHS in (4.2) yields

$$\boldsymbol{V_{out}} = \boldsymbol{V_{in}}\left(\frac{X_C^2}{R^2 + X_C^2} - \frac{-jX_C R}{R^2 + X_C^2}\right).$$

Thus, our complex number is presented in a rectangular form, $\boldsymbol{c} = a + jb$, which can be converted to a phasor form as $\boldsymbol{c} = A\angle\theta = Ae^{j\theta}$, where $A = \sqrt{\left(a^2 + b^2\right)}$ and $\theta = tan^{-1}\left(\frac{b}{a}\right)$. Performing these calculations for our example gives

$$V_{out} = V_{in}\frac{1}{\sqrt{\left(1 + \left(\frac{R}{X_C}\right)^2\right)}}$$

and

$$\Theta = -tan^{-1}\left(\frac{R}{X_C}\right). \tag{4.3}$$

Plugging these amplitude and phase shift in (4.1) results in

$$v_{out}(t) = V_{out}\cos(\omega t + \theta) = V_{in}\frac{1}{\sqrt{\left(1 + \left(\frac{R}{X_C}\right)^2\right)}}\cos\left(\omega t - \tan^{-1}\left(\frac{R}{X_C}\right)\right). \tag{4.4}$$

It seems that we found the response of an RC circuit; so, what else left to discover? In the preceding analysis, we imply that the sinusoidal signal is applied to the circuit forever; that is, from $-\infty$ *to* $+\infty$. If this would be the case, Equation 4.4 resolves the problem in its entirety. But the truth is that it was a time when no signal was applied and then became a moment when the switch was turned on, and the circuit was excited. Refer to Figure 4.3a showing an RC circuit without a switch and compare $v_{in}(t)$ and $v_{out}(t)$. This is a *steady-state* circuit's *regime.* Next, examine Figure 4.3b, where the switch connects the signal source to the RC circuit. Initially, the switch is open, and no signal is applied. In this state, the circuit's response is zero if the initial condition is zero. Then, at one instant, which we consider as $t = 0$, the switch turns on, and the signal $v_{in}(t)$ is applied. At the same moment, the circuit starts responding by $v_{out}(t)$. However, it takes some time for the circuit to develop its output to the steady-state form. This transition time interval is called a *transient interval*, and the *process of transiting from the initial to a steady-state regime is called* **transient response.**

Inspect Figure 4.3b *again to comprehend these crucial definitions fully. Including the transient response into an electrical circuit's complete response distinguishes the* advanced circuit analysis *from a* traditional circuit analysis. *It seems that considering a circuit transition from an initial rest state to an operational state is a minor issue. After all, this transition doesn't take too long to complete compared to a steady-state operation which could last hours, days, months, or years. Indeed,* Figure 4.3b *demonstrates that the transition will end soon. But no, such a point of view would be superficial. First, no operation can start without turning a switch on. We are referring to not only power switches but billions of operational switches inside endless electronic circuits in today's technology. Also, a digi*tal signal, *by its very nature, is a switch that commutes electronic circuits from one state to the other. Our readers know very well that a vast majority of modern technology is based on digital electronics. Thus, switching occurs constantly and everywhere in our everyday technological life. Secondly, not all transient responses would happily end in a steady-state regime. You might be familiar with voltage or current surges after turning the power switch in your numerous electronic devices. If you never encounter such situations, this is because the engineers who designed and built your devices knew about*

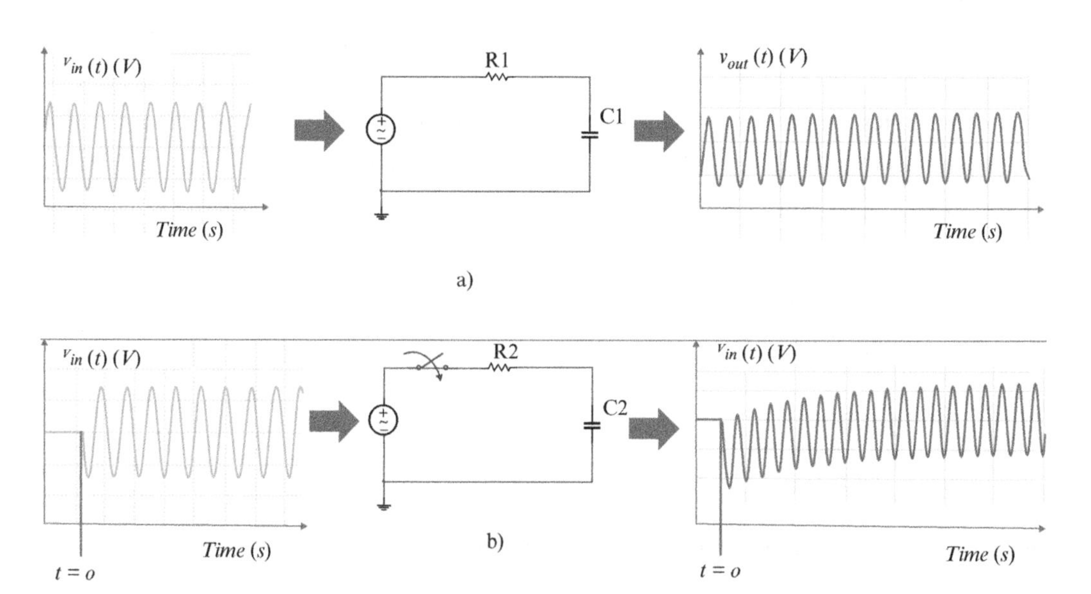

Figure 4.3 Responses of an electrical circuit: a) Steady-state response; b) transient response.

transient responses and took all necessary measures to make these transitions diminish quickly and safely. Reread this paragraph because you cannot overestimate its importance: It describes why we need to study ***the subject of this book.***

4.1.3 Differential Equations–Where They Come from and How to Solve Them

Can we apply the steady-state ac analysis presented by Equations 4.1 through 4.4 to describe a circuit's transient response? No, we cannot because voltages across and currents through all circuit's components are changing over the time in a non-periodic mode. As we know from Chapter 1, resistors react instantaneously to any i or v changes, but reactive components (capacitors and inductors, that is) respond to these changes particularly. Table 4.1 reminds these peculiarities. (We have all this information in Table 3.1; however, here we want to concentrate on the highlighted differential relationships).

Let's focus on the $i - v$ *differential relationship* among the reactive components as highlighted in Table 4.1. Formula $v_L(t) = L\frac{di_L(t)}{dt}$ states that *changing* of an inductor's current causes the appearance and existence of the voltage across the inductor. If $i_L(t)$ *is constant* (dc), then $\frac{di_L(dc)}{dt} = v_L(t) \equiv 0$ by definition of derivative, and there is no voltage across the inductor. The same reasoning holds for $v_C(t) - i_C(t)$ relationship. (Explain to yourself how.) Therefore,

> *At a transient regime, the circuit's voltage and current are changing; therefore, only differential equations can describe the transient response.*

How? Here is the example.

Example 4.1 The response of a series RC circuit—deriving the differential equation.

Problem: Consider a series RC circuit shown in Figure 4.2a whose $R = 1k\Omega$, $C = 0.5mF$, and $v_{in} = v_{in}(t)$. The circuit was initially relaxed, i.e., $v_C(0) = V_0 = 0$. At $t = 0$, the switch is turned on. Derive an equation that relates $v_{out}(t)$ and $v_{in}(t)$.

Solution: Obviously, the circuit will be in an operational state at $t \geq 0$.

What rule in circuit analysis does enable us to derive the required equation? There is Kirchhoff's voltage law (KVL) that puts all circuit voltages in one equation. In our example, the KVL shows how the input voltage is distributed between R and C components:

$$v_R(t) + v_C(t) = v_{in}(t). \tag{4.5}$$

Table 4.1 Current–voltage relationship for R, L, and C components.

Component	Current	Voltage
Resistor	$i_R(t) = \frac{v_R(t)}{R}$	$v_R(t) = i_R(t) \cdot R$
Inductor	$i_L(t) = \frac{1}{L}\int_{-\infty}^{t} v_L(t)dt$	$v_L(t) = L\frac{di_L(t)}{dt}$
Capacitor	$i_C(t) = C\frac{dv_C(t)}{dt}$	$v_C(t) = \frac{1}{C}\int_{-\infty}^{t} i_C(t)dt$

By our circuit's topology, the capacitor's voltage *is* the output voltage (circuit's response); i.e., $v_C(t) \equiv v_{out}(t)$. Substituting $v_{out}(t)$ for $v_C(t)$ in (4.5), we can partly relate $v_{in}(t)$ and $v_{out}(t)$ as

$$v_R(t) + v_{out}(t) = v_{in}(t). \tag{4.6}$$

But all members of LHS of this equation must depend only on $v_{out}(t)$. How can we relate $v_R(t)$ to $v_C(t) \equiv v_{out}(t)$? Examine the given circuit again and ask yourself what these two components, *R* and *C*, have in common during the circuit operation? Obviously, it is the current *i*(*t*). It flows through all the circuit's components, as it must be in a series circuit. Thus,

$$v_R(t) = i(t) \cdot R,$$

according to Table 4.1. This current can be expressed through $v_C(t) \equiv v_{out}(t)$ using the other expression from Table 4.1:

$$i(t) = C\frac{dv_C(t)}{dt}.$$

This is where a derivative enters the play! Therefore, the second member of the LHS of (4.6) can be presented in terms of $v_C(t) \equiv v_{out}(t)$ as

$$v_R(t) = R \cdot i(t) = RC\frac{dv_C(t)}{dt} = RC\frac{dv_{out}(t)}{dt}.$$

Finally, we need to plug this $v_R(t)$ into (4.6) to find that our series RC circuit is described by the following *differential equation*

$$RC\frac{dv_{out}(t)}{dt} + v_{out}(t) = v_{in}(t). \tag{4.7a}$$

Or, using the standard notation $\frac{dy}{dx} = y'(x)$,

$$RCv'_{out}(t) + v_{out}(t) = v_{in}(t). \tag{4.7b}$$

Observe that LHS of (4.7a) or (4.7b) contains only $v_{out}(t)$ and its derivative, whereas RHS of this equation contains only $v_{in}(t)$. Thus, the problem is solved because $v_{out}(t)$ and $v_{in}(t)$ are interrelated.

Strictly speaking, the circuit's operation starts only after the switch connecting the input (excitation) to the circuit gets turned on. To highlight this critical point, we write (4.7b) in the following form,

$$RCv'_{out}(t) + v_{out}(t) = v_{in}(t) \cdot u(t), \tag{4.7c}$$

where *u(t)* is a *unit step (Heaviside) function*. This function is equal to zero before $t = 0$ and becomes equal to 1 after this instant. (See Chapter 7 for a detailed discussion of this topic.) Equation 4.7c is a canonical form of a differential equation describing the operation of an electrical circuit. However, to shorten the writing in future discussions, we will write $u(t)$ explicitly only when the problem's conditions require this.

Therefore, the response of a series RC circuit is described by *a differential equation*. This equation is called *differential* because it contains a derivative. It is also called a *first-order differential equation* because there is only a first-order derivative here. Remember, (4.7a) stems from the fundamental KVL applied to our specific circuit.

Discussion:

- This example demonstrates that a differential equation comes from the need to describe the reactive circuit's transition from one, an initial state, to the other, final state. In other words, the differential equation chracterizes the *transient response* of a circuit.
- The second point is that a differential equation describes the circuit transitive, dynamic regime. This point is general: *Every time we need to describe the changes in a circuit's state, we must use a differential equation.*
- The third point is an origin of a differential equation: It always stems from a fundamental principle. In our example, the differential equation stems from a Kirchhoff's voltage law.
- Finally, can we solve the transient problem by using traditional circuit analysis? No, we can't. Try to follow manipulations performed in Equations 4.1 through 4.4 to verify impossibility of achieving the desired result.

Wait a minute, you might correctly ask, a sinusoidal input is also a time-changing function. How then did we find the circuit's response in (4.4), without employing a differential equation? Good question. The answer is that Equation 4.4 holds only for a steady-state response. In deriving (4.4) we did not consider an instant when a sinusoidal signal was applied to an RC circuit; we implied it was always there. Still, you might rightly object that a sinusoidal function varies with time and therefore needs a differential equation. Excellent observation. We owe Dr. *Steinmetz* the solution to this seems to be a contradiction. He realized that a steady-state circuit's operation could be described by complex numbers, thus avoiding differential equations. (See Section 3.1 for the detailed explanations of this achievement and the brief biography of Dr. Charles Proteus Steinmetz.) It's worth noting that Dr. Steinmetz also developed theory of transient processes in electrical circuits, where differential equations were the main mathematical tool.

How can we *solve a differential equation*? Sidebar 4S considers this and other appropriate questions by reviewing the basics of circuit's differential equations.

Sidebar 4S.1 Basics of Differential Equation

Algebraic equation vs. differential equation

What is *equation*? This is a mathematical expression containing an equality sign. Thus, $ax^2 + bx + c$ is a *polynomial* but $ax^2 + bx + c = 0$ is the equation. The expression 3 = 3 also contains an equality sign, but this obvious equality is called *identity*. The equation, in contrast to the identity, has a question. For example, what value of x turns the equation $ax^2 + bx + c = 0$ into identity? Answering this question means solving the equation.

What is the difference between algebraic and differential equations? An *algebraic equation* is the equation about a quantity. For example,

$$3x + 7 = 13. \tag{4S1.1}$$

The goal here is to find the value of x which will turn this equation into an identity. Simple manipulations,

$$3x = 13 - 7 = 6 \;\Rightarrow\; x = \frac{6}{3} = 2,$$

give us the value of x which is the solution to (4S1.1). Indeed,

$$3 \cdot 2 + 7 = 13.$$

The solution to an algebraic equation is unique; there is no other value of x which could satisfy the given equation. (Referring to our example, try for yourself any other value of x to verify this statement.)

However, (4S.1) operates with numbers, and because of that, it should be called arithmetic equation. Algebra, as we know, generalizes arithmetic methods by using the letters instead of numbers. For example, (4S1.1) can be written as

$$ax + b = c. \tag{4S1.2}$$

The solution to this equation is

$$x = \frac{c - b}{a}. \tag{4S1.3}$$

Now, we can plug into (4S.2) any required numbers and receive the answer by substituting letters in (4S1.3) with the given numbers. Another example is the notorious quadratic equation,

$$ax^2 + bx + c = 0. \tag{4S1.4}$$

Its solution is

$$x_{1,2} = \frac{-b \pm \sqrt{b^2 - 4ac}}{2a}. \tag{4S1.5}$$

Thus, we need to plug the given numbers into (4S1.4) and compute the answer using (4S1.5). (Create a numerical example of quadratic equation for yourself.) We will use (4S1.5) extensively in this book.

A *differential equation* is the equation about a *function*. For example, if $y = f(x)$, that is, $y(x)$, then

$$\frac{dy}{dx} + 3y = 0, \tag{4S1.6}$$

is the differential equation because it contains derivative of y with respect to x. We can't solve this equation by the same manipulations we used for solving an algebraic equation simply because we must obtain not a value but a function.

A differential equation whose RHS is equal to zero, is called *homogenous*. Thus, (4S1.6) is the example of such equation.

Solving a homogeneous differential equation

How then can we solve a differential equation? Differentiation "takes" our function $y = f(x)$ to the other domain; by *integrating* a differential equation, we return our function to its original domain. (Differentiation and integration are two mutually inverse operations, remember?) This is why the process of solving a differential equation is often called integrating the equation.

Back to our example, we rearrange (4S1.6) and separate the variables as

$$\frac{dy}{dx} = -3y \Rightarrow \frac{dy}{y} = -3dx$$

Taking the integrals from the both sides of the equation results in

$$\int_{y_0}^{y} \frac{dy}{y} = -3\int_{x_0}^{x} dx \Rightarrow \ln(y)\Big|_{y_0}^{y} = -3x\Big|_{x_0}^{x}$$

where $y_0 = y(x_0)$. Performing the needed manipulations produces

$$lny(x) - lny_0 = -3(x - x_0) \Rightarrow \ln\left(\frac{y(x)}{y_0}\right) = -3(x - x_0).$$

Taking the power of *e*, we attain

$$\frac{y(x)}{y_0} = e^{-3(x-x_0)} \Rightarrow y(x) = y_0 e^{-3(x-x_0)} \quad (4S1.7)$$

To finish the solution, we need to know the values of y_0 and x_0. These values constitute *initial conditions*; they are necessary for solving any differential equation. Let's assume $x_0 = 8$, and $y_0 = y(0) = 5$. Then the solution to the given differential equation is

$$y(x) = y_0 e^{-3(x-x_0)} = 5e^{-3(x-8)}. \quad (4S1.8)$$

Let's verify our solution: Plugging the obtained $y(x)$ into the given differential equation, we calculate

$$\frac{dy}{dx} + 3y = 0 \Rightarrow \frac{d\left(5e^{-3(x-8)}\right)}{dx} + 3\left(5e^{-3(x-8)}\right) = 0 \Rightarrow -15e^{-3(x-8)} + 15e^{-3(x-8)} = 0.$$

Therefore, our solution is correct because it converts the given equation into identity. (Refer to any textbook in mathematics to refresh your memory on these straightforward manipulations.)

Now, consider a general case of our example: Let the differential equation be given by

$$\frac{dy}{dx} + \alpha y = 0, \quad (4S1.9)$$

where $y_0 = y(0)$, x_0, and α are given numbers. The solution to this equation is provided by (4S1.8), which now takes the following form:

$$y(x) = y_0 e^{-\alpha(x-x_0)} \quad (4S1.10)$$

Plugging into y_0, x_0, and α a set of values, we obtain a numerical solution to (4S1.9). However, if we use the other set of y_0, x_0, and α values, we get the other solution. This means that, in contrast to an algebraic equation, *one differential equation may have many solutions determined by its parameters and initial conditions.*

It's worth knowing that we demonstrate just one method of solving a homogenous differential equation called the *separation of variables*. The other popular method is the use of a *trial solution*, where we assume the form of a possible solution. In our example, we postulate that the solution to a differential equation holds the following form

$$y(x) = Ae^{mx}. \quad (4S1.11)$$

This assumption is based on the fact that most of first- and second-order differential equations do indeed have the solutions in this form. Also, it employs the following peculiarity of an exponential function's derivative,

$$\frac{de^{mx}}{dt} = me^{mx}. \quad (4S1.12)$$

To verify our assumption, we need to plug the proposed solution into (4S1.9) and calculate

$$\frac{dAe^{mx}}{dt}+A\alpha e^{mx}=0 \Rightarrow Ae^{mx}(m+\alpha)=0 \Rightarrow \quad m=-\alpha \tag{4S1.13}$$

because e^{mx} cannot be equal to zero, and $A=0$ is a trivial solution. Therefore,

$$y(x)=Ae^{-\alpha x}. \tag{4S.14}$$

To finalize, we need to evaluate constant A. This can be done using an *initial condition*, $x(0)=x_0$, as follows:

$$y(0)=Ae^{-\alpha x_0}$$

Denoting $y(0)=y_0$, we obtain $A=y_0e^{\alpha x_0}$, which gives

$$y(x)=y_0e^{-\alpha(x-x_0)},$$

as in (4S.10).

Therefore, our trial solution (4S1.11),

$$y(x)=Ae^{mx},$$

is indeed the true solution to (4S1.9) provided that

$$m=-\alpha$$

and

$$A=y_0e^{\alpha x_0}.$$

The solution to a nonhomogeneous differential equation

What if the RHS of a differential equation is not zero? In this case, the equation is called *nonhomogeneous*. How can we solve it? Let's modified (4S1.9) be the example of a nonhomogeneous differential equation.

$$\frac{dy}{dx}+\alpha y=\alpha K, \tag{4S1.15}$$

where α and K are given constants, and x_0 is the initial condition. Since this equation is linear, we can use the principle of superposition and state that its *complete solution*, $y_{cmp}(x)$, is a sum of a *general solution* to its homogeneous version and a *particular solution* to its nonhomogeneous version. Since a general solution is the *homogeneous solution*, we accept this less ambiguous term and denote this solution as $y_h(x)$. Accordingly, we denote a *particular solution* as $y_p(x)$. Then,

$$y_{cmp}(x)=y_h(x)+y_p(x), \tag{4S1.16}$$

where $y_{cmp}(x)$ is the complete (total) solution, which explains the meaning of the subscript *cmp*. Borrowing the homogeneous (general) solution from (4S1.14), we can write

$$y_h(x)=Ae^{-ax}. \tag{4S1.17}$$

The *particular solution is always in the form of the equation's RHS*; that is,

$$y_p(x)=B,$$

where B is constant. To find B, we substitute it into the given nonhomogeneous differential equation and obtain

$$\frac{dy_{p(x)}}{dx}+\alpha y_{p(x)}=\alpha K \Rightarrow \frac{dB}{dx}+\alpha B=\alpha K \Rightarrow B=K. \tag{4S1.18}$$

Combining (4S1.17) and (4S1.18) yields

$$y_{cmp}(x)=y_h(x)+y_p(x) \Rightarrow y_{comp}(x)=Ae^{-\alpha x}+K. \tag{4S1.19}$$

To evaluate A, we turn to the initial condition and find

$$y_{cmp}(x_0)\equiv y_0=y_h(x_0)+y_p(x_0)=Ae^{-\alpha x_0}+K.$$

Hence,

$$A=(y_0-K)e^{\alpha x_0}. \tag{4S1.20}$$

Finally, the complete solution is

$$y_{cmp}(x)=(y_0-K)e^{\alpha x_0}e^{-\alpha x}+K=y_0e^{-\alpha(x-x_0)}+K\left(1-e^{-\alpha(x-x_0)}\right). \tag{4S1.21}$$

If $x_0=0$, which is often the case, the complete solution is simplified to

$$y_{cmp}(x)=y_0e^{-\alpha x}+K\left(1-e^{-\alpha x}\right). \tag{4S1.22}$$

Thus, two important points that distinguish the complete solution to a nonhomogeneous equation from that to the homogeneous are:

- The complete solution to a given nonhomogeneous differential equation is the sum of a general solution to a homogeneous equation and a particular solution to the nonhomogeneous one.
- The particular solution is caused by the RHS of the given differential equation; therefore, it always takes the form of the specific RHS.

This book explores mainly the first- and second-order differential equations. This is why the described two methods of solving of their homogeneous versions –separating the variables and using a trial solution–are sufficient for the time-domain advanced analysis of the proper electrical circuits. Using the superposition principle is satisfactory for finding the nonhomogeneous complete solution.

Nevertheless, we should know that there is the most powerful method to solve differential equations of any form and order; it is called the *Laplace transform*. Part 3 of this book is completely devoted to this method. For now, we leave the discussion of many exciting aspects of differential equations to specialized textbooks in mathematics.

A word of caution regarding the notations: In this sidebar, we designate an independent variable as x and dependent variable as $y(x)$. These designations are agelong mathematical tradition. Even so, in electrical engineering, the independent variable is time, and it is denoted as $t(s)$, whereas the dependent variables are mainly voltage, $v(t)[V]$, and current, $i(t)[A]$. All other parameters describing the processes in electrical circuits or electromagnetic fields (energy, power, the strength of electrical or magnetic fields, etc.) are also used as the variables depending on time.

4.1.4 Finding the Complete Response of an Electrical Circuit-Examples

Now, we are well equipped to find the response of an electrical circuit. The best way to demonstrate this process is to finish the work started in Example 4.1.

Example 4.2 Finding the step response of a series RC circuit.

Problem: Find the step response of the series RC circuit shown in Figure 4.4a. Use the differential equation of this circuit derived in Example 4.1. The circuit's parameters are: $R = 1k\Omega, C = 0.5mF$, and $V_{in} = 12V$. The circuit was initially relaxed, i.e., $v_C(0) = V_0 = 0$. The switch is turned on at $t = 0$. The input signal is a step (Heaviside) function depicted in Figure 4.4b.

Solution: We must realize that we cannot apply phasor-based approach to solve the circuit's differential equation derived in (4.7c) because the input is NOT a sinusoidal signal. Hence, the other approach has to be used to fulfill the task.

The given equation is

$$RCv'_{out}(t) + v_{out}(t) = v_{in}(t) \cdot u(t). \tag{4.7Cr}$$

Here, $R \cdot C$ is a *time constant* of an RC circuit usually denoted as $\tau(s)$,

$$\tau(s) = RC. \tag{4.8}$$

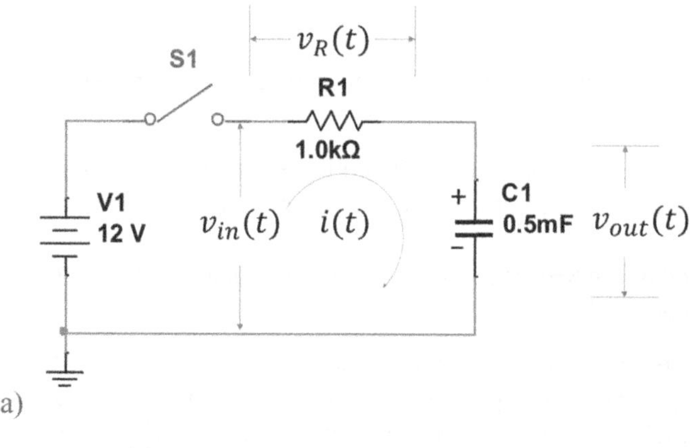

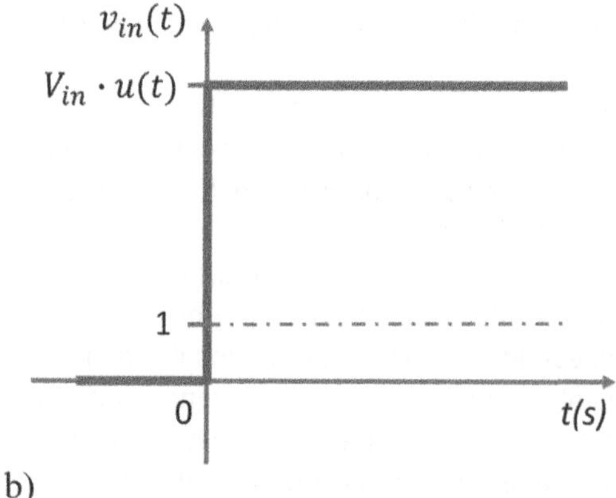

Figure 4.4 Finding the step response of the series RC circuit: a) Circuit's schematic; b) the step input signal for Example 4.2.

Consider the inverse member,

$$\alpha\left(\frac{1}{s}\right)=\frac{1}{\tau}=\frac{1}{RC}, \tag{4.9}$$

which is called a *damping constant (coefficient).*

(**A word of caution**: Chapter 4 uses notations τ and α as they are defined for the first-order circuits in (4.8) and (4.9). Chapters 5 and 6 will use the same notations to introduce similar coefficients for the second-order circuits. When such use might confuse, we will use subscripts to distinguish them. Hence you might see such designations as α_{RC} or α_{RLC}.)

If we divide (4.7a) by RC, the equation takes the form

$$v'_{out}(t)+\alpha v_{out}(t)=\alpha v_{in}(t). \tag{4.10}$$

We omit $u(t)$ in (4.10), as agreed previously, but always bear in mind its presence in the circuit's differential equation. The complete solution to (4.10) is the sum of a *homogeneous (general) solution*, $v_{out-h}(t)$, and a *particular solution*, $v_{out-p}(t)$; that is,

$$v_{out-cmp}(t)=v_{out-h}(t)+v_{out-p}(t) \tag{4.11}$$

The *homogeneous solution* is found by solving (4.10) when its RHS is zero. That is,

$$v'_{out-h}(t)+\alpha v_{out-h}(t)=0. \tag{4.12}$$

The solution to such an equation is given in (4S.10) as $y(x)=y_0e^{-\alpha(x-x_0)}$. Borrowing this solution to our example, we get

$$v_{out-h}(t)=Ae^{-\alpha t} \tag{4.13}$$

because here $y_0\equiv V_0=0$, $x\equiv t(s)$, and $x_0\equiv t_0=0$.

*The homogeneous solution to the circuit's differential equation is called a **natural** (source-free) **response** because it describes the circuit's behavior, in terms of $i-v$ processes, without external excitation.*

(More on the initial condition and natural response are in Section 4.2 and Appendix 4A.)

The *particular solution* must be in the form of $v_{in}(t)$, which is a constant. Thus,

$$v_{out-p}(t)=B\,(V). \tag{4.14}$$

Plugging (4.14) into (4.10) yields

$$B'+\alpha B=\alpha V_{in}.$$

Therefore,

$$B=V_{in}\,(V). \tag{4.15}$$

*The **particular solution** to the circuit differential equation is called a **forced response** because it describes the circuit's reaction to an external excitation.*

Thus, the ***complete solution*** is attained as

$$v_{out-cmp}(t)=v_{out-h}(t)+v_{out-p}(t)=Ae^{-\alpha t}+V_{in}. \tag{4.16}$$

By letting $t=0$ in (4.16), we obtain the *initial condition* that enables us to evaluate A. We get

$$V_0=A+V_{in}\Rightarrow A=V_0-V_{in}$$

where $V_0 = v_{out-cmp}(0)$. Finally, the complete solution to (4.16) is given by

$$v_{out-cmp}(t) = (V_0 - V_{in})e^{-\alpha t} + V_{in} = V_0 e^{-\alpha t} + V_{in}\left(1 - e^{-\alpha t}\right). \quad (4.17a)$$

Observe that the complete response is composed of a *natural response*, $V_0 e^{-\alpha t}$, generated by the circuit's initial condition, V_0, plus the *forced response*, $V_{in}(1-e^{-\alpha t})$, induced by the external independent source, V_{in}.

Plugging in the given numbers, we compute, $RC = \tau = 0.5(s)$ and $\alpha = \frac{1}{\tau} = 2\left(\frac{1}{s}\right)$. So, the numerical answer is

$$v_{out-cmp}(t) = 12\left(1 - e^{-2t}\right). \quad (4.17b)$$

The problem is solved.

Discussion:

- The first thing we must do upon obtaining the solution to a given problem is to verify our answer. And the first method of verification is checking the initial and the final state of the solution and compare them with the given conditions of the problem. An *initial state* is the solution at $t = 0, i.e., v_{out-cmp}(0)$, and the *final state* is $v_{out-cmp}(t)$ when $t \Rightarrow \infty, i.e., v_{out-cmp}(\infty)$. Plugging $t = 0$ into (4.17b) results in $v_{out-cmp}(0) = V_0 - V_{in} + V_{in} = V_0 = 0$, which satisfies the initial condition. Considering $t \Rightarrow \infty$, we find $v_{out-cmp}(\infty) = V_{in} = 12(V)$, which correctly describes the series RC circuit operation: Eventually, the capacitor voltage will reach V_{in} and remain at this value. Therefore, our solution (4.17b) looks reasonable. (Remember, an *initial condition* is provided to a circuit, and an *initial state* refers to the status of the circuit. The same is true for a *final value* and the *final state*.)
- What if a capacitor was initially charged up to $V_0 = 4V$? Solution (4.17a) includes such an initial condition. Thus, (4.17a) remains the same, but the numerical solution becomes

  $$v_{out-cmp}(t) = 12 - 8e^{-\alpha t}.$$

  Also, now $v_{out-cmp}(0) = V_0 = 4V$.
- Figures 4.5a and 4.5b depict (4.17) for two cases, $V_0 = 0V$ and $V_0 = 4V$.
  Carefully review this circuit's response, paying particular attention to the transient and steady-state intervals, and the initial and final values.
- In our example, the series RC circuit's response (output) is given in (4.17a) as a voltage across the capacitor. The current through this capacitor can be readily calculated using the formula from Table 4.1. Specifically,

$$i_c(t)[A] = C\frac{dv_C(t)}{dt} = C\frac{d\left(V_0 e^{-\alpha t} - V_{in}e^{-\alpha t} + V_{in}\right)}{dt}$$

$$= \alpha C\left(V_{in} - V_0\right)e^{-\alpha t} = \frac{1}{R}\left(V_{in} - V_0\right)e^{-\alpha t} \quad (4.18a)$$

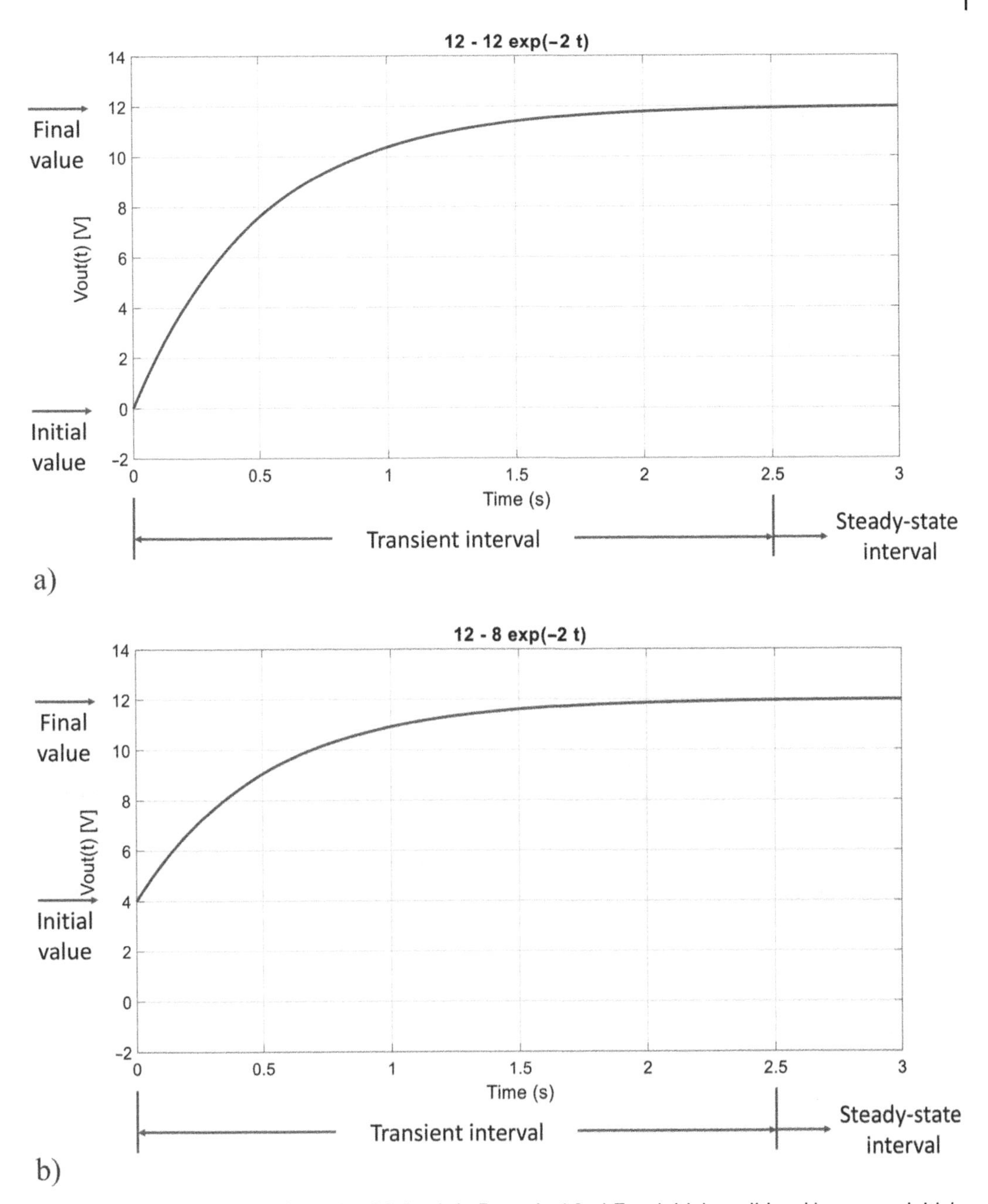

Figure 4.5 Step responses of a series RC circuit in Example 4.2: a) Zero initial condition; b) non-zero initial condition, $V_0 = 4V$.

Plugging the given values into (4.18a) produces for $V_0 = 0$

$$i_c(t)[A] = \alpha C(V_{in} - V_0)e^{-\alpha t} = 0.012e^{-2t}. \tag{4.18b}$$

The units in (4.18a), $LHS[A] = RHS\left[F \cdot \frac{V}{s}\right]$, are correct. Figures 4.6a and 4.6b show the graphs of $i_c(t)$ for $V_0 = 0V$ and $V_0 = 4V$. Review and analyze these figures. When $t = 0$, Equation 4.18b gives the initial value of $i_c(t)$ for $V_0 = 0$ as $i_c(0) = 0.012A$. Letting $t \to \infty$, we compute the final value of $i_c(t)$ for $V_0 = 0V$ as $i_c(\infty) \to 0A$. Compare these calculations with the values shown in the MATLAB-built Figure 4.6a. Perform similar calculations for $V_0 = 4V$ and compare them with Figure 4.6b.

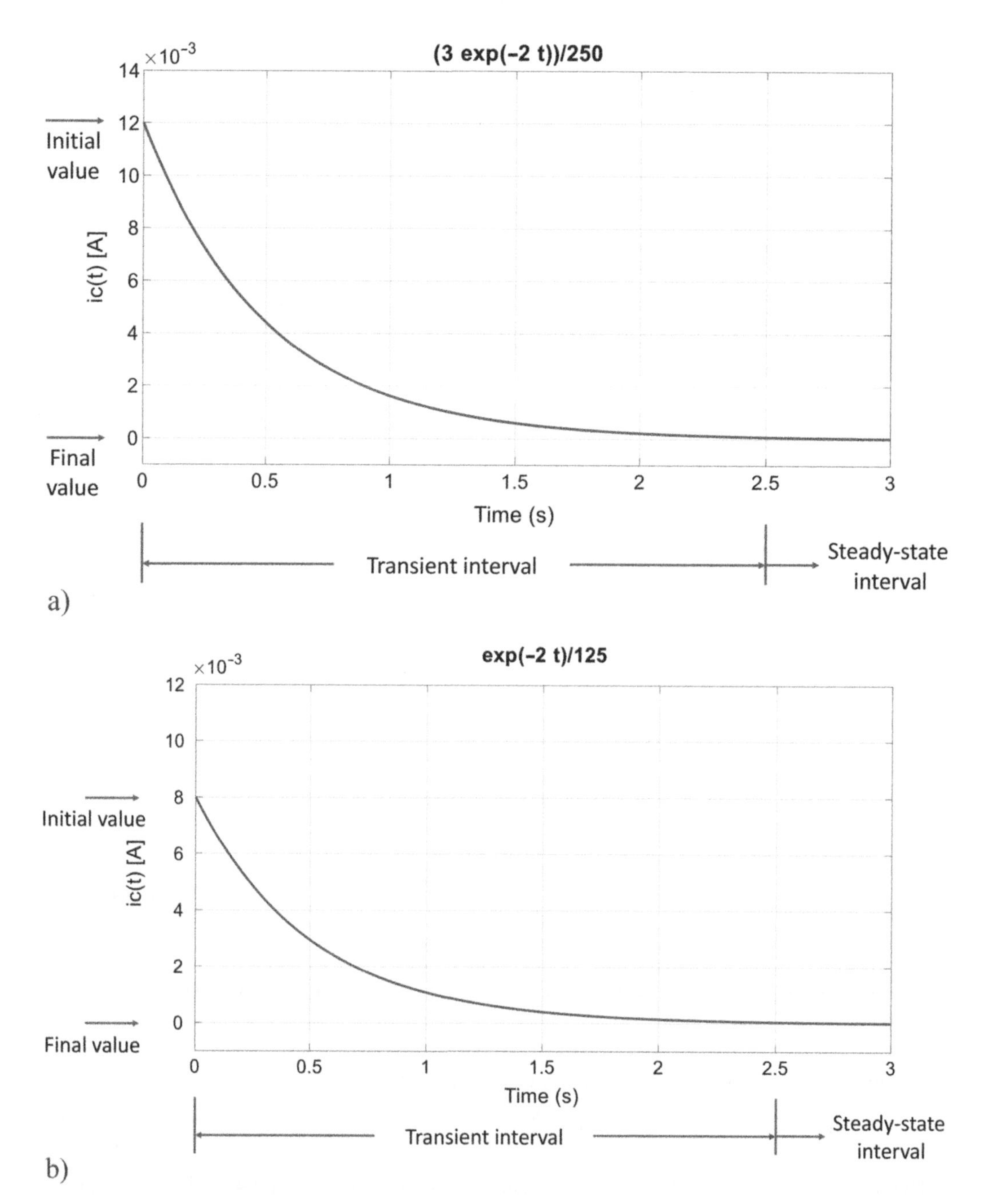

Figure 4.6 The current step response of a series RC circuit: a) $V_0 = 0$; b) $V_0 = 4V$.

Questions

1) Current flowing through a series RC circuit must be the same for R and C. Does Equation 4.18a shows this?
2) Why do step responses of a series RC circuit behave so differently for voltage and current? Namely, whereas $v_{out-cmp}(t)$ increases from V_0 to V_{in}, but $i_c(t)$ declines from $i_{in}(0)$ to zero?

Sidebar 4S.2 How to Measure the Duration of a Transient Interval

Examine Figure 4S2.1: This is the graph of the exponentially decaying function, $f(t) = V_0 e^{-\alpha t} = 12e^{-2t}$. The function's initial value at $t = 0$ is unambiguously determined as $V_0 = f(0)$, which is $V_0 = 12(V)$ in our example. But how can measure its final value, mathematically defined as the function value when $t \to \infty$? Indeed, when t goes to infinity, a decaying exponential function $e^{-\alpha t}$ tends to but never reaches zero, whereas we need a finite value to be measurable! In such a case, a convention always comes to the rescue in engineering practice. There are two types of convention here: The first is based on defining the duration of a transient interval, and the second relies on assuming the final amplitude.

For the circuits containing R, C, and L combinations, it's possible to determine the circuit's time constant as $\tau(s) = RC$ or $\tau(s) = \frac{L}{R}$. Consequently, the transient interval's duration measured in $\tau(s)$ is a natural choice for these circuits. Today, 5τ criterion is considered a gold standard by the industry. The meaning of this criterion is visualized in Figure 4S2.1. If we set $t_f = 5\tau$, where t_f being the assigned final time, then

$$V_f = V_0 e^{-\frac{t_f}{\tau}} = V_0 e^{-\frac{5\tau}{\tau}} = V_0 e^{-5} = 0.0067V_0. \tag{4S2.1}$$

Thus, *the assigned value of the transient interval* $t_f = 5\tau$ *determines the final value of the output's amplitude*. In our example, $\tau = 0.5(s)$, $5\tau = 2.5(s)$, and $V_f = 0.0067V_0 = 0.081(V)$. All these values and their relationships are shown in Figure 4S2.1.

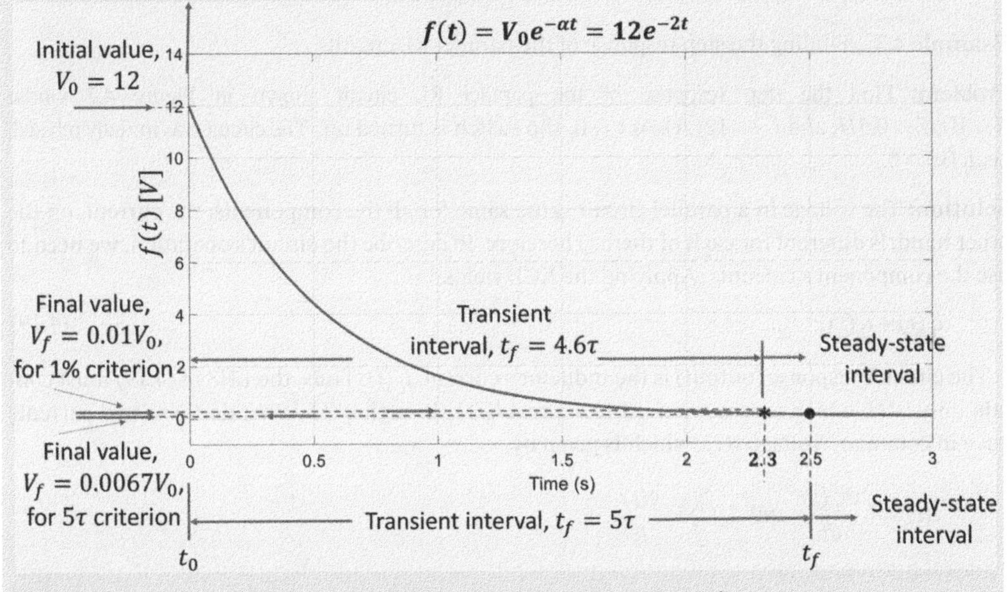

Figure 4S2.1 Two criteria for setting the final values of a transient interval.

The other, more general criterion sets the final amplitude value as 1% of the initial one. In this case, the final amplitude is given by

$$V_f = 0.01V_0 = 0.12(V), \tag{4S2.2}$$

where the numerical value is taken from our example. It's not difficult to find the final time for this criterion. Indeed,

$$V_f = V_0 e^{-\frac{t_f}{\tau}} \Rightarrow 0.01 = e^{-\frac{t_f}{\tau}} \Rightarrow t_f = 4.6\tau. \tag{4S2.3}$$

In our example, 1% criterion gives $t_f = 2.3(s)$. In general, we can set any value of the final amplitude and compute the final time as

$$V_f = V_0 e^{-\frac{t_f}{\tau}} \Rightarrow t_f = -\tau \ln\left(\frac{V_f}{V_0}\right). \tag{4S2.4}$$

What value of any criterion to choose? It depends on the nature of your application. If you were doing an academic laboratory exercise, may be 3τ or 3% work well, but if your job is to perform high-precision scientific measurements, then even 11τ or 0.0001% might not be enough. In any event, always specify the chosen criterion while describing your measurements or calculations.

Know that the above consideration is fully applicable to a growing exponential function. As a practice, determine the final values t_f and V_f for the step response of a series RC circuit in Example 4.2. (See Figures 4.5a and 4.5b.)

Our discussion is based on the properties of an exponential function. This is a fascinating topic, which, unhappily, lies outside of the mainstream of this book. Refer to engineering and mathematics manuscripts.

Let's consider the other example of finding response of a first-order circuit.

Example 4.3 Finding the step response of the parallel RL circuit.

Problem: Find the step response of the parallel RL circuit shown in Figure 4.7 whose $R = 1\Omega$, $L = 0.5H$, and $I_{in} = 12(A)$. At $t = 0$, the switch is turned on. The circuit was initially relaxed, i.e., $i_L(0) = 0$.

Solution: The voltage in a parallel circuit is the same for all the components; the current, on the other hand, is different for each of them. Therefore, to describe the circuit's operation, we need to use the component's currents. Applying the KCL yields

$$i_R(t) + i_L(t) = I_{in}. \tag{4.19}$$

The circuit's response (output) is the inductor's current, $i_L(t)$. Thus, the LHS of (4.19) must contain only $i_L(t)$, which means we need to express $i_R(t)$ through $i_L(t)$. What do these two currents have in common? Voltage $v(t)$, which is given by

$$v(t) = L\frac{di_L(t)}{dt} \text{ and } i_R(t) = \frac{v(t)}{R}.$$

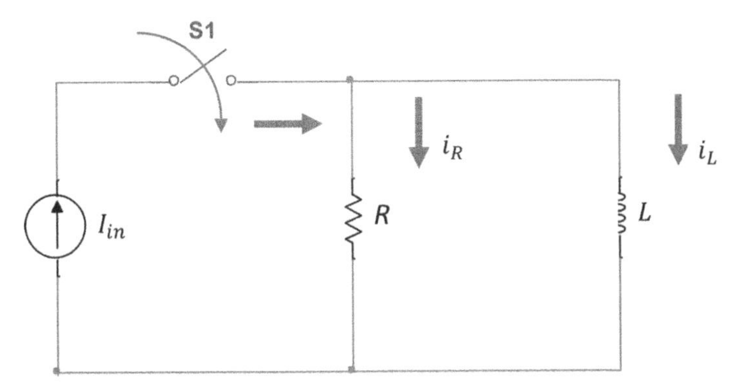

Figure 4.7 Parallel RL circuit for Example 4.3.

Therefore, (4.19) can be expressed as

$$\frac{L}{R}\frac{di_L(t)}{dt}+i_L(t)=I_{in}. \tag{4.20a}$$

or

$$\frac{L}{R}i'_L+i_L(t)=I_{in}. \tag{4.20b}$$

This differential equation describes the operation of a parallel RL circuit in terms of its response, $i_L(t)$.

Introducing the *time constant* $\tau(s)$ and *damping coefficient* $\alpha\left(\frac{1}{s}\right)$ of a parallel RL circuit,

$$\tau(s)=\frac{L}{R} \text{ and } \alpha\left(\frac{1}{s}\right)=\frac{1}{\tau}=\frac{R}{L}, \tag{4.21}$$

we can rewrite (4.20) as

$$i'_L+\alpha i_L(t)=\alpha I_{in}. \tag{4.22}$$

How can we solve this equation? Revisit (4.10), $\frac{dv_{out}(t)}{dt}+\alpha v_{out}(t)=\alpha v_{in}(t)$, to see how we solved this equation in Example 4.2. To remind, we considered two methods: the separation of variables and trial solution. Let's use the latter one for solving (4.22).

The complete solution is a sum of two solutions: a homogeneous, when $I_{in}=0$, and a particular, when the RHS of (4.22) is the input signal. Assuming the homogeneous solution in the standard form,

$$i_{Lh}(t)=Ae^{mt},$$

we plug it into the homogeneous version of (4.22). The result is

$$\frac{dAe^{mt}}{dt}+\alpha Ae^{mt}=0 \;\Rightarrow\; Ae^{mt}(m+\alpha)=0 \;\Rightarrow\; m=-\alpha.$$

Thus,

$$i_{Lh}(t)=Ae^{-\alpha t}.$$

Constant A will be determined based on the initial condition. To remind, $i_{Lh}(t)$ is a *natural response* of a parallel RL circuit.

The particular solution carries the form of the RHS of (4.22), which is a constant B. Plugging $i_{Lp}(t) = B$ into (4.22) yields

$$B = I_{in}.$$

You will recall that $i_{Lp}(t)$ is called a *forced response*, for it describes how the circuit responds to an applied input signal.

Therefore, the complete solution to (4.22) takes the form

$$i_L(t) = Ae^{-\alpha t} + I_{in}. \tag{4.23}$$

At $t = 0$, by the problem's condition, $i_L(0) = 0$. Thus, (4.23) gives

$$i_L(0) = A + I_{in} \;\Rightarrow\; A = -I_{in}$$

Finally, the complete solution to Equation 4.22 is

$$i_L(t) = I_{in}\left(1 - e^{-\alpha t}\right). \tag{4.24}$$

Computation with given numbers produces

$$i_L(t) = 12\left(1 - e^{-2t}\right)(A).$$

(Know that the given component values are not realistic but serve well for illustrative purpose.)

The initial value of (4.24) is zero, and the final value is 12 mA, as expected.

The problem is solved.

Discussion:

- Let's revisit Example 4.1. Analyze the derivation of the differential equation for series RC circuit and notice the close similarity between that reasoning for derivations of (4.7a) and (4.20). If we were to replace in Example 4.1 words "series circuit" by "parallel circuit," interchange words "voltage" and "current," and notations C and L, we'd arrive to (4.20). This interchangeability is called the *principle of duality* discussed in Section 3.3. Accurately using this principle can save time for composing differential equations. Since the differential equations for the different circuits have the same forms, their solution methods are the same.
- What if an inductor has an initial flux resulting in the current $i_L(0) = I_0$? In other words, what if the initial condition is non-zero? Equations 4.16 and 4.17 show how to approach this situation. Taking the complete solution (4.23), $i_L(t) = Ae^{-2\alpha t} + I_{in}$, and plugging in the new initial value of $i_L(t)$, we find new value of constant A

  $$i_L(0) \equiv I_0 = Ae^{-\alpha 0} + I_{in} \;\Rightarrow\; A = I_0 - I_{in}$$

  Therefore, Equation 4.24 becomes

  $$i_L(t) = \left(I_0 - I_{in}\right)e^{-\alpha t} + I_{in} = \; I_0 e^{-\alpha t} + I_{in}\left(1 - e^{-\alpha t}\right). \tag{4.25}$$

  Assume $I_0 = 4mA$ and compute

$$i_L(t) = I_0 e^{-\alpha t} + I_{in}\left(1 - e^{-\alpha t}\right) = 4e^{-2t} + 12\left(1 - e^{-2t}\right).$$

This solution highlights that the complete response is a sum of a ***natural response***, $I_0 e^{-2\alpha t}$, caused only by the circuit's internal source, I_0, and the *forced response*, $I_{in}\left(1 - e^{-\alpha t}\right)$, brought about by the external input, I_{in}. Compare (4.25) with (4.17) and see the duality of RC and RL circuits descriptions again.

- The parallel RL circuit's response can be readily presented in terms of voltage.

Just use the formula, $v_L(t) = L\frac{di_L(t)}{dt}$, and calculate

$$v_L(t) = L\frac{di_L(t)}{dt} = \alpha L\left(I_{in} - I_0\right)e^{-\alpha t}. \tag{4.26}$$

In numbers,

$$v_L(t) = \alpha L\left(I_{in} - I_0\right)e^{-\alpha t} = 8e^{-2t}$$

The initial value of $v_L(t)$ is $v_L(0) = \alpha L\left(I_{in} - I_0\right) = 8(V)$, and its final value is zero, which satisfies the problem conditions. Compare this equation with (4.18).

- The output current and voltage of the parallel RL circuit at non-zero initial condition are shown in Figures 4.8a and 4.8b. Compare them with Figures 4.6b and 4.7b and explain why all these figures are different.

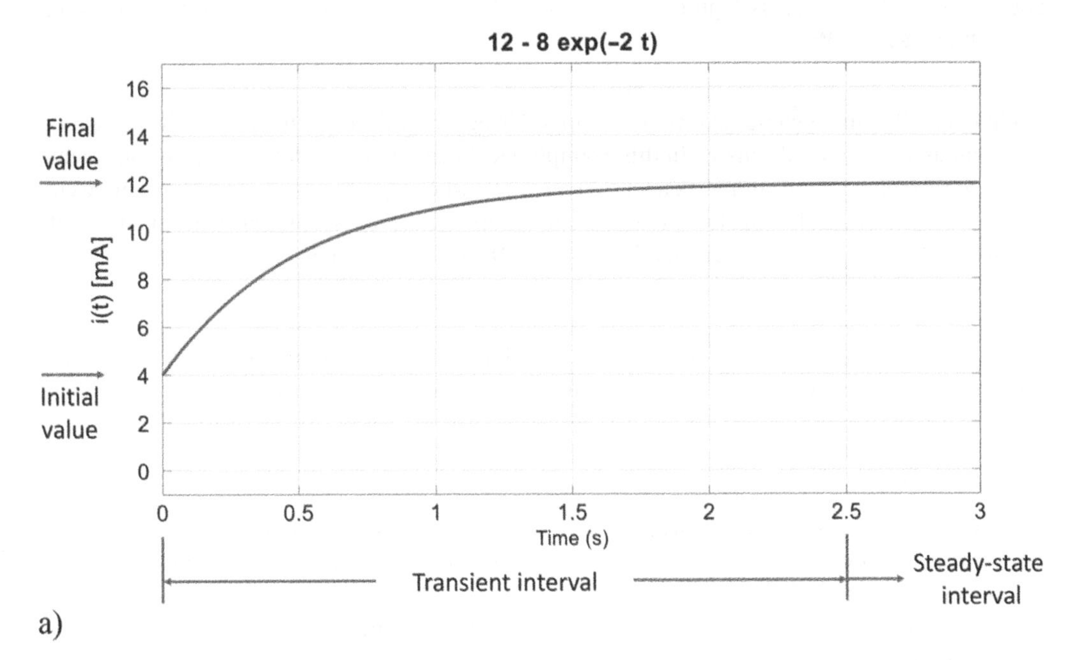

Figure 4.8 The step response of the parallel RL circuit in Example 4.3: a) Current response; b) voltage response.

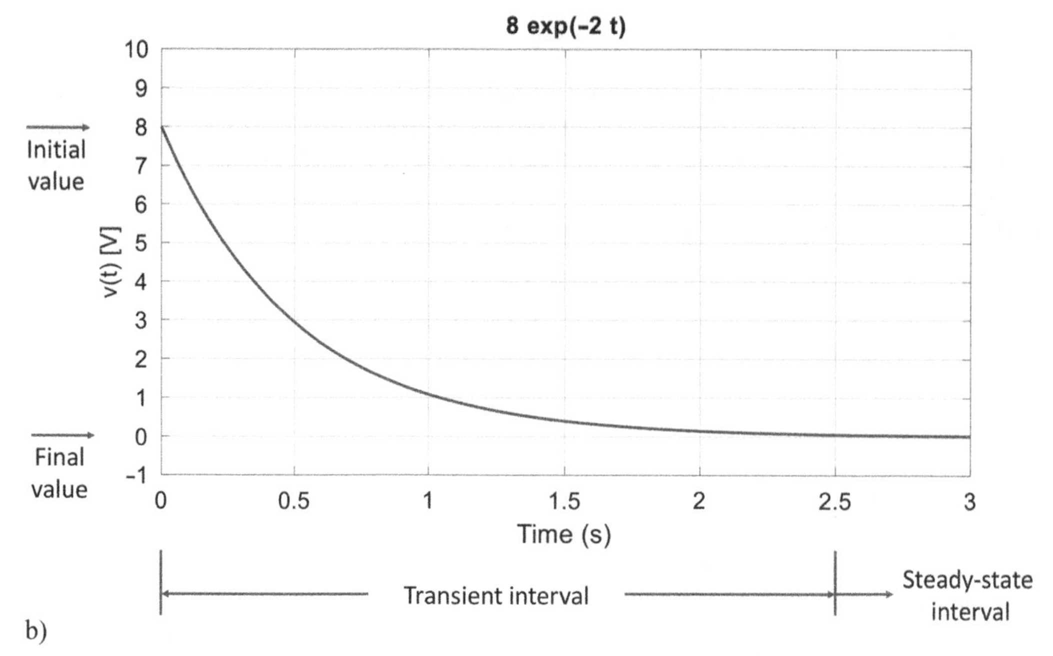

Figure 4.8 **(Cont'd)**

Question The voltage response of a series RC circuit exponentially grows from the initial value to the final, whereas the current response of this circuit exponentially diminishes, as Figures 4.5b and 4.6b show. Nevertheless, Figures 4.8a and 4.8b show the opposite results—current increases and voltage lessens. Why?

- What if an RL circuit contains several resistors? Using Norton's theorem, we can always reduce the circuit to the form discussed in this example. (Refer to Section 2.3 for refreshing your memory on this topic.) For example, consider Figure 4.9a, where I_{in}, R_1, and R_2 form the source circuit and the inductor L is the load. When the given current source is replaced by the open circuit, resistors R_1 *and* R_2, are in series, and therefore the Norton resistance is

$$R_N = R_1 + R_2.$$

Also, the short-circuit current for the circuit to the left of $A - B$ interface is the current source for the Norton equivalent circuit, I_N. It is calculated by the current-division rule as

$$I_N = \frac{R_2}{R_1 + R_2} I_{in}.$$

Thus, we obtain the *Norton equivalent circuit* shown in Figure 4.9b. (See Sections 2.2 and 3.3 to refresh your memory on Norton's theorem.) Now, compare Figure 4.7 and 4.9b and see that the circuits in both figures are identical. Thus, all the results of Example 4.3 are applied to the Norton equivalent circuit. Be aware of the time constant and the damping coefficient: For the circuit in Figure 4.8b, they are

$$\tau_N(s) = \frac{L}{R_N} \text{ and } \alpha_N\left(\frac{1}{s}\right) = \frac{R_N}{L}.$$

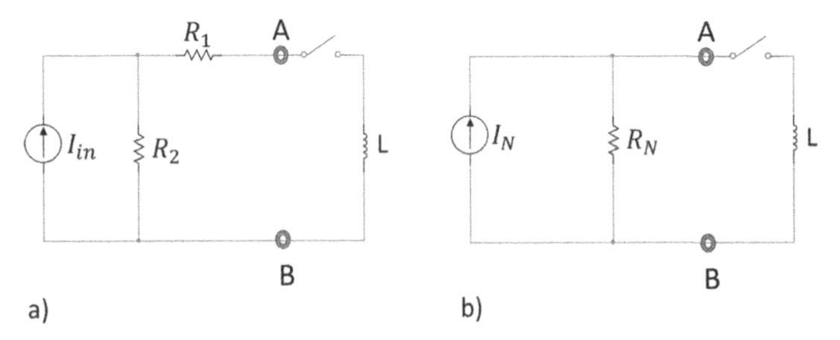

Figure 4.9 Norton resistance in a parallel RL circuit: a) Original circuit; b) equivalent circuit.

Using (4.26), we can write the step response of the circuit shown in Figure 4.9a as

$$i_L(t) = I_0 e^{-\alpha_N t} + I_N\left(1 - e^{-\alpha_N t}\right). \tag{4.27}$$

This brief diversion exemplifies the role of *Norton's theorem* in a parallel circuit.

If you need to solve the problem with a parallel circuit but have to use a voltage source, refer to Section 1.3, where the *source transformation* is discussed.

4.2 Complete First-Order Circuit's Responses and Initial and Final Conditions in Time Domain

In Section 4.1, the basics of the meaning, origin, and applications of differential equations were introduced. Also, that section demonstrates how to use differential equations for solving real-life circuit-analysis problems. This section will discuss additional and essential details of the time-domain circuit analysis. For the readers interested in a deeper look at the natural response of first-order, Appendix 4A offers an additional discussion of this topic.

4.2.1 Search for a Sinusoidal Response of a Series RC Circuit

We've learned in this chapter how a first-order circuit will response to its initial condition and a step input. The third most ubiquitous input is the sinusoidal, which we study in this subsection.

Consider the circuit shown in Figure 4.10a. When the switch is open, the circuit might have the voltage V_0 accumulated by the capacitor. When the switch turns on, the input sinusoidal signal drives the circuit. Our task is to describe the circuit operation mathematically.

We modify a series RC circuit's differential equation given in (4.10) as

$$v'_{out}(t) + \alpha v_{out}(t) = \alpha V_{in} cos(\omega t) \tag{4.28}$$

and remember that the solution of this equation consists of two parts

$$v_{out-cmp}(t) = v_{out-h}(t) + v_{out-p}(t). \tag{4.11R}$$

The homogeneous solution is the circuit's response to its initial condition, which we obtained in Section 4.1 as

$$v_{out-h}(t) = Ae^{-\alpha t}, \tag{4.13R}$$

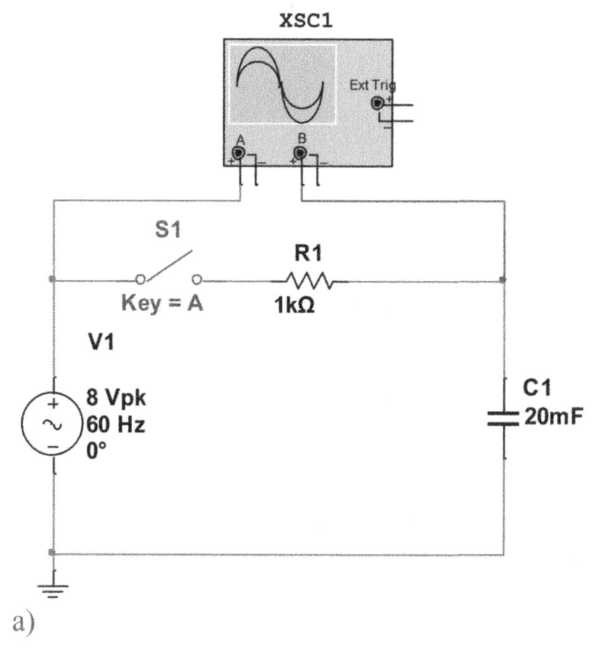

a)

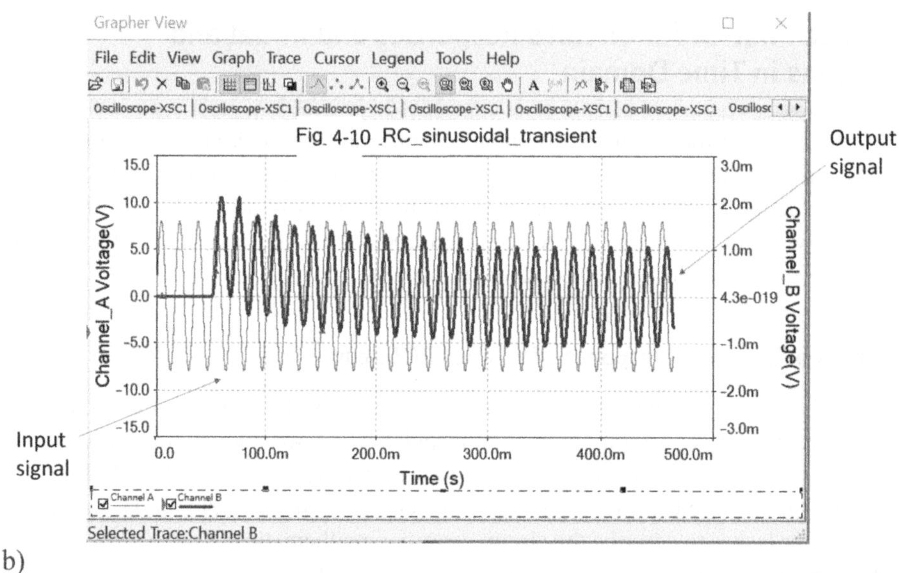

b)

Figure 4.10 Sinusoidal transient response in a series RC circuit: a) Circuit's schematic; b) Multisim simulation of the circuit operation.

where $\alpha = \frac{1}{\tau} = \frac{1}{RC}$ and A is a constant to be found. Thus, (4.13) determines the form of a *natural* (*source-free* or *zero-input*) response of the given circuit.

The *particular solution*, $v_{out-p}(t)$, defines the *forced response*, and it always takes the form of the input signal. Therefore, in this problem, we have

$$v_{out-p}(t) = F cos(\omega t + \theta), \tag{4.29}$$

where F and θ are parameters to be found. The procedure of finding these parameters is presented in Sidebar 4S.3. The results of that manipulations are:

$$F = \frac{V_{in}}{\sqrt{1+\left(\frac{\omega}{\alpha}\right)^2}}, \tag{4.30a}$$

$$\theta = -tan^{-1}\left(\frac{\omega}{\alpha}\right). \tag{4.30b}$$

*Therefore, (4.29) gives us **the forced response to a sinusoidal signal** provided that F and θ are determined by (4.30a) and (4.30b).*

Sidebar 4S.3 Forced Response of a Series RC Circuit to a Sinusoidal Signal: Derivation of Formulas for the Amplitude and Phase Shift of the Output Signal

Consider (4.28),

$$v'_{out}(t) + \alpha v_{out}(t) = \alpha V_{in} cos(\omega t). \tag{4.28R}$$

If (4.29),

$$v_{out-p}(t) = Fcos(\omega t + \theta), \tag{4.29R}$$

is the solution to this equation, then this $v_{out-p}(t)$ must turn the differential equation into the identity. Based on this reasoning, we find F and θ by plugging (4.29) into (4.28), which results in

$$-F\omega sin(\omega t + \theta) + \alpha F cos(\omega t + \theta) = \alpha V_{in} cos(\omega t). \tag{4S3.1}$$

The trigonometric identities for the sine and cosine of the sum of two angles are

$$sin(\omega t + \theta) = sin(\omega t)cos(\theta) + cos(\omega t)sin(\theta)$$

$$cos(\omega t + \theta) = cos(\omega t)cos(\theta) - sin(\omega t)sin(\theta)$$

Plugging these expressions into (4S3.1) and equating the coefficients of the sin(ωt) and cos(ωt) of the LHS and RHS yields

$$-F(\omega cos\theta + \alpha sin\theta)sin(\omega t) = 0$$

$$F(-\omega sin\theta + \alpha cos\theta)cos(\omega t) = \alpha V_{in} cos(\omega t)$$

The first equation results in

$$tan\theta = -\frac{\omega}{\alpha}, \tag{4S3.2}$$

as neither F nor $sin(\omega t)$ can be constantly equal to zero.

The second equation comes down to the configuration:

$$F(-\omega sin\theta + \alpha cos\theta) = \alpha V_{in}$$

We perform the set of the following straightforward manipulations

$$F\left(-\frac{\omega}{\alpha} sin\theta + cos\theta\right) = V_{in} \Rightarrow F(tan\theta sin\theta + cos\theta) = V_{in}$$

$$\Rightarrow F(sin^2\theta + cos^2\theta) = V_{in} cos\theta$$

$$\Rightarrow F = V_{in} cos\theta.$$

But we know from trigonometry that $cos\theta = \frac{1}{\sqrt{1+tan^2\theta}}$, which in our case means $cos\theta = \frac{1}{\sqrt{1+\left(\frac{\omega}{\alpha}\right)^2}}$.

Thus, we arrive at the following expression for F:

$$F = \frac{V_{in}}{\sqrt{1+\left(\frac{\omega}{\alpha}\right)^2}} \tag{4S3.3}$$

Finally, the particular solution of the nonhomogeneous differential equation of a series RC circuit is attained by plugging (4S4.2) and (4S4.3) in (4.29), namely

$$v_{out-p}(t) = \frac{V_{in}}{\sqrt{1+\left(\frac{\omega}{\alpha}\right)^2}} cos\left(\omega t - tan^{-1}\frac{\omega}{\alpha}\right). \tag{4S3.4}$$

Therefore, the forced sinusoidal response of the circuit shown in Figure 4.12 is indeed given by (4.29), (4.30a), and (4.30b).

A curious reader is invited to plug (4S3.4) in (4.28) to confirm that (4S3.4) is a true solution to (4.28).

4.2.2 The Complete (Total) Response of a Series RC Circuit to a Sinusoidal Input

The general form of the complete response of a series RC circuit is given by (4.11) as

$$v_{out-cmp}(t) = v_{out-h}(t) + v_{out-p}(t). \tag{4.11R}$$

Having obtained (4.29), (4.30a), and (4.30b) and taking into account (4.13), we can write the complete response as

$$v_{out-cmp}(t) = Ae^{-\alpha t} + Fcos(\omega t + \theta), \tag{4.31}$$

where F and θ are given in (4.30a) and (4.30b). To find the only unknown member, constant A, we usually refer to the initial condition, but *this time we must use the complete solution, Equation 4.31,* as follows: Denote

$$v_{out-cmp}(0) = V_0,$$

plug $t = 0$ into (4.31), and obtain

$$V_0 = A + Fcos(\theta).$$

This equation yields

$$A = V_0 - Fcos(\theta). \tag{4.32}$$

Thus, the *complete solution* to the series RC circuit's differential equation with sinusoidal input takes the form

$$\boldsymbol{v}_{out-cmp}(\boldsymbol{t}) = \left(\boldsymbol{V}_0\boldsymbol{e}^{-\alpha t} - \left(\boldsymbol{Fcos\theta}\right)\boldsymbol{e}^{-\alpha t}\right) + \boldsymbol{Fcos}\left(\omega \boldsymbol{t} + \boldsymbol{\theta}\right), \tag{4.33}$$

where $F(V) = \frac{V_{in}}{\sqrt{1+\left(\frac{\omega}{\alpha}\right)^2}}$ and $\theta\left(degree\ or\ radian\right) = -tan^{-1}\left(\frac{\omega}{\alpha}\right)$.

Observe that the complete response of a series RC circuit to a sinusoidal input (4.47) includes three members:

- *Natural, source-free response, $V_0e^{-\alpha t}$, which is due to the voltage accumulated by a capacitor before the input is applied (initial condition).*
- *Member $Fe^{-\alpha t}cos(\theta)$, a part of the natural response, which is excited by the applied input sinusoidal signal in the RC circuit.*
- *Forced response $Fcos(\omega t + \theta)$, which induces the permanent oscillations with the input signal frequency but with different amplitude and a phase shift.*

The first two members diminish with time; the forced response lasts as long as the input is applied. For more analysis of circuit's responses see the Summary section that follows. Figure 4.10b shows the input and output signals obtained by the Multisim simulation. We can readily identify the natural (diminishing) and steady-state parts of the circuit's response. Pay attention to the amplitude scales of both graphs: The input amplitude is $8V$, whereas the output amplitude is about $1.2mV$.

Here is a good ***exercise*** for you: Take the component and input values from the circuit in Figure 4.10a, do all necessary calculations, and plug the computed numbers into (4.33) to attain the complete sinusoidal response of this circuit. Then, use MATLAB to build the graphs of the input and output signals, and quantitatively compare your calculated answer with the results of Multisim simulation given in Figure 4.10b. (More on measurements in Multisim graphs can be found in Example 4.4 that follows immediately.)

Now, let's consider the example.

Example 4.4 Sinusoidal response of a series RL circuit.

Problem: Find the complete response of the series RL circuit shown in Figure 4.11, where the circuit's schematic and the component values are presented. The circuit's initial condition is $V_0 = 0V$.

Solution:

We follow the methodology demonstrated for a series RC circuit. Observing that $v_{out} \equiv v_R$, applying KVL, and expressing v_L through v_R as $v_L = L\frac{di}{dt} = \frac{L}{R}\frac{dv_R}{dt}$ enable us to write the series RL circuit's differential equation as

$$\frac{L}{R}\frac{dv_R}{dt} + v_R = v_{in}. \tag{4.48}$$

Recalling from (4.21) the following designations

$$\tau(s) = \frac{L}{R}\ and\ \alpha\left(\frac{1}{s}\right) = \frac{1}{\tau} = \frac{R}{L}, \tag{4.21R}$$

we rewrite (4.48) in a traditional form:

$$v'_{out}(t) + \alpha v_{out}(t) = \alpha V_{in} cos(\omega t). \quad (4.49)$$

The complete solution to this linear equation is given by

$$v_{out-cmp}(t) = v_{out-h}(t) + v_{out-p}(t). \quad (4.11R)$$

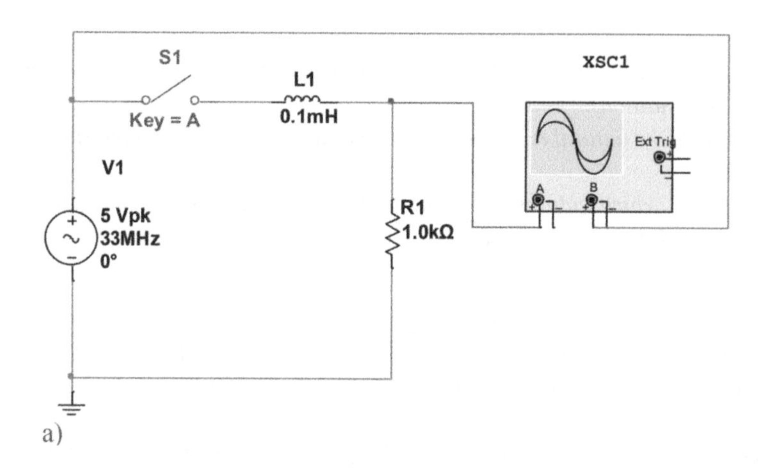

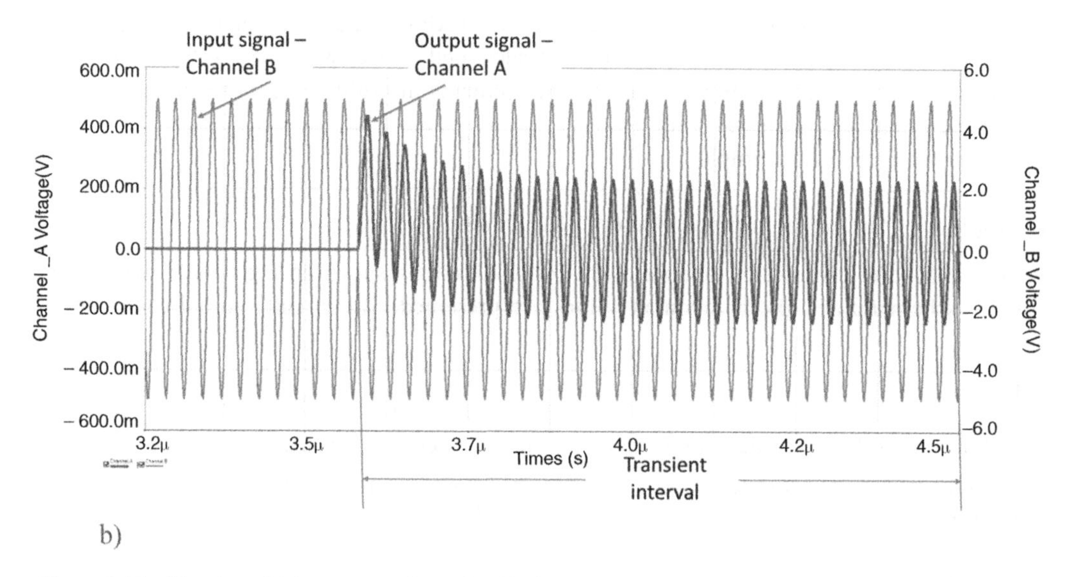

Figure 4.11 The sinusoidal response of a series RL circuit in Example 4.4: a) Circuit's schematic; b) general view at the graphs produced by Multisim simulation; c) detailed view at input and output (response) signals; d) cursors' reading in an enlarged scale.

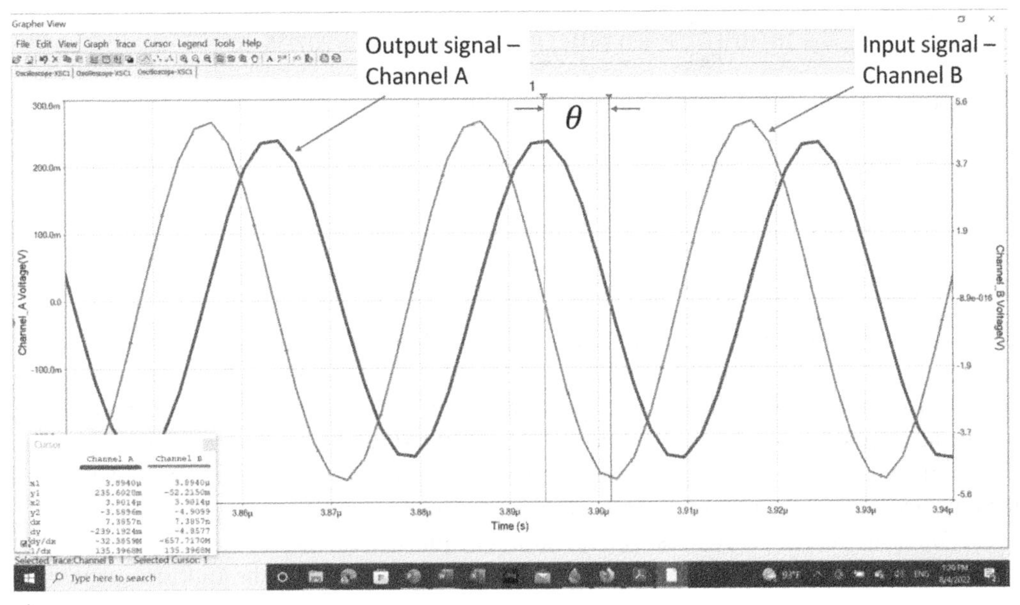

c)

Cursor

	Channel A	Channel B
x1	3.8940µ	3.8940µ
y1	235.6028m	-52.2150m
x2	3.9014µ	3.9014µ
y2	-3.5896m	-4.9099
dx	7.3857n	7.3857n
dy	-239.1924m	-4.8577
dy/dx	-32.3859M	-657.7170M
1/dx	135.3968M	135.3968M

Selected Trace:Channel B | Selected Cursor: 1

d)

Figure 4.11 (Cont'd)

When we repeat all the operation done in Equations from 4.42 through 4.47, we arrive at the familiar result:

$$v_{out-cmp}(t) = \left(V_0 e^{-\alpha t} - \left(F cos\theta\right) e^{-2\alpha t}\right) + F cos\left(\omega t + \theta\right), \tag{4.47R}$$

where $F(V) = \dfrac{V_{in}}{\sqrt{1+\left(\dfrac{\omega}{2\alpha}\right)^2}}$, $\theta(degree) = -tan^{-1}\left(\dfrac{\omega}{2\alpha}\right)$, and $\alpha\left(\dfrac{1}{s}\right) = \dfrac{1}{\tau(s)} = \dfrac{R}{L}$. Taking the given component and input values enables us to compute

$V_0 = 0(V)$, $\alpha = 10 \cdot 10^6 \left(\frac{1}{s}\right)$, $F = 0.24(V)$, $\theta = 87.24^0$, and $\omega = 207.35 \cdot 10^6 \left(\frac{rad}{s}\right)$.

Plugging these computed numbers into (4.47) results in

$$v_{out-cmp}(t) = -0.011e^{-\left(10 \cdot 10^6 t\right)} + 0.24\cos\left(207.35 \cdot 10^6 t + 87.24^0\right).$$

The problem is solved.

Discussion:

To verify our solution, we run the *Multisim simulation* of the circuit shown in Figure 4.11a. The results are presented in Figures 4.11b through 4.11d. Figure 4.11b delivers the general presentation of the input and output (response) signals and shows the *transient interval* within which the output signal passes from the initially excited stage to the steady-state process. Note that even if the circuit did not store initial energy, the response signal starts from a not-zero position because the input excites the circuit's natural response. See the analysis of Equation 4.47.

The *Multisim simulation* enables us to perform quantitative analysis of the results by *making measurements* on the graphs. Here is how to make the measurements on the Multisim graphs:

- Transfer the graphs from the oscilloscope screen to the Grapher by clicking the *Grapher* icon in the Multisim menu.
- In Grapher's menu, click *Cursor*. In the popup menu choose Cursor 1 and/or Cursor 2. Click *Show cursor* and the other popup menu (at the bottom of the screen) with the measurement options will show up.
- Choose a cursor you want.
- Move the cursor to the desired point and see the value in the bottom popup menu.

Here is how we implement the above procedure in our example to measure the phase shift between two sinusoids of the same frequency:

- As Figure 4.11b shows, we choose one cursor to measure Channel A values, the other one does the same for Channel B values. We place cursors at zero points of their sinusoids, as Figure 4.11c demonstrates, to obtain the most accurate result.
- Then we click *Show Cursors* and see the measurements in the other popup menu, which is displayed in Figure 4.11d.
- Next, we measure time interval $\Delta t(s)$, which is the "distance" between points $x1 = 3.8940\,\mu s$ (output signal) and $x2 = 3.9014\,\mu s$ (input signal) in either channel. The difference is $\Delta x \equiv \Delta t = 0.0074\,\mu s$.
- Then we divide $\Delta t(s)$ by period $T(s)$ to learn what portion of the period this interval occupies. The period can be measured or computed. Here, $T = 0.0303\,\mu s$ computed or $T = 0.0304\,\mu s$ measured.
- Next, the quotient $\frac{\Delta t}{T} = 0.2434$ is multiplied by 2π or 360^0 to translate the time shift into angle shift. In our example, the calculations give $0.2434 \cdot 360^0 = 87.63^0$. *This is the measured phase shift*. Note how close this value is to the computed one, $\theta = 87.63^0$. We receive the solid validation of our answer in Example 4.4.

By analyzing the data presented in Figure 4.13d, a curious reader can obtain additional useful information regarding the amplitudes of the signals. However, it's advisable to make the amplitude measurements separately by placing cursors at the peaks or zeroes of both sinusoids.

Table 4.2 The responses of the first-order circuits to the three inputs.

Input signal	Output signal	Graph
Zero input, $v_{in}(t) = 0$	$v_{out}(t) = V_0 e^{-\alpha t}$ Natural (source-free) response	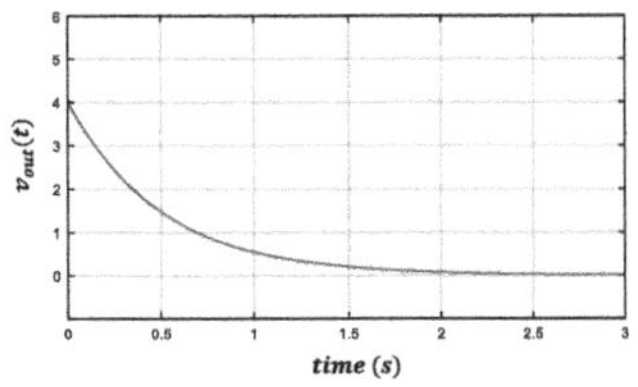
Step input, $v_{in}(t) = u(t)$	$v_{out}(t) = V_0\left(1 - e^{-\alpha t}\right)$ Step response	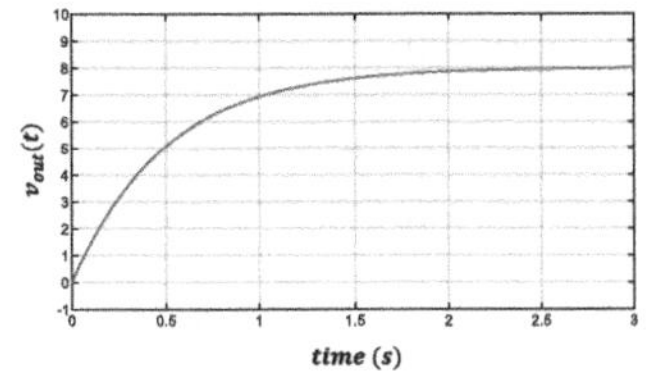
Sinusoidal input, $v_{in}(t) = V_{in}\cos(\omega t)$	$v_{out}(t) = V_0 e^{-\alpha t} - (F\cos\theta)e^{-\alpha t}$ $+ F cos(\omega t + \theta)$, where $F(V) = \frac{V_{in}}{\sqrt{1 + \left(\frac{\omega}{2\alpha}\right)^2}}$, $\theta = tan^{-1}\left(\frac{\omega}{2\alpha}\right)$.	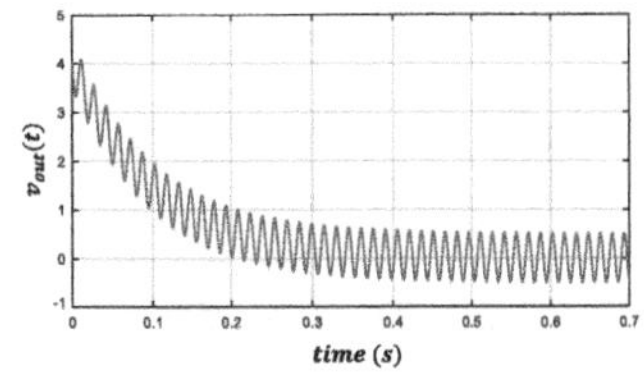

Bear in mind that the measurements on Multisim graphs always deliver approximate values. Nevertheless, as this example shows, these measurements can firmly verify the theoretical results.

We notice a similarity between the responses of RC and RL circuits to the same inputs, which is caused by the commonality of their differential equations. Consequently, their responses to the same inputs are practically identical. Table 4.2 digests the responses of these first-order circuits to the three inputs—zero, step, and sinusoidal.

4.2.3 Power and Energy in the First-Order Reactive Circuits

How can we compute power and energy in the first-order RC and RL circuits? Several examples of such calculations were given in this chapter; here, we offer a summary of this topic. To remind you, the general formula for *instantaneous power* is

$$p(t)\left[W\right] = v(t)i(t), \tag{4.50}$$

where $+p(t)$ means power absorbed into a component (dissipated) and $-p(t)$ means supplied by the element (produced, delivered). *Instantaneous energy* is computed as

$$E(t)[J]=\int_{t_0}^{t_f} p(t)dt=\int_{t_0}^{t_f} v(t)i(t)dt, \tag{4.51a}$$

or

$$E(t)[J]=\int v(t)i(t)dt+E_0. \tag{4.51b}$$

Here, t_0 and t_f are the initial and final moments of the process, respectively, and E_0 is the initial energy stored in the circuit element before its operation starts.

Watt, W, is the unit of power, joule, J, is the energy unit, and $W=\frac{J}{s}$. In words, one watt of power is the amount of energy consumed (dissipated) by a circuit element per second.

There are four following particular cases for calculations of power and energy:

1) For calculations of power and energy on a *single capacitor or inductor*, refer to Table 4.1. Specifically, power produced by a capacitor is given by

$$p_C(t)[W]=v_C(t)i_C(t)=C\frac{dv_C(t)}{dt}v_C(t)=\frac{d}{dt}\left[\frac{1}{2}Cv_C^2(t)\right]. \tag{4.52}$$

Energy stored by a capacitor can be calculated applying 4.51b, justifiably letting $W_0=0$, and using 4.52; thus,

$$E_C(t)[J]=\int v(t)i(t)dt=\frac{1}{2}Cv_C^2(t). \tag{4.53}$$

Create an example for yourself: Set a $v_C(t)$ waveform (for example, $v_C(t)=V_0e^{-bt}$), plug it into (4.52) and (4.53), derive the specific formulas, insert numbers by your choice, calculate the answer, and build the graphs for all four parameters, $v_C(t)$, $i_C(t)$, $p_C(t)$, and $E_C(t)$.

You, our thoughtful reader, can readily write similar formulas for an inductor.

2) Next case is the natural response of a series RC, considered in Example 4.2 and particularly in Appendix 4A. We can find its power and energy using $v_C(t)$ and $i_C(t)$ from (4A.3) and (4A.4), namely,

$$v_c(t)=V_0e^{-\alpha t}(V) \tag{4A.3R}$$

and

$$i_c(t)=C\frac{dv_{out}(t)}{dt}=-\frac{V_0}{R}e^{-\alpha t}(A) \tag{4A.4R}$$

Then

$$p_C(t)=v_C(t)i_C(t)=-\frac{1}{R}V_0^2e^{-2\alpha t}(W). \tag{4.54}$$

The negative sign in (4.54) indicates that the capacitor supplies (delivers) power to the circuit, which is evident for a source-free RC circuit. The graph of $p_C(t)$ depicts that the capacitor starts delivering its power from $p_C(0)=-\frac{1}{R}V_0^2$ until its stored power vanishes completely. If you were to find power consumed by the circuit's resistor, you will see that it mirrors $p_C(t)$ with the positive sign, as it must be. (Do you know why?)

Equation 4.51a enables us to calculate energy delivered by a capacitor as

$$E_C(t)[J] = \int_{t_f}^{t_0} v_C(t) i_C(t) dt = -\frac{V_0^2}{R} \int_{t_0}^{t_0} e^{-2\alpha t} dt = \frac{1}{2C}\left(\frac{V_0}{R}\right)^2 \left[e^{-2\alpha t_f} - e^{-2\alpha t_0}\right]. \tag{4.55a}$$

Letting $t_f = 5\tau$ and $t_0 = 0$, (4.55a) results in

$$E_C(t) = -\frac{1}{2C}\left(\frac{V_0}{R}\right)^2 \left(1 - e^{-10}\right) \approx -\frac{1}{2C}\left(\frac{V_0}{R}\right)^2. \tag{4.55b}$$

Validate the dimensions of LHS and RHS in (4.55b).

Make a numerical example for calculations of $p_C(t)$ and $E_C(t)$, and do similar derivations for a parallel RL zero-source circuit.

3) Power and energy for an RC step-response case is, in essence, the case with the applied dc signal. (Can you validate this statement?) Indeed, for $t \geq 0$, the circuit is under dc excitation with amplitude V_{in}, for which case the initial condition isn't a factor and can be set to zero. Hence, for a series RC circuit we can obtain from (4.17a) the following equation:

$$v_C(t)[V] = V_{in}\left(1 - e^{-\alpha t}\right) \tag{4.56}$$

and

$$i_c(t)[A] = \frac{V_{in}}{R} e^{-\alpha t}. \tag{4.57}$$

Therefore, the RC circuit output power for the step response appears to be

$$p_C(t) = \frac{V_{in}^2}{R}\left(1 - e^{-\alpha t}\right) e^{-\alpha t}. \tag{4.58}$$

The graph of (4.58) is given in Figure 4.14 for $\frac{V_{in}^2}{R} = 10\ (W)$ and $\alpha = 3\left(\frac{1}{s}\right)$. This graph shows how the capacitor (output) power changes during the transient process in an RC circuit step response: After the switch gets closed, the voltage across the capacitor starts accumulating. Thanks to the voltage change, the current through the capacitor begins flowing, and, as a result, power consumed by the capacitor begins increasing. However, the capacitor voltage quickly reaches the limit determined by the input power and becomes constant; consequently, current diminishes to zero. Thus, during the transient process, power reaches its maximum and then vanishes. Figure 4.12 shows this picture. What's more, we can verify that power starts from zero at $t = 0(s)$ and returns to its near-zero value at $5\tau = \frac{5}{3} \approx 1.7(s)$.

Calculating energy in this case gives

$$E_C(t)[J] = \int_{t_0}^{t_f} v_C(t) i_C(t) dt = E_C(t)[J] = \int_{t_0}^{t_f} V_{in}\left(1 - e^{-\alpha t}\right) \frac{V_{in}}{R} e^{-\alpha t} dt. \tag{4.59a}$$

If we set $t_f = 5\tau$ and $t_0 = 0$, then (4.59a) produces

$$E_C(t)[J] = -\frac{CV_{in}^2}{2R}. \tag{4.59b}$$

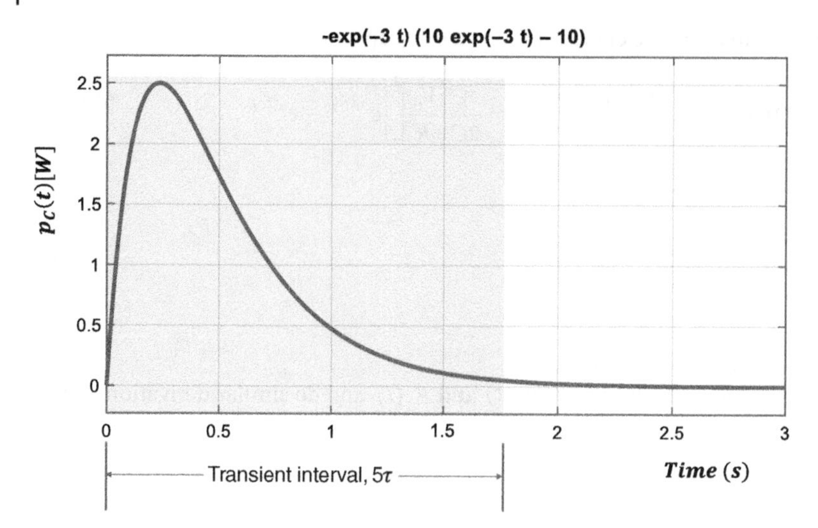

Figure 4.12 Power consumed by a series RC circuit for the step input.

Create a numerical example for $p_C(t)$ and $E_C(t)$.

4) The final case is the sinusoidal response of a series RC circuit. Suppose the sinusoidal voltage signal, $V_m \cos(\omega t)$, and appropriate current, $I_m \cos(\omega t+\theta)$, where $\theta = tan^{-1}\left(\frac{X_C}{R}\right)$, are in use. Obviously, $I_m = \frac{V_m}{Z_{Total}}$. Then, the *instantaneous power* consumed by the circuit is given by

$$p(t) = v(t)i(t) = V_m \cos(\omega t) I_m \cos(\omega t + \theta). \tag{4.60}$$

We resort to the well-known trigonometric identity, $cosA\,cosB = \frac{1}{2}(\cos(A+B) + \cos(A-B)$, and obtain from (4.60) the following formula

$$p(t) = \frac{1}{2} V_m I_m \cos(\theta) + \frac{1}{2} V_m I_m \cos(2\omega t + \theta). \tag{4.61}$$

The constant member of (4.61) is the always-present average power, whereas the average value over a period of the sinusoidal member is zero. Hence, the power for an RC circuit is

$$p(t)[W] = \frac{1}{2} V_m I_m \cos(\theta). \tag{4.62}$$

To find the energy, we apply (4.51a) and calculate, letting $t_f = 5\tau$ and $t_0 = 0$,

$$E(t)[J] = \int_{t_0}^{t_f} p(t)\,dt = \frac{1}{2}\int_0^{5\tau} V_m I_m \cos(\theta) dt = \frac{5\tau}{2} V_m I_m \cos(\theta). \tag{4.63}$$

Interestingly, energy can theoretically increase without bound because it is simply proportional to $\tau(s)$.

Thus, this subsection explains how to deal with power and energy in four typical cases encountered with first-order electrical circuits.

4.2.4 The Circuit Responses and the Initial and Final Values—A Summary

Considering the responses of various circuits in the time domain, we introduced *natural and forced responses* and *transient and steady-state responses*. Though all these terms were defined where they appear first, it's time to sum up all this information. What's more, it's necessary to discuss the linkage among these responses and *initial and final values* and the role of these values in shaping the *complete time-domain response* of a circuit. Additionally, relationship between the *initial values* and *initial conditions* is clarified.

As we learned in the preceding sections, the complete (total) circuit's response can be presented as a sum of natural and forced responses. To reiterate, a *natural response* is the circuit's response to its internally stored energy, without any external input. This is why it is often called a *source-free* or *zero-input response (zir)* and designated as $v_{out}^{zir}(t)$.

Natural response ≡ source-free response

Circuit's response to internally stored energy

Thanks to its origin, a natural response reveals the nature of the circuit in question. It shows all circuit's distinguish features, in terms of $i-v$ processes, that exist regardless of external excitation. Depending on the circuit's design and component's values, the response can be a damping exponential or damping oscillations or something else. In any event, the natural response always diminishes because circuit's stored energy dies out. The diminishing rate is ascertained by a time constant, $\tau = \frac{1}{\alpha}$ (s) determined by the circuit's component values.

A circuit's *stored energy at $t=0$ determines the initial condition or conditions*. For example, in an RC circuit, an initial condition is a voltage across the capacitor, $v_C\big|_{t=0} = V_C(0) = V_{C0}$, whereas an RL circuit's initial condition is current through the inductor at $t=0$, i.e., $i_L\big|_{t=0} = I_L(0) = I_{L0}$. See Appendix 4A *Natural (source-free) response of the first-order circuits*.

To reiterate,

> a *natural response is caused by the energy stored in a capacitor and/or an inductor, that is, by the initial conditions.*

In formulas, a natural response for RC or RL circuit is given by

$$v_{out}^{zir-RC}(t) = V_0 e^{-\alpha_{RC} t} \tag{4.64a}$$

or

$$i_{out}^{zir}(t) = I_0 e^{-\alpha_{RL} t}, \tag{4.64b}$$

respectively, where $\alpha_{RC}\left(\frac{1}{s}\right) = \frac{1}{RC}$ and $\alpha_{RL}\left(\frac{1}{s}\right) = \frac{R}{L}$ are the corresponding rates of decay.

We know that a natural response is due to stored energy at $t=0$, but what exactly is the instant $t=0$? Admittedly, it's up to us how we determine this moment. Typically, we count $t=0$, when a circuit changes its state, for example, it becomes operational or idle. These changes in the circuit states are usually achieved by turning the switch on or off. For example, Figure 4A.1q below shows that a source-free circuit's response starts when the switch disconnects the energy source from the circuit. Figure 4.4a demonstrates that a circuit's step response begins when the switch applies an external step input to the circuit. (Note that an electromechanical switch is a rarity in modern circuits, and digital electronics do a vast majority of switching operations.)

When analyzing experimental graphs of natural responses, students often confuse the origin of the rectangular coordinates, (0, 0), with the starting point of a transient process, $t = 0$. Be aware that, in many cases, these two points do not coincide, as Example 4.4 and its Figure 4.11b demonstrate. Here, Figure 4.13 visualizes these explanations.

Finally, it's worth knowing that since a natural response shows how the circuit's output transits from its initial value to zero, it is also often called *transient response*.

Even if there is no energy stored in the circuit, but external excitation is applied at $t = 0$, the circuit exposes its natural properties in response to this extraneous input. This transient response component, $v_{out}^{tr}(t)$, also diminishes at the rate of a source-free response. *Thus, there is a dying-up, transient member of a circuit's response caused by the external excitation containing V_{in} or I_{in}.* For example, for the step response of a series RC circuit this member is given by

$$v_{out}^{tr}(t) = -V_{in}e^{-\alpha t}, \tag{4.65}$$

whereas a sinusoidal response of such a circuit contains this member

$$v_{out}^{tr}(t) = \left(-\frac{\alpha^2 V_{in}}{\left(\alpha^2 + \omega^2\right)}\right)e^{-\alpha t}. \tag{4.66}$$

A *forced response*, $v_{out}^{fr}(t)$, you will recall, is a circuit's response to external excitation. It includes a transient member mentioned above and a steady-state response that lasts as long as an external input applies. This response shows that, due to the linearity of the circuit, the steady-state member preserves the nature of the input signal (dc, exponential, sinusoidal, or any other function) but carries the affected parameters of an output signal. For instance, a series RC circuit forced response to a dc signal V_{in} (step forced response) is

$$v_{out}^{fr}(t) = V_{in}\,(V), \tag{4.67}$$

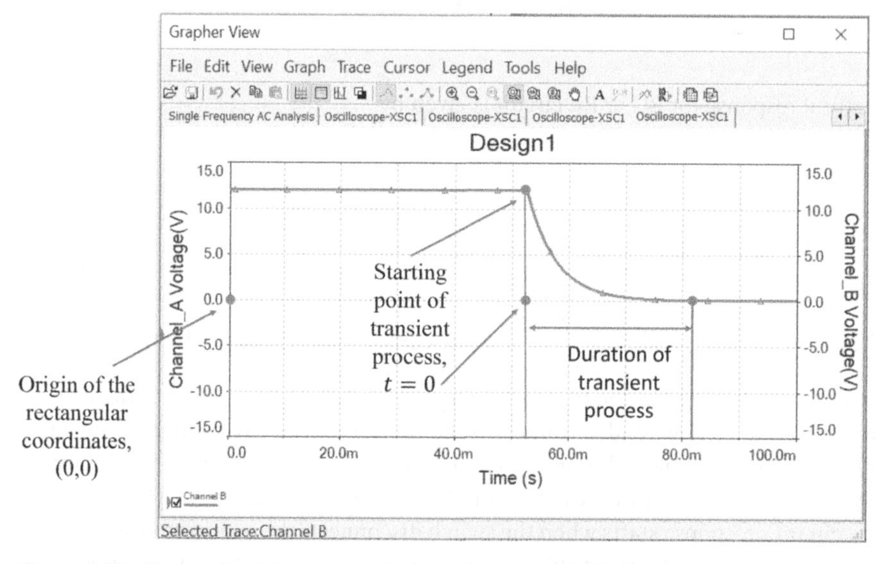

Figure 4.13 The origin of the rectangular coordinate system (0,0) and the starting point of a transient process $t = 0$.

whereas the response of such a circuit to a sinusoidal input $V_{in}\cos(\omega t)$ is

$$v_{out}^{fr}(t)=\frac{V_{in}}{\sqrt{\left(1+(\omega RC)^2\right)}}\cos(\omega t+\theta)\,(V), \tag{4.68}$$

where $\theta = tan^{-1}(-\omega RC)$. Hence, the input's sinusoidal nature and its constants V_{in} and ω remain, but the amplitude and a phase shift of the output sinusoid will change due to the circuit's influence. Since $v_{out}^{fr}(t)$ remains when $t \Rightarrow \infty$ (in contrast to a natural response), this output is a response's *final value*, $v_{out}(\infty)$.
Therefore,

the complete (total) response, which is an algebraic sum of the natural and forced responses, can also be presented as a sum of the initial, transient, and final members of the circuit's response:

$$v_{out}(t)=v_{out}^{zr}(t)+v_{out}^{tr}(t)+v_{out}^{fr}(t). \tag{4.69}$$

For example, the complete response of a series RC circuit to a step input is given by

$$v_{out}(t)[\mathrm{V}]=V_0e^{-\alpha t}-V_{in}e^{-\alpha t}+V_{in} \tag{4.70}$$

and to a sinusoidal input is

$$v_{out}(t)=V_0e^{-\alpha t}-\frac{\alpha V_{in}}{\left(\alpha^2+\omega^2\right)}\alpha e^{-\alpha t}+\frac{\alpha V_{in}}{\sqrt{\left(\alpha^2+\omega^2\right)}}\cos(\omega t-\theta)\,(V). \tag{4.71}$$

Therefore, the complete response is a sum of a natural response, caused by the circuit's stored energy, and forced response, caused by the external input. This statement is shown graphically in Figure 4.14.

Now, it's clear that the initial condition is responsible for natural response only. If the initial condition is zero, then there is no natural response at all.

Analysis of equations shown in Figure 4.14 brings us to the conclusion that

a complete response can also be presented as a sum of transient and steady-state responses.

$$v_{out}(t) = V_0e^{-\alpha t} + V_{in} - V_{in}e^{-\alpha t}$$

Complete response = natural response + forced response

a)

$$v_{out}(t) = V_0e^{-\alpha t} + \frac{\alpha V_{in}}{\sqrt{(\alpha^2+\omega^2)}}\cos(\omega t-\theta) - \frac{\alpha V_{in}}{(\alpha^2+\omega^2)}\alpha e^{-\alpha t}$$

Complete response = natural response + forced response

b)

Figure 4.14 Examples of the complete response as a sum of natural and forced responses: a) Step response of a series RC circuit; b) sinusoidal response of this circuit.

Indeed, two diminishing components of the complete response constitute the transient response, and one remaining regardless of the time flow is the steady-state member. This statement is visualized in Figure 4.15, which is achieved by regrouping the equations shown in Figure 4.14.

Next, let's clarify two more terms, *initial condition*, and *initial value*. An initial condition, as defined early in this section, is the energy level stored by the circuit in the form of I_0 at $t = 0$. As Figure 4.15 shows, when the initial condition is zero, there is no natural response. However, a transient response still exists even when $V_0 = 0$ or $I_0 = 0$, as Figure 4.16 demonstrates. Why? Because this component of a transient response is caused by an external input applied at $t = 0$.

Let's define an *initial value* as the value of $v_{out}(t)$ computed at $t = 0$; that is $v_{out}(0)$. For examples shown in Figures 4.15a and 4.15b respectively, these values are

$$v_{out}(0) = (V_0 - V_{in}) + V_{in} = V_0\,(V) \tag{4.72}$$

and

$$v_{out}(0) = \left(V_0 - \frac{\alpha^2 V_{in}}{(\alpha^2 + \omega^2)}\right) + \frac{\alpha V_{in}}{\sqrt{(\alpha^2 + \omega^2)}}\cos(-\theta). \tag{4.73}$$

Using the trigonometric identity $\cos(-\theta) = \frac{1}{\sqrt{(1 + tan^2(-\theta))}}$ and remembering that $\tan(-\theta) = \frac{\omega}{\alpha}$, the latter equation can be reduced to

$$v_{out}(0) = V_0 - \frac{\alpha^2 V_{in}}{(\alpha^2 + \omega^2)} + \frac{\alpha^2 V_{in}}{(\alpha^2 + \omega^2)} = V_0\,(V). \tag{4.74}$$

Equations 4.72 *and* 4.74 *enable us to conclude that an initial value of the circuit's response—which is, again,* $v_{out}(t)\big|_{t=0} = v_{out}(0)$*—coincides with the initial condition.*

At last, let's revisit the *final value* of a circuit's response, which is $v_{out}(t)\big|_{t \Rightarrow \infty} = v_{out}(\infty)$. Consider complete responses of a series RC circuit shown in Figures 4.4a and 4.12b. When t tends to infinity, the transient members of the complete approaches zero, and $v_{out}(\infty)$ goes to

$$v_{out}(\infty) = V_{in}\,(V)$$

for a step response and to

$$\frac{\alpha V_{in}}{\sqrt{(\alpha^2 + \omega^2)}}\cos(\omega t + \theta).$$

$$\underbrace{v_{out}(t)}_{\text{Complete response}} = \underbrace{(V_0 - V_{in})e^{-\alpha t}}_{\text{transient response}} + \underbrace{V_{in}}_{\text{steady-state response}}$$

a)

$$\underbrace{v_{out}(t)}_{\text{Complete response}} = \underbrace{\left(V_0 - \frac{\alpha^2 V_{in}}{(\alpha^2 + \omega^2)}\right)e^{-\alpha t}}_{\text{transient response}} + \underbrace{\frac{\alpha V_{in}}{\sqrt{(\alpha^2 + \omega^2)}}\cos(\omega t - \theta)}_{\text{steady-state response}}$$

b)

Figure 4.15 Examples of the complete response as a sum of transient and steady-state responses: a) Step response of a series RC circuit; b) sinusoidal response of this circuit.

for a sinusoidal response.

In other words, the *final values* are the steady-state responses, which are excited by and lasted thanks to *external inputs*.

This statement finalizes the discussion of *natural response, final response, transient response, forced response, initial condition, initial value, and final value.*

Appendix 4A More about Natural Response

Natural Response – A Closer Look

The circuit in Figure 4A.1b has no input. What is its output? When you ask this question in a classroom, most answers would be zero. This answer seems to be reasonable and almost obvious. However, it is wrong. The correct answer is that it depends on an *initial condition*: If a capacitor has stored energy before the circuit is put in operation, we will see an output (response). If the capacitor hasn't been charged before the switch turns on, the output will be zero. As said previously, the response in question is called *a natural, source-free, or zero-input* response. It is caused by the energy stored in a circuit before $t = 0$. This energy serves as a source that pushes the circuit to respond. Only *reactive components—capacitors and inductors*—can store energy. Figure 4A.1a shows the charging process of a series RC circuit; R_{in} is the *Thevenin equivalent resistor* seen by the capacitor.

Description of this natural, source-free response of an RC circuit needs the differential equation. Recalling that

$$v_C(t) \equiv v_{out}(t) \text{ and } i_R(t) = i_C(t) = i(t) \tag{4A.1}$$

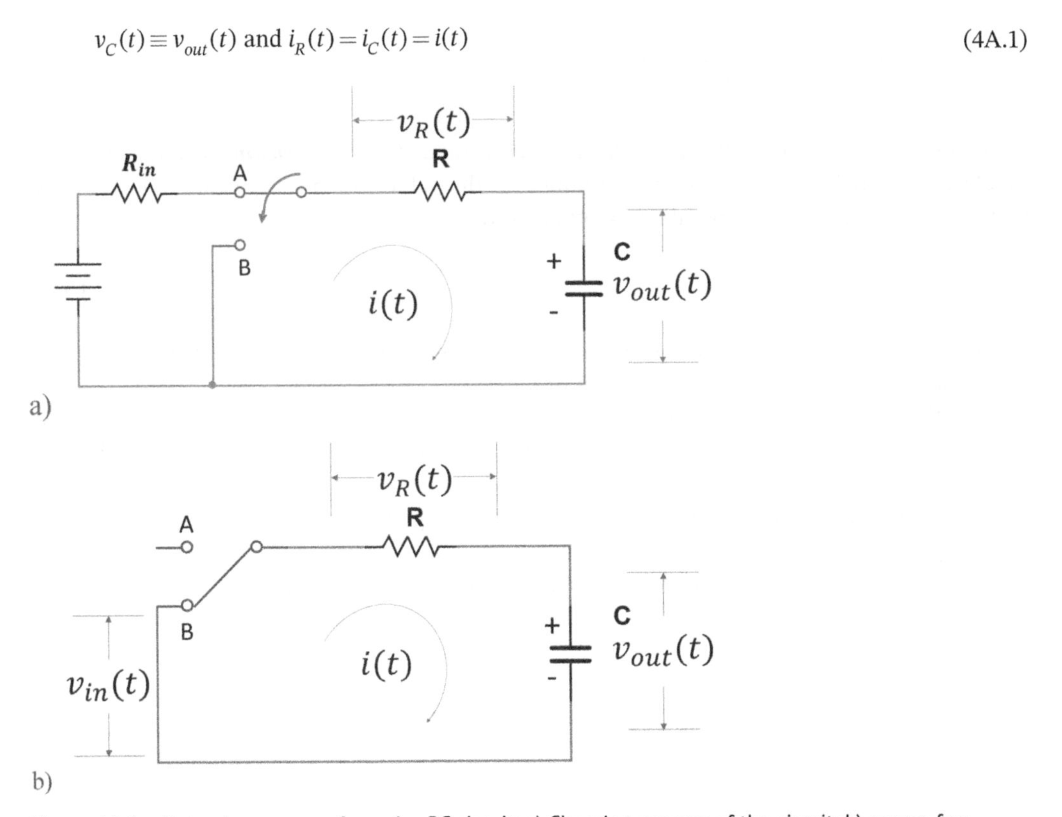

Figure 4A.1 Natural response of a series RC circuit: a) Charging process of the circuit; b) source-free (zero-input) RC circuit's operation.

and applying the KVL to a series RC circuit for $t \geq 0$ produces

$$v_R(t)+v_C(t)=0 \Rightarrow Ri(t)+v_{out}(t)=0 \Rightarrow RC\frac{dv_{out}(t)}{dt}+v_{out}(t)=0$$

$$\Rightarrow v'_{out}(t)+\alpha v_{out}(t)=0. \quad (4A.2)$$

Here we use traditional notations $\tau(s)=RC$ and $\alpha\left(\frac{1}{s}\right)=\frac{1}{\tau}=\frac{1}{RC}$. Note that the change of capacitor's voltage is considered only for $t \geq 0$, that is, only after the switch is turned on.

We know from (4.13) and (4.17) that the solution to this homogeneous equation is

$$v_{out}(t)=V_0 e^{-\alpha t}\left(\text{V}\right). \quad (4A.3)$$

Here

$$V_0 = v_{out}(t)_{t=0} \equiv v_{out}\left(0\right)$$

is the *initial condition*, as before. The graph of $v_{out}(t)$ shows that the output voltage will diminish to zero after the switch is turned on because the finite amount of stored energy will be dissipated, and nothing will support further circuit operation. The course of this graph is similar to that shown in Figure 4.5a.

What current will flow through the capacitor in this case? We use the $i_C - v_C$ relationship and find

$$i_{out}(t)=C\frac{dv_{out}(t)}{dt}=-\frac{V_0}{R}e^{-\alpha t}\left(\text{A}\right). \quad (4A.4)$$

It's important to remember that the *voltage across a capacitor, $v_C(t)$,* ***cannot*** *change instantly;* therefore, its initial value, V_{C0}, is predetermined by the value which a capacitor collects at the moment before the switch is turned on. In other words,

$$v_C\left(0^-\right)=v_C\left(0^+\right)=V_{C0}\equiv V_0, \quad (4A.5)$$

Where $t(s)=0^-$ is the last instant when the switch in Figure 4A.1b was in position A, and $t(s)=0^+$ is the first instant when the switch turns to position B.

Due to this capacitor's property, determining its initial voltage is a simple process. This is why in analyzing an RC circuit, it's always recommended to find the capacitor voltage first, as it's done in our consideration here.

Incidentally, the current through a capacitor **can** change instantaneously. Thus, at the last instant when the switch in Figure 4A.1b still was in position A, current $i_R(0^-)=i_C(0^-)=0$ provided that the switch was in position A long enough to fully charge the capacitor to V_0. Just after the switch turns to position B, the current through a capacitor starts flowing, and its value becomes

$$i_R\left(0^+\right)=i_C\left(0^+\right)=I_0=-\frac{V_0}{R}, \quad (4A.6)$$

as (4A.4) shows at $t=0$.

We've mentioned several times that natural response is caused by the energy stored in a capacitor of a series RC circuit. What is the amount of this energy? Let's start with power delivered to a resistor from a capacitor; it is equal to

$$P_{RC}(t) = \left(v_C(t)\right)^2 R = V_0^2 Re^{-2\alpha t}\left(\text{W}\right). \tag{4A.7}$$

The energy accumulated by the RC circuit at $t = 0$ is precisely the amount of energy dissipated by the resistor during the circuit operation, $t \geq 0$; it is given by

$$E_{RC}(t) = \int_0^t P_R(t)dt = \frac{1}{2}CV_0^2\left(1 - e^{-2\alpha t}\right)(\text{J}). \tag{4A.8}$$

To find the *natural (source-free) response of an RL parallel circuit*, consider Figure 4A.2a, where all values are taken from Example 4.3; that is, $R = 1\Omega$, $L = 0.5H$, and $I_{in} = 12(mA)$. Notice that R_{in} is the Thevenin equivalent resistor seen by the inductor. For $t < 0$, as Figure 4A.2a shows, the switch had been closed for a long time, so the circuit was in a steady state. Assume the inductor's ohmic resistance is $R_L = 1m\Omega$, that is, $R_L = 0.001R$. Hence, the initial inductor's current is practically equal to the input current; i.e.,

$$i_{L0} = I_{in}\left(A\right).$$

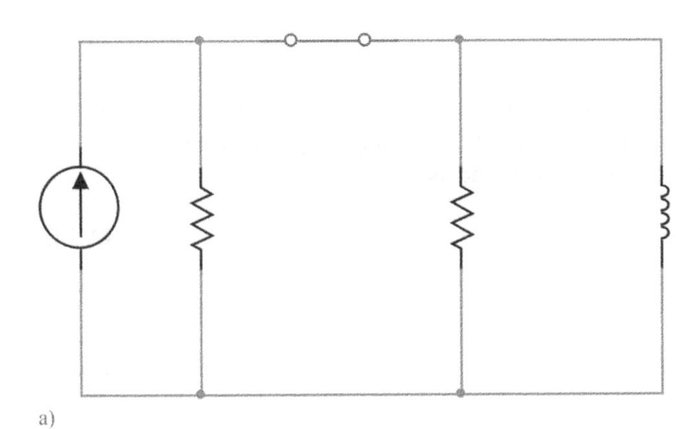

a)

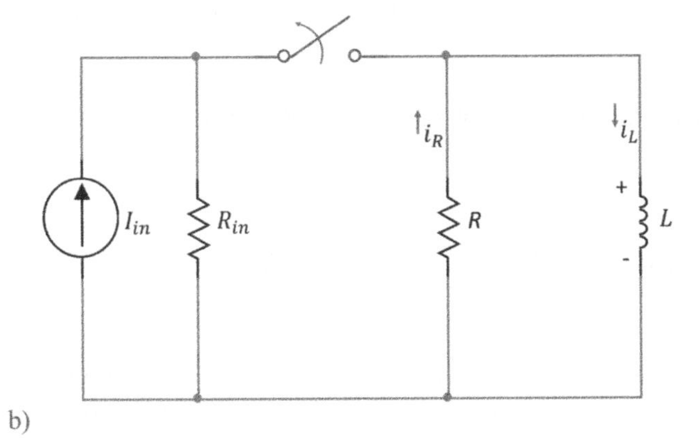

b)

Figure 4A.2 Natural response of an RL parallel circuit: a) Establishing the initial current through an inductor; b) source-free RL parallel circuit.

Then, at $t = 0$, the switch is open, as shown in Figure 4A.2b. The RL circuit becomes source-free and operates under the initial condition only.

Recall that the current through an inductor cannot change instantaneously, and therefore the RL circuit's initial condition is readily defined as

$$i_L(0^-) = i_L(0^+) = i_{L0} \equiv I_0. \tag{4A.9}$$

Refer to Example 4.3 from Section 4.1. Putting the RHS of (4.22) equal to zero, we obtain the following differential equation, describing the source-free parallel RL circuit shown in Figure 4A.2b,

$$\frac{di_{Lh}(t)}{dt} + \alpha i_L(t) = 0, \tag{4A.10}$$

where $\alpha\left(\frac{1}{s}\right) = \frac{1}{\tau} = \frac{R}{L}$. The answer to (4A.10) is the homogeneous solution to the differential equation of (4.22), i.e.,

$$i_{Lh}(t) = Ae^{-\alpha t}.$$

At $t = 0,\ A = i_{Lh}(0) = I_0$. Thus, the current in the source-free RL circuit is given by

$$i_L(t) = I_0 e^{-\alpha t} \tag{4A.11}$$

This is a decaying exponential function whose initial value is I_0 and the final value approaches zero. A curious reader is encouraged to build the $i_L(t)$ graph, using the given numbers.

Power collected in a source-free RL circuit can be found as

$$P_{RL}(t) = (i_L(t))^2 R = I_0^2 Re^{-2\alpha t}\,(\text{W}). \tag{4A.12}$$

Energy stored by the inductor at $t = 0$ is equal to

$$W_{RL}(t) = \int_0^t P_{RL}(t)dt = \frac{1}{2}LI_0^2\left(1 - e^{-2\alpha t}\right)(\text{J}). \tag{4A.13}$$

Assume $t_{final} = 5\tau = \frac{5}{\alpha}$ and compute power and energy stored at the given source-free RL circuit.

Question What is the power dissipated by the resistor of a source-free RL circuit shown in Figure 4A.2b?

Application of a Natural-Response Theory: Charging and Discharging Processes

How does an initial voltage across the capacitor in a series RC circuit appear? Due to charging the circuit's capacitor before the circuit starts its operation at $t = 0$. Technically, it is done by connecting an external dc source to the circuit, as Figure 4A.1a shows. As long the switch connects a dc source to the RC circuit, the capacitor is charging. The value of V_0 depends on the charging duration, but it cannot exceed the voltage, V_{dc}, supplied by a charger, $V_0 \le V_{dc}$, regardless of how long the capacitor is charging.

Comparing Figures 4A.1a and 4.4a shows that they are identical. Therefore, the charging process is described by the forced part of a step response of a series RC circuit given in (4.17). In the self-explanatory notations, this equation takes the form

$$V_{charge} = V_{dc}\left(1 - e^{-\alpha t_{charge}}\right). \tag{4A.14}$$

The course of a graph described by (4A.14) is similar to that shown in Figure 4.5a. This graph shows that V_0 can reach V_{dc} only when $t_{charge} \to \infty$. Therefore, the charger designer must set the final charging time, t_{final}, by setting the final value of V_{dc}. For example, using 5τ criterion, $t_{final} = 5\tau = \frac{5}{\alpha}$, results in

$$V_{charge} = V_{dc}\left(1 - e^{-\alpha t_{final}}\right) = V_{dc}\left(1 - e^{-5}\right) = 0.9933V_{dc}.$$

The whole charging duration, t_{final}, is the interval from $-\infty$ to 0, when $t = 0(v)$ is the final instant of charging and the starting point of discharging. The value of this interval depends on the current generated by the source. Since the charger voltage is fixed, its current determines the charger power. Interestingly, *Tesla*, a manufacturer of electric vehicles (EVs), built a network of charging stations called Tesla Supercharges that are distinct from other charging stations by the higher charging power, which reduces the charging time. In general, the problem of charging EVs becomes so ubiquitous that *IEEE* (Institute of Electrical and Electronics Engineers) even had to standardize this technology. See IEEE 2030.1.1 "DC Quick Charger."

These are the examples of the implementation of the abovementioned theory.

When the switch in Figure 4A.1b turns to position B—that is, disconnecting the charger—the circuit starts dissipating the accumulated energy. The discharging duration depends on V_0 but is determined by the circuit's time constant, $\tau(s) = RC$. Thus, to increase $\tau(s)$, we must increase either R or C or both.

Schematically, this is how our mobile devices are charging and discharging. Of course, neither these devices nor their batteries resemble basic RC circuits; however, the concept of these processes is similar.

Question How does the Charging Duration Depend on the Charger Power?

We will further discuss a natural circuit response from different points of view in the chapters that follow.

Questions and Problems for Chapter 4

Section 4.1 Circuit's Response–Statement of Problem

1. Explain the difference between a system and electrical circuit.
2. How can we analyze the system (circuit) performance and predict the results of its operation theoretically?
3. Suppose you receive the schematic of an unknown electrical circuit. What is the first characteristic of this circuit will you look for?

4 Is any difference between an input and excitation of a circuit? Explain.

5 How can you distinguish output and response?

6 Identify input and output for the circuits shown in Figures 4.P6.

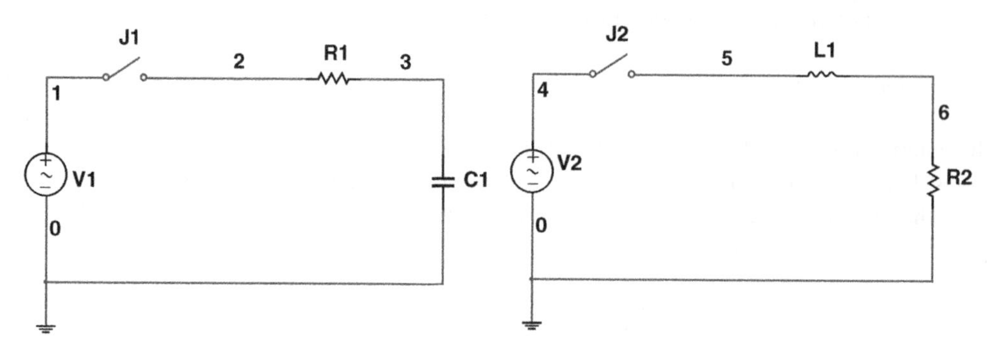

Figure 4.P6 Circuits for Problem 4.6: a) Series RC circuit; b) series RL circuit.

7 Determine input and output of the circuit shown in Figure 4.P7.

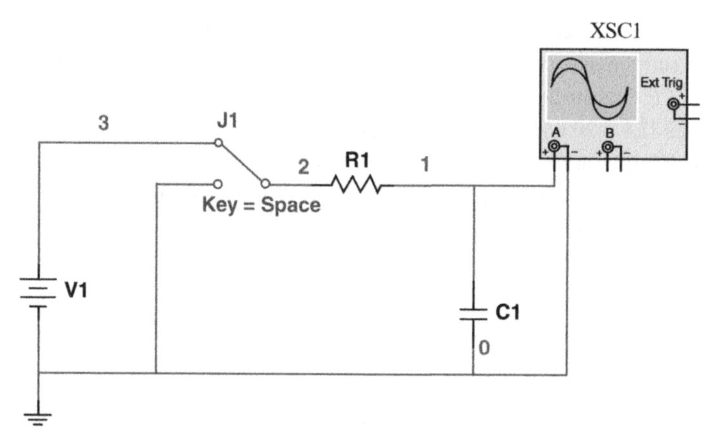

Figure 4.P7 Circuit for Problem 4.7.

8 Calculate the output voltage in circuit shown in Figure 4.P8.

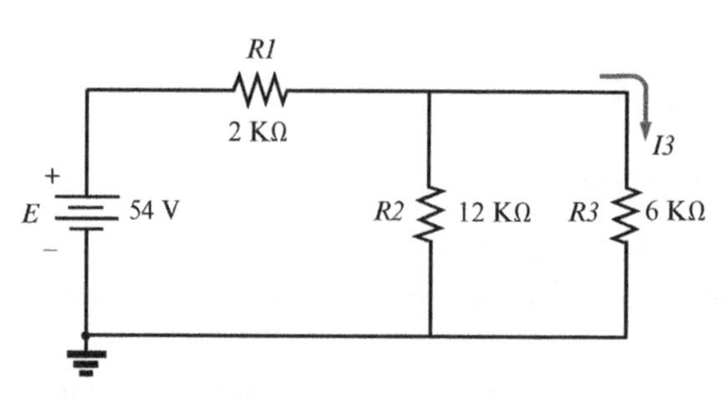

Figure 4.P8 Circuit for Problem 4.8.

9 Consider Figure 4.2c: Why does the output voltage is shown across the capacitor but not across the inductor or resistor?

10 Switch in the circuit shown in Figure 4.P10 is thrown from *a* to *b* position: Can you obtain the output voltage $\boldsymbol{v_{out}(t) \equiv v_C(t)}$ by applying the voltage-divider rule like that given in Equation 4.2? Explain.

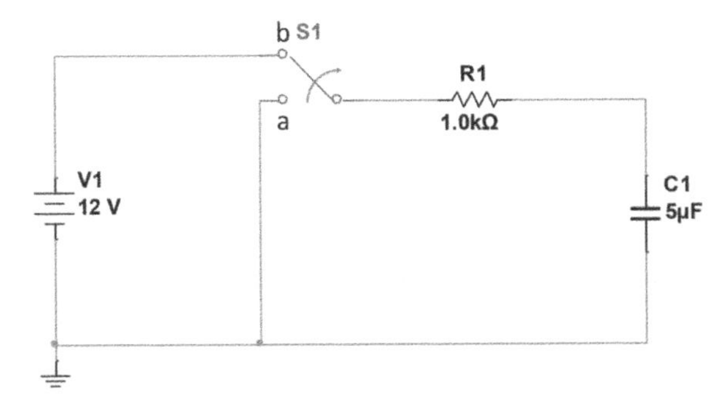

Figure 4.P10 Applicability of voltage-divider rule to RC circuit for Problem 4.10.

11 Examine the most right sides of Figure 4.3a and 4.3b and explain a transient response differs from a steady-state response.

12 Why does the description of an RC or RL circuit's transient response require the use of a differential equation?

13 Refer to Table 4.1: Find the voltage drop across the inductor whose L = 2 mH if $i_L(t) = 3\sin(8t)$.

14 Consider Table 4.1: What is the inductor's current if $v_L(t) = -60e^{-3t}$?

15 Based on Table 4.1, consider a series RL circuit whose $R = 1k\Omega$ and $L = 2mH$. What will be the inductor's current if $V_{in} = 12V$?

16 *What will be the inductor's current in a parallel RL circuit if $V_{in} = 12V$, $R = 1k\Omega$, and $L = 2mH$?

17 Consider a series RC circuit whose $R = 1k\Omega$ and $C = 2mF$. What will be the capacitor's current if $V_{in} = 12V$?

18 In reference to Table 4.1, what will be the capacitor's voltage in a parallel RC circuit if $V_{in} = 12V$, $R = 1k\Omega$, and $C = 2mF$?

19 *Consider the series RC circuit shown in Figure 4.P19 whose $V_0 = 5(V)$.

a) Derive its differential equation. Show all mathematical manipulations.
b) Find $v_{out}(t)$, sketch its graph, and show the transition interval based on 5τ criterion.
c) Find and show the initial and final values of $v_{out}(t)$.
d) Confirm your answers by Multisim simulation.

20 Consider the RC circuit given in Figure 4.P19 whose $V_0 = 5(V)$:

a) Find $i_C(t)$ for the circuit shown in Figure 4.P19 and sketch its graph.
b) Why do voltage across and current through the capacitor behave so differently?
c) Obtain the value of the transient interval duration using 1% criterion.
(Hint: Use the results obtained in Problem 4.17)

21 Consider the RL circuit shown in Figure 4.P21, which is a reproduction of Figure 4.7:

a) Find its response provided that $I_{in} = 3(A), R = 2(\Omega)$, L = 0.3(H), and $I_0 = 0.5(A)$.
b) Sketch the graph $i_L(t)$ and quantitatively show its transient interval.

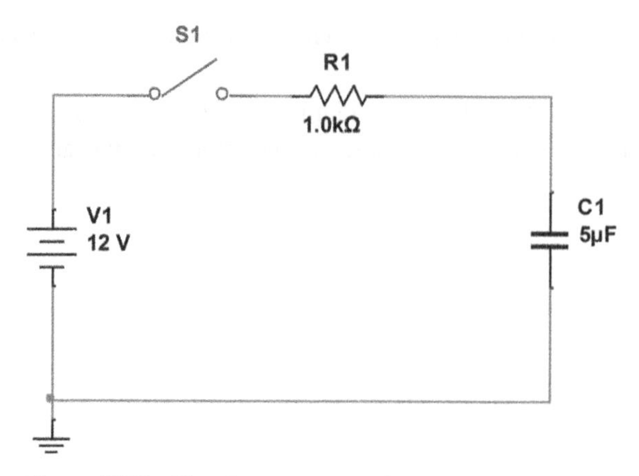

Figure 4.P19 Transient response of a series RC circuit for Problem 4.19.

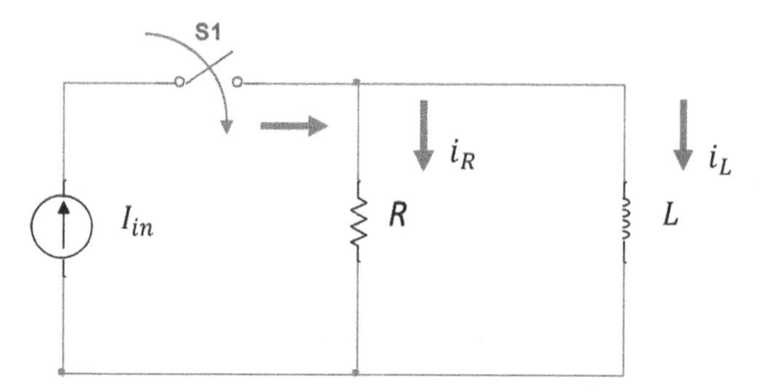

Figure 4.P21 Finding the parallel RL circuit's response for Problem 4.21.

Section 4.2 Complete First-Order Circuit's Responses and Initial and Final Conditions in Time Domain

22 The circuit in Figure 4.P22 demonstrates how to obtain a sinusoidal response of a series RC circuit. Find this response if $V_0 = 0(V)$. Verify your answer by Multisim simulation.

23 Find the complete response of a series RC circuit shown in Figure 4.P22 if $V_0 = 12(V)$.

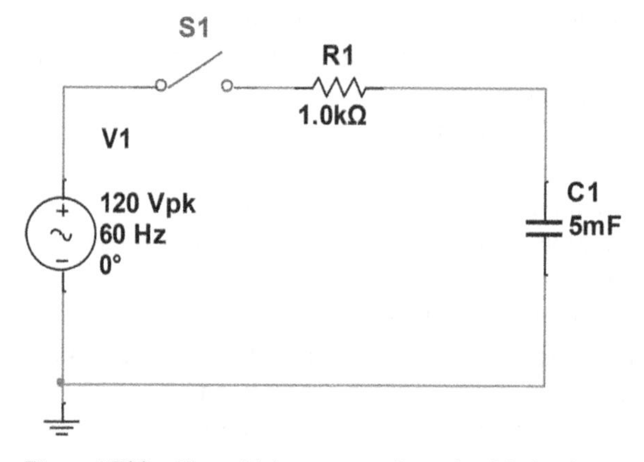

Figure 4.P22 Sinusoidal response of a series RC circuit.

24 Confirm your answer to Problem 23 by running Multisim simulation and quantitatively comparing the results.

25 Why do the first-order circuit responds similarly to the same excitations even when they contain different components?

26 Is a circuit's natural response the same as a transient response? Explain.

27 What is the main feature of a circuit's natural response? What other terms are used for this response?

28 Figure 4.15 shows that a complete circuit response is a sum of the natural and forced responses. In contrast, Figure 4.16 demonstrates that the same complete response consists of transient and steady-state responses. Which figure depicts the correct answer? Explain.

29 Revisit Problem 4.28 and find the initial and final values of the circuit complete response.

Appendix 4A More about Natural Response

30 Figure 4.P30 shows a series RC circuit. What will be its output signal, $v_{out}(t) \equiv v_C(t)$, after the switch will be turned on?

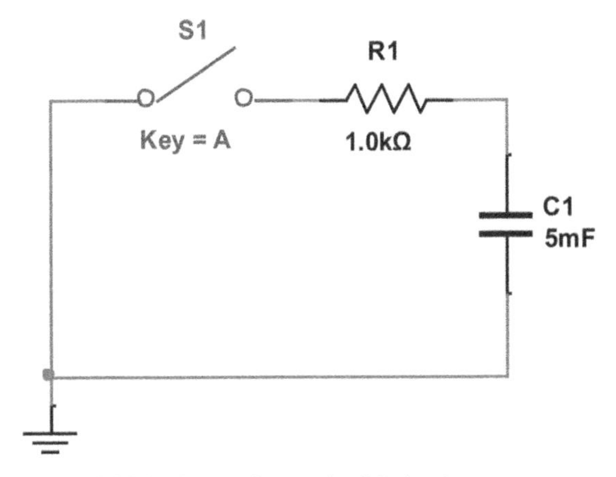

Figure 4.P30 Source-free series RC circuit.

31 What will be the voltage drop across resistor $R1$ and capacitor $C1$ in the circuit shown in Figure 4.P31 immediately after the switch will be thrown on? How about current flowing through these elements? The circuit was initially relaxed.

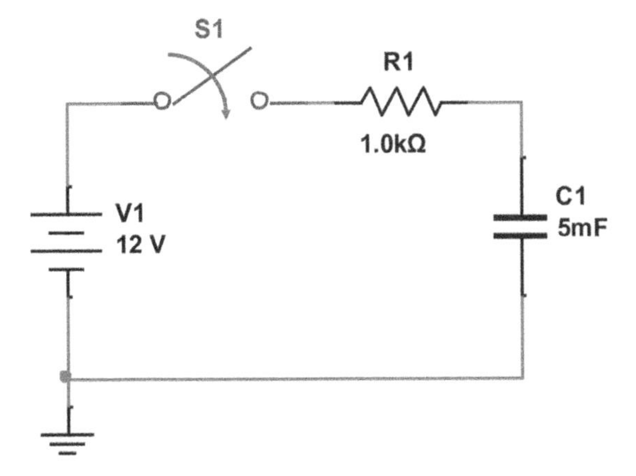

Figure 4.P31 Step response of RC circuit.

32 For the circuit shown in Figure 4.P32, find the voltage across and current through the resistor and inductor immediately after the switch gets closed. Assume $I_0 = 0.2(A)$.

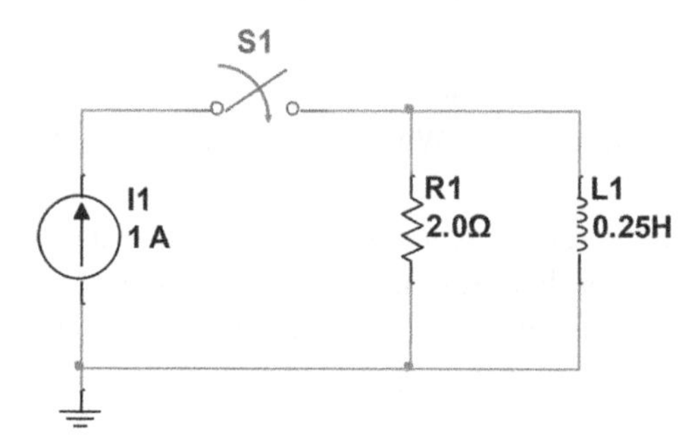

Figure 4.P32 Instantaneous changes of voltages and current in parallel RL circuit.

33 Calculate power, $P_{RC}(t)(W)$, and energy, $E_{RC}(t)(J)$, of a source-free series RC circuit.

34 For a source-free parallel RL circuit, find power, $P_{RL}(t)(W)$, and energy, $E_{RL}(t)(J)$.

35 The charger of a laptop receives 1(A) input current. What power does it deliver to the laptop? What can be done to make the laptop charging faster?

5

Advanced Circuit Analysis in Time-Domain—II

5.1 Natural Responses of RLC Circuits

5.1.1 Natural, Forced, and Complete Responses of RLC Circuit

Consider the circuit shown in Figure 5.1a: What is its response (output)? Most of the answers in a classroom are typically *zero* because the circuit has no input or source. However, this answer is wrong, as we learned in Chapter 4. Indeed, two reactive circuit components—inductor and capacitor—can store energy in this circuit. If they did so before a circuit was put in operation, the circuit response, $v_{out}(t)$, to switching of its status won't be zero.

We refer you to Section 4.1 and especially to Appendix 4A, where this topic has been studied for the first-order circuits. The only difference between that circuits and this one is that the *RLC circuit* can simultaneously store energy in two components, whereas RC and RL circuits have only one reactive element. To remind you, the C and L can *store energy* because they preserve their voltage or current at the instant when the circuit changes its state. Specifically, when the switch closes the loop in an RC circuit, the capacitor will save its voltage as described by (4A.5),

$$v_C\left(0^-\right)=v_C\left(0^+\right)=V_{C0}\equiv V_0, \qquad \text{(4A.5R)}$$

where 0^- is the last instant before the switch flips, and 0^+ is the first instant after the switch turns on. An inductor has a similar property but for current; hence for an RL circuit the genuine relationship is

$$i_L\left(0^-\right)=i_L\left(0^+\right)=I_L(0)=I_0. \qquad \text{(4A.6R)}$$

These two equations display vital peculiarities of the reactive components: *The inductor current and the capacitor voltage cannot change instantly*, and therefore, these elements store the energy they have accumulated before the circuit is switched. In other words, if a voltage across a capacitor was, for example, 12 (V), when the switch was thrown, then these 12 volts will be there immediately after the switch flipped. (It doesn't mean 12 volts will stay across the capacitor forever!) Similar explanations work for an inductor's current. Again, refer to Appendix 4A to refresh your memory on this topic.

Essentials of Advanced Circuit Analysis: A Systems Approach, First Edition. Djafar K. Mynbaev.

Companion Website: www.wiley.com/go/Mynbaev/AdvancedCircuitAnalysis

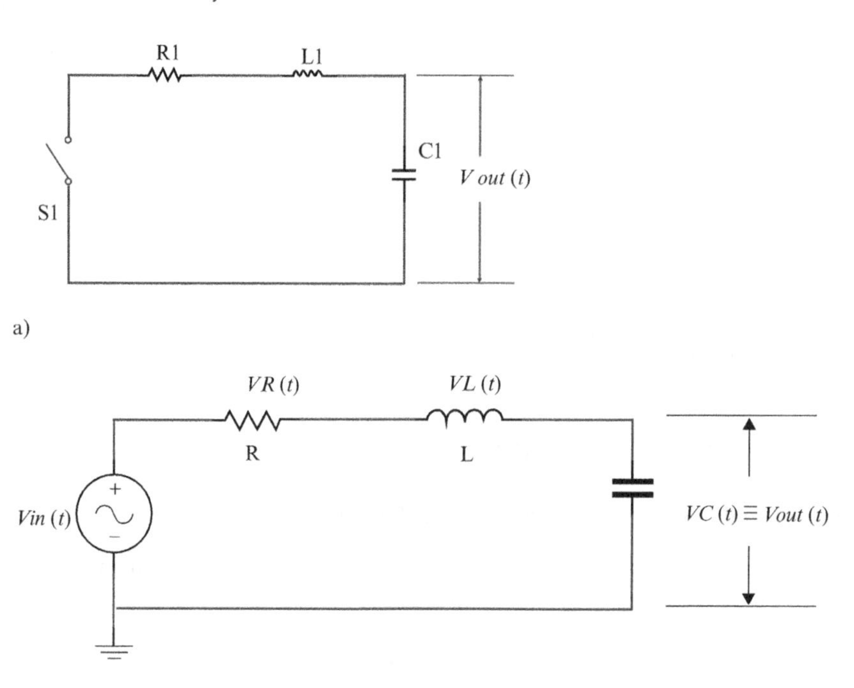

Figure 5.1 Series RLC circuit: a) Source-free (zero-input) version; b) the circuit with the variable input.

The *natural response is the circuit's reaction to its initial condition*, V_{C0} and $I_L(0)$. When the source of the signal is connected to a circuit, as in Figure 5.1b, the circuit *response* to the input only is called the *forced*, and combined output to the initial conditions and the input is termed *complete response*.

5.1.2 Derivation of Second-Order Circuit's Differential Equation–Series RLC Circuit

Consider the RLC series circuit shown in Figure 5.1b. To analyze this circuit, we need to fulfill two tasks:

1) Derive an equation fully describing this circuit.
2) Solve the derived equation to find the circuit response, $v_{out}(t)$, to the external excitation, $v_{in}(t)$.

To derive the RLC circuit equation, we recall that a traditional circuit analysis course introduces the $i-v$ relationship for each circuit's component. The formulas summarizing these rules are given in Table 4.1R, reproduced here for convenience.

To describe the RLC circuit, we apply KVL to our circuit as

$$v_{in}(t) = v_L(t) + v_R(t) + v_C(t). \tag{5.1}$$

This equation is vital from a circuit analysis standpoint. However, from a system's point of view, we need to relate the *output voltage* of our circuit, $v_{out}(t) \equiv v_C(t)$, to the *input voltage*, $v_{in}(t)$. In other words, we need to find the circuit *response*, $v_{out}(t)$, to its *excitation*, $v_{in}(t)$.

Table 4.1R Current–voltage relationship for R, L, and C components.

Component	Current	Voltage
Resistor	$i_R(t) = \frac{v_R(t)}{R}$	$v_R(t) = i_R(t) \cdot R$
Inductor	$i_L(t) = \frac{1}{L}\int_{-\infty}^{t} v_L(t)dt$	$v_L(t) = L\frac{di_L(t)}{dt}$
Capacitor	$i_C(t) = C\frac{dv_C(t)}{dt}$	$v_C(t) = \frac{1}{C}\int_{-\infty}^{t} i_C(t)dt$

Referring again to (5.1), we realize that the problem requires expressing $v_L(t)$ and $v_R(t)$ through $v_C(t) \equiv v_{out}(t)$. How can we do this? The hint is that all these voltages are produced by one current, and therefore,

$$i(t) \equiv i_C(t) = C\frac{dv_{out}(t)}{dt} \; and \; v_L(t) = L\frac{di_L(t)}{dt} \tag{5.2}$$

because for a series circuit, the current is the same for all components,

$$i_R(t) = i_L(t) = i_C(t) = i(t). \tag{5.3}$$

Therefore, (5.1) can be rewritten as

$$\begin{aligned} v_{in}(t) &= v_L(t) + v_R(t) + v_C(t) = L\frac{di(t)}{dt} + Ri(t) + v_{out}(t) \\ &= LC\frac{d^2v_{out}(t)}{dt^2} + RC\frac{dv_{out}(t)}{dt} + v_{out}(t). \end{aligned} \tag{5.4}$$

Dividing (5.4) through by LC and introducing notations

$$\omega_0^2\left(\frac{rad}{s}\right) = \frac{1}{LC} \tag{5.5a}$$

and

$$\alpha\left(\frac{1}{s}\right) = \frac{R}{2L}, \tag{5.5b}$$

enable us to present this equation as

$$\frac{d^2v_{out}(t)}{dt^2} + 2\alpha\frac{dv_{out}(t)}{dt} + \omega_0^2 v_{out}(t) = \omega_0^2 v_{in}(t). \tag{5.6a}$$

Using the common designations,

$$\frac{dy}{dx} = y' and \frac{d^2y}{dx^2} = y'',$$

we can rewrite (5.6a) as

$$v''_{out}(t)+2\alpha v'_{out}(t)+\omega_0^2 v_{out}(t)=\omega_0^2 v_{in}(t). \tag{5.6b}$$

This (5.6a or 5.6b) is the second-order differential equation describing a series RLC circuit when exploring a system's approach.

Thus, task number 1 is complete.

Note well: Similarly to the first-order circuits, description of the second-order circuit in its entirety requires the use of a differential equation even though this equation stems from the algebraic Kirchhoff's voltage law.

A word of caution: For a second-order *series RLC circuit*, a *damping coefficient* (constant) $\alpha\left(\frac{1}{s}\right)$ is defined in (5.5b) as $\alpha\left(\frac{1}{s}\right)=\frac{R}{2L}$, whereas for the first-order circuit RC or RL, this constant was defined as $\alpha\left(\frac{1}{s}\right)=\frac{1}{RC}$ or $\alpha\left(\frac{1}{s}\right)=\frac{R}{L}$. We help you to avoid the confusion over the course of this book.

5.1.3 Three Possible Solutions to the Differential Equation of a Source-Free RLC Circuit

Let's turn to task number 2: How can we solve this equation? We start with considering the *natural, source-free,* or *zero-input* response of a series RLC circuit. In this case, $v_{in}(t)=0$, and (5.6b) becomes

$$v''_{out}(t)+2\alpha v'_{out}(t)+\omega_0^2 v_{out}(t)=0. \tag{5.7}$$

Next, recall that the first-order circuit differential equation was successfully solved in Sidebar 4A by the trial method; namely, by supposing that

$$v_{out}(t)=Ae^{mt}, \tag{5.8}$$

where constants A and m have to be determined. Plugging this trial solution in (5.7) results in

$$\left(m^2+2\alpha m+\omega_0^2\right)Ae^{mt}=0. \tag{5.9}$$

Since Ae^{mt} cannot be equal to zero, (5.9) reduces to

$$m^2+2\alpha m+\omega_0^2=0. \tag{5.10}$$

We know that (5.10) is called the *characteristic equation* of (5.7).

The solution to this quadratic equation is known well:

$$m_{1,2}=-\alpha\pm\sqrt{\left(\alpha^2-\omega_0^2\right)} \tag{5.11}$$

Hence, our trial solution takes the form

$$v_{out}(t)=A_1e^{m_1t}+A_2e^{m_2t}. \tag{5.12}$$

(Prove that (5.7) requires the solution consisting of the sum of two members, as we get in (5.12) by considering the solution for m_1 and m_2 in sequence and using the *linearity* of (5.7) and applying the *superposition principle.*)

To verify the trial solution, we need to plug it into the differential equation and confirm that (5.12) turns (5.7) in identity. Taking the derivatives yields

$$v'_{out}(t) = A_1 m_1 e^{m_1 t} + A_2 m_2 e^{m_2 t}$$

and

$$v''_{out}(t) = A_1 m_1{}^2 e^{m_1 t} + A_2 m_2{}^2 e^{m_2 t}$$

and plugging these derivatives and $v_{out}(t)$ into (5.7) produces

$$A_1 e^{m_1 t}\left(m_1{}^2 + 2\alpha m_1 + \omega_0^2\right) + A_2 e^{m_2 t}\left(m_2{}^2 + 2\alpha m_2 + \omega_0^2\right) = 0.$$

Neither $A_1 e^{m_1 t}$ nor $A_2 e^{m_2 t}$ can be zero. (Explain why.) Thus, the process boils down to the verification of the following equation:

$$\left(m_1{}^2 + m_2{}^2\right) + 2\alpha\left(m_1 + m_2\right) + 2\omega_0^2 = 0.$$

We substitute $m_1 = -\alpha + \sqrt{\left(\alpha^2 - \omega_0^2\right)}$ and $m_2 = -\alpha - \sqrt{\left(\alpha^2 - \omega_0^2\right)}$ and perform the straightforward algebraic manipulations to obtain the following equation

$$2\alpha^2 + 2\alpha^2 - 2\omega_0^2 - 4\alpha^2 + 2\omega_0^2 = 0,$$

which indeed is the identity. Therefore, our *trial solution is the real solution to* (5.7).

Which particular form $v_{out}(t)$ will assume depends on the relationship between α^2 and ω_0^2, as (5.11) shows:

1) If $\alpha^2 > \omega_0^2$, then $\sqrt{\left(\alpha^2 - \omega_0^2\right)}$ is a real number, so m_1 and m_2 are also the real numbers r_1 and r_2, and $v_{out}(t) = A_1 e^{r_1 t} + A_2 e^{r_2 t}$ becomes the sum of two real exponential functions. This case is called ***overdamped***; we will explain the meaning of this and the following terms later in this section.
2) If $\alpha^2 = \omega_0^2$, then $\sqrt{\left(\alpha^2 - \omega_0^2\right)} = 0$, so $m_{1,2} = -\alpha$, and $v_{out}(t)$ again turns to be a real function, whose form must be determined. This case is termed ***critically damped***.
3) If $\alpha^2 < \omega_0^2$, then $\sqrt{\left(\alpha^2 - \omega_0^2\right)}$ is an imaginary, and m_1 and m_2 are the complex numbers. The response $v_{out}(t)$ gets a sum of two complex exponential functions; they, as we know, represent the oscillatory waveforms, so in this case, we will see oscillations in the response. This case is called ***underdamped***.

Note that the roots (5.11) are called *natural frequencies* for they carry frequency dimension. Indeed,

$$m_{1,2}\left(\frac{\text{rad}}{\text{s}}\right) = -\alpha \pm \sqrt{\left(\alpha^2 - \omega_0^2\right)}. \tag{5.11R}$$

However, $\alpha\left(\frac{1}{s}\right)$ is termed as a *damping coefficient*, as introduced early, and the square root member of (5.11R) gets the frequency meaning only in an underdamped case.

But all the above explanations seem to be the mathematical exercises. What is the significance of $\alpha^2 - \omega_0^2$ relationship for an RLC circuit? Well, we need to recall that α and ω_0 are just the notations, and, according to (5.5a), and (5.5b), they mean

$$\alpha = \frac{R}{2L} \text{ and } \omega_0^2 = \frac{1}{LC}. \tag{5.5R}$$

Equation 5.5 enables us to bring the above mathematical manipulations in the circuit-analysis realm by substituting the circuit parameters into $\sqrt{(\alpha^2 - \omega_0^2)}$. This yields:

$$\text{When, } \alpha^2 > \omega_0^2 \text{ then} \left(\frac{R}{2L}\right)^2 > \frac{1}{LC}, \text{ which means } \boldsymbol{R_{od}} > 2\sqrt{\frac{\boldsymbol{L}}{\boldsymbol{C}}} \text{ for an } \textbf{overdamped case,} \tag{5.13a}$$

$$\text{when } \alpha^2 = \omega_0^2, \text{ then } \boldsymbol{R_{cd}} = 2\sqrt{\frac{\boldsymbol{L}}{\boldsymbol{C}}} \text{ for a } \textbf{critically damped case,} \tag{5.13b}$$

$$\text{when } \alpha^2 < \omega_0^2, \text{ then } \boldsymbol{R_{ud}} < 2\sqrt{\frac{\boldsymbol{L}}{\boldsymbol{C}}} \text{ for an } \textbf{underdamped case.} \tag{5.13c}$$

Among the above relationships, the crucially important is (5.13b) because it *determines* the *exact value of the resistor, R_{cd}*, when L and C are given. This is why we should always start the RLC circuit analysis and design by considering Equation 5.13b. The values of R_{od} and R_{ud} are arbitrary; they need to meet the conditions set by (5.13a) and (5.13c), respectively. Simply put, the requirements are $R_{od} > R_{cd}$ and $R_{od} < R_{cd}$. In other words, Condition 5.13a means that R_{od} must be big enough to suppress any oscillations, and Condition 5.13b implies that R_{Ud} is to be small enough to allow for oscillations.

Remember, *family of Equations* 5.13 *holds only for a series RLC circuit!*

Now, the roots of the quadratic equation, after combining (5.5a), (5.5b) and (5.11), come as

$$m_{1,2} = -\alpha \pm \sqrt{(\alpha^2 - \omega_0^2)} = -\frac{R}{2L} \pm \sqrt{\left(\left(\frac{R}{2L}\right)^2 - \frac{1}{LC}\right)} = -\frac{R}{2L} \pm \frac{1}{2L}\sqrt{(R^2 - \left(4\frac{L}{C}\right)} \\ = \frac{-R \pm \sqrt{(R^2 - \left(4\frac{L}{C}\right)}}{2L}. \tag{5.14}$$

Thus, the solution to the series RLC circuit's differential equation, (5.12), appears in the following general form:

$$v_{out}(t) = A_1 \exp(m_1 t) + A_2 \exp(m_2 t) = A_1 \exp\left(\frac{-R + \sqrt{(R^2 - \left(4\frac{L}{C}\right)}}{2L} t\right) + \\ A_2 \exp\left(\frac{-R - \sqrt{(R^2 - \left(4\frac{L}{C}\right)}}{2L} t\right). \tag{5.15}$$

Therefore, task number 2 is formally complete. We say "formally" because there are a lot of important things that must be communicated to clarify the meaning of (5.15).

Analyzing (5.15), we can see that the relationship between a circuit resistance R and the reactive components L and C determines the behavior of a series RLC circuit. Why? To answer this question, we need to recall the operation of an *ideal LC circuit* shown in Figure 5.2.

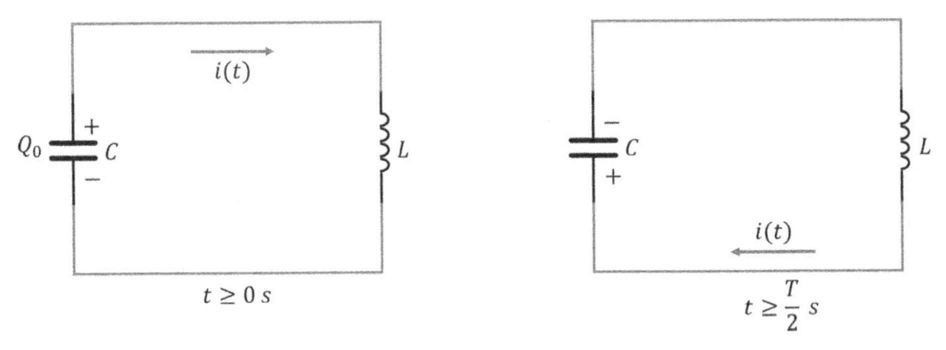

Figure 5.2 Consideration of an LC circuit oscillations.

It's assumed that the capacitor in Figure 5.2 contains charge $Q_0(F)$ before the circuit starts operating. Then the capacitor begins to discharge, thus producing current $i(t)$ that flows through the inductor. This process continues until the capacitor becomes fully discharged, which takes half of the whole period of action. During this time, the flowing current creates the magnetic field, which, in turn, induces the current that charges the capacitor over the second half of the cycle. At $t = T(s)$, the capacitor becomes fully recharged, and the cycle repeats. Thus, the circuit creates the positive halfwave during the first half of the period and the negative half wave over the second half. The underlying process is transferring the capacitor electric energy to the inductor magnetic energy and back.

The ideal LC circuit has no ohmic resistance; this is why there is no energy loss in this circuit. Nevertheless, when frequency of this process is low, the capacitor works like an open circuit and the inductor's impedance, $\boldsymbol{X_L}$, is very low, whereas for the high frequency the situation is the opposite. Thus, it should be the frequency at which both impedances will be acceptable. To find this frequency, we need to recall that (1) $\boldsymbol{X_C} = -j\frac{1}{\omega C}$ and $\boldsymbol{X_L} = j\omega L$ and (2) they are in opposite directions (see Figure 3.7). Therefore, if we set

$$\boldsymbol{X_C} = \boldsymbol{X_L}, \tag{5.16a}$$

these impedances cancel each other, and the ideal LC circuit will have zero impedance. But (5.16a) can be met if and only if

$$|\boldsymbol{X_C}| = |\boldsymbol{X_L}| \Rightarrow \frac{1}{\omega C} = \omega L. \tag{5.16b}$$

Equations 5.16a and 5.16b determine the condition of *resonance* in an LC circuit. Therefore, (5.16b) gives the *resonant frequency* ω_0 of an LC circuit as

$$\omega_0\left(\frac{rad}{s}\right) = \sqrt{\frac{1}{LC}}. \tag{5.17}$$

Now, here is the main point of this consideration: What if we do not have an LC circuit but a series RLC circuit? Well, the resistor dissipates the energy transferring between the capacitor and inductor causing the oscillation process diminish or *damp*. If the resistance is high, as in (5.13a), the LC part of the circuit is unable to generate oscillations, and the process damps exponentially. This is why this case is called *overdamped*. If the resistance is low, as in (5.13c), the LC part can generate oscillations, but the process will eventual damp due to the resistor. This case of diminishing oscillations is justifiably called *underdamped*. The *critically damped* case sits at the border between the abovementioned cases.

Table 5.1 summarizes these classifications and presents the typical graphs of all three natural responses of a series RLC circuit. Derivation and discussion of the response formulas, also shown in Table 5.1, follow.

Table 5.1 Three possible natural responses of a zero-input series RLC circuit.

Differential equation $v''_{out}(t)+2\alpha v'_{out}(t)+\omega_0^2 v_{out}(t)=0$	**Graphs**
Solution $v_{out}(t)=A_1 \exp\left(\frac{-R+\sqrt{\left(R^2-\left(4\frac{L}{C}\right)\right.}}{2L}t\right)$ $+A_2 \exp\left(\frac{-R-\sqrt{\left(R^2-\left(4\frac{L}{C}\right)\right.}}{2L}t\right)$	
Overdamped condition: $R>2\sqrt{\frac{L}{C}}$ Response formula (5.26): $v_{out-od}(t)=\frac{-V'_0-(\alpha-k)V_0}{2k}e^{-(\alpha+k)t}+\frac{V'_0+(\alpha+k)V_0}{2k}e^{-(\alpha-k)t}$, where $k=\sqrt{\left(\alpha^2-\omega_0^2\right)}$.	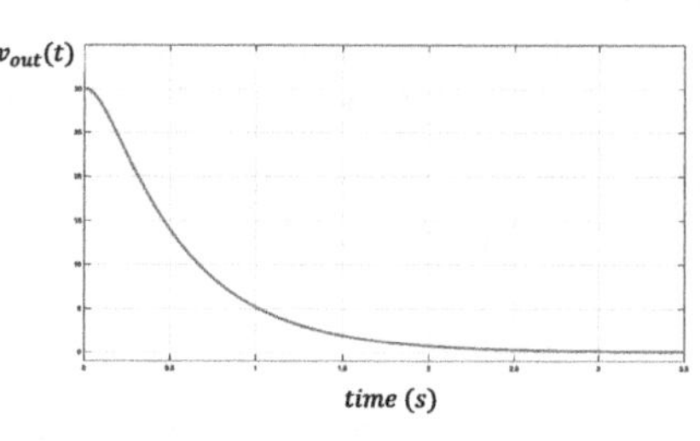
Critically-damped condition: $R=2\sqrt{\frac{L}{C}}$ Response formula (5.30): $v_{out-cd}(t)=V_0e^{-\alpha t}+(V'_0+\alpha V_0 t)\,e^{-\alpha t}$, where $\alpha=\frac{R}{2L}$.	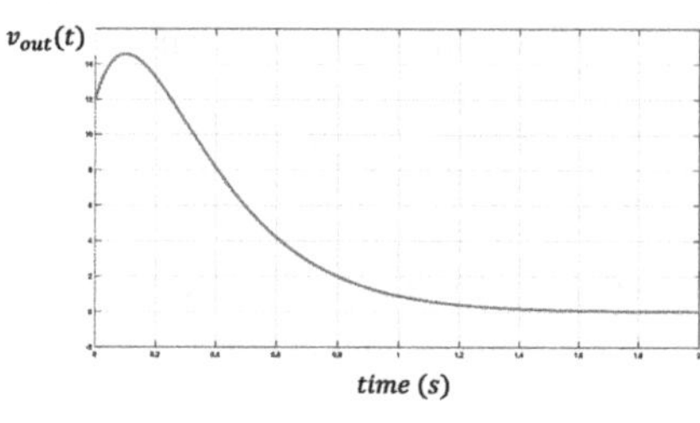
Underdamped condition: $R<2\sqrt{\frac{L}{C}}$ Response formula (5.38): $v_{out-ud}(t)=Ae^{-\alpha t}\cos(\omega_d t-\theta)$, where $\omega_d=\sqrt{\omega_0^2-\alpha^2}$.	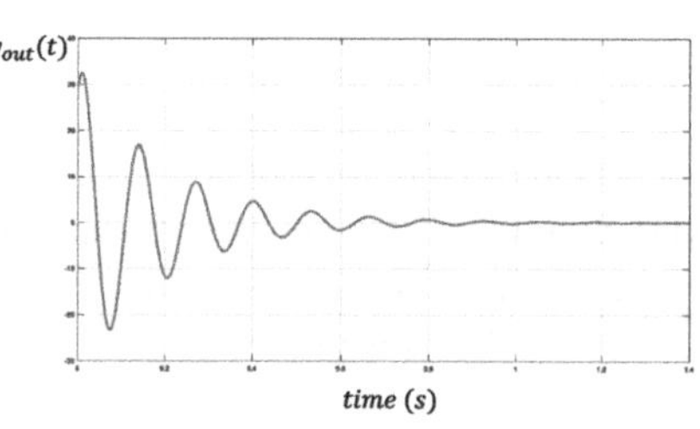

One needs to comprehend that Table 5.1 presents *typical* graphs of the natural responses of a series RLC circuit; in reality, we can meet a variety of their forms. However, several features of these graphs are fixed (which ones?), whereas others can vary. We will meet these varieties in further study. It must be clearly understood that these graphs show the *property of an RLC circuit itself*; the other issue is how this circuit will respond to various inputs.

Having understood the physics underlying the RLC circuit operation, we can turn to the main task of the advanced circuit analysis: describing the circuit operation in mathematical terms. We must derive and solve the circuit's differential equations for the various inputs, and we will do this in the extant chapter for a series and parallel RLC circuit.

We remember that the *complete solution* of a differential equation consists of two parts: a general *solution* to a *homogeneous* (zero-input) equation and a *particular solution* to a nonhomogeneous (non-zero input) equation. Thus, we must start our investigation with a natural (source-free or zero-input) case. *As soon as we obtain the natural response, we can add to it the response of the circuit to any new input.* This is the plan for both sections of Chapter 5.

5.1.4 Solutions to the Second-Order Circuit's Differential Equation with Zero Input

One has to remember that reference to the *response* implies that the circuit operation starts at $t \geq 0$.

5.1.4.1 Initial Conditions for a Series RLC Circuit

A series RLC circuit is described by the second-order differential equation derived in (5.7) for zero input as

$$v''_{out}(t) + 2\alpha v'_{out}(t) + \omega_0^2 v_{out}(t) = 0. \tag{5.7R}$$

Its solution is given by

$$v_{out}(t) = A_1 e^{m_1 t} + A_2 e^{m_2 t}. \tag{5.12R}$$

Here

$$\alpha = \frac{R}{2L} \text{ and } \omega_0^2 = \frac{1}{LC}, \tag{5.5R}$$

and

$$m_{1,2} = -\alpha \pm \sqrt{\left(\alpha^2 - \omega_0^2\right)} = \frac{-R \pm \sqrt{\left(R^2 - \left(4\frac{L}{C}\right)\right.}}{2L}. \tag{5.14R}$$

Table 5.1 shows that the circuit response can be overdamped, or critically damped, or underdamped, depending on the relationship between R and $\frac{L}{C}$. Our goal in this subsection is to derive the formulas for the circuit's response in all three cases.

Since m_1 and m_2 are known, finding the explicit forms for (5.12) reduces to the determination of constants A_1 and A_2. They should be ascertained from the *initial conditions*, but it's not a trivial task.

We must realize that Solution (5.12) requires two initial conditions because we need to determine two unknown constants. Thus, we have A_1 and A_2—not just A, as in the case of RC or RL

circuits—because an RLC circuit is described by the second-order differential equation. There is also a physical reason for the existence of two initial conditions in an RLC circuit: Initial energy can be stored in both capacitor and inductor. Thus, let's assume the following initial capacitor's voltage and inductor's current:

$$v_C(0) = V_0 \; and \; i_L(0) = I_0. \tag{5.18}$$

However, in a series RLC circuit, the current is the same for all components; hence, I_0 is the current through the capacitor also, and it can be expressed through the capacitor voltage as

$$I_0 = Cv'_C(0) = CV'_0 \Rightarrow V'_0 = \frac{I_0}{C}. \tag{5.19}$$

Therefore, *two initial conditions* of a *series RLC circuit are* V_0 *and* V'_0. It's crucial to realize that the formula $V'_0 = \frac{I_0}{C}$ allows for calculating the value of V'_0 because the inductor initial current must be given. However, to find the formula for $V'_0 = v'_{out}(0)$, we need to differentiate $v_{out}(t)$, obtain $v'_{out}(t)$, and then set $t = 0$.

Bear in mind that V'_0 is the change in the value of V_0 at the initial instant, the "thrust," of the output voltage at $t = 0$. The *dimension* of this member, *V/s*, highlights this fact. Contrary to the impression you might have, the value of V'_0 is quite arbitrary; very often it is far from zero. The meaning of this condition is clear from (5.19): It is determined by the current flowing through an inductor at $t = 0$. Note that we use capital letters for V_0 and V'_0 designations because they are numbers, not variables.

The obtained initial conditions enable us to find the solutions for all three responses of a series RLC circuit. It's worth reminding that these solutions differ by only one member,

$$k = \sqrt{\left(\alpha^2 - \omega_0^2\right)} = \sqrt{\left(\left(\frac{R}{2L}\right)^2 - \frac{1}{LC}\right)}, \tag{5.20}$$

that defines which specific case is implemented.

Combining (5.14) and (5.20) gives

$$m_{1,2} = -\alpha \pm \sqrt{\left(\alpha^2 - \omega_0^2\right)} = -\alpha \pm \text{k}. \tag{5.21}$$

These root expressions will play a crucial role in the discussion that follows.

5.1.5 Three Possible Solutions–Overdamped Case

For the *overdamped* case $(R > 2\sqrt{\left(\frac{L}{C}\right)}$, or $(\alpha^2 > \omega_0^2)$, (5.12) takes the following form:

$$v_{out-od}(t) = A_1 e^{m_1 t} + A_2 e^{m_2 t} = A_1 e^{-(\alpha+k)t} + A_2 e^{-(\alpha-k)t}. \tag{5.22}$$

Differentiation of (5.21) gives V'_0 as

$$v'_{out-od}(t) = -(\alpha + k)A_1 e^{-(\alpha+k)t} - (\alpha - k)A_2 e^{-(\alpha-k)t}. \tag{5.23}$$

At $t = 0$, Equations 5.22 and 5.23 give

$$\begin{cases} V_0 = A_1 + A_2 \\ V_0' = -(\alpha + k)A_1 - (\alpha - k)A_2 \end{cases}. \tag{5.24}$$

Straightforward manipulations produce

$$\begin{cases} A_1 = \dfrac{-V_0' - (\alpha - k)V_0}{2k} \\ A_2 = \dfrac{V_0' + (\alpha + k)V_0}{2k} \end{cases}. \tag{5.25}$$

Thus, in the overdamped case, the natural response of a series RLC circuit is

$$v_{out-od}(t) = \frac{-V_0' - (\alpha - k)V_0}{2k} e^{-(\alpha+k)t} + \frac{V_0' + (\alpha + k)V_0}{2k} e^{-(\alpha-k)t}, \tag{5.26}$$

where $k = \sqrt{(\alpha^2 - \omega_0^2)}$ and $\alpha^2 > \omega_0^2$ or $k = \sqrt{\left(\left(\frac{R}{2L}\right)^2 - \frac{1}{LC}\right)}$ and $\left(\frac{R}{2L}\right)^2 > \frac{1}{LC}$.

The graph in Table 5.1 shows that the output voltage, $v_{out-od}(t) = v_C(t)$, tends to zero over time. The physics behind this circuit behavior is clear: Capacitor's initial voltage and inductor's initial current will dissipate as time goes by. The rate of $v_{out-od}(t)$ decay is determined by the exponential functions $e^{-(\alpha+k)t}$ and $e^{-(\alpha-k)t}$, which means that it depends on α and k. Initial values V_0 and V_0' cannot change the decay rate but affect the total process duration. As (5.26) shows, graph of $v_{out-od}(t)$ never increases and never crosses the zero line. (Explain why.) Refer to sidebar 4S.2, where the properties of a decaying exponential function are briefly discussed.

Question The second member of (5.26) contains factor $e^{(-\alpha+k)t}$. If $k > \alpha$, then this exponent will be raised to a positive power and this member will grow. It means that the output voltage of a zero-input series RLC circuit will increase. (To understand this condition better, plug any numbers that will implement it.) Can this situation happen in reality? If no, what relationship between *k and* α must be met? Does an *RLC* circuit always maintain this relationship?

5.1.5.1 Critically Damped Case

In a *critically damped* case, $R = 2\sqrt{\left(\frac{L}{C}\right)}$, or $\alpha^2 = \omega_0^2$, which makes the solution to the RLC circuit's differential equation undetermined. Indeed,

$$v_{out-cd}(t) = A_1 e^{m_1 t} + A_2 e^{m_2 t} = A_1 \exp\left(\frac{-R + \sqrt{\left(R^2 - \left(4\frac{L}{C}\right)\right)}}{2L} t\right) + A_2 \exp\left(\frac{-R - \sqrt{\left(R^2 - \left(4\frac{L}{C}\right)\right)}}{2L} t\right) \tag{5.27}$$

$$= A_1 \exp\left(-\frac{R}{2L}t\right) + A_2 \exp\left(-\frac{R}{2L}t\right) = (A_1 + A_2)\exp\left(-\frac{R}{2L}t\right) = (A_1 + A_2)e^{-\alpha t} = Ae^{-\alpha t}.$$

Thus, (5.27) delivers only one solution, whereas the second-order differential equation requires two. This problem is resolved if the solution in question takes the following form

$$v_{out-cd}(t) = (A_1 + A_2 t)e^{-\alpha t}. \tag{5.28}$$

(See Alexander and Sadiku 2020, pp. 321–322 for the proof of the above statement.) The derivative of (5.28) gives

$$v'_{out-cd}(t) = (-\alpha A_1 + A_2 - \alpha A_2 t)e^{-\alpha t}. \tag{5.29a}$$

At $t = 0$, we find from (5.28) and (5.29a):

$$\begin{cases} A_1 = V_0 \\ A_2 = V'_0 + \alpha V_0 \end{cases}. \tag{5.29b}$$

Thus, the solution for a critically damped case is

$$v_{out-cd}(t) = V_0 e^{-\alpha t} + (V'_0 + \alpha V_0 t)\, e^{-\alpha t}. \tag{5.30}$$

Equation 5.30 contains the member, $\alpha V_0 t$, which increases over time; however, the decaying exponential $e^{-\alpha t}$ will eventually bring the process to zero. The interplay between these two members defines the shape of the response graph. If αV_0 is big compare with V_0 and V'_0, this member can cause the initial increase in the course of $v_{out-cd}(t)$. Such a case is shown in Table 5.1. The $v_{out-cd}(t)$ graph can even "dive" into the negative territory, depending on the relationship between αV_0 and other parameters. (Can you find the condition for such a case?) In most practical cases, the critically damped graph resembles the overdamped one but differs by the decay rate.

5.1.5.2 Underdamped Case

In the *underdamped* case ($R < 2\sqrt{\left(\frac{L}{C}\right)}$, or $\alpha^2 < \omega_0^2$), we have two distinct roots, and therefore, we can utilize the general results of the overdamped case, specifically (5.26).

$$v_{out-od}(t) = \frac{-V'_0 - (\alpha - k)V_0}{2k} e^{-(\alpha+k)t} + \frac{V'_0 + (\alpha + k)V_0}{2k} e^{-(\alpha-k)t}. \tag{5.26R}$$

Remember that k is defined in (5.20) as

$$k = \sqrt{(\alpha^2 - \omega_0^2)} = \sqrt{\left(\left(\frac{R}{2L}\right)^2 - \frac{1}{LC}\right)}, \tag{5.20R}$$

which means that in the underdamped case $(\alpha_{ud}{}^2 - \omega_0^2)$ or $\left(\left(\frac{R}{2L}\right)^2 - \frac{1}{LC}\right)$ is a *negative number* because in this case $\left(\frac{R}{2L}\right)^2 < \frac{1}{LC}$ or $\alpha_{ud}{}^2 < \omega_0^2$. Consequently, in the underdamped case, k_{ud} is a *imaginary number* and must be defined as

$$k_{ud} = j\sqrt{(\omega_0^2 - \alpha_{ud}{}^2)} = j\omega_d \tag{5.31}$$

where $j = \sqrt{-1}$. The new parameter, ω_d, is termed *damping frequency* and is defined as

$$\omega_d\left(\frac{\text{rad}}{\text{s}}\right) = \sqrt{\omega_0^2 - \alpha^2}. \tag{5.32}$$

As (5.32) shows, the damping frequency ω_d is always smaller than the *resonant frequency* ω_0, and the *damping coefficient*, $\alpha = \frac{R}{2L}$, determines their difference.

Hence, the roots m_1 and m_2 assume the forms

$$m_{1,2} = -\alpha \pm j\omega_d, \tag{5.33}$$

and (5.12) can be transformed for the underdamped case as

$$v_{out-ud}(t) = A_1 e^{(-\alpha + j\omega_d)t} + A_2 e^{(-\alpha - j\omega_d)t} = e^{-\alpha t}\left(A_1 e^{(j\omega_d)t} + A_2 e^{(-j\omega_d)t}\right). \tag{5.34}$$

Using Euler's identities,

$$e^{jx} = cosx + jsinx \text{ and } e^{-jx} = cosx - jsinx,$$

we can turn exponential functions into trigonometric as

$$\begin{aligned}\left(A_1 e^{(j\omega_d)t} + A_2 e^{(-j\omega_d)t}\right) &= A_1\left(cos\omega_d t + jsin\omega_d t\right) + A_2\left(cos\omega_d t - jsin\omega_d t\right)\\ &= \left(A_1 + A_2\right)cos\omega_d t + j\left(A_1 - A_2\right)sin\omega_d t.\end{aligned} \tag{5.35}$$

Taking A_1 and A_2 from (5.25) and inserting k from (5.31) brings (5.35) to the following form

$$\begin{aligned}\left(A_1 e^{(j\omega_d)t} + A_2 e^{(-j\omega_d)t}\right) &= B_1 cos\omega_d t + B_2 sin\omega_d t,\\ &= V_0 cos\omega_d t + \frac{\alpha V_0 + V_0'}{\omega_d} sin\omega_d t,\end{aligned} \tag{5.36}$$

where

$$B_1 = V_0 \text{ and } B_2 = \frac{\alpha V_0 + V_0'}{\omega_d}$$

are the constants of an underdamped case equation $v_{out-ud}(t)$.

Finally, putting (5.36) into (5.34) produces the formula for the *natural (zero-input)* response of a series RLC circuit in the *underdamped case* as

$$\begin{aligned}v_{out-ud}(t) &= e^{-\alpha t}\left(B_1 cos\omega_d t + B_2 sin\omega_d t\right)\\ &= e^{-\alpha t}\left(V_0 cos\omega_d t + \frac{\alpha V_0 + V_0'}{\omega_d} sin\omega_d t\right).\end{aligned} \tag{5.37}$$

A clearer presentation of this result can be achieved by performing the following transformation of (5.37): Let's use the trigonometric identity

$$acosx + bsinx = Acox(x - \theta),$$

where $A = \sqrt{a^2 + b^2}$ and $\theta = tan^{-1}\left(\frac{b}{a}\right)$. (You can readily verify this identity by letting $a = Acos\theta$ and $b = Asin\theta$.) Then (5.37) can be rearranged as

$$v_{out-ud}(t) = e^{-\alpha t}\left(V_0 cos\omega_d t + \frac{\alpha V_0 + V_0'}{\omega_d} sin\omega_d t\right) = Be^{-\alpha t}\cos\left(\omega_d t - \theta\right), \tag{5.38}$$

where $B=\sqrt{(V_0)^2+\left(\frac{\alpha V_0+V_0'}{\omega_d}\right)^2}$ and $\theta=tan^{-1}\left(\frac{\alpha V_0+V_0'}{(V_0\omega_d}\right)$. Equation 5.38 explicitly shows that an underdamped response is *damping (decaying) oscillations* whose amplitude B and phase shift θ are determined by the *initial conditions* V_0 and V_0' and the circuit's parameters α and ω_d, and whose decay rate is defined by the *damping coefficient* $\alpha=\frac{R}{2L}$.

It's worth mentioning that we have several members carrying frequency dimensions in the received formulas. First, it is a *resonant frequency*, $\omega_0\left(\frac{rad}{s}\right)=\sqrt{\frac{1}{LC}}$, whose meaning is discussed early in this subsection. Secondly, we introduce a *damping frequency*, $\omega_d\left(\frac{\text{rad}}{\text{s}}\right)=\sqrt{\omega_0^2-\alpha^2}$, whose sense is clear from (5.38): It is the frequency of damping (decaying) oscillations. Third, this is a *damping coefficient*, $\alpha\left(\frac{1}{s}\right)=\frac{R}{2L}$, which is called the *neper*[1] *frequency* because it carries the dimension of frequency. Note that α is measured in $\left(\frac{1}{S}\right)=\left(\frac{\Omega}{H}\right)$, whereas ω_0 and ω_d have the unit $\left(\frac{rad}{s}\right)$. Nevertheless, since *radian* is a dimensionless unit, we accept this discrepancy.

Therefore, the explicit formulas for all three cases of the series RLC circuit natural response are obtained. Now, we are well equipped to consider an example.

Example 5.1 Natural (source-free) response of a series RLC circuit: three cases.

Problem: The series RLC circuit is shown in Figure 5.3. The circuit has been connected to a source before the switch turns on, so that $I_L(0^-)=0.04A$ *and* $V_C(0^-)=8V$. These values are shown in the figure as *initial conditions* (IC). Component values $L=8mH$ and $C=2mF$ are fixed, whereas the resistor assumes the following numbers: (a) 40 Ω, (b) 4 Ω, and (c) 0.4 Ω.

Find the circuit's natural responses for every value of R.

Solution:

Let's start with the problem interpretation: We understand that three resistor values refer to the three cases of the natural circuit response. Nevertheless, it's worth calculating the circuit equation's fixed coefficients to put our understanding on a quantitative basis. Thus,

$$2\sqrt{\frac{L}{C}}=2\sqrt{\frac{8\cdot 10^{-3}}{2\cdot 10^{-3}}}=4\ \Omega=R_{cd}.$$

Therefore, $R_{cd}=4\ \Omega$ constitutes a *critically damped* case, $R_{od}=40\ \Omega$ creates an *overdamped* case, and $R_{ud}=0.4\ \Omega$ is responsible for the *underdamped* case. We can check our understanding by applying the other criterion for a critically damped case, namely $\omega_0^2=\alpha_{cd}^2$. Calculating the *resonant frequency* gives

$$\omega_0^2=\frac{1}{LC}=\frac{10^6}{16}=6.25\cdot 10^4\left(\frac{rad}{s}\right)$$

1 **John Neper** (1550--1617), was a Scottish intellectual who made significant contribution in theology, technology, and mathematics. For engineers and mathematicians, his name is forever associated with logarithms that Neper developed as a mathematical tool to facilitate calculations, especially multiplications. His concept, $\ln(x\cdot y)=lnx+lny$, along with the later developed slide ruler, worked well for over 300 years until mechanical and then electronic calculators arrived. Neper's natural logarithms used the base *e*; the modern logarithms on base 10 were developed later.

and

$$\alpha_{cd}^2 = \left(\frac{R_{cd}}{2L}\right)^2 = \left(\frac{4}{16 \cdot 10^{-3}}\right)^2 = 6.25 \cdot 10^4 \left(\frac{rad}{s}\right),$$

we confirm that $R_{cd} = 4\Omega$ indeed creates the critically damped case.

Obtaining these clarifications enables us to apply the theory developed in this section to solve the problem and compute the numbers.

a) Overdamped case:

Compute the coefficients

$$\alpha_{od} = \frac{R_{od}}{2L} = \frac{40}{16 \cdot 10^{-3}} = 2.5 \cdot 10^3 \left(\frac{1}{s}\right),$$

and

$$k_{od} = \sqrt{\left(\alpha_{od}^2 - \omega_0^2\right)} = \sqrt{\left(6.25 \cdot 10^6 - 6.25 \cdot 10^4\right)} \approx 2.487 \cdot 10^3 \left(\frac{1}{s}\right).$$

Calculate the output voltage using (5.26)

$$v_{out-od}\left(t\right) = \frac{-V_0' - \left(\alpha - k\right)V_0}{2k} e^{-\left(\alpha + k\right)t} + \frac{V_0' + \left(\alpha + k\right)V_0}{2k} e^{-\left(\alpha - k\right)t}$$
$$= \frac{39916}{4974} e^{-13t} - \frac{84}{4974} e^{-4987t}.$$

The MATLAB graph built using the above formula is shown in Figure 5.3b. This case is resolved.

Discussion of the overdamped case:

- How to verify the answer? MATLAB calculates the formulas based on our entry values and builds the corresponding graph. Thus, MATLAB's results are not independent solutions and can't be used for the answer validation. But Multisim performs an independent action—simulates the circuit operation—and as such, it verifies our result.
- To quantitatively compare our answer with Multisim simulation, we need to make the measurements of both MATLAB and Multisim results. But what results should we compare? The most manageable quantity to equate is the duration of a transient process (*transient interval*) because this process is described by a decaying exponential in all three cases. What's more, the result of Multisim's measurement is the graph of a decaying exponential function. We studied how to measure the parameters of such a function in Sidebar 4S.2. Here, we apply this knowledge to our example as follows:
  - Find the initial amplitude, V_0, by setting $t = t_0$ in $v_{out-od}(t)$. We got $V_0 = 8(V)$ as must be.
  - Compute the final amplitude V_{tf} at $t = t_f$, where $t_f = 5\tau$, as $V_{tf} = V_0 \cdot e^{-\frac{t_f}{\tau}} = V_0 \cdot e^{-5} = V_0 \cdot 0.0067(V) = 0.0536(V)$. Here, τ is the time constant of an RLC circuit.
  - The time measured between V_{tf} and V_0, that is, $\Delta t(s) = t_f - t_0$, is the *transient interval.*

We make these measurements with the MATLAB graph given in Figure 5.3b: Here, $V_0 = \frac{39916}{4974} - \frac{84}{4974} = 8V$. Thus, $V_{tf} = V_0 \cdot 0.0067 = 0.0536V$. Clicking "Tools" and using "Data Tips" icons enable us to place the mouse pointer on the MATLAB graph by the left click. The small rectangle with the current coordinates of this poiint will appear. Moving the mouse pointer along the MATLAB graph, we find the point where $Y \approx 0.0536V$. The corresponding X gives the transient interval in search. The data tips box in Figure 5.3b, where X stands for t_f, and Y stands for V_{tf}, reads

$X0.3854$
$Y0.0535228.$

Here, $X = Time(s)$ and $Y = v_{out}(t)$. Don't forget that in general the transient time interval is defined as $\Delta t = t_f - t_0$, but in this example $t_0 = 0$.

To make the *measurements with* Multisim, we transfer the graph from the oscilloscope screen to the Grapher; the latter is displayed in Figure 5.3c. Next, we click "Show cursor" in the Cursor window, place one cursor at the starting spot, where the amplitude reaches its V_0 value, and the second cursor at the transient final point, where $V_{tf} = 0.0067V_0$. It is done by looking at their amplitude values, $y1$ and $y2$, in the cursor popup window, which appears with the click on "Show cursor" icon. For our example, this window is shown in Figure 5.3d, where $y1 = V_0 = 7.9475V$, and $y2 = V_{tf} = 54.0563mV$. The corresponding time points are $x1 = t_0 = 760\,\mu s \approx 0$ and $x2 = t_f = 399.0135ms$; these values are boxed in Figure 5.3d. The difference, $x2 - x1 = t_f - t_0 = \Delta t \approx 398.2535ms$, is the duration of the transient process, as shown in Figure 5.3c.

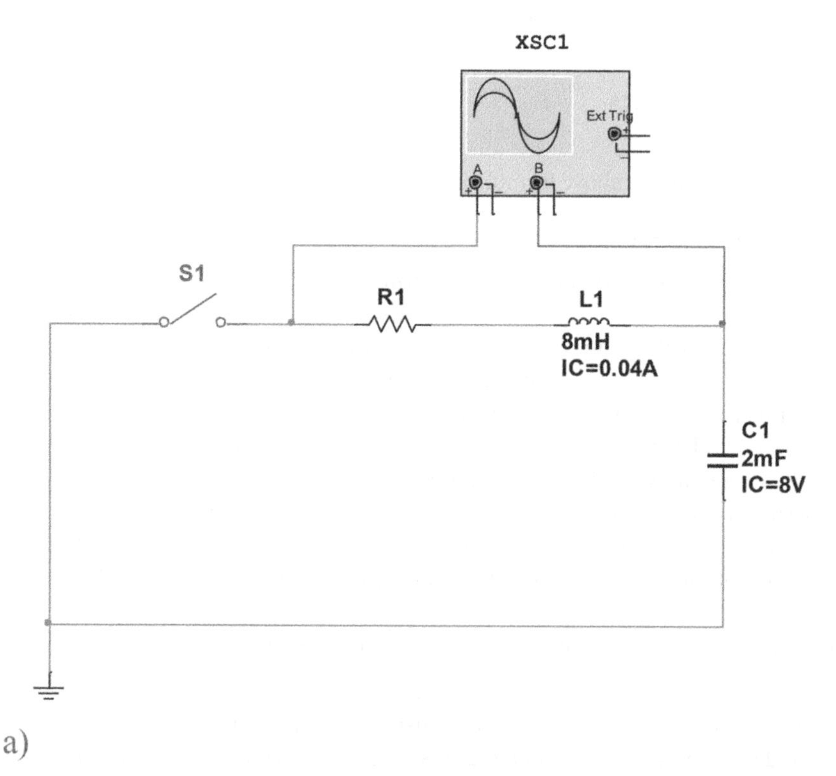

Figure 5.3 MATLAB and Multisim presentations of the series RLC circuit for Example 5.1: a) Circuit's schematics; b) graph of the overdamped response $v_C(t) \equiv v_{out}(t)$ build with MATLAB; c) graph of the overdamped response build by Multisim; d) Multisim cursor measurement table for the overdamped case; e) Multisim graph of the critically damped case; f) Multisim graph of underdamped case.

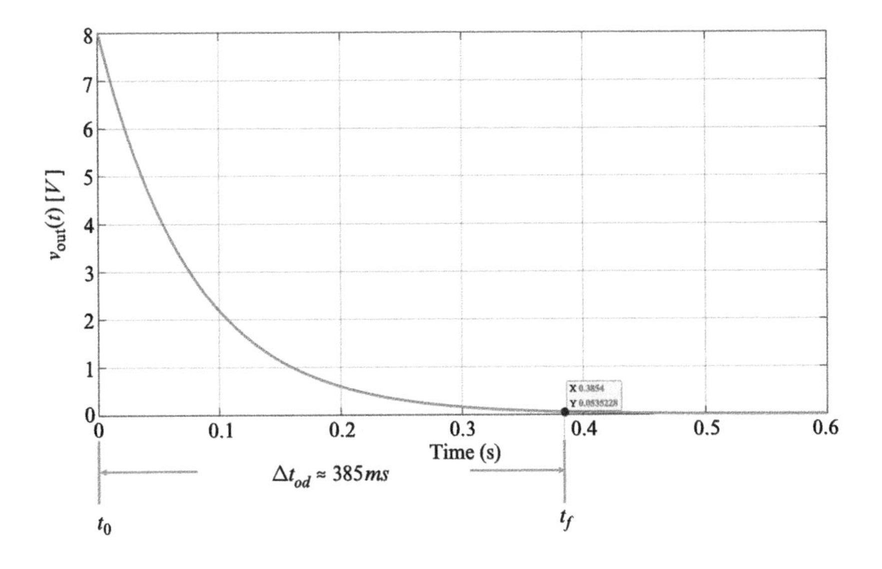

b)

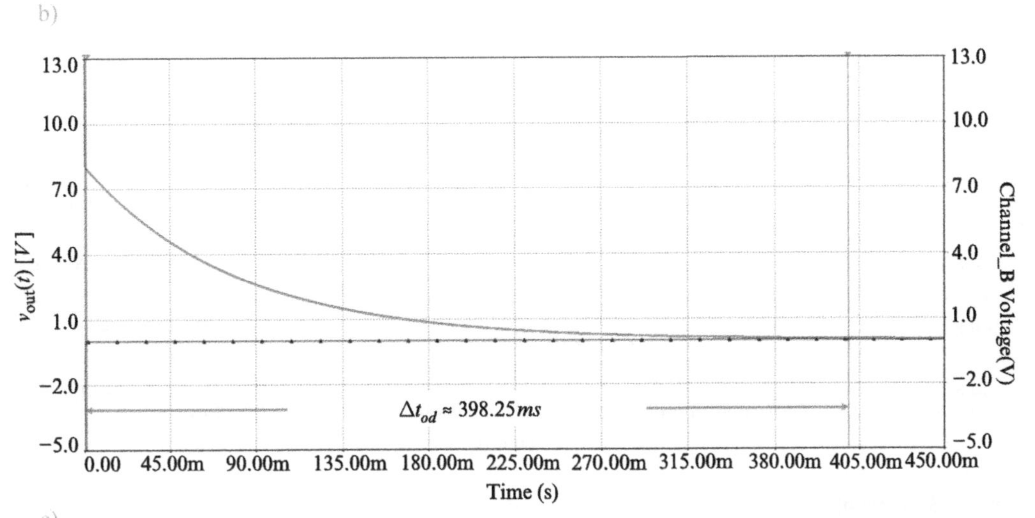

c)

Cursor

	Channel A	Channel B
x1	760.9921µ	760.9921µ
y1	19.3784µ	7.9475
x2	399.0135m	399.0135m
y2	135.4918n	54.0563m
dx	398.2525m	398.2525m
dy	-19.2429µ	-7.8934
dy/dx	-48.3183µ	-19.8201
1/dx	2.5110	2.5110

Grapher measurements – overdamped case, Channel A – V_{in} (V), Channel B – V_{out} (V), x – *time axis* (ms), y – *magnitude axis*(V)

d)

Figure 5.3 (Cont'd)

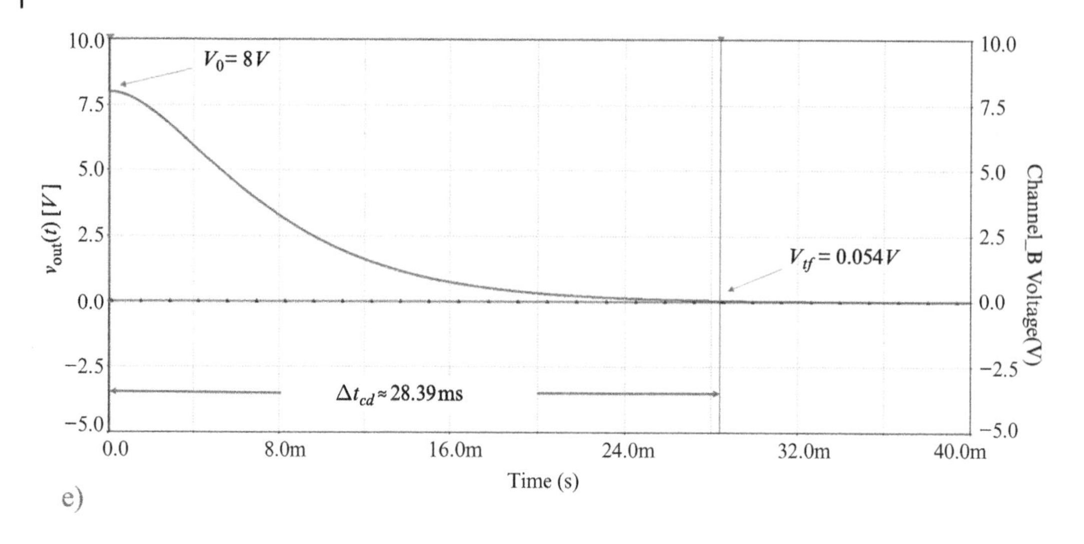

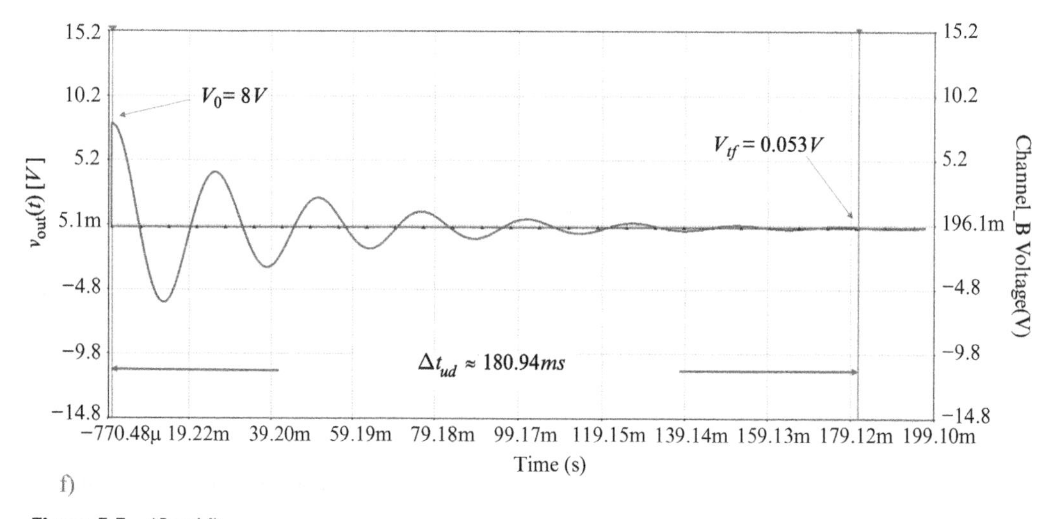

Figure 5.3 (Cont'd)

We see the small—less than 15 ms—time discrepancy between the MATLAB and Multisim measurements, which is inevitable due to the difference between these two platforms' techniques and approximation caused by manual pointing to the measured positions. This difference is less than 4% and is acceptable.

b) Critically damped case

First, we compute

$$\alpha_{cd} = \frac{R}{2L} = \frac{4}{0.016} = 250\left(\frac{1}{s}\right)$$

Next, plugging this and all other values into (5.30) yields

$$v_{out-cd}(t) = V_0 e^{-\alpha_{cd}t} + \left(V_0' + \alpha_{cd} V_0 t\right) e^{-\alpha_{cd}t} = 8e^{-250t} + \left(250 \cdot 8t\right) e^{-250t} = 8e^{-250t} + 2000te^{-250t}.$$

Relegating computational work to MATLAB, we receive from its graph $\Delta t = t_f - t_0 = 0.0284s$at $V_{tf} \approx 0.0534V$.

The case is resolved.

Discussion of the critically damped case:

To compare MATLAB and Multisim results, we conduct the Multisim simulation and obtain the result shown in Figure 5.3e, where $V_{tf} = 0.054V$ and $\Delta t = t_f = 0.02839s$, which practically coincide with the MATLAB calculation.

Observe that the graph of the critically damped transition smoothly decays to zero, and its course is very similar to that of the overdamped graph. The difference—and a big one—is in the values: The overdamped graph needs almost 400 ms to decay, whereas the critically damped chart descends to V_f much faster, for 28.4 ms only.

c) Underdamped case:

Equation 5.38 describes the RLC circuit underdamped natural response as

$$v_{out-ud}(t) = Be^{-\alpha_{ud}t}\cos(\omega_d t - \theta), \quad (5.38R)$$

where $B = \sqrt{(V_0)^2 + \left(\frac{\alpha_{ud}V_0 + V_0'}{\omega_d}\right)^2}$ and $\theta = tan^{-1}\left(\frac{\alpha_{ud}V_0 + V_0'}{V_0\omega_d}\right)$. For this example, we compute the following parameters:

$$\alpha_{ud} = \frac{R_{ud}}{2L} = \frac{0.4}{16 \cdot 10^{-3}} = 25\left(\frac{1}{s}\right),$$

$$\omega_d\left(\frac{\text{rad}}{\text{s}}\right) = \sqrt{\omega_0^2 - \alpha_{ud}^2} = \sqrt{625 \cdot 10^2 - 625} = 248.75\left(\frac{rad}{s}\right),$$

$$B = \sqrt{(12)^2 + \left(\frac{25 \cdot 12}{248.75}\right)^2} = \sqrt{144 + \left(\frac{300}{248.75}\right)^2} = 12.05V,$$

and

$$\theta = tan^{-1}\left(\frac{\alpha_{ud}V_0 + V_0'}{V_0\omega_d}\right) = tan^{-1}\left(\frac{\alpha_{ud}V_0}{(V_0\omega_d}\right) = tan^{-1}\left(\frac{\alpha_{ud}}{(\omega_d}\right) = tan^{-1}\left(\frac{25}{248.75}\right) = 5.74^0.$$

Inserting these values into (5.38) yields

$$v_{out-ud}(t) = Ae^{-\alpha_{ud}t}\cos(\omega_d t - \theta) = 12.05e^{-25t}\cos(248.75t - 5.74^0).$$

The case is resolved.

Discussion of the underdamped case:

- Figure 5.3f shows the Multisim graph for this case. As (5.38) predicts, the underdamped natural response of a series RLC circuit is a decaying sinusoid. Since the decay of this sinusoid is determined by only one exponential function, we can compute its *transient interval* as

$$5\tau_{ud} = \frac{5}{\alpha_{ud}} = 200ms.$$

The measurements of the Multisim graph deliver a slightly different result, $\Delta t = 180.9ms$, and the MATLAB calculations show $\Delta t = 192.3ms$. The difference doesn't exceed 9% and it demonstrates that the tools used in our academic exercise are not as perfect as desired; however, the result is still acceptable to validate our solution.

- How many cycles of the damping oscillations do we want to see? This, in fact, is a design question. Do we need to bring our circuit to the final (rest) state fast or is our goal to make the circuit continue to oscillate for a long time? It depends on the designer's task. But our analysis of an underdamped case enables us to quickly evaluate the oscillation issue. Indeed, the *transient interval* is $\Delta t(s) = 5\tau_{ud}(s) = \frac{5}{\alpha_{ud}}$. The period of damping oscillations is $T_d(s) = \frac{2\pi}{\omega_d} \approx \frac{2\pi}{\omega_0}$ because $\omega_d\left(\frac{\text{rad}}{\text{s}}\right) = \sqrt{\omega_0^2 - \alpha_{ud}{}^2}$ and $\alpha_{ud} < \omega_0$ in an underdamped case. Thus, the ratio of the process duration to one oscillation period gives the number of cycles of damping oscillations, N_d, as

$$N_d = \frac{\Delta t(s)}{T_d(s)}. \tag{5.39}$$

In Example 5.3, $\Delta t = 180.94ms$ and $T_d = \frac{2\pi}{\omega_d} = 25.26ms$, which results in $N_d = 7.2$ cycles.

The Multisim graph shown in Figure 5.3f counts slightly more than seven cycles, close to our calculations.

If we accept the transient process duration as $\Delta t(s) = 3\tau_{ud}(s)$—which means that $V_f = 0.005V_0$, another decaying criterion—and $\omega_0 \approx \omega_d$—a typical case in practice—then (5.39) can be transformed as

$$N_d = \frac{\Delta t(s)}{T_d(s)} = \frac{3\omega_0}{2\alpha\pi} \approx \frac{\omega_0}{2\alpha} = Q, \tag{5.40}$$

where

$$Q = \frac{\omega_0}{2\alpha} \tag{5.41}$$

is a *quality factor*. Thus, the quality factor is approximately equal to the number of cycles in the underdamped case for a series RLC circuit. The greater the quality factor, the more cycles for a transient interval. Also, the greater Q gets closer to N_d. (More about quality factor in a series RLC circuit can be found in Mynbaev and Scheiner 2020, pp. 405–406.)

Interestingly that a *damping frequency*, ω_d, is sometimes called a *ringing frequency*. Do you guess why?

- There is the other way to find constants A_1 and A_2 in an underdamped case. We know that the natural RLC circuit response has the following form

$$v_{out-ud}(t) = e^{-\alpha_{ud}t}(B_1 cos\omega_d t + B_2 sin\omega_d t). \tag{5.42}$$

Taking the derivative of (5.42) and setting $t = 0$ in both equations results in

$$\begin{cases} B_1 = V_0 \\ -\alpha_{ud}B_1 + B_2\omega_d = V_0' \end{cases}. \tag{5.43}$$

Solving this equation system yields

$$\begin{cases} B_1 = V_0 \\ B_2 = \dfrac{V_0' + \alpha_{ud} V_0}{\omega_d}, \end{cases} \tag{5.44}$$

as in (5.36).

Sidebar 5S.1 How to configure the circuit's initial conditions with Multisim

Ever wondering how to create the desired initial conditions performing the Multisim simulation? There are two main steps to do this:

1) In the *user-defined approach, assign* the initial values to the components. For this, double-click on an inductor or capacitor, go to **Value** tab, find **Initial conditions** checkbox, enter the desired value, and click **OK** to confirm. (Why only inductor or capacitor? Explain.)
2) Command Multisim to use those initial conditions that you have set by performing the following operations:
   a) Select **Simulate.**
   b) From **Analyses and Simulation** list, select the **Transient** section.
   c) Choose **Analysis parameters.**
   d) Click **Initial Condition.** From this list, you can choose one of the following options:
      i) Set to zero.
      ii) User-defined.
      iii) Calculate DC operating point.
      iv) Automatically determine initial condition.
   e) After making a choice, click **OK**.
   f) Click **Simulate** to start the simulation.

Figures 5S1.1a, 5S1.1b, and 5S1.1c provide examples of setting the initial conditions in an RLC circuit.

The series RLC circuit in Figure 5.4a has the initial conditions *assigned* according to the above procedure: The inductor's initial current is $I_{L0} = 0.04A$, and the capacitor's initial voltage is $V_{C0} = 8V$. (Note that the acronym *IC* in Figures 5.4a and 5.4c stands for *Initial Condition.*) When the switch is turned on, the simulation starts. It seems that the response, the output voltage across the capacitor, should be equal to the applied voltage, 12 volts. However, as the oscilloscope (Grapher) shows in Figure 5.4b, the initial voltage across the capacitor is 8 volts. This is because the capacitor voltage cannot change instantaneously, as we know, and it obeys the rule $v_C(0^-) = v_C(0^+)$. Since we assigned $v_C(0^-) = 8V$, voltage $v_C(0^+)$ must be of the same value. This is how the *user-defined initial condition* works in practice. In the graph in Figure 5.4b, you might notice a slight increase in the $v_C(t)$ value immediately after the switch has been thrown. This increase reflects the attempt of $v_C(t)$ to reach 12 volts, the attempt that is immediately suppressed by the decaying exponential functions describing the natural response.

Figure 5.4c demonstrates the *parallel RLC circuit*, where the initial conditions are determined automatically. Indeed, as long as the switch is open, the *series circuit* is formed by the 12-volt DC source and resistors R1 and R2. This is because the inductor is replaced by a shorten circuit

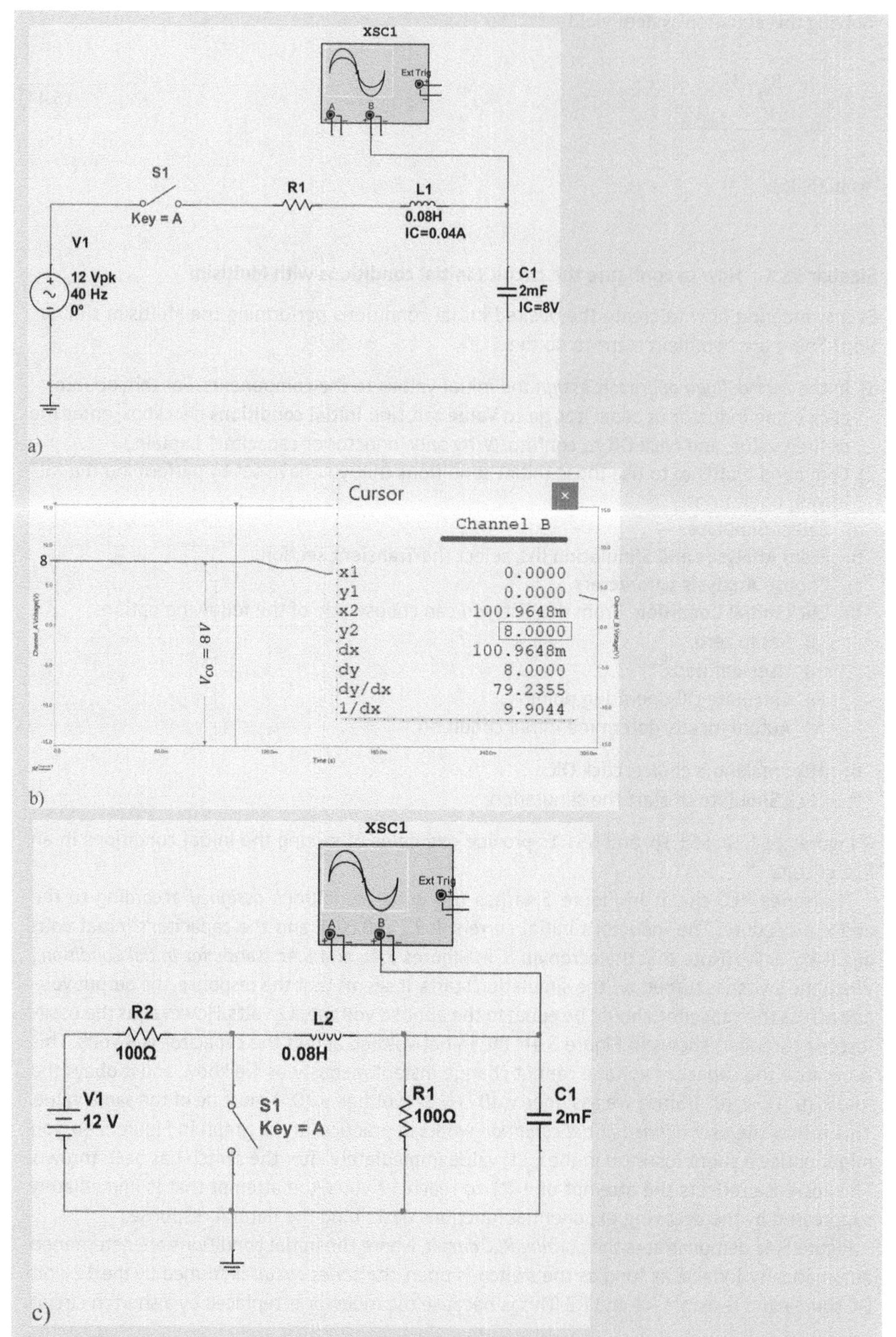

Figure 5S1.1 Examples of configuring initial conditions for transient analysis of a series RLC circuit: a) Series RLC circuit with user-defined initial conditions; b) the series circuit response showing 8-volts initial point; c) parallel RLC circuit with the automatically determined initial condition; d) the parallel circuit response showing 6-volts starting point.

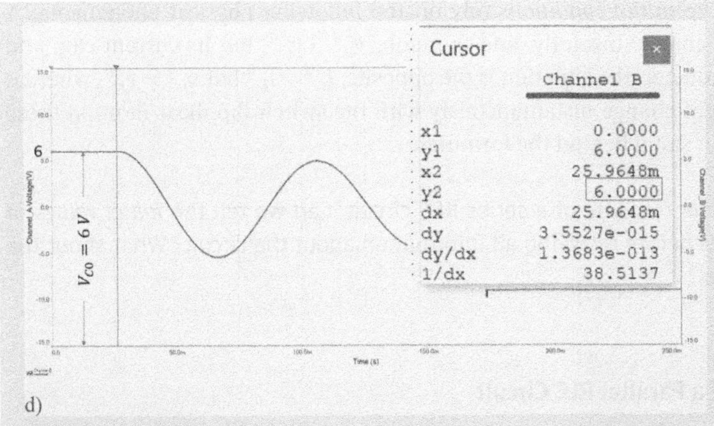

Figure 5S1.1 (Cont'd)

and the capacitor by the open circuit. (Sketch this equivalent circuit for yourself.) The current flowing through these two resistors and the zero-resistance inductor is

$$I_{L0} = \frac{V1}{R1 + R2} = 0.06A.$$

This is the initial inductor's current. The voltage across the open capacitor's terminals equals the voltage drop across resistor R2, which is

$$V_{C0} = I_{L0} \cdot R2 = 6V.$$

Therefore, this circuit automatically generates the initial conditions after the switch is thrown. We restrict ourselves to a conceptual review and skip the details of measuring the initial currents. However, this discussion clarifies configuring the initial conditions in Multisim operations and hints at how to develop a laboratory exercise on this topic.

5.1.6 The Role of Initial Conditions in the Natural Response of a Series RLC Circuit

Example 5.1 in general and Solutions 5.26, 5.30, and 5.37 in particular show that the circuit's natural response depends on the relationship between R and $2\sqrt{\frac{L}{C}}$, or equivalently, $\omega_0^2 = \frac{1}{LC}$ and $\alpha = \frac{R}{2L}$. This is why we have three different cases. However, in each case, the course of the natural response is determined by the *initial conditions*. Consider, for example, the *critically damped* solution given in (5.30). An analysis of this solution, which involves some calculus, shows that we might obtain three possible graphs:

i) The output voltage initially increases but, after reaching the maximum value, starts to diminish, tending to zero. This graph is shown in Table 5.1.
ii) The graph smoothly transits from the initial point, V_0, to zero. It is shown in Figure 5.3c
iii) The $v_{out}(t)$ graph goes down sharply and intersects the time axis. Then, after reaching the minimum value, it approaches zero. (No oscillations, just one intersection of the time axis.)

The actual course of the $v_{out}(t)$ graph depends on the interplay between the values of the initial conditions, V_0 *and* V_0', and α. Regardless of the possible first trends, the *natural circuit response* eventually tends to zero because the capacitor—the only circuit component capable of storing dc energy—will inevitably discharge. You are encouraged to conduct a similar analysis of two other RLC circuit natural response cases.

We must remember that the *initial conditions* rely on the following physical phenomena: A capacitor's voltage, v_C, can't change instantly, and therefore $v_C^{-0} = v_C^{+0}$, but its current can, and therefore $i_C^{-0} \neq i_C^{+0}$. For the inductor, the situation is the opposite: $i_L^{-0} = i_L^{+0}$ but $v_L^{-0} \neq v_L^{+0}$, whereas the resistor voltage and current change instantaneously with the switch flip. Bear in mind these fundamental physics laws that stand behind the formulas.

Question Consider the *natural response* of a series RLC circuit. Can we tell the *initial values* of these circuit's three responses before receiving all information about the circuit? What about the *final values* of these responses?

5.1.7 Natural Response of a Parallel RLC Circuit

5.1.7.1 Application of the Duality Principle to Deriving the Differential Equation of a Parallel RLC Circuit

Suppose we need to find the response of a parallel RLC circuit. Does it mean we must follow the routine developed for a series RLC circuit in this section? Do we need to repeat all the preceding mathematical manipulations? Not so. For one, we can save time deriving the parallel RLC circuit differential equation by applying the *duality principle* discussed in Section 3.3. Excerpts from Table 3.6R are reproduced here to remind us about duality meaning and specific dual electric pairs. Based on these pairs, the differential equations of series and parallel RLC circuits have been related in Section 3.3 as follows: The differential equation of a *series RLC circuit* is given by

$$v_{in}(t)(V) = v_R(t) + v_L(t) + v_C(t) = Ri(t) + L\frac{di(t)}{dt} + \frac{1}{C}\int_T^0 i(t)dt. \tag{5.45}$$

Exchanging currents and voltages, resistance and admittance, capacitance and inductance, and differentiation and integration, as Table 3.6R requires, results in the following differential equation for a *parallel RLC circuit*:

$$i_{in}(t)(A) = i_R(t) + i_L(t) + i_C(t) = \frac{v(t)}{R} + C\frac{dv(t)}{dt} + \frac{1}{L}\int_0^T v(t)dt. \tag{5.46}$$

Table 3.6R Dual electrical pairs (excerpts).

Current	Voltage
KCL	KVL
Conductance	Resistance
Inductance	Capacitance
Admittance	Impedance
Susceptance	Reactance
Parallel circuit	Series circuit
Current division	Voltage division

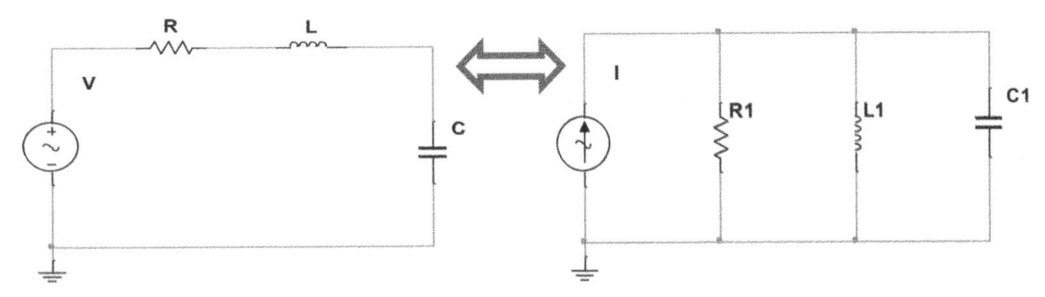

Figure 3.23bR Series and parallel RLC circuits.

Figure 3.23b, reproduced here for convenience, visualizes the duality of these RLC circuits. (Do you understand why Equation 5.46 is the *integro-differential equation of a* ***parallel*** *RLC circuit*? Explain.) A close look at (5.46) and RHS of Figure 3.23b reveals that this equation is simply KCL for a parallel RLC circuit.

5.1.7.2 Three Cases of the Natural Response of a Parallel RLC Circuit

Equation 5.46 for a natural (source-free or zero-input) case accepts the following form:

$$\frac{v_{out}(t)}{R} + C\frac{dv_{out}(t)}{dt} + \frac{1}{L}\int_0^T v_{out}(t)dt = 0.$$

Differentiating this equation once, we arrive at the familiar format of a second-order circuit equation:

$$v''_{out} + \frac{1}{RC}v'_{out} + \frac{1}{LC}v_{out} = 0. \tag{5.47}$$

Let's recall the formula for the resonant frequency,

$$\omega_0^2\left(\frac{rad}{s}\right) = \frac{1}{LC}, \tag{5.5aR}$$

and introduce new formula for a *damping coefficient (constant) for a parallel RLC*,

$$\alpha_{par}\left(\frac{1}{s}\right) = \frac{1}{2RC}. \tag{5.48}$$

To highlight the sameness and difference between notations of a series and parallel RLC circuit, we collect all the notations in Table 5.2.

Using one letter α to designate two different subjects might be confusing. However, the same physical meaning of both quantities—it is a *damping coefficient*—justifies this notation. Usually, the text content clarifies which α implies in an equation. If necessary, we will use an additional subscript to specify which circuit a damping coefficient describes, as we have done in Table 5.2 and (5.48).

The introduced designations transform (5.47) into the following equation:

$$v''_{out} + 2\alpha_{par}v'_{out} + \omega_0^2 v_{out} = 0. \tag{5.49}$$

Table 5.2 Coefficients of the differential equations of series and parallel RLC circuit.

	Series RLC circuit		Parallel RLC circuit	
Damping coefficient (constant)	$\alpha_{ser}\left(\frac{1}{s}\right)=\frac{R}{2L}$	(5.5b)	$\alpha_{par}\left(\frac{1}{s}\right)=\frac{1}{2RC}$	(5.48)
Resonant frequency	$\omega_0\left(\frac{rad}{s}\right)=\sqrt{\frac{1}{LC}}$	(5.5a)	The same as for series RLC circuit.	
Critically damped criterion	$\alpha_{ser}=\omega_0 \Rightarrow R_{cd}(\Omega)=2\sqrt{\frac{L}{C}}$		$\alpha_{par}=\omega_0 \Rightarrow R_{cd}(\Omega)=\frac{1}{2}\sqrt{\frac{L}{C}}$	

By form, (5.49) is undistinguishable from (5.7); the only difference is the content of α, as shown in Table 5.2. Therefore, *all the manipulations applied to (5.7) for the series RLC circuit analysis can be readily used in studying a parallel RLC circuit*. This deduction enables us to turn our consideration to an example. It's worth reminding that the solution to (5.49) takes the form

$$v_{out}(t)=A_1e^{m_1t}+A_2e^{m_2t} \tag{5.12R}$$

whose exponential constants are

$$m_{1,2}=-\alpha\pm\sqrt{\left(\alpha^2-\omega_0^2\right)}=-\alpha\pm \mathrm{k}. \tag{5.11R}$$

This introduction prepares us to consider an example.

Example 5.2 Natural responses of a parallel RLC circuit.

Problem: Consider the parallel RLC circuit demonstrated in Figure 5.4a. The component and the initial condition (IC) values are: $L1=8H, C1=2mH,\ I_{L0}=0.04A,\ and\ V_{C0}=8V$. They are fixed and shown on the circuit schematic. The resistor R value varies to provide (a) overdamped case, (b) critically damped case, and (c) underdamped case. Determine the resistor required values and find the circuit's natural responses for each case.

Solution: Refer to Example 5.1 for the general routine of solving such a problem. Here, we explore only the peculiarities of a parallel RLC circuit and avoid repeating the points common with a series circuit.

Our problem boils down to considering three cases: overdamped, critically damped, and underdamped. We need to start with a critically damped case to determine R_{cd}.

The value of a resonant frequency for this circuit is

$$\omega_0^2\left(\frac{rad}{s}\right)^2=\frac{1}{LC}=\frac{1}{8\cdot 2\cdot 10^{-3}}=62.5.$$

For critically damped case, the damping coefficient $\alpha_{cd}\left(\frac{1}{s}\right)$ must be equal to $\omega_0\left(\frac{rad}{s}\right)$, which enables us to compute the value of R_{cd} as:

$$\alpha_{cd}=\omega_0=\sqrt{62.5}=7.905\left(\frac{1}{s}\right)=7.9\left(\frac{1}{s}\right)$$

and

$$\alpha_{cd} = \frac{1}{2R_{cd}C} \Rightarrow \boldsymbol{R_{cd}} = \frac{1}{2\alpha_{cd}C} = 31.6(\Omega).$$

Observe that α and R are inversely proportional in a parallel RLC circuit, whereas in a series RLC, they are directly proportional.

We assume $R_{od} = 4(\Omega)$ and $R_{ud} = 400(\Omega)$. Then the values of a damping constant for two other cases become

$$\alpha_{od}\left(\frac{1}{s}\right) = \frac{1}{2R_{od}C} = \frac{1}{2 \cdot 4 \cdot 2 \cdot 10^{-3}} = 62.5 \text{ and } \alpha_{ud}\left(\frac{1}{s}\right) = \frac{1}{2R_{ud}C} = \frac{1}{2 \cdot 400 \cdot 2 \cdot 10^{-3}} = 0.625.$$

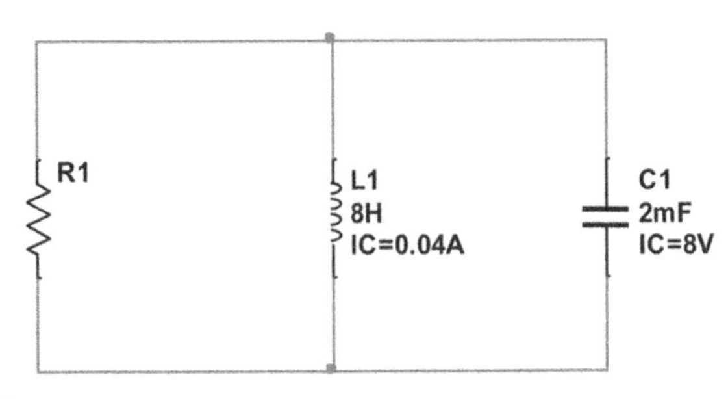

a)

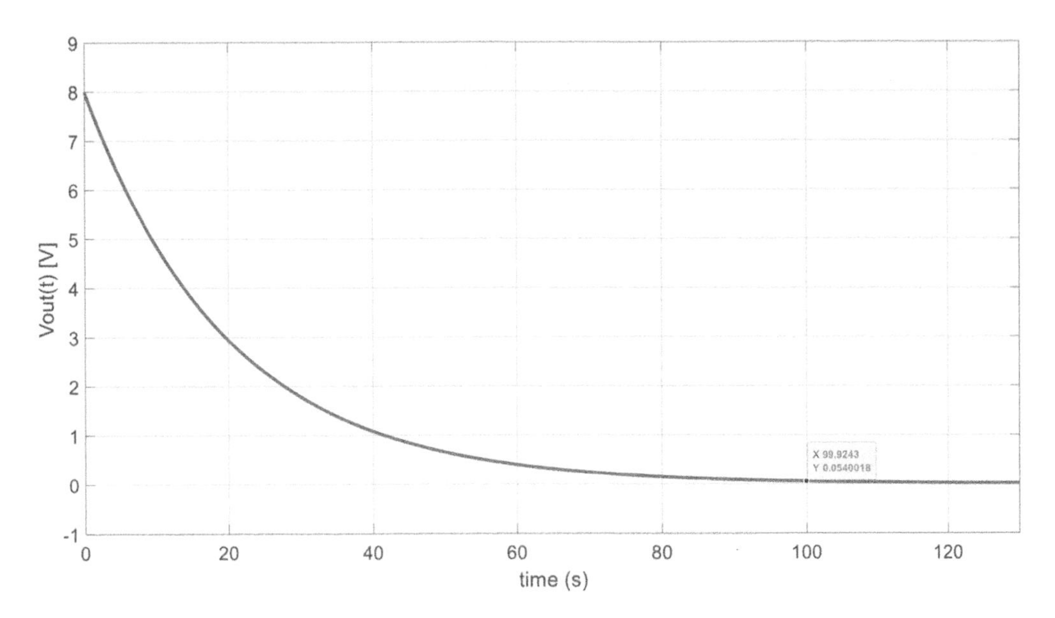

b)

Figure 5.4 Parallel RLC circuit for Example 5.2: a) The circuit's schematic; b) overdamped case; c) critically damped case; d) underdamped case.

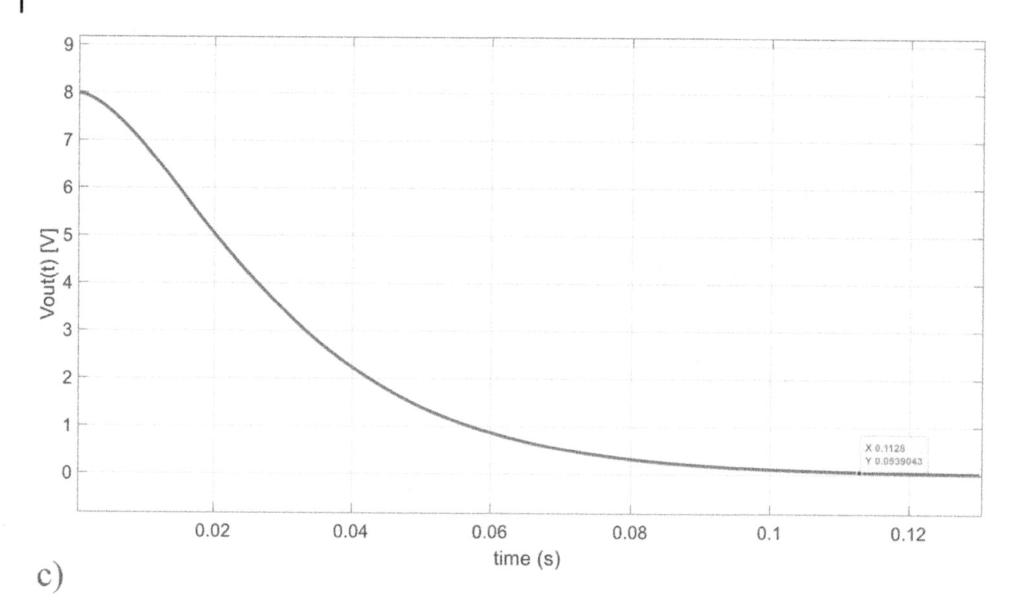

c)

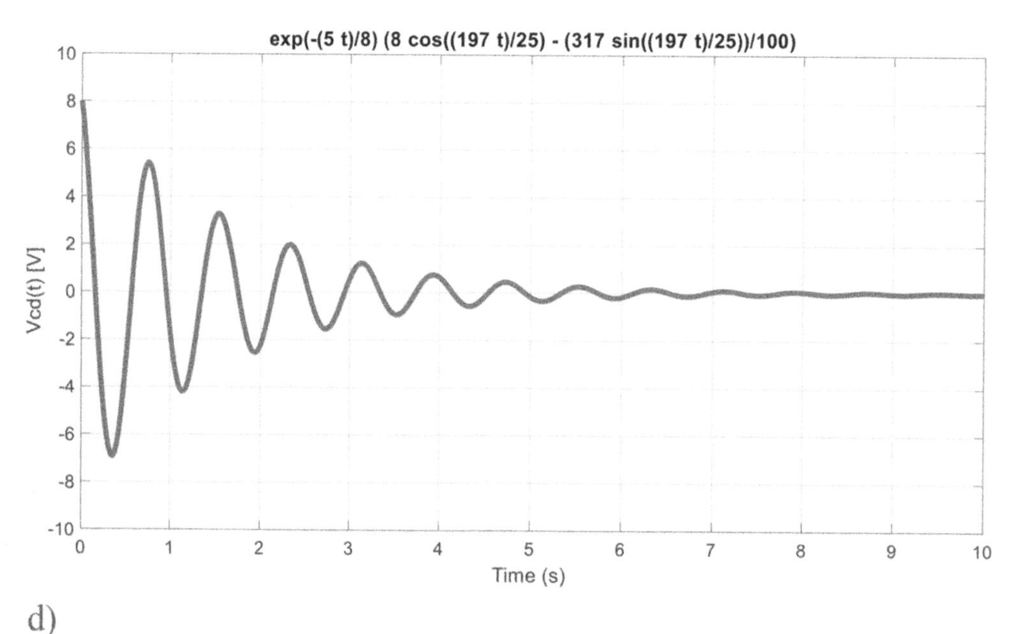

d)

Figure 5.4 **(Cont'd)**

Now, the values of m_1 and m_2 can be computed as

$$m_{1\text{-od}} = -\alpha_{od} + \sqrt{\left(\alpha_{od}^2 - \omega_0^2\right)} = -\alpha_{od} + \mathrm{k}_{od} = -62.5 + 61.99 = -0.502 = 0.5$$

$$\text{and } \; m_{2\text{-od}} = -\alpha_{od} - \mathrm{k}_{od} = -124.5.$$

Also,

$$m_{1\text{-cd}} = m_{2\text{-cd}} = -\alpha_{cd} = -7.9.$$

And

$$m_{1-\mathrm{ud}} = -\alpha_{ud} + \mathrm{j}\sqrt{\left(\omega_0^2 - \alpha_{ud}^2\right)} = -\alpha_{ud} + \mathrm{j}\omega_d = -0.625 + j7.88$$
$$\text{and } m_{2-\mathrm{ud}} = -\alpha_{ud} - \mathrm{j}\omega_d = -0.625 - j7.88.$$

These computations have been prepared for considering the three cases:

a) Overdamped case

The solution to this case is obtain by plugging $m_{1-\mathrm{od}}$ and $m_{2-\mathrm{od}}$ in (5.12R), which gives

$$v_{out-od}(t) = A_1 e^{-0.05t} + A_2 e^{-124.5t}. \tag{5.12R}$$

To find constants A_1 and A_2, we set $t = 0$ in (5.12) and attain

$$V_0 = A_1 + A_2 \Rightarrow 8 = V_0 = A_1 + A_2. \tag{5.50a}$$

The second equation can be found by differentiating (5.12) once and evaluating the derivative at $t = 0$. This operation gives

$$v'_{out-od}(0) = -0.05A_1 - 124.5A_2, \tag{5.50b}$$

where $v'_{out-od}(0) = V'_{C0} = V'_0$. But solving the equation system (5.50a) and (5.50b) requires finding the value of $v'_{out-od}(0) = V'_0$ in (5.50b) independently, which can be done by applying KCL to the parallel RLC and evaluating it at $t = 0$. So,

$$I_{R0} + I_{L0} + I_{C0} = 0.$$

Here, $I_{R0} = 0$ because there is no initial current through $R1$ due to zero initial voltage across it, $I_{L0} = I_0 = 0.04A$ by the problem condition, and $I_{C0} = Cv'_C(0) = Cv'_{out-od}(0) = CV'_0$ by definition. Thus, $I_{C0} = CV'_0$, and KCL at $t = 0$ yield

$$I_0 + I_{C0} = 0 \Rightarrow I_{C0} = -I_0 \Rightarrow V'_0 \equiv \frac{I_{C0}}{C} = -\frac{I_0}{C} = -\frac{0.04}{2 \cdot 10^{-3}} = -20\left(\frac{V}{s}\right).$$

Inserting this value of V_0' into (5.50b) produces the following equation system from (5.50a) and (5.50b):

$$\begin{cases} A_1 + A_2 = 8 \\ 0.05A_1 + 124.5A_2 = 20 \end{cases}. \tag{5.51}$$

Relegating the solution of (5.51) to MATLAB yields

```
>>>>>>>>>>>>>>>>>>>>>>>>>>
>> A = [1 1;
0.05 124.5];
B = [8; 20];
V=linsolve(A,B)
V =
7.8425
0.1575
>>>>>>>>>>>>>>>>>>>>>>>>>>
```

Thus, $A_1 = 7.84$ and $A_2 = 0.16$.

Therefore, the parallel RLC circuit operation in the *overdamped case* is described by the following equation:

$$v_{out-od}(t) = 7.84e^{-0.05t} + 0.16e^{-124.5t}. \tag{5.52}$$

The corresponding graph built with MATLAB is shown in Figure 5.4b. The graph is a smoothly damping exponential function. For making the measurements on such a graph, let's use 5τ criterion: If $t_f = 5\tau$, then $e^{-\frac{t_f}{\tau}} = e^{-5}$, and $V_{tf} = e^{-5} \cdot V_0$. Hence, $\Delta t \approx 99.9\,(s)$ at $V_{tf} = e^{-5} \cdot V_0 = 0.0539V$. The measurements shown in the data-tips box on the graph in Figure 5.4b validate these results. Analysis of (5.52) also supports this result: Since the second member of (5.52) diminishes very quickly (do you know why?), and its initial amplitude is negligibly small, it plays practically no role in the whole process. Thus, the time constant of $v_{out-od}(t)$ is determined by its first member, $\tau \approx \frac{1}{0.05} = 20s$, which gives $\Delta t = 5\tau = 100s$. This number coincides with the MATLAB result.

The case is resolved.

b) Critically damped case

The solution to (5.49) is given by (5.28) as

$$v_{out-cd}(t) = (A_1 + A_2 t)e^{-\alpha_{cd}t}.$$

Its derivative is

$$v'_{cd} = (A_1 + A_2 t)e^{-\alpha_{cd}t} = (-\alpha_{cd}A_1)e^{-\alpha_{cd}t} + A_2 e^{-\alpha_{cd}t} - \alpha_{cd}A_2 te^{-\alpha_{cd}t}$$
$$= (-\alpha_{cd}A_1 + A_2 - \alpha_{cd}A_2 t)e^{-\alpha_{cd}t}.$$

Evaluation of these equations at $t = 0$ produces

$$\begin{cases} A_1 = V_0 \\ -\alpha_{cd}A_1 + A_2 = V'_0 \end{cases}.$$

Since V_0 and V'_0 don't depend on α, we can use their values obtained in the preceding case; hence,

$$\begin{cases} A_1 = 8 \\ -7.9A_1 + A_2 = -20 \end{cases}.$$

Thus, $A_1 = 8V$ and $A_2 = 43.2V$. Finally, the solution for this case takes the form

$$v_{out-cd}(t) = (A_1 + A_2 t)e^{-\alpha_{cd}t} = (8 + 43.2t)e^{-7.9t}. \tag{5.53}$$

The MATLAB graph is shown in Figure 5.4c. Measurements on the graph show $\Delta t = 0.8508s$ *at* $V_{tf} = 0.0536V$.

Two notes:

- If we take into consideration only exponential function of (5.53) and use 5τ criterion, the *transient interval* would be $\Delta t' = 5\tau = \frac{5}{\alpha} = \frac{5}{62.5} = 0.08s$. In reality, it's 1.34 times longer due to presence of $43.2t$ factor.

- As we noted previously, the transient process in a critically damped case diminishes much faster than in an overdamped case. In this case, this difference is enormous: $\Delta t_{od} = 100s$ *vs* $\Delta t_{cd} < 0.1s$. In other words, in this example, the critically damped process reduces more than a thousand times faster than the overdamped response.

The case is resolved.

c) Underdamped case

In this case, the solution to an RLC circuit's differential equation (5.28) is a damping sinusoid in the following form:

$$v_{out-ud}(t) = e^{-\alpha_{ud}t}\left(B_1 cos\omega_d t + B_2 sin\omega_d t\right)$$
$$= e^{-\alpha_{ud}t}\left(V_0 cos\omega_d t + \frac{-\alpha_{ud}V_0 + V_0'}{\omega_d} sin\omega_d t\right). \quad (5.37R)$$

All values have been computed in the preceding parts of this example; hence,

$$v_{out-ud}(t) = e^{-0.625t}\left(8cos7.88t - \frac{0.625 \cdot 8 + 20}{7.88} sin7.88t\right)$$
$$= e^{-0.625t}\left(8cos7.88t - 3.17sin7.88t\right). \quad (5.54)$$

The $v_{out-ud}(t)$ graph built with MATLAB is shown in Figure 5.4d. The transient interval is calculated from (5.54) as $\Delta t(s) = 5\tau = \frac{5}{0.625} = 8(s)$, and the final voltage at 5τ is $V_f \cdot e^{-5} = 0.05$. Data tips on the graph in Figure 5.4d reads $X = 7.99(s)$ and $Y = 0.05(V)$.

Using (5.39), the number of cycles of damping oscillations, N_d, can be computed as

$$N_d = \frac{\Delta t(s)}{T_d(s)} = \frac{8 \cdot 4.7.88}{2\pi} \approx 10.3.$$

Figure 5.4d confirms these calculations by showing 10 full five cycles of $v_{out-ud}(t)$ oscillations. The underdamped case is resolved.

Question Equations 5.40 and 5.41 show that it's possible to estimate the number of cycles of damping oscillations by using a quality factor Q. Which value—Q or N_d—will be greater in this example? Prove your answer. Is your result true in general, or does it hold only for this example?

This is the end of Example 5.2.

5.1.8 Brief Summary of Section 5.1

Let's summarize the technique for finding the solution to a homogeneous second-order differential equation, the solution delivering the natural (source-free or zero input) response of an RLC circuit. This technique includes the following main steps:

- Assuming the solution is a sum of two exponential functions,

$$v_{out}(t) = A_1 e^{m_1 t} + A_2 e^{m_2 t}.$$

- Plugging the assumed solution into the circuit differential equation, deriving the characteristic equation, and finding its roots, m_1 *and* m_2. The roots can be two distinct real numbers, one real number, or two complex conjugate numbers, depending on the values of the circuit's parameters,

- Finding the actual solution for each case by inserting the circuit component values. What natural response will be obtained—underdamped, overdamped, or critically damped—hinges on the nature of the roots.
- Determining constants A_1 *and* A_2 for each case based on the circuit's topology (series or parallel) and its initial conditions.

Taking a broader view at all the manipulations performed in this section, we can summarize them as follows:

- The main goal of our operation is to obtain the output signal if the input signal and parameters of a series RLC circuit are known.
- For a natural response, the initial conditions are the "internal inputs" to our circuit, the differential equation is the description of this circuit, and the solution to this equation is the output, the natural response of the source-free circuit.
- All mathematical manipulations are done in the time domain; the result is the formula for the output signal *waveform*, which is our investigation's primary goal.

To set this summary in action, let's derive a series RLC *circuit differential equation in current* in contrast to such equations in voltage previously presented in this section.

KVL for a series, zero-input RLC circuit is

$$v_L(t)+v_R(t)+v_C(t)=0. \tag{5.55}$$

Using Table 4.1R enables us to rewrite (5.55) as

$$L\frac{di(t)}{dt}+Ri(t)+\frac{1}{C}\int_{-\infty}^{t}i(t)dt=0. \tag{5.56}$$

Differentiating this equation once results in

$$L\frac{d^2i(t)}{dt^2}+R\frac{di(t)}{dt}+\frac{1}{C}i(t)=0. \tag{5.57}$$

Dividing (5.57) through by L, introducing $\alpha=\frac{R}{2L}$, and $\omega_0^2=\frac{1}{LC}$, and using the standard designations $\frac{d^2i(t)}{dt^2}=i(t)''$ and $\frac{di(t)}{dt}=i(t)'$, we rewrite (5.57) as

$$i(t)''+2\alpha i(t)'+\omega_0^2 i(t)=0. \tag{5.58}$$

Equation 5.58 is the differential equation of a series RLC circuit in terms of current. Assuming its solution is in the form

$$i_{out}(t)=A_1e^{m_1t}+A_2e^{m_2t},$$

and inserting this solution into (5.58) yields a *characteristic equation*

$$m_{1,2}^2+2\alpha m_{1,2}+\omega_0^2=0. \tag{5.10R}$$

The solution to this quadratic equation is

$$m_{1,2} = -\alpha \pm \sqrt{\left(\alpha^2 - \omega_0^2\right)} = -\alpha \pm \mathrm{k}. \tag{5.11R}$$

If $\alpha^2 > \omega_0^2$, then $k > 0$ and the *overdamped natural response* of an RLC circuit takes the form

$$i_{out-od}(t) = A_1 e^{(-\alpha+k)t} + A_2 e^{(-\alpha-k)t}, \tag{5.59a}$$

when $\alpha^2 = \omega_0^2$, then $k = 0$ and the *critically damped natural response* is

$$i_{out-cd}(t) = (A_1 + A_2 t)e^{-\alpha t}, \tag{5.59b}$$

and having $\alpha^2 < \omega_0^2$ results in $k = j\omega_d$, where $\omega_d = \sqrt{(\omega_0^2 - \alpha^2)}$. This case is called the *underdamped*, and its formula is

$$i_{out-ud}(t) = \left(A_1 \cos(\omega_d t) + A_2 \sin(\omega_d t)\right)e^{-\alpha t}. \tag{5.59c}$$

The task is fulfilled.

Question Would you be able to write (5.58) without the tedious derivation, but using one of the principles discussed in Chapter 3?

Note: A parallel RLC circuit's differential equation in Example 5.2 was derived and solved in terms of the output (that is, capacitor) *voltage*. In a parallel circuit, the obtained $v_{out}(t)$ is the same for all three circuit elements, which enables us to find each individual *current* $i_R(t)$, $i_L(t)$, and $i_C(t)$ by applying the integro-differential current–voltage relationships shown in Table 4.1R.

Sidebar 5S.2 How the RLC circuit's parameters affect its natural response

We want to consider how the series RLC circuit's parameters affect its natural response using an *underdamped solution* because it is the most complicated case. The typical graphs of the damping oscillations are shown in Table 5.1 and Figure 5.3f, and the formula for the output voltage is given in (5.38) as

$$V_{out-ud}(t) = Ae^{-\alpha t}\cos(\omega_d t - \theta). \tag{5.38R}$$

The *amplitude* and the *phase shift* of these *damping oscillations* are given by

$$A(V) = \sqrt{(V_0)^2 + \left(\frac{\alpha_{ud}V_0 + V_0'}{\omega_d}\right)^2}, \tag{5S2.1}$$

and

$$\theta(rad) = tan^{-1}\left(\frac{\alpha_{ud}V_0 + V_0'}{V_0\omega_d}\right). \tag{5S2.2}$$

Clearly, these parameters are determined by the initial capacitor voltage, V_0 and the initial inductor current, $V_0' = \frac{I_0}{C}$, though constants α_{ud} and ω_d also play a role. Since only the initial voltage and its derivative set off these oscillations, it is clear that the greater the initial "push," the more significant the response, i.e., the greater the amplitude of the damping oscillations.

The *damping frequency* ω_d is given in (5.32), which can be explicitly expressed through the circuit parameters as

$$\omega_d\left(\frac{\text{rad}}{\text{s}}\right) = \sqrt{\omega_0^2 - \alpha_{ud}^2} = \sqrt{\frac{1}{LC} - \left(\frac{R}{2L}\right)^2}. \tag{5S2.3}$$

Equation 5S2.3 shows that the very existence of this frequency, $\omega_0^2 = \frac{1}{LC}$, is due to the presence of both an inductor and a capacitor that support the circuit oscillations. This fact was reflected in the discussion of Figure 5.12 and (5.17). But the actual frequency also depends on R through the parameter α_{ud}. We called this frequency *damping*, but it is also called *natural* to stress its determination by the circuit's parameters, not by any external input.

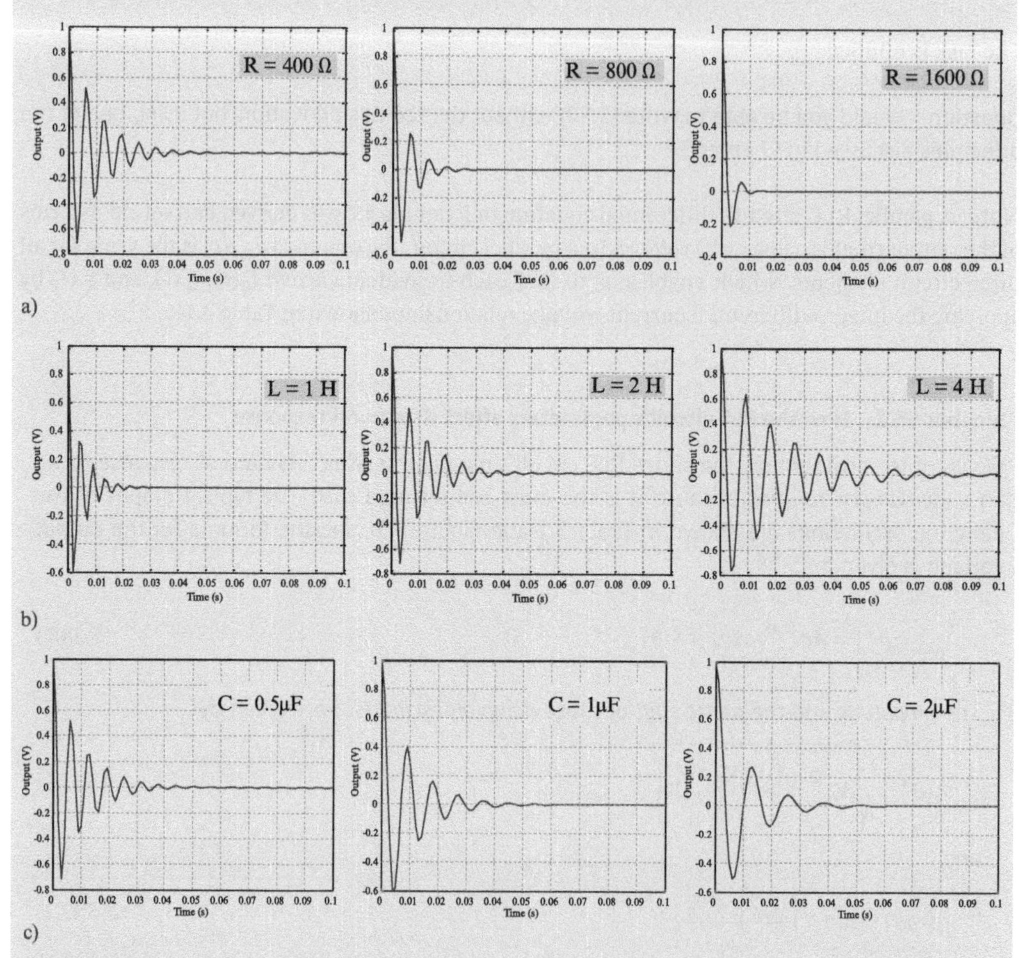

Figure 5S2.1 A series *RLC* circuit: Effect of changing values of (a) *R*, (b) *L*, and (c) *C* on the damping oscillations.

Question Consider the relationship between ω_d and ω_0: What happens with this relationship if R goes to zero? R tends to infinity? Prove your answers.

The *damping process* is determined by the parameter α called the *damping constant or coefficient or damping ratio;* it's also called the *decaying coefficient or ratio* of these oscillations. This coefficient is responsible for damping because α contains the resistance of the circuit, the only element that causes power dissipation, that is, the signal decay.

We assume that all circuit components are ideal, which means that L and C do not exhibit internal resistance. Clearly, if $R = 0$, then α turns to zero, too, and we will have a pure LC resonance circuit, which would indefinitely support sinusoidal oscillations. Nevertheless, we must remember that any *real circuit components inevitably have internal resistance, and therefore, in practice, the circuit's natural response always diminishes.*

This brief review shows that a specific form of the underdamped natural response depends on the values of the circuit's components, *R, L,* and *C.* Therefore, by varying R, L, and C, we can change the amplitude, frequency, and the phase shift of the oscillations, and the damping rate of the entire response.

To demonstrate how the circuit's parameters affect the underdamped natural response, we change the values of *R, L,* and *C* components of a *series RLC circuit.* Figure 5S.1 displays the resulting graphs; all numbers are shown in the figures.

How does the change in *R* influence the underdamped natural RLC circuit response? Discussion of (5.5) shows that an increase in *R* results in a faster oscillation decrease due to the increase in α. Figure 5S2.1a confirms this conclusion. A raise in R will also decrease ω_d, as (5S.23) shows, but this effect is insignificant in Figure 5S2.1a.

An increase in *L* results in a decrease in α; it also boosts ω_0. The increase in *L* also decreases the difference between ω_0 and ω_d. Figure 5S2.1b confirms these statements by showing that the decay process is getting slower due to increasing in *L*.

Changing *C* doesn't affect the damping ratio, but varies the resonant frequency, ω_0. Figure 5S2.1c shows that an increase in *C* results in a decrease in the number of oscillations over the same time interval. This observation is in line with (5.5),

$$\omega_0^2\left(\frac{rad}{s}\right)=\frac{1}{LC}.$$

These comments are helpful tips for solving academic problems and designing real-life circuits and devices.

5.2 Complete Responses of an RLC Circuit

As we know from Chapter 4 and Section 5.1, an *RLC circuit's complete response consists of a natural* (source-free or zero-input) and *forced (driven or excited) response.* Section 5.1 considers the natural response of this circuit in depth; in this section, we concentrate on its *forced (driven) and complete responses.* For the latter topics, we will consider the responses to step and sinusoidal inputs, the two cases mostly encountered in engineering practice. We urge you to review Chapter 4, where such responses of the first-order circuits were considered. The methodology and technique of that study will be exploited here.

The differential equation of an *RLC circuit* in general form is given as

$$v''_{out}(t)+2\alpha v'_{out}(t)+\omega_0^2 v_{out}(t)=\omega_0^2 v_{in}(t). \tag{5.6bR}$$

This equation form works for both series and parallel RLC circuits as (5.6b) and (5.49) attest. For a series RLC circuit, $v_{out}(t)$ is the voltage across the capacitor because $v_C(t)$ is the *state variable* for this circuit. See Section 5.3.

The formula for a *resonant frequency*,

$$\omega_0^2=\frac{1}{LC}, \tag{5.5bR}$$

holds for both circuits, but the *damping coefficient* is defined as

$$\alpha=\frac{R}{2L} \tag{5.5aR}$$

for a *series RLC* and as

$$\alpha=\frac{1}{2RC} \tag{5.48R}$$

for a *parallel RLC circuit.*

A step input is described as

$$v_{in}(t)=V_{in}\cdot u(t), \tag{5.60a}$$

where V_{in} is the input amplitude, and $u(t)$ is a *unit step (Heaviside) function*, and a sinusoidal input is given by

$$v_{in}(t)=V_{in}sin(\omega t)u(t), \tag{5.60b}$$

where $V_{in}(V)$ is the signal's amplitude and $\omega\left(\frac{rad}{s}\right)$ is its frequency.

Since (5.6b) is a linear equation, we can apply the *superposition principle*, which enables us to solve this equation by pursuing the methodology shown below.

The methodology of finding the complete response of an RLC circuit

1) Assume that the natural response—the solution of (5.6b) for $v_{in}(t)=0$—has the following form:

$$v_{out-nat}(t)=A_1e^{m_1t}+A_2e^{m_2t}. \tag{5.12R}$$

2) Presume that the forced response always takes the form of an input signal, which is a constant for a step input,

$$v_{out-frc}(t)=A_3, \tag{5.61a}$$

or a sinusoid for a sinusoidal signal,

$$v_{out-frc}(t)=Fsin(\omega t+\Phi), \tag{5.61b}$$

or in the other form

$$v_{out-frc}(t)=D_1\sin(\omega t)+D_2\cos(\omega t). \tag{5.61c}$$

3) Insert (5.61b) or (5.61c) into (5.6b) and find the constants F and Φ or D_1 and D_2. This part involves the relatively sophisticated manipulations.

4) Write the complete solution to (5.60) as

$$v_{out-cmp}(t) = v_{out-nat}(t) + v_{out-frc}(t). \tag{5.62}$$

5) Determine the constants A_1 *and* A_2, which is the most intriguing point of this methodology.
5.1. For a *step input*, it can be done as follows:
a) Evaluate (5.62) at $t = 0$ and obtain $A_1 + A_2 + A_3 = V_0$, where an initial value V_0 is given.
b) Differentiate (5.62) once, evaluate the result at $t = 0$, and receive $A_1 m_1 + A_2 m_2 = V_0'$, where the value of V_0' is given.
c) Insert A_3 into (5.12) and find for a *step response*, $A_3 = V_{in}$.
d) Collect all the results and get an equation system, which for a step excitation is

$$\begin{cases} A_1 + A_2 + A_3 = V_0 \\ A_1 m_1 + A_2 m_2 = V_0' \\ A_3 = V_{in} \end{cases}. \tag{5.63}$$

e) Solve (5.63), find A_1, A_2, and A_3, and obtain $v_{out-cmp}(t)$ in an explicit form as

$$v_{out-cmp}(t) = v_{out-nat}(t) + v_{out-frc}(t) = \left(A_1 e^{m_1 t} + A_2 e^{m_2 t}\right) + V_{in}. \tag{5.64}$$

6) For a *sinusoidal input*, the response constants must be determined for each case—overdamped, critically damped, and underdamped—that will be discussed shortly in the appropriate subsection.

In this section, we will employ the above methodology for finding step and sinusoidal responses of an RLC circuit. We studied similar manipulations in Chapter 4 and Section 5.1, and now we are capable of considering an example.

5.2.1 Step Response of Series RLC circuit

Example 5.3 Step responses of series RLC circuit.

Problem: The series RLC circuit is shown in Figure 5.5a, where IC stands for "Initial Conditions." They are $I_0 = 0.04A$ for the inductor and $V_0 = 8V$ for the capacitor. The inductor and capacitor numbers, $L = 8mH$ and $C = 2mF$, are fixed, but the resistor's value varies from (a) 40 Ω to (b) 4 Ω to (c) 0.4 Ω. At $t = 0$, the switch connects the circuit to the dc source, $V_{in} = 12V$. Find the complete response of this circuit for every value of R.

Solution: You will recall that Example 4.2 considers the step response of an RC circuit, and Figure 4.4b demonstrates a step function. It would be a wise move to review that material.

The interpretation of this problem is similar to that given in Example 5.1. The main difference is the initial conditions: Here, they are set to $V_0 = 8V$ and $I_0 = 0.04A \Rightarrow V_0' = \frac{I_0}{C} = 20\frac{V}{s}$. The values of the *resonant frequency*, ω_0, and the overdamped damping constant, α_{od} remain the same as in Example 5.1.

1) RLC circuit step response in overdamped case:

We borrow from Example 5.1 three parameters,

$$\omega_0^2 = \frac{1}{LC} = \frac{10^6}{16} = 6.25 \cdot 0^4 \left(\frac{rad}{s}\right)^2,$$

$$\alpha_{od}^2 = \left(\frac{R_{od}}{2L}\right)^2 = 6.25 \cdot 10^6 \left(\frac{1}{s}\right)^2,$$

and

$$k_{od} = \sqrt{\left(\alpha_{od}^2 - \omega_0^2\right)} \approx 2.487 \cdot 10^3 \left(\frac{1}{s}\right).$$

To find constants A_1, A_2, and A_3, we calculate $m_1 = -13$ and $m_2 = -4987$ and plug these and all other known values into (5.63), which yields

$$\begin{cases} A_1 + A_2 + A_3 = 8 \\ -13A_1 - 4987A_2 = 20. \\ A_3 = 12 \end{cases} \tag{5.65}$$

MATLAB automates our calculations. (See Sidebar 2S for refreshing your memory on this topic.) The equation system 5.65 must be written in the matrix form as

$$\begin{vmatrix} 1 & 1 \\ -13 & -4987 \end{vmatrix} \begin{vmatrix} A_1 \\ A_2 \end{vmatrix} = \begin{vmatrix} -4 \\ 20 \end{vmatrix}.$$

Let's denote the *coefficient matrix* as A, the *matrix of variables* as V, and the *matrix of constants* as B. Then MATLAB, executing the attached code, produces the following result for A_1 and A_2:

```
>>>>>>>>>>>>>>>
>> A = [1 1;
-13 -4987];
>> B = [-4; 20];
>> V = linsolve(A,B)
>> V =
-4.0064
0.0064
>>>>>>>>>>>>>>>>>>
```

(To validate the result, plug the obtained A_1 *and* A_2 into (5.65) and confirm that they turn this equation system into identity.)

Therefore, the step overdamped response of a series RLC circuit is attained from (5.64) as

$$\begin{aligned} v_{out-od}(t) &= A_3 + A_1 e^{m_1 t} + A_2 e^{m_2 t} \\ &= 12 - 4.0064\, e^{-13t} + 0.0064 e^{-4987t}. \end{aligned} \tag{5.66}$$

The case is resolved.

Discussion of the step overdamped case:

- We can determine the *transient interval,* $\Delta t(s)$, by building a MATLAB graph visualizing (5.66). The example of such an approach is given in Figure 5.3b and its discussion. For this example, the MATLAB graph shows $\Delta t \approx 300 ms$.
- To verify our answer, we run a Multisim simulation, whose results are shown in Figure 5.5b and 5.5c. The graph course in Figure 5.5b shows that the output signal approaches the input value 12 V, as it must be. The Multisim measurements displayed in Figure 5.5c show that $\Delta t \approx 300 ms$, which coincide with the MATLAB measurements. Therefore, our solution is validated.

2) RLC circuit step response in critically damped case:
To obtain the step response of a series RLC circuit for a critically damped case, we modify the natural response, (5.28), by adding constant member A_3, which represent the input. Thus,

$$v_{out-cd}(t) = A_3 + (A_1 + A_2 t)e^{-\alpha_{cd} t}. \tag{5.67}$$

Parameters calculated for this case are as follows:

$$\omega_0^2 = \frac{1}{LC} = \frac{10^6}{16} = 6.25 \cdot 10^4 \left(\frac{rad}{s}\right)^2,$$

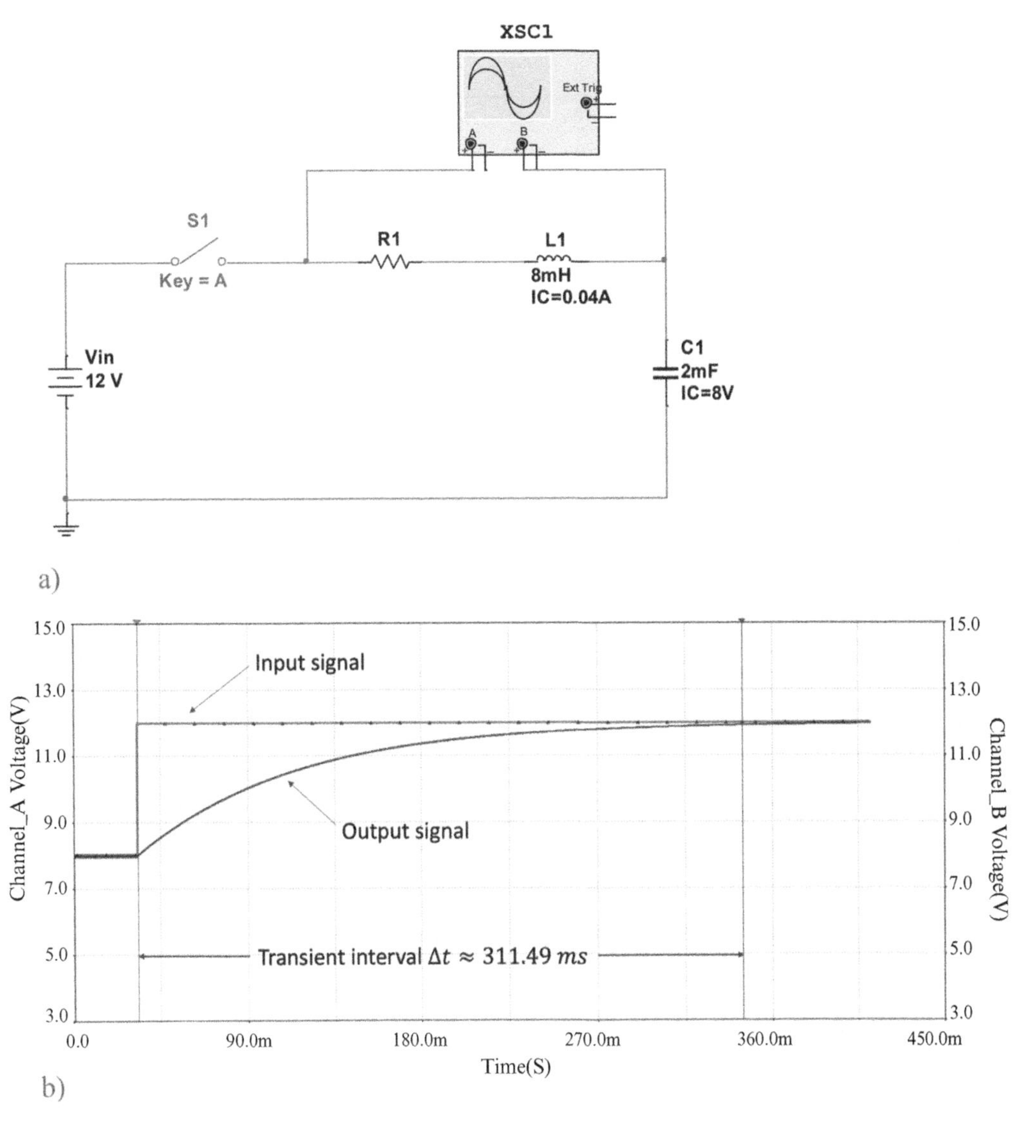

Figure 5.5 Multisim simulation of a series RLC circuit with the step input for Example 5.3: a) Circuit's schematic; b) Multisim circuit response in the overdamped case; c) Multisim measurements for the overdamped case; d) response in the critically damped case; e) circuit output in the underdamped case.

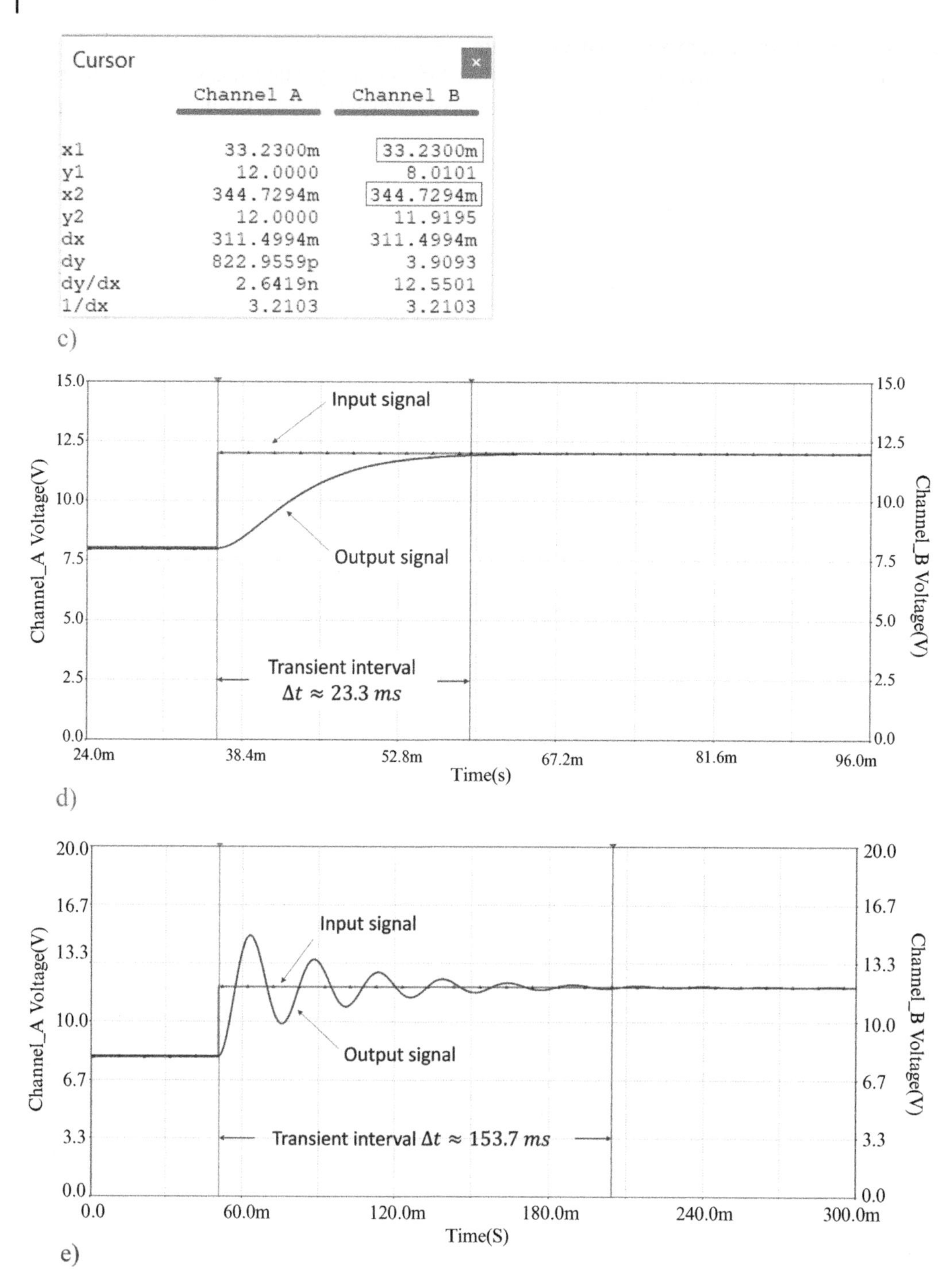

Cursor

	Channel A	Channel B
x1	33.2300m	33.2300m
y1	12.0000	8.0101
x2	344.7294m	344.7294m
y2	12.0000	11.9195
dx	311.4994m	311.4994m
dy	822.9559p	3.9093
dy/dx	2.6419n	12.5501
1/dx	3.2103	3.2103

Figure 5.5 (Cont'd)

$$\alpha_{cd} = \frac{R_{cd}}{2L} = 2.5 \cdot 10^2 \left(\frac{1}{s}\right), \textit{ and } \alpha_{cd}^2 = 6.25 \cdot 10^4 \left(\frac{1}{s}\right)^2$$

and

$$k_{od} = \sqrt{\left(\alpha_{cd}^2 - \omega_0^2\right)} = 0\left(\frac{1}{s}\right).$$

To find constants A_1, A_2, and A_3 for (5.67), we repeat the procedure employed in the preceding case, namely:

- Compute $m_1 = m_2 = -\alpha_{cd} = -250\left(\frac{1}{s}\right)$ and plug these and V_{in} values into (5.56), which yields $A_1 + A_3 = V_0$. at $t = 0$.
- Differentiate (5.67) once and set $t = 0$ again to find $V_0' = A_2$.
- Plug A_3 into (5.6b) to find $A_3 = V_{in}$.
- Collect these results in an equation system

$$\begin{cases} A_1 + A_3 = 8 \\ -250A_1 + A_2 = 20. \\ A_3 = 12 \end{cases} \tag{5.68}$$

Hence, $A_1 = -4$ and $A_2 = -980$, and (5.56) takes the form

$$v_{out-cd}(t) = A_3 + (A_1 + A_2 t)e^{-\alpha_{cd}t} = 12 - 4e^{-250t} - 980te^{-250t}. \tag{5.69}$$

From (5.69), MATLAB determines the transient interval as 23.3 ms. The Multisim measurements shown in Figure 5.5d coincide with the MATLAB results. Thus, the case is resolved and its solution is verified.

Discussion of the critically damped case:

- Though the graph of the critically damped case in Figure 5.5d resembles that of the overdamped case in Figure 5.5b, the transient interval of the former is more than ten times shorter than that of the latter. See discussion of this point in Section 5.1.

3) RLC circuit step response in underdamped case:

Follow the routine developed in the preceding step response cases, we present the RLC circuit output as a sum of the natural response, (5.37), and the force response, A_3, that is,

$$v_{out-ud}(t) = A_3 + e^{-\alpha_{ud}t}\left(B_1 cos\omega_d t + B_2 sin\omega_d t\right), \tag{5.70}$$

where the *damping coefficient* is

$$\alpha_{ud} = \frac{R_{ud}}{2L} = \frac{0.4}{16} \cdot 10^3 = 25\left(\frac{1}{s}\right),$$

and the *damping frequency* is

$$\omega_d\left(\frac{\text{rad}}{\text{s}}\right) = \sqrt{\omega_0^2 - \alpha_{ud}^2} = \sqrt{62500 - 625} = 248.75\left(\frac{1}{s}\right).$$

In Equation 5.70, B_1 and B_2 are constants. When $t = 0$, (5.70) turns to

$$A_3 + B_1 = V_{0.}$$

Differentiating (5.70) once produces

$$-\alpha_{ud} B_1 - \omega_d B_2 = V_0'.$$

Also, $A_3 = V_{in} = 12V$, as in the two preceding cases. Plugging the known values, we obtain the following equation system:

$$\begin{cases} B_1 = -4 \\ -25B_1 - 248.75B_2 = 20 \end{cases}$$

from which we compute $B_2 = -0.48$.
Therefore, (5.70) takes the following form for the underdamped case

$$v_{out-ud}(t) = 12 - e^{-25t}(4cos248.75t + 0.48sin248.75t). \tag{5.71}$$

From (5.71), MATLAB calculates the transient interval as $\Delta t \approx 153.82ms \ at \ V_{tf} = 11.9222V$. Multisim measurements (not shown) results in $\Delta t \approx 153.72ms \ at \ V_{tf} = 11.9262V$. Excellent validation of our manual solution (5.71).

Discussion of underdamped case:

- When measuring the *transient final amplitude* value, V_{tf}, of a sinusoidal signal, the amplitude varies and can exceed the final value, $V_f = 12V$. In our example, even after we pass $V_{tf} = 11.9262V$, we still can measure the values greater than 12 V. This might be a point of confusion. Nevertheless, the V_{tf} value is unique, and as soon as it is found, there is no need for a further search.
  The problem for Example 5.3 is fully solved.

5.2.2 Sinusoidal Response of a Series RLC Circuit

A sinusoidal response of a series RLC circuit can be found by following **The methodology of finding the complete response of an RLC circuit,** which has been developed in this section for complete response in general.

Let's consider a series RLC circuit first. Its differential equation for a sinusoidal input is given by

$$v''_{out}(t) + 2\alpha v'_{out}(t) + \omega_0^2 v_{out}(t) = \omega_0^2 V_{in} sin(\omega t). \tag{5.72}$$

We know that a *complete response* (output) is a sum of a *natural response*,

$$v_{out-nat}(t) = A_1 e^{m_1 t} + A_2 e^{m_2 t}, \tag{5.73}$$

and a *forced response*,

$$v_{out-frc}(t) = D_1 \sin(\omega t) + D_2 \cos(\omega t), \tag{5.61cR}$$

as

$$v_{out-cmp}(t) = A_1 e^{m_1 t} + A_2 e^{m_2 t} + D_1 \sin(\omega t) + D_2 \cos(\omega t). \tag{5.74}$$

To find the *constants* D_1 *and* D_2, we insert (5.61c) into (5.72) and obtain:

$$-D_1\omega^2 sin(\omega t) - D_2\omega^2 cos(\omega t) + 2\alpha\omega D_1 cos(\omega t) - 2\alpha\omega D_2 sin(\omega t) \\ +\omega_0^2 D_1 sin(\omega t) + \omega_0^2 D_2 cos(\omega t) = \omega_0^2 V_{in} sin(\omega t). \tag{5.75}$$

Collecting the coefficients of cosine and sine at both sides of (5.75), we find

$$sin(\omega t) \Rightarrow -D_1\omega^2 - 2\alpha\omega D_2 + \omega_0^2 D_1 = \omega_0^2 V_{in}$$

and

$$cos(\omega t) \Rightarrow -D_2\omega^2 + 2\alpha\omega D_1 + \omega_0^2 D_2 = 0,$$

from which we obtain the following system of two equations:

$$\begin{cases} D_1\left(\omega_0^2 - \omega^2\right) - 2\alpha\omega D_2 = \omega_0^2 V_{in} \\ 2\alpha\omega D_1 + D_2\left(\omega_0^2 - \omega^2\right) = 0 \end{cases}. \tag{5.76}$$

Solving (5.76) for D_1 and D_2 yields

$$D_1 = \frac{\omega_0^2\left(\omega_0^2 - \omega^2\right)}{\left[\left(\omega_0^2 - \omega^2\right)^2 + (2\alpha\omega)^2\right]} V_{in}$$

and

$$D_2 = \frac{-2\alpha\omega\omega_0^2}{\left[\left(\omega_0^2 - \omega^2\right)^2 + (2\alpha\omega)^2\right]} V_{in}. \tag{5.77}$$

We want to present the forced sinusoidal response in the compact form given in (5.61b),

$$v_{out-frc}(t) = F sin(\omega t + \Phi), \tag{5.61bR}$$

where $F(V)$ is the amplitude and Φ(rad) is the constant phase shift acquired by the sinusoidal forced response. For this, we relate (5.61b) and (5.61c) as

$$v_{out-frc}(t) = D_1 \sin(\omega t) + D_2 \cos(\omega t) = F \sin(\omega t + \Phi). \tag{5.78}$$

Letting

$$D_1 = F\, cos\Phi$$

and

$$D_2 = F\, sin\Phi \tag{5.79}$$

brings (5.61c) to the following form

$$D_1 \sin(\omega t) + D_2 \cos(\omega t) = F\left[\sin(\omega t) cos\Phi + \cos(\omega t) sin\Phi\right]. \tag{5.80}$$

The member in brackets can be presented as a sine function according to the well-known trigonometric identity, namely

$$\sin(\omega t)cos\Phi + \cos(\omega t)sin\Phi = \sin(\omega t + \Phi).$$

Amplitude F and the phase shift Φ of a forced response can be readily found from (5.79) as

$$F = \sqrt{\left(D_1^2 + D_2^2\right)} = \frac{\omega_0^2 V_{in}}{\sqrt{\left(\omega_0^2 - \omega^2\right)^2 + (2\alpha\omega)^2}} \tag{5.81}$$

and

$$\Phi = tan^{-1}\left(\frac{C_2}{C_1}\right) = -tan^{-1}\left(\frac{2\alpha\omega}{\omega_0^2 - \omega^2}\right). \tag{5.82}$$

Thus, we receive the *forced response* in the explicit form as

$$v_{out-frc}(t) = Fsin(\omega t + \Phi) = \frac{\omega_0^2 V_{in}}{\sqrt{\left(\omega_0^2 - \omega^2\right)^2 + (2\alpha\omega)^2}} \sin\left(\omega t - tan^{-1}\left(\frac{2\alpha\omega}{\omega_0^2 - \omega^2}\right)\right). \tag{5.83}$$

Performing further manipulations requires recalling that $m_1 = -\alpha + k$ and $m_2 = -\alpha - k$, where $\alpha = \frac{R}{2L}$, $k = \sqrt{\alpha^2 - \omega_0^2}$, and $\omega_0^2 = \frac{1}{LC}$.

Now, the *complete sinusoidal response* of a series RLC circuit, Equation 5.74, can be written as

$$\begin{aligned} v_{out-cmp}(t) &= A_1 e^{m_1 t} + A_2 e^{m_2 t} + Fsin(\omega t + \Phi) \\ &= A_1 e^{(-\alpha + k)t} + A_2 e^{(-\alpha - k)t} + \frac{\omega_0^2 V_{in}}{\sqrt{\left(\omega_0^2 - \omega^2\right)^2 + (2\alpha\omega)^2}} \sin\left(\omega t - tan^{-1}\left(\frac{2\alpha\omega}{\omega_0^2 - \omega^2}\right)\right). \end{aligned} \tag{5.84}$$

To find A_1 and A_2, we need to consider three cases of a sinusoidal response.

1) RLC circuit sinusoidal response in overdamped case.

In this case, m_1 and m_2 are real and distinct roots, and the evaluation of (5.84) and its derivative at $t = 0$ enables us to obtain the following equation system

$$\begin{cases} A_{1od} + A_{2od} + F_{od}sin(\Phi_{od}) = V_0 \\ A_{1od}m_{1od} + A_{2od}m_{2od} + F_{od}cos(\Phi_{od}) = V_0' \end{cases}. \tag{5.85}$$

Solving (5.85) for A_{1od} and A_{2od} results in

$$A_{1od} = \frac{m_{2od}V_0 - m_{2od}F_{od}sin(\Phi_{od}) + \omega F_{od}cos(\Phi_{od}) - V_0'}{m_{2od} - m_{1od}}$$

and

$$A_{2od} = \frac{-m_{1od}V_0 + m_{1od}F_{od}sin(\Phi_{od}) - \omega F_{od}cos(\Phi_{od}) + V_0'}{m_{2od} - m_{1od}}. \tag{5.86}$$

All members in the right-hand sides of (5.86) are known. Therefore, we obtained the complete solution shown in (5.84), where α_{od} is specific for an overdamped case; i.e.

$$\begin{aligned} v_{out-cmp}^{od}(t) &= A_{1od}e^{m_{1od}t} + A_{2od}e^{m_{2od}t} + F_{od}sin(\omega t + \Phi_{od}) \\ &= \left(\frac{m_{2od}V_0 - m_{2od}F_{od}sin(\Phi_{od}) + \omega F_{od}cos(\Phi_{od}) - V_0'}{m_{2od} - m_{1od}} \right) e^{m_{1od}t} \\ &+ \left(\frac{-m_{1od}V_0 + m_{1od}F_{od}sin(\Phi_{od}) - \omega F_{od}cos(\Phi_{od}) + V_0'}{m_{2od} - m_{1od}} \right) e^{m_{2od}t} \\ &+ \frac{\omega_0^2 V_{in}}{\sqrt{\left((\omega_0^2 - \omega^2)^2 + (2\alpha_{od}\omega)^2\right)}} \sin\left(\omega t - tan^{-1}\left(\frac{2\alpha\omega}{\omega_0^2 - \omega^2}\right)\right). \end{aligned} \tag{5.87}$$

2) RLC circuit sinusoidal response in critically damped case

Here, $m_{1cd} = m_{2cd} = -\alpha_{cd}$, and (5.28) gives the *natural response* as

$$v_{out-cd}(t) = (A_{1cd} + A_{2cd}t)e^{-\alpha_{cd}t}. \tag{5.28R}$$

Thus, the *complete sinusoidal response for critically damped case* can be written as

$$v_{out-cmp}^{cd}(t) = (A_{1cd} + A_{2cd}t)e^{-\alpha_{cd}t} + F_{cd}sin(\omega t + \Phi_{cd}). \tag{5.88a}$$

Its derivative is

$$v_{out-cmp}'(t) = (-\alpha_{cd}A_{1cd} + A_{2cd} - \alpha_{cd}A_{2cd}t)e^{-\alpha_{cd}t} + \omega cF_{cd}cos(\omega t + \Phi_{cd}). \tag{5.88b}$$

The evaluation of (5.88a) and (5.88b) at $t = 0$ gives the following equation system

$$\begin{cases} A_{1cd} + F_{cd}sin(\Phi_{cd}) = V_0 \\ -\alpha_{cd}A_1 + A_2 + \omega F_{cd}cos(\Phi_{cd}) = V_0' \end{cases}. \tag{5.89}$$

Hence, in the critically damped case, constants A_{1cd} and A_{2cd} are determined as

$$A_{1cd} = V_0 - F_{cd}sin(\Phi_{cd})$$

and

$$A_{2cd} = \alpha_{cd}V_0 - \alpha_{cd}F_{cd}sin(\Phi_{cd}) - \omega F_{cd}cos(\Phi_{cd}) + V_0'. \tag{5.90}$$

Equations 5.83a and 5.85, along with Equations 5.81 and 5.82, enable us to find the complete response in the critically damped case as

$$\begin{aligned} v_{out-cmp}^{cd}(t) &= (A_{1cd} + A_{2cd}t)e^{-\alpha_{cd}t} + F_{cd}sin(\omega t + \Phi_{cd}) \\ &= (V_0 - F_{cd}sin(\Phi_{cd}))e^{-\alpha_{cd}t} + (\alpha_{cd}V_0 - \alpha_{cd}F_{cd}sin(\Phi_{cd}) - \omega F_{cd}cos(\Phi_{cd}) + V_0')te^{-\alpha_{cd}t} \\ &+ \frac{\omega_0^2 V_{in}}{\sqrt{(\omega_0^2 - \omega^2)^2 + (2\alpha_{cd}\omega)^2}} sin\left(\omega t - tan^{-1}\left(\frac{2\alpha_{cd}\omega}{\omega_0^2 - \omega^2}\right)\right). \end{aligned} \tag{5.91}$$

3) RLC circuit sinusoidal response in underdamped case

The *natural underdamped response* of the series RLC circuit response is given in (5.37) or (5.38) as

$$v_{out-ud}(t) = e^{-\alpha_{ud}t}\left(B_1 cos\omega_d t + B_2 sin\omega_d t\right) \tag{5.37R}$$

or

$$v_{out-ud}(t) = Be^{-\alpha_{ud}t}\cos(\omega_d t - \theta), \tag{5.38R}$$

where $B = \sqrt{B_1^2 + B_2^2}$ and $\theta = tan^{-1}\left(\frac{B_2}{B_1}\right)$. Thus, *the complete sinusoidal response* in this case is given by

$$v_{out-cmp}^{ud}(t) = e^{-\alpha_{ud}t}\left(B_1 cos\omega_d t + B_2 sin\omega_d t\right) + F_{ud} sin\left(\omega t + \Phi_{ud}\right). \tag{5.92}$$

Note that (5.92) contains two frequencies: the *damping frequency of the natural response*, $\omega_d = \sqrt{\omega_0^2 - \alpha_{ud}^2}$ and the *input frequency*, ω. Differentiating (5.92) one time produces

$$\begin{aligned} v'_{out-cmp}(t) = &-\alpha_{ud} e^{-\alpha_{ud}t}\left(B_1 cos\omega_d t + B_2 sin\omega_d t\right) + e^{-\alpha_{ud}t} \\ &\left(-B_1\omega_d sin\omega_d t + B_2\omega_d cos\omega_d t\right) + \omega F_{ud} cos\left(\Phi_{ud}\right). \end{aligned} \tag{5.93}$$

Assessing (5.92) and (5.93) at $t = 0$ results in the following equation system:

$$\begin{cases} B_1 + F_{ud} sin\left(\Phi_{ud}\right) = V_0 \\ -\alpha_{ud} B_1 + B_2\omega_d + \omega F_{ud} cos\left(\Phi_{ud}\right) = V_0' \end{cases}. \tag{5.94}$$

Solving (5.94) yields

$$B_1 = V_0 - F_{ud} sin\left(\Phi_{ud}\right)$$

and

$$B_2 = \frac{\alpha_{ud} V_0 - \alpha_{ud} F_{ud} sin\left(\Phi_{ud}\right) - \omega F_{ud} cos\left(\Phi_{ud}\right) + V_0'}{\omega_d}. \tag{5.95}$$

Therefore, the complete sinusoidal response in the underdamped case can be found by gathering B_1 and B_2 from (5.95) and F_{ud} and Φ_{ud} from (5.81) and (5.82), and inserting all these formulas into (5.92), i.e.,

$$\begin{aligned} v_{out-cmp}^{ud}(t) &= e^{-\alpha_{ud}t}\left(B_1 cos\omega_d t + B_2 sin\omega_d t\right) + F_{ud} sin\left(\omega t + \Phi_{ud}\right) \\ &= e^{-\alpha_{ud}t}\left(\left(V_0 - F_{ud} sin\Phi_{ud}\right)cos\omega_d t\right. \\ &\quad + \left.\left(\frac{\alpha_{ud} V_0 - \alpha_{ud} F_{ud} sin\left(\Phi_{ud}\right) - \omega F_{ud} cos\left(\Phi_{ud}\right) + V_0'}{\omega_d}\right) sin\omega_d t\right) \\ &\quad + \frac{\omega_0^2 V_{in}}{\sqrt{\left(\omega_0^2 - \omega^2\right)^2 + \left(2\alpha_{cd}\omega\right)^2}} sin\left(\omega t - tan^{-1}\left(\frac{2\alpha\omega}{\omega_0^2 - \omega^2}\right)\right). \end{aligned} \tag{5.96}$$

Therefore, Equations 5.87, 5.91, and 5.96 provide the explicit formulas for the responses of a series RLC circuit to a sinusoidal input for overdamped, critically damped, and underdamped cases. These equations deliver the theoretical basis for consideration of a sinusoidal transient response of a series RLC circuit, which enables us to move to an example.

Example 5.4 Sinusoidal response of a series RLC circuit.

Problem: The series RLC circuit with a sinusoidal input is shown in Figure 5.6a. All values shown in Figure 5.6a are fixed. The resistor value changes from (a) 126.5Ω to (b) 12.65Ω to (c) 1.26Ω. Find the circuit complete response for every value of R after the switch turns on at $t = 0$ and connects the circuit to the source $v_{in}(t)[V] = 12\sin(2\pi \cdot 40 \cdot t) = 12\sin(251.33t)$.

Solution: We split the solution into two parts: *theory and experiment.*

Theory for Example 5.4:

The theoretical solution reduces to inserting the values of the RLC circuit's parameters into Equations 5.87, 5.91, and 5.96. Though these equations look intimidatingly complex, they require only straightforward but meticulously performing calculations. We can relegate this tedious work to MATLAB. The MATLAB codes and the results of computations are collected in Sidebar 5S.3 in this section.

The only computation to be performed manually is finding a critically damped resistor, R_{cd}; its value must satisfy the following condition:

$$\alpha_{cd}^2 = \omega_0^2.$$

(Do you understand why? Check (5.27).) From this condition we find

$$\alpha_{cd}^2 = \omega_0^2 \Rightarrow \left(\frac{R_{cd}}{2L}\right)^2 = \frac{1}{LC} \Rightarrow \boldsymbol{R_{cd}} = 2\sqrt{\frac{L}{C}}.$$

Inserting the given values, we compute:

$$R_{cd} = 2\sqrt{\frac{L}{C}} = 2\sqrt{\frac{0.08}{0.002}} = 2\sqrt{40} = 12.6\ (\Omega).$$

Based on this R_{cd} value, we assign

$$R_{od} >> R_{cd} = 126(\Omega)$$

and

$$R_{ud} << R_{cd} = 1.26(\Omega).$$

Applying these resistance values and using the codes presented in Sidebar 5S.3, MATLAB builds the graphs shown in Figures 5.6b, 5.6c, and 5.6d. The amplitude measurements are made on the graphs using *the Data Tips* command from the MATLAB *Tools* figure window. These three graphs visualize Equations 5.87, 5.91, and 5.96 and thus present *theoretical predictions* for the transient responses of the series RLC circuit for overdamped, critically damped, and underdamped cases.

Equally importantly to building the graphs, MATLAB enables us to calculate all the *values of this transient response*. Sidebar 5S.3 presents the codes that readily permit computations of the amplitudes, phase shifts, and all other values needed to answer the problem of Example 5.4. The commentaries and explanations in the sidebar help in deeper understanding of all the operations and results.

Collecting all the values calculated by MATLAB and referring to Equations 5.87, 5.91, and 5.96, we obtain *the **answers** to the problem posed in this example* as follows:

Overdamped case:

$$\begin{aligned} v_{out-cmp}^{od}(t) &= A_{1od}e^{m_{1od}t} + A_{2od}e^{m_{2od}t} + F_{cd}sin(\omega t + \Phi_{cd}) \\ &= 8.22e^{-3.98t} - 0.04e^{-1571.02t} + 0.19sin(251.3t - 98.2^0). \end{aligned} \tag{5.97}$$

Critically damped case:

$$\begin{aligned} v_{out-cmp}^{cd}(t) &= (A_{1cd} + A_{2cd}t)e^{-\alpha_{cd}t} + +F_{cd}sin(\omega t + \Phi_{cd}) \\ &= (8.62 + 921.86t)e^{-78.75t} + 1.08\sin(251.3t - 145.2^0). \end{aligned} \tag{5.98}$$

Underdamped case:

$$\begin{aligned} v_{out-cmp}^{ud}(t) &= (B_1 cos\omega_d t + B_2 sin\omega_d t)e^{-\alpha_{ud}t} + F_{ud}sin(\omega t + \Phi_{ud}) \\ &= (8.09cos78.66t + 4.75sin78.66t)e^{-7.88t} + 1.31\sin(251.3t - 176.0^0). \end{aligned} \tag{5.99}$$

One of the first methods to validate our answers is checking the values of $v_{out}(t)$ amplitudes at $t = 0$. For all three cases, they must be equal to $8(V)$, as required by the initial condition. Overdamped amplitude, $v_{out-cmp}^{od}(t)$, at $t = 0$ gives

$$v_{out-cmp}^{od}(0) = 8.22 - 0.04 + 0.19sin(-98.2^0) \approx 8.22 - 0.04 - 0.19 \approx 8.0\ (V).$$

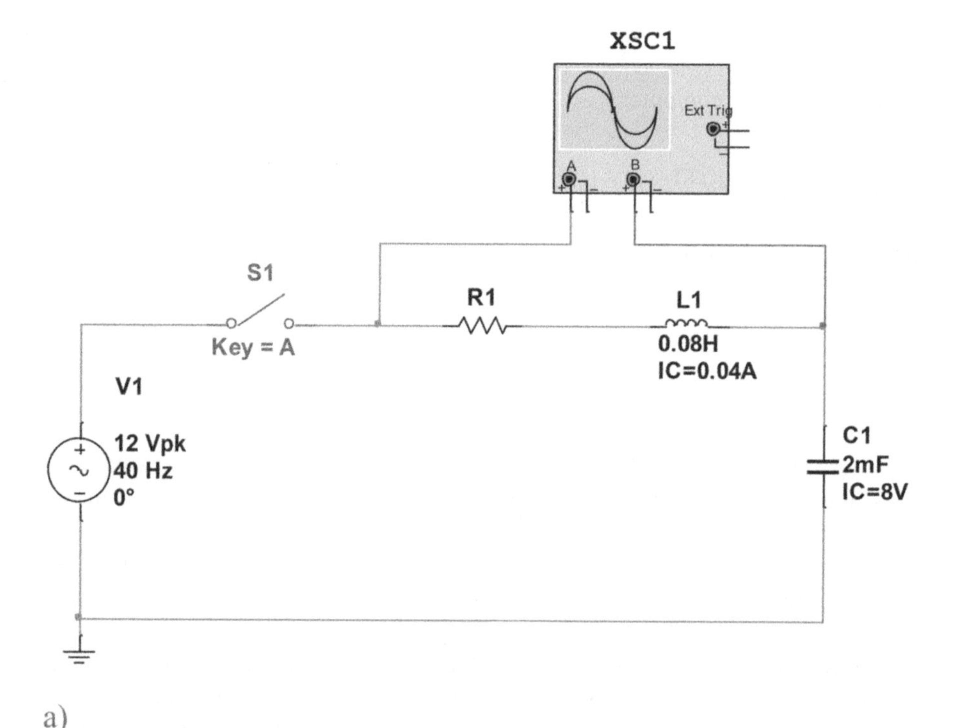

Figure 5.6 MATLAB calculations of the sinusoidal responses of a series RLC circuit: a) Circuit's schematic; b) overdamped case; c) critically damped case; d) underdamped case.

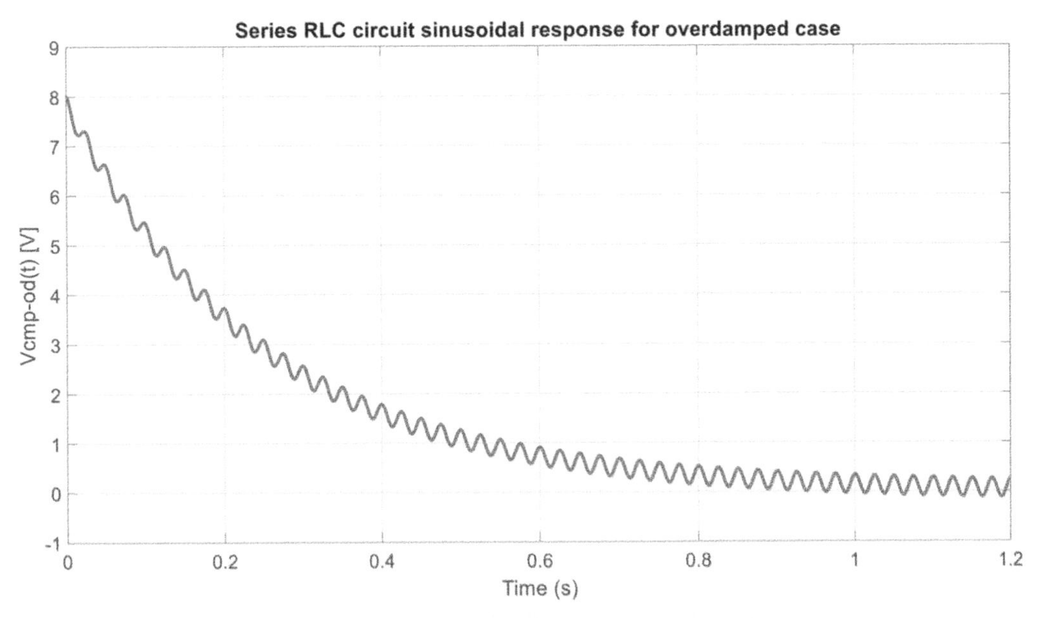

MATLAB calculation of sinusoidal overdamped case for Example 5.4 - measured amplitude: $F_{od} = \frac{-Y1+Y2}{2} = \frac{0.372\,V}{2} = 0.186\,V$. (Data tips are not shown.)

b)

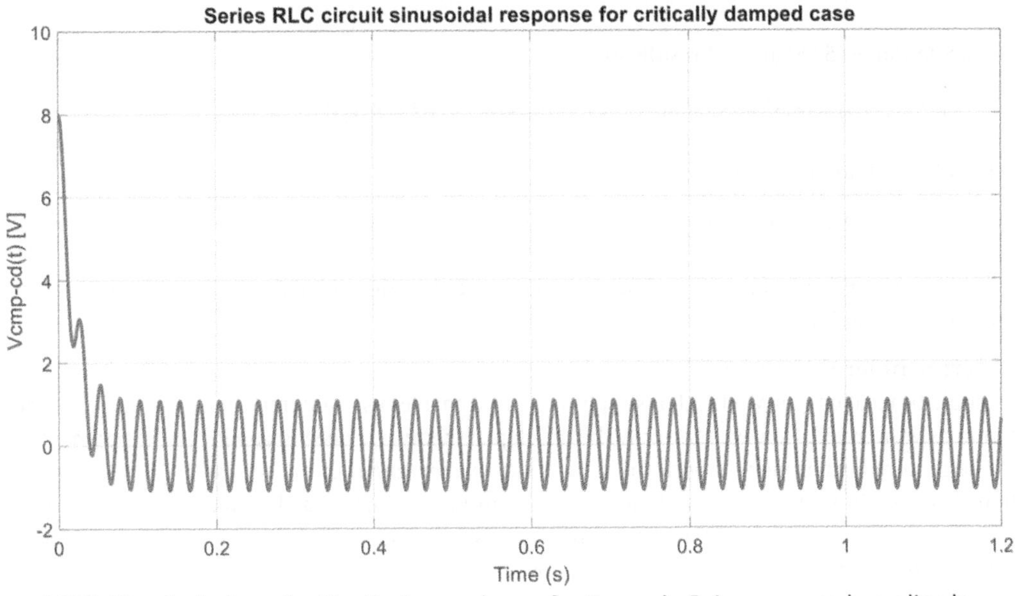

MATLAB calculation of critically damped case for Example 5.4 - measured amplitude: $F_{od} = \frac{-Y1+Y2}{2} = \frac{2.210}{2} = 1.105\,V$. (Data tips are not shown.)

c)

Figure 5.6 (Cont'd)

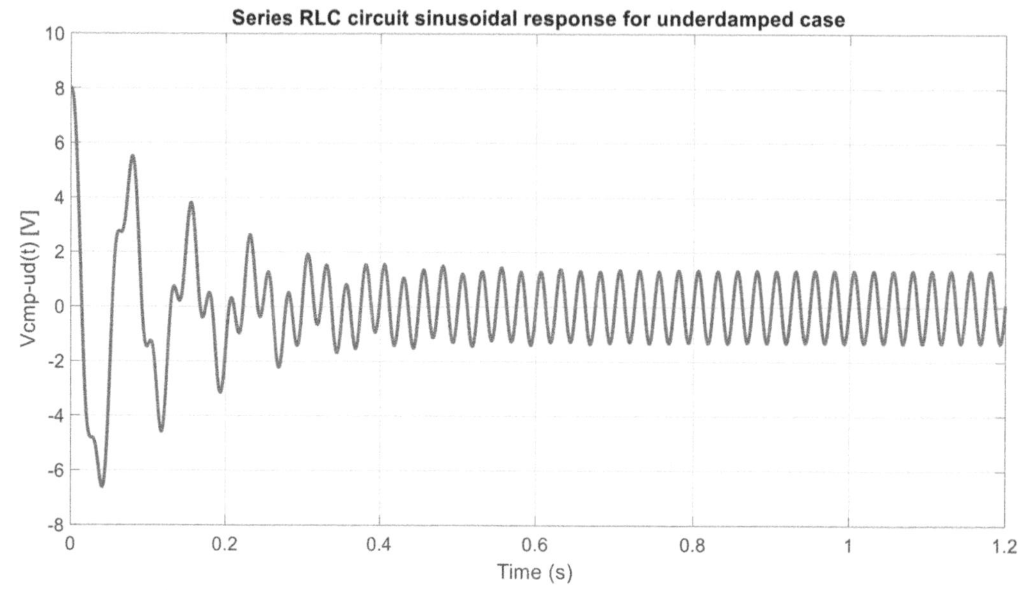

MATLAB calculation of critically underdamped case for Example 5.4 - measured amplitude: $F_{od} = \frac{-Y1+Y2}{2} = \frac{2.624}{2} = 1.312\ V$. (Data tips are not shown.)

d)

Figure 5.6 (Cont'd)

Calculation of (5.98) at $t=0$ results in

$$v_{out-cmp}^{cd}(0)=(8.62)+1.08\sin\left(-145.2^{0}\right)=8.62-0.62=8.0\,(V).$$

Finally, for Equation 5.99 we have

$$v_{out-cmp}^{ud}(0)=8.09+1.31\sin\left(-176.0^{0}\right)=8.09-0.09=8.0\ (V).$$

Therefore, our results have passed this critical test. This is the end of the theory part of the solution for Example 5.4.

Experiment for Example 5.4:
It should be recalled that MATLAB is simply a tool that enables us to automate the calculations of the formulas we derived. However, to verify the developed theory, we need to conduct an experiment.

The required experiment is performed by *Multisim simulation* of the series RLC circuit operation. The circuit used for the simulation is shown in Figure 5.6a, and obtained graphs are shown in Figures 5.7a, 5.7b, and 5.7c, respectively. The visual comparison of the MATLAB and Multisim graphs confirm their resemblance. Moreover, the quantitative comparison of the calculated amplitude and phase shift values shown in the sets of Figures 5.6 and 5.7 demonstrates that they are close enough to validate our theory. This comparison is presented in Table 5.3.

Here are some insights into the Multisim simulations and measurements. To perform simulations of all three cases, we consequently change the resistor values from 126Ω to 12.65Ω to 1.26Ω. We get input and output sinusoids on one screen by connecting the circuit input and output to the oscilloscope, as shown in Figure 5.6a. The scales of the amplitude and time axes retain the same for

Table 5.3 Comparison of theoretical calculations and experimental measurements of the transient processes in Example 5.4.

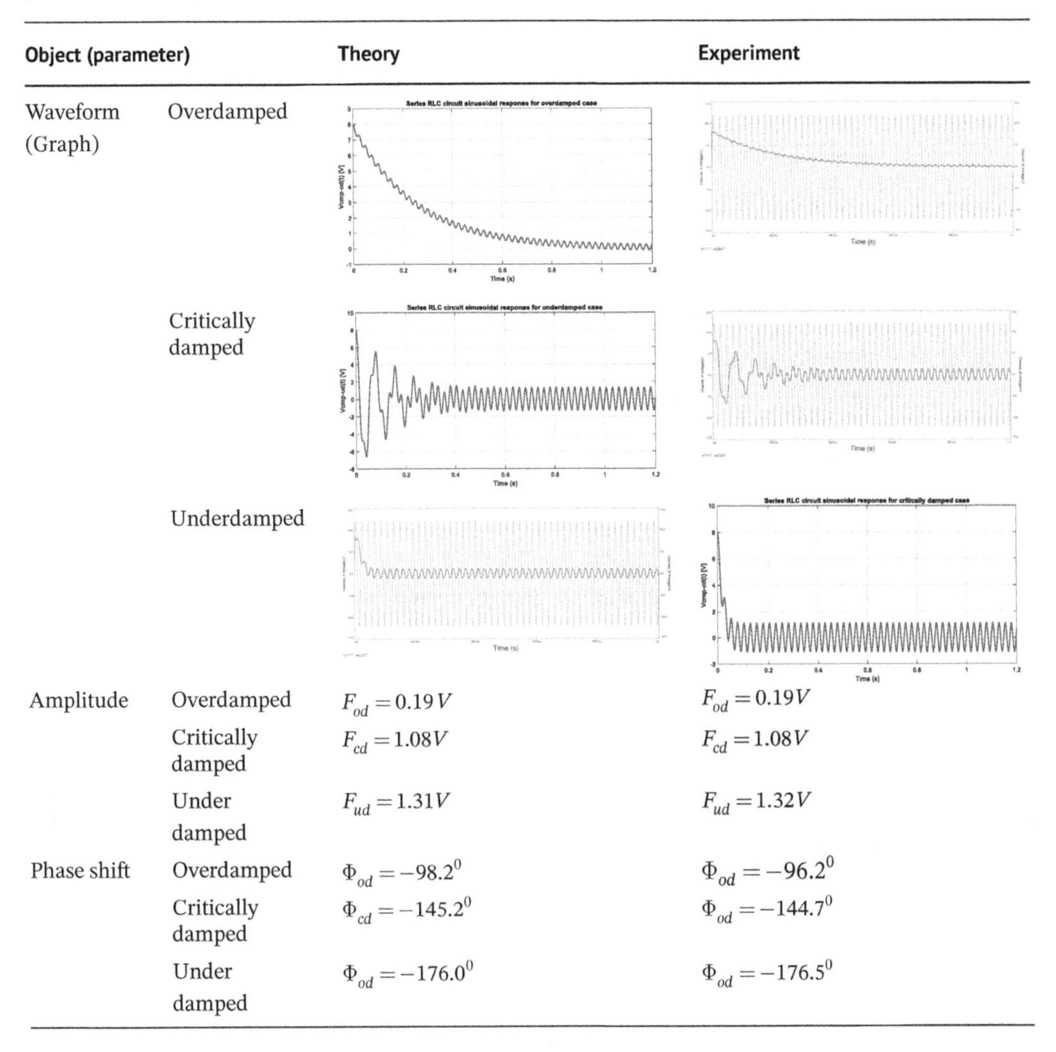

Object (parameter)		Theory	Experiment
Waveform (Graph)	Overdamped		
	Critically damped		
	Underdamped		
Amplitude	Overdamped	$F_{od} = 0.19\,V$	$F_{od} = 0.19V$
	Critically damped	$F_{cd} = 1.08V$	$F_{cd} = 1.08V$
	Under damped	$F_{ud} = 1.31V$	$F_{ud} = 1.32V$
Phase shift	Overdamped	$\Phi_{od} = -98.2^0$	$\Phi_{od} = -96.2^0$
	Critically damped	$\Phi_{cd} = -145.2^0$	$\Phi_{od} = -144.7^0$
	Under damped	$\Phi_{od} = -176.0^0$	$\Phi_{od} = -176.5^0$

all three graphs. This approach enables us to observe the difference between all three waveforms and assess their parameters as follows:

- The *responses* in the *overdamped* and *critically damped* cases are sinusoidal signals sitting on the decaying exponential functions. In contrast, the *underdamped* case transient demonstrates a sophisticated combination of two sinusoidal signals (natural oscillations of an RLC circuit and the input sinusoid) controlled by the decaying exponential function.
- The *transient interval* is the longest in the overdamped case and the shortest in the critically damped case. (Do you understand why? Compare the exponents in (5.87) and (5.91).)
- The measurements on the Multisim graphs are made by using the cursors. See Sidebar 4S.2 for the explanations of this technique.

The routine of these measurements is like that discussed in Figure 5.3, but the details are slightly different, and they are shown in Figures 5.7d and 5.7e. The amplitude measurements must be made at the time when the transient process is almost diminished. At this stage, the amplitudes become constants; obviously, they are final (forced) amplitudes. To measure an amplitude, we place two cursors at the highest (maximum) and the lowest (minimum) points of the sinusoid. The sum of these two measurements gives us the peak-to-peak value, from which the amplitude of the transient sinusoidal response is computed as $F_{od} = \frac{-Y1+Y2}{2} = 0.189\ (V)$. The measured values Y1 and Y2 are boxed in the cursor table in the bottom left part of the figure. Similarly, for measuring the phase shift in Figure 5.7e, two cursors mark the closest-to-zero points of the input and output sinusoids, and their difference along the time axis delivers the time interval, $\Delta t(s) = X1 - X2$, between these two graphs. The values of X1 and X2 are shown in the cursor box. This interval translates to the phase shift as $\Phi(degrees) = \frac{\Delta t(s)}{T(s)} \cdot 360^0 = \frac{-10.5(ms)}{25(ms)} \cdot 360^0 = -144.72^0$. The negative sign appears due to shifting the output (response) signal to the right from the input signal.

The described technique clarifies that these measurements depend on the manual placements of the cursors, which implies that such measurements are approximate. Nonetheless, for our exercises, the used measurement method is solid enough to verify the obtained theoretical results.

The analysis of Table 5.3 convinces that our theory derivations are confirmed by the experimental measurements.

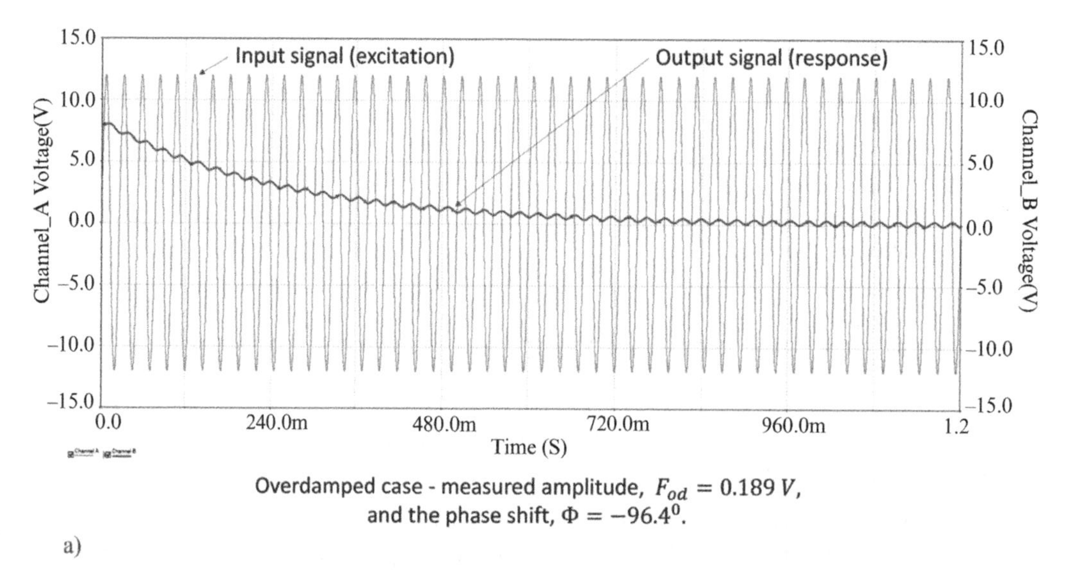

Figure 5.7 Multisim simulation of the series RLC circuit sinusoidal transient response for Example 5.4: a) Overdamped case; b) critically damped case; c) underdamped case; d) measuring amplitude on the Multisim graph; e) measuring phase shift on the graph.

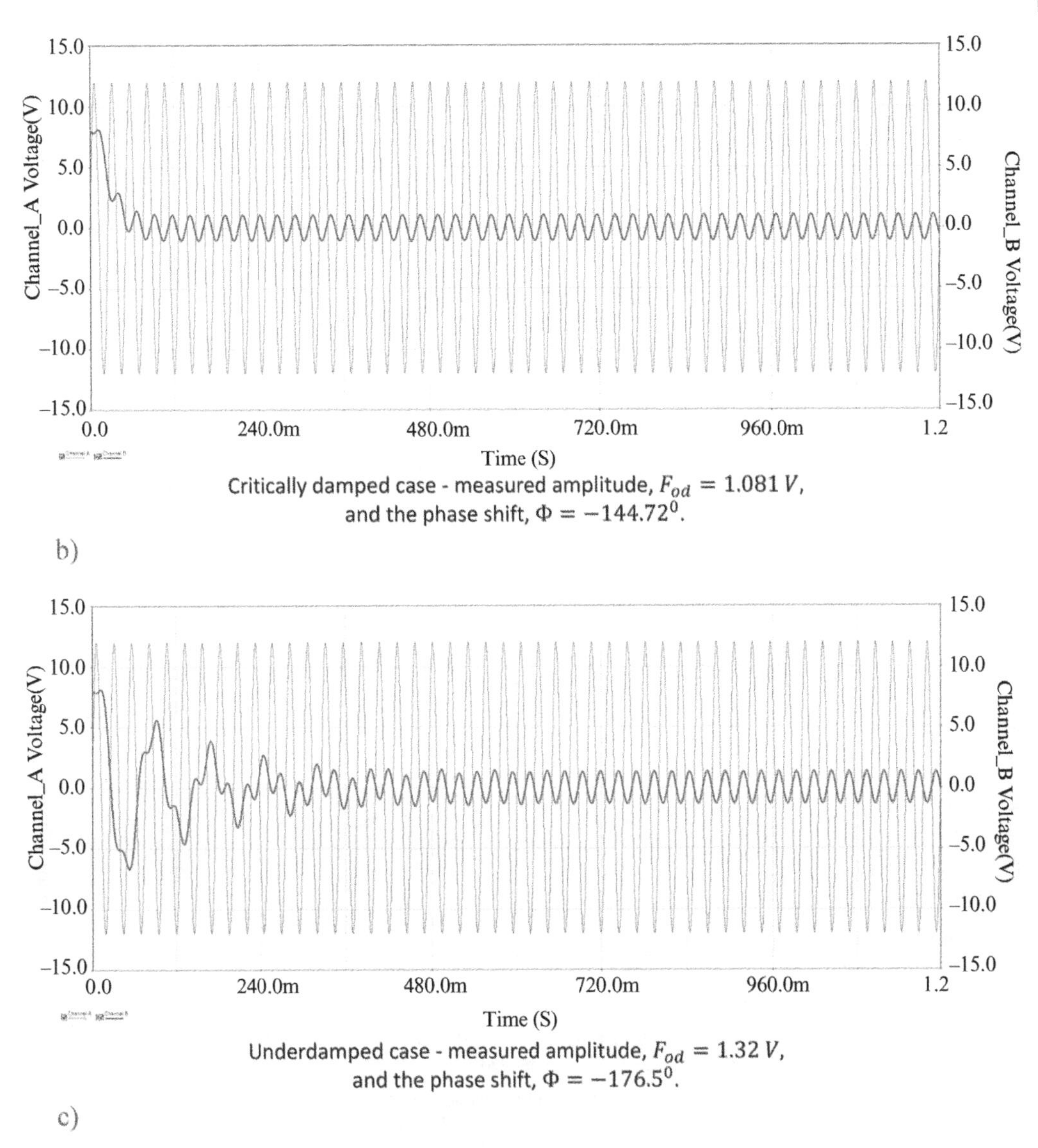

Critically damped case - measured amplitude, $F_{od} = 1.081\ V$, and the phase shift, $\Phi = -144.72^0$.

b)

Underdamped case - measured amplitude, $F_{od} = 1.32\ V$, and the phase shift, $\Phi = -176.5^0$.

c)

Figure 5.7 (Cont'd)

Sidebar 5S.3 MATLAB calculations for Example 5.4

This sidebar presents the MATLAB scripts and the results of calculations for Example 5.4, along with necessary explanations. To satisfy the MATLAB requirements for the code syntax and simplify writing the code, we rename several variables shown in Table 5S3.1 from mathematical to MATLAB formats.

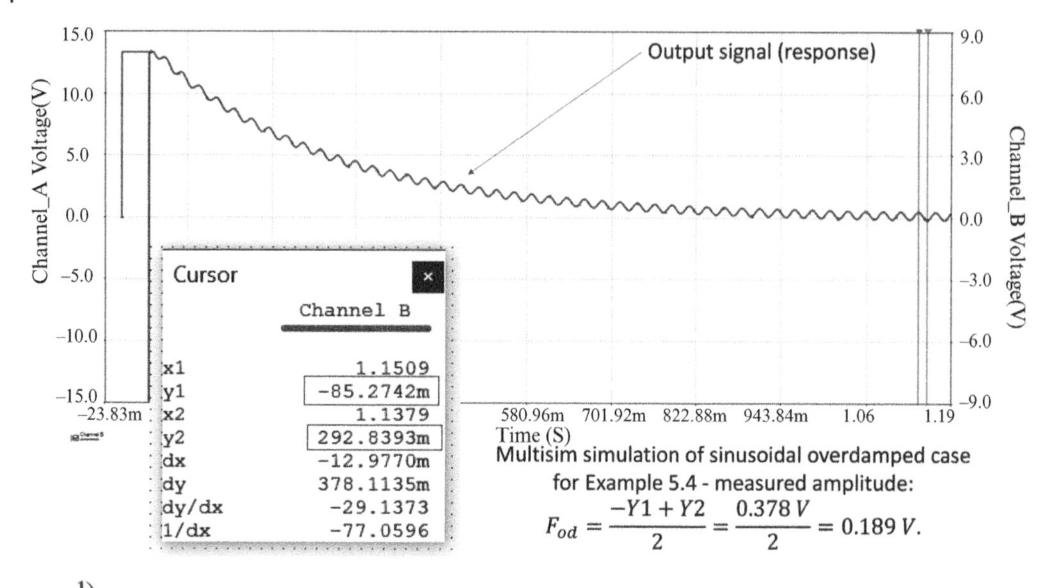

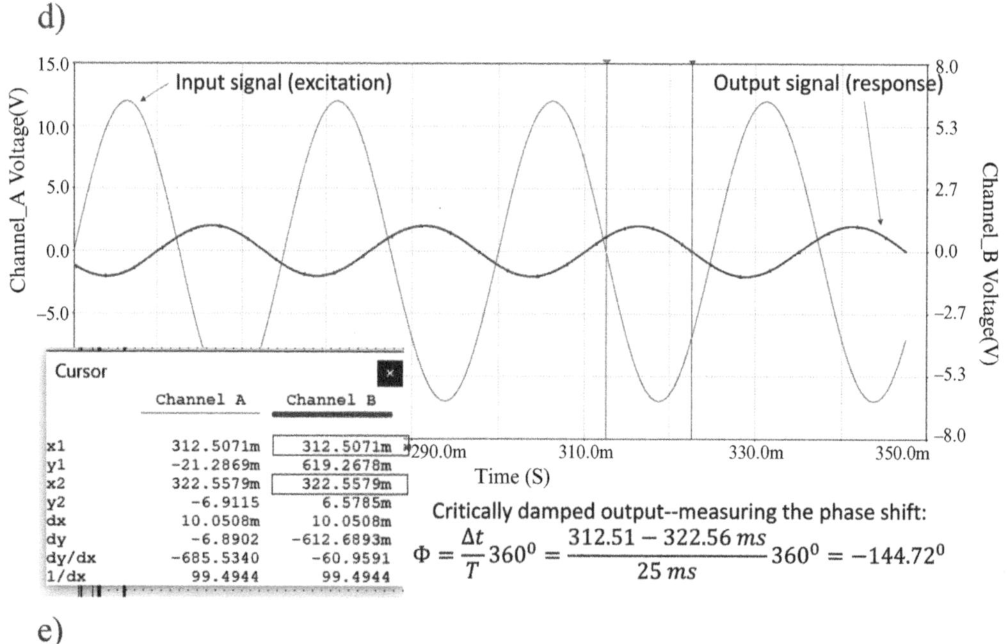

Figure 5.7 (Cont'd)

Below are the MATLAB scripts that enable us *to build the graph* shown in Figure 5.6b. Using the similar scripts, we have made the graphs for the critically damped and underdamped cases. They can be seen in Figures 5.6c and 5.6d, respectively.

⋙⋙⋙⋙⋙⋙⋙⋙⋙⋙⋙⋙⋙⋙⋙

MATLAB code for building the graph of the overdamped transient response in Example 5.4:

```
%RLC series sinusoidal complete overdamped response
%RLC sinusoidal forced overdamped
```

Table 5S3.1 Conversion of mathematical to MATLAB notations for the scripts in this sidebar.

Mathematical notations	MATLAB notations
V_0	V0
V_0'	V01
ω_0	w0
ω	w1
α_{od} or α_{cd} or α_{ud}	aod *or* acd *or* aud
F_{od} or F_{cd} or F_{ud}	Fod *or* Fcd *or* Fud
Φ_{od} or Φ_{cd} or Φ_{ud}	Qod *or* Qcd *or* Qud
m_{1od} and m_{2od} *or* m_{1od} and m_{2od} *or* m_{1od} and m_{2od}	M1od and M2od *or* M1cd and M2cd *or* M1ud and M2ud
A_{1od} and A_{2od} *or* A_{1od} and A_{2od} *or* A_{1od} and A_{2od}	A1od and A2od *or* A1cd and A2cd *or* A1ud and A2ud
B_{1od} and B_{2od} *or* B_{1od} and B_{2od} *or* B_{1od} and B_{2od}	B1od and B2od *or* B1cd and B2cd *or* B1ud and B2ud
ω_d	wd

```
Vin = 12;
R = 126;
L = 0.08;
C = 0.002;
w0 = sqrt(1/(L*C));
p = pi;
w1 = 2*p*40;
aod = R/(2*L);
t = (0:0.0001:1.2);
Fod = (w0^2*Vin)/sqrt((w0^2-w1^2)^2 + (2*aod*w1)^2);
Qod = -atan((2*aod*w1)/(w0^2-w1^2));
Vfrcod = Fod*sin(w1*t + Qod);
%RLC series circuit natural overdamped response to sinusoidal input
M1od = -aod + sqrt(aod^2-w0^2);
M2od = -aod-sqrt(aod^2-w0^2);
V0 = 8;
V01 = 20;
A1od = (M2od*V0-M2od*Fod*sin(Qod)+w1*Fod*cos(Qod)-V01)/(M2od-M1od);
A2od = (-M1od*V0 + M1od*Fod*sin(Qod)-w1*Fod*cos(Qod)+V01)/(M2od-
M1od);
Vnod = A1od*exp(M1od*t) + A2od*exp(M2od*t);
Vod = Vfrcod + Vnod;
figure
```

```
h = plot(t,Vod)
grid
xlabel('Time (s)')
ylabel('Vcmp-od(t) [V]')
title('Series RLC circuit sinusoidal response for overdamped case')
set(gca,'FontSize',18)
```

≫≫≫≫≫≫≫≫≫≫≫≫≫≫≫≫≫≫≫≫≫>

Besides building the graphs, MATLAB enables us to calculate the *forced (final) values* of this transient response. Below is the code that readily permits computations of all three sets of the final amplitude, $V_f \equiv F_{od}$, and phase shift, $\Phi_f \equiv \Phi_{od}$, by changing the value of just one circuit parameter, *R*, from $R_{od} = 126\Omega$ to $R_{cd} = 12.6\Omega$ to $R_{ud} = 1.26\Omega$.

Pay attention to the line for calculating the phase shift, "`atan2`." You recall that $tan\Phi = \frac{sin\Phi}{cos\Phi}$, and the command "atanΦ" gives the correct answer only for the first and third quadrants, where both sine and cosine are either positive or negative. In other words, the validity of "atanΦ" command ranges from $-\pi/2$ to $\pi/2$ or -90^0 to 90^0. In our example, however, $sin\Phi = 2\alpha\omega$ is positive but $cos\Phi = \omega_0^2 - \omega^2$ is negative, which require the command covering $-\pi$ to π or -180^0 to 180^0 range. This command is "atan2Φ," and the values returned by MATLAB are true angles.

Nonetheless, if you know the signs of both $sin\Phi$ and $cos\Phi$ functions, you can still use atanΦ code, but the correct result will be attained by subtracting the obtained angle from π or 180^0, i.e., $\Phi_{correct}(rad) = \pi - \Phi_{calculated}$ $\Phi^0_{correct} = 180^0 - \Phi_{calculated}$. (Verify this consideration by using both codes, atanΦ and atan2Φ, for this example.) The bottom line is to use "atan2Φ" command to avoid possible errors.

≫≫≫≫≫≫≫≫≫≫≫≫≫≫≫≫≫≫≫≫≫≫≫

MATLAB code for calculating the forced (final) values of the RLC circuit sinusoidal response in Example 5.4.

```
%RLC sinusoidal response - final values
Vin = 12;
R1 = 126;
R2 = 12.6;
R3 = 1.26;
L = 0.08;
C = 0.002;
w0 = sqrt(1/(L*C));
w1 = 2*pi*40;
aod1 = R1/(2*L);
aod2 = R2/(2*L);
aod3 = R3/(2*L);
t = (0:0.0001:1.2);
Fod1 = (w0^2*Vin)/sqrt((w0^2-w1^2)^2 + (2*aod1*w1)^2)
Fod2 = (w0^2*Vin)/sqrt((w0^2-w1^2)^2 + (2*aod2*w1)^2)
Fod3 = (w0^2*Vin)/sqrt((w0^2-w1^2)^2 + (2*aod3*w1)^2)
Qod1rad = -atan2((2*aod1*w1),(w0^2-w1^2))
Qod2rad = -atan2((2*aod2*w1),(w0^2-w1^2))
Qod3rad = -atan2((2*aod3*w1),(w0^2-w1^2))
```

```
Qod1deg=Qod1rad*180/pi
Qod2deg=Qod2rad*180/pi
Qod3deg=Qod3rad*180/pi
%MATLAB returns
Fod1 = 0.1875
Fod2 = 1.0818
Fod3 = 1.3146
Qod1rad = -1.7136
Qod2rad = -2.5339
Qod3rad = -3.0722
Qod1deg = -98.1821
Qod2deg = -145.1818
Qod3deg = -176.0215
```

≫≫≫≫≫≫≫≫≫≫≫≫≫≫≫≫≫≫≫≫≫≫≫≫≫

The forced amplitudes and phase shifts calculated by MATLAB in the above script are close enough to the values obtained from the MATLAB graphs, which justifies using MATLAB graphs for quantitative assessments of the theoretical derivations.

Below is the last set of MATLAB codes for calculations of *the natural responses*. Since the response equations for all three cases—overdamped, critically damped, and underdamped—differ in their structures and parameters, it's necessary to write a unique code for each case. The commentaries make the codes self-explanatory.

≫≫≫≫≫≫≫≫≫≫≫≫≫≫≫≫≫>

MATLAB codes for calculations of the parameters of the natural responses for Example 5.4.

≫≫≫≫≫≫≫≫≫≫≫≫≫

Overdamped case

```
% Calculations RLC series circuit natural overdamped %response to
sinusoidal input Example 5.4
Vin = 12;
V0 = 8;
V01 = 20;
R1 = 126;
L = 0.08;
C = 0.002;
aod1 = R1/(2*L);
w0 = sqrt(1/(L*C));
w1 = 2*pi*40;
%t = (0:0.0001:1.2);
Fod1 = (w0^2*Vin)/sqrt((w0^2-w1^2)^2 + (2*aod1*w1)^2);
Qod1rad = -atan2((2*aod1*w1),(w0^2-w1^2));
M1od = -aod1+sqrt(aod1^2-w0^2);
M2od = -aod1-sqrt(aod1^2-w0^2);
A1od = (M2od*V0-M2od*Fod1*sin(Qod1rad)+w1*Fod1*cos(Qod1rad)-V01)/
(M2od-M1od);
```

```
A2od = (-M1od*V0+M1od*Fod1*sin(Qod1rad)-w1*Fod1*cos(Qod1rad)+V01)/
(M2od-M1od);
vpa(M1od)
vpa(M2od)
vpa(A1od)
vpa(A2od)
%MATLAB returns
ans =
-3.9783027892590325791388750076294
```

%*This is* M1od $\Rightarrow m_{1od} \approx -3.98\left(\frac{1}{s}\right)$.

```
ans =
-1571.0216972107409674208611249924
```

$\approx -1571.02\left(\frac{1}{s}\right)$

%*This is* M2od $\Rightarrow m_{2od} \approx -1571.02\left(\frac{1}{s}\right)$.

```
ans =
8.2234572372860554878570837900043
```

$\approx 8.22(V)$ %*This is* A1od $\Rightarrow A_{1od} \approx 8.22(V)$.

```
ans =
-0.037824790414189957132862218713854
```

$\approx -0.04(V)$

%*This is* A2od $\Rightarrow A_{2od} \approx -0.04(V)$.

>>

Critically damped case

```
% Calculations RLC series circuit natural critically damped
%response to sinusoidal input for Example 5.4
Vin = 12;
V0 = 8;
V01 = 20;
R2 = 12.6;
L = 0.08;
C = 0.002;
acd = R2/(2*L);
w0 = sqrt(1/(L*C));
w1 = 2*pi*40;
%t = (0:0.0001:1.2);
Fcd = (w0^2*Vin)/sqrt((w0^2-w1^2)^2 + (2*acd1*w1)^2);
Qod1rad = -atan2((2*acd*w1),(w0^2-w1^2));
M1cd = -acd;
M2cd = -acd;
A1cd = V0-Fcd*sin(Qod1rad);
A2cd = (acd*V0-acd*Fcd*sin(Qcdrad)-w1*Fcd*cos(Qcdrad)+V01);
vpa(acd)
vpa(A1cd)
vpa(A2cd)
```

```
%MATLAB returns
ans =
78.75
```

%*This is* M1cd = M2cd ⇒ $-\alpha_{1cd} \approx -78.75\left(\frac{1}{s}\right)$.

```
ans =
8.6176943305463886702000309014693
```

%*This is* A1cd⇒ $A_{1cd} \approx 8.62\,(V)$.

```
ans =
921.85842540120711419149301946163
```

%*This is* A2cd⇒ $A_{2cd} \approx 921.86\,(V)$.

⋙⋙⋙⋙⋙⋙⋙⋙⋙

Underdamped case[2]

```
% Calculations RLC series circuit natural underdamped %response to
sinusoidal input for Example 5.4
Vin = 12;
V0 = 8;
V01 = 20;
R3 = 1.26;
L = 0.08;
C = 0.002;
aud = R3/(2*L);
w0 = sqrt(1/(L*C));
w1 = 2*pi*40;
wd = sqrt(w0^2-aud^2);
%t = (0:0.0001:1.2);
Fud = (w0^2*Vin)/sqrt((w0^2-w1^2)^2 + (2*aud*w1)^2);
Qud = -atan2((2*aud*w1),(w0^2-w1^2));
B1 = V0-Fud*sin(Qud);
B2 = (aud*V0-aud*Fud*sin(Qud)-w1*Fud*cos(Qud)-V01)/wd;
vpa(aud)
vpa(B1)
vpa(B2)
%MATLAB returns
ans =
7.875
```

%*This is* aud ⇒ $-\alpha_{ud} \approx -7.875\left(\frac{1}{s}\right)$.

```
ans =
8.0912064218293959783068203250878
```

%*This is* B1 ⇒ $B_1 \approx 8.09\,(V)$.

```
ans =
4.7456313253469861734856749535538
```

%*This is* B2 ⇒ $B_2 \approx 4.75\,(V)$.

⋙⋙⋙⋙⋙⋙⋙⋙⋙⋙⋙⋙⋙⋙⋙

---

2 All MATLAB codes in Sidebar 4S.3 were developed by Ms. Ina Tsikhanava.

5.2.3 Parallel RLC Circuit

This section investigates the *complete response* of a parallel RLC circuit to various inputs. Consider Figure 5.8a, whose input can be either DC or sinusoidal voltage source. When the switch is thrown, the RLC circuit receives the input, and the problem turns to finding its complete response.

In Section 5.1, we investigated the natural response of a parallel RLC circuit. The voltage differential equation (5.46) for a parallel RLC circuit has been derived using the *duality principle*. Here, we want to derive a *current equation* based on KCL as

$$i_L(t) + i_R(t) + i_C(t) = i_{in}(t)(A) \tag{5.97a}$$

or

$$i_L + \frac{v}{R} + C\frac{dv}{dt} = i_{in}(t). \tag{5.97b}$$

Since

$$v = L\frac{di_L}{dt},$$

(5.97b) can be transformed into the following required differential equation of a parallel RLC circuit:

$$i_L''(t) + 2\alpha i_L'(t) + \omega_0^2 i_L(t) = \omega_0^2 i_{in}(t), \tag{5.98}$$

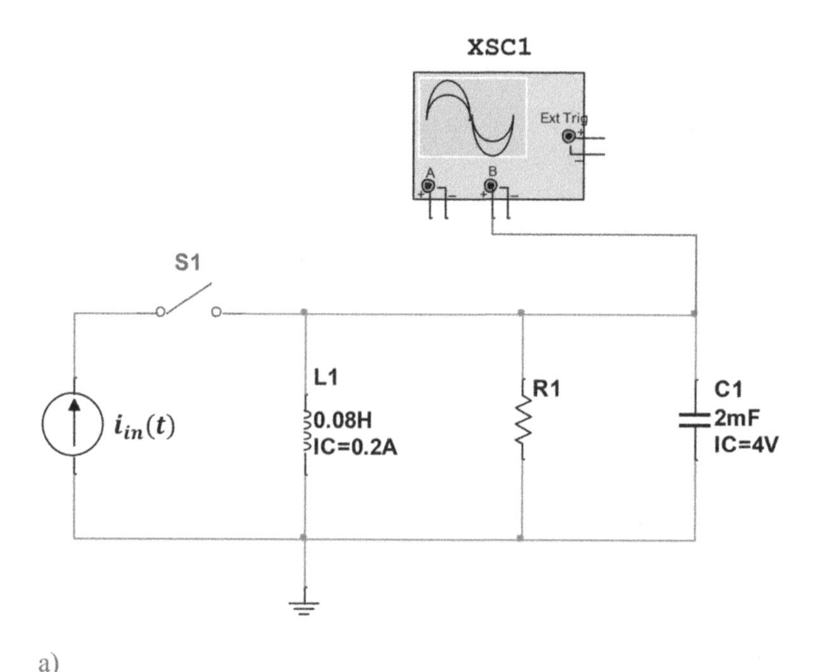

Figure 5.8 Complete response of a parallel RLC circuit to step input in Example 5.5: a) The circuit schematic; b) the graph of the overdamped case; c) critically damped case; d) underdamped case.

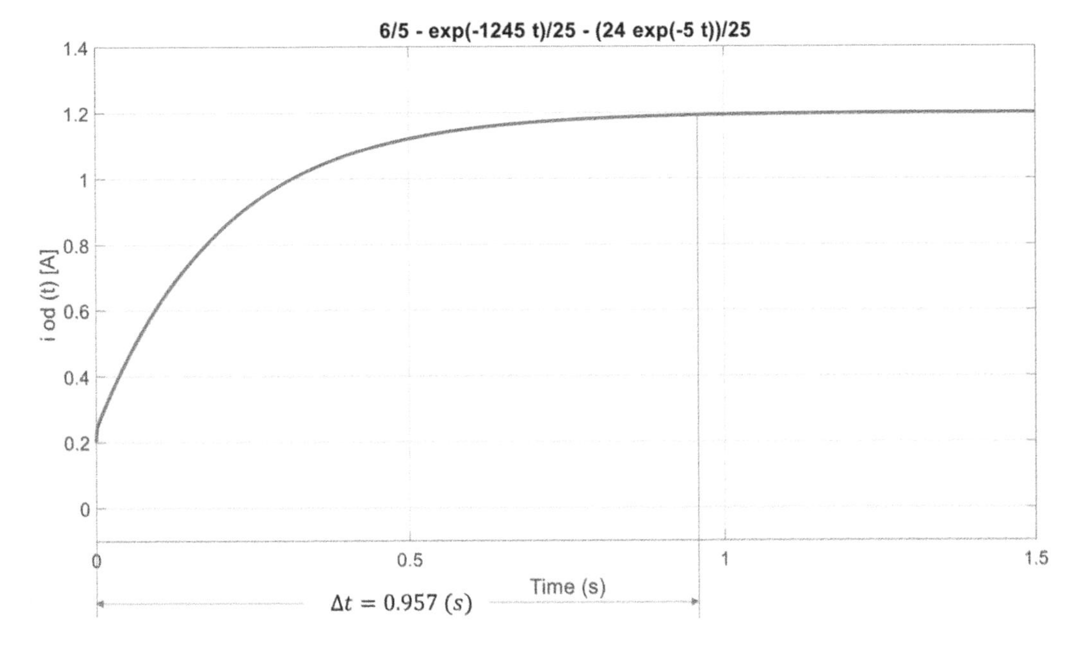

b)

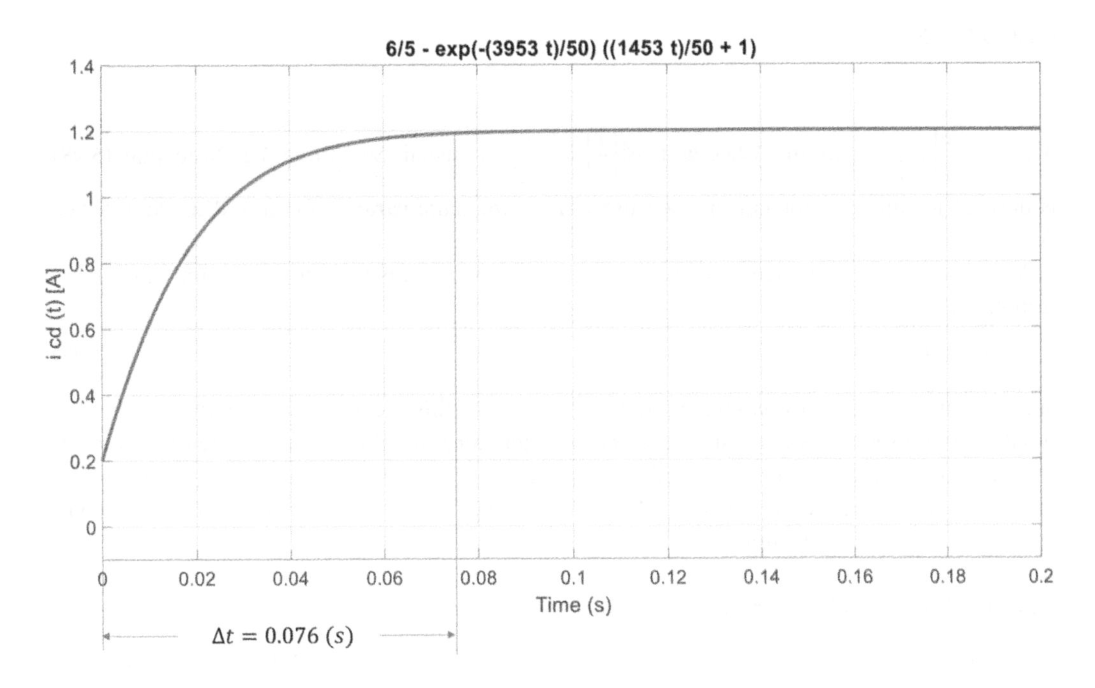

c)

Figure 5.8 (Cont'd)

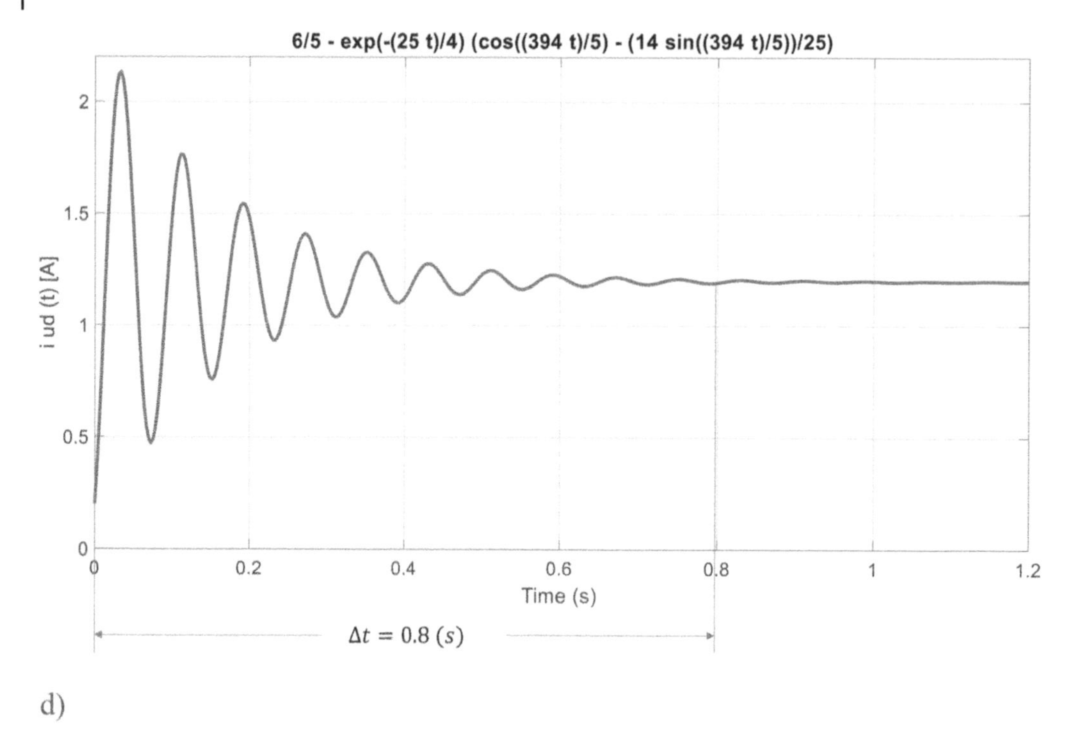

Figure 5.8 **(Cont'd)**

where $\alpha\left(\frac{1}{s}\right)=\frac{1}{2RC}$, as in (5.48), and $\omega_0^2\left(\frac{1}{s}\right)^2=\frac{1}{LC}$, as usual. See Table 5.2. Note that (5.98) is derived for an inductor current because $i_L(t)$ is the *state variable* for this RLC circuit. See Section 5.3.

We know that the complete response, the solution to (5.98), is the sum of natural and forced responses,

$$i_{out}=i_{nat}+i_{frc}. \tag{5.99}$$

Since the natural response is known, and we learned in this section how to find the complete (total) solution the series RLC circuit, further mathematical manipulations will be repetitions of what we already performed in this section. However, the practical implementation of the general procedure for finding the solutions to (5.101), (5.102), and (5.103) have some peculiarities that will be best discussed in an example.

Example 5.5 Complete response of a parallel RLC circuit.

Problem: Consider a parallel RLC circuit shown in Figure 5.8a. The input is a step function, $i_{in}^{step}(t)=1.2\cdot u(t)(A)$. The values of L1 and C1, along with their user-defined initial conditions (ICs), are shown and fixed; the values of R1 change to set up the overdamped, critically damped, and underdamped cases for each input. Find the complete responses for all three cases.

Solution: We follow **"The methodology of finding the complete response of an RLC circuit"**

discussed at the beginning of this section.

a) The input for this case is given by

$$i_{in}(t) = I_{in} \cdot u(t) = 12 \cdot u(t), \tag{5.100}$$

where $u(t)$ is the *unit step (Heaviside) function.* The three cases of a natural response are thoroughly discussed in Section 5.1. For the current flowing through the inductor, the complete responses of a parallel RLC circuit for $t \geq 0(s)$ take the form:

$$\text{For overdamped case: } i_{cmp-od}(t) = A_1 e^{m_1 t} + A_2 e^{m_2 t} + I_{in} \tag{5.101}$$

$$\text{For critically damped case: } i_{cmp-cd}(t) = (A_1 + A_2 t) e^{-\alpha_{cd} t} + I_{in} \tag{5.102}$$

$$\text{For underdamped case: } i_{cmp-ud}(t) = (A_1 \sin(\omega_d t) + A_2 \cos(\omega_d t)) e^{-\alpha_{ud} t} + I_{in}. \tag{5.103}$$

Here,

$$m_{1,2} = -\alpha \pm \sqrt{(\alpha^2 - \omega_0^2)}. \tag{5.11R}$$

Let's compute the values to be used in all three cases:

The square of the resonant frequency for this circuit is

$$\omega_0^2 \left(\frac{rad}{s}\right)^2 = \frac{1}{LC} = \frac{1}{0.08 \cdot 2 \cdot 10^{-3}} = 6250.$$

The value of a *damping coefficient* for the *critically damped case* is

$$\alpha_{cd}\left(\frac{1}{s}\right) = \omega_0 = 79.06.$$

Therefore,

$$R_{cd}(\Omega) = \frac{1}{2\alpha_{cd} C} = 3.16.$$

The values of a damping constant for the remaining cases are

Overdamped case: $\alpha_{od}\left(\frac{1}{s}\right) = \frac{1}{2R_{od}C} = \frac{1}{2 \cdot 0.4 \cdot 2 \cdot 10^{-3}} = 625$

Underdamped case: $\alpha_{ud}\left(\frac{1}{s}\right) = \frac{1}{2R_{ud}C} = \frac{1}{2 \cdot 40 \cdot 2 \cdot 10^{-3}} = 6.25.$

Similarly, the values of m_1 and m_2 are

$$m_{1-od}\left(\frac{1}{s}\right) = -\alpha_{od} + \sqrt{(\alpha_{od}^2 - \omega_0^2)} \approx -625 + 620 = -5$$

$$\text{and } m_{2-od}\left(\frac{1}{s}\right) = -\alpha_{od} - \sqrt{(\alpha_{od}^2 - \omega_0^2)} \approx -1245.$$

Also,

$$m_{1-\text{cd}}\left(\frac{1}{s}\right)=m_{2-\text{cd}}=\alpha_{cd}=-79.06.$$

And

$$m_{1-\text{ud}}=-\alpha_{ud}+\text{j}\sqrt{\left(\omega_0^2-\alpha_{ud}^2\right)}=-\alpha_{ud}+\text{j}\omega_d=-6.25+j78.8$$
$$\text{and } m_{2-\text{ud}}=-\alpha_{ud}-\text{j}\omega_d=-6.25-j78.8.$$

Now, the problem reduces to finding the constants A_1 and A_2, which can be done by finding $I_{cmp}^{step}(0)$ and its derivative.

b) For the **step overdamped case**, we have

$$I_{od}(0)=A_1+A_2+I_{in}.$$

Taking the derivative of (5.101) and setting $t=0$ yields

$$\frac{dI_{od}(0)}{dt}=m_1A_1+m_2A_2.$$

Since $v_L(t)=L\frac{di(t)}{dt}$, we can write

$$\frac{dI_{od}(0)}{dt}=\frac{V(0)}{L}.$$

Using the given values of the initial conditions produces the following equation system

$$\begin{cases} 0.2(A)=A_1+A_2+1.2 \\ \frac{4}{0.08}(A)=-5A_1-1245A_2 \end{cases}. \tag{5.104}$$

Presenting (5.104) in a matrix form,

$$\begin{vmatrix} 1 & 1 \\ -5 & -1245 \end{vmatrix}\begin{vmatrix} A_1 \\ A_2 \end{vmatrix}=\begin{vmatrix} -1 \\ 50 \end{vmatrix},$$

and denoting the *coefficient matrix* as A, the *matrix of variables* as V, and the *matrix of constants* as B enable us to employ MATLAB for calculating A_1 and A_2 as

```
>>>>>>>>>>>>>>>>>>>>>>>>>
>> A = [1 1;
-5 -1245];
B = [-1; 50];
V = linsolve(A,B)

V =
-0.9637
-0.0363
>>>>>>>>>>>>>>>>>>>>>>>>>>
```

Therefore, the step overdamped response is attained as

$$\begin{aligned} i_{cmp-od}(t)[A] &= A_1 e^{m_1 t} + A_2 e^{m_2 t} + I_{in} \\ &= -0.96e^{-5t} - 0.04e^{-1245t} + 1.2. \end{aligned} \tag{5.105}$$

c) The **step critically damped case** has the following initial conditions:

$$I_{in}(0) = A_1 + 1.2 \Rightarrow A_1 = -1.0.$$

and

$$\frac{dI_{od}(0)}{dt} = -\alpha_{cd} A_1 + A_2 \Rightarrow 50 = -79.06A_1 + A_2 \Rightarrow A_2 = -29.06.$$

Thus, (5.102) takes the following numerical presentation

$$i_{cmp-cd}(t) = (A_1 + A_2 t)e^{-\alpha_{cd} t} + I_{in} = -(1 + 29.06t)e^{-79.06t} + 1.2. \tag{5.106}$$

d) For the **step underdamped case**, the solution is obtained by the similar actions, namely: Equation (5.103) gives the following initial conditions

$$I_{in}(0) = A_2 e^{-\alpha_{ud} t} + I_{in} \Rightarrow 0.2 = A_2 + 1.2 \Rightarrow A_2 = -1.0$$

and

$$\frac{dI_{od}(0)}{dt} = \omega_d A_1 - \alpha_{ud} A_2 \Rightarrow A_1 = \frac{43.75}{78.8} = 0.56.$$

Therefore, the answer to this case is

$$\begin{aligned} i_{cmp-ud}(t)[A] &= \left(A_1 \sin(\omega_d t) + A_2 \cos(\omega_d t)\right)e^{-\alpha_{ud} t} + I_{in} \\ &= \left(0.56 sin(78.8t) - \cos(78.8t)\right)e^{-6.25t} + 1.2. \end{aligned} \tag{5.107}$$

Figures 5.8b, 5.8c, and 5.8d depict the graphs visualizing (5.105), (5.106), and (5.107), respectively.

The problem is solved.

- The first step to verify the result is checking our solutions' initial and final values. The initial value of $i_{cmp}(t)$ for every complete solution attained in (5.105), (5.106), and (5.107) must be equal to $I_0 = 0.2A$ at $t = 0$. The final value of every $i_{cmp}(t)$ must tend to $I_f = 1.2A$ at $t \to \infty$. Checking the results, we find:

*Equation 5.105 for overdamped case gives: $i_{cmp-od}(t) = -0.96e^{-5t} - 0.04e^{-1245t} + 1.2$.

At $t = 0 \Rightarrow i_{cmp-od}(t) = 0.2A$
At $t \to \infty \Rightarrow i_{cmp-od}(\infty) = 1.2A$

*Equation 5.106 for critically damped case: $i_{cmp-cd}(t) = -(1 + 29.06t)e^{-79.06t} + 1.2$.

At $t = 0 \Rightarrow i_{cmp-cd}(t) = 0.2A$.
At $t \to \infty \Rightarrow i_{cmp-cd}(\infty) = 1.2A$.

*Equation 5.107 for underdamped case: $i_{cmp-ud}(t) = \left(0.56 sin(78.8t) - \cos(78.8t)\right)e^{-6.25t} + 1.2$.

At $t=0 \Rightarrow i_{cmp-ud}(t)=0.2A$.
At $t \to \infty \Rightarrow i_{cmp-ud}(\infty)=1.2A$.

- Thus, our answers meet the requirements of the initial and final values. However, experimental verification must be done to acquire the comprehensive validation of the theoretical results. We leave this step as an exercise for you.
- The graphs presented in Figures 5.8a, 5.8b, and 5.8c give an additional measure for checking the correctness of our results, the transient interval $\Delta t(s)$. As Equations 5.101, 5.102, and 5.103 suggest, the critically damped case must have the shortest Δt, and the overdamped case must have the longest. The measurements shown in the figures confirm these predictions. (You will recall that we consider Δt equals 5τ, where $\tau(s)=\frac{1}{\alpha\left(\frac{1}{s}\right)}$.) However, two different exponential functions involved in the *overdamped case* prevent us from direct calculations of Δt, and we need to measure Δt on the graph. For this, we compute the final amplitude as $I_f=12(1-e^{-5})=1.191$ and place *Data Tips* box on the graph point where its amplitude is $Y=1.191$, and get $X=0.957$. This reading gives $\Delta t=5\tau=0.957(s)$.)

Applying the methodology used in Example 5.5, we can find the complete response of a parallel RLC circuit to a sinusoidal input. Example 5.4, where the sinusoidal response of a series RLC circuit is thoroughly considered, makes this task quite manageable. We leave this problem as an exercise for you.

5.3 State Variables in Time Domain

5.3.1 Application of State Variables in Circuit Analysis

What is the state of a circuit? It is extant status, the situation in which the circuit exists at a given moment. How can we describe the circuit's state? By the circuit's parameters that recite its condition from the beginning to the current situation and able to predict how this state will evolve in the future. For example, what is the initial state of an RLC circuit? It is described by the initial values of its inductor current, $I_L(0)$, and the capacitor voltage, $V_C(0)$. What is the state of an RLC circuit at $t=23(s)$? It is still defined by the value of its inductor current, I_L and V_C accumulated from $t=0(s)$ to $t=23(s)$. What will be the circuit's state at $t \to \infty$? Of course, it will be determined by the values of I_L and V_C stored from $t=0(s)$ and projected to $t \to \infty$.

Naturally, the parameters that describe the circuit's state are called *state variables*. For an RLC circuit, the *inductor current* and the *capacitor voltage* are the state variables.

Can you guess why the resistance is not involved in the description of an RLC circuit's status, and is not a state variable? Your guess is correct: The current through and the voltage across a resistor change instantly, regardless of their preceding values. In contrast, the current through an inductor and the voltage across the capacitor cannot change immediately; we remember that $I_L(0^-)=I_L(0^+)$ and $V_C(0^-)=V_C(0^+)$, and these rules are valid for any instant, not only for $t=0$. In other words, the inductor and capacitor carry a memory of the circuit's past. (Interestingly, modern electronics developed an electrical component termed memory-resistor, *memristor*, whose extant resistance depends on the value of the current that flowered through it previously. The memristor's operation, however, is based on the *feedback* meaning that this component is *active* in contrast to *passive* resistor, inductor, and capacitor.)

Strictly speaking, the *fundamental state variable* for an inductor is its *flux linkage* $\Lambda(Wb)$ whose relationship with the inductor current is given in (1.36) as

$$\Lambda(Wb) = L(H)i_L(t)[A], \tag{1.36R}$$

where $L(H)$ is the inductance. Similarly, for the capacitor, the fundamental state variable is the *charge*, $Q(C)$, which relates to the capacitor voltage as

$$Q(C) = C(F)v_C(t)[V], \tag{1.22R}$$

where $C(F)$ is the *capacitance*. The parameters $\Lambda(Wb)$ and $Q(C)$ accumulate all the relevant preceding activities of an inductor and capacitor, respectively, and thus present the *actual state* of these components. The latter can also serve as the state variables thanks to their linear relationships with the current and voltage.

Why do we need to know the circuit state and all these state variables if we have the description of a circuit and its operation delivered by a circuit analysis, which we are studying?

Here are the reasons:

- A circuit analysis under our systems approach is focused on the input–output relationship. Solving the circuit's differential equation enables us to find the input–output formula and, with additional manipulations, determine the state of each circuit element, and, therefore, the state of the circuit. For example, in a series RLC circuit, obtaining $v_{C-out}(t) = f(v_{in}(t), R, L, C, I_{L0}, \text{and } V_{C0})$ allows for calculating $i_c(t) = i(t)$, $v_L(t)$, *and* $v_R(t)$. In contrast, the state-variables approach is based on defining the state variables' rate of change. The two differential equations for an RLC circuit stem from $\frac{di_L(t)}{dt}$ and $\frac{dv_C(t)}{dt}$ definitions, and as soon as we solve these two equations, all other circuit's values can be readily found.
- For example, consider an RLC circuit's differential equation that can be presented in a general form as

$$\ddot{x} + 2\alpha\dot{x} + \omega_0^2 x = \omega_0^2 x_{in}, \tag{5.108}$$

where x can be either $v(t)$ or $i(t)$, $\dot{x} = \frac{dx}{dt}$, and $\ddot{x} = \frac{d^2x}{dt^2}$. (Clearly, $x = x(t)$, but we omit the argument (t) for the notational simplicity.) Let's introduce the following notations:

$$x_1 = x \tag{5.109a}$$

$$\dot{x}_1 = x_2 \tag{5.109b}$$

$$\dot{x}_2 = \ddot{x}_1 \tag{5.109c}$$

$$x_{in} = y. \tag{5.109d}$$

Then, presenting (5.108) as

$$\ddot{x} = -\omega_0^2 x - 2\alpha\dot{x} + \omega_0^2 x_{in},$$

we attain

$$\dot{x}_1 = x_2 \tag{5.110a}$$

$$\dot{x}_2 = -\omega_0^2 x_1 - 2\alpha x_2 + \omega_0^2 y. \tag{5.110b}$$

Therefore, one second-order Equation 5.108 is transformed into the system of two first-order differential equations 5.110. Also, we need to consider the initial conditions when writing the whole equation system.

In fact, (5.110a) and (5.110b) are the *matrix differential equation* to be written as

$$\begin{vmatrix} \dot{x}_1 \\ \dot{x}_2 \end{vmatrix} = \begin{vmatrix} 0 & 1 \\ -\omega_0^2 & -2\alpha \end{vmatrix} \begin{vmatrix} x_1 \\ x_2 \end{vmatrix} + \begin{vmatrix} 0 \\ \omega_0^2 \end{vmatrix} y. \tag{5.111}$$

Any matrix equation can be readily solved by using MATLAB, as we did previously and will demonstrate again shortly.

- *The main advantage of using the state-variables approach is its ability to find the state of the higher-order circuits.* Our discussion has been restricted by the second-order circuits. Why? Because their *characteristic equations* are quadratic, whose roots can be easily determined, as we've seen in this chapter. The differential equation of a circuit containing three reactive elements is the third-order one, and its characteristic equation will be cubic. We can find the roots of a cubic equation, but the procedure is cumbersome, and all the required manipulations become prohibitively sophisticated. There is no reasonable way to solve the fourth and higher-order differential equations by applying the traditional method discussed in this chapter. Here, the state-variables approach comes to the rescue because it offers three first-order differential equations instead of a single third-order one. If we have a circuit with n reactive elements, then using the state variables enables us to build the n-th order matrix resembling (5.111). In addition, the *output matrix* $\boldsymbol{z}$ might be added. Then, the concise form of these equations becomes

$$\dot{\boldsymbol{x}} = \boldsymbol{Ax} + \boldsymbol{By} \tag{5.112a}$$

$$\boldsymbol{z} = \boldsymbol{Cx} + \boldsymbol{Dy}. \tag{5.112b}$$

Matrix equations 5.112a and 5.112b are called *state matrix equations*; they describe n-th order circuit with k inputs and l outputs. They use the following notations: $\boldsymbol{x}$ is n-th *state vector* whose components are the state variables, $\boldsymbol{y}$ is the k-th *input vector*, and $\boldsymbol{z}$ is the l-th *output vector*. Also, $\boldsymbol{A}$ is *nxn circuit matrix*, $\boldsymbol{B}$ is *nxk input matrix*, $\boldsymbol{C}$ and $\boldsymbol{D}$ are respective *lxn* and *lxk* output matrixes.

We discuss these new ideas in the examples that follow.

Example 5.6 Analysis of the third-order circuit using the state-variables approach.

Problem: The third-order series-parallel RLC circuit is shown in Figure 5.9. The switch has been in its initial position at $V2 = -1\ (V)$ for a long time, and then, at $t = 0\ (s)$, it was thrown to connect $V1 = 1\ (V)$ to the circuit. All element values are shown in the figure. Find all three state variables. The initial conditions are not assigned.

Solution: The solution consists of two parts: The first is deriving the *state matrix equations*. In the second part, we must solve these equations to find out how the circuit state variables change over time.

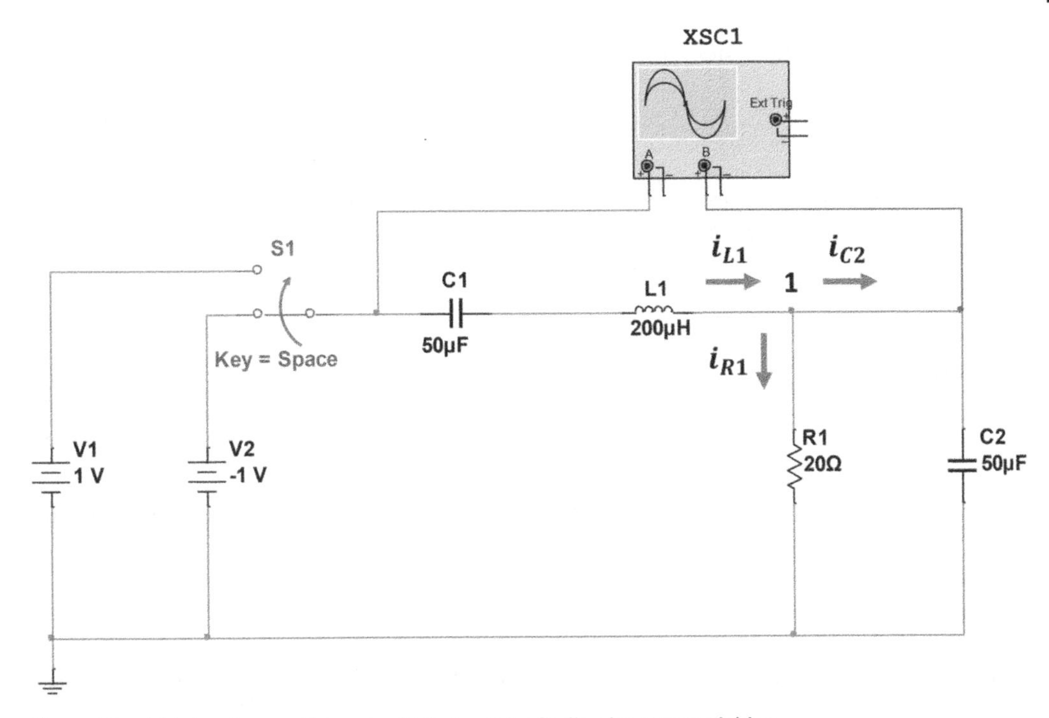

Figure 5.9 Third-order parallel-series RLC circuit for finding its state variables.

- ***Part 1—deriving the state matrix equations.***

The *state variables* of the circuit in Figure 5.9 are voltage $v_{C1}(t)$ across capacitor $C1$, the current $i_{L1}(t)$, through the inductor $L1$, and voltage $v_{C2}(t)$ across capacitor $C2$. We denote them as

$$\begin{aligned} x_1 &= v_{C1}(t) \\ x_2 &= i_{L1}(t) \\ x_3 &= v_{C2}(t). \end{aligned} \tag{5.113}$$

In a vector form, Equations 5.113 are given by

$$\boldsymbol{x} = \begin{vmatrix} x_1 \\ x_2 \\ x_3 \end{vmatrix} = \begin{vmatrix} v_{C1} \\ i_{L1} \\ v_{C2} \end{vmatrix}, \tag{5.114}$$

where $\boldsymbol{x}$ is a *state vector*. Now, we need to obtain equations containing the following derivatives

$$\begin{aligned} \frac{dv_{C1}(t)}{dt} &= \frac{dx_1}{dt} = \dot{x}_1, \\ \frac{di_{L1}(t)}{dt} &= \frac{dx_2}{dt} = \dot{x}_2, \\ \frac{dv_{C2}(t)}{dt} &= \frac{dx_3}{dt} = \dot{x}_3. \end{aligned} \tag{5.115}$$

The first of 5.115 equations stems from $i_{C1}(t)=C1\frac{dv_{C1}(t)}{dt}$, which gives

$$\dot{x}_1=\frac{x2}{C1}. \tag{5.116a}$$

The second equation involves $L1\frac{di_{L1}(t)}{dt}=v_{L1}(t)$; it can be found if we apply KVL to the loop $V1-C1-L1-C2$ in Figure 5.9 as

$$-V1+x1+L1\frac{dx_2}{dt}+x_3=0,$$

which results in

$$\dot{x}_2=-\frac{(x_1+x_3)}{L1}+\frac{V1}{L1}. \tag{5.116b}$$

Finally, the third equation concerning with $\frac{dv_{C2}(t)}{dt}=\frac{dx3}{dt}$ can be derived by applying KCL to Node 1:

$$i_{L1}(t)=i_{C2}(t)+i_{R1}(t).$$

Knowing that $i_{C2}(t)=C2\frac{dv_{C2}(t)}{dt}\equiv C2\frac{dx}{dt}$, $i_{L1}(t)=x_2$, and $i_{R1}(t)=\frac{v_{C2}(t)}{R1}\equiv\frac{x_3}{R1}$ enables us to write

$$\dot{x}_3=\frac{x_2}{C2}-\frac{x_3}{R1C2}. \tag{5.116c}$$

Collecting all Equations 5.116, we get the *state matrix equation* of our RLC circuit as

$$\dot{\boldsymbol{x}}=\begin{vmatrix}\dot{x}_1\\ \dot{x}_2\\ \dot{x}_3\end{vmatrix}=\begin{vmatrix}0 & \frac{1}{C1} & 0\\ -\frac{1}{L1} & 0 & -\frac{1}{L1}\\ 0 & \frac{1}{C2} & -\frac{1}{R1C2}\end{vmatrix}\begin{vmatrix}x_1\\ x_2\\ x_3\end{vmatrix}+\begin{vmatrix}0\\ \frac{1}{L1}\\ 0\end{vmatrix}V1. \tag{5.117a}$$

To complete Part 1, we need to identify the *initial conditions*. As usual, they can be found based on the physical property of the reactive components: The voltage across a capacitor and the current through an inductor can't change instantaneously. Thus, for the RLC circuit in Figure 5.9a, we have

$$\begin{aligned}v_{C1}(0^-)&=v_{C1}(0^+)=V2,\\ i_{L1}(0^-)&=i_{L1}(0^+)=0,\\ v_{C2}(0^-)&=v_{C2}(0^+)=0.\end{aligned} \tag{5.117b}$$

These initial conditions can be rewritten in a vector form as

$$\boldsymbol{x_0}=\begin{vmatrix}x_{10}\\ x_{20}\\ x_{30}\end{vmatrix}=\begin{vmatrix}V2\\ 0\\ 0\end{vmatrix}. \tag{5.118}$$

Therefore, Part 1 of the solution is completed. We can put down our results in the concise form (5.112a) as

$$\dot{\boldsymbol{x}} = \boldsymbol{A}\boldsymbol{x} + \boldsymbol{B}\boldsymbol{y}, \tag{5.112aR}$$

where $\boldsymbol{x}$ is given in (5.114), $\dot{\boldsymbol{x}}$ is defined in (5.117a), $\boldsymbol{x}_0$ is determined in (5.118), and matrixes $\boldsymbol{A}$ and $\boldsymbol{B}$ can be collected from (5.117a) as

$$\boldsymbol{A} = \begin{vmatrix} 0 & \frac{1}{C1} & 0 \\ -\frac{1}{L1} & 0 & -\frac{1}{L1} \\ 0 & \frac{1}{C2} & -\frac{1}{R1C2} \end{vmatrix} \tag{5.119}$$

and

$$B = \begin{vmatrix} 0 \\ \frac{1}{L1} \\ 0 \end{vmatrix}. \tag{5.120}$$

- ***Part 2—solving the matrix equations using the MATLAB code.***

The MATLAB code for solving the matrix equations and plotting the first state variable is presented below. The codes for two other variables are almost the same. See **Comments** at the end of the code. The graphs are displayed in Figure 5.10.

The codes are prepared by Dr. Vitaly Sukharenko.

```
%% MATLAB CODE for state variable VC1
clear all;
clc;

t0 = 0; %Starting time
tf = 10e-3; %End time

tspan = [t0 tf];%Time span

x0 = [-1 0 0]'; %Initial conditions for Vc1 = -1; i1 = 0; V2 = 0
[t,x] = ode45(@ckt, tspan, x0); %Call ODE function

plot(t*1000,x(:,1),'-b') %Plot Scope response
xlabel('Time (ms)');
ylabel('Voltage');
grid on; hold on;

%ODE FUNCTION
function xdot = ckt(t,x)
vg = 1; %SOURCE
C1 = 50e-6; %Cap1
C2 = 50e-6; %Cap2
R=20; %R
L=200e-6; %L
```

```
xdot(1) = x(2)/C1; % X1 = Vc1
xdot(2) = vg/L-x(1)/L-x(3)/L; % X2 = i1
xdot(3) = x(2)/C2 - x(3)/(R*C2);% X3 = V2
xdot = [xdot(1); xdot(2); xdot(3)]; %XDOT array for ODE
end
%END OF CODE
```

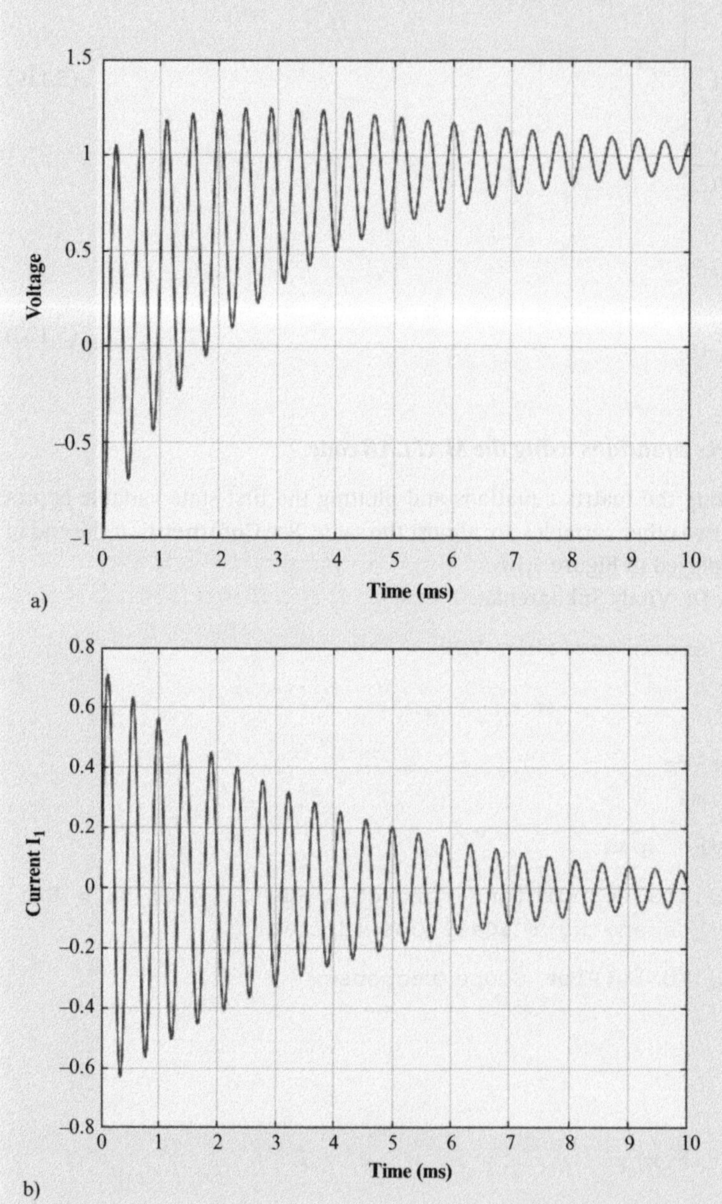

Figure 5.10 Graphs of the three state variables for Example 6.6: a) Variable $VC1$; b) variable $I1$; c) variable $VC2$.

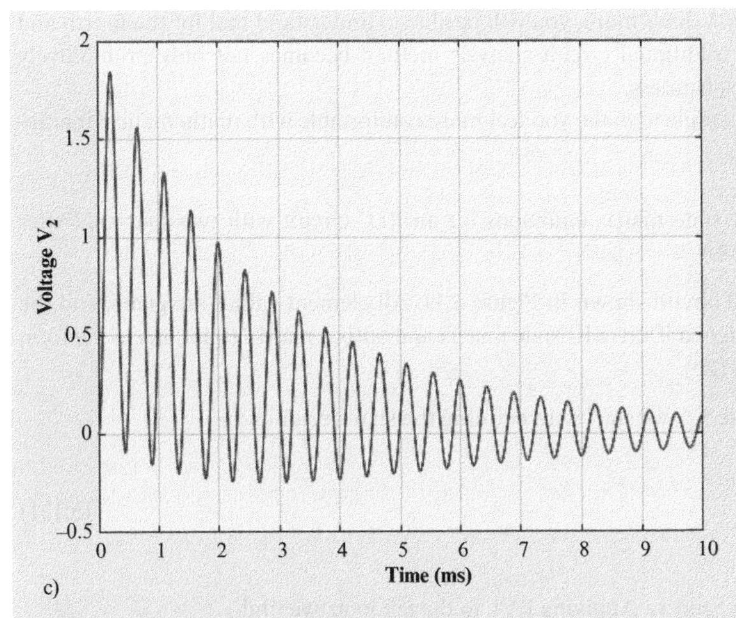

Figure 5.10 (Cont'd)

Comments:

1) The codes for $I1$ and $VC2$ are practically identical to the given code. The only difference is in the plot commands, which for these variables must be written as

```
plot(t*1000,x(:,2),'-b') %Plot Scope response
xlabel('Time (ms)');
ylabel('Current I_1');
and
plot(t*1000,x(:,3),'-b') %Plot Scope response
xlabel('Time (ms)');
ylabel('Voltage V_2');
```

2. To run the code, you must open the NEW SCRIPT window of MATLAB, paste the code into the EDITOR window, and press RUN (green triangle button).

Discussion:

- Carefully analyze the three graphs in Figure 5.10 regarding the circuit in Figure 5.9. Understand that these graphs present signals seen by the B terminal of the oscilloscope. Answer the following questions concerning the initial and final values: Why does state variable $VC1$ start at −1 volt and tend to 1 volt? Why does current $I1$ jump from 0 (A) to 0.8 (A), but still tend to zero? Why does the voltage drop across capacitor $C2$ begin from zero, increase to almost 2 (V) and still go to zero? Why do these three variables oscillate but not transit smoothly from the initial to final values? Can we change the behavior of $VC1$, $I1$, *and* $VC2$, and if we can, then how?
- If you think using the state variables is a sophisticated task, try solving the problem presented in this example using a classical circuit analysis method. You immediately find out how good the

state-variables approach is. What's more, you will be able to understand that for the fourth and higher-order circuits the traditional circuit-analysis method becomes not only prohibitively complex but, in most cases, fruitless.

Let's consider the other example to make you feel more comfortable with mathematical manipulations of the state variables.

Example 5.7 Deriving the state matrix equations for an RLC circuit with two sources. (After Glyn James 2015, pp. 487–488.)

Problem: Consider the RLC circuit shown in Figure 5.11. All element values are given, and the initial conditions are not assigned. Derive its state matrix and output matrix equations in the form provided in (5.112a) and (5.112b).

Solution: Following Example 5.6, we start with defining the state vector as

$$\begin{aligned} x_1 &= i_{L1}(t), \\ x_2 &= i_{L2}(t), \\ x_3 &= v_{C1}(t). \end{aligned} \tag{5.121}$$

Next, we need to find $\dot{x}_1$, $\dot{x}_2$, and $\dot{x}_3$. Applying KVL to the left loop, we find

$$-V1 + i_{L1}R1 + L1\frac{di_{L1}}{dt} + v_{C1} = 0,$$

which gives

$$\dot{x}_1 = \frac{x_1 R1}{L1} + \frac{x_3}{L1} + \frac{V_1}{L1}. \qquad 5.122$$

Employing KVL for the right loop produces

$$-V_2 + L2\frac{di_{L2}}{dt} + v_{C1} = 0.$$

Using notations given in Equations 1.121 results in

$$\dot{x}_2 = \frac{x_3}{L2} + \frac{V_2}{L2}. \tag{5.123}$$

Considering KCL for Node 1 brings

$$i_{L1} + i_{L2} = C1\frac{dv_{C1}}{dt},$$

from which we deduct

$$\dot{x}_3 = \frac{x_1}{C1} + \frac{x_2}{C1}. \tag{5.124}$$

Therefore, the *state matrix equation* of the RLC circuit in Figure 5.11 is given by

$$\begin{vmatrix} \dot{x}_1 \\ \dot{x}_2 \\ \dot{x}_3 \end{vmatrix} = \begin{vmatrix} \frac{R1}{L1} & 0 & \frac{1}{L1} \\ 0 & 0 & \frac{1}{L2} \\ \frac{1}{C1} & \frac{1}{C1} & 0 \end{vmatrix} \begin{vmatrix} x_1 \\ x_2 \\ x_3 \end{vmatrix} + \begin{vmatrix} \frac{1}{L1} & 0 \\ 0 & \frac{1}{L2} \\ 0 & 0 \end{vmatrix} \begin{vmatrix} V1 \\ V2 \end{vmatrix}. \tag{5.125}$$

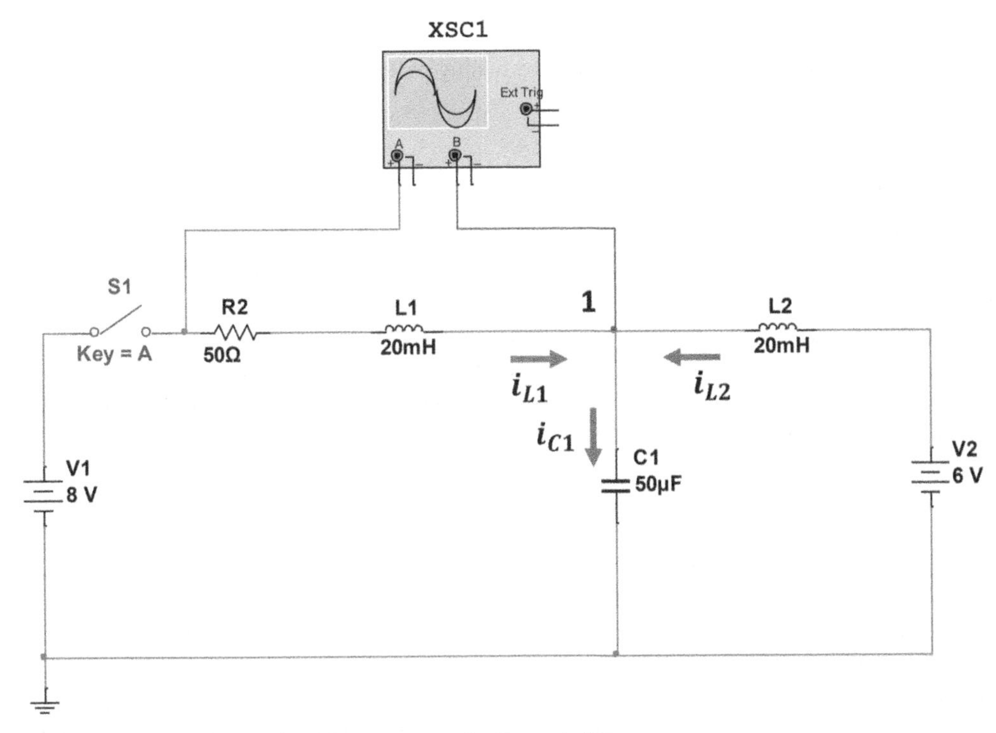

Figure 5.11 The RLC circuit with two sources for Example 5.7.

Concerning the *output state matrix*, we can choose any circuit's reactive element as an output and calculate the voltage across it. For example, let's consider v_{L2}. KVL of the right circuit loop in Figure 5.11 gives

$$-V2 + v_{L2} + v_{C1} = 0.$$

Thus,

$$v_{L2} = -v_{C1} + V2.$$

Using the notations of (5.112b),

$$\boldsymbol{z} = \boldsymbol{Cx} + \boldsymbol{Dy}, \tag{5.112Br}$$

we can write

$$\boldsymbol{z}_{L2} = \begin{bmatrix} 0 0 -1 \end{bmatrix} \begin{vmatrix} x_1 \\ x_2 \\ x_3 \end{vmatrix} + \begin{bmatrix} 0 1 \end{bmatrix} \begin{vmatrix} V1 \\ V2 \end{vmatrix}, \tag{5.126}$$

where $\boldsymbol{z}_{L2} = v_{L2}$, $\boldsymbol{C} = \begin{bmatrix} 0 0 -1 \end{bmatrix}$, and $\boldsymbol{D} = \begin{bmatrix} 0 1 \end{bmatrix}$.

Performing the similar operations, we can write the *output matrix equation* for v_{L1} as

$$\boldsymbol{z}_{L1} = \begin{bmatrix} -R 0 -1 \end{bmatrix} \begin{vmatrix} x_1 \\ x_2 \\ x_3 \end{vmatrix} + \begin{bmatrix} 1 0 \end{bmatrix} \begin{vmatrix} V1 \\ V2 \end{vmatrix}, \tag{5.127}$$

where $\boldsymbol{z}_{L1} = v_{L1}$, $\boldsymbol{C} = [-R0-1]$, and $\boldsymbol{D} = [10]$.

Likewise, the output matrix equation v_{C1} can be readily written.

The problem for Example 5.7 is solved.

5.3.2 ZIR and ZSR

The state-variables approach offers another advantage in transient circuit analysis: It enables us to solve problems using the *superposition principle.* Specifically, we consider the circuit response to its initial conditions when any external input (excitation or drive, or source) is zero. Naturally, such a response is called *a zero-input response, ZIR.* Secondly, we deliberate the circuit response to the external input assuming the circuit's *initial state* is zero, that is, all capacitor voltages and inductor currents are zero. Commonly, this is *a zero-state response, ZSR.* The circuit's *complete response* is a sum of ZIR and ZSR. Needless to say, this approach works only for the linear time-invariant (LTI) circuits.

Wait a minute, you might correctly ask, isn't it the same method we employed in the preceding and the current chapters by finding the complete response as a sum of transient and steady-state responses or by adding up the natural and forced responses? Yes and no. **Yes**, because the preceding summations also exploited the superposition principle, but **no**, because there is a difference between a natural-plus-forced response and a ZIR+ZSR response. As we highlighted many times in our discussions and examples in Chapters 4 and 5, the natural response doesn't satisfy the initial conditions in traditional transient circuit analysis. This is because *the constants of a complete solution can be found only from the sum of the natural and forced responses.* In contrast, the zero-input response alone meets the initial conditions, and the complete response with the state-variables approach becomes correct as the zero-state response itself assumes zero initial conditions. In other words, the complete response, in this case, is the authentic sum of ZIR and ZSR.

To clarify our discussion, let's consider a short example.

Example 5.8 Comparison of circuit-analysis and state-variables approaches.

Problem: Consider a series RC circuit shown in Figure 5.12. Find its step response by (1) the traditional circuit-analysis method and (2) by the state-variables approach. Compare both techniques.

Solution:

Part 1: Traditional circuit-analysis solution:

Refer to Chapter 4, where the step response of a series RC circuit is discussed in depth by a traditional circuit-analysis method. The *differential equation* of a series RC circuit with a step input V_{in} is given by

$$\frac{dv_C(t)}{dt} + \alpha v_C(t) = \alpha V_{in}, \tag{5.128}$$

where $\alpha\left(\frac{1}{s}\right) = \frac{1}{RC}$ and $v_C(t) \equiv v_{out}(t)$. The complete response is

$$v_{out-cmp}(t) = Ae^{-\alpha t} + V_{in}, \tag{5.129}$$

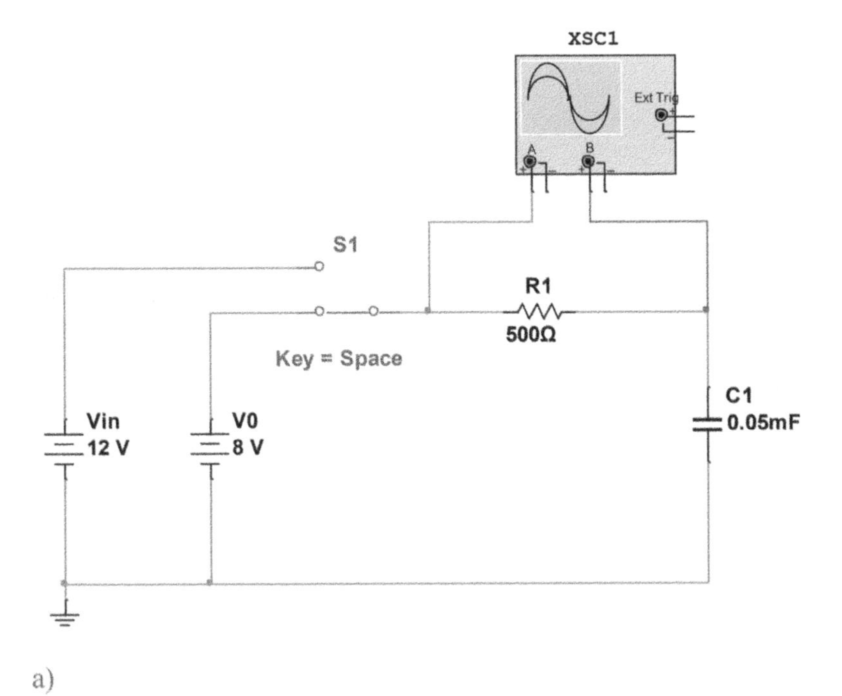

a)

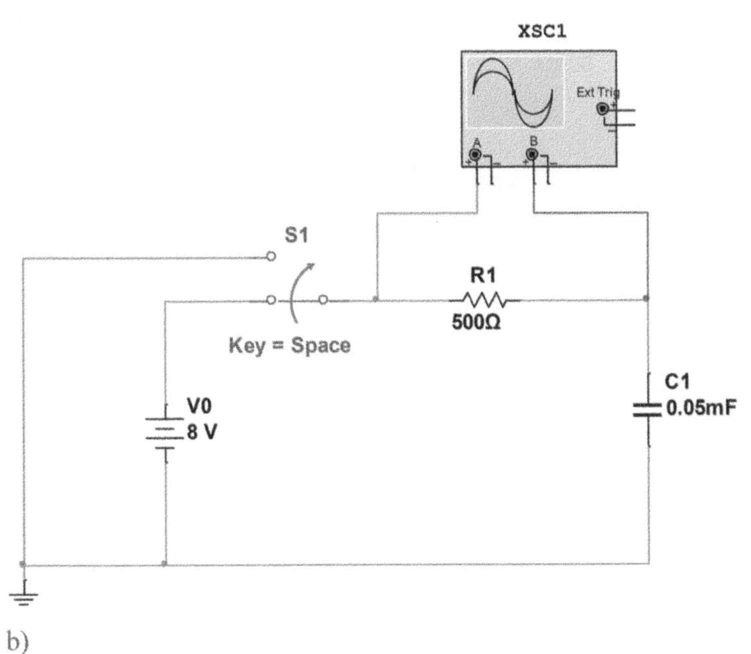

b)

Figure 5.12 Series RC circuit for ZIR and ZST analysis in Example 5.8: a) Circuit's schematic; b) ZIR circuit; c) ZIR response; d) ZSR circuit; e) ZSR response; f) complete (total) response, ZIR + ZSR.

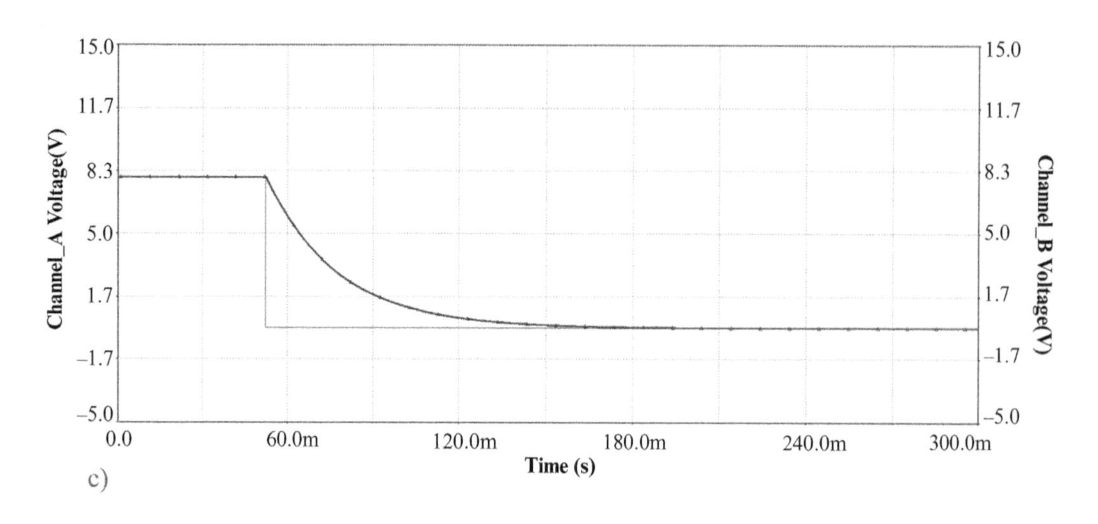

c)

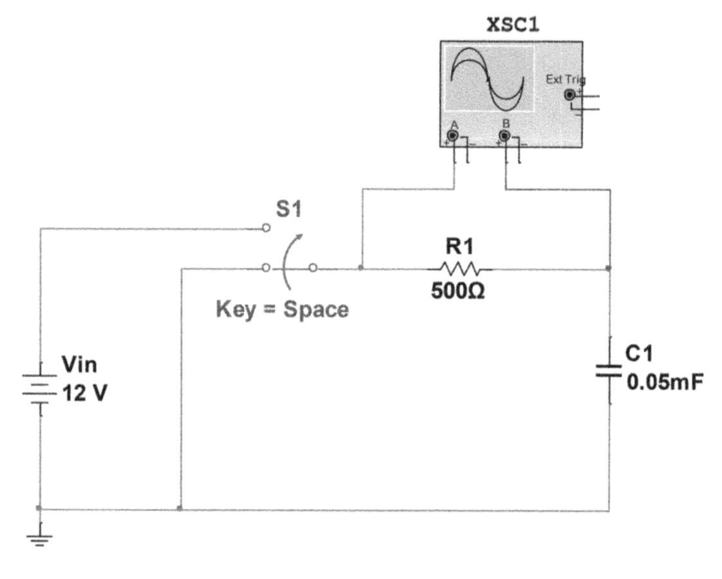

d)

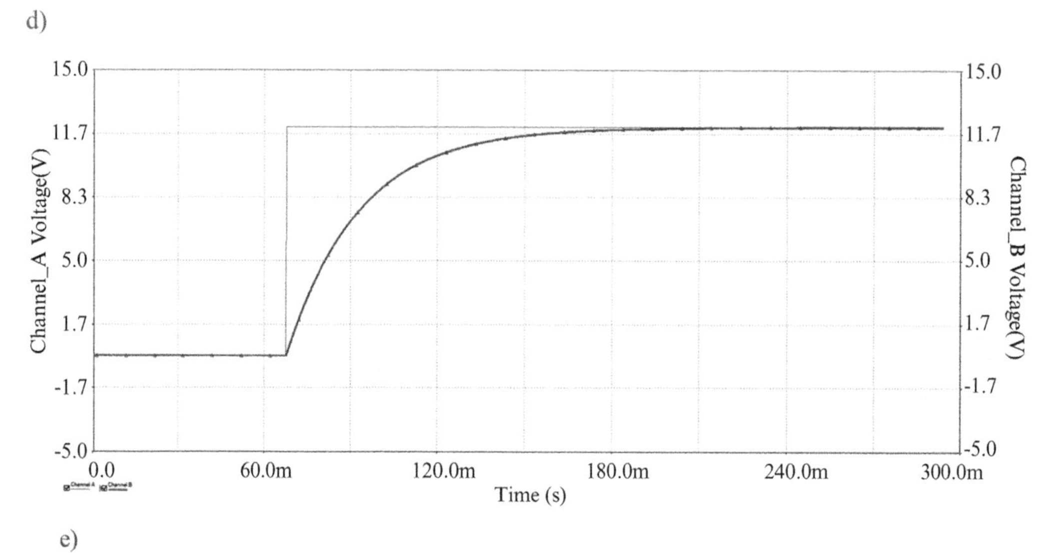

e)

Figure 5.12 (Cont'd)

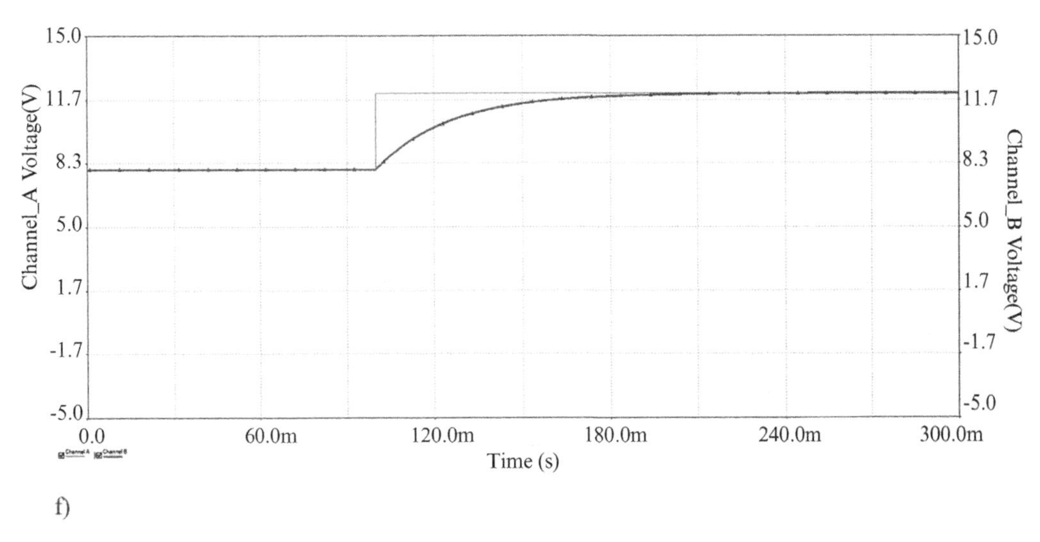

f)

Figure 5.12 (Cont'd)

where $Ae^{-\alpha t}$ is a *natural response*, and V_{in} is a *forced response*. To find constant A, we must—and this is the key point of the actual discussion—evaluate (5.129) at $t=0$. Therefore, *to find the constant of a natural response, we must involve the complete—the sum of natural and forced—response*. Denoting $v_{out-cmp}(0)=V_0$, we get from (5.129)

$$V_0 = A + V_{in},$$

which gives

$$A = V_0 - V_{in}. \tag{5.130}$$

Therefore, the answer to Part 1 is

$$v_{out-cmp}(t) = (V_0 - V_{in})\, e^{-\alpha t} + V_{in}. \tag{5.131}$$

To highlight it one more time, the complete response is a sum of natural and forced responses, but the complete solution—specifically, finding the constant—must use both responses simultaneously.

Part 2: The state-variables approach—ZIR and ZSR

In this approach, the complete solution can be found as a true sum of two independent entities, ZIR and ZSR.

The ZIR circuit is shown in Figure 5.12b, and its equation is obtained from (5.128) when $V_{in}=0$, that is,

$$\frac{dv_C(t)}{dt} = -\alpha v_C(t).$$

The solution to this equation is

$$v_{out-ZIR}(t) = Ae^{-\alpha t}.$$

Letting $t = 0$ gives

$$A = v_{out-ZIR}(0) \equiv V_0.$$

Thus, the ZIR solution is

$$v_{out-ZIR}(t) = V_0 e^{-\alpha t}. \tag{5.132}$$

The graph of (5.132) is shown in Figure 5.12c.

The ZSR circuit is shown in Figure 5.12d, and its equation stems from (5.128) as

$$\frac{dv_C(t)}{dt} = -\alpha v_C(t) + \alpha V_{in}.$$

The solution to this equation is given in (5.129) as

$$v_{out-ZSR}(t) = Ae^{-\alpha t} + V_{in}. \tag{5.129R}$$

At $t = 0$, we have $v_{out-cmp}(0) \equiv V_0 = 0$, and $A = -V_{in}$.

Thus, the ZSR solution is

$$v_{out-ZSR}(t) = V_{in}\left(1 - e^{-\alpha t}\right), \tag{5.133}$$

and its graph is demonstrated in Figure 5.12e.

Combining ZIR and ZSR solutions yields the complete solution as

$$\begin{aligned} v_{out-cmp}(t) &= v_{out-ZIR}(t) + v_{out-ZSR}(t) \\ &= V_0 e^{-\alpha t} + V_{in}\left(1 - e^{-\alpha t}\right). \end{aligned} \tag{5.134}$$

Figure 6.12f visualizes the complete solution, ZIR+ZSR. (Pay attention to the starting (initial) and the final points of this graph. Explain why they are of such value.)

Part 3: Comparison of both solutions.

- Comparing the solution obtained by the traditional circuit-analysis method (5.130) and the state-variables approach (5.133), we see that they are identical, as expected. This fact validates our solutions.
- The real question is, which method is better? Example 5.8 demonstrates how to apply both methods, though it is too simple to highlight the real advantage of the state-variables technique. Nonetheless, this example shows that ZIR+ZSR approach enables us to fully separate two states of the circuit: (1) with initial condition but without the source and (2) with the source but without the initial conditions. This separation allows for varying ZIR conditions for the circuit whose ZSR state has already been determined, and vice versa. Such flexibility becomes a great advantage for the circuit design.
- If you try to solve Examples 5.6 and 5.7 by a traditional circuit-analysis method, you'd appreciate the advantages of ZIR+ZSR approach at a much higher degree.

The problem is solved.

Question 1 Compare these results with those discussed in subsection "The circuit responses and the initial and final values–a summary in Section 4.2." What similarity and difference between these results do you see?

Question 2 Measurements on the graphs in Figures 5.12b, 5.12c, 5.12d, and 5.12e show that the transient interval of every case is different. Specifically, $\Delta t_{ZIR} = 114.97\,ms$, $\Delta t_{ZSR} = 125.65\,ms$, and $\Delta t_{ZIR+ZSR} = 97.74\,ms$. Why do they vary?

5.4 Summary of Chapter 5

1) The second-order circuit contains two reactive elements, inductor L and capacitor C. The $i-v$ relationship of each element is governed by a differential equation; hence, the circuit containing L and C components needs the second-order differential equation. The typical form of such an equation is

$$x''(t) + 2\alpha x'(t) + \omega_0^2 x(t) = \omega_0^2 y(t),$$

where variable $x(t)$ is either $v(t)$ or $i(t)$, the damping coefficient α is either $\alpha\left(\frac{1}{s}\right) = \frac{1}{2L}$ for a series RLC circuit or $\alpha\left(\frac{1}{s}\right) = \frac{1}{2RC}$ for a parallel RLC circuit, the resonant frequency is $\omega_0^2\left(\frac{1}{s}\right)^2 = \frac{1}{LC}$, and $y(t)$ is the input (source or excitation).

2) The second-order differential equation describing the behavior of an RLC circuit is still linear. Based on a superposition principle, its solution is the sum of a natural (source-free) response, $x_{out-nat}(t)$, the circuit's reaction to its initial conditions, and the forced response, $x_{out-frc}(t)$, the circuit's reply to the input (source). That is,

$$x_{out}(t) = x_{out-nat}(t) + x_{out-frc}(t).$$

Depending on the relationship between $R, L,$ and$C$, there are three possible solutions to the circuit's differential equation:

Overdamped case for $\alpha > \omega_0$: $x_{out-od}(t) = A_1 e^{m_1 t} + A_2 e^{m_2 t} + x_{out-frc}(t)$
Critically damped case for $\alpha = \omega_0$: $i_{cmp-cd}(t) = (A_1 + A_2 t)e^{-\alpha_{cd} t} + x_{out-frc}(t)$
Underdamped case for $\alpha < \omega_0$:

$$i_{cmp-ud}(t) = \left(A_1 \sin(\omega_d t) + A_2 \cos(\omega_d t)\right)e^{-\alpha_{ud} t} + x_{out-frc}(t).$$

Here, $m_{1,2} = -\alpha \pm \sqrt{\left(\alpha^2 - \omega_0^2\right)}$. The natural response $x_{out-nat}(t)$ describes the *transient process*, and it always diminishes with time, whereas the forced response $x_{out-frc}(t)$ presents the *transient part* and the *steady-state output* of the circuit, and its form depends on the input. For a step input, $x_{out-frc}(t) = Y_{in} \cdot u(t)$, and for a sinusoidal input, $x_{out-frc}(t) = D_1 \sin(\omega t) + D_2 \cos(\omega t) + F\sin(\omega t + \Phi)$.

3) To finish, all constants of the complete solutions, Y_{in}, A_1, A_2, D_1, $and\ D_2\ (or\ F\ and\ \Phi)$ must be determined. Finding the forced solution constants requires plugging $x_{out-frc}(t)$ into the circuit's differential equation. Since its RHS, $y(t)$, is given, this insertion allows for finding Y_{in}, D_1, and $D_2\ (or\ F and \Phi)$. However, describing A_1 and A_2 is trickier. For overdamped case, we assess the solution, $x_{out-od}(t) = A_1 e^{m_1 t} + A_2 e^{m_2 t} + x_{out-frc}(t)$, at $t = 0$, which gives one equation because $x_{out-od}(0)$ and $x_{out-frc}(0)$ are known. The second required equation is obtained by differentiating both sides of $x_{out-od}(t)$ equation once, and evaluating the result at $t = 0$. The problem is that $x'_{out-od}(t)$ is not given, and it must be expressed through $I_L(0)$ or $V_C(0)$. This can be done by using the

element's $i-v$ relationship, namely either $i_c(t)=C\frac{dv_C(t)}{dt}$ or $v_L(t)=L\frac{di_L(t)}{dt}$, and evaluating it at $t=0$. The obtained system of two equations enables us to calculate A_1 and A_2.

All constants for two other cases—critically damped and underdamped—are found in a similar manner. See Examples 5.3, 5.4, and 5.5.

4) *State variables* are the parameters that provide the comprehensive information regarding the circuit's extant status, which is formed by the circuit's initial conditions and the applied source (input). For an RLC circuit, the state variables are the inductor current $i_L(t)$ and capacitor voltage $v_C(t)$. Employing the state-variables approach for the advanced analysis of the circuit containing n reactive components enables us to build the system of the n first-order differential equations instead of one n-th order equation. This system can be readily presented in the form of *state matrix equations* as

$$\dot{\boldsymbol{x}} = \boldsymbol{A}\boldsymbol{x} + \boldsymbol{B}\boldsymbol{y} \tag{5.112aR}$$

$$\boldsymbol{z} = \boldsymbol{C}\boldsymbol{x} + \boldsymbol{D}\boldsymbol{y}, \tag{5.112bR}$$

where $\boldsymbol{x}$ is n-th *state vector* whose components are the state variables, $\boldsymbol{y}$ is the k-th *input vector*, and $\boldsymbol{z}$ is the l-th *output vector*, $\boldsymbol{A}$ is *nxn circuit matrix*, $\boldsymbol{B}$ is *nxk input matrix*, $\boldsymbol{C}$ and $\boldsymbol{D}$ are, respectively, *lxn* and *lxk* output matrixes.

The advantage of using the state-variables approach is its ability to provide analysis of the higher-order and multiple inputs circuits. However, this approach becomes helpful even for analysis of the second-order RLC circuit by using ZIR + ZSR method. It enables us to apply the superposition principle in its pure form by finding the circuit's complete response first only to the initial conditions, secondly to the input alone, and summing up both responses. Furthermore, ZIR + ZSR technique allows for varying the initial conditions while keeping ZSR constant or changing ZSR leaving the initial conditions intact. This ability becomes greatly beneficial in circuit design. See Examples 5.6, 5.7, and 5.8.

Chapter 5 Questions and Problems

5.1 Natural responses of RLC circuits

1 Refer to Figure 5.1a: Why does a circuit without any driving source could possibly deliver any output (response)?

2 Is there any difference between a source-free and zero-input circuits? Which one generates natural response?

3 Which circuit—RC, RL, or RLC—can store more energy?

4 What circuit's element—R, L, or C—can store energy? Why?

5 Why L and C are called the *reactive* components?

6 Explain the difference between natural, forced, and complete responses.

7 Why does an equation, describing a circuit with a reactive element, must be a differential one?

8 Why do RL and RC circuit are described by the first-order differential equation but an RLC circuit requires the second-order one?

9 *Find the natural response of the series RLC circuit shown in Figure 5.P9. Show all your mathematical operations.

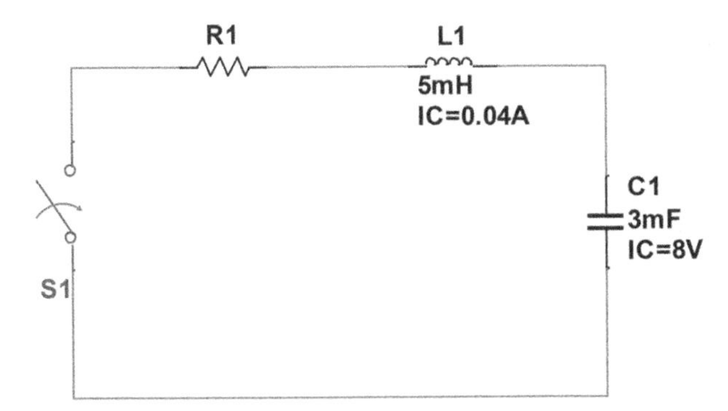

Figure 5.P9 Finding the natural response of series RLC circuit.

10 Do KVL and KCL play any role in finding the mathematical description of an RLC circuit? Explain and give examples.

11 Deriving the differential equation for an RLC circuit seems to be an easy task. But how can you solve it for a source-free circuit? Explain by considering an initially relaxed zero-input series RLC circuit whose $R = 7(\Omega)$, $L = 2(mH)$, and $C = 4mF$.

12 *Solve Problem 11 if $V_0 = 12(mV)\ and\ I_0 = 8(mA)$.

13 Why do the response of an RLC circuit breaks into three cases, whereas the responses of RC or RL circuit do not?

14 Consider three responses of an RLC circuit:

a) By what criterion do these responses break down to overdamped, critically damped, and underdamped?

b) Why do these cases are called overdamped, critically damped, and underdamped? Why do they contain the word "damped"?

c) If $\alpha^2 > \omega_0^2$, then we have overdamped case, and when $\alpha^2 = \omega_0^2$, the case is called critically damped. But what if $\alpha^2 \geq \omega_0^2$?

15 Consider damping oscillations in the natural response of an RLC circuit:

a) How an RLC circuit can generate oscillations, but RC and RL circuits cannot?

b) The resonance frequency of these oscillations, $\omega_0^2\left(\frac{rad}{s}\right) = \frac{1}{LC}$, does not depend on alpha, but damping frequency, $\omega_d = \sqrt{\omega_0^2 - \alpha^2}$, does. How? Why?

16 The complete response of an RLC circuit consists of two parts, but its natural response includes only one. Why?

17 Consider an RLC circuit's initial conditions: How can these condition be created?

18 Consider the circuit in Figure 5.P18, where switch 1 is in A position and switch 2 has been closed for a long time. At $t = 0$, switch 2 gets open, but switch 1 remains in position A. Find:

a) $I_{L0} \equiv i_L(0^+)$ and $V_{C0} \equiv v_C(0^+)$.

b) $I'_{L0}(0^+)$ and $V'_{C0}(0^+)$.

c) $i_L(\infty)$ and $v_C(\infty)$.

(Hint: Build the equivalent circuits for each stage of your consideration, namely, for $t = 0^-$, $t = 0^+$, and $t \to \infty$.)

19 Find $v_C(t) \equiv v_{out}(t)$ for the circuit in Figure 5.P18 whose switch 1 turns to position B at $t = 0$, and switch 2 remains open. Assume that the circuit retains the initial conditions determined in Problem 5.18. Additionally:

a) Using MATLAB, plot the graph of $v_{out}(t)$ which you will find. On the MATLAB graph, determine the duration of a transient interval employing "Data Tips" icon.
b) Run Multisim simulation of the circuit and measure the transient interval. (See Sidebar 5S.1.) Compare the experimental results by Multisim with theoretical ones by MATLAB.
c) Discuss your answers and their verification.

20 What can be done with the circuit in Problem 5.18 to make its response the overdamped? Do it, and work out all assignments given in Problem 5.19 with the new condition.

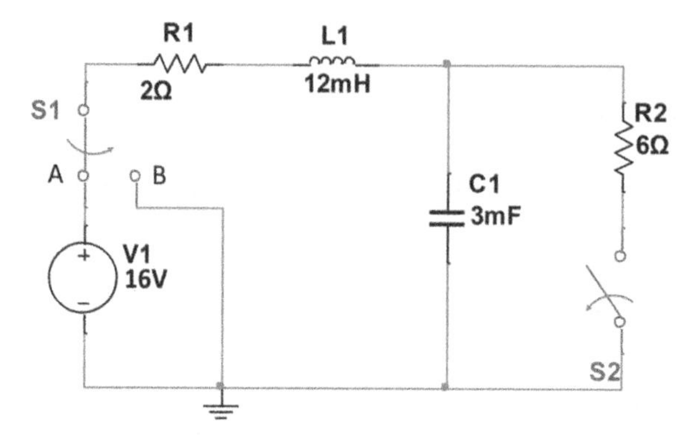

Figure 5.P18 Creating the initial conditions for a series RLC circuit.

21 The switch in a parallel RLC circuit in Figure 5.P21 gets closed at $t=0$ so that the circuit's RHS turns to the source-free. Calculate all the initial conditions. (After Alexander and Sadiku 2020 pp. 330–331.)

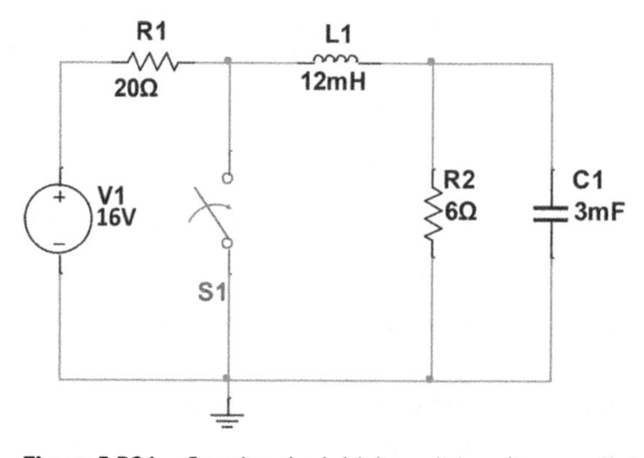

Figure 5.P21 Creating the initial conditions in a parallel RLC circuit.

22 Find $v_C(t) \equiv v_{out}(t)$ for the circuit in Figure 5.P21 for $t \geq 0^+$, that is, immediately after the switch gets closed. Use MATLAB to build the appropriate graph; use Multisim to validate your answer. See Problem 5.18 for the assignment clarifications.

23 Consider Figure 5.4b in Example 5.2:

a) What specific point the "Data Tips" box indicates on the graph?

b) How do you interpret the numbers in that MATLAB box:
   X 99.9243
   Y 0.05400
c) Why do this graph decaying smoothly, whereas the graphs in Figures 5.4c and 5.4d exhibit some increases and even oscillations?

24 Consider the natural response of an RLC circuit: Does an underdamped response always results with oscillations? Explain.

25 How do the circuit components' values affect its output?

26 Figure 5.P18 shows a series RLC circuit. Apply the duality principle and construct corresponding parallel RLC circuit.

27 Greek letter α is used throughout the text to designate various quantities for RC, RL, series RLC and parallel RLC. What are these notations? Why do we use so confusing designation?

28 Why does a solution to an RLC circuit's differential equation include two exponential functions, $A_1e^{m_1t}$ and $A_2e^{m_2t}$?

29 Why do we pay so much attention to the RLC circuit's initial conditions? What do we need them for?

30 The overdamped and critically damped cases of the RLC circuit responses contain the resonant-frequency parameter, whereas their solutions don't include any oscillations. Why is it so?

31 The underdamped case of the RLC circuit response include two frequencies—resonant and damping. Does this response oscillate on two frequencies? If no, at which frequency does the response oscillate?

32 What is a quality factor, Q? How Q is used in analysis of the responses of an RLC circuit?

5.2 Complete responses of an RLC circuit

33 Consider the complete response of an RLC circuit:
a) What response of an RLC circuit is called *complete*?
b) What other terms we use to denote this type of response?
c) Does it mean that there is an incomplete response? Explain.

34 **Consider the series RLC circuit in Figure 5.P34 whose switch is closed at $t = 0$. All circuit's values except of $R1$ are fixed, and the resistor can be (a) 1Ω, (b) 4Ω, and (c) 16Ω. Find the circuit's step response for all three cases. Validate your answers.

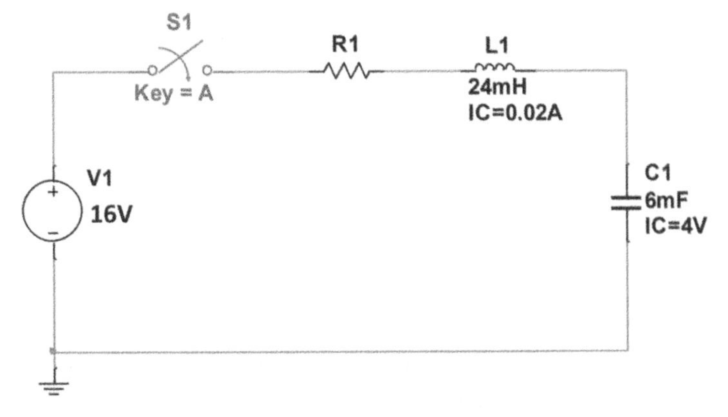

Figure 5.P34 Step response of a series RLC circuit.

35 The Multisim-simulated response of the series RLC circuit operation is shown in Figure 5.P35:

a) Which signal is the input and which is the output? Explain your answer.
b) How do you know whether it is natural or complete response?
c) If this is a forced response, what signal drives the circuit? Explain your answer.

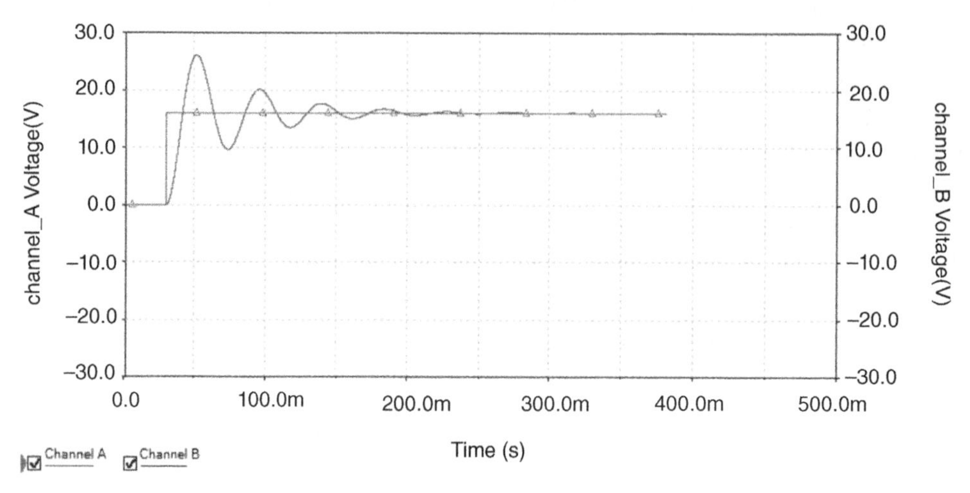

Figure 5.P35 Multisim-simulated response of a series RLC circuit.

36 **The dc source for the series RLC circuit in Figure 5.P34 is replaced by the ac input, $v_{in}(t) = 12\sin(628t)$. With closing the switch at $t = 0$, the circuit operation starts. Find all circuit's three sinusoidal responses and verify the solutions.

37 Consider the Multisim-simulated sinusoidal response of a series RLC circuit in Figure 5.P37:

a) Which of two sinusoidal signals is the input one? Why?
b) Show approximately the transient interval in Figure 5.P37. How do you determine it?
c) It seems that both signals eventually have the same frequency. Approve or disapprove this statement.
d) The figure shows that at the steady-state regime the signals have a 180^0 phase shift. Is this the occasional event or a rule for a series RLC circuit with a sinusoidal excitation?

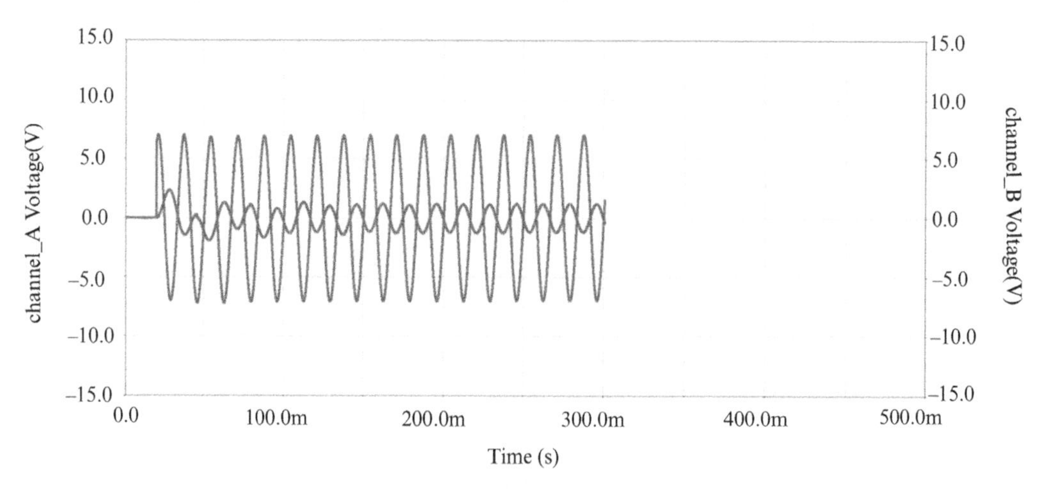

Figure 5.P37 Sinusoidal response of a series RLC circuit.

38 The text affirms, "The *transient interval* is the longest in the overdamped case and the shortest in the critically damped case." Why so?

39 **Circuit in Figure 5.P39 gets to action at $t = 0$, when the switch turns on:

a) Find the value R1 which makes critically damped case and determine the circuit's response in this case. Using MATLAB, build the response graph and find the point for $t_f - V_f$ point on the graph, employing the "Data Tips" tool.
b) Repeat the task of the preceding point for the overdamped case.
c) Find the underdamped response of this circuit and present the answer in the format shown in point a).
d) Verify all three answers by checking their initial and final values.
e) Verify whether the critically damped case has the shortest transient interval compared with the other two.

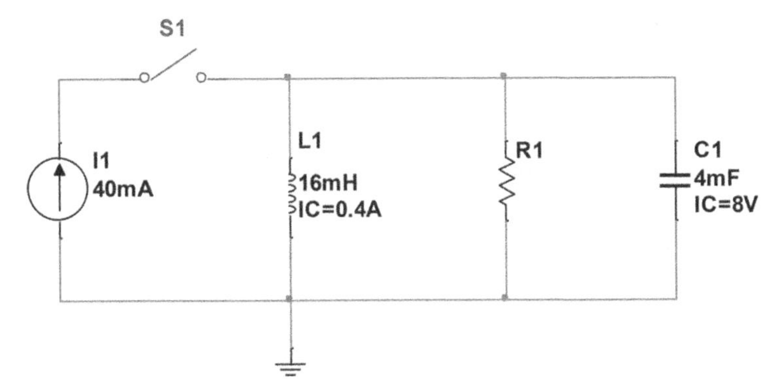

Figure 5.P39 Step response of a parallel RLC circuit.

40 **Perform all assignments of Problem 5.39 using $i_{in}(t) = 0.12\cos(cos120\pi t)$ instead of $i_{in} = 40\ (mA)$.

41 Does MATLAB calculations validate the results of your solution to a problem? Explain. What other methods of result verification can you offer?

5.3 State variables in time domain

42 Explain and give examples:

a) What is the state of a circuit?
b) What are the state variables?
c) What are the state variables in RLC circuits?

43 Why do we need to use the state variables if the advanced circuit analysis provide comprehensive information about the circuit in question?

44 What is the main advantage of the state-variables approach over the advanced circuit analysis approach?

45 Discuss Example 5.6:

a) Why the circuit in Figure 5.9 requires the use of three state variables?
b) What are those state variables?
c) After all the manipulations, we arrive at the matrix equations (5.112a) and (5.112b), but why is this equation of the first-order, whereas there are three variables here?

46 Consider Example 5.7:

a) Examples 5.6 and 5.7 discuss the third-order circuits, each introducing three state variables. However, Example 5.6 includes two voltage and one current variable, but Example 5.7 contains two current and one voltage variable. Why?

b) Compare the output matrix equations in Examples 5.6 and 5.7. Why are they different?

47 What are *zero-input response* (ZIR) and *zero-state response (ZSR)?*

48 Is there any difference between ZIR and a natural response of the same circuit? Explain.

49 How does ZIR + ZSR approach differ from using the superposition principle?

50 Consider Example 5.8, where the network for executing ZIR and ZSR approaches to the advanced analysis of the RC circuit is shown. What circuits for finding the complete response of this circuit by the traditional advanced-analysis method would you need. Sketch them.

51 Compare the ZIR+ZSR method with the advanced circuit analysis approach and explain what advantage the former has for circuit design.

52 Why does a damping coefficient differ for all reactive circuits discussed in this chapter?

53 Why do RLC circuits have three response cases but RC and RL circuits have only one for each of them?

54 Summarize all used methods for finding the constants defining the solutions of all differential equations.

55 Which approach—ZIR + ZSR or differential equations—is better for the circuit analysis? For circuit design? Explain.

6

Advanced Circuit Analysis in Time Domain—Convolution-Integral Technique

6.1 Convolution Integral—The Other Technique for Finding Circuit's Response

6.1.1 The Objective of Chapter 6

You are reminded that the objective of this book is to find the system's (circuit's) response when the circuit description and the input are known. We reproduce here the modified version of Figure 4.1a that illustrates this objective.

Within the *traditional circuit analysis* (reviewed in Part 1), this goal is achieved by solving the algebraic equations that relate the input and the output via the circuit's description. This approach works only for a steady-state circuit regime and for dc and sinusoidal signals. The *advanced circuit analysis* in the time domain (subject of the current Part 2) requires working out the differential equations, much more demanding mathematical manipulations. As a reward, this analysis enables us to find the response for any input signal and both transient and steady-state operations; still, complex mathematics is involved.

Is there any way to directly relate circuit's input $v_{in}(t)$ and output $v_{out}(t)$ signals, as Figure 4.1RM suggests? The answer is yes, and Equation 6.1 shows such a relationship,

$$v_{out}(t) = h(t) \star v_{in}(t). \tag{6.1}$$

Here, $\star$ is the symbol of the mathematical operation called *convolution integral*, and $h(t)$ is a circuit's descriptor that presents all circuit's properties needed to relate the circuit's response to the excitement. This descriptor is called an *impulse response*. Why impulse response, you might ask? It appears that the circuit's response to a single impulse comprehensively carries all circuit's features necessary to maintain the input–output operation. That's great, but (6.1) must work for any input signal, not just for a single impulse, you can object. True, and the answer is that an impulse—more accurately, the set of impulses—can represent any possible signal. How? The discussion of this question and the entirely new approach to finding a circuit's response is the subject of this chapter.

Here is the *outline of Chapter 6*:

- In Section 6.1, we collect the necessary background to understand what a convolution integral is and what it does, what are its main constituents—impulse and impulse response—and how to find them.

Essentials of Advanced Circuit Analysis: A Systems Approach, First Edition. Djafar K. Mynbaev.

Companion Website: www.wiley.com/go/Mynbaev/AdvancedCircuitAnalysis

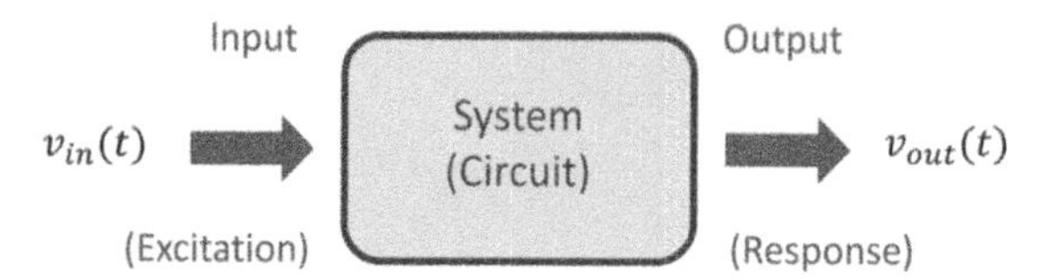

Figure 4.1RM General view at system (circuit) response. (R stands for "Reproduced" and M—for "Modified").

- This knowledge brings us to necessity of studying of circuit's pulse response, which is the objective of Section 6.2.
- Finally, Section 6.3 teaches how to apply the convolution-integral technique to finding the circuit's output.

6.1.2 Continuous Signal, Pulse, and Impulse

A continuous signal, in layman's terms, is what can be sketched without lifting a pen from a paper. The vast majority of signals we use in practice for a circuit excitement are continuous.

The convolution-integral presents a continuous signal as an assembly of pulses or impulses, passes this assembly through a circuit, and reassembles them at the output to obtain the circuit's response. Thus, we must start our discussion by introducing pulse and impulse functions.

6.1.2.1 Pulse and Impulse

Pulse function, P(t), is defined as

$$P(t) = \begin{cases} 0 & for\ t < -\frac{\Delta T}{2} \\ \frac{A}{\Delta T} & for\ -\frac{T}{2} < t < \frac{\Delta T}{2} \\ 0 & for\ t > \frac{\Delta T}{2} \end{cases} \tag{6.2a}$$

or

$$P(t) = \begin{cases} \frac{A}{\Delta T} & for - \frac{\Delta T}{2} < t < \frac{\Delta T}{2} \\ 0 & elsewhere \end{cases} \tag{6.2b}$$

The pulse described by (6.2a) and (6.2b) is centered at $t' = 0$; it is shown in Figure 6.1a. Figure 6.1b. displays the pulse for arbitrary t`. In words, a *pulse is a function equal to zero everywhere except at the time interval, ΔT, centered at t'. Pulse amplitude (height), H, equals $\frac{A}{\Delta T}$, where A is the area under a specific pulse.* Mathematically speaking, we need to integrate the function $P(t)$ to find A, *i.e.*,

$$\int_{-\infty}^{\infty} P(t)dt = \int_{t'-\Delta T/2}^{t'+\Delta T/2} \left(\frac{A}{\Delta T}\right) dt = A. \tag{6.2c}$$

It is essential to realize that the amplitude of this pulse, $H = \frac{A}{\Delta T}$, is closely related to its area, A.

If we assume $A = 1$, then we get a *unit pulse, p*(t), whose amplitude is $\frac{1}{\Delta T}$ and whose area is equal to 1. Regarding the *dimensions*, the pulse parameters in electrical-circuit applications are measured as follows:

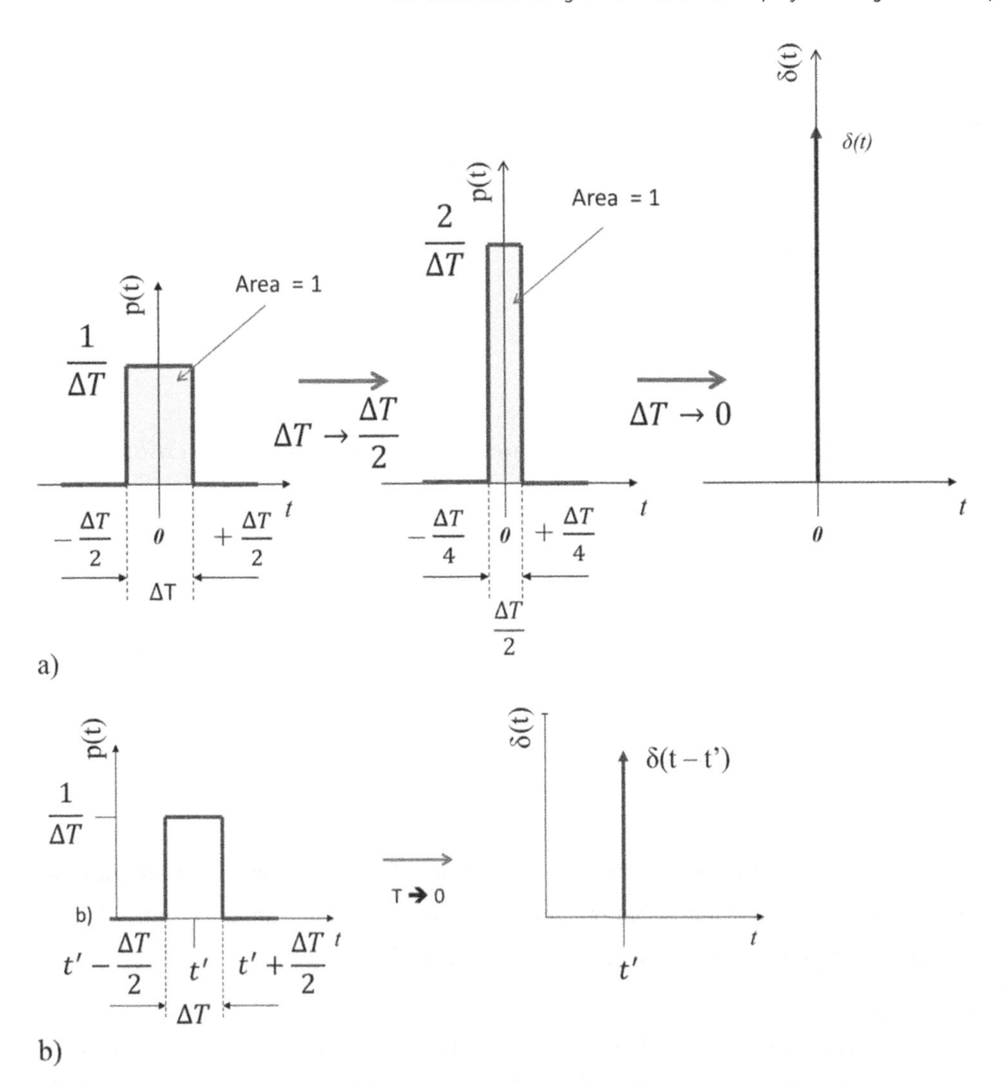

Figure 6.1 Pulse and impulse functions: a) Pulse transformation for ΔT → 0 leads to the definition of the impulse function; b) the time-shifted pulse and impulse functions defined at $t = t'$.

Duration, ΔT —in seconds (s),
Area, A – in volt times seconds (Vs),
Amplitude (height), H – in volts (V) as $A(V \cdot s)/\Delta T(s)$.

6.1.2.2 Impulse Function

To understand the origin of the *impulse function*, we make the *duration of the unit pulse, ΔT, approach zero, whereas keeping its area equal to 1.* (See Figure 6.1a.) Clearly, to keep the area constant with the decrease of ΔT, the unit pulse amplitude, $\frac{1}{\Delta T}$, must increase.

Figure 6.1a shows that when ΔT decreases by half, the pulse amplitude doubles and becomes equal to $\frac{2}{\Delta T}$, whereas the area covered by the pulse remains equal to 1; when the pulse duration becomes $\frac{\Delta T}{4}$, the pulse amplitude becomes equal to $\frac{4}{\Delta T}$, and so on. In the limit, when $\Delta T \to 0$, the unit pulse amplitude, $\frac{1}{\Delta T}$, approaches infinity.

This is how we can obtain the *impulse function*, $\delta(t)$, defined by the following properties:

$$\delta(t) = \begin{cases} 0 & when\, t \neq t` \\ \to \infty & when\, t = t` \end{cases} \tag{6.3a}$$

and

$$\int_{-\infty}^{\infty} \delta(t)dt = 1. \tag{6.3b}$$

Therefore, *the impulse function is equal to zero everywhere except at the point where it exists; its value at this point tends to infinity to keep the area of the impulse equal to 1.* The impulse function is known as the *Dirac*[1] *delta function* (or simply the *delta function*). *Equations* 6.3a and 6.3b *are, in fact, the definition of the delta function,* $\delta(t)$.

The delta function can be shifted in time by interval t′; the definition given in *Equation* 6.3a and 6.3b holds true for $\delta(t - t')$. We depict this delta function in Figure 6.1b. The delta function can be scaled in amplitude by factor M called the *strength* of the impulse, $M\ \delta(t)$. Factor M is the *area* of the scaled impulse; clearly,

$$\int_{-\infty}^{\infty} M\delta(t)dt = M\int_{-\infty}^{\infty} \delta(t)dt = M \tag{6.4}$$

by (6.3b).

The dimension of delta function, $\delta(t)$, is 1/s because the units of the product, $\delta(t).dt$, must be dimensionless. Indeed, (6.4) requires that integral $\int_{-\infty}^{\infty} M\delta(t)dt = M$ must produce units of area, which implies that the integral $\int_{-\infty}^{\infty} \delta(t)dt$ must be dimensionless.

The impulse function cannot be realized practically (how could you possibly build a function whose area is equal to 1 but whose width is zero?). Still, it is a handy mathematical tool in practical engineering work because some signals (very short pulses, for example) can be closely approximated by the delta function. Strictly speaking, the delta function is not a function at all because calculus requires that for one value of an independent variable—time, in our case—the function must give one value, which should be $\delta(t)$ in our example. The impulse function doesn't satisfy this requirement, which is why it is called a *generalized function* or *distribution*.

6.1.3 Convolution Integral—The First Encounter

Consider the arbitrary input signal $v_{in}(t)$ shown in Figure 6.2a. This is a *continuous signal*. Now, let's *approximate this continuous signal,* $v_{in}(t)$, *by a set of impulses*, as in Figure 6.2b. The *strength* of

---

1 Paul Dirac (1902–1984) was a British physicist who made crucial contributions to quantum mechanics development. It's sufficient to refer to his theoretical discovery of positron—an elementary particle equivalent to an electron but having a positive electrical charge—which states antimatter discovery. The other outstanding Dirac's achievement was his wave equation, which reconciles relativity and quantum theories, the fields considered before as opposing each other. He was awarded Nobel Prize in 1933; since 1930, he was a Lucasian Professor at Cambridge, the chair occupied by such figures as Newton and Babbage. He was buried in Westminster Abbey alongside the most distinguished British scientists.

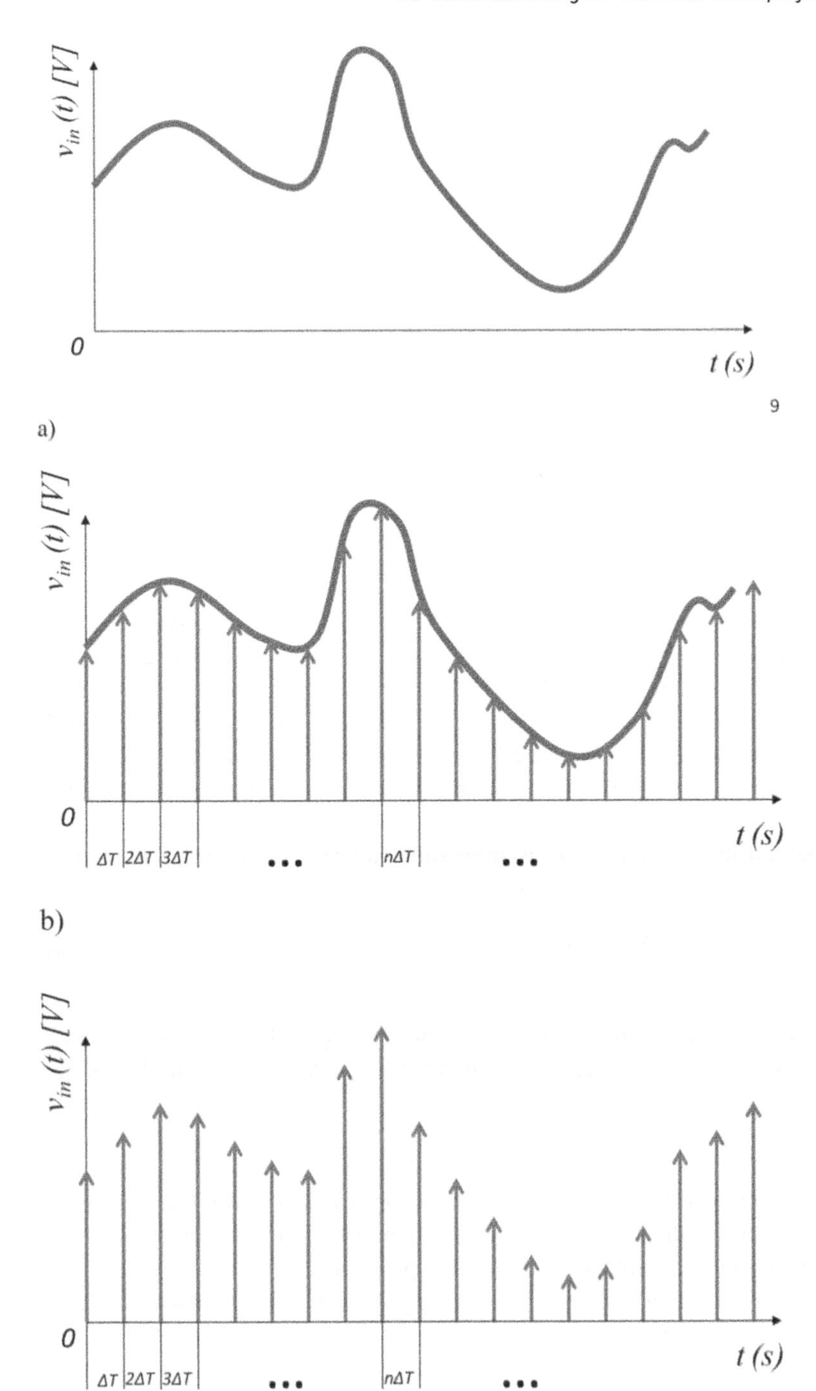

Figure 6.2 Approximation of a continuous signal by a set of impulses: a) The continuous signal; b) a set of impulses and the signal; c) the set of impulses approximating the signal; d) better approximation of the signal by a set of impulses with reduced spacing.

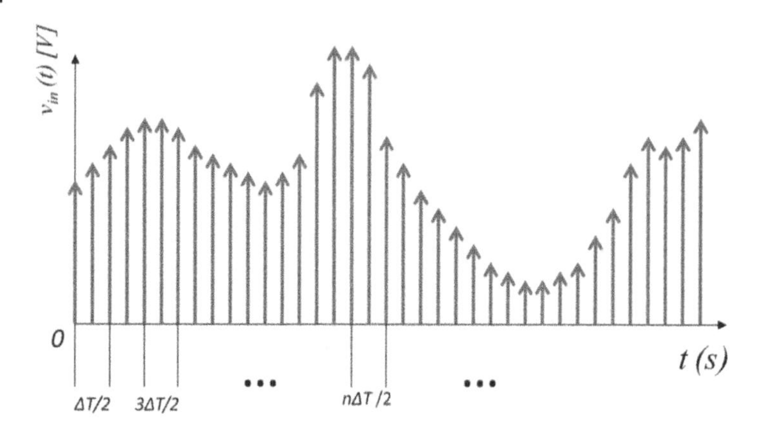

d)

Figure 6.2 (Cont'd)

each impulse is given by $V_{inm}\Delta T = V_{in}(\mathrm{m}\Delta T)\Delta T = V_{in}(t_m)\Delta T$, where ΔT is the interval (spacing) between two adjacent impulses, $m = 0, 1, 2, 3, \ldots,$, is the number of the impulse, and $\mathrm{m}\Delta T = t_m$ is the time shift of an individual m-th impulse. This definition means that the strength of an individual pulse varies depending on t_m. The set of impulses that approximates the continuous signal is shown in Figure 6.2c. Note that if we decrease ΔT (or, equivalently, increase the number of impulses within the same interval), we will obtain a better approximation of the signal, as Figure 6.2d shows.

Now, we can write that our input signal, $v_{in}(t)$, is approximated by a set of impulses in the following mathematical form:

$$v_{in}(t) \approx \sum_{m=0}^{m \to \infty} V_{inm}\delta(\mathrm{t} - t_m)\Delta T = \sum_{m=0}^{m \to \infty} V_{in}(\mathrm{m}\Delta T)\delta(\mathrm{t} - \mathrm{m}\Delta T)\Delta T. \tag{6.5}$$

Here, $V_{in}(\mathrm{m}\Delta T)\delta(\mathrm{t} - \mathrm{m}\Delta T)\Delta T$ is an individual impulse whose strength is $V_{inm}\Delta T = V_{in}(\mathrm{m}\Delta T) \times \Delta T = V_{in}(t_m)\Delta T$. It is located at $t_m = \mathrm{t} - \mathrm{m}\Delta T$, and separated from an adjacent impulse by the interval ΔT.

6.1.3.1 Impulse Response

A single impulse, δ(t), causes an individual response by our circuit, which we denote as h(t). Figure 6.3 shows this relationship. Physically, the circuit's response to an individual impulse involves all the system's dynamics, and this response displays all the circuit's properties.

In general, an impulse, δ(t), causes a response, h(t); on the other hand, a continuous signal is approximated by a set of individual impulses V_{inm} taken at different times, t_m. Therefore, we can determine the circuit's output by (1) finding the circuit's individual response, V_{outm}, to a single input impulse, V_{inm} and (2) summarizing these individual responses.

The system's response to an individual input is described as

$$V_{outm} = V_{inm}h(t_m) \tag{6.6}$$

because the strength of individual impulse, V_{inm}, is a coefficient (number) that passes through a linear system without change, and $h(t_m)$ is the response to $\delta(t_m)$. In other words, (6.6) states that

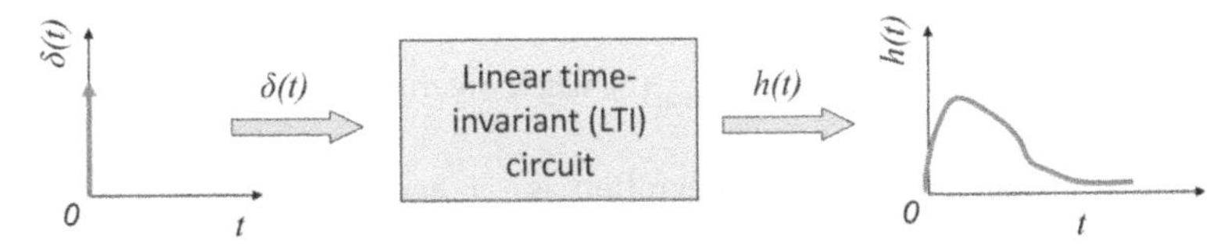

Figure 6.3 A circuit's typical response to an individual impulse.

an individual system's response to each impulse of various strengths is equal to its amplitude-scaling factor, V_{inm}, times the response to the individual impulse, $\delta(t_m)$.

To obtain the complete (total) response, we need to summarize these responses, for which a *convolution integral* is required.

6.1.3.2 Convolution Integral–Introduction

To obtain the total system's response to the entire set of the input impulses (which approximates our actual input signal, remember), we need to sum up the individual outputs given in (6.6); that is,

$$v_{out}(t) \approx \sum_{m=0}^{\infty} V_{inm} h(t_m) = \sum_{m=0}^{\infty} V_{in}(m\Delta T)\text{h}(\text{t} - \text{m}\Delta T)\Delta T. \tag{6.7a}$$

It is crucial to realize that we can present the output as a sum of individual responses to the sum of individual input impulses only because we are dealing with a linear time-invariant circuit, for which the superposition principle holds.

To better approximate our actual continuous input, $v_{in}(t)$, we need to take the impulses more often, which implies decreasing the interval ΔT, or—equivalently—taking a greater number of impulses over the same time span. (Compare Figures 6.2c and 6.2d.) Thus, the smaller the ΔT, the more accurate approximation of a signal by the set of impulses. When $\Delta T \to 0$, we will be nearing the exact reproduction of the input signal. Then $m\Delta T$, the time shift of the m-th individual impulse, tends to a continuous variable, which we denote as τ. In other words, since $\Delta T \to 0$ and $m \to \infty$, the discrete time shifts, $m\Delta T$, merge into a continuum of time shifts, τ. Obviously, the discrete interval, ΔT, goes to the differential, $d\tau$. In the limit, the sum must be replaced by an integral. In formulas, all these operations can be shown as

$$\begin{cases} m\Delta T \Rightarrow \tau \\ \Delta T \Rightarrow d\tau \\ \sum \Rightarrow \int \end{cases} \tag{6.7b}$$

By exercising all these approaches in (6.5), we arrive at the following formula for the *input signal*:

$$v_{in}(t) \approx \sum_{m=0}^{m\to\infty} V_{in}(\text{m}\Delta T)\delta(\text{t} - \text{m}\Delta T)\Delta T = \int_0^{\infty} v_{in}(\tau)\delta(t-\tau)d\tau. \tag{6.7c}$$

Accordingly, the expression for the circuit's *output signal* takes the form

$$v_{out}(t) \approx \sum_{m=0}^{m\to\infty} V_{in}(\text{m}\Delta T)\text{h}(\text{t} - \text{m}\Delta T)\Delta T \Rightarrow v_{out}(t) = \int_0^{\infty} v_{in}(\tau)h(t-\tau)d\tau. \tag{6.8}$$

This formula is the objective of the entire operation:

Equation 6.8 enables us to calculate the circuit's output, $v_{out}(t)$, when the input, $v_{in}(t)$, and the system's impulse response, $h(t)$, are known.

Strictly speaking, we need to consider the limits of integration as $-\infty$ and $+\infty$; the appropriate integral

$$\int_{-\infty}^{\infty} v_{in}(\tau)h(t-\tau)d\tau \tag{6.9}$$

is called the ***convolution integral***. The *common notation for convolution operation is* ☼. Hence, (6.9) can be written as

$$v_{out}(t)=\int_{-\infty}^{\infty} v_{in}(\tau)h(t-\tau)d\tau \equiv v_{in}(t)☼h(t). \tag{6.10}$$

Equation 6.10 is the main point of this discussion. It shows that if we know the impulse response of a circuit, $h(t-\tau)$, and its input, $v_{in}(\tau)$, then the circuit's response to an arbitrary input signal, $v_{in}(t)$, can be obtained by applying (6.10).

This formula is valid provided that the circuit's initial conditions are zero; that is, the circuit is in an initial quiescent (relaxed) state.

(Don't be confused with using two letters, $t(s)$ and $\tau(s)$, for time functions. This point will be carefully discussed in Section 6.3. For now, understand they both are different time variables.)

Convolution integral calculates the overlapping area of two functions, $v_{in}(t)$ and $h(t-\tau)$, when one shifts (scans) over the other. When these functions do not overlap, their convolution integral is zero; when their overlapping area reaches the maximum, the integral value is the utmost. We'll discuss the meaning of a convolution integral in Section 6.3.

Three properties of the convolution integral:

1) The impulse response, $h(t)$ is a causal function. This means that $h(t)=0$ when $t<0$. In addition, we are interested in the circuit's response within a finite time interval, t. Thus, because the future components (those appear after t) do not contribute to the sum, we can change the limits of integration as follows:

$$v_{out}(t)=\int_{-\infty}^{\infty} v_{in}(\tau)h(t-\tau)d\tau=\int_{0}^{t} v_{in}(\tau)h(t-\tau)d\tau \tag{6.11}$$

2) Since the convolution is a commutative operation, we can exchange the places of $v_{in}(\tau)$ and $h(t-\tau)$:

$$v_{out}(t)=\int_{0}^{t} v_{in}(\tau)h(t-\tau)d\tau=\int_{0}^{t} h(t-\tau)v_{in}(\tau)d\tau \tag{6.12a}$$

or

$$v_{in}(t)☼h(t)d\tau=h(t)☼v_{in}(t). \tag{6.12b}$$

3) The convolution integral allows for the following symmetrical time presentation:

$$v_{out}(t)=\int_{0}^{t} v_{in}(\tau)h(t-\tau)d\tau=\int_{0}^{t} v_{in}(t-\tau)h(\tau)d\tau. \tag{6.12c}$$

Here, $v_{in}(t-\tau)$ shows the input value, which was τ seconds ago, and $h(\tau)$ displays how much the extant output depends on the input value that existed τ seconds ago.

These properties help us in our manipulations with the convolution integral.

The convolution-integral technique calls for the following steps:

- When an arbitrary continuous input signal, $v_{in}(t)$, is approximated by a set of impulses and when this set goes to infinity, this approximation approaches the original signal, $v_{in}(t)$, and
- when the system's response, h(t), to the individual impulse, δ(t), is determined, then the output, $v_{out}(t)$, or the system's response to the input, can be found through the convolution integral, $v_{out}(t) = \int_0^t v_{in}(\tau)h(t-\tau)d\tau = v_{in}(t) \star h(t)$.

6.1.4 Analysis of an Impulse Response

If we analyze (6.10) closely, we immediately ask ourselves an obvious question: What exactly is the impulse response, $h(t)$? It is the solution to the system's nonhomogeneous differential equation, whose RHS is the delta function. To put these words into a formula, we need to refer to the general circuit's differential Equation 6.13a,

$$a_n y^n(t) + a_{n-1}y^{n-1}(t) + \ldots + a_1 y'(t) + a_0 y(t) = \\ b_m x^m(t) + b_{m-1}x^{m-1}(t) + \ldots + b_1 x'(t) + b_0 x(t), \tag{6.13a}$$

where we have to replace (1) the output, $y(t)$, by the impulse response, $h(t)$, and (2) the input signal, $x(t)$, by the delta function, $\delta(t)$, as follows:

$$a_n h^n(t) + a_{n-1}h^{n-1}(t) + \ldots + a_1 h'(t) + a_0 h(t) = \delta(t). \tag{6.13b}$$

To find the impulse response, $h(t)$, Equation 6.13b must be solved for it under zero initial conditions.

For instance, consider the differential equation, describing a series RLC circuit,

$$v''_{out}(t) + 2\alpha v'_{out}(t) + \omega_0^2 v_{out}(t) = \omega_0^2 v_{in}(t). \tag{5.6bR}$$

Now, let's replace $v_{out}(t)$ and $v_{in}(t)$ with $h(t)$ and $\delta(t)$, respectively. We obtain an equation similar to (5.6bR), but describing the *RLC circuit impulse response*:

$$h''(t) + 2\alpha h'(t) + \omega_0{}^2 h(t) = \omega_0{}^2 \delta(t). \tag{6.14}$$

The solution of (6.14) is the impulse response, $h(t)$, of a series RLC circuit.

*Thus, the **impulse response** of a system, h(t), is the solution to* (6.13b) *whose δ(t) is the input and initial conditions are zero.*

Note well: The main difference—and it's a huge one—between the classical solution (integration) and the impulse response is that the former requires finding a new solution for every new input signal, whereas with the convolution-integral approach we solve most of the problem, as soon as we find the output signal for an individual impulse. This is because (1) any new input signal can be presented as a set of impulses, (2) we can find the response to an individual impulse, and (3) the total response can be attained as an integral of the individual impulse responses. However, finding an impulse response by directly integrating the system's differential equation is not easy.

6.1.4.1 Step (Heaviside) Function

To continue our discussion, we need to elaborate on the *step (Heaviside) function* which was mentioned several times in the preceding chapters. What is a unit step function? When we turn the power in any device on, we abruptly (step-like) change this device's state from one mode (off) to the other (on). This operation is mathematically described by *a step (or Heaviside[2]) function*.

The particular type, the *unit* step function is defined as

$$u(t) = \begin{cases} 0 & for\ t < 0 \\ 1 & for\ t > 0 \end{cases}. \tag{6.15}$$

This function is shown in Figure 6.4a.

The step function can be shifted in time and starts at an instant τ. It can also be scaled in amplitude; in this case, the step function is not a unit function, obviously. Such a step function is shown in Figure 6.4b; it is described as

$$Au(t-\tau) = \begin{cases} 0 & for\ t < \tau \\ A & for\ t > \tau \end{cases}. \tag{6.16}$$

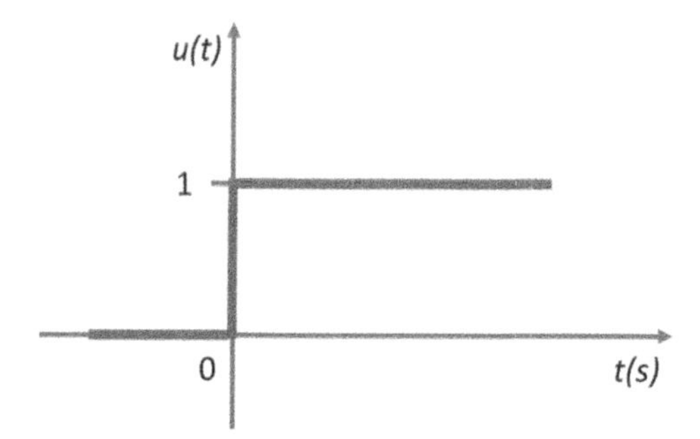

a)

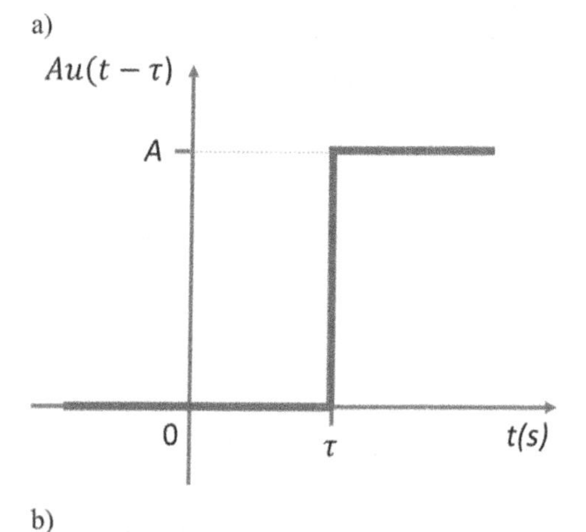

b)

Figure 6.4 Step function: a) Unit step function; b) time-shifted and amplitude-scaled step function.

---

2 Oliver Heaviside (1850–1925) was a British scientist who made significant contributions to mathematics, physics, and electrical engineering. A self-taught genius, lacking formal higher education, he reformulated Maxwell's equation to the form we use today, developed vector analysis, and invented a method of solving differential equations (similar to the modern Laplace method). Heaviside also made significant inventions in the theory and practice of electrical communications. He suggested an ionized atmospheric layer (called the Kennelly-Heaviside layer) used today for long-range radio transmission, and Heaviside introduced a step function now named after him. A reclusive, eccentric man, he was mostly self-employed and chronically poor despite being a fellow of The Royal Society and an Honorary Doctor of The University of Göttingen.

As mentioned, a step function describes the circuit switching from "off" to "on" state. It can also present the reverse operation. (Sketch the graph of a step function given by $Bu(t-\tau)=\begin{cases} B & for\ t<\tau \\ 0 & for\ t>\tau \end{cases}$.)

We must know that the first derivative of a unit-step function is the delta function, as (6.17) shows. This property follows from the sense of a unit-step function; for the formal proof, see Section 7.3.

$$u`(t) = \delta(t). \tag{6.17}$$

The unit-step function is a standard signal used for investigating a system's transient response, as we saw in Sections 4.2 and 5.2.

6.1.4.2 Causal Systems and Functions

The unit-step function helps us to introduce an important class of systems and functions called *causal*.

A circuit (system) is called **causal** if its reaction always comes after excitement of the circuit; in other words, its output' $y(t')$, at any given time, t', is caused by the inputs applied at any time before the instant t'. *If* $t'=0$, then the output of a causal system, $y(t)$, is equal to zero for $t \leq 0$. This definition also implies that future input components do not contribute to current output.

Why, you may ask, do we need to introduce such an obvious definition? Indeed, any real system is a causal one because the real system can't start operating before the input forces it to do so. In other words, practical systems respond to present or past input signals only; they can't respond to future inputs. Theoretically, however, we can imagine systems that don't obey this definition; such systems are called *noncausal*. They are helpful in developing the mathematical tools that are applied to the investigation of practically realizable systems. Therefore, it's necessary to clarify what kind of system is under investigation in the extant problem.

The definition of causal systems can be applied, obviously, to functions, too.

The unit-step function, $u(t)$, enables us to put our discussion of causal systems in mathematical language. Consider, for example, an arbitrary continuous function, $f(t)$; we can find its causal equivalent by multiplying this function by the unit-step function as $f(t)u(t)$. Thus, we understand that

$$f(t)=\begin{cases} f(t)\ at\ t>0 \\ 0\ at\ t<0 \end{cases} \Leftrightarrow f(t)u(t). \tag{6.18}$$

An arbitrary continuous function, *f(t)*, and its causal equivalent, *f(t)u(t)*, are depicted in Figure 6.5.

Note: From now on, we will consider only the *causal functions*; however, instead of writing *f(t)* $\times$ *u(t)* every time, we will continue to write $f(t)$ with the understanding that $f(t)$ is the causal function, unless otherwise specified.

Now we are ready to consider an example of how to find the impulse response.

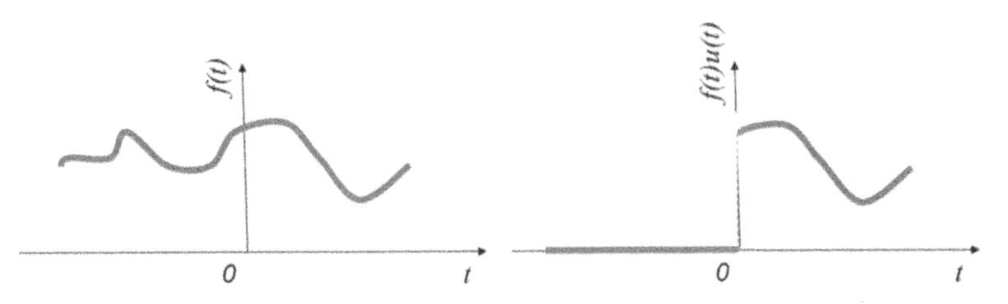

Figure 6.5 Application of the unit-step function: a) Arbitrary continuous function, $f(t)$; b) its causal equivalent, $f(t)u(t)$.

Example 6.1 Impulse response of a series RC circuit.

Problem: Find the impulse response of the series RC circuit shown in Figure 6.6 with $R = 2.0\ k\Omega$ and $C = 0.1$ mF. (You will recall that an RC circuit is thoroughly considered in Sections 4.1 and 4.2.)

Solution:
You will recall that the first-order differential equation describing this RC circuit is derived in Section 4.1 as

$$v`_{out}(t) + \alpha\ v_{out} = \alpha\ v_{in}(t), \tag{6.19a}$$

where $\alpha = \frac{1}{RC}$. We need to modify (6.19a) to involve the impulse response, $v_{out} = h(t)$, and the input impulse $v_{in} = \delta(t)$, that is,

$$h`(t) + \alpha h(t) = \alpha\delta(t). \tag{6.19b}$$

To solve this equation, let's execute the following approach: We know that this system is *causal*, which means that $h(t)$ must be equal to zero for $t < 0$. On the other hand, for $t > 0$, the circuit is not excited because the impulse is applied at $t = 0$ only. (See Figure 6.1 and Equation 6.3.) Therefore, for $t < 0$ and $t > 0$, our circuit is undriven and we should consider the zero-input response as a solution to (6.19). We know that the natural response of the RC circuit is given by

$$v_{out}(t) = Ae^{-\alpha t};$$

hence,

$$h(t) = Ae^{-\alpha t} \qquad for\ t > 0. \tag{6.20}$$

This solution becomes a general one if it satisfies this requirement,

$$h(t) = 0\ for\ t < 0, \tag{6.21a}$$

as it must be for a causal system. To meet this requirement, the solution $h(t)$ must be multiplied by the unit-step function. Thus, we get

$$h(t) = Ae^{-\alpha t}u(t). \tag{6.21b}$$

This is the complete (total) solution for $t > 0$. Therefore, the impulse response is

$$h(t) = \begin{cases} 0\ for t < 0 \\ Ae^{-\alpha t}u(t)\ t > 0 \end{cases}. \tag{6.22}$$

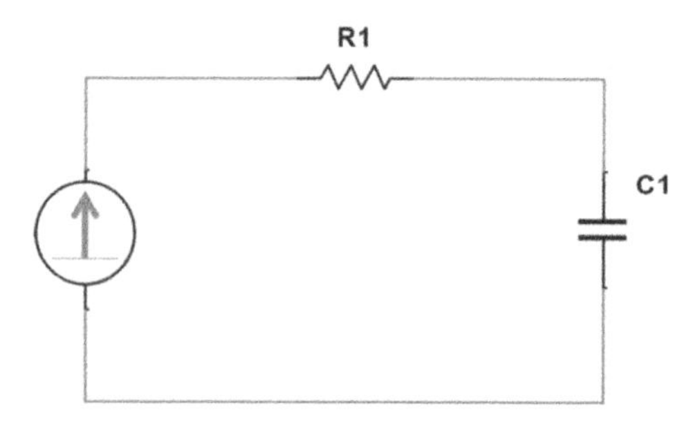

Figure 6.6 Circuit diagram for Example 6.1.

To find constant A, we determine response $h(t)$ at $t=0$, as we did in Section 4.1. Since (6.19) must be satisfied at all times, we can obtain the solution at $t=0$ by integrating this equation from $t=0^-$ into $t=0^+$, where 0^- and 0^+ are infinitesimal times just before and after the zero instant, $t=0$. Thus,

$$\int_{0^-}^{0^+} h'(t)dt + \alpha\int_{0^-}^{0^+} h(t)dt = \alpha\int_{0^-}^{0^+} \delta(t)dt. \tag{6.23}$$

You will recall that $h'(t) = dh/dt$ and

$$\int_{0-}^{0+} h'(t)dt = \int_{0-}^{0+}\left(dh(t)/dt\right)dt = \int_{0-}^{0+} dh(t) = (h(0^+) - h(0^-)).$$

Thus, integrating (6.23) and remembering that $\int_{0-}^{0+}\delta(t)dt = 1$, as (6.3) states, we obtain

$$(h(0^+) - h(0^-)) + \alpha\int_{0^-}^{0^+} h(t)dt = \alpha. \tag{6.24}$$

Here, $h(0^-)=0$ because this system is causal. The integral $\int_{0^-}^{0^+} h(t)dt$ must be equal to zero because it is the area covered by the finite-value function, $h(t) = Ae^{-\alpha t}u(t)$, over an infinitesimal increment, $(0^+ - 0^-)$. Hence, (6.24) reduces to

$$(h(0^+) = \alpha. \tag{6.25}$$

To find A, we recall that $h(t) = Ae^{-\alpha t}u(t)$, as (6.21b) shows, and obtain

$$(h(0^+) \equiv (Ae^{-\alpha\, 0+})u(0^+) = \alpha. \tag{6.26}$$

Thus, the constant A is found as

$$A = \alpha \equiv 1/RC \tag{6.27}$$

because $e^0 = 1$ and $u(0^+) = 1$ by definition.

Therefore, *the* ***solution*** *to (6.19)* is given by

$$h(t) = \alpha e^{-\alpha t}u(t). \tag{6.28}$$

The problem is solved.

To *validate our solution*, we substitute (6.28) into the original Equation 6.17. We find that its LHS becomes

$$h`(t) + \alpha h(t) = [\alpha e^{-\alpha t}u(t)]` + \alpha[\alpha e^{-\alpha t}u(t)]$$

$$= \alpha[-\alpha e^{-\alpha t}u(t) + e^{-\alpha t}u`(t) + \alpha e^{-\alpha t}u(t)] = \alpha\, e^{-\alpha t}(u(t))`.$$

Thus, (6.17) turns into

$$\alpha\, e^{-\alpha t}u`(t) = \alpha\delta(t), \tag{6.29}$$

which holds true for $t=0$ because of (1)$u`(t) = \delta(t)$ as (6.14) states, and (2) at $t=0$ (6.29) becomes

$$e^0\delta(t) = \delta(t).$$

But (6.18) satisfies (6.17) not only at t = 0 but at any time because (6.19) reduces to (6.29), which holds true since derivative $u`(t)$ is zero for $t<0$ and $t>0$. Also, at $t=0$, this equation turns to u`(t) = δ(t)). The solution is verified.

Turning to the calculations, we compute

$$\alpha = \frac{1}{RC} = \frac{1}{\left(2.0 \cdot 10^3 \cdot 0.1 \cdot 10^{-3}\right)} = 5\left(\frac{1}{s}\right),$$

and (6.28) becomes

$$h(t) = \alpha e^{-\alpha t}u(t) = 5e^{-5t} \quad \text{for } t > 0.$$

Figure 6.7 depicts the graph of this response.

Discussion:

- This example demonstrates the concept of the **impulse response**: *When the RHS of (6.19b) is δ(t), the solution is h(t).* Analyze our manipulations from start to finish—from (6.19b) to (6.28)—and review Figure 6.3, which illustrates this definition in general case. (It is reproduced here for your convenience in Figure 6.8a.) Examine, too, Figure 6.8b, which shows what the impulse response of an RC circuit is a decaying exponential function.
- The *dimension* of the output signal can be determined from (6.28) as follows:

$$h(t)\left[\frac{1}{s}\right] = \left(\frac{1}{RC}\left[\frac{1}{s}\right]\right)e^{-\frac{1}{RC}t}[1]u(t)[1]$$

Why does the output signal of an RC circuit, which is the voltage across the capacitor, carry dimension $\frac{1}{s}$? The answer is that $h(t)$ is not an actual physical output signal, which an instrument like an oscilloscope could measure; it's just a mathematical response to the generalized mathematical function $\delta(t)$. The precise response is given by (6.7),

$$v_{out}(t)[V] \approx \sum_{m=0}^{m\to\infty} V_{in}(\mathrm{m}\Delta T)\mathrm{h}(\mathrm{t}-\mathrm{m}\Delta T)\Delta T \Rightarrow \int_0^\infty v_{in}(\tau)h(t-\tau)d\tau. \tag{6.7R}$$

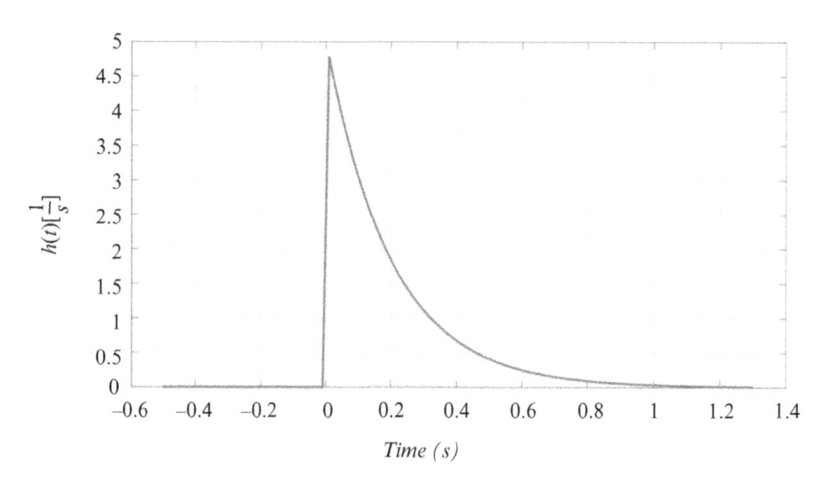

Figure 6.7 Impulse response of a series RC circuit.

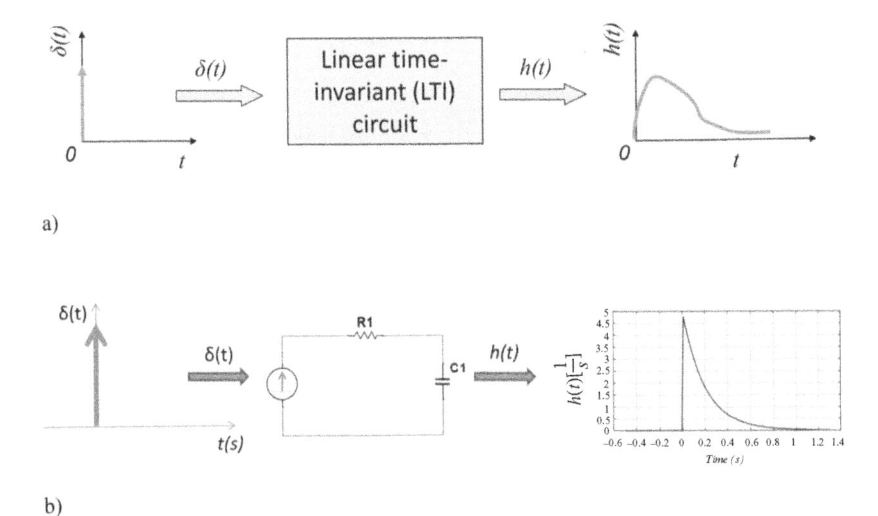

Figure 6.8 Conceptual view of the impulse response: a) The system's response to an individual impulse (reproduction of Figure 6.3); b) the impulse response of an RC circuit.

which shows that the *actual output signal* is the result of the integration of $v_{in}(\tau)\left[V\right]xh(t)\left[\frac{1}{s}\right]xdt\left[s\right]$, which gives $v_{out}(\tau)\left[V\right]$.

As an exercise, analyze the dimension of (6.19), which can be done based on Example 4.1.

Questions

1) Does this form of an impulse response of an RC circuit—the decaying exponent—look reasonable to you?
2) Based on this example, can you find the impulse response of a series RLC circuit by using Equation 6.13b?
3) Why does the convolution-integral technique work only for zero initial conditions?

6.1.5 Finding the Total Output Signal by Using an Impulse Response

Example 6.1 shows that finding an impulse response is not a simple task; nonetheless, as explained previously, we need to do it only once for a given circuit. Having the impulse response of an electrical circuit, we can find the total response to any signal that a set of impulses can approximate. Figures 6.9a and 6.9b demonstrate this idea graphically. The first three top graphs of Figure 6.9a show the individual amplitude-scaled and time-shifted input impulses (left) and their corresponding responses (right). The fourth graphs in Figure 6.9a demonstrate how the sum of the three input impulses approximates a triangle input signal and how the sum of the responses approximates an actual output signal. Figure 6.9b illustrates that adding more input impulses results in a better approximation of an output signal after the summation of all responses. Figure 6.9b is a graphical demonstration of Equation 6.7: When ΔT tends to zero, an approximation of the output by the summing up of the scaled impulse responses gets closer to the actual signal; eventually, when the integration replaces summation, we will obtain the exact waveform of the output signal.

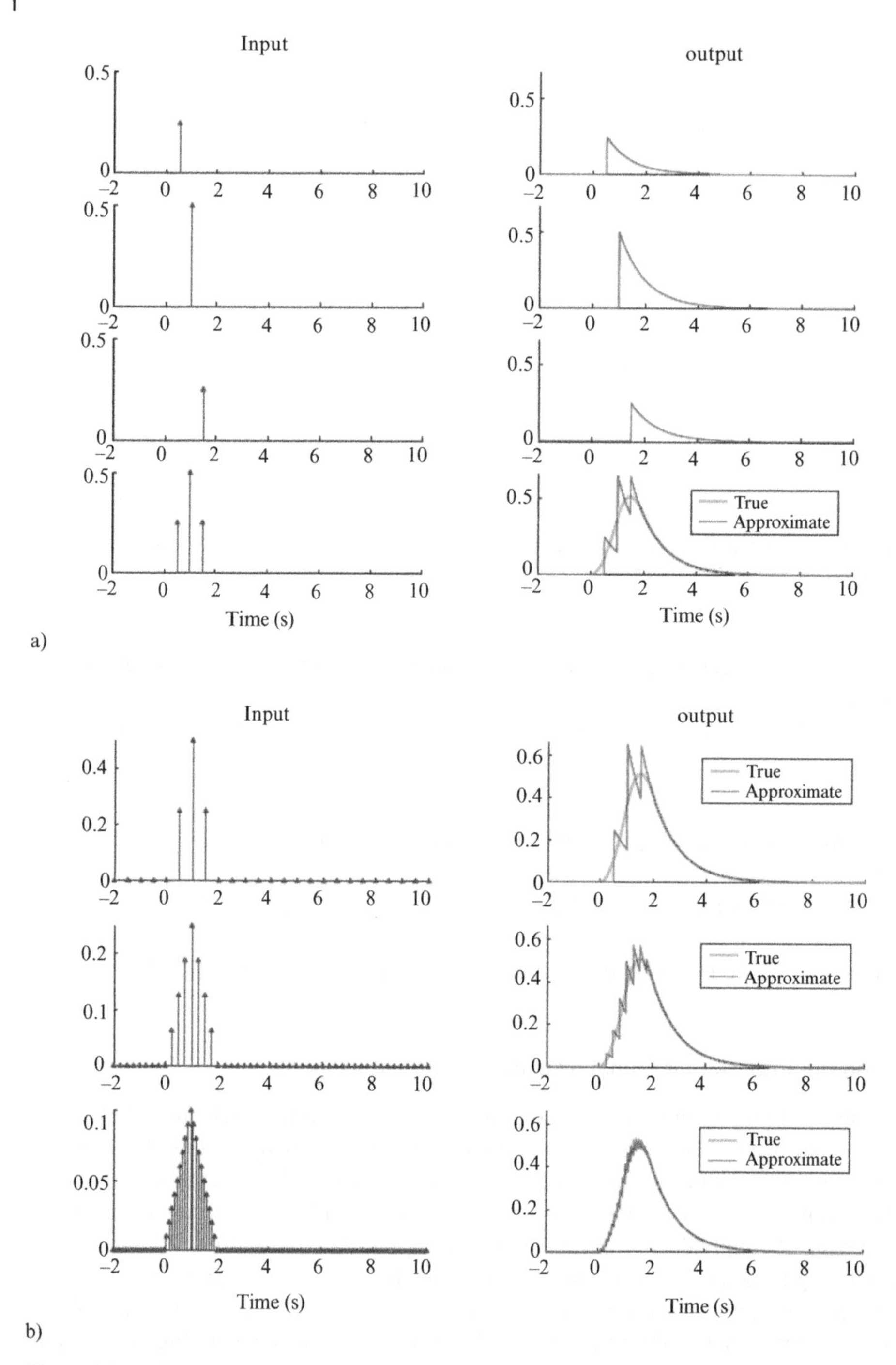

Figure 6.9 a) Responses to amplitude- and time-scaled impulses and the total response to the sum of the impulses; b) improving the approximation of an input triangle signal by decreasing the time interval between input impulses and refining the approximation of the output signal by a summation of the increasing number of the responses. (From: James McNames, *web.cecs.pdx.edu/~**ece**2xx/**ECE222**/Slides/**ConvolutionIntegral**.pdf*).

6.2 Convolution-Integral Technique—The Pulse-Response Approach

6.2.1 Statement of the Problem

The convolution-integral method discussed in Section 6.1 looks attractive, but how can we implement the technique based on an equation member, an impulse function $\delta(t)$, that cannot be materialized? The answer is using a real-life signal that approximates the impulse, a *pulse function*. We devote this section to discussing the pulse-response approach.

The definition of *pulse function* is given in (6.2a) and (6.2b); Figure 6.1 shows the relationship between pulse and impulse. Now, we want to use a pulse to solve the same problem: Given an input signal and the circuit's differential equation, find the circuit's response. To solve the problem, we will apply the steps used for the impulse function in Section 6.1. These steps are:

1) Approximate a continuous input signal by pulses and present these pulses to a circuit one at a time.
2) Find the system's response $h_{pm}(t)$ to a single input pulse $p_m(t)$.
3) Sum up (and eventually integrate) the individual responses to obtain the circuit's total response.

Remember that we can exercise this approach as long as we deal with *linear, time-invariant (LTI)* circuits, for which the *superposition principle* holds. It is essential to realize that we can apply each pulse to the system one at a time, obtain the system's individual response, and then add up all these responses to receive the total response.

6.2.2 Implementing the Pulse-Response Approach

Let's take these three steps to implement the pulse-response approach. First, we approximate the continuous input signal (Figure 6.10a) by the set of contiguous pulses, $P_n(t)$, of the same duration (width), ΔT, but of various amplitudes (Figure 6.10b). The result is the pulses approximating the original signal (Figure 6.10c).

Mathematically, this process is described as

$$v_{in}(t) = \sum_{-\infty}^{\infty} P_m(t), \tag{6.30}$$

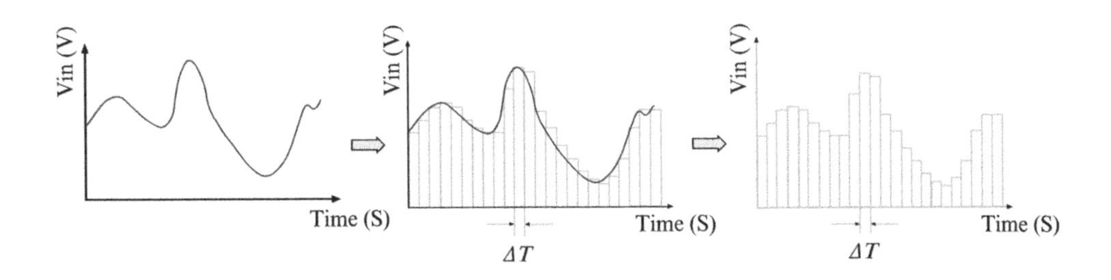

Figure 6.10 Arbitrary continuous input signal and its approximation by pulses: a) The original continuous signal; b) the continuous signal approximated by pulses; c) the set of pulses approximating the original signal.

where $v_{in}(t)$ is an arbitrary continuous signal and $P_m(t)$ is the m-th pulse defined in (6.1). Note that $P_m(t)$ can be presented as

$$P_m(t) = \begin{cases} V_{in}(mT)p(t) & for\ mT \le t < (m+1)\Delta T \\ 0 & elsewhere \end{cases}, \tag{6.31}$$

with $V_{in}(mT) = V_{inm}$ being the amplitude of the m-th pulse and $p(t) = \frac{1}{\Delta T}$ being a *unit pulse.*

It's intuitively clear that the smaller the duration of an individual pulse, the better the approximation of a continuous signal by pulses.Figure 6.11b, 6.11c, and 6.11d demonstrate how much the approximation quality improves when the pulse duration decreases from ΔT to $\frac{\Delta T}{2}$ to $\frac{\Delta T}{3}$.

Continuing to implement the *first step*, we present to our circuit every pulse individually, one at a time, as shown in Figure 6.12. We introduce the first pulse at t; the second then gets shifted by ΔT; that is, it is located at $t - \Delta T$; the third is at $t - 2\,\Delta T$, the fourth at $t - 3\,\Delta T$, and so on.Figure 6.12 shows the first pulse, an arbitrary pulse at $t - mt$, and the last n-th pulse. In addition to time shifting, every pulse is scaled in amplitude.

To implement the *second step*, we need to find the system's response, $H_p(t)$, to an individual pulse.Figure 6.13 illustrates how the circuit transforms an individual input pulse, $P_m(t)$, into a unique output $H_{pm}(t)$ (Refer to Figure 6.1 for comparison with the impulse use.)

To describe the *second step* mathematically, we need to find the pulse response, *Hpm(t)*. How? In the same way how we find any *system's response: solving the system's differential equation with the given input.* In the pulsed case, we need to solve the following equation:

$$a_n H_{pm}{}^n(t) + a_{n-1} H_{pm}{}^{n-1}(t) + \ldots + a_1 H_{pm}{}^`(t) + a_0 H_{pm}(t) = P_m(t), \tag{6.32}$$

where $P_m(t)$ is the input and $H_{pm}(t)$ is the output. This equation is analogous to the impulse response given in (6.12), and it implicitly describes the system's response, $H_{pm}(t)$ to every individual pulse, $P_m(t)$ According to (6.31), we can express $H_{pm}(t)$ through a unit-pulse response $h_p(t)$ as

$$H_{pm}(t) = V_{in}(mT)h_p(t)T. \tag{6.33}$$

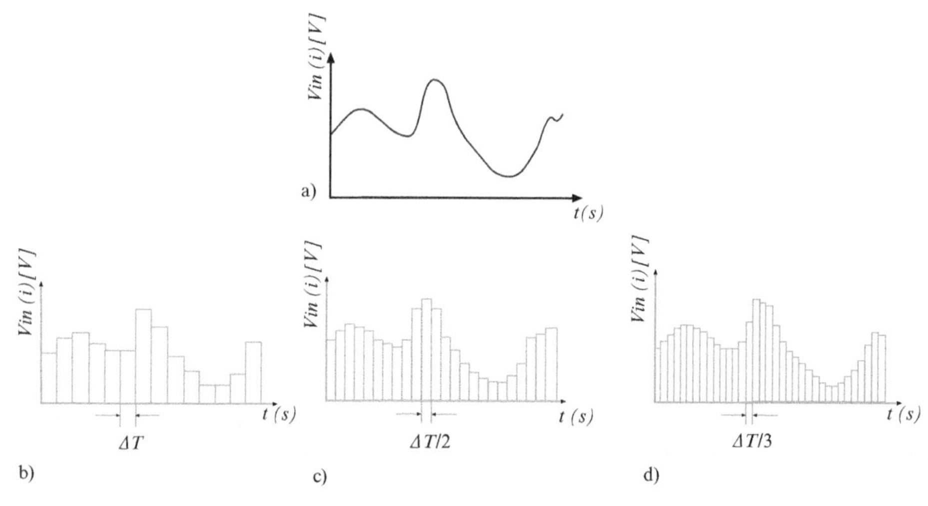

Figure 6.11 The quality of a continuous signal's approximation improves when the pulse duration decreases: a) Original continuous signal; b) the same signal approximated by pulses whose duration is ΔT; c) pulses duration is $\Delta T/2$; d) pulses duration is $\Delta T/3$.

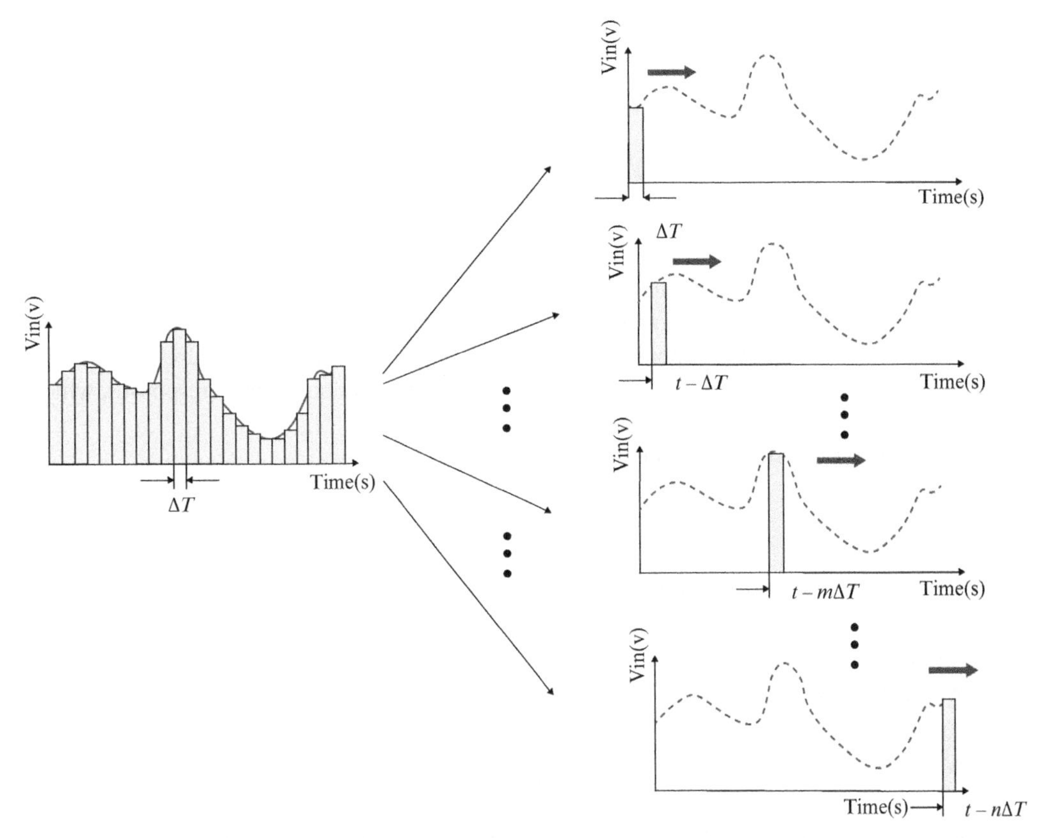

Figure 6.12 Presenting the input pulses to the system.

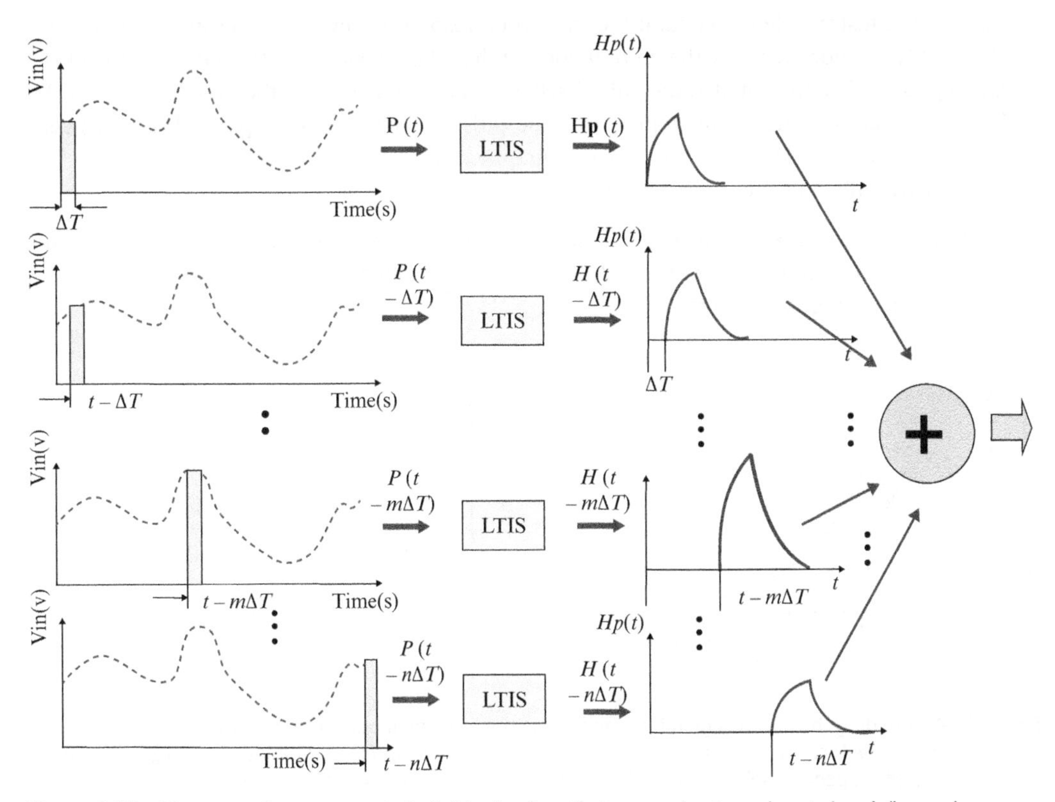

Figure 6.13 The system's responses to individual pulses that approximate an input signal. (Legend: LTIS—linear time-invariant system).

The unit-pulse response, $h_p(t)$, can be found by solving the following differential equation,

$$a_n h_p^{\,n}(t) + a_{n-1} h_p^{\,n-1}(t) + \ldots + a_1 h_p`(t) + a_0 h_p(t) = p(t), \tag{6.34}$$

which is analogous to (6.12) or (6.32).

To implement the *third step*, we need to sum up the individual circuit's pulse responses, $H_{pm}(t)$, and—by tending the pulse duration to zero—arrive at the convolution integral.Figure 6.14 demonstrates building the output signal by adding up the individual pulse responses. The graphic on the left shows the individual pulse responses (the solid lines) along with the input pulses (the dashed lines), and the graph on the right shows the result of the summation of these responses, which produces the approximated output signal.

We can describe the *third step* mathematically as follows:

$$v_{out}(t) \approx \sum\nolimits_{m=-\infty}^{\infty} H_{pm}(t). \tag{6.35}$$

Compare this equation with (6.30): You can see that they are quite similar and, in fact, (6.35) can be obtained from (6.30) by simply replacing the input signal, $v_{in}(t)$, with its output counterpart, $v_{out}(t)$, and the input pulse, $P_m(t)$, with the system's response to this pulse, $H_{pm}(t)$. Using (6.33), we can rewrite (6.35) in the following form:

$$v_{out}(t) \approx \sum\nolimits_{m=-\infty}^{\infty} H_{pm}(t) = \sum\nolimits_{m=-\infty}^{\infty} V_{in}(mT) h_p(t) \Delta T \tag{6.36}$$

(See reference [28], where the MATLAB codes for executing (6.36) and drawing of Figure 6.14 are presented.)

Figure 6.15 illustrates how to obtain the exact total response: Obviously, the smaller the pulse duration, ΔT, the more accurate the reproduction of the total response, $v_{out}(t)$. But when the pulse duration gets smaller and smaller, the pulse itself gets closer and closer to the impulse, as the LHS of Figure 6.1 demonstrates. This implies that the pulse response should approach the impulse response with $\Delta T \to 0$. The RHS of Figure 6.15 graphically proves this point.

To summarize the pulse-response approach:

- A continuous signal is approximated by a set of pulses. Each input pulse, $P(t)$, is transformed by the system into an individual output, $H_p(t)$. After being added up, these outputs give the system's complete approximated response.

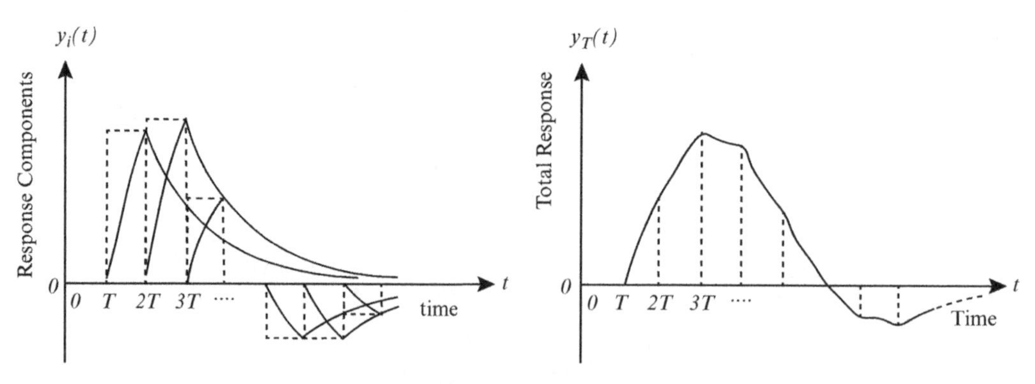

Figure 6.14 Building the output signal by summing up the individual pulse responses. (Adapted from Trumper, David. *2.14 Analysis and Design of Feedback Control Systems, Spring 2007*. (Massachusetts Institute of Technology: MIT OpenCourseWare), http://ocw.mit.edu. License: Creative Commons BY-NC-SA.)

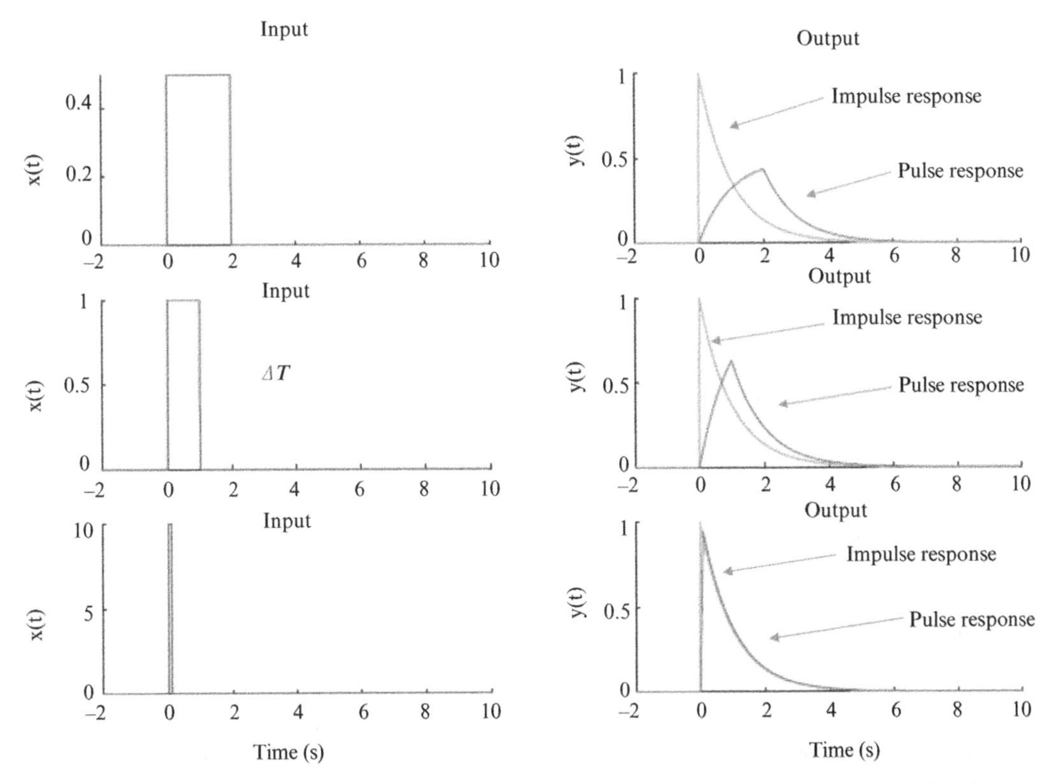

Figure 6.15 Comparison of pulse and impulse responses: As an input pulse approaches an impulse, the pulse response approaches the impulse response. (After James McNames, ECE 212, Convolution Integral ver. 1.68, Portland State University http://web.cecs.pdx.edu/~ece2xx/ECE222/Slides/ConvolutionIntegral.pdf).

- By letting the pulse duration, ΔT, go to zero (or by allowing the total number of pulses to go to infinity), we obtain the system's exact response.
- Mathematically, this means that approaching $\Delta T \to 0$, we go from pulses to impulses and from the summation of pulse responses to the integration of the impulse responses, thus arriving at the *convolution integral*.

Bottom line: *To obtain the impulse response using a convolution integral, we need to start with the pulse response.*

6.2.2.1 Constructing a Pulse as a Combination of Two Step Functions

Before proceeding to the example of finding the pulse response, we need to derive the pulse formula. To do so, we construct an individual pulse as a combination of two step functions. One function is an amplitude-scaled step (Heaviside) function, $V_{in} \cdot u(t)$, and the second is the amplitude-scaled and time-shifted step function, $V_{in} \cdot u(t - T_p)$. Such step functions are defined by (6.15) and shown in Figure 6.4b. When we take the difference of these two functions, we get a pulse, $P(t)$, whose duration is ΔT. Note that we imply $\Delta T = T_p - 0$.

$$P(t) = \begin{cases} 0 \; for \; t < 0 \\ V_{in}\left(u(t) - u\left(t - T_p\right)\right) for \; 0 < t < T_p. \\ 0 \; for \; t > T_p \end{cases} \tag{6.37}$$

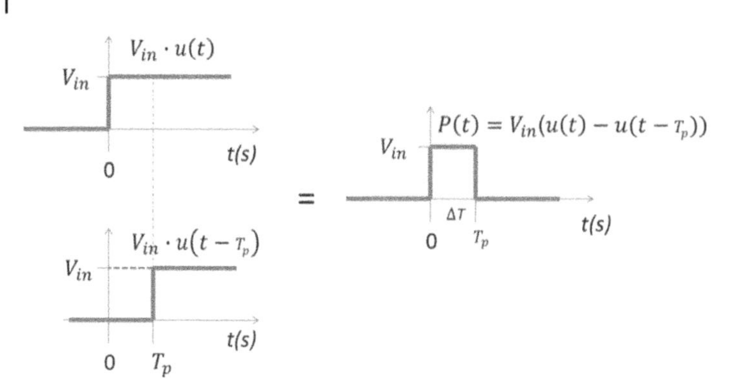

Figure 6.16 Constructing a unit pulse from two step functions.

This construction is illustrated in Figure 6.16.

Now, we are well equipped to consider the example of finding the circuit's response to a pulse input.

Example 6.2 Finding the pulse response of a series RC circuit.

Problem: Refer to the RC circuit shown in Figure 6.6 for Example 6.1, where $R = 2(k\Omega)$ and $C = 0.1(mF)$. Replace the impulse input with the pulse, and find the pulse response of this modified RC circuit. Assume the pulse duration is $\Delta T = 2(s)$, $V_{in} = 12(V)$, and the initial condition is zero.

Solution:

We offer two approaches to finding the solution: qualitative and quantitative.

Qualitative solution The qualitative construction of the response in question is shown in Figure 6.17. The physics behind this construction is as follows:

- When we apply a step-up input $V_{in} \cdot u(t)$ at instant $t = 0^+$, an RC circuit responds with an exponential increase in the output voltage, $v_{out}(t)$, from zero to V_{out} because the capacitor will be charging. Here, $V_{out} \le V_{in}$ is an unknown constant whose value depends on *pulse duration*, ΔT, and the circuit *damping coefficient*, $\alpha = 1/RC$. See Figure 6.17a.
- After we apply at instant $t = T^-$ the time-shifted Heaviside function, $V_{in} \cdot u(t - T_p)$ and subtract it from $V_{in} \cdot u(t)$, the circuit becomes undriven, and its output voltage, $v_{out}(t)$, exponentially decays from V_{out} to zero, as demonstrates.
- Finally, we construct the input pulse, $P(t)$, by adding up the step-up and step-down signals, and find the RC circuit's response, $H_p(t)$, to this pulse as a sum of the two output voltages. Figure 6.17c shows this result. In all Figure 6.17 family we omit designations 0^+ and T^- to avoid overloading the figures by the additional details and make them better observable.

The *key point* to come away with from this discussion is that the rising edge of the input pulse causes a forced response (a growing exponent) and the pulse amplitude from the rising to falling edges charges the RC circuit capacitor whose accumulated voltage causes a natural (source-free) response (a decaying exponent) for $t > T$. Sections 4.1 and 4.2 discussed each constituent response thoroughly.

Quantitative solution To present this qualitative consideration in a mathematical form, we need to describe the RC circuit and the input signal, and find the output as a solution to the circuit's differential equation. (Make yourself comfortable with the notation $\Delta T = t - T_p$ shown in Figure 6.17c.)

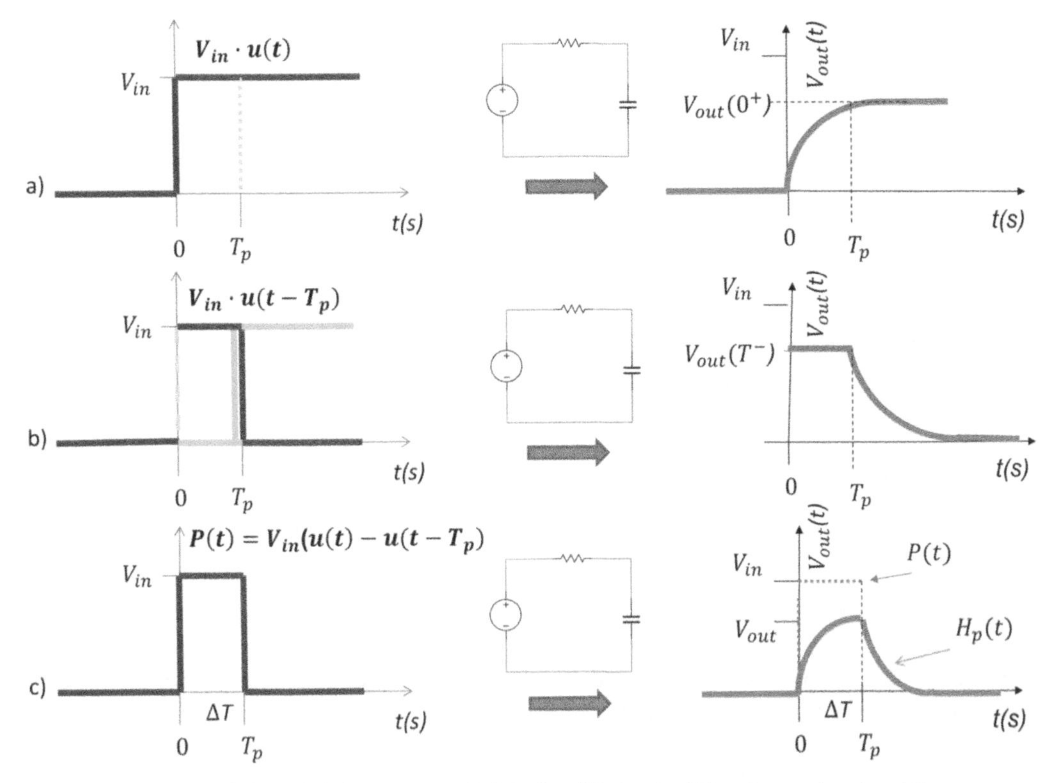

Figure 6.17 Inputs (left) and their responses (right) of an RC circuit: a) The step-up input and its response; b) step-down input and its response; c) input pulse (sum of two step inputs) and its response (sum of two step responses).

We will find the pulse response of the RC circuit in three steps shown in Figures 6.17: for the input step function (Step 1 shown in Figure 6.17a), for the source-free RC circuit (Step 2 in Figure 6.17b), and for the complete input pulse (Step 3 in Figure 6.17c).

Step 1: The input step function is given by

$$v_{in}(t) = V_{in} \cdot u(t). \tag{6.38}$$

The well-discussed RC circuit's differential equation is

$$v'_{out}(t) + \alpha v_{out}(t) = \alpha v_{in}(t) \tag{6.39}$$

with $\alpha = 1/RC$. Its solution is the *complete response* of the RC circuit.

Here is a quick reminder of the procedure of finding the RC circuit complete response thoroughly discussed in Sections 4.1 and 4.2:

- The complete response of any circuit is a sum of the natural, $v_{out-nat}(t)$, and the forced, $v_{out-frc}(t)$, solutions:

$$v_{out-cmp}(t) = v_{out-nat}(t) + v_{out-frc}(t). \tag{6.40}$$

- A natural (source-free) response of an RC circuit is given by

$$v_{out-nat}(t) = Ae^{-\alpha t}, \tag{6.41}$$

and the step forced response is described as

$$v_{out-frc}(t) = V_{in} \text{ for } t > 0. \tag{6.42}$$

- Constant A can be found from (6.41), by setting $t = 0$. It gives

$$v_{out-cmp}(0) = A + V_{in}. \tag{6.43}$$

As $v_{out-cmp}(0) = 0$ in our example, we obtain

$$A = -V_{in}. \tag{6.44}$$

- Finally, the complete *(total) response of an RC circuit to the step-up function* takes the form

$$v_{out-cmp}(t) = V_{in}(1 - e^{-\alpha t}) \cdot u(t) \text{ for } t > 0. \tag{6.45}$$

We are confident that you recognize (6.45) as a step response of a series RC circuit studied in detail in Section 4.2.

Step 2: The response to the $-V_{in} \cdot u(t - T_p)$ input can be found by following the same pattern, which results in

$$v_{out-cmp}(t - T_p) = V_{in}(-u(t - T_p))(1 - e^{-\alpha(t - T_p)}) \; for \; t > T_p. \tag{6.46}$$

Step 3: Summing up (6.45) and (6.46), we determine the complete response, $H_p(t)$, of the RC circuit to an individual pulse with amplitude Vin as

$$H_p(t) = v_{out-cmp}(t) + v_{out-cmp}(t - T_p)$$
$$= \begin{cases} V_{in}(u(t)(1 - e^{-\alpha t}) \; for \; 0 \le t \le T_p \\ -V_{in}u(t - T_p)\left(1 - e^{-\alpha(t - T_p)}\right) for \; t \ge T_p \end{cases}. \tag{6.47}$$

Equation 6.47 is the pulse response of an RC circuit.

Inserting the given values $(R = 2.0 \; k\Omega, \; C = 0.1 \; mF, \; \alpha = 1/RC = 5(1/s), \; and \; T = 2(s) \; into (6.47),$ we get

$$H_p(t) = V_{in}(u(t)((1 - e^{-\alpha t}) - u(t - T_p)(1 - e^{-\alpha(t - T_p)})$$
$$= 12(u(t)(1 - e^{-5t}) - u(t - 2)(1 - e^{-5(t-2)})). \tag{6.48}$$

Figure 6.18a shows the RC circuit's pulse response plot as given in (6.48).Figure 6.18b demonstrates how the pulse response changes if the pulse duration shortens to 0.5 (s). The graphs support the response we qualitatively constructed in Figure 6.17c based on the physics of the RC circuit.

For convenience, we show how (6.48) is written in the MATLAB code[3] to plot this figure:

MATLAB code for

```
t=linspace(0,4);
H = 12.*(heaviside(t).*(1-exp(-5*t))-heaviside(t-2).*(1-exp(-
5.*(t-2))));
plot(t,H)
    grid
```

3 This code was written by Dr. Vitaly Sukharenko.

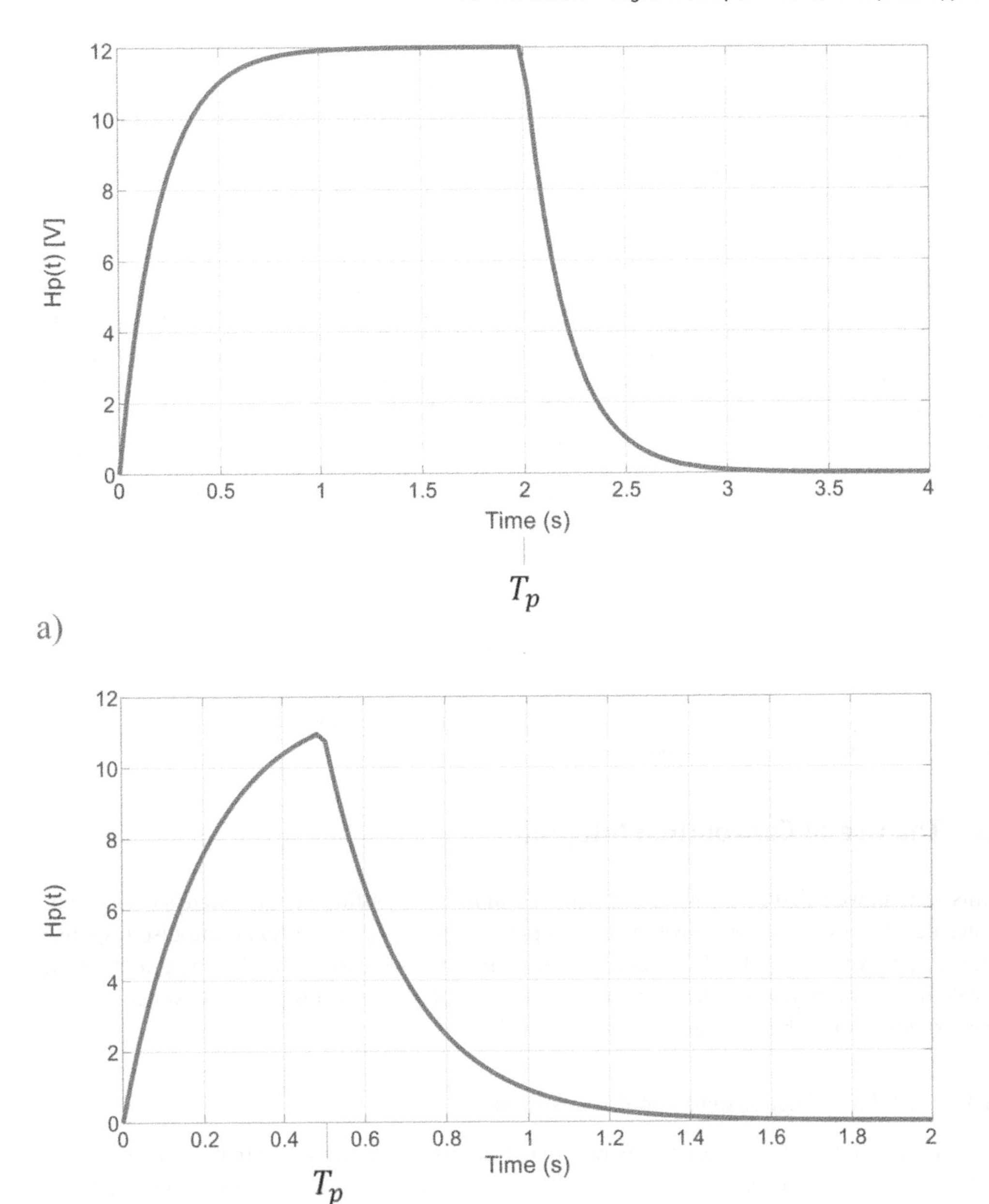

Figure 6.18 Example 6.2: Pulse response of the RC circuit whose R = 2.0 kΩ, C = 0.1 mF, and α = 1/RC = 5 (1/s) when a) T_p = 2 (s) and b) T_p = 0.5 (s).

Discussion:

- Finding the pulse response is not easy for the most practical systems; it's very similar to finding the impulse response discussed above. But—again—we need to do it only once for the same circuit; changing the values is not a problem, as Figure 6.18b demonstrates.

- Does (6.48) look familiar to you? It should be because this equation describes the *complete transient response of an RC circuit* to the step input, as we studied in Section 4.2.
- Why does the response in Figure 6.18a differ from that in 6.18b? It is due to the difference in the ratio $\frac{\Delta T(s)}{\tau(s)} \equiv \frac{\Delta T(s)}{RC(s)}$. For the graph in Figure 6.18a, $\frac{\Delta T(s)}{RC(s)} = \frac{2}{0.2} = 10$, and for Figure 6.18b, $\frac{\Delta T(s)}{RC(s)} = \frac{0.5}{0.2} = 2.5$ because of shortening $\Delta T(s)$. This is why the capacitor, whose $\tau(s) = RC$ retains, doesn't have enough time for the full charging. As a result, the RC circuit's response amplitude is smaller than the input. In Figure 6.18a, $V_{out} = V_{in} = 12(V)$, whereas in Figure 6.18b, $V_{in} = 12(V)$ and $V_{out} \approx 11(V)$. Compare graphs in Figure 6.17c and 6.18b and see their similarity. This observation has vital practical implications concerning the transmission of a digital signal through a reactive circuit, the point we'll discuss in Section 6.3.
- It's essential to understand that the results shown in Figure 6.18 hold regardless of what parameter, T *or* RC, changes because they depend on the ratio $\frac{T(s)}{RC(s)}$, not on the individual parameter.

6.2.2.2 Conclusion of Section 6.2

The set of pulses approximates an input signal, each pulse produces an individual pulse response, and the total response is found by adding up the individual pulse responses. In this section, we learned how to find a pulse response for an RC circuit; see (6.47) and Figure 6.17c and 6.18a. Following this routine, we can find the pulse responses of other circuits.

We must always remember that the pulse-response approach approximates an output signal; we need to turn to the convolution integral to obtain the exact output.

6.3 The Use of Convolution Integral

In this section, we will discuss the use of convolution integral for finding the system's response in greater detail. To start, we'll transition from a pulse-response approach to an impulse-response process. Then we will find the RC circuit's response by employing a convolution integral. Finally, we will discuss—applying our findings and using a convolution integral—how the system's complete response to an arbitrary signal can be obtained.

6.3.1 From Pulse Response to Impulse Response

We are looking for a circuit's exact response to a given input, and we know that such a response can be obtained using a convolution integral. But first, we must prove the several times declared statement: *When the pulse width, ΔT, goes to zero, the pulse response, $H_p(t)$, in the limit becomes the impulse response, $h(t)$*; i.e.,

$$h(t) = \lim_{\Delta T \to 0} H_p(t). \tag{6.49}$$

Applying (6.49) to the result of Example 6.2, we find

$$h(t) = \lim_{\Delta T \to 0} H_p(t) = \lim_{\Delta T \to 0} \left[\frac{u(t)}{\Delta T}\left(1 - e^{-\alpha t}\right) - \frac{u(t - T_p)}{\Delta T}\left(1 - e^{-\alpha(t - T_p)}\right) \right]$$

$$= \alpha e^{-\alpha t} \; for \; t > 0. \tag{6.50}$$

Q.E.D.

***Sidebar 6S* Proof of Equation 6.50**

Here is the proof of Equation 6.50: Recall that $\Delta T = t - T_p$. Then, for $t > T > 0$, we find $u(t) = 1$ *and* $u(t-) = 1$. *Then* $H_p(t)$ can be simplified as

$$H_p(t) = \frac{u(t)}{\Delta T}(1 - e^{-\alpha t}) - \frac{u(t - T_p)}{\Delta T}\left(1 - e^{-\alpha(t - T_p)}\right)$$

$$= \frac{1}{\Delta T}\left(1 - e^{-\alpha t}\right) - \frac{1}{\Delta T}\left(1 - e^{-\alpha(t - T_p)}\right) = \frac{1}{\Delta T}\left(e^{-\alpha(t - T_p)} - e^{-\alpha t}\right).$$

But

$$\frac{1}{\Delta T}\left(e^{-\alpha(t - T_p)} - e^{-\alpha t}\right) = \frac{1}{\Delta T}\int_t^{t-T_p} \alpha e^{-\alpha z} dz,$$

where z is a dummy variable. You can easily verify this equality by direct integration of the RHS. This function is the continuous on the closed interval $[t, t-T]$; therefore, we can denote it as

$$\frac{1}{\Delta T}\int_t^{t-T_p} \alpha e^{-\alpha z} dz = f(y),$$

where y is the point somewhere between t and $t-T_p$, *i.e.*, $t < y < t-T_p$. Thus, we find

$$H_p(t) = \frac{1}{\Delta T}\left(e^{-\alpha(t - T_p)} - e^{-\alpha t}\right) = \frac{1}{\Delta T}\int_t^{t-T_p} \alpha e^{-\alpha z} dz = f(y).$$

In the limit, when the pulse duration goes to zero, $\Delta T \to 0$, we find that

$$t < y < t - T_p \to t < y < t - 0,$$

which means that

$$y = t.$$

(Remember, the condition $\Delta T \to 0$ is equivalent to $T_p \to 0$ because $\Delta T = t - T_p$ by definition.) Finally, we arrive at

$$\lim_{\Delta T \to 0} H_p(t) = \lim_{\Delta T \to 0} f(y) = f(t) = \alpha e^{-\alpha t} = h(t),$$

as in (6.50). Refer to Figure 6.17 for a graphical presentation of this proof.

6.3.2 From Summation to Integration—Ultimate Presentation of the Convolution-Integral Technique

Finding the system's complete output for the pulse response comes down to the application of (6.36),

$$v_{out}(t) \approx \sum_{m=-\infty}^{\infty} H_{pm}(t) = \sum_{m=-\infty}^{\infty} V_{in}(mt) h_p(t)\Delta T. \tag{6.36R}$$

Allowing $\Delta T \to 0$ in this equation, we approach the exact reproduction of both the input and output signals, and the summation will be replaced by integration, as we pointed out in the derivation of the convolution integral. Simply put, letting $\Delta T \to 0$, (6.36) arrives at (6.10) because $H_p(t)$ approaches $h(t)$, as (6.50) states. Therefore,

$$v_{out}(t) \approx \sum_{m=-\infty}^{\infty} H_{pm}(t) = \sum_{m=-\infty}^{\infty} V_{in}(mt)h_p(t)\Delta T \underset{\Delta T \to 0}{\Rightarrow} v_{out}(t) = \int_{-\infty}^{\infty} v_{in}(\tau)h(t-\tau)d\tau. \quad (6.51)$$

Now, it's time to collect all our preceding discussions regarding the pulse and impulse responses and the convolution integral for the *ultimate presentation of the convolution-integral technique.* To make this presentation more applicable, we consider the *pulse response* of a series RC circuit. Let's invoke (6.10),

$$v_{out}(t) = \int_{\infty}^{-\infty} v_{in}(\tau)h(t-\tau)d\tau = v_{in}(t)h(t). \quad (6.10R)$$

Here, $v_{in}(t)$ is a pulse determined in (6.47), and $h(t)$ is the impulse response of a series RC circuit found in (6.28) as

$$h(t) = \alpha e^{-\alpha t}u(t) \quad (6.28R)$$

with $\alpha = \frac{1}{RC}$. The graphs of both $v_{in}(t)$ and $h(t)$ are reminded in Figure 6.19a and 6.19b, respectively.

In the convolution integral, 6.10, the *integrand* is the product of two functions, $v_{in}(\tau)$ and $h(t-\tau)$. Let's consider the latter: There are four possible $h(t)$ transformations; they are shown in Figures 6.20a through 6.20d, and it's critical to understand their meaning. The impulse response, initially given in (6.10) as $h(t)$, now can take the form $h(-\tau)$ called *time-reversed (folded)* and shown in Figure 6.20b. A *time-reversed and positively time-shifted impulse response* is presented in Figure 6.20c; it is formed as a sequence of (1) time reversing from $h(\tau)$ to $h(-\tau)$ and (2) time-shifting from $h(-\tau)$ to $h(-(\tau-t) = h(t-\tau)$. Finally, Figure 6.20d demonstrates *time-reversed and negatively time-shifted* $h(t+\tau)$.

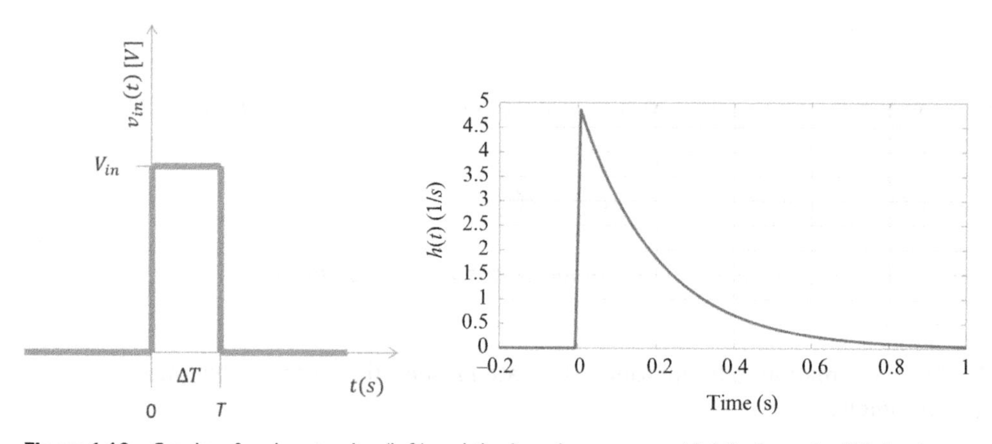

Figure 6.19 Graphs of an input pulse (left) and the impulse response (right) of a series RC circuit.

For convenience, below is the sample of the MATLAB code for Figure 6.20d:

% Figure 6.20d – reversed and left-shifted impulse response $(t+\tau)$

```
>> t=linspace(-1.2, 1.2);
h = (5.*heaviside(-t-0.2).*exp(-5.*-t-0.2));
plot (t,h)
grid
xlabel('Time (s)')
ylabel('h(t) [1/s]')
```

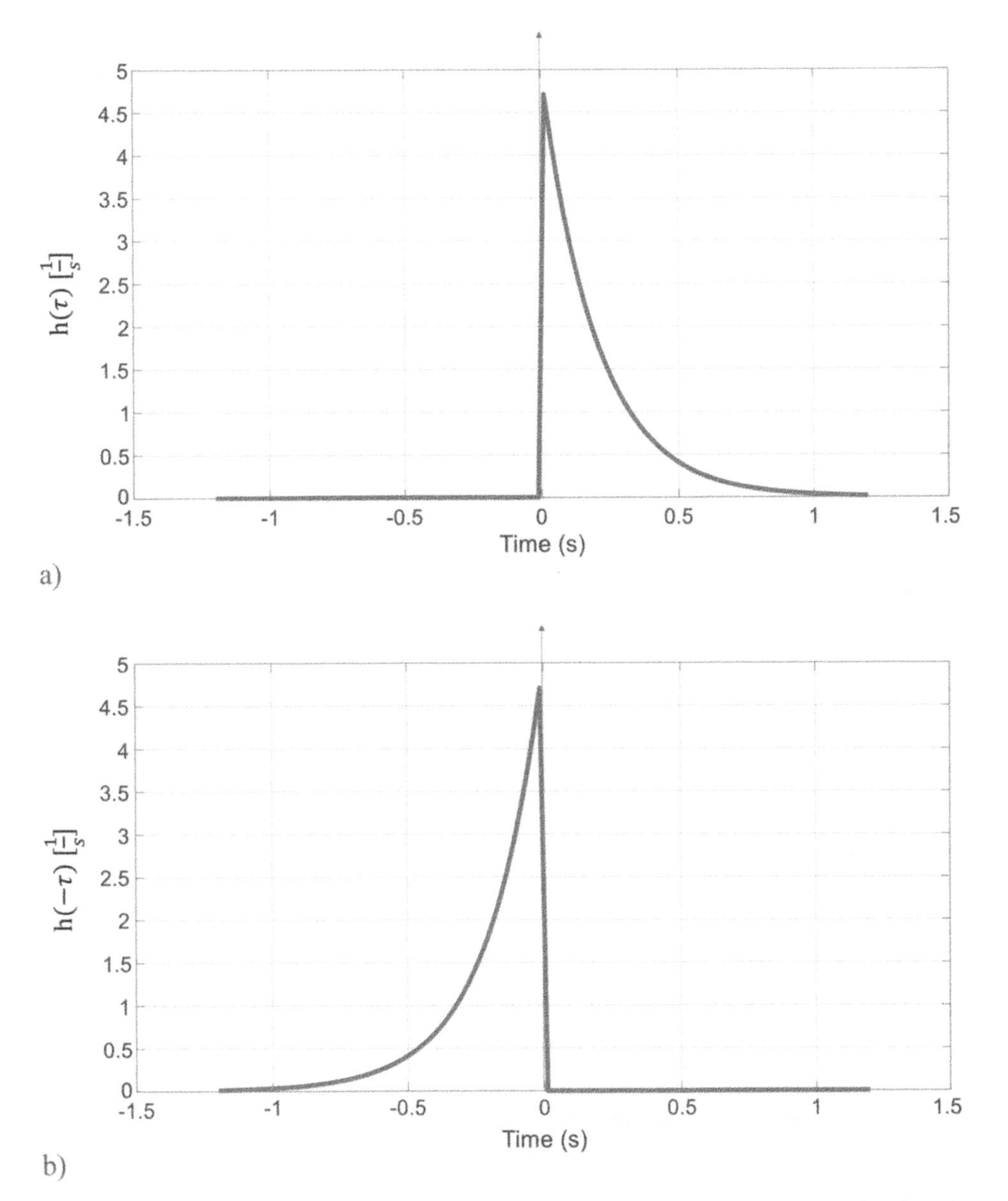

Figure 6.20 Impulse response and its transpositions: a) The original impulse response, $h(\tau)$; b) a time-reversed impulse response, $h(-\tau)$; c) a time-reversed and positively time-shifted impulse response, $h(t - \tau)$; d) a time-reversed and negatively time-shifted impulse response, $h(t + \tau)$.

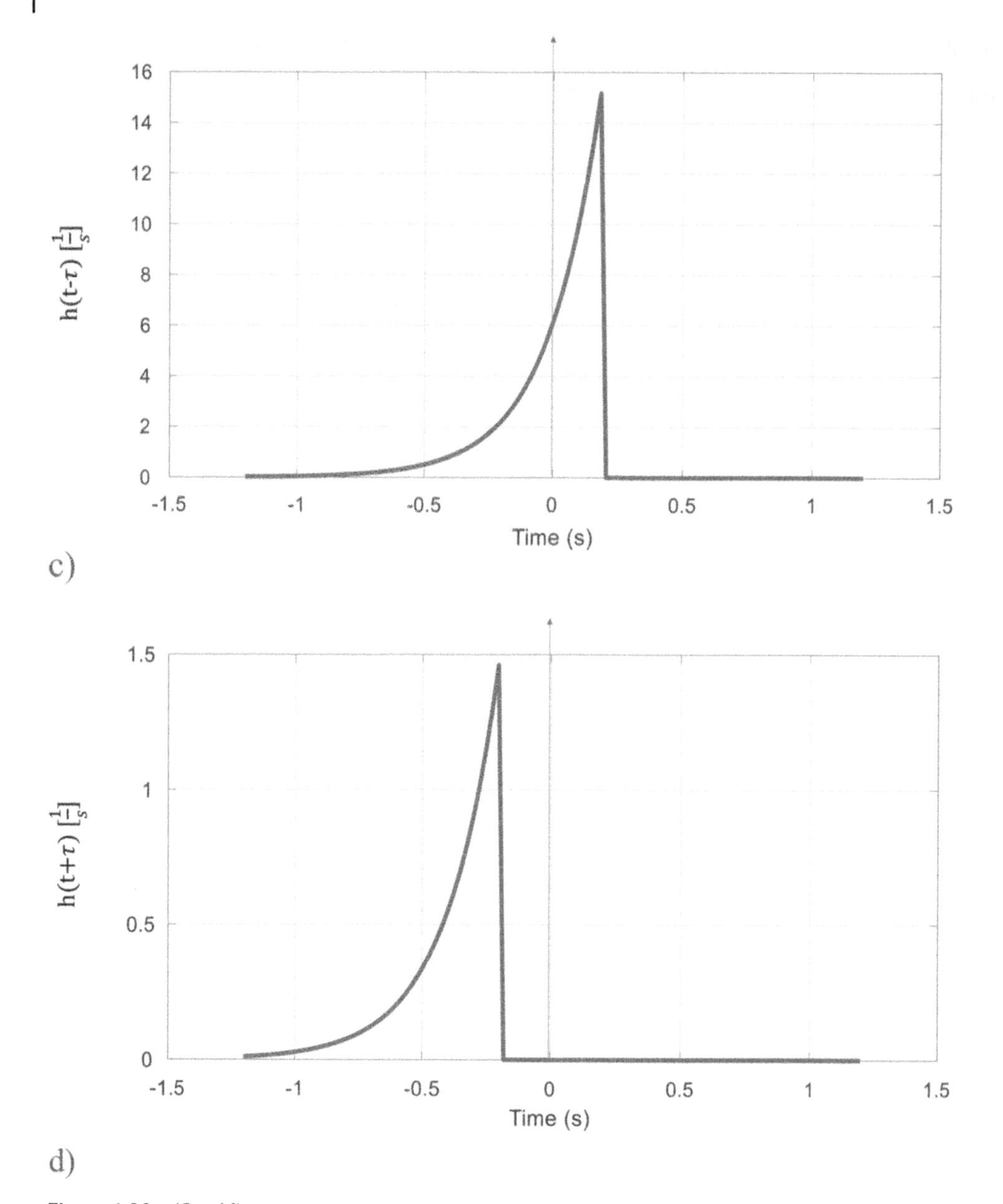

Figure 6.20 **(Cont'd)**

Now, let's consider how the *integrand* of the convolution integral, $\int_{-\infty}^{\infty} v_{in}(\tau)h(t-\tau)d\tau$, is forming: As t changes, the impulse response, $h(t-\tau)$, slides through the input signal, $v_{in}(\tau)$, as the arrow in Figures 6.21a and 6.21b visualize. (We take $h(t-\tau)$ as an example; any other transformation of $h(\tau)$ can be considered too.) Figure 6.21a exemplifies the situation where, at $t=-3(s)$, the impulse response $h(-3+\tau)$ and input signal $v_{in}(\tau)$, located at $0<t<2$, do not overlap; therefore, the integrand $h(-3+\tau)\cdot v_{in}(\tau)$ is zero everywhere. In contrast, at $t=2(s)$, both functions fully overlapped, and their product (the integrand, that is) reaches its maximum value, as Figure 6.21b elucidates.

Therefore, for each instant, $t(s)$, the value of the integrand, $v_{in}(\tau)\cdot h(t-\tau)$, depends on the area of the overlapping parts of $v_{in}(\tau)$ and $h(t-\tau)$. To calculate this area, we need to evaluate the convolution integral, $\int_{-\infty}^{\infty} v_{in}(\tau)h(t-\tau)d\tau$, over the whole range of τ from $-\infty$ *to* $+\infty$. In our example, the result is the value of the shaded area in Figure 6.21b. Note firmly that this number is obtained at fixed t (here, $t=2(s)$), and it is the circuit response, $v_{out}(t)$, at this instant. Thus, the value of the shaded area in Figure 6.21b is $V_{out}(t=2(s))$.

Next step: As t progresses from $-\infty$ to $+\infty$, the calculations of the overlapping areas $v_{in}(\tau)\cdot h(t-\tau)$ repeat again and again for every new t value. The result of this two-step integration—first over τ for each t, and then for all values of t—is the total area of $v_{in}(\tau)\cdot h(t-\tau)$ overlapping parts for $-\infty < t < +\infty$. This result is $v_{out}(t)$, whose $V_{out}(t_n)$ is the area of the overlapping parts at a specific instant t_n.

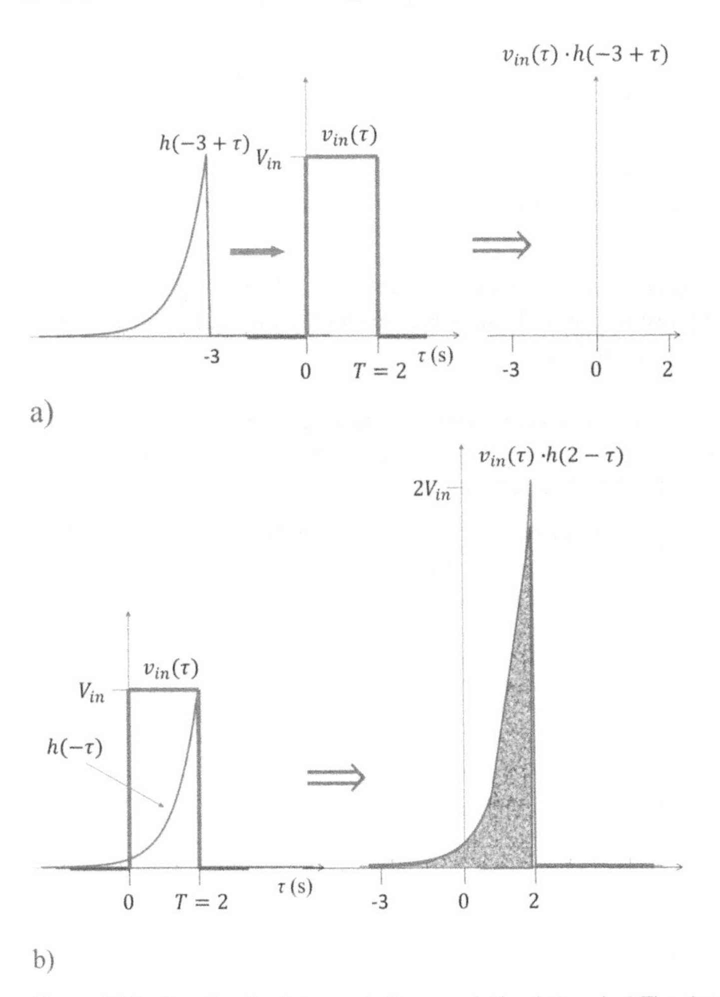

Figure 6.21 Forming the integrand of a convolution integral: a) The time-reversed and time-shifted impulse response, $h(t-\tau)$, slides toward the input pulse, $v_{in}(\tau)$, but doesn't overlap it yet; b) the impulse response and the input function overlaps.

To sum up, the evaluation of the convolution integral,

$$v_{out}(t) = \int_{-\infty}^{\infty} v_{in}(\tau)h(t-\tau)d\tau \equiv v_{in}(\tau) \ast h(t-\tau), \tag{6.10R}$$

produces the circuit response, $v_{out}(t)$, for the whole range of t. This evaluation, however, is performed in two steps:

1) Integration over the entire range of τ for each fixed t,
2) Integration over the total range of t.

Eventually, the circuit output, $v_{out}(t)$, is the response to all past and present values of the input signal, $v_{in}(t)$.

Concerning the specifics of the convolution integral evaluation, let's refer to the example shown in Figure 6.21. Now, consider the three situations for sliding the impulse response, $h(t-\tau)$, through the input pulse, $v_{in}(\tau)$, with the increase in time. They are shown in Figures 6.22.

a) The two integrand functions don't touch yet, and their overlapping area is zero, as Figure 6.22a shows. Here, $t_a < 0$.
b) The two functions of the integrand overlap, and their common area enlarges with increasing t, but $h(t-\tau)$ doesn't cross T yet, the last point of the input pulse. This situation is depicted in Figure 6.22b, where $0 < t_b < T_p$.
c) The two integrand functions overlap on the whole interval from 0 to T_p; the impulse response, $h(t-\tau)$, has crossed the point T_p; and the common area decreases with increasing t. Figure 6.22c shows that in this situation $t_c > T_p$.

To comprehend the subsequent explanations, it's vitally important to follow the position of the point t, the front point of the impulse response. Figures 6.22 exemplify this statement by three specific positions of the front point, t_a, t_b, and t_c.

Question For $t > T_p$, the input pulse, $v_{in}(\tau)$, becomes zero as Figures 6.22 show. Why then do we need to consider the range $t > T_p$ for the convolution integral evaluation? After all, we don't include the time interval $t < 0$ in calculating the convolution integral.

The value of the convolution integral starts increasing when $h(t-\tau)$ and $v_{in}(\tau)$ begin to overlap; this value reaches its maximum when the overlapping area becomes the biggest; after that, the value of the overlapping area—and therefore the convolution integral—begins to decrease.

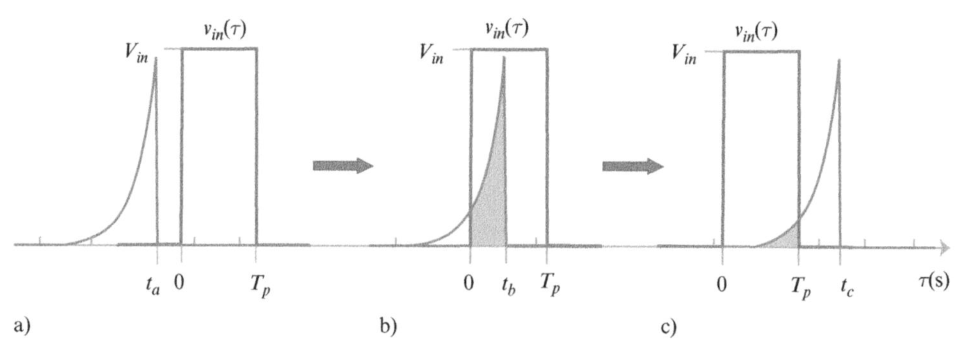

Figure 6.22 Three situations in evaluating the convolution integral: a) The overlapping area of the two integrand functions is zero, $t_a < 0$; b) the overlapping area of the two functions increases for $0 < t_b < T_p$; c) the overlapping area of the two functions decreases since $t_c > T_p$.

To put the above statement into quantitative assessment, we need to evaluate the convolution integral (6.10) for three cases shown in Figure 6.22. This can be done as follows:

a) For $t = t_a < 0$, the integrand is obviously zero, and the circuit output is

$$v_{out}(t) = \int_{-\infty}^{\infty} v_{in}(\tau)h(t-\tau)d\tau = 0, \tag{6.52}$$

which is true for any $t < 0$.

b) For $0 < t = t_b < T_p$, the *input pulse* can be considered a step function. This is so because, on the one hand, a pulse is a sum of two step functions, as (6.37) shows, but, on the other hand, its second part is zero for all t in the range $0 < t < T_p$. See Figure 6.16. Therefore, $v_{in}(\tau) = V_{in}u(t)$ for $0 < t < T_p$.

The folded and positively time-shifted *impulse response* of an RC circuit is given as $h(t-\tau) = \alpha e^{-\alpha(t-\tau)}$.

We remember that $u(t) = 1$ and V_{in} is constant, which allows for changing the convolution integral limits from infinities to $0 \to t$. Thus, the integration of (6.10) for this interval produces

$$V_{in}e^{-\alpha t})\int_0^{T_p} \alpha e^{\alpha(\tau)}d\tau v_{out}(t) = \int_{-\infty}^{\infty} v_{in}(\tau)h(t-\tau)d\tau = \int_0^t V_{in}u(t)\alpha e^{-\alpha(t-\tau)}$$

$$d\tau = V_{in}\int_0^t \alpha e^{-\alpha(t-\tau)}d\tau = (V_{in}e^{-\alpha t})e^{\alpha\tau}\Big|_0^t = V_{in}(1-e^{-\alpha t}) \; for \; 0 < t < T_p. \tag{6.53}$$

This is how the output value changes while the impulse response glides through the input pulse, and its front point t_k $(k = a,b,c)$ remains within the pulse width.

c) For $t = t_c > T_p$, the front point of the impulse response leaves the pulse, but the impulse's tail still slides through the input pulse. Evaluating the convolution integral for this interval results in

$$\begin{aligned} v_{out}(t) &= V_{in}\int_0^{T_p} \alpha e^{-\alpha(t-\tau)}d\tau = (V_{in}e^{-\alpha t})\int_0^{T_p} \alpha e^{\alpha(\tau)}d\tau = (V_{in}e^{-\alpha t})e^{\alpha(\tau)}\Big|_0^{T_p} \\ &= (V_{in}e^{-\alpha t})(e^{-\alpha T_p} - 1) \; for \; t > T_p. \end{aligned} \tag{6.54}$$

We have to understand that the parameter T_p in (6.54) is just a number and therefore the member $(e^{\alpha}T_p - 1)$ is merely a coefficient. Thus, (6.54) shows that these two functions still overlap, but as t increases, the area of the overlapping parts decreases due to member $e^{-\alpha t}$.

In reviewing the process of evaluating a convolution integral in Equations 6.52 through 6.54, pay attention to the fact that we calculated $v_{out}(t)$ by integrating over the whole range of τ for a fixed value of t—such as t_a, t_b, and t_c in Figures 6.22 —the sequence stressed many times in the preceding discussions.

Finally, the total response of an RC circuit to the pulse input obtained by the convolution integral is as follows:

$$v_{out}(t) = \begin{cases} 0 \; for \; t < 0 \\ V_{in}\left(1 - e^{-\alpha t}\right) for \; 0 \le t < T_p \\ -V_{in}\left(1 - e^{\alpha T_p}\right)e^{-\alpha t} for \; t \ge T_p \end{cases} . \tag{6.55}$$

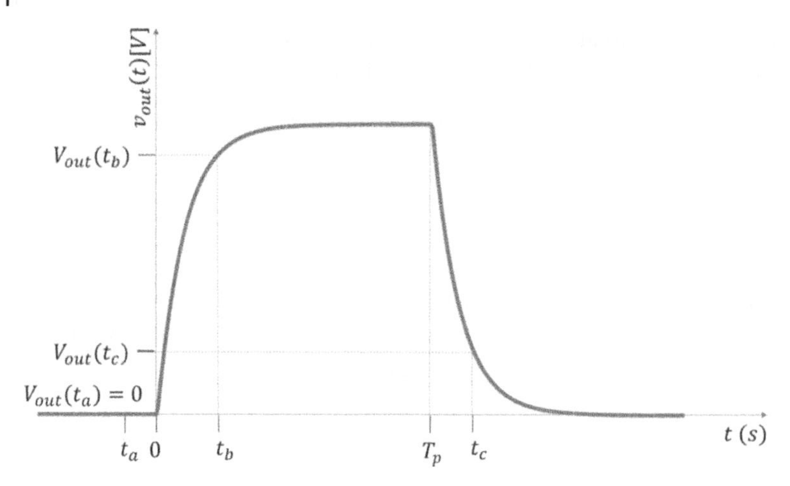

Figure 6.23 The RC circuit's output voltage, a response to the pulse input.

Figure 6.23 is a graphical presentation of Equation 6.55. The figure shows three positions of the impulse response, t_a, t_b, and t_c, with respect to the input pulse, and corresponding values of the output voltages, $V(t_a)$, $V(t_b)$, and $V(t_c)$.

It's no surprise that this graph has the same shape as the one demonstrated in Figures 6.15 and 6.19; this is because all of them present the response of an RC circuit to a pulse input.

The integral $v_{out}(t) = \int_0^t h(\tau)u(t-\tau)d\tau$ gives the actual circuit's output. Here, the integrand member $u(t-\tau)$ shows input's status back τ seconds ago, and $h(\tau)$ presents how much the extant output depends on the input mode τ seconds ago.

Example 6.3 shows how to apply the convolution integral to find the response of an RC circuit to an input pulse in practice.

Example 6.3 Finding the pulse response of an RC circuit by the convolution-integral technique.

Problem: Consider the series RC circuit discussed in Examples 6.1 and 6.2, whose $R = 2.0(k\Omega)$, $C = 0.1(mF)$, and $\alpha = 1/RC = 5(1/s)$. The initial condition is zero, and the input pulse whose $\Delta T = 2$ (s) is given by

$$v_{in}(t) = \begin{cases} V_{in} = 12(V) \; for \; 0 \leq t \leq T_p \\ V_{in} = 0 \; for \; t < 0 \; and \; t > T_p \end{cases}. \tag{6.56}$$

Find the output of this circuit using the convolution-integral technique.

Solution:

Though the answer was obtained in Equation 6.2, we want to receive the result using the convolution integral.

To achieve this goal, we need to invoke (6.10),

$$v_{out}(t) = \int_{-\infty}^{\infty} v_{in}(\tau)h(t-\tau)d\tau = v_{in}(t) \ast h(t), \tag{6.10R}$$

where, $v_{in}(t)$ is a pulse determined by (6.56), and $h(t)$ is the impulse response of the RC circuit found in (6.28) as

$$h(t) = \alpha e^{-\alpha t} u(t). \tag{6.28R}$$

The graphs of $v_{in}(t)$ and $h(t)$ are shown in Figure 6.19. The solution to this problem boils down to plugging the numbers into (6.55), which produces

$$v_{out}(t) = \begin{cases} 0 \ for \ t < 0 \\ 12u(t)\left(1 - e^{-5t}\right) \ for \ 0 \le t < 2 \\ -12u(t-2)\left(1 - e^{10}\right)e^{-5t} \ for \ t \ge 2 \end{cases}. \tag{6.57}$$

Figure 6.24 visualizes (6.57).

For convenience, below is the MATLAB code for building Figure 6.24.

```
≫ syms t
Vout = 12.*(heaviside(t).*(1-exp(-5*t))-heaviside(t-2).*(1-exp(-
5.*(t-2))));
ezplot(Vout,[-1, 4.0])
axis([-1 4.0 -1 14])
grid
xlabel('Time (s)')
ylabel('Vout(t)(t) [V]')
```

Figure 6.24 shows that for $t < 0$, $v_{out}(t) = 0$, as expected. For $t = 0.5(s)$, $V_{out}(0.5) = 11(V)$; for $1 \le t(s) \le 2$, $V_{out}(1 \le t \le 2) = 12(V)$, and for $t = 2.5(s)$, $V_{out}(2.5) = 1(V)$. These results are the numerical illustrations of the concept discussed in regard to Figure 6.23.

The problem is solved.

Discussion:

This example serves two purposes: It introduces the use of the convolution-integral technique and shows how to find the response to a pulse input of a specific circuit by applying the convolution integral.

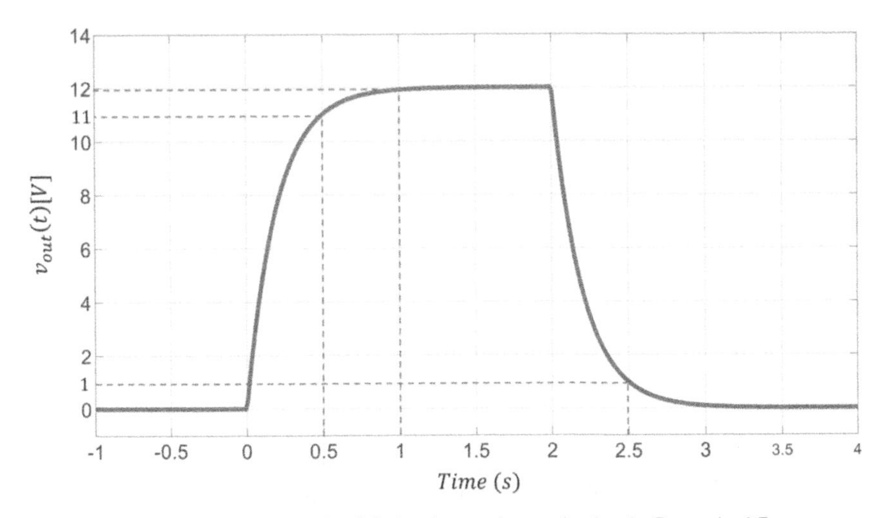

Figure 6.24 Response of a series RC circuit to pulse excitation in Example 6.3.

The pulse response of an RC circuit finds many practical applications; we will concentrate here on its use for digital transmission.

6.3.3 Bit Rate and the Output Waveform

A series RC circuit works as a lowpass filter (LPF) in electrical and telecommunications networks. This is why its response to a pulse—which represents a digital signal—is an essential issue in engineering practice.

The graphs Figure 6.25 in show the input pulses and corresponding responses of a series RC circuit whose $\tau = RC = 0.2(s)$. We considered this circuit and its pulse response in Examples 6.2 and 6.3. In fact, Figure 6.25 partly reproduces Figure 6.18.

Figure 6.25 demonstrates that (1) the RC circuit's response does not reproduce an input pulse in general, and (2) the shorter the input pulse, the larger the response deviation from the input.

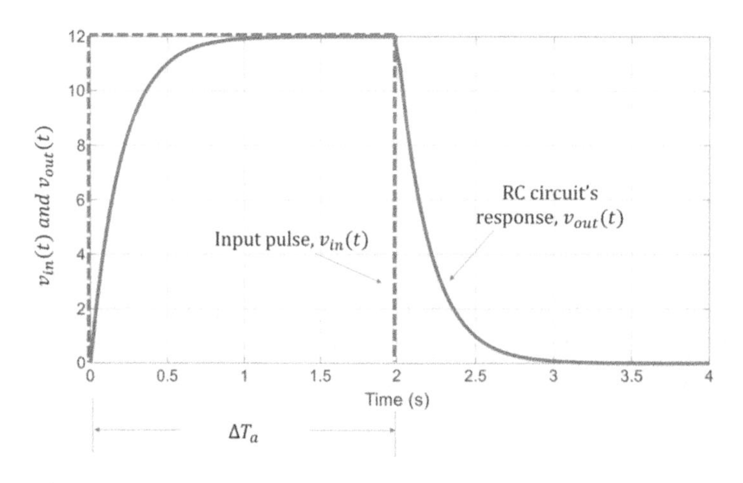

a)

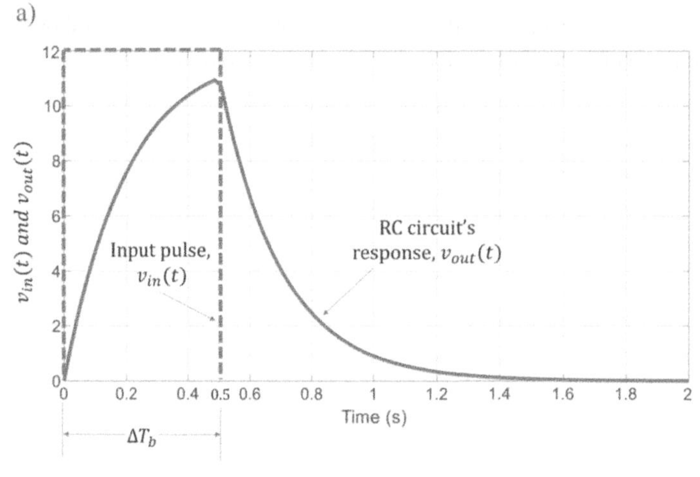

b)

Figure 6.25 The RC circuit's responses to the input pulses of different durations: a) Input pulse is $\Delta T_a = 2(s)$; b) $\Delta T_b = 0.5(s)$. The RC circuit's time constant remains unchanging and equal to 0.2(s).

The latter observation is vitally important for the application of RC circuits in telecommunications. Why? Let's assume that each pulse carries one bit. Then, how many pulses (bits, that is) can a telecommunication system transmit for one second? If the pulse width is $\Delta T = 1(s)$, then the answer is one bit. If $\Delta T = 0.5(s)$, then—two bits, and so on. Now, permit the *number of bits per second* be a *bit rate*, $BR\left(\frac{b}{s}\right)$. Then, the bit rate of a telecommunication system is inversely proportional to the pulse width, $\Delta T(s)$, that is,

$$BR\left(\frac{b}{s}\right) = \frac{1}{\Delta T(s)}. \tag{6.58}$$

Understand that $BR\left(\frac{b}{s}\right)$ is the measure of the *transmission capacity* of a telecommunication system. Realize, too, that the unit $\frac{b}{s}$ denotes the *number of bits per second* and can be written as $\frac{1}{s}$ because a number carries no dimension.

Now, Figure 6.25 enables us for the other interpretation: We can say that the smaller the pulse width ΔT—which means the higher the bit rate of a digital signal according to (6.58)—the smaller the amplitude of the output voltage and the shorter the duration of the output signal. But the shorter the pulse duration, the more the response deviates from the input pulse, which causes difficulty for a receiver correctly interpret the incoming signal. Indeed, when the pulse duration is $\Delta T = 2$ seconds, as in Figure 6.25a, the output voltage resembles the input very closely, which means that a receiver can readily recognize this pulse, and, therefore, count bit 1. But when the pulse duration shortens to 0.5 seconds, the output signal deviates from the input significantly, confusing the receiver with apprehending the sense of the received signal. Compare Figures 6.25a and 6.25b to see that the response to a shorter pulse in Figure 6.25b has a smaller amplitude and a shorter duration, which means that this signal carries less energy than the output in Figure 6.25a.

Eventually, the RC circuit will not reproduce an input digital signal at all. This is why we can't transmit a high-speed digital signal through a traditional electrical wire line, which is, in essence, a lowpass filter.

Figures 6.26 shows an experimental setup and the results of the Multisim simulation of such a transmission. These figures display how the changes in the input and output signals depend on an increasing bit rate. To visualize those changes, the oscilloscope settings remain the same for all three measurements. At the low bit rate, $BR_b = \frac{1}{\Delta T_b} = 1\left(\frac{kb}{s}\right)$, as in Figure 6.26b, a receiver can easily distinguish between bit 1 delivered by the signal of 5-V amplitude and bit 0 brought by practically-zero signal. Figure 6.26c demonstrates transmission at the intermediate bit rate, $BR_c = \frac{1}{\Delta T_c} = 10\left(\frac{kb}{s}\right)$. Here, bit 1 is delivered by 3-V amplitude, whereas bit 0 is provided by 2-V amplitude. The difference between the amplitudes is still significant, and such a telecommunication system is operational. Eventually, when a bit rate becomes high, $BR_d = \frac{1}{\Delta T_d} = 100\left(\frac{kb}{s}\right)$, as in Figure 6.26d, the difference between the highest and the lowest amplitudes becomes $2.56 - 2.42 = 0.14(V)$, and not every receiver can correctly decipher such a small voltage fluctuation to distinguish whether it is bit 1 or 0.

6.3.4 Δ*T*/RC Parameter

Analyses of Figure 6.25 and 6.26 bring us to another critical parameter describing a pulse response of a series RC circuit:

$$\frac{\Delta T}{RC} = \alpha \Delta T = \frac{\Delta T(s)}{\tau(s)} \tag{6.59}$$

Parameter $\frac{\Delta T}{RC}$ describes how well an RC LPF reproduces the input pulse or, putting it another way, how severely the RC circuit distorts the input signal. Consider Figure 6.26: Since $RC = \tau = 0.2(ms)$ is constant, the *parameter* $\frac{\Delta T}{RC}$ *changes here only due to the change in* ΔT. In numbers, this statement shows

Figure 6.26b: $\frac{\Delta T_b}{RC} = \frac{1(ms)}{0.2(ms)} = 5,$

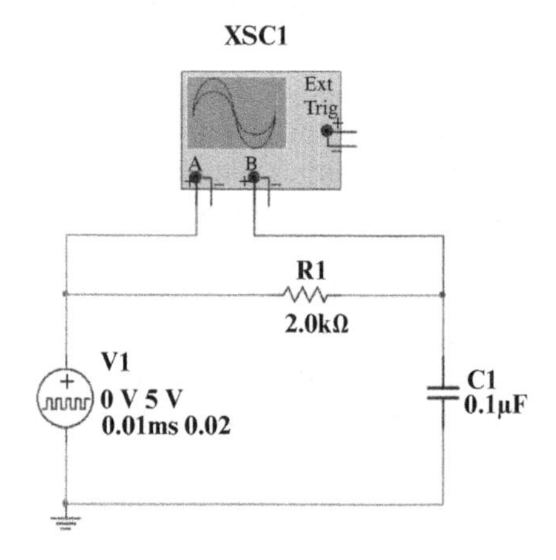

a)

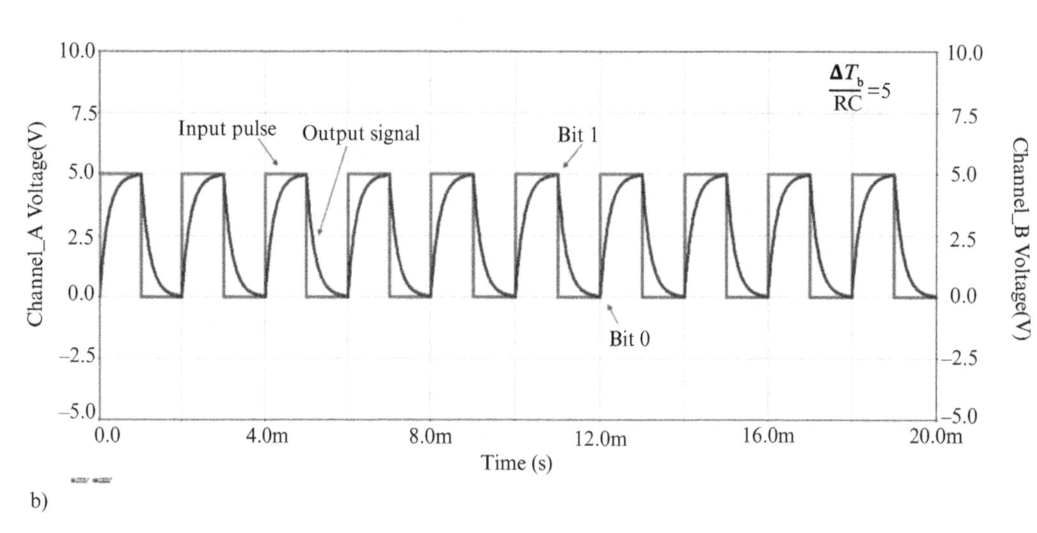

b)

Figure 6.26 Simulation of a series RC circuit operation showing the input pulses and the output signals: a) Experimental setup; b) input pulse width, $\Delta T_b = 1ms$; c) $\Delta T_c = 0.1ms$; d) $\Delta T_d = 0.001ms$.

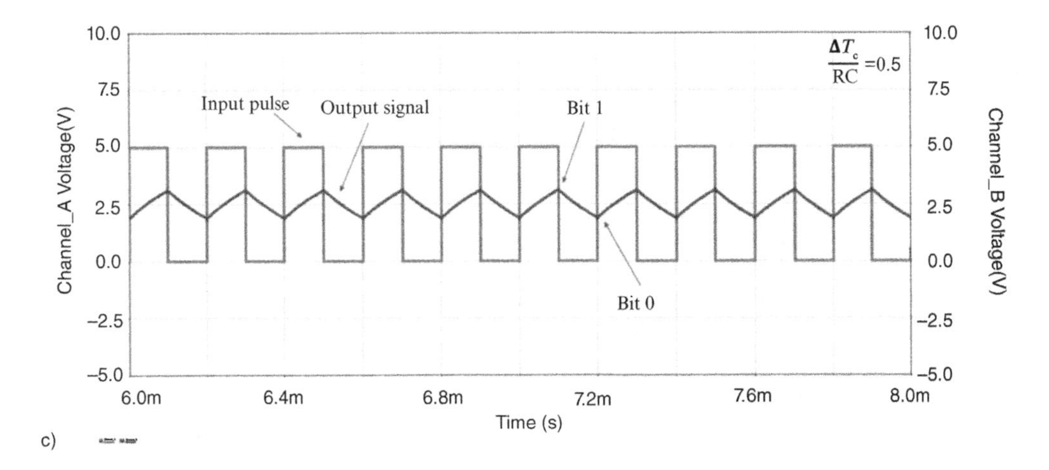

c)

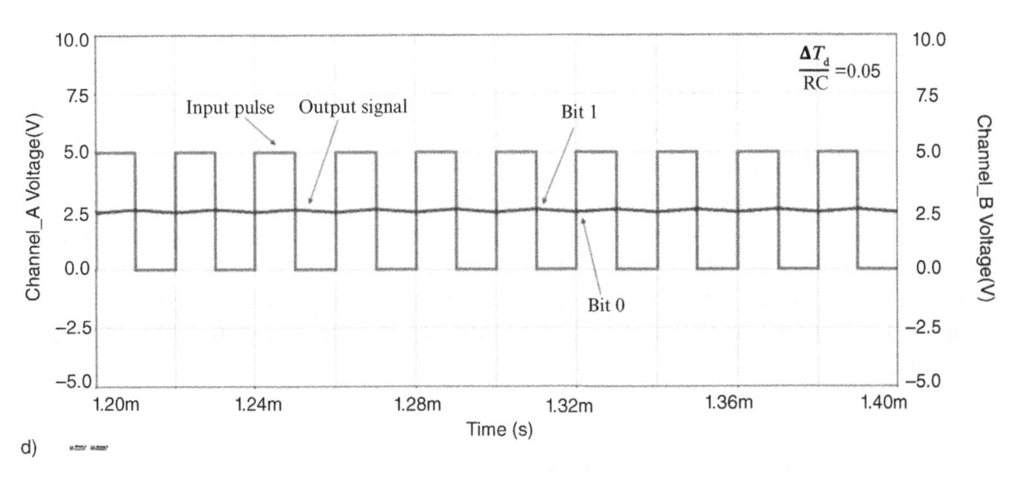

d)

Figure 6.26 (Cont'd)

Figure 6.26c: $\frac{\Delta T_c}{RC} = 0.5$,

Figure 6.26d: $\frac{\Delta T_d}{RC} = 0.05$.

These numbers are shown in each graph of Figure 6.26. Analyzing these graphs, we can conclude that the smaller $\frac{\Delta T}{RC}$, the greater the distortion of the output signal.

Question Figures 6.26 show the output signal's degradation with the bit rate increase. What physical phenomena stand behind such degradation?

We could develop the other view at the meaning of $\frac{\Delta T}{RC}$ parameter by referring to the *spectral analysis* of the process, which calls for turning to the *frequency domain*. We will pursue this approach in Chapter 9.

This subsection discussion demonstrates the significance of the output signal waveform in practice. The ability to determine this waveform beforehand, which can be achieved by using the convolution-integral technique, enables a designer to take all necessary measures to guarantee the adequate operation of the circuit in construction.

6.4 Brief Summary of Chapter 6

Our study's objective is to find an electrical circuit's response to a given input. In this chapter, we learned that this goal can be directly achieved by evaluating the convolution integral, if we know only an input function and the circuit's impulse response.

The main advantage of the convolution-integral approach, compared with classical (direct-solution) and state-variables methods, is that it requires solving the circuit's differential equation only once to obtain the impulse response. As soon as this response is known, we can find this circuit's output for any input because the convolution integral essentially scans the input over the circuit's mathematical model. The only task left is to perform the integration. However, finding the impulse response and performing integration are, in reality, challenging undertakings.

Modern computational technology enables us to execute mathematical operations with convolution integrals much faster and more efficiently than using manual manipulations. Nonetheless, knowing the mathematical basis of the convolution-integral technique is necessary for applying this newest tool to perform specific calculations.

There are other critical motivations for studying the convolution-integral technique. For one, some mathematical models of circuits and systems can be treated only by the convolution-integral technique, as no other methods, such as Laplace or Fourier transforms, can't be applied. Also, the convolution integral serves today as a powerful tool in digital signal processing (DSP).

All in all, knowing the convolution integral method greatly enhances your engineering skills and background.

Questions and Problems for Chapter 6

6.1 Convolution integral–the other technique for finding the circuit's response

1 What is the objective of this book and what methods for achieving this objective do you know?

2 Consider a convolution integral:

a) What is it?

b) What is the role of the convolution integral in finding the circuit's input-output relationship compared with the traditional and advanced circuit analyses?

c) Why an *impulse response*, $h(t)$, is the key member of Equation 6.1?

3 An advanced circuit analysis claims to be able to work with any input signal, though only the step and sinusoidal ones were considered so far. The convolution-integral method also declares its ability to work with any type of a circuit's excitation, but its claim seems to be more plausible. Why?

4 Consider the definition of pulse:

a) Sketch the pulse $P(t) = \begin{cases} \frac{5}{3} \; for -\frac{3}{2} < t < \frac{3}{2} \\ 0 \quad elsewhere \end{cases}$.

b) Repeat the plot for $t' = 7(s)$.

c) What is the pulse's height?
d) Calculate the area under this pulse. Show all your manipulations.
e) Convert this pulse into a unit pulse and writes the new formula for the latter.
f) What is the dimension of $P(t)$?

5 Consider the Dirac delta function $\delta(t)$:
a) Draw the function $11\delta(t-4)$.
b) What is the dimension of the function $f(t)=11\delta(t-4)$? Indicate the dimensions of all constituents.
c) The text states that an impulse function can be obtain from an appropriate pulse function by tending pulse's width ΔT to zero. Write the formula of the original pulse from which the impulse $11\delta(t-4)$ was obtained.
d) In formula $M\delta(t)$, the coefficient M is called *strength* of delta function. Why M isn't called an *amplitude*?
e) Indicate the dimension of each constituent in formula $f(t)=11\delta(t-4)$. Explain your reasoning.
f) Why an impulse function cannot be realized in practice?

6 The convolution-integral technique approximates of a continuous input signal by the set of impulses:
a) What requirements these impulses must meet?
b) Write the formula for the set of impulses that approximates the pulse, $P(t)=\begin{cases}\frac{5}{3} \; for \; -\frac{3}{2}<t<\frac{3}{2} \\ 0 \quad elsewhere\end{cases}$.

7 Consider Equation 6.10, $v_{out}(t)=\int_{\infty}^{-\infty} v_{in}(\tau)h(t-\tau)d\tau \equiv v_{in}(t) \ast h(t)$:
a) Why is it called the main point of Chapter 6 discussion?
b) What part of (6.10) is the convolution integral?
c) This discussion starts with approximating an input continuous signal by the discrete set of impulses, as shown in Figure 6.2. Explain how we can progress from the discrete set and its summation to a continuum and integration.
d) In Equation 6.10, the integrand consists of two members. Explain their meaning.

8 What exactly is the impulse response, $h(t)$? How it can be obtained?

9 The text states that finding an impulse response requires solving a circuit's differential equation. But we do the same in the classical circuit analysis! What's the difference?

10 Consider step (Heaviside) function:
a) Sketch the function $3u(t-\tau)=\begin{cases}0 \; for \; t<2 \\ 3 \; for \; t>2\end{cases}$.
b) What is the difference between a unit step function and a step function? Plot the unit step function.
c) Draw the derivative of the function given in 6.10a.
d) What is the relationship between the unit step function and a causal function?

11 *Consider the parallel RL circuit in Figure 6.P11:
a) Find its impulse response.
b) What is the dimension of the output signal?
c) What is the initial condition for this circuit?
d) Plot the graph.

(Hint: Refer to Example 6.1.)

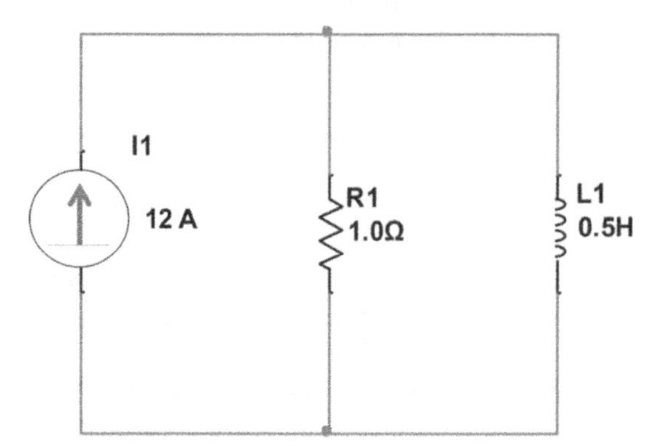

Figure 6.P11 Finding the impulse response of a parallel RL circuit.

12 Use Figure 6.9 to explain how the set of individual impulse responses of an electrical circuit forms the circuit complete response to the input signal.

6.2 Convolution-integral technique—the pulse-response approach

13 Section 6.1 tried to convince us that the *impulse response* solves the problem of finding the circuit total response. Now, Section 6.2 turns back to the *pulse response*. Why?

14 Both impulse-response and pulse-response methods rely on the superposition principle. Under what condition this principle can be applied? Explain.

15 **Consider the use of the pulse response for finding the circuit's output: Figure 6.10 through 6.13 show approximation of an input signal by the set of pulses, one-by-one passage of these pulses through a circuit, and appearance the individual pulse responses at the circuit's output. However, neither of these figures show the total output signal. Can you plot it? (Consider this problem as a mini-project. Hint: See Figure 6.14.)

16 **Figure 6.15 proves that when the pulse width $\Delta T \rightarrow 0$, a pulse approaches an impulse, and their responses tends to coincide. Prove that statement mathematically, starting at Equation 6.36. (Hint: Review Equations 6.7 through 6.10.)

17 Figure 6.16 and Equation 6.37 show how to construct and describe a pulse as a combination of two step functions. Can you build a pulse located in the negative amplitude half-plane by performing a similar operation?

18 Consider Figure 6.17 and Equation 6.47:

a) What parameter of the input pulse $-V_{in}$ or $\Delta T = t - T_p$ – determines the response amplitude V_{out}? Explain.

b) Prove it mathematically.

6.3 The use of convolution integral

19 Is Equation 6.50 universal, or it can be applied for a specific circuit only? Explain.

20 Discussing the integrand of the convolution integral, $\int_{-\infty}^{\infty} v_{in}(\tau)h(t-\tau)d\tau$, the text states that "the impulse response, $h(t-\tau)$, slides through the input signal, $v_{in}(\tau)$... ." Why does the text say so? The integrand looks like a simple product of two functions.

21 The convolution integral, $\int_{-\infty}^{\infty} v_{in}(\tau)h(t-\tau)d\tau$, contains two variables, t and τ. Over which one does the integration occur?

22 *Examine Figures 5.21b and 5.22b: Both show the situation when two integrand constituents overlap. However, Figure 5.21b shows the impulse response drastically increased in size and is fully shaded, whereas the impulse response in Figure 5.22b retains its original size and is shaded partly. Why is a difference?

23 **Using the convolution integral, find the pulse response of a parallel RL circuit in Figure 6.P23 if the switch turns on at $t_{on} = 2(s)$ and off at $t_{off} = 3(s)$. Sketch the graphs of $v_{in}(t)$ and $v_{out}(t)$ on the same set of axes. Discuss the result. All parameters are given in the figure. (Consider as a mini-project. Hint: See Examples 4.3 and 6.3.)

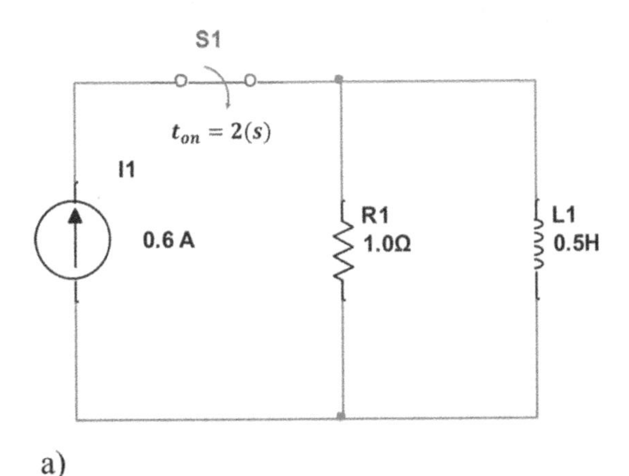

a)

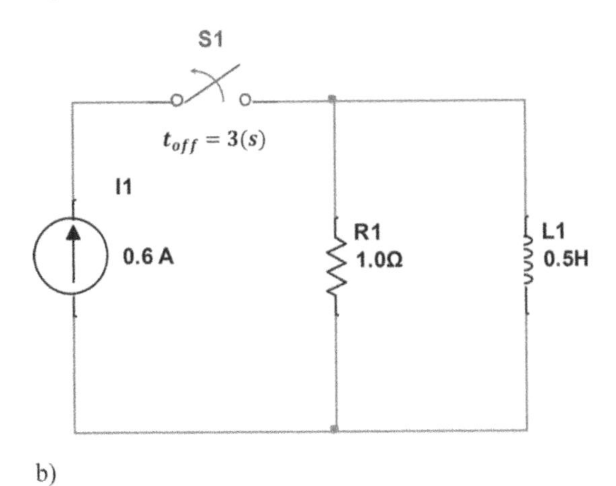

b)

Figure 6.P23 Finding the pulse response of a parallel RL circuit with convolution integral: a) The switch is on; b) the switch is off.

24 Consider Figure 6.26d:

a) What is the physical reason for such degradation of the output signal?

b) What parameters of the input signal and the whole circuit determine the level of the output degradation?

Brief summary of Chapter 6

25 Why do we need to use a convolution integral to meet the objective of this book?

26 What is the main advantage of using the convolution integral compared with differential-equations and state-variables methods?

27 The convolution integral application requires laborious manual mathematical manipulations, as the study of this chapter demonstrates. Is there any other approach to achieve the results provided by the convolution integral application? Explain.

28 Are there any situations when the convolution integral is the only method to solve an input–output problem? Explain.

29 **The text states that the convolution-integral technique is especially useful in the digital signal processing (DSP) area, but did not give an example. Volunteer to make a project on this subject.

Part 3

Frequency-Domain Advanced Circuit Analysis

7

What and Why of the Laplace Transform

The *Laplace transform* is the main tool that we use in advanced circuit analysis. Naturally, we need to know what this tool is, how it works, its strengths and limitations, and how to apply it for our purposes. This chapter answers these needs.

7.1 Function and Transform—Overview

To understand the term "transform," we need to recall the meaning of the term "function" discussed in Section 3.2:

> A ***function*** *is a rule* that assigns to each element of an input set of numbers a unique element in the output set of numbers. The input set is called the *domain*, and the output set is called the *range of the function*. The range is sometimes called *codomain*.

If x is an arbitrary element of the input set X and y is its unique assigned element in the output set Y, then we say that *y is the function of x* and denote this as

$$y = f(x). \tag{7.1}$$

This definition is illustrated in Figure 7.1a. We must recall that if y is the function of x, $y = f(x)$, x *may or may not* be an *inverse function* of y denoted as $x = f^{-1}(y)$. (Refer to Sidebar 3.2S *Waveform and Function* to refresh your memory on this topic.)

Consider, for example, the function $y = 2^x$. Here, the *domain* is a set of any possible numbers that x can assume. We can restrict this domain by requesting that x can only be a real number or a real positive number. When x takes on a specific value (e.g. $x = 3$), then y will be assigned the value $y = 2^3 = 8$, according to the rule (function!) defining the relationship between y and x. All possible values of y make up the *range* of this function. Since x can take on any value in its domain, but y must assume the value dictated by the function (rule!), x is called an *independent variable* and y is a *dependent variable*. A function, as it follows from this discussion, is a relation between y and x. This relationship is denoted by letter f shown in (7.1).

Essentials of Advanced Circuit Analysis: A Systems Approach, First Edition. Djafar K. Mynbaev.

Companion Website: www.wiley.com/go/Mynbaev/AdvancedCircuitAnalysis

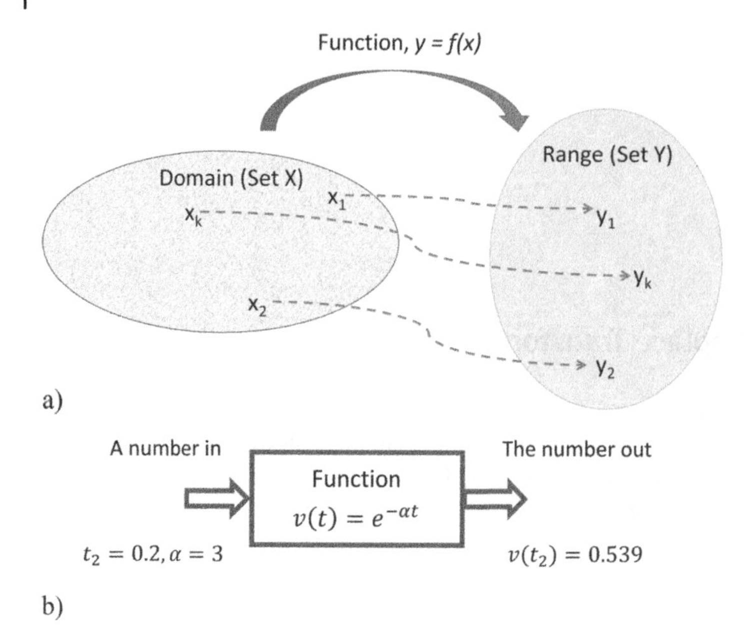

Figure 7.1 Definition of a function: a) Function is the relation between each domain element and uniquely corresponding range element; b) function is a rule that produces a unique y output value in response to a single x input value.

By definition, a function is a relation (rule) that produces a unique y value in response to a single x value, as exemplified in Figure 7.1b. But if we take $y = 4^x$ instead of $y = 2^x$, then for $x = 3$ the rule results in $y = 4^3 = 64$. Does this mean, you might ask, we have a different y in response to the same x, which violates the function definition? The answer is no because $y = 2^x$ and $y = 4^x$ are two different functions. They differ in their constants (parameters). The general formula for this type of function (called *exponential function*) is $y = a^x$, where a is a constant (parameter). Here, the term *function* denotes the general dependence y from x, but the relationship $y = a^x$ becomes a function if and only if a assumes a particular value.

Question If k is a constant, what is the inverse function for $y = ka^x$? Under what condition does this inverse function exist?

In engineering, we deal with the functions that depend on time, *t(s)*. These functions are called *signals*; they are mathematical descriptions of the real processes in our circuits, devices, and systems occurring in time. Time changes independently while current, voltage, power, energy, or any other system output measure depend on *t(s)*, as the function definition requires.

Figure 7.1b symbolically shows and exemplifies a *function operation*. It takes a time-domain function, $v(t) = e^{-\alpha t}$, sets its constant, $\alpha = 3\left(\frac{1}{s}\right)$, assumes $t_2 = 0.2(s)$, and computes the output number $v(t_2) = 0.539$. Note we use the other common notation for a function, $v(t)$. You can take, for another example, a sinusoidal function, $v(t) = a\cos(\omega t)$, set constants as $a = 2(V)$ and $\omega = 4\left(\frac{rad}{s}\right)$, assume $t_5 = 0.7(s)$, and compute $v(t_5) = 1.9976$.

Question In this example, $v(t) = e^{-\alpha t}$, what are x and y in notations of Figure 7.1?

A *transform* is also a mathematical operation but this operation produces the *function* in response to an input *function*, as Figure 7.2 demonstrates. For instance, the *Laplace transform* produces s-domain function $F(s) = \frac{1}{(s+\alpha)}$ when the time-domain function $f(t) = e^{-\alpha t}$ is presented to this transform. This example is shown symbolically in Figure 7.2a as $F(s) = L\{f(t)\} = L\{e^{-\alpha t}\} = \frac{1}{(s+\alpha)}$, where $F(s)$ is the s-domain function corresponding to the time-domain function $f(t)$, and $L\{\}$ stands for the *Laplace operator*. Another example is a time-domain sinusoidal function $f(t) = cos(\omega t)$, whose Laplace transform is $F(s) = L\{f(t)\} = L\{cos\omega t\} = \frac{s}{(s^2+\omega^2)}$.

Here, you naturally want to ask, (1) what does mystical s stands for and (2) where do these s-domain functions come from? The answer to (1) is that s is a *complex frequency* defined as $s = \sigma + j\omega$, where σ is a real and $j\omega$ is complex parts of s. As for the other question, you need to wait until the next subsection explains it.

In contrast to function, a transform is a reversible operation; that is, the original function can be obtained from the transformed-by function by the inverse transform. For example, the inverse Laplace transform of s-domain function $\frac{1}{(s+\alpha)}$ produces $e^{-\alpha t}$ in time domain. This is shown in Figure 7.2b.

Note: Unfortunately, the industry uses the same letter s to designate seconds and the complex frequency $s = \sigma + j\omega$, which is the variable in the s-domain. Fortunately, their applications make clear what is what in most cases. You cannot confuse $F(s)$ and $f(t)[s]$. If any ambiguity still arises, we will clarify the notations.

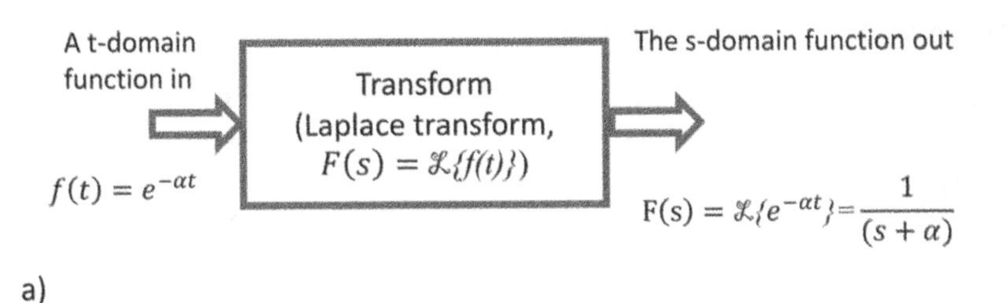

a)

The t-domain function out

Inverse transform (Inverse Laplace transform, $f(t) = \mathcal{L}^{-1}\{F(s)\}$)

An s-domain function in

$f(t) = e^{-\alpha t}$

$F(s) = \mathcal{L}\{e^{-\alpha t}\} = \frac{1}{(s+\alpha)}$

b)

Figure 7.2 The transform operations: a) The (forward) Laplace transform; b) the inverse Laplace transform.

7.2 The Laplace Transform–Concept, Table, and Examples

7.2.1 Table of the Laplace Transform Pairs

Following the conceptual view of a transform, we can present our study's subject in this section, the Laplace transform, as a table that relates time-domain functions to corresponding s-domain functions. See Table 7.1, which is called the *Laplace transform pairs* table because it provides the pairs of the related time-domain and s-domain functions. (See Section 7.3 for all the mathematical manipulations lead to this table results.)

Notes:

- We denote the (forward) Laplace transform operation as $F(s) = L\,\{f(t)\}$, and the inverse Laplace transform operation as $f(t) = L^{-1}\,\{F(s)\}$.

Table 7.1 The Laplace transform pairs.

		f(t) **⇒** ***L{f(t)}*** **⇒** ***F(s)*** **(Forward) Laplace transform**	
	Function	**Time domain function, *f(t)***	**s-domain function, *L{f(t)}* = *F(s)***
Designation		***f(t)*** **⇐** ***L⁻¹{F(s)}*** **⇐** ***F(s)*** **Inverse Laplace transform**	
LT-1	Constant	A	$\frac{A}{s}$
LT-2	Unit step (Heaviside)	1 or $u(t)$	$\frac{1}{s}$
LT-3	Unit ramp $(t > 0)$	t	$\frac{1}{s^2}$
LT-4	Ramp of n-th power $(n\ is\ an\ integer,\ t > 0)$	t^n	$\frac{n!}{s^{n+1}}$
LT-5	Delta	$\delta(t)$	1
LT-6	First derivative of $\delta(t)$	$\frac{d\delta(t)}{dt}$	s
LT-7	Exponential decay $(\alpha\ is\ constant,\ t > 0)$	$e^{-\alpha t}$	$\frac{1}{s+\alpha}$
LT-8	Exponentially decaying (damped) unit ramp $(t > 0)$	$t \cdot e^{-\alpha t}$	$\frac{1}{(s+\alpha)^2}$
LT-9	Exponentially decaying (damped) ramp of n-th power $(t > 0)$	$t^n \cdot e^{-\alpha t}$	$\frac{n!}{(s+\alpha)^{n+1}}$
LT-10	Sine $(\omega\ is\ real\ constant)$	$\sin(\omega t)$	$\frac{\omega}{(s^2+\omega^2)}$
LT-11	Cosine $(\omega\ is\ real\ constant)$	$\cos(\omega t)$	$\frac{s}{(s^2+\omega^2)}$
LT-12	Exponentially decaying (damped) sine $(\alpha\ and\ \omega\ are\ real\ constants,\ t > 0)$	$e^{-\alpha t}\sin(\omega t)$	$\frac{\omega}{((s+\alpha)^2+\omega^2)}$
LT-13	Exponentially decaying (damped) cosine $(\alpha\ and\ \omega\ are\ real\ constants, t > 0)$	$e^{-\alpha t}\cos(\omega t)$	$\frac{(s+\alpha)}{((s+\alpha)^2+\omega^2)}$

- Entries from LT-1 through LT-6 require $Re(s) \equiv \sigma > 0$. Entries LT-7, LT-8, LT-9, LT-12, and LT-13 require $Re(s) \equiv \sigma > \alpha$. These conditions determine the *regions of conversion* of the Laplace transform integral for each entry. The meaning of these conditions becomes apparent in the Example 7.1 discussion.
- Section 7.4 explicitly shows how to obtain each entry in Table 7.1 by applying the Laplace transform definition (7.5).

7.2.2 MATLAB Application to Finding the Laplace and Inverse Laplace Transforms

MATLAB enables us to find the forward and inverse Laplace transform by using its symbolic Toolbox and command *laplace* and *ilaplace*. For example, to find $L\{e^{-\alpha t}\}$, we write the following code:

```
syms s a t; % Refer to the symbolic toolbox and define the variables
f = exp(-a*t); % write down the time-domain function
laplace(f,s) % command MATLAB to perform the Laplace transform of f
```

% MATLAB returns the following answer:

```
ans = 1/(a + s)
```

To find the inverse Laplace transform, we need to apply *ilaplace* command. Consider, for instance, the MATLAB code for finding $L^{-1}\left\{\dfrac{s+3}{(s+3)^2+4^2}\right\}$:

```
syms s t % Refer to the symbolic toolbox and define the variables
c = (s+3)/((s+3)^2 + 4^2); %Introduce a variable c describing the
%given s-domain %function
v = ilaplace(c,t) %command MATLAB to perform the inverse Laplace
%transform of ((s + 3)/((s + 3)^2 + 4^2))
```

% MATLAB returns the following answer:

```
ans = e^(-3t)cos4t as in LT-13
```

If you wish to plot the graph of the obtained time-domain function, add the following line:

```
ezplot(c, [0,3]); %command MATLAB to plot the time-domain function and
%display the formula; choose the proper time span by varying [0,3]
%numbers
```

Know that everything written after the percentage sign % is your comments, which can't "seen" and won't be executed by MATLAB. You don't need to write these comments unless you wish to do so. Without the comments, the MATLAB script looks pretty simple:

```
syms s t
c = ilaplace((s+3)/((s+3)^2 + 4^2));
ezplot(c, [0,3]);
..............................................
ans = e^(-3t)cos4t
```

as in LT-13.

If the given $F(s)$ is a sophisticated formula, you break the inverse Laplace transform command into several lines of code for easy control of the entire script. Then the MATLAB code will look as follows:

```
syms s t
a = s + 3; %write down the numerator
b = (s+3)^2 + 4^2; %write down the denominator
c = ilaplace(a/b,t) %command MATLAB to perform the inverse Laplace
%transform
```

Bear in mind that MATLAB offers a variety of options for plotting your graphs; just check the "Plots" entry in the main menu.

```
ezplot(c,[0,3]); %command MATLAB to plot the time-domain
%function and display %the formula
```

7.2.3 Concept of the Laplace Transform Method

What is the meaning of this table? A constant member in the time domain, A, is represented by $\frac{A}{s}$ in the s-domain; an exponentially decaying function in the time domain, $e^{-\alpha t}$, is $\frac{1}{s+\alpha}$ in the s-domain; a signal $\sin(\omega t)$ in the time domain corresponds to $\frac{\omega}{(s^2+\omega^2)}$ in the s-domain, and so on. For example, applying the Laplace transform, we obtain $\frac{4}{s}$ from number 4, $\frac{1}{s+5}$ from e^{-5t}, and $\frac{6}{(s^2+6^2)}$ from $\sin(6t)$. Inversely, we find $\cos(\omega t)$ from $\frac{s}{(s+\alpha)^2}$ and $t \cdot e^{-\alpha t}$ from $\frac{1}{(s+\alpha)^2}$.

Formally speaking, the Laplace[1] *transform is an integral operation that converts a time-domain function, f(t), into a function, F(s), in the s-domain, where s is a complex frequency.* We will study this mathematical approach soon.

At this point, two questions may come to our mind: (1) What is this mystical s-domain and (2) why do we need all these transformations? Well, the s-domain is the domain where all functions depend on the variable s; this situation is similar to the time domain, where all functions depend on the variable t. As we proceed with this study, we will feel more comfortable with the s-domain; for now, we need to accept this simple explanation. Regarding the second question, the short answer is this: *The Laplace transform provides an easy method for solving differential equations.* How? Using the Laplace transform, we (1) transform *time-domain differential equations* into *s-domain algebraic equations, (2) solve these equivalent algebraic equations and obtain the circuit's output (response) in the s-domain, and (3) applying the inverse Laplace transform, we obtain the circuit's output in the time domain, thus achieving the goal of the entire operation.*

The concept and the example of the Laplace transform application are illustrated in Figure 7.3. It shows that by using a classical method—integration of differential equations—we can directly obtain the solutions. This statement is depicted as the move from Box 1 to Box 2. However, Part 2 of this textbook should convince you how difficult to solve even second-order differential equation by integrating it directly. Figure 7.3 shows this arduous path by shaded banner-shaped box and

---

1 Marquis de Pierre Simon Laplace (1749–1827) was born in Normandy, France. By the end of his life, though a humble origin, he had won every academic honor available, was awarded by the highest civil titles and was considered the greatest scientist of his time. He was elected in Paris Academy at the age of 24. He contributed to many areas of mathematics and astronomy. Still, he is most famous for his work in (1) astronomy, where he used calculus in the application of Newton's theory to celestial problems, and (2) in statistics, where he is considered a father of the probabilities theory. Among his many other achievements is the method of solving the differential equations (which is a generalization of Euler's idea) bearing his name, the method whose its modern interpretation we discuss in this chapter.

curved arrows to stress its complexity. The Laplace-transformed method briefed above is visualized in Figure 7.3 by Boxes 3 and 4 and the straight arrows; we discuss this method below.

In Figure 7.3, $v_{in(t)}$ is a circuit's input signal, $v(t)$ is the output signal, and $V_0 \equiv V(0)$ *and* $V_0' \equiv V_0'(0)$ are the initial conditions.

We start applying the Laplace transform by converting the obtained differential equation into the s-domain. This operation is symbolically shown as $L\{v(t)\} = V(s)$. The rules of this transformation, mandated by the Laplace transform properties, will be considered shortly; the result is shown in Box 3. We can see that the differential (mean, the time-domain) equation, $v''(t) + 5v'(t) + 6v_{out}(t) = 0$, is transformed into the algebraic s-domain equation, $s^2V(s) + 5sV(s) + 6V(s) = 2s + 16$. We solve this equation for V(s), as shown in Box 4. To convert $V(s)$ back into the time-domain formulas, we apply the inverse Laplace transform, $L^{-1}\{V(s)\} = v(t)$, and use Table 7.1. Thus, $L^{-1}\left\{\frac{12}{(s+2)}\right\} = 12e^{-2t}$ and $L^{-1}\left\{-\frac{10}{(s+3)}\right\} = -10e^{-3t}$. Their sum gives us the required solution of the given differential equation in the time domain, $v(t) = 12e^{-2t} - 10e^{-3t}$, as shown in Box 2. Now, we can see that obtaining the system's response by employing the Laplace transform is much easier than resorting to the integration of the system's differential equation.

(Note that Figure 7.3 shows $v_{in}(t) = 0$. In fact, $v_{in}(t)$ could be any; we've chosen the simplest case to not overshadowing the main concept of the Laplace transform by extraneous mathematical manipulations.)

Unhappily, no benefit comes for free: The price for gaining this advantage is the need for additional operations: direct and inverse Laplace transforms. Nevertheless, overall, the use of the Laplace transform is well worth undertaking these additional operations. In short, remember this: *The Laplace transform method is a powerful tool for obtaining a circuit's response—a task we carried out throughout this textbook.*

7.2.4 Examples of the Laplace Transform Operations

Let's consider the examples of Laplace transform operations. Applying the Laplace transform to time-domain functions to obtain their s-domain equivalents is a straightforward operation: We just need to use the rules in Tables 7.1. To demonstrate this operation, several examples are assembled

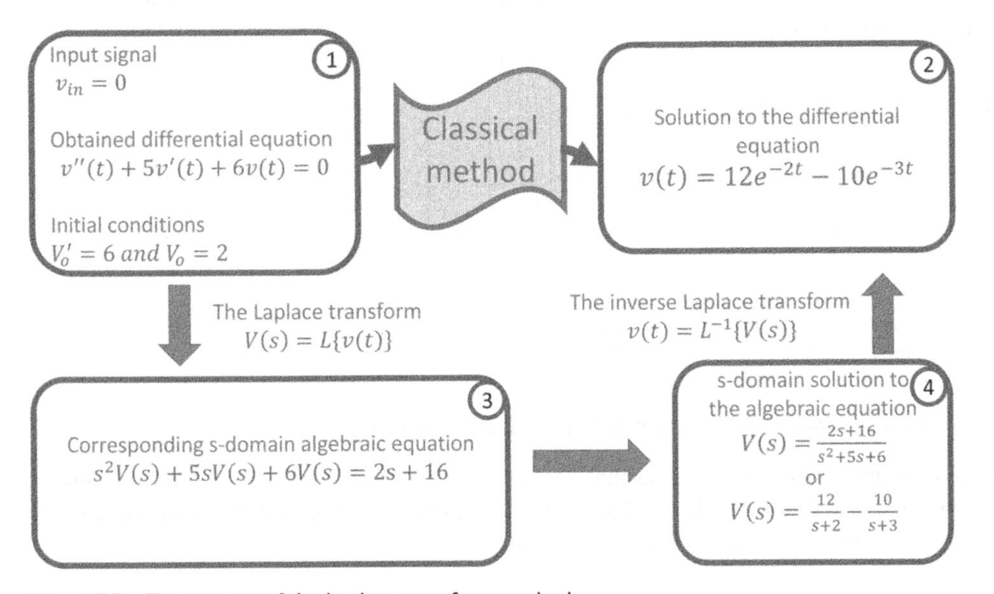

Figure 7.3 The concept of the Laplace transform method.

in Table 7.2. In this table, at the Laplace transform side, we first show the general rule of the Laplace transform to be used and then demonstrate this rule's application to a specific example. For instance, consider Example 7.1, $f(t) = 3$. The general Rule LT-1 in Table 7.1 says $L\{A\}=\frac{A}{s}$, applying which, we find $L\{3\}=\frac{3}{s}$. Consider Example 7.2, $f(t) = 3t$. To perform the Laplace transform of this function, we first need to factor out the coefficient, $B = 3$, to obtain $L\{3t\}=3L\{t\}$. We can do this thanks to the Laplace transform properties discussed in Section 7.3. Secondly, we apply Rule LT-5 and find $L\{t\}=\frac{1}{s^2}$. Thus, $L\{3t\}=\frac{3}{s^2}$.

Pay special attention to the difference between constant member A as a stand-alone function ($A = 3$ in Example 7.1) and constant B as a time function's coefficient ($B = 3$ in Example 7.2). Now, we apply the above procedure to the time-domain functions and build Table 7.2.

These examples cover the essential time-domain functions presented in Table 7.1. We urge you to analyze these examples, create your examples to gain confidence in transforming time-domain functions into their s-domain equivalents, and—of course—do the homework problems given at the end of this chapter.

Table 7.2 Examples of the Laplace transforms of time-domain functions.

Example number (Rule No. in Table 7.1)	Time-domain function $f(t)$	Laplace transform of $f(t)$, $F(s) = L\{f(t)\}$	
		Rules to be applied	**Results of the Laplace transforms**
1 (LT-1)	$f(t) = 3$	$L\{A\}=\frac{A}{s}$	$L\{A\}=\frac{3}{s}$
2 (LT-3)	$f(t) = 3t$	$L\{Bt\}=LB\{t\}$ and $BL\{t\}=\frac{B}{s^2}$	$L\{3t\}=3L\{t\}$ and $3L\{t\}=\frac{3}{s^2}$
3 (LT-4)	$f(t)=5t^2$	$L\{Bt^n\}=B\frac{n!}{s^{n+1}}$	$L\{5t^2\}=\frac{10}{s^3}$
4 (LT-7)	$f(t)=4e^{-6t}$	$L\{Be^{-\alpha t}\}=B\frac{1}{(s+\alpha)}$	$L\{4e^{-6t}\}=\frac{4}{(s+6)}$
5 (LT-9)	$f(t)=2e^{-5t}t^3$	$L\{Be^{-\alpha t}t^n\}=B\frac{n!}{(s+\alpha)^{n+1}}$	$L\{2e^{-5t}t^3\}=2\frac{6}{(s+5)^4}$
6 (LT-10)	$f(t)=4\ sin8t$	$L\{Bf(t)\}=BL\{f(t)\}$ and $BL\{sin\omega t\}=B\frac{\omega}{(s^2+\omega^2)}$	$L\{4sin8t\}=4L\{sin8t\}$ and $4L\{sin8t\}=\frac{32}{(s^2+8^2)}$
7 (LT-11)	$f(t)=5\ cos6t$	$L\{Bcos\omega t\}=B\frac{s}{(s^2+\omega^2)}$	$L\{5cos\omega t\}=\frac{30}{(s^2+6^2)}$
8 (LT-12)	$f(t)=7e^{-6t}sin4t$	$L\{Be^{-\alpha t}sin\omega t\}=B\frac{\omega}{((s+\alpha)^2+\omega^2)}$	$L\{7e^{-6t}sin4t\}=\frac{28}{((s+6)^2+4^2)}$
9 (LT-13)	$f(t)=4e^{-2t}cos6t$	$L\{Be^{-\alpha t}cos\omega t\}=B\frac{(s+\alpha)}{((s+\alpha)^2+\omega^2)}$	$L\{4^{-2t}cos6t\}=4\frac{(s+2)}{((s+2)^2+6^2)}$

Notes:

- Recall a factorial rule: $n!= n\cdot(n-1)\cdot(n-2)\cdot\ldots\cdot1$. E.g. $5! = 5\cdot4\cdot3\cdot2\cdot1 = 120$.
- Pay attention to the designations used in Table 7.2.
- Perform all these operations using MATLAB observing Notes to Table 7.1.

7.2.5 The Mathematical Foundation of the Laplace Transform

The preceding discussion demonstrates that all Laplace transform operations rely on the Laplace transform pairs displayed in Table 7.1. But where do these pairs come from? The answer comes from the explanation of the mathematics behind this table.

Formally, the Laplace transform is defined as follows:

If f(t) is a function in the time domain, then its ***Laplace transform****, F(s), is given by the following two-sided integral:*

$$F(s) = L\{f(t)\} = \int_{-\infty}^{\infty} e^{-st} f(t) dt \tag{7.2}$$

where $F(s)$ is the result of the Laplace transform, $L\{\cdot\}$ stands for the Laplace transform operator, and e^{-st} is the kernel of transformation. Also, s is the complex frequency commonly written in the form

$$s = \sigma + j\omega. \tag{7.3}$$

Equation 7.3 relates s to reality through a growing or damping sinusoid in the time domain as

$$s = \sigma + j\omega \Rightarrow f(t) = e^{\sigma t}\cos(\omega t), \tag{7.4}$$

where the real part, $\sigma\left(\frac{1}{s}\right)$, is an exponent (power) of e, which determines the rate of an exponential growth or decay in time, and $\omega\left(\frac{rad}{s}\right)$ is an angular frequency of a sinusoid.

For physically realizable time-domain systems, whose responses are equal to zero before $t = 0$, we introduce the *one-sided* Laplace transform as

$$F(s) = \int_{0}^{\infty} e^{-st} f(t) dt. \tag{7.5}$$

Throughout this book, we will use the one-sided Laplace transform. It's important mentioning that in this case, *f(t) is called a causal function*, and it is formally defined as

$$causal\ f(t) = f(t) \cdot u(t), \tag{7.6}$$

where *u(t)* is a *unit-step (Heaviside) function* given by

$$u(t) = \begin{cases} 1\ for\ t > 0 \\ 0\ for\ t < 0 \end{cases}. \tag{7.7}$$

(Refer to (6.15) and Figure 6.4a.) In this text, we always refer to a *causal* time-domain function, *f(t)*, and normally omit *u(t)* for simplicity's sake. *Many entries in* Table 7.1 *contain the condition t > 0; now, we understand that these entries refer to causal functions.*

The Laplace transform exists if and only if Integrals (7.2) *and* (7.5) *converge*. These conditions put some restrictions on the functions *f(t)*, though it's not a concern for the vast majority of functions (signals) encountered in engineering practice. All complex frequencies s for which this integral converges form *the region of convergence* in the s-domain. They are introduced in the Notes to Table 7.1; see more on this point in Example 7.1 below. We omit these notations in Table 7.1 to avoid overloading it with subsidiary information.

Example 7.1 The Laplace transform of a real constant.

Problem: Find the Laplace transform of *f(t)* = *A*, where *A* is a real constant (number).

Solution: Applying the definition of the Laplace transform (7.5) produces

$$F(s) = L\{ft\} = \int_0^{\infty} e^{-st} A dt = A \int_0^{\infty} e^{-st} dt. \tag{7.8}$$

This is an improper integral because its upper limit is infinity. This integral converges if the integral $\int_0^T e^{-st} dt$ and the limit of $T \to \infty$ exist and $\lim_{T \to \infty}$ is finite. Thus,

$$\int_0^{\infty} e^{-st} dt = \lim_{T \to \infty} \int_0^T e^{-st} dt. \tag{7.9}$$

Therefore, the problem boils down to (1) integrating RHS of Equation 7.9 and (2) finding the limit of the result for $T \to \infty$. The integration yields

$$\lim_{T \to \infty} A \int_0^T e^{-st} dt = \lim_{T \to \infty} \left(-\frac{A}{s} e^{-st} \right) \Big|_0^T = \frac{A}{s} \left(1 - \lim_{T \to \infty} e^{-sT} \right) \tag{7.10}$$

because $e^0 = 1$. Recalling that $s = \sigma + j\omega$, where σ and ω are real, and using *Euler's identity*, $e^{j\omega T} = cos\omega T + jsin\omega T$, we arrive at

$$\lim_{T \to \infty} e^{-sT} = \lim_{T \to \infty} e^{-(\sigma + j\omega)T} = \lim_{T \to \infty} (e^{-\sigma T} e^{-j\omega T}) = \lim_{T \to \infty} e^{-\sigma t} (cos\omega T - jsin\omega T). \tag{7.11}$$

Since sine and cosine are finite, *this limit exists and is equal to* zero (finite, that is) if the real part, $Re(s) \equiv \sigma$, of the complex frequency, $s = \sigma + j\omega$, is a positive number; that is, $Re(s) \equiv \sigma > 0$. This condition makes $e^{-\sigma T}$ tend to zero when T goes to infinity. Thus, from (7.10) we get

$$F(s) = L\{A\} = \frac{A}{s} \tag{7.12}$$

as (LT-1) in Table 7.1 shows.

The problem is solved.

Discussion:

- Again, the result presented in (7.12) holds true (mean, Integral 7.5 converges) only under condition

$$Re(s) \equiv \sigma > 0, \tag{7.13}$$

which determines the conversion region for this integral. Now we understand where the convergence condition, $Re(s) \equiv \sigma > 0$, comes from. Indeed, if $\sigma < 0$ in our example, then $e^{-\sigma t}$ goes to infinity, and Integral 7.5 doesn't converge.

Putting this consideration on a mathematical footing, we need to recall that Integrals 7.2 and 7.5 must converge to provide the Laplace transform existence. All complex frequencies s for which this integral converges form the *region of convergence*, $\boldsymbol{C}$, in the s-plane. In Example 7.1, the region of convergence is all the positive values of σ, as proved above. Hence, (7.12) holds true for all s values that belong to $\boldsymbol{C}$ (denoted as $s \in \boldsymbol{C}$), where $\boldsymbol{C}$ is $Re(s) \equiv \sigma > 0$. This is why we use notation $\sigma > 0$, *not* $\sigma \geq 0$. (Does $\boldsymbol{C}$ include $s = 0$? If not, why?)

- This example shows how to obtain the Laplace transform pair of a time-domain function. Fortunately, we don't need to perform such operations every time; the Laplace transforms of the essential functions that we use in the analysis of practical circuits are readily available in the form of a table similar to Table 7.1. Such a table can be found in all the textbooks on electrical engineering. Section 7.3, which follows, reveals all mathematical manipulations for finding the Laplace transform pairs shown in Table 7.1. Also, MATLAB makes finding the Laplace transforms an easy task.

Question Can you find the Laplace transform of a unit step function, $u(t)$, by following the pattern of this example? Prove your answer by explicitly performing all the mathematical manipulations.

Now, we are ready to discover how all the Laplace transform pairs shown in Table 7.1 were obtained.

7.3 Proofs of the Laplace Transform Pairs–Step and Ramp Functions

Since finding the Laplace transform of LT-1 is covered in Example 7.1, we start with LT-2, unit step (Heaviside) function.

LT-2: The Laplace transform of a unit step (Heaviside) function.
Refer to Section 6.1 to recall the definition and the main properties of a step (Heaviside) function. See Equations 6.15 and 6.16 and Figure 6.4.

You are reminded that the *unit* step function is defined as

$$u(t) = \begin{cases} 0 \ for\ t < 0 \\ 1 \ for\ t > 0 \end{cases}. \tag{6.15R}$$

How can we obtain LT-2 when using (6.15)? Apply the definition of the Laplace transform, as we did in Example 7.1, and find:

$$\begin{aligned} F(s) = L\{u(t)\} = \int_0^{\infty} e^{-st} u(t)dt = \int_0^{\infty} e^{-st} dt = \int_0^{\infty} e^{-st} dt \\ = -\frac{1}{s}\left[e^{-st}\right]_0^{\infty} = -\frac{1}{s}\left[e^{-s\infty} - e^{-s0}\right] = \frac{1}{s}. \end{aligned} \tag{7.14}$$

Thus, $L\{u(t)\} = \frac{1}{s}$, as in LT-2 of Table 7.1. Note that $e^{-s0} = 1$ regardless of s, but $e^{-s\infty} = e^{-\sigma\infty} \cdot e^{-j\omega\infty}$ goes to zero only if $\sigma \equiv Re(s) > 0$. This is the condition for the Laplace transform integral conversion, which we met in Example 7.1. Recall that $(e^{-j\omega t} = cos\omega t - jsin\omega t) \leq 1$ by the cosine and sine definitions.

Question

1) What will be the answer to this problem if a step function is delayed by 4(s) interval? If a step function is multiplied by *f(t)* = 3?
2) What problem with finding *L{u(t)}* would you encounter if $\sigma \equiv Re(s) < 0$?

LT-3 The Laplace transform of a unit ramp function
A dictionary defines the word "ramp" as a sloping way or plane leading from one level to another. In electrical engineering, a ramp signal describes voltage or current linearly changes from one value to the other in contrast to a step signal describing an abrupt value change.

Consider a linear function $r(t)$. Let t_0 and t_f be the initial and final times, and A_0 and A_f be the initial and final values of $r(t)$. Then, the ramp function is defined as

$$r(t) = \begin{cases} A_0 \ for \ t < t_0 \\ mt \quad for \ t_0 \le t \le t_f . \\ A_f \ for \ t > t_f \end{cases} \tag{7.15}$$

Here the ramp slope is given by $m = \dfrac{A_f - A_0}{t_f - t_0}$. The slope can be positive, giving an upward function trend, or negative, leading to a downward trend, as shown in Figures 7.4a and 7.4b. If we set $A_0 = 0$, $A_f = 1$, $t_f = 1$, and $t_0 = 0$, then $m = 1$, which gives a special function known as a *unit ramp*. This signal is shown in Figure 7.4c and used in Table 7.1.

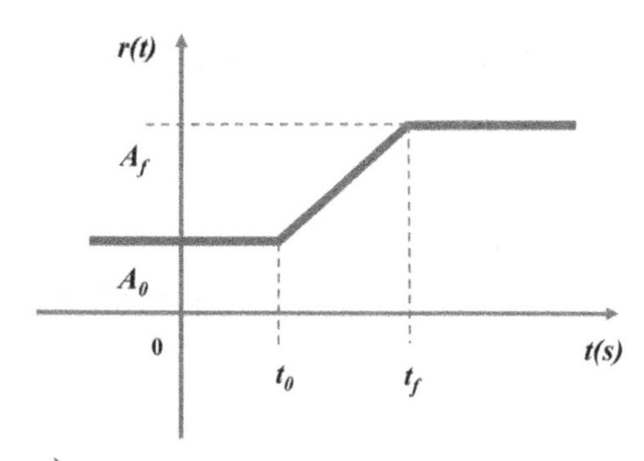

a)

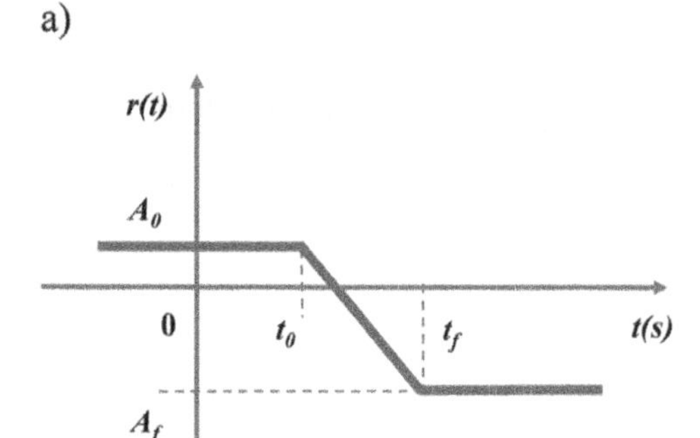

b)

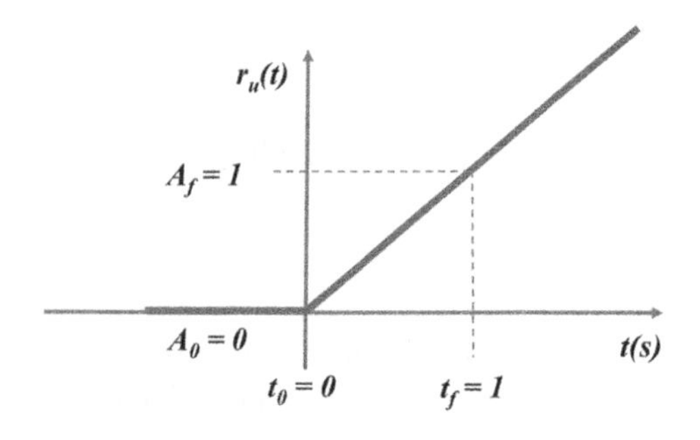

c)

Figure 7.4 Ramp functions: a) With a positive slope and time shift, b) with a negative slope and time shift, and (c) a unit ramp.

We have to realize that (7.15) determines a *piecewise function*, consisting of three parts: two constant members, A_0 and A_f, and the linear member, mt. In fact, the *ramp*—a connection of two levels—is provided by only the linear part of (7.17). This is why more often than not, a ramp function is defined as

$$r_u(t) = \begin{cases} t \ for \ t > 0 \\ 0 \ for \ t < 0 \end{cases}. \quad (7.16a)$$

In other words, the term *ramp* typically refers to a *unit ramp function*, $r_u(t)$. This function exists only from $t > 0$ and has unity slope. Therefore, the other way to define this function is

$$r_u(t) = t \cdot u(t). \quad (7.16b)$$

As well as a step function, the ramp can be scaled in amplitude and in time shift, as shown in Figures 7.4a and 7.4b.

To find the Laplace transform of the unit ramp, we again use the definition:

$$F(s) = L\{r_u(t)\} = \int_0^\infty t \cdot e^{-st} dt \quad (7.17)$$

The rule for the integration of the two-functions product is given by

$$\int_0^\infty u dv = [u \cdot v]_0^\infty - \int_0^\infty v du. \quad (7.18a)$$

Referring to the differential definition, $dv(t) = v' dt$, we can rewrite (7.18a) as

$$\int_a^b u \cdot v' dt = [u \cdot v]_a^b - \int_a^b v \cdot u' dt. \quad (7.18b)$$

Let's introduce $u = t$, $v'(t) = e^{-st}$, $a = 0$, *and* $b = \infty$. Then, $dt = du$ and $v = -\frac{1}{s} e^{-st}$. (You can verify the latest notation by taking the derivative $v'(t) = \frac{dv}{dt} = \frac{d\left(-\frac{1}{s}e^{-st}\right)}{dt} = e^{-st}$.) Applying (7.18b) to the ramp function, we find

$$\int_0^\infty t \cdot e^{-st} dt = \left[t \cdot (-\frac{1}{s} e^{-st} \right]_0^\infty - \int_0^\infty -\frac{1}{s} e^{-st} dt \quad (7.19)$$

as $t' = \frac{dt}{dt} = 1$. The first member of the RHS of (7.19) is zero because

$$\left[t \cdot (-\frac{1}{s} e^{-st} \right]_0^\infty = -\frac{1}{s} \lim_{t \to \infty} \frac{t}{e^{st}} + \frac{1}{s} \lim_{t \to 0} \frac{t}{e^{st}} = 0 - 0. \quad (7.20)$$

Equation 7.20 holds provided that $Re(s) > 0$ and $s \neq 0$. (Can you see why these two conditions must be imposed?)

The second member of the RHS of (7.19) turns to

$$-\int_0^\infty -\frac{1}{s}e^{-st}dt = \frac{1}{s}\int_0^\infty e^{-st}dt = \frac{1}{s}\int_0^\infty e^{-st}dt.$$

We showed above that $\frac{d\left(-\frac{1}{s}e^{-st}\right)}{dt} = e^{-st}$, which means that $e^{-st}dt = d\left(-\frac{1}{s}e^{-st}\right)$. Thus, the second term becomes

$$\frac{1}{s}\int_0^\infty d(e^{-st}) = \frac{1}{s}\left[e^{-st}\right]_0^\infty = \frac{1}{s}\lim_{t\to\infty} e^{-st} - \frac{1}{s}\lim_{t\to 0} e^{-st} = \frac{1}{s}\cdot\frac{1}{s} = \frac{1}{s^2}. \tag{7.21}$$

Therefore,

$$F(s) = L\{t\} = \int_0^\infty t\cdot e^{-st}dt = \frac{1}{s^2}, \tag{7.22}$$

as in Table 7.1.

Question

1) What is the Laplace transform of r(t) = $5t$?
2) What is the Laplace transform of r(t) = t^0?

LT-4 Ramp function of the *n*-th power

Realize that in LT-3, the case $L\{t^1\}$ was discussed. For the LT-4, we need to generalize that derivation. Consider the Laplace transform of a given ramp function of the n-th power

$$F(s) = L\{t^n\} = \int_0^\infty t^n\cdot e^{-st}dt. \tag{7.23}$$

Integration by parts, similar to that in the preceding case, yields

$$\int_0^\infty t^n\cdot e^{-st}dt = \left[t^n\cdot(-\frac{1}{s}e^{-st}\right]_0^\infty + \frac{n}{s}\int_0^\infty t^{n-1}e^{-st}dt. \tag{7.24}$$

Compare this formula with (7.19) and see that the first member of (7.24) is zero, as proved in (7.22). Then use this derivation recursively and obtain

$$F(s) = L\{t^n\} = \int_0^\infty t^n\cdot e^{-st}dt = \frac{n!}{s^{n+1}}. \tag{7.25}$$

Sidebar 7S.1 Step and Ramp Functions and Their Derivatives and Integrals

Relationship between step and ramp functions

To proceed with the justification of Table 7.1, we need to consider the relationship between a step function and a ramp function.

A ramp function changes, and we need to know the rate of its changes. To remind, we have differentiation–a comprehensive mathematical tool enabling us to evaluate this rate. If *y* is a continuous function of *t*, *y(t)*, then its derivative describes the instantaneous rate of change of the function *y(t)* at a specific instant *t*, which is mathematically shown as

$$y`(t) \equiv \frac{dy}{dt} = \lim_{\Delta t \to 0}\left(\frac{y(t+\Delta t) - y(t)}{\Delta t}\right) = \lim_{\Delta t \to 0}\frac{\Delta y}{\Delta t}. \tag{7S1.1}$$

Therefore, to evaluate the rate of change of a ramp function, *r(t)*, we need to take its derivative, *r'(t)*; i.e.,

$$r`(t) = \frac{dr(t)}{dt}. \tag{7S1.2}$$

What kind of function *r'(t)* will be? Let's investigate.

Consider the three-piece ramp function shown at the top of Figure 7S1.1a: Using the definition of a ramp function (7.15), we can describe this function as

$$r(t) = \begin{cases} 0 \ for \ t < 0 \\ mt \quad for \ 0 \le \tau \le t \ and \ m = \dfrac{A}{\Delta t}. \\ A \ for \ \tau > t \end{cases} \tag{7S1.3}$$

Taking the derivative of (7S1.3) by pieces, we find

$$r`(t) = \begin{cases} \dfrac{d(0)}{dt} \\ \dfrac{d(mt)}{dt} \\ \dfrac{d(A)}{dt} \end{cases} = \begin{cases} 0 \ for \ t < 0 \\ m \quad for \ 0 < \tau < t. \\ 0 \ for \ \tau > t \end{cases} \tag{7S1.4}$$

The derivative is zero for the region t < τ because the ramp function doesn't change in this area. Within the interval $0 < \tau < t$, the derivative $r`(t) = m = \frac{A}{\Delta t}$. The graph of the whole *r'(t)*, (7S1.4), fits a profile of a rectangular pulse shown at the bottom of Figure 7S.1.1a. This pulse is zero everywhere, except for the interval $0 < \tau < t$, where its amplitude is $m = \frac{A}{\Delta t}$. Also, the area under the pulse is A, as Figure 7S1.1a demonstrates

$$Area = m \cdot \Delta t = \frac{A}{\Delta t} \cdot \Delta t = A. \tag{7S1.5}$$

Therefore, the *derivative of a three-piece ramp function produces a rectangular pulse.* (In Section 6.1, we introduced a pulse function. See Equations 6.1 and 6.2 and Figure 6.1. Over there, a pulse is denoted as P(t). This new notation stems from the new approach to defining a rectangular pulse. We hope these different designations, $\boldsymbol{P(t) \equiv r'(t)}$, won't confuse you.)

Wait a minute, you might rightly ask, what about the two-piece unit ramp function shown at the top of Figure 7S.1.1b? Writing the formal definition of a unit ramp given in (7.15) and taking its derivative, we find

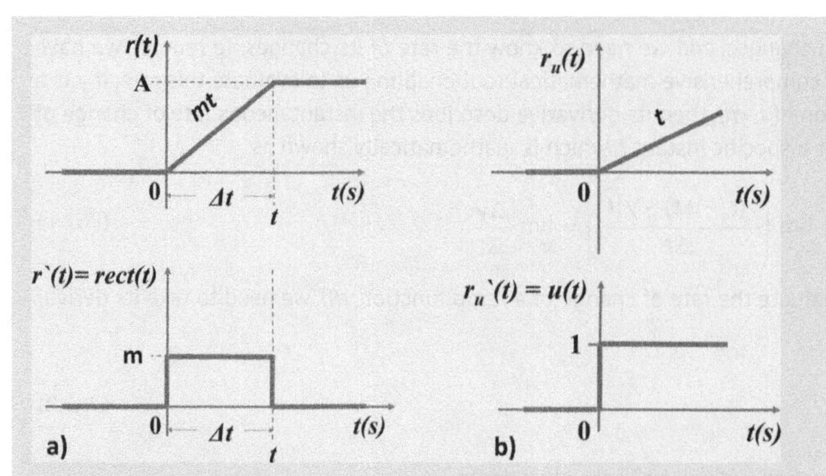

Figure 7S1.1 (a) Derivative of a three-piece ramp function produces a rectangular pulse; (b) derivative of a unit ramp function produces a unit step function.

$$r_u^{\prime}(t) = \begin{cases} \dfrac{d(0)}{dt} \\ \dfrac{d(t)}{dt} \end{cases} = \begin{cases} 0 \ \ for \ t < 0 \\ 1 \ \ for \ t > 0 \end{cases}. \tag{7S1.6}$$

The RHS of (7S1.6) fits the definition of a unit step function given in (7.13) as $u(t) = \begin{cases} 0 \ \ for \ t < 0 \\ 1 \ \ for \ t > 0 \end{cases}$. Thus, the *derivative of a unit ramp function produces a unit step function,*

$$r_u^{\prime}(t) = u(t). \tag{7S1.7}$$

Can you guess what function will be obtained by integrating a unit step function? Integration of both sides of (7S1.7) answers:

$$\int_0^t \frac{dr_u(t)}{dt} dt = \int_0^t u(t)dt. \tag{7S1.8}$$

That is,

$$r_u(t) = \int_0^t u(t)dt. \tag{7S1.9}$$

Given (7S1.7), the result obtained in (7S1.9) is evident because differentiation and integration are the inverse mathematical operations. See Figures 7S1.1b and 7S1.2.

It's worth reminding that an integral over a function *f(t)* provides the area under this function. Hence, At is the area under a step function whose amplitude is *A* and duration is *t*.

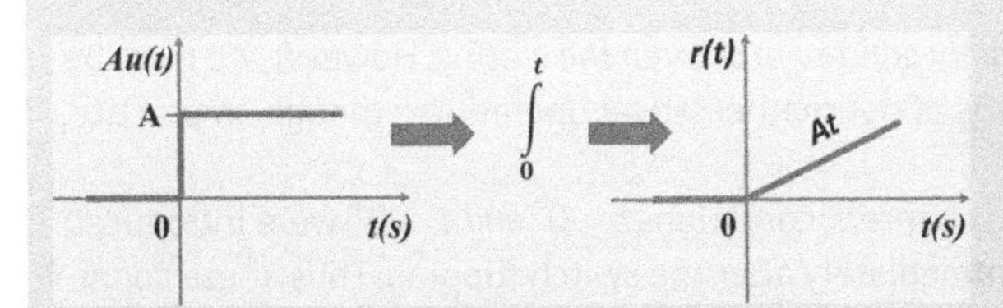

Figure 7S1.2 Integration of a step function gives a ramp function.

Question

1) Can you provide the area's formula under a step function over the interval [0, t] without integration?
2) What is the "area" under a ramp function if $m = 5$ (V/s) and $t = 5$ (s)? What is the dimension of this "area" and what is the meaning of this "area"?

We will encounter these *ramp* and *step* functions very often in advanced circuit analysis, where they serve as input signals; this is why it is essential to know the relationship between them.

Differentiation and integration of step and ramp functions–zero values and their infinitesimally small limits

We should realize that there is a more rigorous view at our mathematical manipulations. Specifically, the definition of a derivative given in (7S.1.1) is valid for a *continuous function*, whereas both step and ramp functions exhibit *discontinuities*. As a result, a ramp signal is not differentiable at $t = 0$, and (7S1.1) must be given as

$$\dot{r_u}(t) = u(t) \text{ for } t \neq 0. \tag{7S1.10}$$

Also, the integration in (7S1.9), $r_u(t) = \int_0^t u(t)dt$, is not a valid operation if one of its limits is zero because the step unit function is not defined at $t = 0$.

The mathematics found the solution to these problems in introducing *infinitesimally small limits*, 0^- being diminutively less than zero and 0^+ being diminutively greater than zero, as we mentioned previously. Then, (7S1.10) is to be written as

$$\dot{r_u}(t) = u(t) \text{ for } t > 0^+, \tag{7S1.11}$$

and (7S1.9) has to be presented as

$$\int_{0^+}^{t} u(t)dt = r_u(t). \tag{7S1.12}$$

Thus, (7S1.11) and (7S1.12) show that we avoid the problem with discontinuities by starting differentiation and integration from infinitesimally small points greater than zero, the points where the ramp and step functions are defined. Table 7.1 should indicate that similar considerations hold for 0^+ in the proper mathematical manipulations. (Can you identify these manipulations?)

In this textbook, these notations are mainly omitted to shorten the writing. However, we must be aware of these conditions and limitations of our mathematical operations; though, in practice, we often neglect to mention them.

In the circuit analysis in the preceding chapters, conditions $t < 0^-$ *and* $t > 0^+$ were introduced to denote the instants just before and immediately after the switch flipping. Thus, these conditions are not merely mathematical abstractions but the realities used in engineering practice.

Question Can you symbolically show these starting points in Figures 7S1.1b and 7S1.2?

7.4 Proofs of the Laplace Transform Pairs—The Rest of Table 7.1

LT-5 Impulse or Dirac delta function, δ(t).

You will recall that the impulse function is introduced in Section 6.1. See Equations 6.3a, 6.3b, and 6.4 and Figure 6.1. Here, we consider this function from different perspectives.

We saw that a step function, a ramp function, and a rectangular pulse are all interconnected. Here we'd like to show how these functions' relationship leads to the *impulse* or *Dirac delta function*.

Consider a *ramp function* again: What if we decrease its linear interval, $(0, t_i)$, as keeping its amplitude constant and equal to 1? Figure 7.5 depicts the whole picture as a sequence of steps.

We can see in Figure 7.5 that by decreasing the ramp interval, $\Delta t_i \equiv (0, t_i)$, we increase the ramp slope's values; that is, $m_3 > m_2 > m_1$ for $t_3 < t_2 < t_1$ because $m_i = \frac{1}{\Delta t_i}$. Eventually, when the interval tends to zero, $\Delta t \to 0$, the slope of the ramp function approaches infinity, $m \to \infty$; that is,

$$m = \lim_{\Delta t \to 0} \frac{1}{\Delta t} \to \infty. \tag{7.26}$$

This is where a *ramp function* turns into a *step function*.

What happens with the *rectangular pulses* obtained by differentiating each of these changing ramp functions? Discussion of Figure 7S.1.1a shows that each of these rectangular pulses has amplitude $m_i = \frac{1}{\Delta t_i}$, duration Δt_i, and area $A = m_i \cdot \Delta t_i = 1$. As the ramp interval, Δt_i, becomes smaller, the amplitude of a rectangular pulse, m_i, becomes greater to keep the area under the pulse constant and equal to 1. Eventually, *when pulse width,* Δt*, approaches zero, its amplitude tends to infinity. The area under the resulting impulse, however, remains equal to 1.* Such an impulse whose width goes to zero and amplitude tends to infinity is called *impulse* or *Dirac delta function*. This consideration is visualized in Figure 7.6.

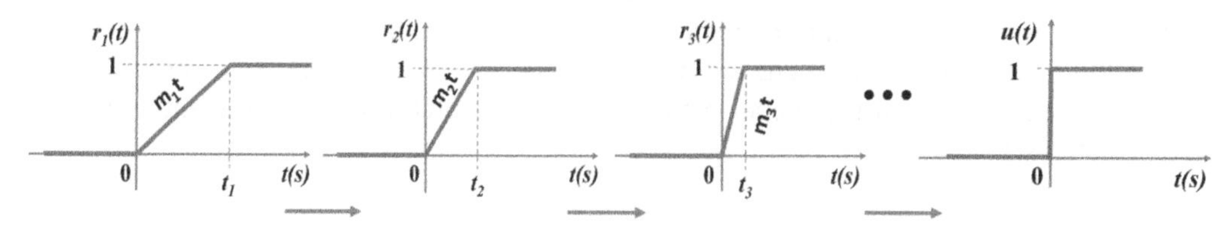

Figure 7.5 From ramp to a step function.

Each pair of drawings in Figure 7.6 related by a vertical line constitutes a function and its derivative; for example, beneath the ramp m_3t, its derivative—the rectangular pulse with amplitude m_3 —is shown. Thus, we can conclude that the rightmost pair in Figure 7.6 states that *a step function's derivative is an impulse function*. Indeed, at $t = 0$, the step function instantaneously changes from one value to the other. The ramp change rate tends to infinity as the width of the rectangular pulse approaches zero, which produces an impulse function.

It must be reminded that an *impulse function (Dirac delta function)* denoted as δ(t), was introduced in subsection "Impulse function" of Section 6.1. Its definition,

$$\delta(t) = \begin{cases} 0 \ for \ t \neq 0 \\ \rightarrow \infty \ for \ t = 0' \end{cases} \tag{6.3aR}$$

and main property

$$\int_{-\infty}^{\infty} \delta(t)dt = 1 \tag{6.3bR}$$

were discussed in that section too. Unfortunately, we cannot avoid some repetition to keep the current narrative's logic continuous. Here, we add new material stems from Figure 7.6:

- Definition 6.3a is supported by the graph of an impulse function (bottom rightmost figure) in Figure 7.6, which shows that the delta function is equal to zero everywhere except for the point of its origin. Here, this point is $t = 0$, where δ(t) goes to infinity.
- The bottom row of drawings in Figure 7.6 shows that the impulse function is the limiting case of a rectangular pulse when the pulse's width ΔT tends to zero. The area under the pulse, and therefore under the impulse function, remains equal to 1; thus, (6.3b).
- The magnitude (strength) of an impulse function can be scaled to any value A as $A\delta(t)$, which results in

$$\int_{-\infty}^{\infty} A\delta(t)dt = A. \tag{6.4R}$$

This case is shown in Figure 7.7.

- After examining the rightmost vertical pair in Figure 7.6, we conclude that the delta function is

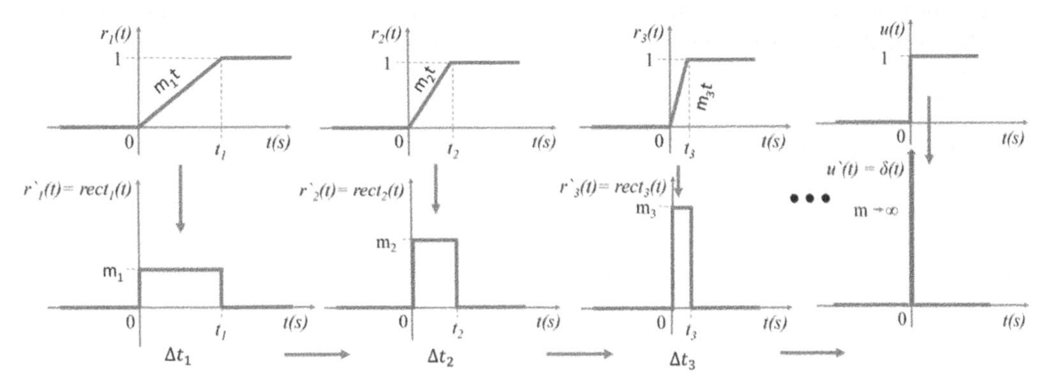

Figure 7.6 As the ramp interval tends to zero, the ramp function becomes a step function (top row), and the rectangular pulse becomes an impulse (bottom row).

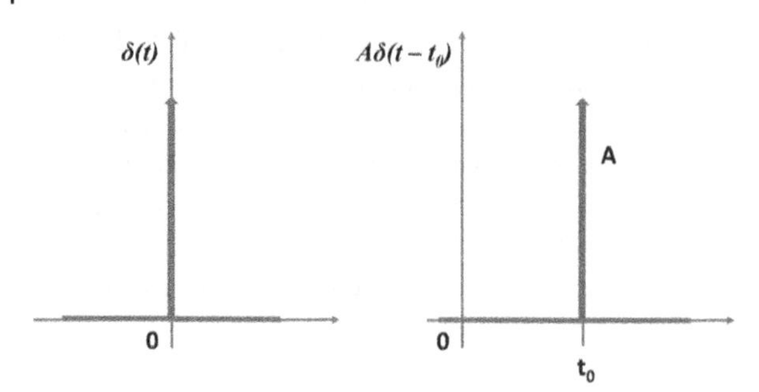

Figure 7.7 Impulse or Dirac delta function at $t = 0$ (left) and at $t = t_0$ (right, with amplitude scaled to A).

the derivative of a unit step function; i.e.,

$$\delta(t) = u`(t), \tag{7.27}$$

as in (6.17).

- What is an indefinite integral of a delta function? The answer follows from (7.27) as

$$\int \delta(t)dt = u(t). \tag{7.28}$$

(Constant C must be shown in (7.28), but we can always assume C = 0.)

- A delta function can exist at any point in time (horizontal) axis, as shown in Figure 7.7. In this situation, (6.3a) takes the form

$$\delta(t - t_0) = \begin{cases} 0 \; for \; t \neq t_0 \\ \to \infty \; for \; t = t_0. \end{cases} \tag{7.29}$$

Therefore,

$$\int_{-\infty}^{\infty} \delta(t - t_0)dt = 1 \tag{7.30}$$

because the point where the delta function exists, t_0, will be certainly within the integration interval. However, if the integration limits are finite, the delta function integral is evaluated as follows:

$$\int_a^b \delta(t - t_0)dt = \begin{cases} 1 \; for \; a < t_0 < b \\ 0 \; otherwise \end{cases}. \tag{7.31}$$

- The delta-function *sifting property* is of practical significance because it enables us to compute the value of any function *f(t)* at any point:

$$\int_{-\infty}^{\infty} f(t)\delta(t - t_0)dt = f(t_0). \tag{7.32}$$

Consider, for example, $f(t) = t^2$. At $t_0 = 3$, $f(t)$ assumes the value 9 because

$$\int_{-\infty}^{\infty} t^2\delta(t - 3)dt = 3^2 = 9.$$

The rigorous proofs of all the formulas presented here exist, of course, but we leave them to specialized textbooks in mathematics.

Questions

1) Can you prove (7.29)? Show your proof explicitly.
2) What form will Equation 7.32 take if $t_0 = 0$?
3) What is the value of integral $\int_{-\infty}^{\infty} f(t)\delta(t)dt = f(t)$ be if $f(t) = t^2$?

Dirac delta function is not a traditional function shown in Figure 7.1; this is why it is often called a *generalized function* or *distribution*.

The impulse function describes a spike that happens at an instant. Think of the impulse function as infinitely large in magnitude and infinitely small in time duration and have an area of one, as symbolized by an arrow in Figures 7.7a and 7.7b. It is practically impossible to generate a perfect impulse in the real world; therefore, a very brief pulse is used to approximate an impulse. Some physical phenomena come very close to being modeled with impulse functions. *An example of an impulse is lightning*. An impulse function can model lightning because the latter is a sudden burst of energy and lasts for a very short time. In electric circuits, spikes in voltage, current, or energy caused by various external or internal factors are examples of phenomena described by the impulse functions.

This time, we find the **Laplace transform of a delta function** by using two-sided integral (7.2) because the delta function can exist in both negative and positive sides of time axis. Thus,

$$F(s) = L\{\delta(t)\} = \int_{-\infty}^{\infty} e^{-st} f(t)dt = \int_{-\infty}^{\infty} e^{-st}\delta(t)dt.$$

Let's multiply both sides of (6.3b), $\int_{-\infty}^{\infty} \delta(t)dt = 1$, by e^{-st} and get

$$\int_{-\infty}^{\infty} (e^{-st} \cdot \delta(t))dt = e^{-st} \cdot 1.$$

Comparing two last expressions, we conclude

$$F(s) = L\{\delta(t)\} = e^{-st} \cdot 1.$$

Since the impulse function exits only at t = 0, we get

$$e^{-s0} \cdot 1 = 1.$$

Therefore, the Laplace transform of a delta function is given by

$$L\{\delta(t)\} = \int_{-\infty}^{\infty} e^{-st} \cdot \delta(t)dt = e^{-s0} \cdot 1 = 1, \tag{7.33}$$

as in Table 7.1.

LT-6 The Laplace transform of a delta function derivative

To find the Laplace transform of the delta function derivative, we need to resort to the rule applied to such a mathematical subject as *distribution*, Δ, to which a delta function belongs. The rule is

$$\int_{-\infty}^{\infty} \Delta` f(t)dt = -\int_{-\infty}^{\infty} \Delta f`(t)dt. \tag{7.34}$$

Using (7.34) for a delta function and referring to (7.5), we get

$$\int_{-\infty}^{\infty} e^{-st}\delta`(t)dt = -\int_{-\infty}^{\infty} (e^{-st})`\delta(t)dt = s\int_{-\infty}^{\infty} e^{-st}\delta(t)dt = s. \tag{7.35}$$

Thus,

$$L\{\delta`(t)\} = s, \tag{7.36}$$

as can be seen in (LT-6) of Table 7.1.

It's worth mentioning that step, ramp, pulse, and impulse (delta) functions are often called *singularity* or *generalized* or *switching functions*. Each of them has a point of singularity (discontinuity); also, their derivatives might have such points. As mentioned above, they model switching processes in electrical circuits and therefore play an essential role in advanced circuit analysis. For more on singularity functions see, for example, (Alexander and Sadiku 2020, pp. 265–273).

LT-7 The Laplace transform of an exponentially decaying function

An exponentially decaying function (or damped exponential) and its properties are extensively discussed in Section 5.1. You will recall that it is given by

$$v(t) = V_0 e^{-\alpha t}, \tag{7.37}$$

where $V_0 \equiv V(0)$ is the initial value of $v(t)$ calculated at $t = 0$, and $\alpha\left(\frac{1}{s}\right) = \frac{1}{\tau}$ is a *decay constant* (*damping coefficient*) with $\tau(s)$ being a *time constant*. (Equation 7.37 presents a time-domain signal commonly found in circuit transient responses. This sense of $v(t)$ implies that $\tau(s)$ and therefore $\alpha\left(\frac{1}{s}\right) = \frac{1}{\tau}$ are positive numbers.) Constants V_0 and $\alpha\left(\frac{1}{s}\right)$ play different roles in the function behavior. An initial value V_0 ascertains the starting point of the exponential function, whereas the value of an exponent α describes the rate of the exponential function decay.

The Laplace transform of an exponentially decaying function e^{-at} can be found by applying (7.5) and evaluating the resulting integral

$$\begin{aligned} F(s) &= \int_0^{\infty} e^{-st} f(t)dt = \int_0^{\infty} e^{-st}e^{-at}dt = \int_0^{\infty} e^{-(s+a)t}dt = \\ &\left[-\frac{1}{(s+a)}e^{-(s+a)t}\right]_0^{\infty} = -\frac{1}{(s+a)}\left[e^{-\infty} - e^{0}\right] = \frac{1}{(s+a)}. \end{aligned} \tag{7.38}$$

Now we understand where (LT-7) in Table 7.1 came from.

Compare (LT-7) and (LT-1) of Table 7.1: The Laplace transform of the *f(t)* = 1 is $\frac{1}{s}$, but the Laplace transform of the $f(t)=1e^{-at}$ is $\frac{1}{s+a}$. Thus, in general, *multiplying f(t) by* e^{-at}, *will result in replacing s by* $(s+\alpha)$ *in the Laplace transform of the f(t)*; that is,

$$L\{f(t)e^{-at}\}=\int_0^\infty e^{-st}f(t)e^{-at}dt=\int_0^\infty e^{-(s+a)t}f(t)dt=F(s+a). \tag{7.39}$$

We will use this rule shortly.

Question Can we apply this result to an exponentially growing function? What will be the Laplace transform of the *f(t)* = $5e^{3t}$?

LT-8 The Laplace transform of an exponentially decaying unit ramp function (damped ramp)

This function is defined in Table 7.1 as $t\cdot e^{-at}$, which is the product of a unit ramp, *t*, and exponential functions, e^{-at}. It is depicted in Figure 7.8.

The first step in finding the Laplace transform of this function is straightforward: Apply the Laplace transform integral given in (7.5).

$$L\left\{te^{at}\right\}=\int_{-\infty}^{\infty} e^{-st}te^{-at}dt=\int_{-\infty}^{\infty} e^{-(s+a)t}tdt. \tag{7.40}$$

Here, the integrand is a product of two functions, and we need to use the integration-by-parts formula given in (7.20). This would be the second step. The shortest way to solve this problem, however, is to use Rule (7.39), which enables us to write immediately

$$L\{te^{-at}\}=F(s)=\frac{1}{(s+a)^2}. \tag{7.41}$$

Indeed, initially $L\{t\}=F_1(s)=\frac{1}{(s)^2}$; since *f(t)* = *t* is multiplied by $-e^{at}$, we replace *s* in the $F_1(s)$ by $(s+a)$ and obtain the result, $F_2(s)=\frac{1}{(s+a)^2}$. This result is shown in (LT-8) of Table 7.1.

Figure 7.8 Exponentially decaying unit ramp function.

Questions

1) Can you verify the obtained result by integrating (7.40) by parts? Show all your mathematical manipulations.
2) At what value of t (s) is the peak of the graph in Figure 7.8 located?

LT-9 The Laplace transform of an exponentially decaying unit ramp function of n-th order

The Laplace transform in question is the transform of the two functions product; that is, $L\{t^n \cdot e^{-\alpha t}\}$. From (7.25) we know that

$$L\{t^n\} = \int_0^{\infty} t^n \cdot e^{-st} dt = \frac{n!}{s^{n+1}}. \tag{7.25R}$$

Rule (7.39) says that the Laplace transform $L\{t^n \cdot e^{-\alpha t}\}$ is the same as $L\{t^n\}$ whose s must be replaced by $(s+\alpha)$. That is,

$$L\{t^n \cdot e^{-\alpha t}\} = \frac{n!}{(s+\alpha)^{n+1}}, \tag{7.42}$$

as Table 7.1 shows.

LT-10 and LT-11 The Laplace transform of sine and cosine functions

We thoroughly discussed both sine and cosine functions (collectively referred to as sinusoidal functions or simply sinusoids) in the previous chapters. Without the repetition, we turn immediately to finding the Laplace transform of these functions.

To find the Laplace transform of sinusoids, recall Euler's identity:

$$e^{j\omega t} = \cos(\omega t) + j sin(\omega t). \tag{7.43}$$

When we find the Laplace transform of $e^{j\omega t}$, then its real part will give us the Laplace transform of cosine, and the imaginary part represents that of a sine function. Thus, applying the Laplace transform definition (7.5), we find

$$\int_0^{\infty} e^{-st} e^{j\omega t} dt = \int_0^{\infty} e^{-st} cos\omega t dt + j \int_0^{\infty} e^{-st} sin\omega t dt. \tag{7.44}$$

The Laplace transform of $e^{j\omega t}$ is

$$\int_0^{\infty} e^{-st} e^{j\omega t} dt = \int_0^{\infty} e^{-(s-j\omega)t} dt = -\frac{1}{(s-j\omega)} \left[e^{-(s-j\omega)t} \right]_0^{\infty} = \frac{1}{(s-j\omega)} \tag{7.45}$$

because $e^{-\infty} \to 0$ and $e^{-0} = 1$. Now, we need to present $\frac{1}{(s-j\omega)}$ in $a + jb$ format. Then, a will be $L\{cos\omega t\}$ and b will be $L\{sin\omega t\}$. Jump to Section 3.1, where the operations with complex numbers are discussed, to find

$$\frac{1}{(s-j\omega)} \cdot \frac{(s+j\omega)}{(s+j\omega)} = \frac{s}{(s^2+\omega^2)} + j\frac{\omega}{(s^2+\omega^2)}. \tag{7.46}$$

Therefore,

$$L\{cos\omega t\} = \frac{s}{\left(s^2 + \omega^2\right)} \tag{7.47}$$

and

$$L\{sin\omega t\} = \frac{\omega}{\left(s^2 + \omega^2\right)}, \tag{7.48}$$

as Table 7.1 shows.

Question What will be the Laplace transform of 4cos(6t)? 4sin(6t + 30^0)? (Hint: for the latter question, we introduce without the proof the following formula: $L\{\sin(\omega t + \theta)\} = \frac{s\,sin\theta + \omega\,cos\theta}{s^2 + \omega^2}$.)

LT-12 and LT-13 The Laplace transform of exponentially decaying sine and cosine functions

An exponentially decaying sine function is described as

$$f(t) = e^{-\alpha t} sin\omega t. \tag{7.49}$$

It is depicted in Figure 7.9. Its Laplace transform can be obtained by applying the Laplace transform definition, as we did in the preceding examples, namely

$$L\{e^{-\alpha t} sin\omega t\} = \int_0^{\infty} e^{-st}\left(e^{-\alpha t}\right) sin\omega t dt. \tag{7.50}$$

Compare left-hand sides of (7.50) and (7.48): The former differs from the latter by only one member, $e^{-\alpha t}$. But we know from (7.39) that multiplication of a time-domain function by $e^{-\alpha t}$ results in changing s to $(s + \alpha)$ in s-domain. Therefore,

$$L\{e^{-\alpha t} sin\omega t\} = \frac{\omega}{\left((s+\alpha)^2 + \omega^2\right)}. \tag{7.51}$$

Figure 7.9 Exponentially decaying sine function.

Similarly,

$$L\{e^{-\alpha t}\cos\omega t\} = \frac{s}{\left((s+\alpha)^2+\omega^2\right)}. \quad (7.52)$$

Questions

1) Build the graph of $f(t) = 3e^{-0.2t}\cos1.26t$. What is the role of $e^{-0.2t}$ member in this graph? Mark its traces on the graph.
2) Can you find the Laplace transform of $3e^{-5t}\cos7t$? $2e^{4t}\sin6t$?

Summary: This section explicitly shows how all entries in Table 7.1 are found. It's essential to understand all members of this table's origin because they are the most popular input signals encountered in the advanced circuit in practice. Mastering the technique of determining the Laplace transform of any time-domain function helps solving the problems you can meet in your professional activity.

7.5 Properties of the Laplace Transform—Table and Example

To apply the Laplace transform to the circuit analysis, we need to know its properties. They are presented in Table 7.3.

Table 7.3 Properties of the Laplace transform.

Designation	Property	Time domain	s-domain
PLT-1	Signal designation	$f(t)$	$F(s)$
PLT-2	Units	$Time, t\ (second)$	$Complex\ frequency,\ s = \sigma + j\omega\left(\frac{1}{second}\right)$
PLT-3		$Voltage,\ v(t)[volt]$	$V(s)[volt - second]$
PLT-4		$Current,\ i(t)[amer]$	$I(s)[amper - second]$
PLT-5	Linearity	$Af_1(t) + Bf_2(t)$	$AF_1(s) + BF_2(s)$
PLT-6	Scaling (k—constant)	$f(kt)$	$\frac{1}{k}F\left(\frac{s}{k}\right)$
PLT-7	Time shift $\tau(second) - constant)$	$f(t-\tau)$	$e^{\tau s}F(s)$
PLT-8	Frequency shift $(\alpha - constant)$	$e^{-\alpha t}f(t)$	$F(s+\alpha)$
PLT-9	Time differentiation	$f'(t)$	$sF(s) - f(0)$
PLT-10		$f''(t)$	$s^2F(s) - sf(0) - f'(0)$
PLT-11		$f^n(t)$	$s^nF(s) - \sum_{i=1}^{n} s^{n-i}f^{i-1}(0)$
PLT-12	Frequency differentiation	$f(t)$	$-\frac{dF(s)}{ds}$
PLT-13	Time integration	$\int_0^t f(\tau)d\tau$	$\frac{F(s)}{s}$
PLT-14	Initial value	$f(0)$	$\lim_{s\to\infty}(sF(s)) = f(0)$
PLT-15	Final value	$f(t)$	$\lim_{s\to 0} sF(s)) = f(\infty)$
PLT-16	Convolution	$g(t)\ast h(t)$	$G(s)\cdot H(s)$

Note: Entries PLT-9, PLT-10, and PLT-11 should show $f(0^-)$instead of $f(0)$, and PLT-14 should be $f(0^+)$ instead*of* $f(0)$, where 0^- is infinitesimally less than zero and 0^+ is infinitesimally greater than zero. (See Sidebar 7.1S for more on this point.) We omit these notations to shorten the writing.

We will discuss these properties in detail shortly. For now, let's consider an example to put this introductory presentation into practical terms.

Example 7.2 Finding an RC circuit's free (natural) response by using the Laplace transform.

Problem: Consider the RC circuit presented in Figure 7.10a whose R = 220 kΩ and C = 20 nF. For $t < 0$, the switch connected the circuit to the battery long enough to fully charge the capacitor up to $V_0 = 12V$. At $t = 0$, the switch connects the circuit to the ground leaving it source free for $t > 0$. Find the circuit's *natural response.*

Solution: The *differential equation of an RC circuit* was derived in Section 5.1 as

$$RCv'_{out}(t) + v_{out}(t) = v_{in}(t). \tag{5.24R}$$

Natural response means response to a zero input; hence, the differential equation takes the form

$$v'_{out}(t) + \alpha v_{out}(t) = 0 \tag{7.53}$$

with $\alpha = \frac{1}{RC}$. Converting this time-domain differential equation into its s-domain equivalent requires the application of the Laplace transform, which gives

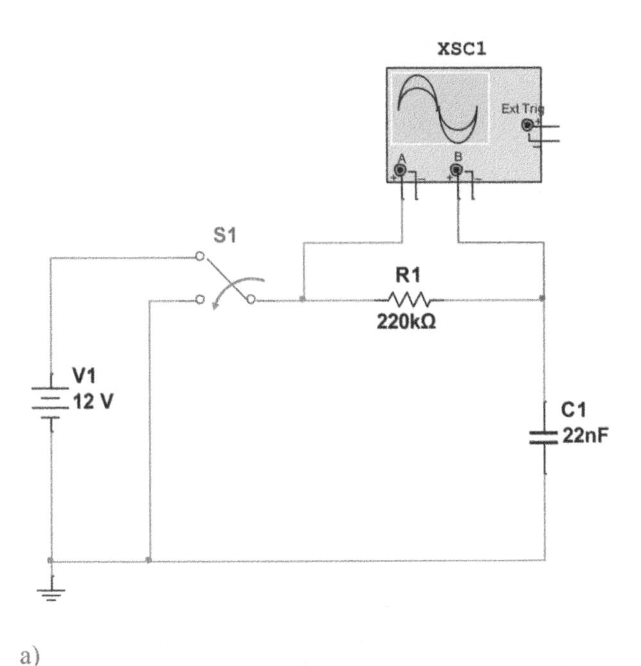

Figure 7.10 Natural response of an RC circuit: a) Circuit's schematic; b) MATLAB calculated $v_{out}(t)$; c) Multisim simulated the circuit's output, $v_{out}(t)$ after the switch flipped.

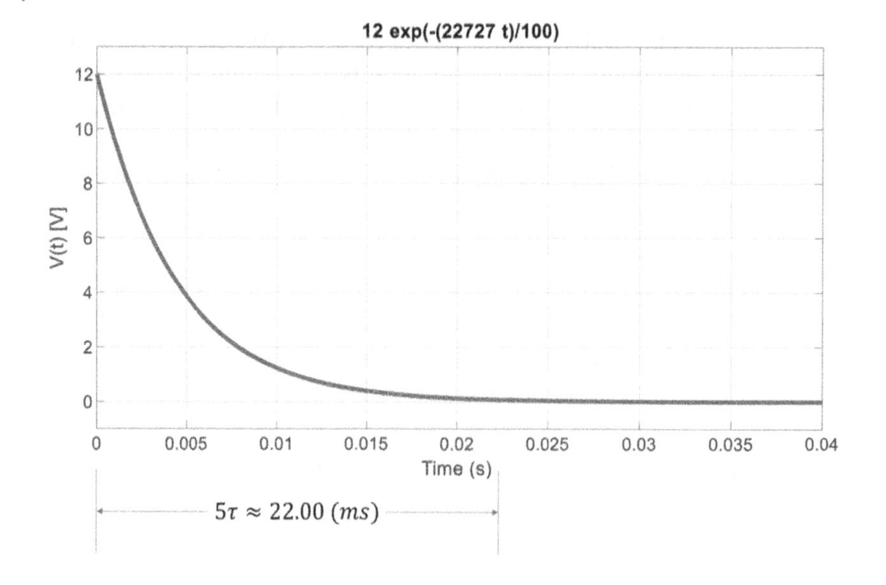

b)

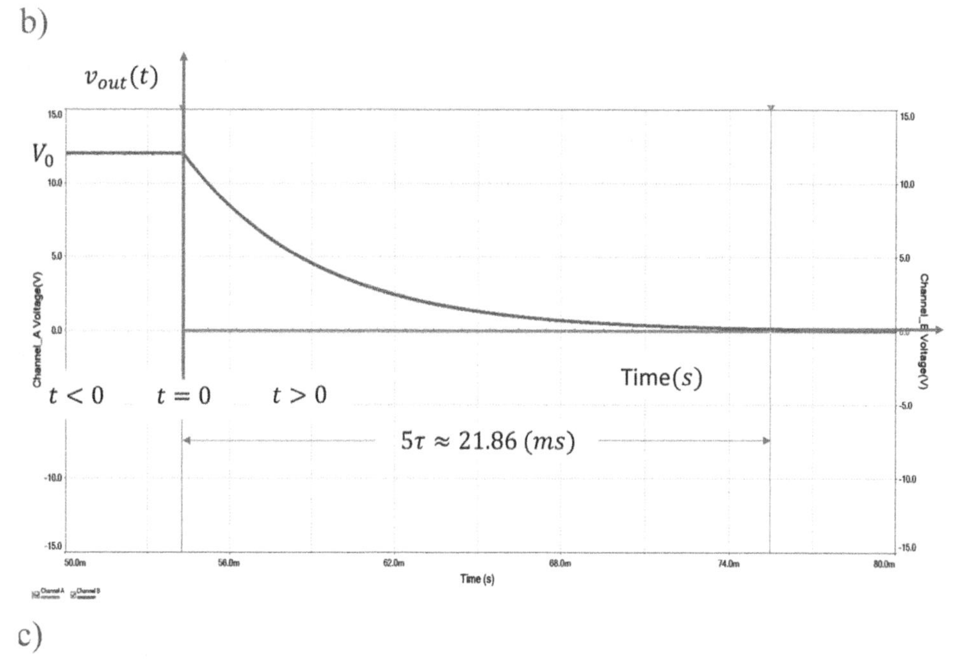

c)

Figure 7.10 (Cont'd)

$$L\{v'_{out}(t) + \alpha v_{out}(t) = 0\} \Rightarrow sV_{out}(s) - V_{out}(0^-) + \alpha V_{out}(s) = 0. \tag{7.54}$$

Here we used (1) linearity property, PLT-5, to apply the Laplace transform to each member of (7.55) and factor out a constant α, (2) signal designation property, PLT-1, to obtain $V_{out}(s) = L\{v_{out}(t)\}$, and (3) time differentiation property, PLT-9, to find $L\{v'_{out}(t)\} = sV_{out}(s) - V_0$.

(Don't be confused with using capital V in the designation V_0; remember that an initial value, $V_0 \equiv V_{out}(0)$, is a number (constant), which in this example is $V_0 = 12(V)$. We use capital V_0 to contrast this constant to a function $v_{out}(t)$.)

Thus, the *equivalent equation* of the relaxed RC circuit in the s-domain is given by

$$sV_{out}(s) + \alpha V_{out}(s) = V_o. \tag{7.55}$$

Solving (7.55) for *the RC circuit's response in the s-domain,* $V_{out}(s)$, we find

$$V_{out}(s) = \frac{V_o}{(s+\alpha)}. \tag{7.56}$$

This solution provides the RC circuit's response, but in the s-domain. To obtain the required response in the time domain, $v_{out}(t)$, we have to perform an *inverse Laplace transform*:

$$v_{out}(t) = L^{-1}\{V_{out}(s)\}. \tag{7.57}$$

Referring to Entries LT-7 in Table 7.1 and PLT-14 in Table 7.3, we find

$$v_{out}(t) = L^{-1}\{V_{out}(s)\} = L^{-1}\left\{\frac{V_o}{(s+\alpha)}\right\} = V_0 L^{-1}\left\{\frac{1}{(s+\alpha)}\right\} = V_0 e^{-\alpha t}. \tag{7.58}$$

Thus, the formula for the RC circuit's natural response is

$$v_{out}(t) = V_0 e^{-\alpha t}. \tag{7.59}$$

Plugging the given numbers into (7.59), we compute

$$\tau = RC = 4.4 \cdot 10^{-3}(s) \text{ and } \alpha = \frac{1}{RC} = 227.27\left(\frac{1}{s}\right),$$

and the solution to our problem takes the form

$$v_{out}(t) = 12e^{-227.27t}\ (\text{V}).$$

Its graph is presented in Figure 7.10b. Note that the graph actually shows $v_{out}(t) \cdot u(t)$, a causal function. (Can you explain why it is $v_{out}(t) \cdot u(t)$ but not merely $v_{out}(t)$?)

Discussion:

- Solution 7.59 is, obviously, the same one that we obtained in the time domain — see Section 5.1. The graph shown in Figure 7.10b, too, is very much familiar to us, particularly in reference to that section. And what will be the output amplitude at $t = 5\tau(s)$?
- To fully understand the similarity and difference between the time-domain and the Laplace transform operations, it's worth comparing side by side finding the free response of an RL circuit by both methods. We encourage you to perform such a comparison, which would highlight the advantages of using the Laplace transform. Specifically, not only do we need to take fewer steps to find the free response of an RL circuit, but we also employ much simpler mathematical manipulations to achieve our goal.

- We can find the solution to this problem with MATLAB by writing the following script for (7.58), $v_{out}(t) = L^{-1}\left\{\frac{V_o}{(s+\alpha)}\right\}$:

```
syms s t % Refer to symbolic toolbox and define the variables
a = 12; %write down the numerator
b = s + 227.27; %write down the denominator
c = ilaplace(a/b,t) %command MATLAB to perform inverse Laplace
%transform
ezplot(c,[0,3]); %command MATLAB to plot the time-domain function
%and display the formula.

axis([0 3 -1 13])
grid
xlabel('Time (s)')
ylabel('V(t) [V]')
```

The graph is displayed in Figure 7.10b.

- We run the *Multisim simulation* of the given RC circuit to validate our result and obtain the graph shown in Figure 7.10c. It's left to you to thoroughly compare both graphs and make the conclusion.

A short overview of Sections 7.1 through 7.5*: Let's reiterate main points of these sections—a summary of the Laplace transform operation and the key reason for using the Laplace transform:*

- *The Laplace transform operation in obtaining a circuit's time-domain response to a given input (which is the goal of advanced circuit analysis) consists of three steps:*
  1. *Deriving a differential (time-domain) circuit's equation and converting it into an s-domain equivalent algebraic equation.*
  2. *Finding the circuit's response in the s-domain by solving the equivalent algebraic equation for $V_{out}(s)$.*
  3. *Obtaining the required response in the time domain $v_{out}(t)$ by performing an inverse Laplace transform on the s-domain circuit response.*
- *The key reason to use the Laplace transform method is its ability to transform a time-domain differential equation into an equivalent s-domain algebraic equation, which enables us to obtain the s-domain circuit's output by performing straightforward algebraic manipulations. The price for this simplification is performing the inverse Laplace transform.*

7.6 Properties of the Laplace Transform—Proofs and Discussions

In this section, we will prove each property and discuss Table 7.3 line by line.

PLT-1 Signal designation

The Laplace transform of a time-domain function $f(t)$ is simply $F(s)$, that is, $L\{f(t)\} = F(s)$, by definition of the Laplace transform.

PLT-2, PLT-3, and PLT-4 Units

Time is measured in seconds, and frequency (even the complex one) is measured in the number of cycles per second. The units of any Laplace transform, as follows from Integral 7.5, are determined by the units of its integrand. We reproduce here the Laplace transform integral with units shown for every member of the integrand:

$$F(s) = \int_0^{\infty} e^{-st}\left[dimensonless\right] f(t)\left[units\,of\;a\;time - domain\;function\right] dt\left[second\right]. \tag{7.60}$$

Therefore, if *f(t)* is the voltage signal, *v(t) [V]*, its Laplace transform results in *V(s) [V·s]*, as shown in PLT-3 of Table 7.3. You are encouraged to analyze the units of all other signals and parameters used in electronic circuits. Bear in mind that the Laplace transform units are intermediate results; eventually, we need to invert all Laplace transforms to the time-domain functions with their units.

PLT-5 Linearity

If $f(t) = Af_1(t) + Bf_2(t)$, where $A\,\text{and}\,B$ are constants, then

$$L\{Af_1(t) + Bf_2(t)\} = AL\{f_1(t)\} + BL\{f_2(t)\} = AF_1(s) + BF_2(s). \tag{7.61}$$

Thus, the Laplace transform exhibits both *homogeneity* and *superposition (additivity)* properties, which constitute its linearity. You are reminded that homogeneity means that an increase in the *f(t)* amplitude will result in a proportional increase in the *F(s)* amplitude, whereas superposition means that the sum of two time-domain signals will result in the sum of two s-domain signals. Refer to Section 5.1, where we discussed the definition of linearity, and to Example 7.2, where we used this property for circuit analysis.

PLT-6 Scaling

Let *F(s)* be the Laplace transform of *f(t)*, and *k* be a positive constant, then we get the scaling property as

$$L\{f(kt)\} = \frac{1}{k}F\left(\frac{s}{k}\right). \tag{7.62}$$

Indeed, if we substitute $t = \frac{x}{k}r$ or $x = kt$ in (7.5), we find

$$F(s) = \int_0^{\infty} e^{-st} f(kt)dt = \frac{1}{k}\int_0^{\infty} e^{-\frac{s}{k}x} f(x)dx = \frac{1}{k}F\left(\frac{s}{k}\right). \tag{7.63}$$

This is because $\int_0^{\infty} e^{-\frac{s}{k}x} f(x)dx = F\left(\frac{s}{k}\right)$ by definition. For example, recalling that $L\{\cos(\omega t)\} = \frac{s}{(s^2 + \omega^2)}$, we find

$$L\{\cos(3\omega t)\} = \frac{1}{3}\frac{\frac{s}{3}}{\left(\frac{s}{3}\right)^2 + \omega^2} = \frac{s}{\left(s^2 + 9\omega^2\right)}. \tag{7.64}$$

PLT-7 Time shift (time-domain translation)
Let $L\{f(t)\} = F(s)$; then the time shift is defined as

$$\mathrm{L}\{f(t-\tau)\} = e^{-s\tau}F(s) \tag{7.65}$$

where constant $\tau \geq 0$. To prove this property, refer again to the definition of the Laplace transform (7.5)

$$L\{f(t) = \int_0^\infty e^{-st}f(t)dt. \tag{7.5R}$$

Before plugging in (7.5) a new function, we need to recall that we agreed to use merely a causal function, $f(t)\cdot u(t)$, which exists only for $t>0$. But in (7.65) we are dealing with the function shifted in time with respect to zero by τ. Therefore, (7.65) must be rewritten as

$$\mathrm{L}\{f(t-\tau)\cdot u(t-\tau)\} = e^{-s\tau}F(s). \tag{7.66}$$

Now, the Laplace transform of $f(t-\tau)$ given in (7.5) takes the form

$$L\{f(t-\tau)\} = \int_0^\infty e^{-st}f(t-\tau)\cdot u(t-\tau)dt. \tag{7.67}$$

A time-shifted unit step function is shown in Figure 7.14 and defined in (7.15) as

$$u(t-\tau) = \begin{cases} 0 \text{ for } t<\tau \\ 1 \text{ for } t>\tau \end{cases}. \tag{7.68}$$

This definition means that the lower limit in Integral 7.67 must be not 0 but τ because for $t<\tau$ the unit step function and the whole integral become zero. Since $u(t-\tau)=1$ for $t>\tau$, we get

$$L\{f(t-\tau) = \int_0^\infty e^{-st}f(t-\tau)\cdot u(t-\tau)dt = \int_\tau^\infty e^{-st}f(t-\tau)dt. \tag{7.69}$$

Changing the variable, $x = t-\tau$, results in the following modifications: $t = x+\tau$ and $dx = dt$; also, when $t \to \tau$, $x \to 0$ and $t \to \infty$, $x \to \infty$. Hence, we can rewrite (7.69) as

$$L\{f(t-\tau) = \int_0^\infty e^{-s(x+\tau)}f(x)dx = e^{-s\tau}\int_0^\infty e^{-sx}f(x)dx = e^{-s\tau}F(s) \tag{7.70}$$

because $\int_0^\infty e^{-sx}f(x)dx = F(s)$, by definition.

Example: Introducing a time delay of 5(s) to a cosine signal, we find $L\{\cos\omega(t-5)\} = e^{-s5}\left(\frac{s}{(s^2+\omega^2)}\right)$.

PLT-8 Frequency shift (s-domain translation)
Let $L\{f(t)\} = F(s)$; then the frequency shift is defined as

$$\mathrm{L}\{e^{-\alpha t}f(t)\} = F(s+\alpha) \tag{7.71}$$

where α is a constant. To prove this property, refer again to the definition of the Laplace transform (7.5) and find

$$L\left\{e^{-\alpha t}f(t)\right\}=\int_0^{\infty}e^{-st}e^{-\alpha t}f(t)dt=\int_0^{\infty}e^{-(s+\alpha)t}f(t)dt=F(s+\alpha). \quad (7.72)$$

As an example, consider $f(t)=e^{-4t}cos9t$ and find $L\left\{e^{-4t}cos9t\right\}=\frac{(s+4)}{\left((s+4)^2+9^2\right)}$.

We learned this property early as a Rule (7.41) when proving the Laplace transform pairs, Table 7.1.

PLT-9, PLT-10, and PLT-11 Time differentiation (differentiation in the time domain)
This is one of the most important properties of the Laplace transform; after all, it is due to this property we can transform a differential equation into an equivalent algebraic equation. We introduced and used this property for the first derivative in Example 7.1 to analyze first-order circuits. Now it's time to prove it and expand it to the higher-order derivatives.

If $L\{f(t)\}=F(s)$, then the first derivative is defined as

$$L\{f'(t)\}=sF(s)-f(0) \quad (7.73)$$

where *f(0)* is the initial value of *f(t)*, that is, the *f(t)* value calculated at *t* = *0*. Remember that *f(0)* is a number; it's often designated as $F_0\equiv F(0)\equiv f(0)$.

We can prove this property by considering, as before, the definition of the Laplace transform (7.5), which can be presented as

$$L\{f'(t)\}=\int_0^{\infty}e^{-st}f'(t)dt=\int_0^{\infty}udv \quad (7.74)$$

with $u=e^{-st}$and $dv=f'(t)dt=\frac{df(t)}{d}dt=df(t)$. Thus, $v=f(t)$ and $du=-se^{-st}dt$. Now we can integrate (7.74) by parts, as shown here. (You may recall that $\int_0^{\infty}udv=\left[u\cdot v\right]_a^b-\int_0^{\infty}vdu$, as (7.20) states.)

$$\begin{aligned}L\{f'(t)\}&=\int_0^{\infty}e^{-st}f'(t)dt=\left[e^{-st}f(t)\right]_0^{\infty}-\int_0^{\infty}f(t)\left(-se^{-st}dt\right)=\\&\left[0-f(0)\right]+s\int_0^{\infty}f(t)e^{-st}dt=sF(s)-f(0).\end{aligned} \quad (7.75)$$

In (7.75), the term $e^{-st}f(t)$ goes to zero when $t\Rightarrow\infty$ provided that e^{-st} goes to zero faster than $f(t)$ goes to infinity. This is the condition for the Laplace integral convergence, and, therefore, it must be met. Also, the initial condition, $f(0)$, must be taken at 0^-, that is, at the instant before a switch is turned on.

Thus, property PLT-9 is proved.

Consider, for example, $f(t)=cos\omega t$. According to PLT-9,

$$L\{(cos\omega t)'\}=s\frac{s}{\left(s^2+\omega^2\right)}-\cos(0)=-\frac{\omega^2}{\left(s^2+\omega^2\right)}\omega=-\omega\left[\frac{\omega}{\left(s^2+\omega^2\right)}\right]. \quad (7.76)$$

Question Does (7.76) make sense as the Laplace transform of the derivative of a cosine function? (Hint: Apply the inverse Laplace transform to (7.76) and recall that $(cos\omega t)' = -\omega sin\omega t$ in the time domain.))

The Laplace transform of a $f(t)$ second derivative can be obtained by considering $f'(t)$ as an original function and applying (7.75) to it. Thus,

$$L\{(f'(t))'\} = s(sF(s) - f(0)) - f'(0) = s^2F(s) - sf(0) - f'(0), \tag{7.77}$$

where $(sF(s) - f(0)) = L\{f'(t)\}$. Repeat this procedure n times and get the Laplace transform for the n-th derivative of a $f(t)$ given in PLT-11.

For the second derivative of $f(t)$, there are two initial conditions, which—from an engineering standpoint—can be considered as an initial "position," $f(0)$, and the initial "velocity," $f'(0)$. For example, in a series RLC circuit, the initial "position," V_0, is the voltage across the output capacitor at $t = 0$, and the initial "velocity," V_0', corresponds to the voltage drop caused by the current through this capacitor at $t = 0$. (Recall that $i_C(t) = Cv_C'(t)$.)

PLT-12 Frequency differentiation
If $F(s)$ is the Laplace transform of $f(t)$, then

$$\frac{dF(s)}{ds} = -L\{tf(t)\} \tag{7.78}$$

To prove this, start with (7.5), $F(s) = \int_0^\infty e^{-st} f(t)dt$, and take the derivative of both sides with respect to s,

$$\frac{dF(s)}{ds} = \int_0^\infty \left(\frac{de^{-st}}{ds}\right) f(t)dt = \int_0^\infty -t\left[e^{-st} f(t)\right]dt = -L\{tf(t)\}. \tag{7.79}$$

As an example, consider $F(s) = \frac{1}{(s+\alpha)} = L\{e^{-\alpha t}\}$. On the one hand, take its derivative, and get

$$-\frac{d\left(\frac{1}{(s+\alpha)}\right)}{ds} = \frac{1}{(s+\alpha)^2}.$$

On the other hand,

$$L\{te^{-\alpha t}\} = \frac{1}{(s+\alpha)^2},$$

according to LT-8. Since each RHS of the above equations is equal to one another, their LHSs are also equal. Therefore,

$$\frac{d\left(\frac{1}{(s+\alpha)}\right)}{ds} = -L\{te^{-\alpha t}\},$$

as in (7.78). You can readily generalize this formula to the n-th frequency derivative.

PLT-13 Integration in the time domain

Let $L\{f(t)\} = F(s)$; then the Laplace transform of the integral of $f(t)$ is given by

$$L\left\{\int_0^t f(\tau)d\tau\right\} = \frac{1}{s}L\{f(t)\} = \frac{1}{s}F(s). \tag{7.80}$$

That said, the integration in the time domain is equivalent to the division by s in the s-domain. This rule looks logical: Since differentiation in the time domain is equivalent to multiplication by s in the s-domain, then integration in the time domain is matched to division by s in the s-domain.

Indeed,

$$L\left\{\int_t^0 f(\tau)d\tau\right\} = \int_0^\infty [\int_0^t f(\tau)d\tau]e^{-st}dt = \int_0^\infty udv \qquad 7.81$$

with $u = \int_0^t f(\tau)d\tau$ and $dv = e^{-st}dt$, which means that $du = f(t)dt$ and $v = -\frac{1}{s}e^{-st}$. Integration by parts yields

$$\begin{aligned} L\left\{\int_0^t f(\tau)d\tau\right\} &= \int_0^\infty udv = [u\cdot v]_0^\infty - \int_0^\infty vdu = \\ &\left[\left(\int_0^t f(\tau)d\tau\right)\cdot\left(-\frac{1}{s}e^{-st}\right)\right]_0^\infty - \int_0^\infty\left(-\frac{1}{s}e^{-st}\right)f(t)dt \\ &= 0 + \frac{1}{s}F(s). \end{aligned} \tag{7.82}$$

The first RHS term is equal to ∞ because it vanishes at $t \to 0$ due to member e^{-st}, and it vanishes at $t = 0$ due to integral $\int_0^0 f(\tau)d\tau$). Hence, property PLT-13 is proved.

As an example, let's take $f(t) = e^{-\alpha t}$. Integration in the time domain produces

$$\int_0^t e^{-\alpha\tau}d\tau = \left[-\frac{1}{\alpha}e^{-\alpha t}\right]_0^\infty = \frac{1}{\alpha}(1 - e^{-\alpha t}). \tag{7.83}$$

And the Laplace transform of this expression yields

$$\mathcal{L}\left\{\frac{1}{\alpha}(1 - e^{-\alpha t}\right\} = \frac{1}{\alpha}\left(\frac{1}{s} - \frac{1}{s+\alpha}\right) = \frac{1}{s(s+\alpha)}. \tag{7.84}$$

On the other hand, using (7.80), we find

$$L\left\{\int_0^t e^{-\alpha\tau}d\tau\right\} = \frac{1}{s}L\{e^{-\alpha t}\} = \frac{1}{s(s+\alpha)} \tag{7.85}$$

which validates (7.80).

PLT-14 Initial Value Theorem

Let $F(s) = \mathcal{L}\{f(t)\}$; then

$$\lim_{s\to\infty} sF(s) = f(0). \tag{7.86}$$

This property enables us to find the initial value of a time-domain function through its Laplace transform. Property PLT-14 is also known as the *initial-value theorem*, and the following manipulations can prove it: Let's rearrange (7.86) and consider its limit at s→∞:

$$\lim_{s\to\infty} sF(s) - f(0) = \lim_{s\to\infty} \int_0^\infty e^{-st} f'(t)dt. \tag{7.87}$$

Under condition $s \to \infty$ the integrand vanishes because of the e^{-st} member, and, according to the time differentiation property, PLT-9, we find

$$\lim_{s\to\infty} (sF(s) - f(0)) = 0.$$

Since $f(0)$ doesn't depend on s, we arrive at (7.86).

Consider, for example, the following time-domain function: $f(t) = 4e^{-2t}cos6t$. Its initial value is equal to $f(0) = 4e^{-2\cdot 0}cos6\cdot 0 = 4$. To verify (7.86), let's find the initial value of this function through its Laplace transform. First, applying LT-13, we get

$$\mathfrak{L}\left\{4e^{-2t}cos6t\right\} = \frac{4s+8}{(s+2)^2+36}.$$

Secondly, the limit in question is

$$\lim_{s\to\infty} \frac{s(4s+8)}{(s+2)^2+36} = \lim_{s\to\infty} \frac{\left(4+\frac{8}{s}\right)}{\left(1+\frac{2}{s}+\frac{40}{s^2}\right)} = 4.$$

Thus, property PLT-14 delivers the correct result.

PLT-15 Final value theorem

If $F(s) = \mathcal{L}\{f(t)\}$, then

$$\lim_{s\to 0} sF(s) = f(\infty). \tag{7.88}$$

This statement is known as the *final-value theorem*; it is sometimes is written as

$$\lim_{s\to 0} sF(s) = \lim_{t\to\infty} f(t). \tag{7.89}$$

To prove this property, we use (7.87) and find

$$\lim_{s\to 0} sF(s) - f(0) = \lim_{s\to 0} \int_0^\infty e^{-st} \left(\frac{df(t)}{dt}\right) dt = \lim_{s\to 0} \int_0^\infty df(t) = f(\infty) - f(0). \tag{7.90}$$

Therefore,

$$\lim_{s\to 0} sF(s) = f(\infty).$$

In general, both initial-value and final-value theorems are useful tools to check the performed Laplace transforms' correctness.

Note: In contrast to the Initial Value theorem, the *Final Value rule can be applied if and only if the real part of the complex frequency, $\sigma = \alpha + j\omega$, is positive.* The Notes to Table 7.1 and Discussion in Example 7.1 explain why: This is because $Re(\sigma) \equiv \alpha$ is the power of an exponentially decaying function, $e^{-\alpha t}$. When $t \to \infty$, this function approaches zero if $\alpha > 0$. If, however, $\alpha < 0$, function $e^{-(-\alpha)t}$ grows to infinity when $t \to \infty$. Since the function final value is determined by (7.89) as $\lim_{t\to\infty} f(t)$, the requirement $Re(\sigma) \equiv \alpha > 0$ must be met, too.

Consider, for example, the time-domain function $f(t) = 4e^{-2t}cos6t$. According to property PLT-14, its Laplace transform is $F(s) = \dfrac{4s+8}{(s+2)^2+36}$, and its final value, according to (7.89) is

$$\lim_{t\to\infty} f(t) = \lim_{s\to 0} sF(s) = \lim_{s\to 0} s\frac{4s+8}{(s+2)^2+36} = 0.$$

On the other hand, the final value of this time-domain function can be determined immediately as

$$\lim_{t\to\infty}(4e^{-2t}cos6t) = 0$$

thanks to the positive real coefficient 2. However, if we have $-2 < 0$, then

$$\lim_{t\to\infty}(4e^{2t}cos6t) \to \infty,$$

whereas (7.89) still gives

$$\lim_{s\to 0} s\frac{4s-8}{(s-2)^2+36} = 0,$$

which is **wrong**.

Thus, in this case, the application of the Final Value theorem is **invalid**. Bear in mind this word of caution.

PLT-16 Convolution

Refer to Chapter 6 to refresh your memory about the convolution integral. In particular, generalizing its definition, we can present the convolution integral as

$$\int_{-\infty}^{+\infty} g(\tau)h(t-\tau)d\tau = g(t)\star h(t), \tag{7.91}$$

where the symbol $\star$ denotes the convolution integration.

Now we introduce the convolution property of the Laplace transform as

$$\mathcal{L}\{g(t)\star h(t)\} = G(s)H(s). \tag{7.92}$$

In Chapter 6, we learned that the convolution integral is the tool to find a system's response to the input in the time domain,

$$v_{out}(t) = \int_{-\infty}^{\infty} g(\tau)h(t-\tau)d\tau = g(t)\star h(t). \tag{7.93}$$

The convolution property of the Laplace transform, PLT-16, allows for replacing the complex, sophisticated mathematical operation in the time domain by multiplication in the s-domain,

$$V_{out}(s) = G(s)\cdot H(s). \tag{7.94}$$

To find the time-domain output, we will need to find the inverse Laplace transform of $V_{out}(s)$, of course. Still, the beauty of Equation 7.94 is that we can determine many essential properties of our system directly from this equation without inverting it into the time domain. We will learn more about this topic in the "Transfer Function" section of Chapter 10. It should be noted that the proof of (7.93) is rather challenging and lies outside the scope of this book.

We will use the Laplace transform properties throughout the rest of this textbook, where many examples of their applications will be found.

Finally, we want to remind you that it's possible to verify the correctness of the applications of the Laplace transform properties in any mathematical manipulations by employing MATLAB for solving the given problems. Below are two examples of the MATLAB usage:

Problems:

1) Find the Laplace transform of the time-domain function: $f(t) = e^{-\alpha t} \cdot u(t)$.

2) What is the inverse Laplace transform of $F(s) = \dfrac{s+0.2}{(s+0.2)^2 + (1.26)^2}$?

Solutions:

1) The following MATLAB script, along with the comments, provides the solution to Problem 1. You will recall that the comments are separated from the script lines by the percentage sign, %. They are not executed by MATLAB and simply explain the meaning of specific commands.

```
syms s a t % Refer to symbolic toolbox and define the variables s
%and t, and constant a
  f = exp(-a*t)*heaviside (t); % Denote the function in question by
%the symbol f and use heaviside(t) for the step unit function
  F = laplace(f,s) % Denote the answer by F and command MATLAB to
%perform the Laplace transform operation F = 1/(s + α)
  simplify (F) % This command rationalizes the answer to the form
%we obtain it manually ans = 1/(s + α)
```

We urge you to verify all the answers obtained in examples in Section 7.6 by using this MATLAB code.

2) Here is the solution to Problem 2 of this example:

```
syms s t
a = s + 0.2;
b = (s + 0.2)^2 + (1.26)^2;
c = ilaplace(a/b, t)
ezplot(c, [0,25]);
```

Run this program and obtain the formula and the graph of the time-domain function in search.

Question What would be the result of the MATLAB operation for this example,

$L^{-1}\left\{\dfrac{s+0.2}{(s+0.2)^2 + (1.26)^2}\right\}$? Can you predict it by analyzing the given code?

We will continue using MATLAB for the Laplace transform operations in the chapters that follow.

Questions and Problems for Chapter 7

1 The text says, *Laplace transform* is a tool. What tool is it?

2 Consider function $y = x^2$. What are its domain and the range?

3 The text's definition of a function states, A *function is a rule* that assigns to each element of an input set of numbers a unique element in the output set of numbers. However, function $y = cosx$ assigns y = 1 to x = 0, x = 2π, x = 4π, and so on. This fact implies that the relationship $y = cosx$ doesn't meet the function's definition. Is this conclusion correct? Explain.

4 Does function $y = cosx$ have an inverse function? Prove your answer.

5 Is there any difference between functions $v_1(t) = e^{-2t}$ and $v_2(t) = e^{-3t}$? Explain.

6 What is the difference between functions and signals?

7 Give definition and examples of a transform.

8 What is the difference between a function and a transform? Give examples.

9 Why does a function may not have the inverse function but a transform always have the inverse transform?

10 Use the attached Laplace transform table to find the Laplace transform of the following functions:

a) $f(t) = 3u(t)$
b) $f(t) = 3t^4$
c) $f(t) = 3e^{-5t}$
d) $f(t) = 3t^4e^{-5t}$
e) $f(t) = 12t^2 + 7cos8t$
f) $f(t) = 6e^{-4t}sin6t + 2e^{-4t}t^2$

11 Use the attached Laplace transform table to find the inverse Laplace transform of the following functions:

a) $F(S) = \dfrac{3}{s+5} + \dfrac{24}{(s+2)^5}$

b) $F(s) = \dfrac{9}{s^2+81} + \dfrac{6}{s^2}$

c) $F(s) = \dfrac{s+4}{(s+4)^2+16}$

d) $F(s) = \dfrac{28}{\left((s+6)^2+16\right)}$

e) $F(s) = \dfrac{5}{s+3} + \dfrac{120}{(s+3)^6} + \dfrac{9}{s^2+81}$

f) $F(s) = \dfrac{6}{s^2} + \dfrac{s+4}{(s+4)^2+16}$.

12 Solve all the problems in Problems 10 and 11 using MATLAB.

13 Two-sided and one-sided integrals defining of the Laplace transform given by (7.2), $F(s) = L\{f(t)\} = \int_{-\infty}^{\infty} e^{-st} f(t)dt$, and (7.5), $F(s) = \int_0^{\infty} e^{-st} f(t)dt$, respectively. In general, when we replace a two-sided integral by its one-sided equivalent, we double the value of the one-sided integral. However, in relationship between (7.2) and (7.5), this logic is broken. Why?

14 Why a unit-step (Heaviside) function is so important in the Laplace transform operation?

15 Derive the Rule LT-2, $L\{u(t)\} = \frac{1}{s}$. Show all the mathematical manipulations. Why does this rule require $Re(s) \equiv \sigma > 0$? What is the region of convergence of this function? Does this region include $s = 0$?

16 Sections 7.3 and 7.4 study the Laplace transform pairs. What is the role of these pairs, particularly their time-domain functions, in the advanced circuit analysis?

17 Find the Laplace transform of $f(t) = 4u(t-3)$. What is its region of conversion?

18 Consider a ramp function:

a) According to (7.18b), a unit ramp function is defined as $r_u(t) = t \cdot u(t)$. What is the Laplace transform of this function, $L\{t \cdot u(t)\}$?

b) Whether $r(t) = t^2$ is still a ramp function? What is the Laplace transform of this function?

c) What is the area under a unit ramp function if $t = 5s$? What is the dimension and meaning of this area?

19 What is the relationship among ramp, step, and delta functions?

20 Evaluate the following integrals containing the delta function:

a) $\int_0^3 \delta(t-5)dt$.

b) $\int_{-\infty}^{\infty} e^{-2t}\delta(t-4)dt$.

21 Find the Laplace transform of the following functions:

a) $f(t) = 3e^{-5t}$.

b) $f(t) = 3t^4 e^{-5t}$.

c) $f(t) = 3e^{-5t} cos8t$.

22 Find the Laplace transforms of two time-domain functions, $f_1(t) = 3e^{-5t} + 3cos8t$ and $f_2(t) = 3e^{-5t}cos8t$. Build their graphs using MATLAB and explain the difference between them.

23 Consider all the entries shown in Table 7.1: Why does Table 7.1 list these time-domain functions but not any others?

24 On the one hand, Problem 18c asks about area under the unit-ramp function, $v(t)V = t$ over $t = 5s$ and its dimension. The answer is $A = 5V \cdot s$. Here, the unit $V \cdot s$ is the measure of the magnetic flux intensity; this measure is known as weber. On the other hand, entry PLT-3 in Table 7.3 shows that Laplace transform of a voltage signal also carries the unit $V \cdot s$. Is this a confusing situation? We know that the System International assigns a unique unit to each individual measurable quantity in this world. Therefore, using the same unit, $V \cdot s$, for two different entities violates this fundamental principle. How would you resolve this contradiction?

25 Find the Laplace transform of the following time-domain functions:

a) $f(t) = 3te^{-5t} + 3t^2 e^{-5t}$.

b) $f(t) = 3e^{-5t} sin7t$.

c) $f(t) = 3e^{-5(t-7)}$.

d) $f(t) = 3e^{-5t} t^3$.

e) $f(t) = (3sin5t)^{`}$.
f) $f(t) = \int_0^t sin5\tau d\tau$.

26 **Find the initial and final values of $f(t) = 3e^{-5t}sin7t$ using the Laplace transform.

27 Solve all the problems in Problems 21, 25, and 26 using MATLAB.

28 Consider a series RL circuit without excitation whose R = 2 Ω and L 0.3 H. Find its response using the Laplace transform if $V_0 = 12V$.

8

Laplace Transform Application to Advanced Circuit Analysis

8.1 The Sufficient Condition of Using the Laplace Transform—The Inverse Laplace Transform

8.1.1 The Inverse Laplace Transform—The Need and the Methods

Chapter 7, particularly Tables 7.1 and 7.2 and Example 7.2, demonstrate why and how to apply the *Laplace transform* to circuit analysis, namely:

1) Derive a circuit's *differential equation* describing the circuit behavior.
2) Convert this differential equation into the equivalent s-domain *algebraic equation* using the (forward) Laplace transform.
3) Solve the s-domain equivalent algebraic circuit equation to find the circuit output in the s-domain.
4) Convert the obtained *s-domain circuit output* into the time domain by using the *inverse Laplace transform.*

Note that steps 1 through 3 constitute the *necessary condition* for the Laplace transform application to circuit analysis. Step 4—converting the obtained s-domain output into the time domain by using the inverse Laplace transform—is the *sufficient condition* of this application.

Figure 7.3 (reproduced here for convenience) visualizes this point: Performing all the Laplace transform manipulations is only the intermediate step toward obtaining the solution to a differential equation, which describes an electrical circuit being analyzed. To attain the actual solution, a time-domain function, we need to perform the inverse Laplace transform.

Example 7.2 demonstrates how to find the inverse Laplace transform for the free response of an RC circuit. Based on this example, it might seem that finding the Laplace transform is an easy matter, but such a conclusion would be too superficial. To clarify, consider the following example.

Example 8.1 Step response of a first-order circuit—Part 1: The problem.

Problem: Using the Laplace transform, find the RC circuit response to a step signal. The circuit's schematics and all the parameters are shown in Figure 8.1. When the switch is turned on, the input signal, $V_{in} = 12V$, is applied to the circuit. Assume zero initial condition, $V_0 = 0\ (V)$ and compute $\alpha = \frac{1}{RC} = 166.67\left(\frac{1}{s}\right)$.

Essentials of Advanced Circuit Analysis: A Systems Approach, First Edition. Djafar K. Mynbaev.

Companion Website: www.wiley.com/go/Mynbaev/AdvancedCircuitAnalysis

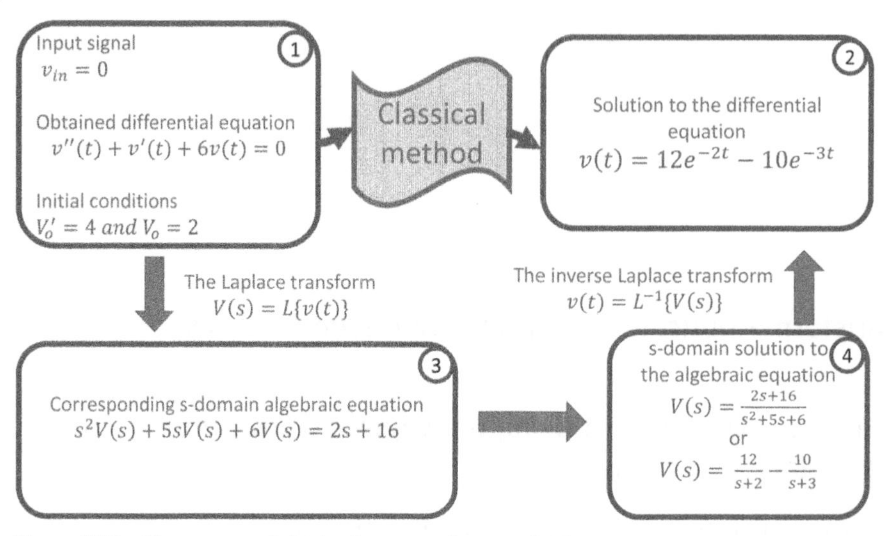

Figure 7.3R The concept of the Laplace-transform method.

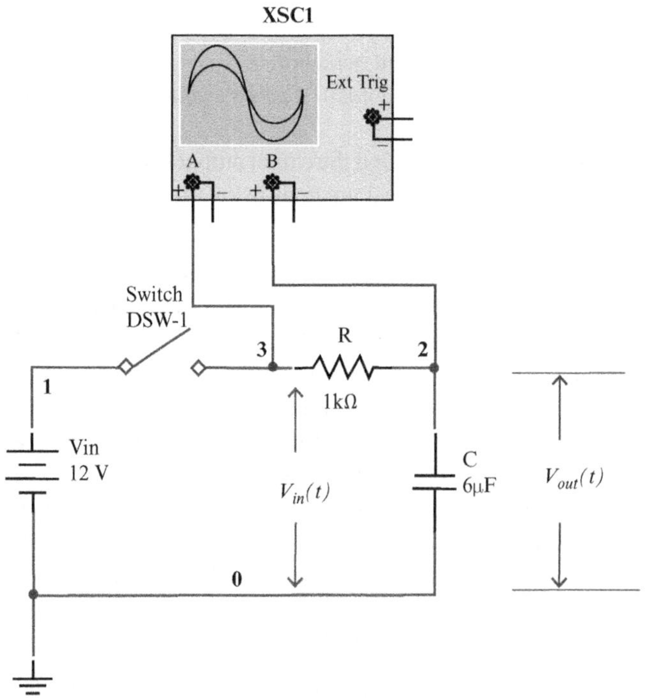

Figure 8.1 Schematic of the experimental setup for finding the step response of an RC LPF.

Solution:

You will recall that this problem was considered in Section 4.1, where the circuit differential equation was derived as

$$v'_{out}(t) + \alpha v_{out}(t) = \alpha V_{in} \cdot u(t). \tag{8.1}$$

Here, $u(t)$ is the *unit step (Heaviside) function* and $V_{in} = 12V$ is the input dc signal.
(Recall how we derive this equation: The *KVL* gives

$$v_R(t) + v_C(t) = v_{in}(t),$$

where $v_C(t) \equiv v_{out}(t)$, and $v_{R(t)} = i(t) \cdot R$. Since $i(t) = C\frac{dv_C(t)}{dt}$, we find (8.1) by inserting $v_{R(t)} = RC\frac{dv_C(t)}{dt}$ into the KVL equation and multiplying it through by $\alpha = \frac{1}{RC}$.)

The Laplace transform of this equation, $L\left\{v'_{out}(t) + \alpha v_{out}(t) = \alpha V_{in} \cdot u(t)\right\}$, yields

$$sV_{out}(s) + \alpha V_{out}(s) = \frac{\alpha V_{in}}{s}. \tag{8.2}$$

The solution for $V_{out}(s)$ is

$$V_{out}(s) = \frac{\alpha V_{in}}{s(s+\alpha)}. \tag{8.3}$$

Following the routine shown in Figure 7.3 and discussed in Example 7.2, we need to find the inverse Laplace transform of (8.3); however, *there is no appropriate entry in Table 7.1*. Therefore, it seems we are unable to find the time-domain response of this circuit. Do you have any suggestions? Thinking logically, we should reconstruct (8.3) in the form that can be found in Table 7.1, but how? Let's investigate. (End of Part 1 of Example 8.1.)

We postpone the solution's search for Part 2 of this example. For now, we have to notice that Example 7.2 demonstrates that the inverse Laplace transform of the s-domain output signal, $V_{out}(s) = \frac{V_o}{(s+\alpha)}$, can be obtained by examining Table 7.1. Indeed, $v_{out(t)} = L^{-1}\left\{V_{out}(s) = \frac{V_o}{(s+\alpha)}\right\} = V_0 e^{-\alpha t}$, according to LT-7 in Table 7.1. However, such a situation is not always the case, as Example 8.1 demonstrates.

Thus, we have to conclude that there are two cases in finding the inverse Laplace transform:

- Case 1, when the inverse Laplace transform can be found by examining Table 7.1.
- Case 2, when the s-domain output signal has no counterpart in Table 7.1.

In Case 2, the inverse Laplace transform might be a sophisticated operation, which requires in-depth investigation. Such an investigation will be provided in the later sections; in Section 8.2, we will concentrate on Case 1.

The inverse Laplace transform comes into play when an s-domain solution to a problem is transformed back into a real-time-domain function (signal). In many cases, it is the most time-consuming and challenging operation in the entire cycle. In fact, this is the price we pay for all the advantages of the Laplace transform method.

8.2 Finding the Inverse Laplace Transform by Examining Table 7.1

This section considers the examples of straightforward inverse Laplace transforms; they are collected in Table 8.1.

Examples 1 to 9 are rather straightforward: We examine Table 7.1, find the corresponding rule, and apply it. Examples 2 through 11 call for the use of the Linearity property, PLT-5 from Table 7.3.

Table 8.1 Examples of the inverse Laplace transforms performed by examination of Table 7.1.

The inverse Laplace transforms of s-domain functions				
		Rules to be applied		
No.	Given s-domain functions	Rules designations in Tables 7.1 and 7.3	The rules per se	Result of the inverse Laplace transform
1	$F(s) = \frac{5}{s}$	LT-1	$\mathcal{L}^{-1}\{\left\{\frac{A}{s}\right\} = A$	$\mathcal{L}^{-1}\left\{\frac{5}{s}\right\} = 5$
2	$F(s) = \frac{4}{s^2}$	PLT-5 and LT-3	$\mathcal{L}^{-1}\left\{\frac{k}{s^2}\right\} = kt$	$4\,\mathcal{L}^{-1}\left\{\frac{1}{s^2}\right\} = 4t$
3	$F(s) = \frac{12}{s^4}$	PLT-5 and LT-4	$k\,\mathcal{L}^{-1}\left\{\frac{n!}{s^{n+1}}\right\} = kt^n$	$2\,\mathcal{L}^{-1}\left\{\frac{6}{s^4}\right\} = 2t^3$
4	$F(s) = 5$	PLT-5 and LT-5	$\mathcal{L}^{-1}\{A\} = A\delta(t)$	$\mathcal{L}^{-1}\{5\} = 5\delta(t)$
5	$F(s) = 3s$	PLT-5 and LT-6	$\mathcal{L}^{-1}\{As\} = A\delta'(t)$	$\mathcal{L}^{-1}\{3s\} = 3\delta'(t)$
6	$F(s) = \frac{3}{s+4}$	PLT-5 and LT-7	$k\,\mathcal{L}^{-1}\left\{\frac{1}{s+\alpha}\right\} = k\,e^{-\alpha t}$	$3\,\mathcal{L}^{-1}\left\{\frac{1}{s+4}\right\} = 3e^{-4t}$
7	$F(s) = \frac{12}{(s+4)^2}$	PLT-5 and LT-8	$k\,\mathcal{L}^{-1}\left\{\frac{n!}{(s+a)^{n+1}}\right\} = k\,te^{-at}$	$2\,\mathcal{L}^{-1}\left\{\frac{3!}{(s+5)^4}\right\} = 2e^{-5t}t^3$
8	$F(s) = \frac{27}{(s^2+9)}$	PLT-5 and LT-10	$k\,\mathcal{L}^{-1}\left\{\frac{\omega}{(s^2+\omega^2)}\right\} = k\sin\omega t$	$9\,\mathcal{L}^{-1}\left\{\frac{3}{(s^2+3^2)}\right\} = 9\sin(3t)$
9	$F(s) = \frac{5s}{(s^2+16)}$	PLT-5 and LT-11	$k\,\mathcal{L}^{-1}\left\{\frac{s}{(s^2+\omega^2)}\right\} = k\cos\omega t$	$5\,\mathcal{L}^{-1}\left\{\frac{s}{(s^2+4^2)}\right\} = 5\cos(4t)$

10	$F(s)=\frac{3s+24}{(s^2+16)}$	PLT-5, LT-10, and LT-11	$\mathcal{L}^{-1}\{k_1F_1(s)+k_2F_2(s)\}$ $=k_1\mathcal{L}^{-1}\left\{\frac{s}{(s^2+\omega^2)}\right\}+k_2\mathcal{L}^{-1}\left\{\frac{\omega}{(s^2+\omega^2)}\right\}$ $=k_1\cos\omega t+k_2\sin\omega t$	$\mathcal{L}^{-1}\left\{\frac{3s+24}{(s^2+16)}\right\}$ $=3\mathcal{L}^{-1}\left\{\frac{s}{(s^2+4^2)}\right\}$ $+6\mathcal{L}^{-1}\left\{\frac{4}{(s^2+4^2)}\right\}$ $=3\cos 4t+6\sin 4t$
11	$F(s)=\frac{3s+36}{s^2+4s+29}$	PLT-5, LT-12, and LT-13	$\mathcal{L}^{-1}\{k_1F_1(s)+k_2F_2(s)\}=$ $+k_1\mathcal{L}^{-1}\left\{\frac{\omega}{(s+\alpha)^2+\omega^2}\right\}$ $+k_2\mathcal{L}^{-1}\left\{\frac{s+\alpha}{(s+\alpha)^2+\omega^2}\right\}$ $=k_1e^{-\alpha t}\sin\omega t+k_2e^{-\alpha t}\cos\omega t$	$\mathcal{L}^{-1}\left\{\frac{3s+36}{s^2+4s+29}\right\}$ $=\mathcal{L}^{-1}\left\{\frac{3s+6+30}{s^2+4s+4+25}\right\}$ $=\mathcal{L}^{-1}\left\{\frac{30}{(s+2)^2+5^2}\right\}$ $+\mathcal{L}^{-1}\left\{\frac{3s+6}{(s+2)^2+5^2}\right\}$ $=6\mathcal{L}^{-1}\left\{\frac{5}{(s+2)^2+5}\right\}$ $+3\mathcal{L}^{-1}\left\{\frac{s+2}{(s+2)^2+5^2}\right\}$ $=6e^{-2t}\sin 5t+3e^{-2t}\cos 5t$

Example 11, however, requires additional attention. Here are the steps we take (along with the reasoning behind them) to obtain the solution to this example:

- Examine the given s-domain function, $F(s)=\frac{3s+36}{s^2+4s+29}$, and then review Table 7.1, where the Laplace transforms are given. Here, there are two possible candidates, $\frac{s+\alpha}{(s+\alpha)^2+\omega^2}$, and $\frac{\omega}{(s+\alpha)^2+\omega^2}$, that can be used in this inverse Laplace transform.
- Realize that the members of the given s-function must be in the forms required by Table 7.1. To attain these forms, we must
  - split the given F(s) into two functions with a common denominator, as $F(s) = \frac{k_1(s+\alpha)}{s^2+4s+29} + \frac{k_2\omega}{s^2+4s+29}$,
  - present the common denominator in a $(s+\alpha)^2 + \omega^2)$ form, and
  - break the numerator into the sum of two members, $k_2\omega$ and $k_1(s+\alpha)$.

- The following steps are performed to execute this plan:
  - To convert the denominator, $(s^2 + 4s+29)$, into the form $(s+\alpha)^2 + \omega^2)$, we reconfigure it as $(s^2 + 4s+29) = (s^2 + 4s+4-4+29) = (s+2)^2 + 25 = (s+2)^2 + 5^2$.
  - To split our numerator and present it as $k_1(s+\alpha)+k_2\omega$, we perform the following manipulations $3s+36=3s+6+30=3(s+2)+30=3(s+2)+6\cdot5$.
  - Thus, we obtain: $F(s) = k_1\frac{(s+\alpha)}{(s+\alpha)^2+\omega^2} + k_2\frac{\omega}{(s+\alpha)^2+\omega^2} = 3\frac{s+2}{(s+2)^2+5^2} + 6\frac{5}{(s+2)^2+5^2}$.
  - Now, we can apply Rules LT-12 and LT-13 from Table 7.1 and get the results as $L^{-1}\{F(s)\} = 3e^{-2t}cos5t + 6e^{-2t}sin5t$.

All these operations are relatively simple, but we must get used to them, which takes exercise.

Unhappily, most of the circuits encountered in practice are described by much more sophisticated equations than those considered in Table 8.1. For these situations, a straightforward application of the Laplace transform table can't help. In such cases, we must apply the main alternative method of finding the inverse Laplace transforms called *partial fraction expansion*.

8.3 Partial Fraction Expansion—The Main Method of Finding the Inverse Laplace Transform

Let's return to **Example 8.1**: We stuck at the point that the s-domain step response of the RC circuit, $V_{out}(s)=\frac{\alpha V_{in}}{s(s+\alpha)}$, cannot be converted into the time-domain solution by examining Table 7.1. How can this problem be solved?

We don't have any other means to find the inverse Laplace transform but Table 7.1. Therefore, this problem's solution boils down to bringing (8.3) to the form given in this table. Here is the clue: If we were able to reformat (8.3) as

$$V_{out}(s)=\frac{\alpha V_{in}}{s(s+\alpha)}=\frac{A_1}{s}+\frac{A_2}{s+\alpha}, \tag{8.4}$$

the problem would be almost solved. Indeed, Table 7.1 gives the inverse Laplace transform of (8.4) as

$$v_{out}(t)=L^{-1}\{V_{out}(s)\}=L^{-1}\left\{\frac{A_1}{s}+\frac{A_2}{s+\alpha}\right\}=A_1+A_2e^{-\alpha t}. \tag{8.5}$$

Thus, to complete the task, we would "merely" need to find constants A_1 and A_2.

But how can we present $\frac{V_{in}}{s(s+\alpha)}$, the member having a product $s\cdot(s+\alpha)$ in the denominator, as a sum of two members? Examine the required mathematical manipulation again:

$$\frac{\alpha V_{in}}{s(s+\alpha)}=\frac{A_1}{s}+\frac{A_2}{s+\alpha}. \tag{8.6}$$

If we bring the RHS to the common denominator, then (8.6) becomes

$$\frac{\alpha V_{in}}{s(s+\alpha)}=\frac{A_1}{s}+\frac{A_2}{s+\alpha}=\frac{A_1(s+\alpha)\cdot A_2 s}{s(s+\alpha)}. \tag{8.7}$$

Equation 8.7 shows that we can attain (8.6) provided that A_1 and A_2 satisfy the following condition:

$$\alpha V_{in}=A_1(s+\alpha)\cdot A_2 s. \tag{8.8}$$

Operation (8.6) is called partial fraction expansion; it is crucial in finding the inverse Laplace transform of any s-domain function.

Nevertheless, (8.8) shows that the "small" step to get A_1 and A_2 is not a straightforward task because it requires determining two unknown members from one equation. Analysis of (8.6) hints that this problem can be solved by the following two-step operation:

1) Multiplying both sides of (8.6) by s releases A_1 as

$$\frac{\alpha V_{in}\cdot s}{s\cdot(s+\alpha)}=\frac{A_1\cdot s}{s}+\frac{A_2\cdot s}{s+\alpha}\Rightarrow\frac{\alpha V_{in}}{(s+\alpha)}=A_1+\frac{A_2\cdot s}{s+\alpha}. \tag{8.9}$$

2) Computing both sides at $s=0$ eliminates A_2 and gives A_1 as

$$\left.\frac{\alpha V_{in}}{(s+\alpha)}\right|_{s=0}=A_1+\left.\frac{A_2\cdot s}{s+\alpha}\right|_{s=0}\Rightarrow V_{in}=A_1. \tag{8.10}$$

Follow the same reasoning, A_2 can be found as

$$\frac{\alpha V_{in}\cdot(s+\alpha)}{s\cdot(s+\alpha)}=\frac{A_1\cdot(s+\alpha)}{s}+\frac{A_2\cdot(s+\alpha)}{s+\alpha}\Rightarrow\frac{\alpha V_{in}}{s}=\frac{A_1\cdot(s+\alpha)}{s}+A_2 \tag{8.11}$$

and

$$\left.\frac{\alpha V_{in}}{s}\right|_{s=-\alpha}=\left.\frac{A_1\cdot(s+\alpha)}{s}\right|_{s=-\alpha}+A_2\Rightarrow -V_{in}=A_2. \tag{8.12}$$

Note that we multiply both sides in (8.11) by $(s+\alpha)$ to release A_2, and in (8.12) we compute both sides at $s=-\alpha$ to eliminate A_1.

Now, we can plug $A_1=V_{in}$ and $A_2=-V_{in}$ into (8.5) and ascertain the solution to our problem as

$$v_{out}(t)=V_{in}-V_{in}e^{-\alpha t}=V_{in}\left(1-e^{-\alpha t}\right). \tag{8.13}$$

Though very specific, this discussion presents the concept of finding the inverse Laplace transform of any s-domain function. Generalization of this approach follows; at this point, let's employ the obtained result to complete Example 8.1.

Example 8.1 Continued Step response of a first-order circuit—Part 2: the solution.

Problem: Find the RC circuit response to a step signal. The circuit's schematics and all the parameters are shown in Figure 8.1. When the switch is turned on, the input signal, $V_{in} = 12V$, is applied to the circuit. Assume zero initial condition, $V_0 = 0V$.

Recall that Part 1 of this example in Section 8.1 gives the circuit's response in the s-domain as

$$V_{out}(s) = \frac{\alpha V_{in}}{s(s+\alpha)}. \tag{8.3R}$$

Solution:

Following the procedure described above in Equations 8.3 through 8.13, we expand (8.3) into the partial fractions as

$$V_{out}(s) = \frac{\alpha V_{in}}{s(s+\alpha)} = \frac{A_1}{s} + \frac{A_2}{s+\alpha} \tag{8.14}$$

and attain

$$v_{out}(t) = L^{-1}\{V_{out}(s)\} = L^{-1}\left\{\frac{\alpha V_{in}}{s(s+\alpha)}\right\} = L^{-1}\left\{\frac{A_1}{s} + \frac{A_2}{(s+\alpha)}\right\} = V_{in} - V_{in}e^{-\alpha t} = V_{in}\left(1 - e^{-\alpha t}\right). \tag{8.15}$$

Using the given values, we get the answer

$$v_{out}(t) = 12\left(1 - e^{-166.67t}\right). \tag{8.16}$$

The graph of $v_{out}(t)$ built with MATLAB is shown in Figure 8.2. Remember that at $t = 5\tau = 5RC(s)$, the time-domain output becomes equal to $v_{out}(t) = 12\left(1 - e^{-\frac{5\tau}{\tau}}\right) = 12(1 - 0.0067) \approx 11.92V$. The problem is solved.

Discussion:

- Does the obtained solution make sense? Considering the initial and final values of the solution usually helps to answer this question. In our example, initially, before the switch was turned on, the circuit was in the relaxed state. Therefore, its output at $t = 0^-$ must be zero. At the final state, when $t \to \infty$, the capacitor will be fully charged, and the voltage drop across it should be equal to the input signal; that is, $v_{out}(t) \to 12V$ as $t \to \infty$. The solution obtained in (8.15) and its visual presentation in Figure 8.2 confirm our propositions.

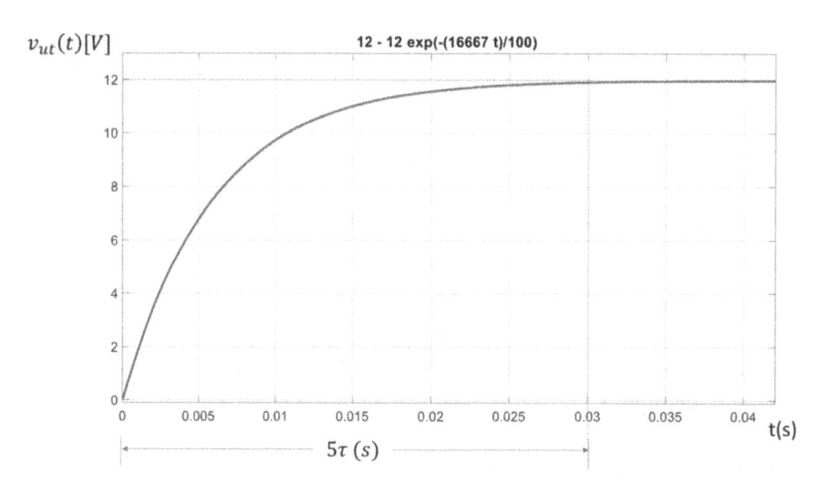

Figure 8.2 Step response of an RC circuit in Example 8.1.

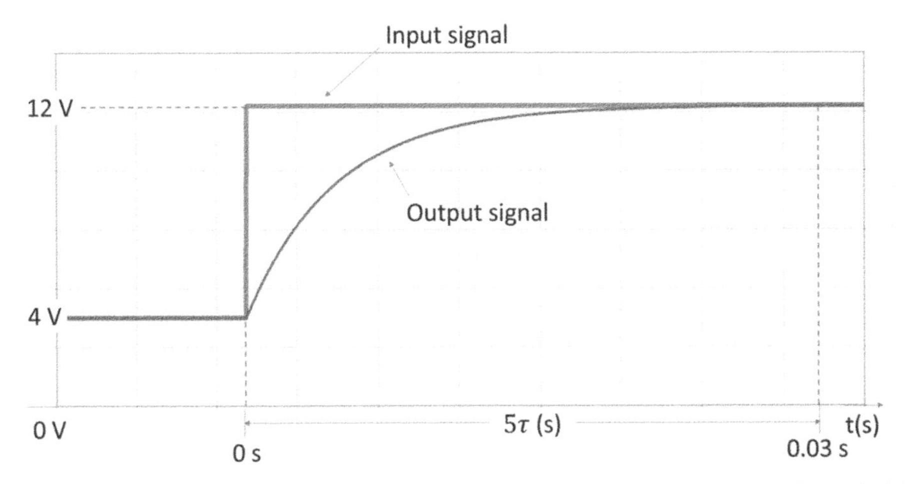

Figure 8.3 The result of Multisim simulation of the RC circuit's step response for Example 8.1.

- What if the initial condition is not zero? Multisim simulation of the RC circuit's step response for $V_0 = 4\ (V)$ is shown in Figure 8.3.
- Can we apply the Laplace transform to an RC circuit with a non-zero initial condition? Certainly. If the capacitor's initial voltage is $v_{out}(0) = V_0 = 4$ V, as in Figure 8.3, then

$$L\{v'_{out}(t)\} = sV_{out}(s) - V_0, \tag{8.17}$$

according to PLT-9. This addition causes the following change in (8.2)

$$sV_{out}(s) + \alpha V_{out}(s) = \frac{\alpha V_{in}}{s} + V_0. \tag{8.18}$$

Now, the circuit's response in the s-domain takes the form

$$V_{out}(s) = \frac{\alpha V_{in}}{s(s+\alpha)} + \frac{V_0}{(s+\alpha)}. \tag{8.19}$$

Remember that V_0 is a constant, and therefore we need to treat the RHS of (8.18) as a whole function in the s-domain. Hence, we apply the technique used in this example and find the step response of the initially charged RC circuit as

$$\begin{aligned} v_{out}(t) &= L^{-1}\{V_{out}(s)\} = L^{-1}\left\{\frac{\alpha V_{in}}{s(s+\alpha)} + \frac{V_0}{(s+\alpha)}\right\} \\ &= V_{in} - V_{in}e^{-\alpha t} + V_0 e^{-\alpha t} = V_{in} - (V_{in} - V_0)e^{-\alpha t}). \end{aligned} \tag{8.20}$$

or

$$v_{out}(t) = 12 - 8e^{-166.67t}.$$

Figure 8.4 shows the MATLAB solution for this example for $V_0 = 4V$. At what point does this graph start? Does this solution make sense? Can we consider (8.20) as a general result?

Compare the graphs displayed in Figures 8.3 and 8.4 and comment on their likeness and difference. Remember that at $t = 5\tau = 0.03s$ and $v_{out}(t)\big|_{t=5\tau} = V_{in} \cdot 0.993V$. Is this rule applied to both cases when $V_0 = 0V$ and $V_0 = 4V$? In consideration of this question, refer to Equations 8.16 and 8.20.

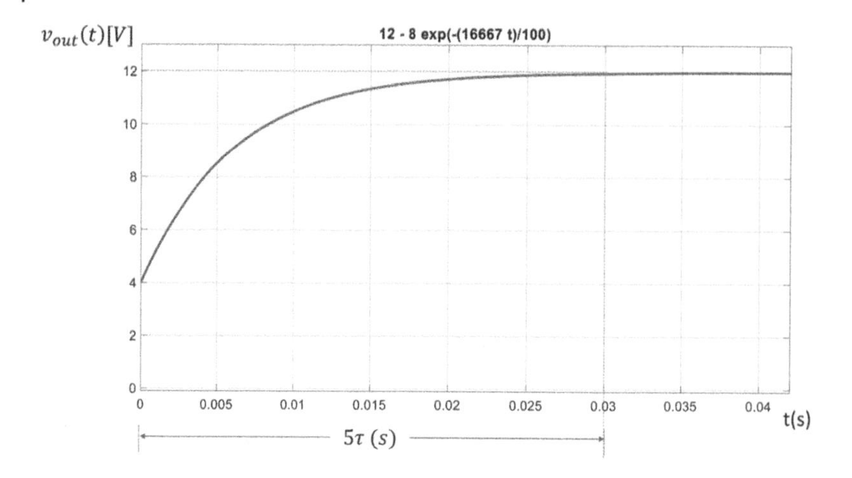

Figure 8.4 MATLAB solutions for the RC circuit's step responses for $V_0 = 4\,(V)$ in Example 8.1.

Here is the MATLAB script for the case $V_0 = 4V$:

```
syms s t
Vin = 12;
V0 = 4;

a = 166.67;
n = a*Vin+s*V0;
d = s*(s+a);
c =ilaplace(n/d,t)

pretty(c)
ezplot(c,[0,0.05])
axis([0 0.042 -0.1 13]);
```

- Rewriting (8.20) as

$$v_{out}(t) = (V_0 - V_{in})e^{-\alpha t} + V_{in}, \tag{8.21a}$$

we arrive at the general format presenting the response of a first-order circuit as

$$v_{out}(t) = (v(0) - v(\infty))e^{-\alpha t} + v(\infty). \tag{8.21b}$$

Thus, we find the other verification of the decisive role of the initial and final conditions in finding the transient response of a circuit under a non-zero initial condition to an excitation.

Exercise 8.1 Consider a parallel RC circuit shown in Figure 8.5a. Its output, $v_{out}(t)$, obtained by Multisim simulation is given in Figure 8.5b. Using the Laplace transform, calculate its step response, and build the graph of $v_{out}(t)$. Compare the calculated and experimental graphs and discuss the result.

Question Why Figure 8.5b doesn't show the input voltage? What graph of $v_{in}(t)$ would you expect to see?

Now, we need to delve into the technique for finding the partial fraction expansions.

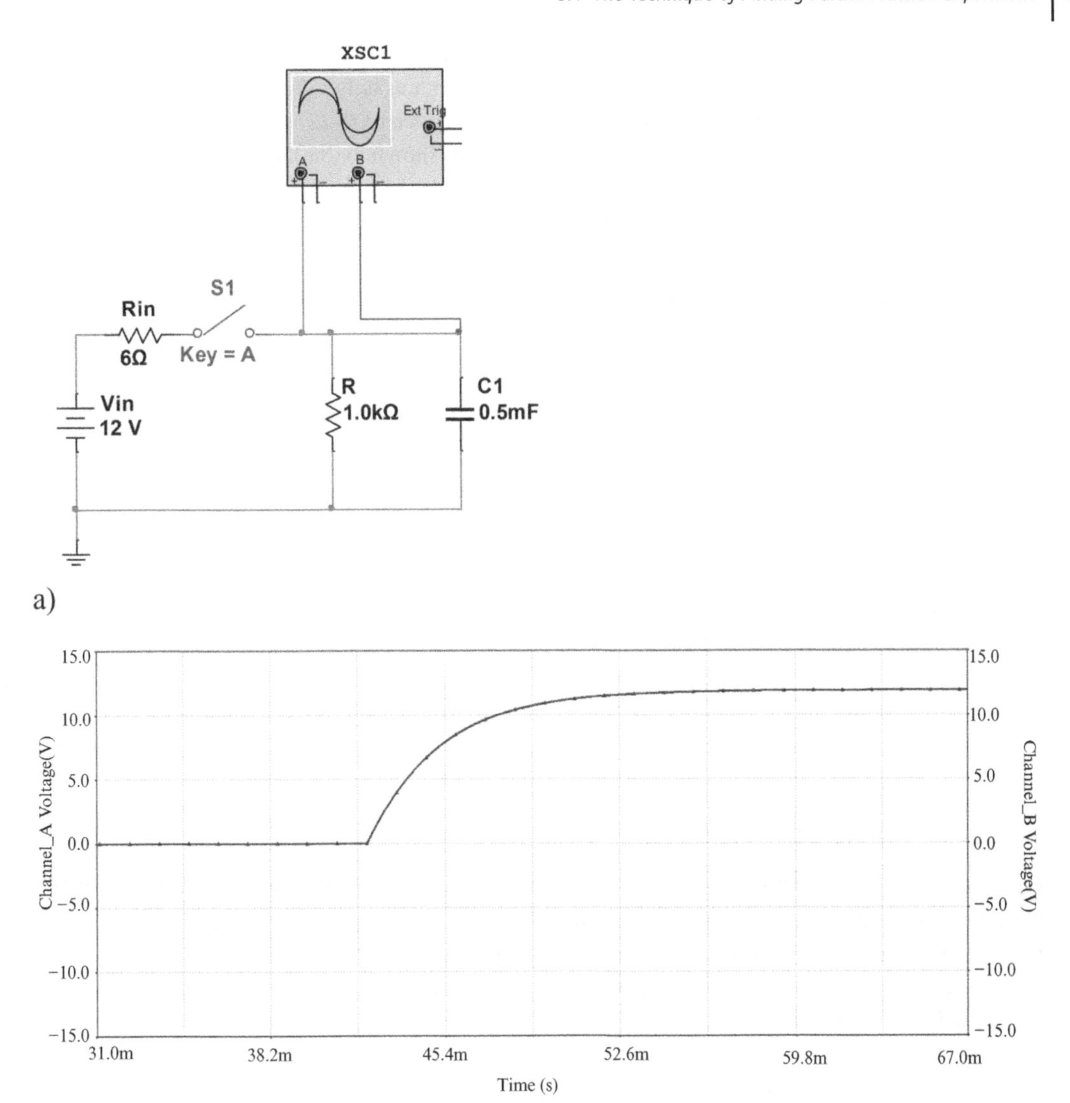

Figure 8.5 Step response of a parallel RC circuit: a) The experimental setup; b) output voltage, $v_{out}(t)$, obtained by Multisim simulation.

8.4 The Technique of Finding Partial Fraction Expansions

8.4.1 Problem and Solution

Let's reiterate the problem we encounter: On the one hand, we will meet the s-domain functions, F(s), whose general form is given by

$$F(s) = \frac{N(s)}{D(s)} = \frac{a_n s^n + a_{n-1} s^{n-1} + \ldots + a_0}{b_m s^m + b_{m-1} s^{m-1} + \ldots + b_0}, \tag{8.22}$$

where F(s) is called a *rational function* since it is the ratio of two polynomials. Here, N(s) and D(s) stand for numerator and denominator, respectively, a_n, and b_m are the constants, n and m are the orders of the numerator and denominator. On the other hand, Table 7.1 of the Laplace transforms—which enables us to find the inverse Laplace transforms—contains only s-domain formulas with binomial denominators. Therefore, *the problem is to present F(s) given as (8.22) as a sum of the ratios with binomial denominators*. The example of the required form is shown in (8.6); it can be generalized as

$$F(s)=\frac{A_1}{s-p_1}+\frac{A_2}{s-p_2}+\frac{A_3}{s-p_3}+\ldots\frac{A_m}{s-p_m}. \tag{8.23}$$

Note that the denominators in (8.23) can take any form given in Table 7.1, including those containing s^2, as shown in (LT-8) through (LT-13). The method enabling us to proceed from (8.22) to (8.23) is called *partial fraction expansion*, as mentioned above.

> *The partial fraction expansion is an algebraic technique that presents a sophisticated rational function as a sum of simple ratios.*

Let's elaborate on this point: Consider, for example, the following F(s) function

$$F(s)=\frac{11s+26}{s^2+5s+6}. \tag{8.24}$$

Compare (8.24) with (8.22) and find that here $a_1=11$, $a_0=26$, and $n=1$. Also $b_2=1$, $b_1=5$, $b_0=6$, and $m=2$.

After the algebraic manipulations, similar to that we use in Example 8.1, this function can be brought to the form

$$F(s)=\frac{11s+26}{s^2+5s+6}=\frac{4}{s+2}+\frac{7}{s+3}. \tag{8.25}$$

(Don't worry about the details of this transformation: We will study them shortly.) Compare (8.25) with (8.23) to understand that here $A_1=4$, $p_1=-2, A_2=7$, and $p_2=-3$. You can verify the correctness of this partial fraction expansion by bringing the RHS to the common denominator.

Equation 8.25 enables us immediately find the required inverse Laplace transform of the given F(s) by using LT-7 of Table 7.1 and PLT-5 of Table 7.3 as

$$f(t)=L^{-1}\{F(s)\}=L^{-1}\left\{\frac{11s+26}{s^2+5s+6}\right\}=L^{-1}\left\{\frac{4}{s+2}+\frac{7}{s+3}\right\}=4e^{-2t}+7e^{-3t}. \tag{8.26}$$

The concept of obtaining the inverse Laplace transform by using partial fraction expansion seems to be clear and easy, but the first question that should cross our mind is, how we can progress from $\frac{11s+26}{s^2+5s+6}$ to $\frac{4}{s+2}+\frac{7}{s+3}$? Indeed, this move is not apparent and requires a thorough investigation.

8.4.2 Factored form and poles

For the incoming mathematical manipulations, we need to recall the following rule of two binomials multiplication,

$$(s+2)(s+3)=s\cdot s+s\cdot 3+2\cdot s+2\cdot 3=s^2+5s+6 \tag{8.27}$$

The first step in moving from the rational s-domain function $\frac{11s+26}{s^2+5s+6}$ to its partial fraction expansion is presenting the given F(s) denominator in a *factored form*, as

$$\frac{11s+26}{s^2+5s+6}=\frac{11s+26}{(s+2)(s+3)}. \tag{8.28}$$

The real question is, why we use $(s+2)$ and $(s+3)$ but not, say $(s+5)$ and $(s+8)$? In other words, how do we know in advance what numbers will work for a particular case? We can't apply here a trial-and-error approach with its infinite number of possible combinations. This is an aged-known problem with a long-known solution, whose procedure we demonstrate below, exemplifying it by (8.28):

Step 1. Equate the denominator, D(s), to zero, and attain a *characteristic equation* for s as

$$s^2+5s+6=0 \tag{8.29}$$

Step 2. Solve the characteristic equation for s and find *roots*.
The general way to solve a quadratic equation, $as^2+bs+c=0$, is by using the well-known formula,

$$s_{1,2}=\frac{-b\pm\sqrt{b^2-c^2}}{2a}. \tag{8.30}$$

Applying this formula to (8.29), we find roots $s_1=-2$ and $s_2=-3$.

For $F(s)=\frac{N(s)}{D(s)}=\frac{a_ns^n+a_{n-1}s^{n-1}+\ldots+a_0}{b_ms^m+b_{m-1}s^{m-1}+\ldots+b_0}$ given in (8.22), the numerator's roots are called *zeros*, z_i, and the roots of the denominator are called ***poles***, $\boldsymbol{p_i}$. Here i = 1, 2, 3, ..., m. Observe that ***the number of poles is equal to a polynomial order.***

Step 3. Present a given denominator in a *factored form*:

$$b_ms^m+b_{m-1}s^{m-1}+\ldots+b_0=(s-p_1)(s-p_2)(s-p_3)\ldots(s-p_m), \tag{8.31}$$

where $p_1,p_2,p_3,\ldots p_m$ are its poles (roots). This is why in our example we have

$$s^2+5s+6=(s+2)(s+3). \tag{8.32}$$

We can verify our operations correctness by multiplying two binomials, as shown in (8.27).

Note that manipulation given in (8.27) offers the following *alternative method of finding the roots (poles) of a quadratic equation*:

For $s^2+bs+c=(s-p_1)(s-p_2)$, we have

$$b=-(p_1+p_2) \text{ and } c=p_1\cdot p_2. \tag{8.33}$$

Solving these two equations, we find p_1 and p_2. Thus, for $s^2+5s+6=0$, we have $5=-(p_1+p_2)$ and $6=p_1\cdot p_2$, which results in $p_1=-2$ and $p_2=-3$. *Pay attention to the signs of the poles.*

Step 4. Present a whole F(s) whose denominator is given in a factored form as

$$F(s)=\frac{N(s)}{D(s)}=\frac{N(s)}{(s-p_1)(s-p_2)(s-p_3)\ldots(s-p_m)}. \tag{8.34}$$

In our example, we find

$$F(s)=\frac{11s+26}{s^2+5s+6}=\frac{11s+26}{(s+2)(s+3)},$$

as shown in (8.28).

Step 5. Expand the obtained factored-form *F(s)* into partial fractions.
Using the basic algebraic rule, we can present a given rational function as a sum of fractions with binomial denominators; that is,

$$F(s)=\frac{N(s)}{D(s)}=\frac{N(s)}{(s-p_1)(s-p_2)(s-p_3)\ldots(s-p_m)}=\frac{A_1}{(s-p_1)}+\frac{A_2}{(s-p_2)}+\frac{A_3}{(s-p_3)}+\cdots+\frac{A_m}{(s-p_m)}. \quad (8.35)$$

For our example, we get

$$F(s)=\frac{11s+26}{s^2+5s+6}=\frac{11s+26}{(s+2)(s+3)}=\frac{A_1}{s+2}+\frac{A_2}{s+3}. \quad (8.36)$$

This expansion enables us to determine the inverse Laplace transform of *F(s)* from Table 7.1 as

$$f(t)=\mathcal{L}^{-1}\{F(s)\}=A_1e^{p_1t}+A_2e^{p_2t}+A_3e^{p_3t}+\ldots+A_me^{p_mt}. \quad (8.37)$$

In the example we are discussing, this step results in

$$f(t)=L^{-1}\{F(s)\}=L^{-1}\left\{\frac{A_1}{s+2}+\frac{A_2}{s+3}\right\}=A_1e^{-2t}+A_2e^{-3t}. \quad (8.38)$$

Thus, our task is almost complete. As we can see, *we need the partial fraction expansion because each member of this expansion has a straight inverse Laplace transform.*

8.4.3 Finding the Residues (Constants) $A_1, A_2, A_3, \ldots, A_m$

To attain a time-domain function $f(t)$ shown in (8.38) explicitly (which is the ultimate goal of these manipulations, remember?), we need to find constants A_1, A_2, A_3,..., A_m. These constants (called *residues*) can be real or complex. The residues, A_i, are the amplitudes of the time-domain functions, as (8.38) shows.

In Section 8.3 and Part 2 of Example 8.1, we develop reasoning how to find residues A_1 and A_2. Extending that approach, we need to take a close look at (8.35) presented here as

$$\frac{N(s)}{(s-p_1)(s-p_2)(s-p_3)\ldots(s-p_m)}=\frac{A_1}{(s-p_1)}+\frac{A_2}{(s-p_2)}+\frac{A_3}{(s-p_3)}+\cdots+\frac{A_m}{(s-p_m)}. \quad (8.39)$$

To find A_3, for example, we can multiply both sides of (8.39) by the binomial $(s-p_3)$ and compute the result at $s=p_3$. That is,

$$A_3=\frac{N(s)}{(s-p_1)(s-p_2)(s-p_3)\ldots(s-p_m)}(s-p_3)\bigg|_{s=p_3}$$
$$=\frac{A_1(s-p_3)}{(s-p_1)}\bigg|_{s=p_3}+\frac{A_2(s-p_3)}{(s-p_2)}\bigg|_{s=p_3}+\frac{A_3(s-p_3)}{(s-p_3)}\bigg|_{s=p_3}+\cdots+\frac{A_m(s-p_3)}{(s-p_m)}\bigg|_{s=p_3}. \quad (8.40)$$

At the LHS, binomials $(s-p_3)$ at the numerator and denominator cancel each other, and, instead of a variable *s*, we plug in a constant p_3. At the RHS, all members, except for A_3, turn to zero. Thus, we obtained

$$\frac{N(s)}{(p_3-p_1)(p_3-p_2)(p_3-p_4)\ldots(s-p_m)} = A_3. \tag{8.41}$$

Generalizing this consideration, we can write

$$A_i = \frac{N(s)}{(s-p_1)(s-p_2)(s-p_3)\ldots(s-p_m)}(s-p_i)\Big|_{s=p_i} = F(s)(s-p_i)\Big|_{s=p_i}. \tag{8.42}$$

Applying (8.40) to our example yields

$$A_1 = \frac{11s+26}{(s+2)(s+3)}(s+2)\Big|_{s=-2} = \frac{-22+26}{-2+3} = 4$$

and

$$A_2 = \frac{11s+26}{(s+2)(s+3)}(s+3)\Big|_{s=-3} = \frac{-33+26}{-3+2} = 7.$$

Thus, the given $F(s) = \frac{11s+26}{s^2+5s+6}$ is expanded into *partial fractions* as

$$F(s) = \frac{11s+26}{s^2+5s+6} = \frac{11s+26}{(s+2)(s+3)} = \frac{4}{s+2} + \frac{7}{s+3}. \tag{8.43}$$

The method shown in (8.40) is called a *cover-up algorithm* because it temporarily covers one unknown residue to find the other. It is also called a *residue method* or a *standard method.* Be aware that there are several other methods of finding the residues, which will be discussed soon.

Before proceeding further, it's worth pausing and summarizing the partial fraction expansion technique in the visual form shown in Figure 8.6. As mentioned above, we can check the accuracy of this expansion by bringing the right-hand sum to the common-denominator form and verify that it gives the original $F(s)$. The other, more robust approach to checking the partial fraction expansion is called a *test-value method.* We substitute any value for s into both LHS and RHS and verify that the obtained numbers are the same. Plug $s = 2$, for example, into both sides of the equation in Figure 8.6 and confirm that they produce $\frac{12}{5} = \frac{12}{5}$. Let $s = 1$ and verify (8.43).

Now, we need to plug the values of residues A_1 and A_2 in (8.36) to ascertain the final answer

$$f(t) = \mathcal{L}^{-1}\{F(s)\} = A_1 e^{-2t} + A_2 e^{-3t} = 4e^{-2t} + 7e^{-3t}, \tag{8.44}$$

as shown previously.

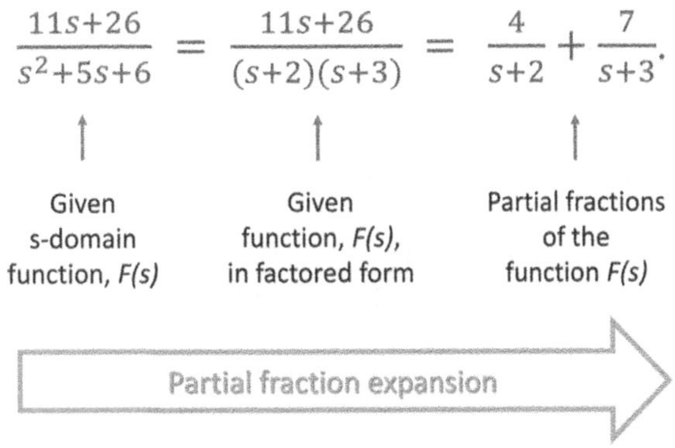

$$\frac{11s+26}{s^2+5s+6} = \frac{11s+26}{(s+2)(s+3)} = \frac{4}{s+2} + \frac{7}{s+3}.$$

Given s-domain function, *F(s)* — Given function, *F(s)*, in factored form — Partial fractions of the function *F(s)*

Partial fraction expansion

Figure 8.6 Example of expanding of an s-domain function into partial fractions.

The techniques and results in finding $f(t)$ vary significantly depending on the nature of poles, which is determined by the nature of a circuit in question. In the next sections, we'll consider all possible cases of poles' nature and the resulting circuits' outputs.

Question The inverse Laplace transforms performed in (8.9) through (8.12) are reminded below:

$$\frac{\alpha V_{in}\cdot s}{s\cdot(s+\alpha)}=\frac{A_1\cdot s}{s}+\frac{A_2\cdot s}{s+\alpha}\Rightarrow\frac{\alpha V_{in}}{(s+\alpha)}=A_1+\frac{A_2\cdot s}{s+\alpha}, \tag{8.9R}$$

$$\left.\frac{\alpha V_{in}}{(s+\alpha)}\right|_{s=0}=A_1+\left.\frac{A_2\cdot s}{s+\alpha}\right|_{s=0}\Rightarrow V_{in}=A_1, \tag{8.10R}$$

$$\frac{\alpha V_{in}\cdot(s+\alpha)}{s\cdot(s+\alpha)}=\frac{A_1\cdot(s+\alpha)}{s}+\frac{A_2\cdot(s+\alpha)}{s+\alpha}\Rightarrow\frac{\alpha V_{in}}{s}=\frac{A_1\cdot(s+\alpha)}{s}+A_2, \tag{8.11R}$$

and

$$\left.\frac{\alpha V_{in}}{s}\right|_{s=-\alpha}=\left.\frac{A_1\cdot(s+\alpha)}{s}\right|_{s=-\alpha}+A_2\Rightarrow -V_{in}=A_2. \tag{8.12R}$$

Why did we determine A_1 by calculating both sides of (8.10) at $s=0$ but to find A_2, we calculate (8.12) at $s=-\alpha$?

Example 8.2 Finding the inverse Laplace transform of an s-domain function.

Problem: Consider $F_1(s)=\frac{8s+17}{s^2+9s+20}$. Find its inverse Laplace transform, $f(t)$.

Solution: This example is a convenient way to summarize the procedure of the *inverse Laplace transform finding*:

- The goal is to present the given $F_1(s)$ in a form shown in the *table of the Laplace transform pairs*, Table 7.1.
- This goal is attained by the *partial fraction expansion* of the $F_1(s)$ with residues A_1 and A_2.
- This expansion requires presenting the $F_1(s)$ in a *factored form.*
- The factored form needs finding the $F_1(s)$ *poles.*

Let's execute this plan, starting with finding the poles.

- Poles of $F_1(s)=\frac{8s+17}{s^2+9s+20}$ are the roots of the following *characteristic equation*

$$s^2+9s+20=0. \tag{8.45}$$

To solve (8.45) for p_1 and p_2, we can use either (8.30), $s_{1,2}\equiv p_{1,2}=\frac{-b\pm\left(\sqrt{b^2-4ac}\right)}{2a}$, or (8.33) which for $s^2+bs+c=(s-p_1)(s-p_2)$ gives $b=-(p_1+p_2)$ and $c=p_1\cdot p_2$. Either way, we find

$$p_1=-4 \text{ and } p_2=-5.$$

- Thus, the $F_1(s)$ *factored form* is

$$F_1(s)=\frac{8s+17}{s^2+9s+20}=\frac{8s+17}{(s+4)(s+5)}. \tag{8.46}$$

- The partial fraction expansion of the $F(s)$ turns to be

$$F_1(s) = \frac{8s+17}{(s+4)(s+5)} = \frac{A_1}{(s+4)} + \frac{A_2}{(s+5)}. \tag{8.47}$$

- *Residues (constants) A_1 and A_2 can be found by one of the three following methods:*
  - *Direct method*: Bring the RHS of (8.47) to the common-denominator form

    $$\frac{8s+17}{(s+4)(s+5)} = \frac{A_1(s+5)+A_2(s+4)}{(s+4)(s+5)},$$

    and equate the numerators of the LHS and RHS

    $$8s+17 = A_1(s+5)+A_2(s+4).$$

    Compute this equation at $s=-4$, and find $A_1 = \frac{-15}{1} = -15$. Repeat this computation at $s=-5$, and find $A_2 = \frac{-23}{-1} = 23$.
    These manipulations yield

    $$F_1(s) = \frac{8s+17}{(s+4)(s+5)} = \frac{-15}{s+4} + \frac{23}{s+5}. \tag{8.48}$$

    Apply the test-value method by plugging, for example, $s=3$ into (8.48), and calculate $\frac{41}{56} = \frac{41}{56}$. Thus, the obtained partial fraction expansion is correct.
  - *Algebraic method* (Alexander and Sadiku 2020, p. 694): Multiply both sides of (8.48) by $(s+4)(s+5)$ and get

    $$8s+17 = A_1(s+5)+A_2(s+4) = (A_1+A_2)s+5A_1+4A_2.$$

    Equating coefficients of like powers of s from both the right- and left-hand sides produces two equations for two unknown variables A_1 and A_2,

    $$\begin{cases} A_1 + A_2 = 8 \\ 5A_1 + 4A_2 = 17 \end{cases}.$$

    Solving these equations for A_1 and A_2, we find $A_1 = -15$ and $A_2 = 23$, as before.
  - Use the *general method*, i.e., the *cover-up algorithm*: Apply (8.42) and obtain

    $$A_1 = F(s)(s-p_1)|_{s=p_1} = \frac{8s+17}{(s+4)(s+5)}(s+4)|_{s=-4} = -15$$

    and

    $$A_2 = F_1(s)(s-p_2)|_{s=p_2} = \frac{8s+17}{(s+4)(s+5)}(s+5)|_{s=-5} = 23.$$

Thus, all three methods produce the same residues A_1 and A_2, as expected.
Therefore, we've obtained the partial fraction expansion of our $F(s)$ as

$$F_1(s) = \frac{8s+17}{(s+4)(s+5)} = \frac{A_1}{(s+4)} + \frac{A_2}{(s+5)} = \frac{-15}{(s+4)} + \frac{23}{(s+5)} \tag{8.49}$$

Finally, the inverse Laplace transform of the given F(s) is determined as

$$f_1(t) = L^{-1}\{F_1(s)\} = L^{-1}\left\{\frac{8s+17}{s^2+9s+20}\right\} = L^{-1}\left\{\frac{-15}{(s+4)} + \frac{23}{(s+5)}\right\} = -15e^{-4t} + 23e^{-5t} \quad (8.50)$$

The problem is solved.

Discussion:

- We can generalize the procedure of finding the inverse Laplace transform as follows: Consider the following s-domain rational function of the second-order,

$$F(s) = \frac{ms+n}{s^2+(a+b)s+ab}, \quad (8.51)$$

whose poles $p_1 = -a$ and $p_2 = -b$ allows for presenting the denominator in a factored form as

$$s^2+(a+b)s+ab = (s-p_1)(s-p_2) = (s+a)(s+b). \quad (8.52)$$

Then the given F(s) can be resolved in partial fractions as

$$F(s) = \frac{ms+n}{(s+a)(s+b)} = \frac{A_1}{s+a} + \frac{A_2}{s+b}. \quad (8.53)$$

We need this partial fraction expansion because its RHS enables us to find the inverse Laplace transform of the given F(s) using LT-7 of Table 7.1. Indeed,

$$L^{-1}\left\{\frac{A_1}{s+a} + \frac{A_2}{s+b}\right\} = A_1 e^{-at} + A_2 e^{-bt}. \quad (8.54)$$

The next question is, what are those coefficients A_1 and A_2? The answer is given by the equation of the partial fraction expansion (8.53): *A_1 and A_2 are numbers that make the RHS of this equation equal to its LHS*. In other words, after bringing the RHS to a common denominator, we can write

$$\frac{ms+n}{(s+a)(s+b)} = \frac{A_1(s+b)+A_2(s+a)}{(s+a)(s+b)}, \quad (8.55)$$

and A_1 and A_2 must turn this equation into identity.

The partial fraction expansion itself suggests the way to find the constants (residues) A_1 and A_2. In fact, if we multiply both sides of this equation by $(s + a)$

$$\frac{(ms+n)(s+a)}{(s+a)(s+b)} = \frac{A_1(s+a)}{s+a} + \frac{A_2(s+a)}{s+b}, \quad (8.56)$$

and compute the obtained expression at $s = p_1 = -a$

$$\left.\frac{(ms+n)}{(s+b)}\right|_{s=-a} = A_1 + \left.\frac{A_2(s+a)}{s+b}\right|_{s=-a}, \quad (8.57)$$

we find

$$\frac{-ma+n}{(-a+b)} = A_1. \quad (8.58)$$

Here a, b, m, and n are known numbers, enabling us to compute A_1. Similarly, we can find A_2 as

$$\frac{-mb+n}{(-b+a)}=A_2. \tag{8.59}$$

Finally, the inverse Laplace transform of the given F(s) is determined as

$$\begin{aligned} L^{-1}\{F(s)\}=L^{-1}\left\{\frac{ms+n}{s^2+(a+b)s+ab}\right\}=A_1e^{-at}+A_2e^{-bt} \\ =\frac{-ma+n}{-a+b}e^{-at}+\frac{-mb+n}{-b+a}e^{-bt}. \end{aligned} \tag{8.60}$$

Therefore, if F(s) is given in the form of (8.51), then its inverse Laplace transform can be found immediately by plugging the given numbers m, n, a, and b into (8.60). Incidentally, by plugging any numbers for n, m, a, and b into (8.51), you can create your own numerical examples.

- What if a *denominator's polynomial has the order higher than 2*? Consider, for example the following *F*(*s*):

$$F_2(s)=\frac{9s^2+34s+29}{s^3+6s^2+11s+6}. \tag{8.61}$$

The three poles (roots of the denominator's characteristic equation), are p_1 = -1, $p_2=-2$, and $p_3=-3$. (We refer you to mathematics textbooks or online sources to find out how the roots of a cubic equation can be determined.) These poles enable us to resolve (8.61) into partial fractions as

$$F_2(s)=\frac{9s^2+34s+29}{s^3+6s^2+11s+6}=\frac{9s^2+34s+29}{(s+1)(s+2)(s+3)}=\frac{A_1}{s+1}+\frac{A_2}{s+2}+\frac{A_3}{s+3}. \tag{8.62}$$

Then we find the residues by using the cover-up algorithm as

$$\begin{aligned} A_1=F(s)(s-p_1)\Big|_{s=p_1}=\frac{9s^2+34s+29}{(s+1)(s+2)(s+3)}(s+1)\Bigg|_{s=-1}=2, \\ A_2=F(s)(s-p_2)\Big|_{s=p_2}=\frac{9s^2+34s+29}{(s+1)(s+2)(s+3)}(s+2)\Bigg|_{s=-2}=3,\} \\ A_3=F(s)(s-p_3)\Big|_{s=p_3}=\frac{9s^2+34s+29}{(s+1)(s+2)(s+3)}(s+3)\Bigg|_{s=-3}=4. \end{aligned} \tag{8.63}$$

Therefore,

$$F_2(s)=\frac{9s^2+34s+29}{s^3+6s^2+11s+6}=\frac{2}{s+1}+\frac{3}{s+2}+\frac{4}{s+3}. \tag{8.64}$$

Finally,

$$f_2(t)=L^{-1}\{F_2(s)\}=L^{-1}\left\{\frac{9s^2+34s+29}{s^3+6s^2+11s+6}\right\}=L^{-1}\left\{\frac{2}{s+1}+\frac{3}{s+2}+\frac{4}{s+3}\right\}=2e^{-t}+3e^{-2t}+4^{-3t}. \tag{8.65}$$

These manipulations show that the higher-order s-domain functions are treated the same way as the two-order *F*(*s*). Of course, the complexity of mathematical (in fact, algebraic) manipulations increases exponentially with the increase in the denominator polynomial order; however, modern technology resolves this issue.

- Below are the MATLAB solutions to Equation 8.46 for Example 8.2 and the discussion of this example given in Equation 8.61:
  - The MATLAB solution to Equation 8.46, $F_1(s)=\frac{8s+17}{s^2+9s+20}$, is as follows:

```
≫ syms s t
a = 8*s+17;
b = s^2+9*s+20;
c = ilaplace(a/b,t)
```

MATLAB gives

```
c =
 23*exp(-5*t) - 15*exp(-4*t),
```

as (8.50) shows.

- For Equation 8.61, $F_2(s)=\frac{9s^2+34s+29}{s^3+6s^2+11s+6}$, the MATLAB script is

```
syms s t
a = 9*s^2+34*s+29;
b = s^3+6*s^2+11*s+6;
c = ilaplace(a/b,t)
```

MATLAB returns

```
c =
 2*exp(-t) + 3*exp(-2*t) + 4*exp(-3*t),
```

as in Equation 8.65.

Exercises for Example 8.2

1. Consider a *series RL circuit* displayed in Figure 8.7a. Find its step response.
2. Consider a *parallel RL circuit* shown in Figure 8.7b. Find its output voltage and explain it.

The circuits and the corresponding graphs are shown, as the hints, in Figures 8.7a and 8.7b.

Question Compare the step response of a parallel RC circuit shown in Figure 8.5 with that of the parallel RL circuit shown in Figure 8.7b. Why are they so different?

8.4.4 Summary of the Finding of the Inverse Laplace Transform

Let's highlight the main steps in finding the inverse Laplace transform:

- Review the given s-domain function, $F(s)$, to see whether it could be reduced to the form shown in the Laplace transform Table 7.1. If it could, we obtained the result. If not, we have to use the partial-fraction-expansion method.

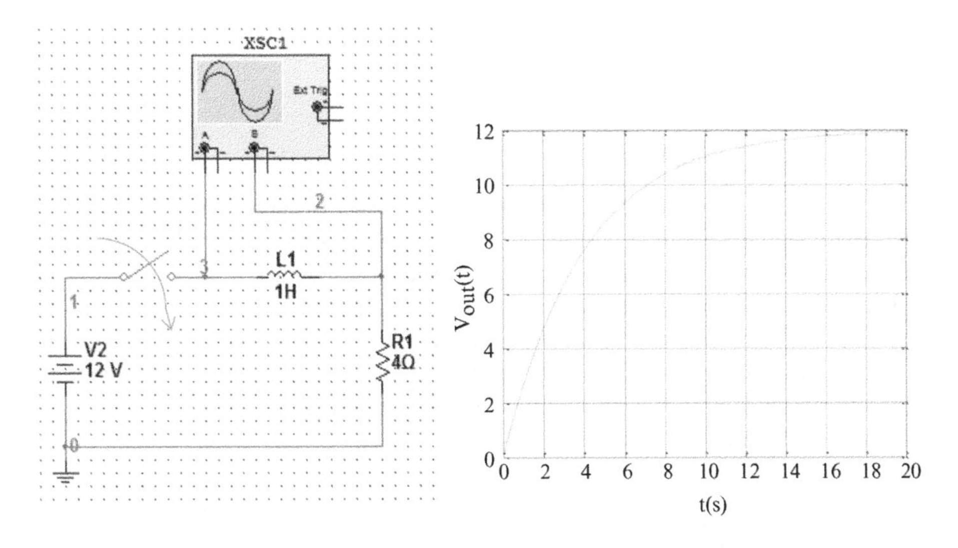

a)

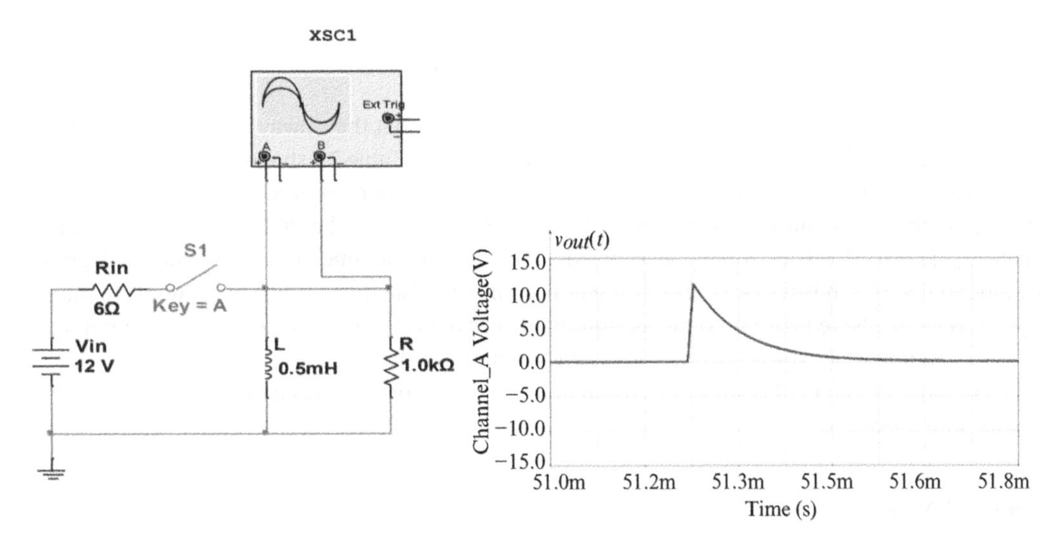

b)

Figure 8.7 Step response of RL circuits: a) Setup for finding the transient response of a series circuit (left) and the step response's graph (right); b) experimental setup for a parallel RL circuit (left) and corresponding graph (right).

- To use the partial-fraction-expansion approach, we need
  - Find the poles, present the $F(s)$ denominator in a factored form, and expand the $F(s)$ function into the partial fractions.
  - Determine the residues (constants) of the partial fraction expansion by using either direct method, or algebraic method, or cover-up method. Don't forget to verify the correctness of the obtained partial fraction expansion by the test-value method.
  - Perform the inverse Laplace transform using Table 7.1 and obtain the required time-domain function, which is the final result of your search.

It seems that we are ready to use the Laplace transform to find the transient response of a first-order circuit. However, some obstacles still need to be addressed.

8.4.5 Real and Distinct (Simple) Poles

Preceding sections considered first-order circuits whose s-domain functions contain only *real and distinct (simple) poles*. The procedure of finding the inverse Laplace transform in this case is the straightforward, as discussed in depth in the antecedent sections. In practice, however, we can meet the circuits whose equations lead to the poles of different nature. These cases we will consider in the sections that follow. For now, it's essential to understand the relationship between the poles' nature and the circuit's outputs. This understanding enables us to link mathematical manipulations with poles in the s-domain to actual circuits' operations in the time domain.

To clarify, combine (8.15) and (8.23) as

$$F(s)=\frac{N(s)}{D(s)}=\frac{a_n s^n + a_{n-1}s^{n-1}+\ldots+a_0}{b_m s^m + b_{m-1}s^{m-1}+\ldots+b_0}=\frac{A_1}{s-p_1}+\frac{A_2}{s-p_2}+\frac{A_3}{s-p_3}+\ldots\frac{A_m}{s-p_m}. \tag{8.66}$$

Refer to (8.17) and write

$$f(t)=\mathcal{L}^{-1}\{F(s)\}=A_1e^{p_1t}+A_2e^{p_2t}+A_3e^{p_3t}+\ldots+A_me^{p_mt} \tag{8.37R}$$

Hence, the inverse Laplace transforms of these poles show that they always become exponential terms in the time-domain functions. For a real pole, as LT-7 in Table 7.1 shows, the picture is clear: If a pole is negative, $p_i<0$, then $v_i(t)$ will be a decaying exponential function, e^{-p_it}. When such $v_i(t)$ describes a circuit's response, the circuit's final state will be stable because the transient $v_{out}(t)\to 0$. However, if the pole is positive, $p>0$, then the time-domain function will grow exponentially, e^{pt}. In this case, the circuit will become unstable because its output tends to infinity, $v_{out}(t)\to\infty$. Note that a binomial denominator has a form $(s-p_i)$, so a negative p_i produces $(s+p_i)$ denominator, which results in e^{-p_it} time-domain function.

To conclude, when the F(s) poles are *real and distinct*, the corresponding *f(t)* is a sum of the *real exponential functions.*

8.4.6 What If...

After all these discussions and examples, the inverse Laplace transform seems a straightforward operation with a well-explained procedure. But what if we would need to find $L^{-1}\left\{\frac{3s+9}{s^2+2s+5)}\right\}$? The attempt in finding the denominator poles of this $F(s)$ would result in

$$p_{1,2}=\frac{-b\pm\sqrt{(b^2-4ac)}}{2a}=\frac{-2\pm\sqrt{(2^2-4\cdot 5)}}{2}=\frac{-2\pm\sqrt{-16}}{2}. \tag{8.67}$$

The procedure of finding the inverse Laplace transform summarized in this section certainly cannot be used with these poles because of $\sqrt{-16}$ member. What to do?

This example shows a case (in fact, there are three such cases in total) when we need to develop a new routine to solve the inverse Laplace transform problem. The next sections will consider these cases and their routines.

Sidebar 8S Proper and Improper Rational Functions

One obstacle for using a cover-up method might stem from the structure of an s-domain function.

Consider a *rational function* $F(s)$ given in (8.23). You are reminded that the integers n and m are the degree or orders of $N(s)$ and $D(s)$, respectively. If $n < m$, then $F(s)$ is called a *proper rational function*, and it can be resolved in partial fractions, as we've done in the preceding section. However, if $n \geq m$, that is, the polynomial numerator's degree, n, is equal or greater than that of the polynomial denominator, m, then $F(s)$ is called an *improper rational function*. In such a case, the function can't be resolved in partial fractions directly.

To illustrate this statement, consider, for example, the following $F(s)$,

$$F(s) = \frac{3s^2+13s+7}{s^2+6s+5} = \frac{3s^2+13s+7}{(s+1)(s+5)}, \tag{8S.1}$$

whose both polynomials have the same order. The partial fraction expansion of this function should be

$$F(s) = \frac{3s^2+13s+7}{(s+1)(s+5)} = \frac{A_1}{s+1} + \frac{A_2}{s+5} (?) \tag{8S.2}$$

But this is an invalid expansion. To prove, let's bring the RHS of (8S.2) to the common denominator as

$$\frac{3s^2+13s+7}{(s+1)(s+5)} = \frac{A_1}{s+1} + \frac{A_2}{s+5} = \frac{A_1(s+5)+A_2(s+1)}{(s+1)(s+5)} (?) \tag{8S.3}$$

We can see that the order of the RHS numerator's polynomial ($n = 1$) is not equal to the order of the LHS numerator's polynomial ($n = 2$).

To resolve the issue with an improper function, we divide $N(s)$ into $D(s)$ to reduce $F(s)$ to the quotient, $Q(s)$, and the remainder, $\frac{N_R(s)}{D(s)}$, as shown in (8S.4),

$$F(s) = Q(s) + \frac{N_R(s)}{D(s)}. \tag{8S.4}$$

The *remainder is now a proper rational function*, and it can be treated the way we studied in the preceding sections. Let's illustrate this idea with $F(s) = \frac{3s^2+13s+7}{s^2+6s+5}$ given in (8S.1). Use the long division as shown,

$$\begin{array}{r} 3 \\ s^2+6s+5\overline{)(3s^2+13s+7)} \\ -\;(3s^2+18s+15) \\ \hline -5s-8 \end{array}$$

and obtain

$$F(s) = 3 + \frac{-5s-8}{(s+1)(s+5)}, \tag{8S.5}$$

where 3 is the quotient and $\frac{-5s-8}{(s+1)(s+5)}$ is the remainder. We can easily verify the following equality:

$$\frac{3s^2+13s+7}{(s+1)(s+5)} = 3 + \frac{-5s-8}{(s+1)(s+5)}. \tag{8S.6}$$

(You will recall the long-division rule:

- Write both polynomials in descending order. Insert the long-division symbol, $\overline{)}$.
- Place the divisor (s^2+6s+5) at the left of the symbol and the dividend $(3s^2+13s+7)$ inside the symbol.
- Divide the dividend's member with the highest power $(3s^2)$ by the similar divisor's member (s^2) and obtain quotient (3). Write the quotient (3) at the top of the long-division symbol.
- Multiply the divisor by the quotient and subtract the result, $3\cdot(s^2+6s+5)$, from the dividend. This operation eliminates the term with the highest power in the difference and gives you the remainder's numerator.

  That is, $(3s^2+13s+7)-3\cdot(s^2+6s+5)=-5s-8$.

  Thus, an improper function shown in (8S.1) as $\frac{3s^2+13s+7}{(s+1)(s+5)}$, is presented as a sum of the quotient and the remainder, $F(s)=3+\frac{-5s-8}{(s+1)(s+5)}$, as symbolically shown in (8S.4).)

Expanding the remainder into partial fractions enables us to find the expansion of the whole s-domain function as

$$F(s)=\frac{3s^2+13s+7}{s^2+6s+5}=3+\frac{-5s-8}{(s+1)(s+5)}=3-\frac{3}{4(s+1)}-\frac{17}{4(s+5)}. \quad (8S.7)$$

Use the test-value method to verify the correctness of this expansion.

Finally, consulting Table 7.1, we find the inverse Laplace transform of given $F(s)$ as

$$f(t)=L^{-1}\{F(s)\}=L^{-1}\left\{3-\frac{3}{4(s+1)}-\frac{17}{4(s+5)}\right\}=3\delta(t)-\frac{3}{4}e^{-t}-\frac{17}{4}e^{-5t}. \quad (8S.8)$$

This example clarifies the technique for working with an improper function. In practice, the circuits whose descriptions lead to improper functions are rarities.

8.5 Repeated (Multiple) Poles and the Circuit's Output

We start considering the case of repeated (multiple) poles in the s-domain function with a real-life problem of a cascaded RC filter.

8.5.1 The Example—Response of the Cascaded RC Filters

To attain the desired characteristics of passive filters, a designer ought to build the high-order filters. Increasing the filter order is typically achieved by cascading the first- and second-order filters. (See, for example, Mynbaev and Scheiner, 2020, pp. 463 and 468.) Let's consider the Laplace transform application to finding the output of cascaded filters.

Example 8.3 Finding the step response of two cascaded RC filters—Part 1 (problem).

Problem: Consider two cascaded identical RC filters shown in Figure 8.8. The first filter has a step input. Find the second filter's output if $R=1\ k\Omega$ and C = 0.05 mF, which gives $\alpha=\frac{1}{\tau}=\frac{1}{RC}=20\frac{1}{s}$, and $V_{in}=4V$, and both circuits are initially relaxed.

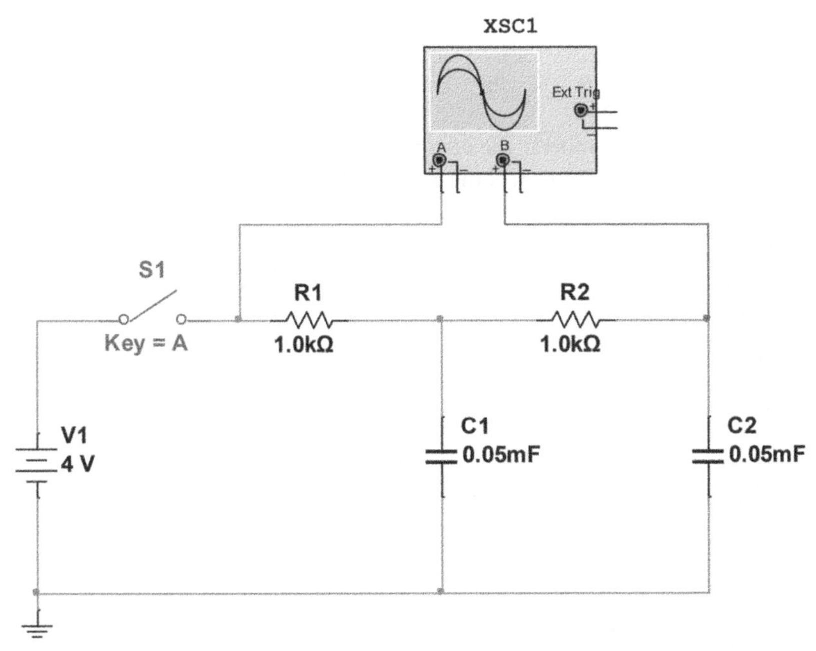

Figure 8.8 Two cascaded RC filters in Example 8.3.

Solution:

From Part 2 of Example 8.2, we know that an RC circuit's step response in the s-domain is given by (8.14). This signal, $V_{out1}(s)$, is now the input for the second RC circuit; hence, we change the notation as

$$V_{in2}(s) = V_{out1}(s) = \frac{\alpha V_{in}}{s(s+\alpha)}, \tag{8.68}$$

where $V_{in} = 4V$ is a constant. If we repeat our reasoning in finding the RC circuit response given by (8.1) and (8.2), we now see the second circuit's differential equation as

$$v'_{out2}(t) + \alpha v_{out2}(t) = \alpha v_{in2}(t). \tag{8.69}$$

After applying the Laplace transform to (8.69), we obtain

$$sV_{out2}(s) + \alpha V_{out2}(s) = \alpha V_{in2}(s). \tag{8.70}$$

Thus, the second circuit's s-domain response becomes

$$V_{out2}(s) = \frac{\alpha V_{in2}(s)}{(s+\alpha)} = \frac{\alpha^2 V_{in}}{s(s+\alpha)(s+\alpha)} = \frac{\alpha^2 V_{in}}{s(s+\alpha)^2}. \tag{8.71}$$

Note that (8.71) is, in fact, the product of two s-domain functions, $V_{in2}(s)$ and $\frac{\alpha}{(s+\alpha)}$, describing the first and the second circuits in the cascade.

Following the established procedure for the partial fraction expansion, we should write

$$V_{out}(s) = \frac{\alpha^2 V_{in}}{s(s+\alpha)(s+\alpha)} = \frac{\alpha^2 V_{in}}{s(s+\alpha)^2} = \frac{A_1}{s-p_1} + \frac{A_2}{s-p_2} + \frac{A_3}{s-p_3} = \frac{A_1}{s} + \frac{A_2}{s+\alpha} + \frac{A_3}{s+\alpha} = ? \tag{8.72}$$

However, this expansion makes no sense. Indeed, the denominator's polynomial is the third-order one, and therefore we need three *distinct* members of the expansion; yet, the two last RHS fractions have the *identical poles*, $p_2 = p_3 = -\alpha$. Such a case is called *repeated or multiple poles*. Clearly, the members with identical denominators can't have different numerators due to the very nature of partial fraction expansion. Refer to Sections 8.3 and 8.4, where this mathematical method is discussed in-depth, and recall specifically (8.40),

$$A_i = \frac{N(s)}{(s-p_1)(s-p_2)(s-p_3)\ldots(s-p_m)}(s-p_i)\bigg|_{s=p_i} = F(s)(s-p_i)\big|_{s=p_i}. \tag{8.73}$$

This formula shows that we can't obtain a different A_i when multiplying an F(s) by the same $(s-p_i)$. For example, to calculate A_2, we need to multiply an F(s) by $(s-p_2)$ and compute the product at $s = p_2$. Therefore, if we multiply F(s) by the same $(s-p_i)$ several times, we repeatedly obtain the same A_i. All this means that the previously developed partial fraction expansion procedure doesn't provide the required solution in the case of *repeated (multiple) poles*. What to do?

To reiterate, here we encounter the situation with two *identical* fractions with *identical (repeated) poles*. We know that the members with identical denominators can't have different numerators due to the very nature of partial fraction expansion. Therefore, we need to develop new technique for solving the problem with *repeated poles*.

We need to postpone solving this problem until the introduction of the necessary theoretical background.

8.5.2 The Procedure of Obtaining Partial Fraction Extension in Case of Repeated (Multiple) Poles

All the preceding cases have one feature in common: One $F(s)$ function contains only distinct poles. There are, however, real-life situations, as shown in Example 8.4 above, when an $F(s)$ has two or more equal (repeated) poles called *multiple-order poles* or *multiple poles*. Consider the $F(s)$ whose factored form is

$$F(s) = \frac{N(s)}{(s-p_1)(s-p_2)^2}. \tag{8.74}$$

The expansion of this F(s) into partial fractions seems to yield

$$F(s) = \frac{N(s)}{(s-p_1)(s-p_2)^2} = \frac{(A_1)}{(s-p_1)} + \frac{(A_2)}{(s-p_2)^2}. \tag{8.75}$$

There is a problem, however, in this equation: *It contains three (though not distinct) poles, p_1, p_2, and p_2, whereas (8.72) includes only two members. To resolve this issue, we state that the member $\frac{A_2}{(s-p_2)^2}$ must be presented as*

$$\frac{(A_2)}{(s-p_2)^2} = \frac{A_{21}}{(s-p_2)} + \frac{A_{22}}{(s-p_2)^2}. \tag{8.76}$$

Why? Here is the proof (Thomas, Rosa, and Toussaint 2019, pp. 476–477):

Factoring one of the repeated polynomials results in

$$F(s)=\frac{N(s)}{(s-p_1)(s-p_2)^2}=\frac{1}{(s-p_2)}\left[\frac{N(s)}{(s-p_1)(s-p_2)}\right]. \tag{8.77}$$

Expanding the member inside the brackets into partial fractions yields

$$F(s)=\frac{1}{(s-p_2)}\left[\frac{N(s)}{(s-p_1)(s-p_2)}\right]=\frac{1}{(s-p_2)}\left[\frac{(A_1)}{(s-p_1)}+\frac{(A_2)}{(s-p_2)}\right]. \tag{8.78}$$

Opening the brackets gives

$$F(s)=\frac{N(s)}{(s-p_1)(s-p_2)^2}=\frac{A_1}{(s-p_2)(s-p_1)}+\frac{A_{22}}{(s-p_2)^2}, \tag{8.79}$$

which means that $(A_2)=A_{22}$. Expanding $\frac{A_1}{(s-p_2)(s-p_1)}$ of (8.79) into the partial fractions results in

$$F(s)=\frac{N(s)}{(s-p_1)(s-p_2)^2}=\frac{A_{11}}{(s-p_1)}+\frac{A_{21}}{(s-p_2)}+\frac{A_{22}}{(s-p_2)^2}, \tag{8.80}$$

which conclude the proof.

The above proof provides us with the method of finding the residues A_1, A_{21}, and A_{22}. Indeed, equating the brackets of RHS and LHS of the (8.78), we can determine A_1 and A_{22} by using the traditional cover-up technique because $N(s)$ is given. Then, expanding member $\frac{A_1}{(s-p_2)(s-p_1)}$ of (8.79) into partial fractions, we get

$$\frac{A_1}{(s-p_2)(s-p_1)}=\frac{A_{11}}{(s-p_1)}+\frac{A_{21}}{(s-p_2)}. \tag{8.81}$$

Note that we can compute A_1 and A_{22} because we know the values of $N(s)$, p_1, and p_2. Further, we can calculate A_{11} and A_{21} because we know the A_1 value. Therefore,

$$F(s)=\frac{A_{11}}{(s-p_1)}+\frac{A_{21}}{(s-p_2)}+\frac{A_{22}}{(s-p_2)^2},$$

as in (8.80).

Finally, the inverse Laplace transform of the given *F(s)* according to LT-7 and LT-8 of Table 7.1 is

$$\begin{aligned} f(t)=\mathcal{L}^{-1}\{F(s)\}&=\mathcal{L}^{-1}\left\{\frac{A_{11}}{(s-p_1)}+\frac{A_{21}}{(s-p_2)}+\frac{A_{22}}{(s-p_2)^2}\right\}=A_{11}e^{p_1t}+A_{21}e^{p_2t}+A_{22}te^{p_2t} \\ &=A_{11}e^{p_1t}+(A_{21}+A_{22}t)e^{p_2t}. \end{aligned} \tag{8.82}$$

This discussion enables us to complete Example 8.3.

Example 8.3 Finding the step response of two cascaded RC filters—Part 2 (solution).

Problem: Consider two cascaded identical RC filters shown in Figure 8.8. The first filter has a step input. Find the second filter's output if $R = 1\ k\Omega$, $C = 0.05\ \text{mF}$, $V_{in} = 4V$, and both circuits are initially relaxed.

Solution:
The circuit's output in the s-domain is obtained in Part 1 of this example as

$$V_{out}(s) = \frac{\alpha V_{in}(s)}{(s+\alpha)} = \frac{\alpha^2 V_{in}}{s(s+\alpha)(s+\alpha)} = \frac{\alpha^2 V_{in}}{s(s+\alpha)^2}. \tag{8.71R}$$

Resolving $V_{out}(s)$ into partial fractions, as (8.69) requires, yields

$$V_{out}(s) = \frac{\alpha^2 V_{in}}{s(s+\alpha)^2} = \frac{A_1}{s(s+\alpha)} + \frac{A_{22}}{(s+\alpha)^2}. \tag{8.83}$$

To obtain the equation in the form given in (8.77), we need to factor the repeated member in (8.83) as

$$\frac{1}{(s+\alpha)}\left[\frac{\alpha^2 V_{in}}{s(s+\alpha)}\right] = \frac{1}{(s+\alpha)}\left[\frac{A_1}{s} + \frac{A_{22}}{(s+\alpha)}\right]. \tag{8.84}$$

Following (8.78), we equate members in the brackets of the RHS and LHS of (8.84) and find

$$\left[\frac{\alpha^2 V_{in}}{s(s+\alpha)}\right] = \left[\frac{A_1}{s} + \frac{A_{22}}{(s+\alpha)}\right]. \tag{8.85}$$

Equation 8.85 enables us to calculate residues A_1 and A_{22} as follows

$$A_1 = \left.\frac{s\alpha^2 V_{in}}{s(s+\alpha)}\right|_{s=0} = \left.\frac{\alpha^2 V_{in}}{(s+\alpha)}\right|_{s=0} = \frac{\alpha^2 V_{in}}{(\alpha)} = \alpha V_{in} \tag{8.86}$$

and

$$A_{22} = \left.\frac{(s+\alpha)\alpha^2 V_{in}}{s(s+\alpha)}\right|_{s=-\alpha} = \left.\frac{\alpha^2 V_{in}}{s}\right|_{s=-\alpha} = \frac{\alpha^2 V_{in}}{(-\alpha)} = -\alpha V_{in}. \tag{8.87}$$

Now, we take $\frac{A_1}{s(s+\alpha)}$ from (8.83), plug $A_1 = \alpha V_{in}$ into this formula, and resolve $\frac{\alpha V_{in}}{s(s+\alpha)}$ into the partial fractions as

$$\frac{\alpha V_{in}}{s(s+\alpha)} = \frac{A_{11}}{s} + \frac{A_{21}}{(s+\alpha)}. \tag{8.88}$$

Residues A_{11} and A_{21} can be found from (8.88) as follows

$$A_{11} = \left.\frac{s\alpha V_{in}}{s(s+\alpha)}\right|_{s=0} = \left.\frac{\alpha V_{in}}{(s+\alpha)}\right|_{s=0} = \frac{\alpha V_{in}}{(\alpha)} = V_{in}, \tag{8.89}$$

$$A_{21} = \left.\frac{(s+\alpha)\alpha V_{in}}{s(s+\alpha)}\right|_{s=-\alpha} = \left.\frac{\alpha V_{in}}{s}\right|_{s=-\alpha} = \frac{\alpha V_{in}}{-\alpha} = -V_{in}. \tag{8.90}$$

Thus, we obtain the needed partial fraction expansion of the given $V_{out}(s)$ as

$$V_{out}(s) = \frac{\alpha^2 V_{in}}{s(s+\alpha)^2} = \frac{A_{11}}{s} + \frac{A_{21}}{(s+\alpha)} + \frac{A_{22}}{(s+\alpha)^2} = \frac{V_{in}}{s} - \frac{V_{in}}{(s+\alpha)} - \frac{\alpha V_{in}}{(s+\alpha)^2}. \tag{8.91}$$

Before we proceed further, we have to verify our partial fraction expansion by the test-value method. Recalling that $V_{in} = 4V$ and $\alpha = \frac{1}{RC} = 20\left(\frac{1}{s}\right)$ and plugging $s = 2$ for the test, we compute $1.653 = 1.653$. Also, you are encouraged to apply a general test by bringing the RHS the common denominator and comparing the RHS and LHS of (8.91).

Finally, the inverse Laplace transform in question is attained as

$$v_{out}(t) = \mathcal{L}^{-1}\{V_{out}(s)\} = \mathcal{L}^{-1}\left\{\frac{V_{in}}{s} - \frac{V_{in}}{(s+\alpha)} - \frac{\alpha V_{in}}{(s+\alpha)^2}\right\} = V_{in} - V_{in}e^{-\alpha t} - \alpha V_{in}te^{-\alpha t}. \tag{8.92}$$

Plugging the given values yields

$$v_{out}(t) = 4 - 4e^{-20t} - 80te^{-20t}. \tag{8.93}$$

The problem is solved.

Discussion:

- Does this solution make sense? To check, we want to consider the initial and final states. At $t = 0$, (8.92) shows that $v_{out}(0) = 0$. This result meets the given circuit's relaxed initial condition. At $t \to \infty$, $v_{out}(t)$ tends to V_{in}, as it should be, because two variable members of (8.92) describing the transient response vanish. Recall that an *exponentially decaying unit ramp function*, $te^{-\alpha t}$, also called *damped ramp*, goes to zero after initial growth, as Figure 7.9 demonstrates. Nevertheless, the initial and final states' correctness is only necessary, but not a sufficient test. In other words, the received initial and final conditions might be correct, but the obtained answer still might be wrong.
- Compare members $\frac{A_{21}}{(s-p_2)} + \frac{A_{22}}{(s-p_2)^2}$ and $A_{21}e^{p_2 t} + A_{22}te^{p_2 t}$ and notice that multiplication the $\frac{1}{(s-p_2)}$ by itself, which yields the $\frac{1}{(s-p_2)^2}$ in the s-domain, results in multiplication of the corresponding member, $e^{p_2 t}$, by t in the time domain. We could come to this conclusion earlier by comparison of LT-7 and LT-8 in Table 7.1.
- The summary of the discussed technique for finding the inverse Laplace transform of $F(s)$ with multiple poles of the second-order is as follows:
  - Present the given s-domain function in a factored form as in (8.74),

$$F(s) = \frac{N(s)}{(s-p_1)(s-p_2)^2}.$$

  - Resolve the $F(s)$ into partial fractions as in (8.79),

$$F(s)=\frac{N(s)}{(s-p_1)(s-p_2)^2}=\frac{A_1}{(s-p_2)(s-p_1)}+\frac{A_{22}}{(s-p_2)^2}.$$

- Factor a repeated-pole member from both sides of (8.79) and obtain (8.78)

$$\frac{1}{(s-p_2)}\left[\frac{N(s)}{(s-p_1)(s-p_2)}\right]=\frac{1}{(s-p_2)}\left[\frac{A_1}{(s-p_1)}+\frac{A_{22}}{(s-p_2)}\right].$$

- Find A_1 and A_{22} from (8.78) by the cover-up method.
- Expand $\frac{A_1}{(s-p_2)(s-p_1)}$ of (8.79) into partial fractions, as shown in (8.79),

$$\frac{A_1}{(s-p_2)(s-p_1)}=\frac{A_{11}}{(s-p_1)}+\frac{A_{21}}{(s-p_2)}.$$

- Find A_{11} and A_{21} from (8.80).
- Finally, attain the required inverse Laplace transform as

- $f(t)=L^{-1}\{F(s)\}=L^{-1}\left\{\frac{A_{11}}{(s-p_1)}+\frac{A_{21}}{(s-p_2)}+\frac{A_{22}}{(s-p_2)^2}\right\}=A_{11}e^{p_1t}+A_{21}e^{p_2t}+A_{22}te^{p_2t}.$
- Can we use this technique in finding the inverse Laplace transform of $F(s)$ with multiple poles of third and the higher order? Analyzing the above summary, we should definitely say yes. We just need to consequently apply this technique, starting from the highest-order (e.g., $(s+\alpha)^3$) member.

Exercise (Stanley 2002, pp. 198–199): Find the inverse Laplace transform of $F(s)=\frac{s^2+4}{s(s+1)(s+2)^3}$.

Answer: $f(t)=0.5-5e^{-t}+\left(4.5+4t+2t^2\right)e^{-2t}$.

- Note that an s-domain function (in this example, $F(s)=\frac{A_{11}}{(s-p_1)}+\frac{A_{12}}{(s-p_2)}+\frac{A_2}{(s-p_2)^2}$) contains fractions depending only on single poles $\left(\frac{A_{11}}{(s-p_1)}\right)$ and fractions depending only on multiple poles $\left(\frac{A_{21}}{(s-p_2)}+\frac{A_2}{(s-p_2)^2}\right)$. Consequently, the inverse Laplace transform of $F(s)$ is the sum of $f_1(t)=A_{11}e^{p_1t}$ and $f_2(t)=(A_{21}+A_{22}t)e^{p_2t}$.
- The result of the Multisim simulation is shown in Figure 8.9. Comment on how this graph confirms (or not?) the answer presented in (8.93).

Question Time constant of an RC circuit is $\tau=RCs$. For the RC filter shown in Figure 8.8, $\tau=0.05s$. Why then does Figure 8.9 show $5\tau=0.5s$, *not* $0.25s$?

- MATLAB solution to this problem is attained by taking the inverse Laplace transform of $V_{out}(s)=\frac{\alpha^2V_{in}}{s(s+\alpha)^2}$ whose $\alpha=20\frac{1}{s}$ and $V_{in}=4V$ as follows:

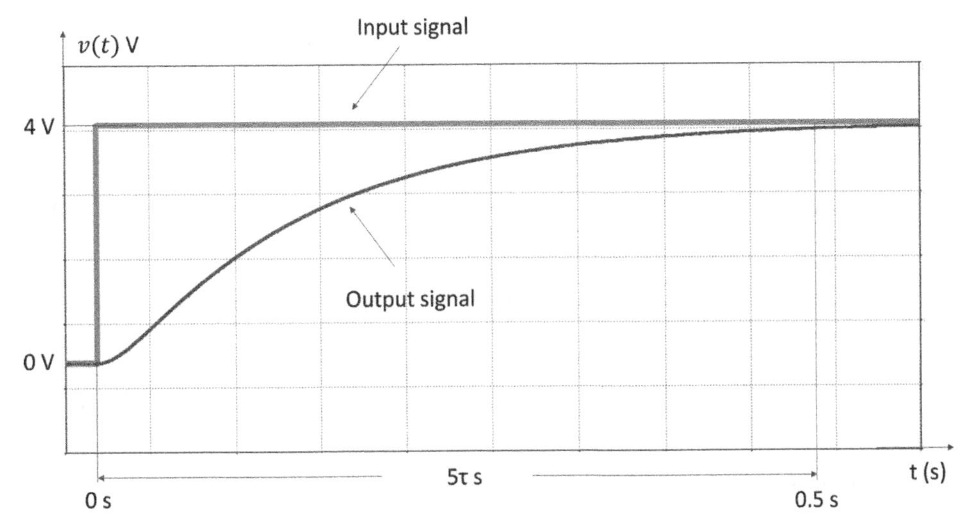

Figure 8.9 The result of Multisim simulation for Example 8.3.

```
>> syms s t
n=20^2*4;
d=s*(s+20)^2;
f=ilaplace(n/d,t)
ezplot(f,[0,3])
```

MATLAB returns the answer and builds the graph shown in Figure 8.10.

It is essential to compare all three results for this example—manual, Multisim, and MATLAB—to better understand the Laplace transform application to solving the real-life problem in advanced circuit analysis.

8.5.3 The Alternative Method of Finding the Residues for the Partial Fraction Expansion with Repeated (Multiple) Poles

Refer again to the $F(s)$ whose factored form is considered in the preceding subsection

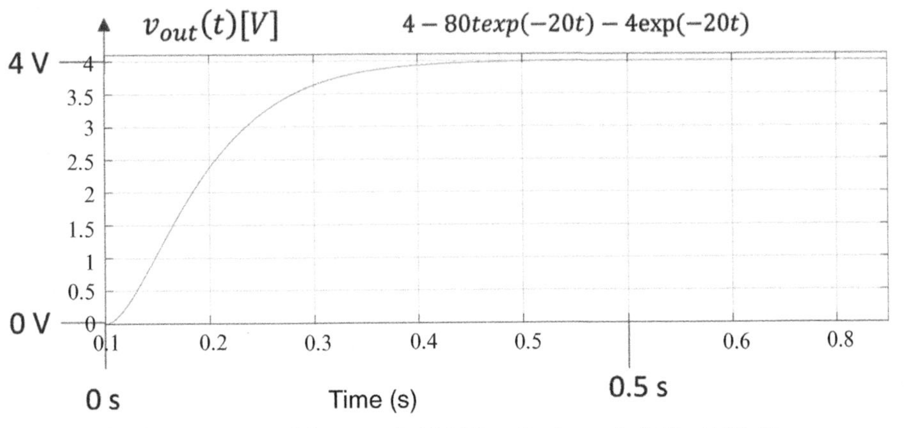

Figure 8.10 Step response of the cascaded RC filters for Example 8.3 by MATLAB.

$$F(s)=\frac{N(s)}{(s-p_1)(s-p_2)^2}. \tag{8.72R}$$

Its partial fractions expansion has the form

$$F(s)=\frac{N(s)}{(s-p_1)(s-p_2)^2}=\frac{A_{11}}{(s-p_1)}+\frac{A_{21}}{(s-p_2)}+\frac{A_{22}}{(s-p_2)^2}, \tag{8.78R}$$

which offers the following *alternative (differentiating) method of finding the residues* A_{11}, A_{21}, and A_{22}. The use of the cover-up algorithm for A_{11} yields

$$A_{11}=(s-p_1)F(s)\big|_{s=p_1}=\frac{(s-p_1)N(s)}{(s-p_1)(s-p_2)^2}\Bigg|_{s=p_1}=\frac{N(s)}{(s-p_2)^2}\Bigg|_{s=p_1}=\frac{N(p_1)}{(p_1-p_2)^2}. \tag{8.94}$$

Following the pattern, we find A_{22} as

$$A_{22}=(s-p_2)^2 F(s)\big|_{s=p_2}=\frac{(s-p_2)^2 N(s)}{(s-p_1)(s-p_2)^2}\Bigg|_{s=p_2}=\frac{N(p_2)}{(p_2-p_1)}. \tag{8.95}$$

You will recall that finding A_{21} by directly applying the cover-up algorithm results in an invalid operation because it requires dividing by zero.

$$A_{21}=(s-p_2)F(s)\big|_{s=p_2}=\frac{(s-p_2)N(s)}{(s-p_1)(s-p_2)^2}\Bigg|_{s=p_2}=\frac{N(s)}{(s-p_1)(s-p_2)}\Bigg|_{s=p_2}=? \tag{8.96}$$

The solution comes from analyzing (8.78): We can separate A_{22} by multiplying both sides of this equation by $(s-p_2)^2$:

$$(s-p_2)^2 F(s)=(s-p_2)^2\frac{N(s)}{(s-p_1)(s-p_2)^2}=(s-p_2)^2\left[\frac{A_{11}}{(s-p_1)}+\frac{A_{21}}{(s-p_2)}+\frac{A_{22}}{(s-p_2)^2}\right]. \tag{8.97}$$

Thus, we obtain

$$(s-p_2)^2 F(s)=\frac{(s-p_2)^2 A_{11}}{(s-p_1)}+(s-p_2)A_{21}+A_{22}. \tag{8.98}$$

Now, we denote the part of the F(s) that doesn't contain a multiple pole p_2 as $F_1(s)$; that is,

$$\frac{A_{11}}{(s-p_1)}=F_1(s). \tag{8.99}$$

Then, (8.98) becomes

$$(s-p_2)^2 F(s)=(s-p_2)^2 F_1(s)+(s-p_2)A_{21}+A_{22}. \tag{8.100}$$

To find A_{21}, we need to eliminate the constant A_{22}. How? Take the derivative of both sides of (8.100) over s, which yields

$$\frac{d}{ds}\left[(s-p_2)^2 F(s)\right]=\frac{d}{dt}\left[(s-p_2)^2 F_1(s)+(s-p_2)A_{21}+A_{22}\right]=\left[2(s-p_2)F_1(s)+(s-p_2)^2\frac{dF_1(s)}{ds}\right]+A_{21} \quad (8.101)$$

because $\frac{d(s-p_2)}{ds}=1$. Evaluating the RHS of (8.101) at $s=p_2$, enables us to find A_{21} as

$$A_{21}=\frac{d}{ds}\left[(s-p_2)^2 F(s)\right]\Bigg|_{s=p_2}=\frac{d}{ds}\left[\frac{(s-p_2)^2 N(s)}{(s-p_1)(s-p_2)^2}\right]\Bigg|_{s=p_2}=\frac{d}{ds}\left[\frac{N(s)}{(s-p_1)}\right]\Bigg|_{s=p_2}. \quad (8.102)$$

Again, to determine A_{21}, we need to *take the derivative first* (as in Equation 8.101) and *evaluate the result at* $s=p_2$ *second* (as in Equation 8.102).

Therefore, the partial fraction expansion of the given *F(s)* is

$$F(s)=\frac{N(s)}{(s-p_1)(s-p_2)^2}=\frac{A_1}{(s-p_1)}+\frac{A_{21}}{(s-p_2)}+\frac{A_{22}}{(s-p_2)^2}, \quad (8.78R)$$

as before. Here A_1, A_{21}, and A_{22} are determined from (8.94), (8.95), and (8.102), respectively.

Let's apply this method for finding the residues in Example 8.3, where $F(s)$ is given in (8.68) as $V_{out}(s)=\frac{\alpha^2 V_{in}}{s(s+\alpha)^2}$. Here, $p_1=0$ and $p_2=\alpha$. Using (8.94) and (8.95) yields

$$A_{11}=(s-p_1)F(s)\big|_{s=p_1}=\frac{(s-p_1)N(s)}{(s-p_1)(s-p_2)^2}\Bigg|_{s=p_1}=\frac{\alpha^2 V_{in}}{(s-\alpha)^2}\Bigg|_{s=0}=\frac{\alpha^2 V_{in}}{(0-\alpha)^2}=V_{in}$$

and

$$A_{22}=(s-p_2)^2 F(s)\big|_{s=p_2}=\frac{(s-p_2)^2 N(s)}{(s-p_1)(s-p_2)^2}\Bigg|_{s=p_2}=\frac{\alpha^2 V_{in}}{(s-0)}\Bigg|_{s=-\alpha}=\frac{\alpha^2 V_{in}}{-\alpha}=-\alpha V_{in}.$$

These residues were obtained in (8.82) and (8.83).

Finding A_{21} requires taking the derivative and evaluating the result at $s=p_2$, as (8.102) shows. Performing these operations, we get

$$A_{21}=\frac{d}{ds}\left[\frac{N(s)}{(s-p_1)}\right]\Bigg|_{s=p_2}=\frac{d}{ds}\left[\frac{\alpha^2 V_{in}}{(s-0)}\right]\Bigg|_{s=\alpha}=\alpha^2 V_{in}\frac{d}{dt}\left(\frac{1}{s}\right)\Bigg|_{s=\alpha}=-\frac{\alpha^2 V_{in}}{s^2}\Bigg|_{s=\alpha}=-\frac{\alpha^2 V_{in}}{\alpha^2}=-V_{in}. \quad (8.103)$$

Therefore, the s-domain circuit's response given in (8.69) takes the same form as obtained in Example 8.3,

$$V_{out}(s)=\frac{\alpha^2 V_{in}}{s(s+\alpha)^2}=\frac{A_{11}}{s}+\frac{A_{21}}{(s+\alpha)}+\frac{A_{22}}{(s+\alpha)^2}=\frac{V_{in}}{s}-\frac{V_{in}}{(s+\alpha)}-\frac{\alpha V_{in}}{(s+\alpha)^2}. \quad (8.85R)$$

To reaffirm this alternative method of finding the residues, let's do the examples.

8.5.4 Two Examples of the Differentiation Application for Finding the Residues in Case of Multiple Poles

Example 8.4 The inverse Laplace transform of an F(s) with multiple poles.

Problem: Find the inverse Laplace transform of

$$F(s)=\frac{8(s+4)}{(s+1)(s^2+6s+9)}.$$

Solution:

Step 1. Determine the poles and present the F(s) in a factored form:
The poles are $p_1=-1$, $p_2=p_3=-3$. (To find p_2 and p_3, we can use either (8.30) or (8.33).) Thus, the given *F(s)* has multiple poles $p_2=p_3$, and its factored form is

$$F(s)=\frac{N(s)}{(s-p_1)(s-p_2)^2}=\frac{8(s+4)}{(s+1)(s+3)^2}.$$

Step 2. Expand the $F(s)$ into partial fractions and find the corresponding constants (residues):
According to (8.78), the partial fraction expansion of the $F(s)$ is given by

$$F(s)=\frac{A_{11}}{(s+1)}+\frac{A_{21}}{(s+3)}+\frac{A_{22}}{(s+3)^2}.$$

Equation 8.94 determines A_{11} by using a cover-up method as

$$A_{11}=(s-p_1)F(s)\Big|_{s=p_1}=\frac{N(p_1)}{(p_1-p_2)^2}=\frac{8(-1+4)}{(-1+3)^2}=6.$$

Similarly, from (8.95), we get

$$A_{22}=(s-p_2)^2F(s)\Big|_{s=p_2}=\frac{N(p_2)}{(p_2-p_1)}=\frac{8(-3+4)}{(-3+1)}=-4.$$

Now, Equation 8.102 turns to

$$A_{21}=\frac{d}{ds}\left[(s-p_2)^2F(s)\right]\Big|_{s=p_2}=\frac{d}{ds}\left[\frac{N(s)}{(s-p_1)}\right]\Bigg|_{s=p_2}=\frac{d}{ds}\left[\frac{8(s+4)}{(s-p_1)}\right]\Bigg|_{s=p_2}.$$

Using the derivative quotient rule, we obtain

$$\frac{d}{ds}\left[\frac{8(s+4)}{(s-p_1)}\right]=\frac{\left[\frac{d}{ds}8(s+4)\right](s-p_1)-8(s+4)\left[\frac{d}{ds}(s-p_1)\right]}{(s-p_1)^2}=\frac{8(s-p_1)-8(s+4)}{(s-p_1)^2}.$$

Evaluating this quotient at $s=p_2$ yields

$$A_{21}=\frac{8(s-p_1)-8(s+4)}{(s-p_1)^2}\Big|_{s=p_2}=\frac{8(-p_1)-8(4)}{(p_2-p_1)^2}=8\frac{(1-4)}{(-3+1)^2}=-6.$$

Thus, the given $F(s)$ is expanded into partial fractions as

$$F(s)=\frac{8(s+4)}{(s+1)(s^2+6s+9)}=\frac{6}{(s+1)}+\frac{-6}{(s+3)}+\frac{-4}{(s+3)^2}.$$

To verify the obtained residues' correctness, we use the test-value method. Plugging for computational simplicity $s=2$ into both LHS and RHS, we find $\frac{48}{75}=\frac{48}{75}$.

Step 3. Finding the inverse Laplace transform of the given F(s):

$$f(t)=\mathcal{L}^{-1}\{F(s)\}=\mathcal{L}^{-1}\left\{\frac{A_1}{(s-p_1)}+\frac{A_{21}}{(s-p_2)}+\frac{A_{22}}{(s-p_2)^2}\right\}=A_1e^{p_1t}+(A_{21}+A_{22}t)e^{p_2t}$$

$$=6e^{-t}-(6+4t)e^{-3t}.$$

The problem is solved.

Question What will be the partial fraction expansion of $F(s)=\frac{8(s+4)}{(s+1)(s+3)^3}$? [Hint: Follow the procedure shown in this example.

Answer: $F(s)=\frac{8(s+4)}{(s+1)(s+3)^3}=\frac{3}{(s+1)}+\frac{-4}{(s+3)^3}+\frac{-6}{(s+3)^2}+\frac{-3}{(s+3)}$.]

Discussion:

- Remember that $F(s)=\frac{8(s+4)}{(s+1)(s^2+6s+9)}$. Then the MATLAB solution to this example can be written as

```
syms s t
a = 8*(s+4);
b = (s+1)*(s^2+6*s+9);
c = ilaplace(a/b,t);
pretty(c)
```

MATLAB returns

```
c = 6 exp(-t) - exp(-3 t) 6 - t exp(-3 t) 4,
```

as obtained by the manual solution.

As mentioned previously, a higher-order electrical circuit is typically built by the cascading of the first-order and the second-order stages. Nevertheless, if we encounter a higher-order circuit that can't be presented as a cascade and whose Laplace transform contains multiple poles, then its analysis can be done as shown in the example that follows. (It helps to review the Discussion section of Example 8.2, where the circuits with the higher-order polynomials are considered.)

Consider the s-domain function, F(s)

$$F(s)=\frac{N(s)}{(s-p_1)(s-p_2)(s-p_3)^4}. \tag{8.104}$$

This F(s) is expanded in the partial fractions as

$$F(s)=\frac{A_1}{(s-p_1)}+\frac{A_2}{(s-p_2)}+\frac{A_{44}}{(s-p_3)^4}+\frac{A_{43}}{(s-p_3)^3}+\frac{A_{42}}{(s-p_3)^2}+\frac{A_{41}}{(s-p_3)}. \tag{8.105}$$

The residues for the distinct poles A_1 and A_2 are found by the cover-up method as

$$A_1=(s-p_1)F(s)|_{s=p_1}=\frac{N(p_1)}{(p_1-p_2)(p_1-p_3)^4} \tag{8.106}$$

and

$$A_2=(s-p_2)F(s)|_{s=p_2}=\frac{N(p_2)}{(p_2-p_1)(p_2-p_3)^4}. \tag{8.107}$$

The same rule determines the residue A_{44},

$$A_{44}=(s-p_3)^4F(s)|_{s=p_3}=\frac{N(p_3)}{(p_3-p_1)(p_3-p_2)}. \tag{8.108}$$

To simplify the further manipulations, we denote the part of the $F(s)$ without repeated poles as $F_1(s)$; that is,

$$F_1(s)=\frac{A_1}{(s-p_1)}+\frac{A_2}{(s-p_2)}. \tag{8.109}$$

Thus,

$$F(s)=F_1(s)+\frac{A_{44}}{(s-p_3)^4}+\frac{A_{43}}{(s-p_3)^3}+\frac{A_{42}}{(s-p_3)^2}+\frac{A_{41}}{(s-p_3)}. \tag{8.110}$$

Now, we recall that the residues of all other repeated poles except A_{44} require differentiation. To remind, multiplication of (8.96) by $(s-p_3)^4$ results in

$$(s-p_3)^4F(s)=(s-p_3)^4F_1(s)+A_{44}+(s-p_3)A_{43}+(s-p_3)^2A_{42}+(s-p_3)^3A_{41}. \tag{8.111}$$

The first derivative from both sides of (8.111) takes the form

$$\frac{d}{ds}\left[(s-p_3)^4F(s)\right]=\frac{d}{ds}\left\{\left[(s-p_3)^4F_1(s)\right]+A_{44}+(s-p_3)A_{43}+(s-p_3)^2A_{42}+(s-p_3)^3A_{41}\right\}. \tag{8.112}$$

Performing derivation (8.112), we eliminate A_{44} because $\frac{d}{ds}A_{44}=0$. Also, $\frac{d}{ds}[(s-p_3)A_{43}]=A_{43}$, which yields

$$\frac{d}{ds}[(s-p_3)^4F(s)]=\frac{d}{ds}[(s-p_3)^4F_1(s)]+A_{43}+2(s-p_3)A_{42}+3(s-p_3)^2A_{41}. \tag{8.113}$$

Now, all members of the RHS in (8.113), except A_{43}, will turn to zero when we evaluate $\frac{d}{ds}[RHS]$ at $s=p_4$. Therefore,

$$A_{43} = \frac{d}{ds}\left[(s-p_3)^4 F(s)\right]\Bigg|_{s=p_3}. \tag{8.114}$$

Continue this procedure, we can reveal A_{42} by differentiating (8.113) again and thus eliminating A_{43} from this equation; thus,

$$\frac{d^2}{ds^2}\left[(s-p_3)^4 F(s)\right] = \frac{d^2}{ds^2}\left\{\left[(s-p_3)^4 F_1(s)\right] + 2A_{42} + 6(s-p_3)A_{41}\right\} = \frac{d^2}{ds^2}\left[(s-p_3)^4 F_1(s)\right] + 6A_{41}. \tag{8.115}$$

Evaluation of (8.115) at $s = p_3$ enables us to find A_{42} as

$$A_{42} = \frac{1}{2}\frac{d^2}{ds^2}\left[(s-p_3)^4 F(s)\right]\Bigg|_{s=p_3}. \tag{8.116}$$

Differentiating (8.115) again brings

$$\frac{d^3}{ds^3}\left[(s-p_3)^4 F_1(s)\right] + \frac{d}{ds}\left[2A_{42} + 6(s-p_3)A_{41}\right] = \frac{d^3}{ds^3}\left[\left[(s-p_3)^4 F_1(s)\right] + 6A_{41}\right]. \tag{8.117}$$

Now, A_{41} is obtained as

$$A_{41} = \frac{1}{6}\frac{d^3}{ds^3}\left[(s-p_3)^4 F(s)\right]\Bigg|_{s=p_3} = \frac{1}{3!}\frac{d^3}{ds^3}\left[(s-p_3)^4 F(s)\right]\Bigg|_{s=p_3}. \tag{8.118}$$

Therefore, all the required residues are attained, and all members of (8.105) expansion are determined.

Based on the above example, we can derive a *general formula for finding the residues of F(s) with multiple poles* as follows:

Let $F(s) = \frac{N(s)}{D(s)}$ be an s-domain function whose denominator contains a set of l distinct poles, $p_1, p_2, p_3, \ldots, p_l$ and a set of m repeated poles p_n. Let q be a distinct pole current number, $n = l+1$ be the current number of a multiple pole, and r be the current order of a multiple pole. (In the above example, $q = 1, 2$ and $l = 2, n = 3, r = 1, 2, 3, 4$ and $m = 4$.) Then, *F(s)* can be written in a factored form as

$$F(s) = \frac{N(s)}{(s-p_1)(s-p_2)(s-p_3)\ldots(s-p_l)(s-p_n)^m}. \tag{8.119}$$

Expanding this F(s) into partial fractions results in

$$F(s) = \frac{A_1}{(s-p_1)} + \frac{A_2}{(s-p_2)} + \frac{A_3}{(s-p_3)} + \ldots + \frac{A_l}{(s-p_l)}$$

$$+ \frac{A_{n,m}}{(s-p_n)^m} + \frac{A_{n,m-1}}{(s-p_n)^{m-1}} + \frac{A_{n,m-2}}{(s-p_n)^{m-2}} + \frac{A_{n,m-3}}{(s-p_n)^{m-3}} + \ldots + \frac{A_{n,1}}{(s-p_n)}$$

$$= \sum_{q=1}^{q=l} \frac{A_q}{(s-p_q)} + \sum_{r=1}^{r=m} \frac{A_{n,r}}{(s-p_n)^r}. \tag{8.120}$$

All residues of the distinct poles, A_{1q}, for $q = 1, 2, 3, ..., l$ can be found by the traditional cover-up method as

$$A_{1q} = \left(s - p_{1q}\right)F(s)\Big|_{s=p_{1q}}. \tag{8.121}$$

The constants for the multiple poles are determined by (8.108) through (8.118) whose generalized formula is

$$A_{n,r} = \frac{1}{(m-r)!}\frac{d^{m-r}}{ds^{m-r}}\left[\left(s - p_n\right)^m F(s)\right]\Bigg|_{s=p_r}, \tag{8.122}$$

where r = 1, 2, 3, ..., m, and the exclamation mark denotes a factorial. Don't forget that $n! = 1\cdot2\cdot3\cdot...\cdot n$ and $0! = 1$ by convention. You are encouraged to verify (8.122) by applying this formula to any residue considered in the above example.

Now, we want to consider the response of an RC circuit to a *ramp input*. This real-life problem enables us demonstrate the use of the alternative (differentiating) method for finding the inverse Laplace transform in case of multiple (repeated) poles.

Example 8.1 considers a step response of an RC circuit. Nevertheless, Figure 7.6 shows that a step function is the limiting case of a ramp function whose ramp interval, Δt, tends to zero. In reality, Δt can never reach zero; thus, an ideal step function can never be materialized. Hence, to refer to a real-life situation, we need to consider a ramp response of an electrical circuit, which Example 8.5 does.

Example 8.5 Ramp response of an RC circuit.

Problem: Consider the RC circuit shown in Figure 8.11 whose input is a ramp signal determined by (7.23) as

$$v_{in}(t) = r(t) = \begin{cases} V_0 \ for\ t < t_0 \\ \quad mt \qquad for\ t_0 \le t \le t_f, \\ V_f \ for\ t > t_f \end{cases} \tag{7.23R}$$

where $V_0 = 0$, $t_0 = 0.01s$, $t_{f1} = 0.02s$, $V_f = 4V, and\ m_1 = 400\frac{V}{s}$. The input signal's graph is shown in Figure 8.12. The circuit's parameter values are R = 1 kΩ and C = 0.5 mF, which means that $\alpha = \frac{1}{RC} = 2\frac{1}{s}$. Find the response of this circuit using the Laplace transform.

Solution: The differential equation of this circuit is

$$v'_{out}(t) + \alpha v_{out}(t) = \alpha v_{in}(t), \tag{8.123}$$

where $\alpha\left(\frac{1}{s}\right) = \frac{1}{RC}$. The input signal is the sum of three components, as (7.23R) describes above. We, however, focus only on its *ramp* part because (1) we know well about the RC circuit's responses to 0 and V_f inputs, and (2) we want to exercise more with multiple-poles cases. Thus,

$$v_{inr}(t) = mt. \tag{8.124}$$

The Laplace transform of this input signal is

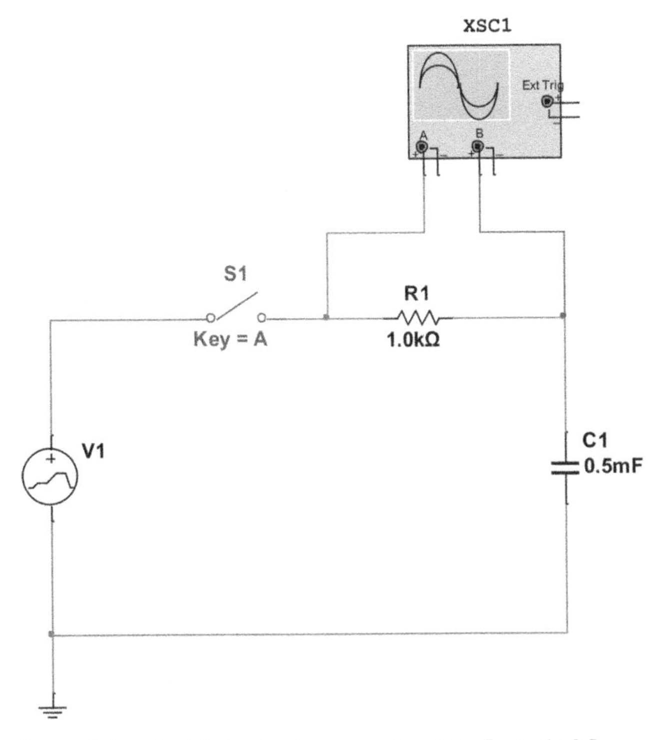

Figure 8.11 An RC circuit with a ramp input for Example 8.5.

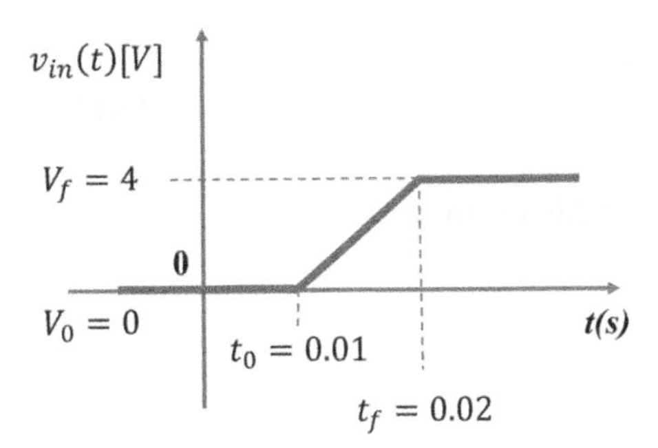

Figure 8.12 Ramp input signal for an RC circuit in Example 8.5.

$$V_{in}(s) = L\{v_{inr}(t)\} = \frac{m}{s^2}. \tag{8.125}$$

The Laplace transform of the RC circuit's Equation (8.60) gives

$$L\{v'_{out}(t) + \alpha v_{out}(t) = \alpha v_{in}(t)\} \Rightarrow sV_{out}(s) + \alpha V_{out}(s) = \alpha V_{in}(s). \tag{8.126}$$

Therefore, the circuit's response in the s-domain is

$$\begin{aligned} sV_{out}(s) + \alpha V_{out}(s) &= \alpha\left(\frac{m}{s^2}\right) \\ \Rightarrow V_{out}(s) &= \frac{\alpha m}{s^2(s+\alpha)} \end{aligned} \tag{8.127}$$

Thus, the task is to find $v_{out}(t)$:

$$v_{out}(t) = L^{-1}\{V_{out}(s)\} = L^{-1}\left\{\frac{\alpha m}{s^2(s+\alpha)}\right\}. \tag{8.128}$$

To find the inverse Laplace transform of $V_{out}(s)$, we need to expand the latter into partial fractions, as (8.83) requires

$$V_{out}(s) = \frac{\alpha m}{s^2(s+\alpha)} = \frac{A_{11}}{s+\alpha} + \frac{A_{21}}{s} + \frac{A_{22}}{s^2}. \tag{8.129}$$

The inverse Laplace transform of (8.129) yields

$$L^{-1}\{V_{out}(s)\} = L^{-1}\left\{\frac{A_{11}}{s+\alpha} + \frac{A_{21}}{s} + \frac{A_{22}}{s^2}\right\} = A_{11}e^{-\alpha t} + A_{21} + A_{22}t. \tag{8.130}$$

This equation gives an idea about the output signal's nature: It is a sum of a damped exponential, constant member, and a linear growing part.

Finding the residues by the alternative (differentiating) method—which is the goal of this example—requires the following steps:

- Using (8.129), we can find A_{11} and A_{22} by the standard cover-up algorithm; namely,

$$A_{11} = \left.\frac{(s+\alpha)(\alpha m)}{(s+\alpha)s^2}\right|_{s=-\alpha} = \frac{m}{\alpha}; \tag{8.131}$$

$$A_{22} = \left.\frac{s^2\alpha m}{s^2(s+\alpha)}\right|_{s=0} = m. \tag{8.132}$$

Thus, the circuit's response to the ramp signal, (8.129), takes the form

$$V_{out}(s) = \frac{\alpha m}{s^2(s+\alpha)} = \frac{m}{\alpha(s+\alpha)} + \frac{A_{21}}{s} + \frac{m}{s^2}. \tag{8.133}$$

- Multiplying (8.133) through by s^2 produces

$$\frac{s^2\alpha m}{s^2(s+\alpha)} = \frac{ms^2}{\alpha(s+\alpha)} + sA_{21} + m. \tag{8.134}$$

- Taking the derivative over s from both sides of (8.134) and evaluating the result at $s=0$ give A_{21} as

$$A_{21} = \left.\frac{d}{ds}\left(\frac{s^2\alpha m}{s^2(s+\alpha)}\right)\right|_{s=0} = \left.\frac{d}{ds}\left(\frac{\alpha m}{(s+\alpha)}\right)\right|_{s=0} = -\left.\frac{\alpha m}{(s+\alpha)^2}\right|_{s=0} = -\frac{m}{\alpha}. \tag{8.135}$$

Now, we attain the full expansion of $V_{out}(s)$ as

$$V_{out}(s) = \frac{\alpha m}{s^2(s+\alpha)} = \frac{m}{\alpha(s+\alpha)} - \frac{m}{\alpha s} + \frac{m}{s^2}. \tag{8.136}$$

You are urged to check the correctness of our partial fraction expansion of $V_{out1}(s)$ by bringing its RHS to the common denominator. We, of course, can use the test-value method.

At last, the inverse Laplace transform of the ramp is attained as

$$v_{out}(t) = L^{-1}\{V_{out}(s) = L^{-1}\left\{\frac{m}{\alpha(s+\alpha)} - \frac{m}{\alpha s} + \frac{m}{s^2}\right\} = \frac{m}{\alpha}e^{-\alpha t} - \frac{m}{\alpha} + mt. \quad (8.137)$$

Finally, let's plug in the given values and obtain the bottommost equation

$$v_{out}(t)V = 200e^{-2t} - 200 + 400t. \quad (8.138)$$

The problem is solved.

Discussion:

- Checking the initial value, we find that $v_{out}(0)V = 0V$, as the problem's condition requires. The final value of $v_{out}(t)$ given in (8.138) cannot be specified because this equation determines only the difference between the ramp—which is the linear part of the whole input signal—and an RC circuit's response to this ramp. However, we can calculate this difference, which is the value of $v_{out}(t)$ when the ramp turns to constant a 4-volts line. Plugging into (8.138) the duration of the ramp, $t_f - t_0 = 0.1s$, gives $v_{out}(t)_{t=0.01} = 0.0397V$.
- MATLAB script and the solution for this example are shown below.

```
syms s t
a = 2;
m = 400;
n = a*m;
d = s^2*(s+a);
v = ilaplace(n/d,t);
ezplot(v, [0,0.02]);
```

Figure 8.13 shows the point $t = 0.01$ s, where the $v_{out}(t)_{t=0.01}$ takes on the value of 0.0397 V. Remember, the correctness of the *obtained solution's initial and final values* is the only necessary condition of the solution's accuracy.

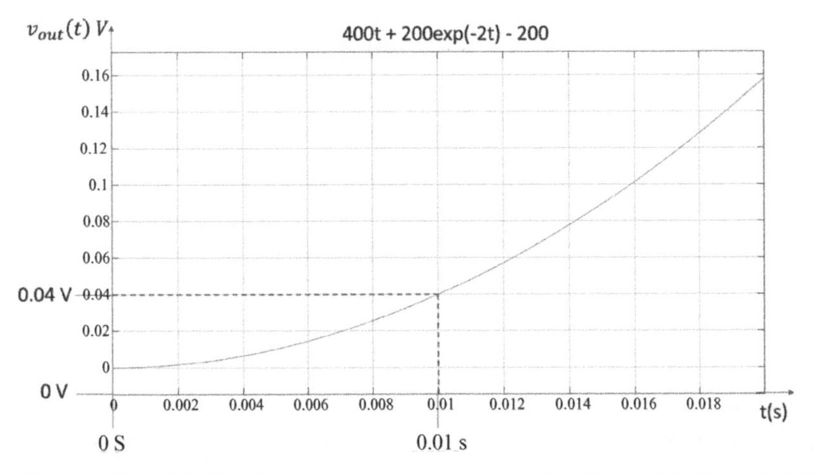

Figure 8.13 MATLAB solution for the ramp response of an RC circuit in Example 8.5.

- The Multisim enables us to simulate the RC circuit's response to the whole ramp signal described by (7.23), and the result is shown in Figure 8.14. To generate the required input signal, we need to adhere to the following instructions:
  - Build the circuit shown in Figure 8.11.
  - Choose voltage source "Piecewise_Linear_Voltage_Source, PWL."
    - Double left-click on the PWL and choose "Value."
    - Mark "Enter data points in the table."
    - Set time and voltage for each piece of the ramp signal as assigned by the given voltage, $v_{in}(t)$, in Equation 7.23.
  - In the main menu, select window "Simulate," then choose "Analyses," and then select "Transient Analysis."
    - In "Analysis Parameters"
      - Initial conditions → "Set to zero"
      - Start time → 0
      - Stop time → Whatever you need (e.g., 9 or 10)
  - In "Outputs," select V(1) and obtain the graph.
  - Play with the graph to obtain the best graphical view. Keep the background white and add the grid. Save the graph.
  - If you want to see the output current, repeat this step by choosing output I(v2). This is your result.

The results of this simulation are shown in Figure 8.14. Compare the input and the output signals and see how the output signal's graph in Figure 8.14a corresponds to Equation 8.138 whose $m_1 = \frac{4}{0.01} = 400\frac{V}{s}$. Figure 8.14b shows the result of additional simulation for $m_2 = 2m_1 = \frac{4}{0.005} = 800\frac{V}{s}$. Also, Figure 8.14c demonstrates the outcome for

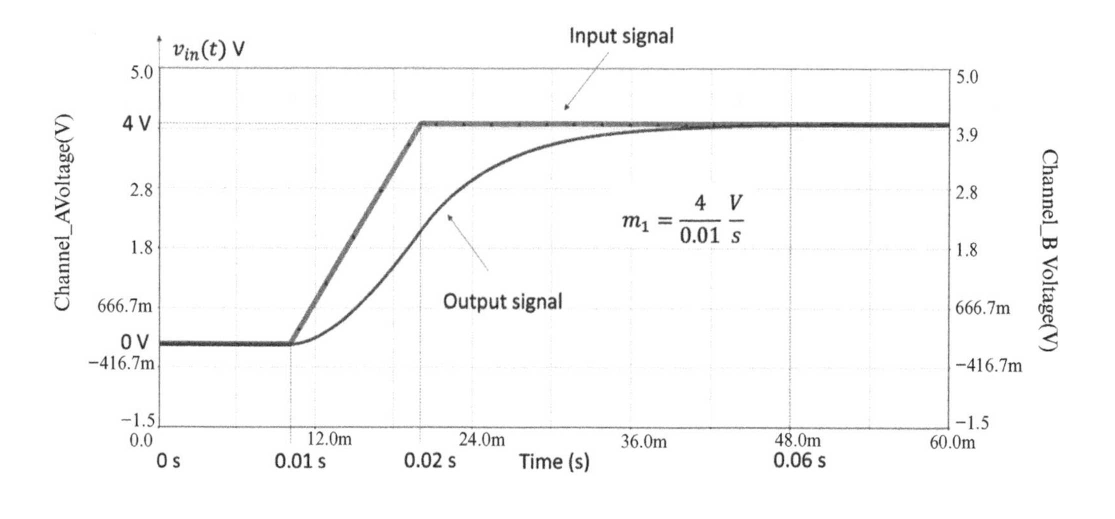

Figure 8.14 Multisim simulation of the RC circuit with a ramp input in Example 8.5–the simulation results: a) Input and output signals whose $m_1 = \frac{4}{0.01}\frac{V}{s}$; b) these signals whose $m_2 = \frac{4}{0.005}\frac{V}{s}$; c) these signals whose $m_3 = \frac{4}{0.02}\frac{V}{s}$.

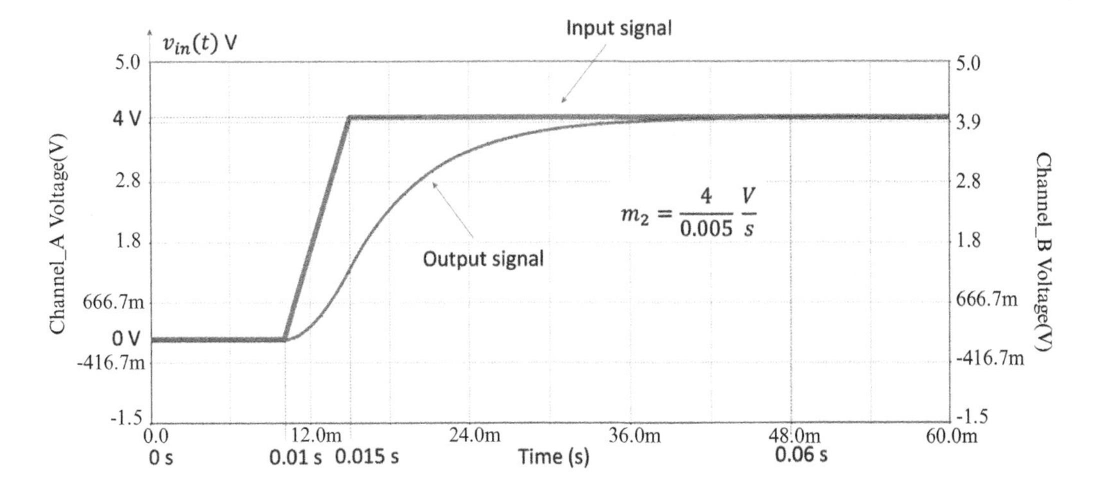

b)

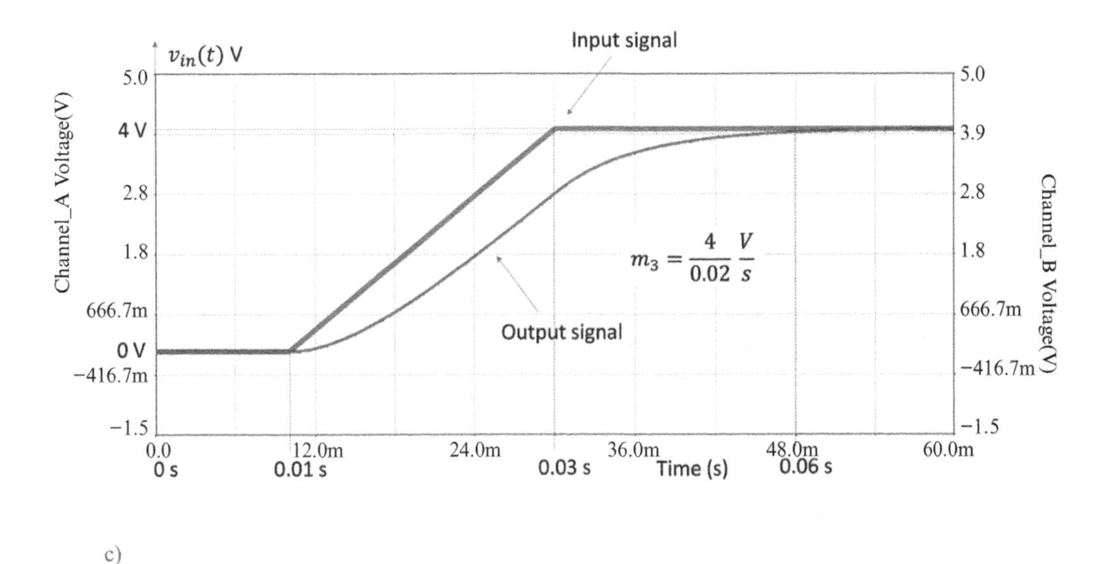

c)

Figure 8.14 **(Cont'd)**

$m_3 = \frac{1}{2}m_1 = \frac{4}{0.02} = 200\frac{V}{s}$. Analyze the difference between Figures 8.14a, 8.14b, and 8.14c and record your comments.

8.5.5 Repeated (Multiple) Poles: A Common-Sense Approach

You certainly notice that we solve all our preceding problems by two methods: First, by reformatting the $F(s)$ for direct application in the table of the Laplace transform pairs, Table 7.1, and secondly, by expanding a given $F(s)$ into partial fractions and finding the residues by a cover-up algorithm or other methods. There are, however, cases where a problem can be solved by using the common-sense approach. The following example demonstrates such a case.

Example 8.6 Finding the inverse Laplace transform of a given $F(s)$ by a common-sense approach.

Problem: Determine the inverse Laplace transform of

$$F(s)=\frac{7(s+5)}{(s^2+6s+9)}.$$

Solution:

Step 1. We find the poles, $p_1 = p_2 = -3$, and expand the $F(s)$ into the partial fractions as

$$F(s)=\frac{7(s+5)}{(s^2+6s+9)}=\frac{7(s+5)}{(s+3)^2}=\frac{A_1}{(s+3)}+\frac{A_2}{(s+3)^2}.$$

Step 2. The constant A_2 is still obtained by employing the cover-up algorithm

$$A_2=(s+3)^2F(s)\Big|_{s=-3}=\frac{(s+3)^2 7(s+5)}{(s+3)^2}\Bigg|_{s=-3}=7(-3+5)=14.$$

To find A_1 this time, we resort to algebraic manipulations. Namely, we substitute $A_2 = 14$ into original expansion of $F(s)$ and find

$$\frac{7(s+5)}{(s+3)^2}=\frac{A_1}{(s+3)}+\frac{14}{(s+3)^2}.$$

Bringing the RHS to the common denominator yields

$$\frac{7(s+5)}{(s+3)^2}=\frac{A_1(s+3)+14}{(s+3)^2}.$$

Compare the coefficients of s at both sides of this equation and attain

$$A_1=7.$$

Therefore, the partial fraction expansion of the given F(s) is

$$\frac{7(s+5)}{(s^2+6s+9)}=\frac{14}{(s+3)^2}+\frac{7}{(s+3)}.$$

Verify the expansion by a test-value method by plugging $s = 2$ (or any other s value) in both sides. For $s = 2$, compute $\frac{49}{25}=\frac{49}{25}$.

Step 3. Performing the inverse Laplace:

Substitute the values of A_1 and A_2 into the expanded $F(s)$ and perform inverse Laplace transform as

$$f(t) = \mathcal{L}(F(s)) = 7\mathcal{L}^{-1}\left(\frac{1}{(s+3)}\right) + 14\mathcal{L}^{-1}\left(\frac{1}{(s+3)^2}\right) = 7e^{-3t} + 14te^{-3t}.$$

The problem is solved.

Discussion:

- Though the demonstrated method provides no significant innovation in finding the residues, this example shows that there is a possibility to avoid differentiation for finding constants in a multiple-poles case. In any event, a common-sense approach is always helpful.

  (Historically, mathematicians have widely used the equating coefficients method for problem solving and theorem proving. The figure no less than Newton extensively employed it in his mathematical manipulations.)
- MATLAB's answer for this example shown in Figure 8.15 confirms our manual solution and presents the graphical view of this solution. Mark the point where $t = 5\tau = \frac{5}{3}s$ and verify that the $f(t)$ amplitude at this point is indeed equal to $0.067 \cdot 7 = 0.047V$.

This concludes the investigation of finding the inverse Laplace transform when the original $F(s)$ contains multiple poles.

8.6 Poles Are Pure Imaginary Numbers

This chapter's preceding sections show that the Laplace transform application drastically mitigates finding a circuit's response. Still, the price for this benefit is the need for finding the inverse Laplace transform, which can be found by either immediate comparison of an *F(s)* with the functions provided in Table 7.1 or by expanding the given *F(s)* into partial fractions and then referring to Table 7.1. Yet, we need to comprehend that a partial fraction expansion is the most general method and can be used for any function. This point's importance becomes evident when we encounter

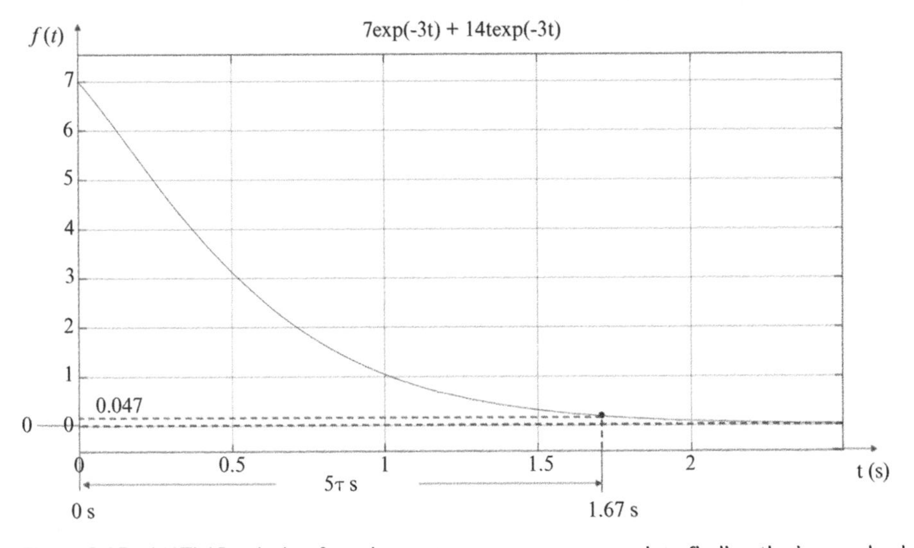

Figure 8.15 MATLAB solution for using a common-sense approach to finding the inverse Laplace transform with multiple poles in Example 8.6.

complex poles, the poles whose nature is different from the real poles considered previously. In this section, we start studying this topic with pure imaginary poles.

8.6.1 Circuit Analysis with Imaginary Poles

Now, we turn to the case of imaginary poles with a real-life problem presented in Example 8.7.

Example 8.7 Finding an RC circuit response to a sinusoidal input.

Problem: Find the response of an RC circuit shown in Figure 8.16 if $v_{in}(t) = 5\cos(2\pi 100t)$, $R = 1\ k\Omega$, and C = 80 μF. Assume $v_{out}(0) \equiv V_0 = 0V$.

Solution:
The circuit's differential equation is derived in (4.21) as

$$v'_{out}(t) + \alpha v_{out}(t) = \alpha V_{in} cos\omega t. \tag{8.142}$$

where $\alpha = \frac{1}{RC} = 12.5 \left(\frac{1}{s}\right)$.

The Laplace transform of this differential equation is

$$sV_{out}(s) + \alpha V_{out}(s) = \frac{\alpha V_{in} s}{\left(s^2 + \omega^2\right)}. \tag{8.143}$$

Solving this algebraic equation for $V_{out}(s)$, we find

$$V_{out}(s) = \frac{\alpha V_{in} s}{(s+\alpha)\left(s^2 + \omega^2\right)}. \tag{8.144}$$

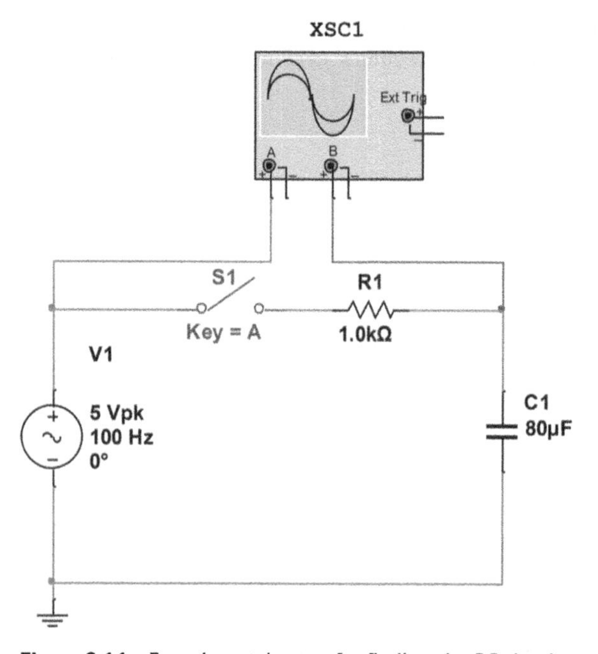

Figure 8.16 Experimental setup for finding the RC circuit response to a sinusoidal signal in Example 8.7.

To find the circuit's response in the time domain, $v_{out}(t)$, we need, of course, obtain the inverse Laplace transform of $V_{out}(s)$, which requires the expansion of the RHS of (8.144) in partial fractions.

The poles of (8.144) are

$$p_1 = -\alpha \text{ and } p_{2,3} = \pm j\omega.$$

The factored form and the partial fraction expansion of the $V_{out}(s)$ is

$$\begin{aligned} V_{out}(s) &= \frac{\alpha V_{in} s}{(s+\alpha)(s^2+\omega^2)} = \frac{\alpha V_{in} s}{(s-p_1)(s-p_2)(s-p_3)} \\ &= \frac{\alpha V_{in} s}{(s+\alpha)(s-j\omega)(s+j\omega)} = \frac{A_1}{(s+\alpha)} + \frac{\mathbf{A}_2}{(s-j\omega)} + \frac{\mathbf{A}_3}{(s+j\omega)}. \end{aligned} \tag{8.145}$$

The residues $\mathbf{A}_2$ and $\mathbf{A}_3$ are boldfaced because we expect them to be complex numbers. We find all the residues by applying the cover-up method:

$$A_1 = (s-p_1)F(s)\Big|_{s=p_1} = (s+\alpha)\left[\frac{\alpha V_{in} s}{(s+\alpha)(s^2+\omega^2)}\right]\Bigg|_{s=-\alpha} = -\frac{\alpha^2 V_{in}}{(\alpha^2+\omega^2)}, \tag{8.146}$$

$$\begin{aligned} \mathbf{A}_2 &= (s-p_2)F(s)\Big|_{s=p_2} = (s-j\omega)\left[\frac{\alpha V_{in} s}{(s+\alpha)(s-j\omega)(s+j\omega)}\right]\Bigg|_{s=+j\omega} \\ &= \frac{\alpha V_{in}(\alpha-j\omega)}{2(\alpha^2+\omega^2)} \end{aligned} \tag{8.147}$$

$$\begin{aligned} \mathbf{A}_3 &= (s-p_3)F(s)\Big|_{s=p_3} = (s+j\omega)\left[\frac{\alpha V_{in} s}{(s+\alpha)(s-j\omega)(s+j\omega)}\right]\Bigg|_{s=-j\omega} \\ &= \frac{\alpha V_{in}(\alpha+j\omega)}{2(\alpha^2+\omega^2)} \end{aligned} \tag{8.148}$$

Note that $\mathbf{A}_3$ is a complex conjugate to $\mathbf{A}_2$ so we can write $\mathbf{A}_3 = \mathbf{A}_2^*$, which means there is no need to perform calculations for $\mathbf{A}_3$ shown in (8.148).

How can we treat the complex residues $\mathbf{A}_2$ and $\mathbf{A}_3$? In general, we can write (8.145) in the following form

$$V_{out}(s) = A_1\frac{1}{(s-p_1)} + \mathbf{A}_2\frac{1}{(s-p_2)} + \mathbf{A}_3\frac{1}{(s-p_3)}. \tag{8.149}$$

Then, the invert $V_{out}(s)$ is given by

$$v_{out}(t) = L^{-1}\{V_{out}(s)\} = L^{-1}\left\{A_1\frac{1}{(s-p_1)} + \mathbf{A}_2\frac{1}{(s-p_2)} + \mathbf{A}_3\frac{1}{(s-p_3)}\right\} = A_1 e^{p_1 t} + \mathbf{A}_2 e^{p_2 t} + \mathbf{A}_3 e^{p_3 t}. \tag{8.150}$$

When we plug all the residues and poles in (8.150), we should obtain $v_{out}(t)$, which would be the solution to our problem.

Plugging all the residues and poles from (8.146), (8.147), and (8.148) into (8.145) yields

$$V_{out}(s) = A_1\frac{1}{(s+\alpha)} + A_2\frac{1}{(s-j\omega)} + A_3\frac{1}{(s+j\omega)}$$
$$= -\frac{\alpha^2 V_{in}}{(\alpha^2+\omega^2)}\frac{1}{(s+\alpha)} + \frac{\alpha V_{in}(\alpha-j\omega)}{2(\alpha^2+\omega^2)}\frac{1}{(s-j\omega)} + \frac{\alpha V_{in}(\alpha+j\omega)}{2(\alpha^2+\omega^2)}\frac{1}{(s+j\omega)}. \tag{8.151}$$

Inserting this $V_{out}(s)$ into (8.150) produces

$$v_{out}(t) = A_1 e^{p_1 t} + A_2 e^{p_2 t} + A_3 e^{p_3 t} = -\frac{\alpha^2 V_{in}}{(\alpha^2+\omega^2)}e^{-\alpha t} + \frac{\alpha V_{in}(\alpha-j\omega)}{2(\alpha^2+\omega^2)}e^{j\omega t} + \frac{\alpha V_{in}(\alpha+j\omega)}{2(\alpha^2+\omega^2)}e^{-j\omega t}? \tag{8.152}$$

We know that $v_{out}(t)$ can contain only the real numbers; hence, (8.152) is not the required solution because it contains the complex numbers. However, the fact that the second and the third members of (8.152) are the complex-conjugate pair assures us that this equation can be brought to the form with real numbers only. The tool enabling us to achieve this goal is the Euler's formula, $e^{j\theta} = cos\theta + jsin\theta$, applying which to (8.152) yields the 2nd member:

$$\frac{\alpha V_{in}(\alpha-j\omega)}{2(\alpha^2+\omega^2)}e^{j\omega t} = \frac{\alpha V_{in}}{2(\alpha^2+\omega^2)}\left(\alpha e^{j\omega t} - j\omega e^{j\omega t}\right)$$
$$= \frac{\alpha V_{in}}{2(\alpha^2+\omega^2)}(\alpha cos\omega t + j\alpha sin\omega t - j\omega cos\omega t + \omega sin\omega t)$$

and
the 3rd member:

$$\frac{\alpha V_{in}(\alpha+j\omega)}{2(\alpha^2+\omega^2)}e^{-j\omega t} = \frac{\alpha V_{in}}{2(\alpha^2+\omega^2)}\left(\alpha e^{-j\omega t} + j\omega e^{-j\omega t}\right)$$
$$= \frac{\alpha V_{in}}{2(\alpha^2+\omega^2)}(\alpha cos\omega t - j\alpha sin\omega t + j\omega cos\omega t + \omega sin\omega t). \tag{8.153}$$

The sum of these two members,

$$\frac{\alpha V_{in}(\alpha-j\omega)}{2(\alpha^2+\omega^2)}e^{j\omega t} + \frac{\alpha V_{in}(\alpha+j\omega)}{2(\alpha^2+\omega^2)}e^{-j\omega t} = \frac{\alpha V_{in}}{(\alpha^2+\omega^2)}(\alpha cos\omega t + \omega sin\omega t), \tag{8.154}$$

indeed, contains only real numbers.

Therefore, the solution to our problem (the inverse Laplace transform of $V_{out}(s)$, that is) is given by

$$v_{out}(t) = L^{-1}\{V_{out}(s)\} = -\frac{\alpha^2 V_{in}}{(\alpha^2+\omega^2)}e^{-\alpha t} + \frac{\alpha V_{in}}{(\alpha^2+\omega^2)}(acos\omega t + \omega sin\omega t). \tag{8.155}$$

Plugging the given numbers in (8.155), we compute the manual solution as

$$v_{out}(t) \approx -\ 0.001978e^{-12.5t} + 0.001978\cos(628.32t) + 0.099432\sin(628.32t). \tag{8.156}$$

The problem is solved.

Discussion:

- Does this solution make sense? Let's check its initial and final states of (8.155): At $t = 0$, $v_{out}(0) = 0$, which meets the initial condition of this problem. At $t \to \infty$, $v_{out}(t) \to (acos\omega t + \omega sin\omega t)$. This result is correct because the transient member containing the factor $e^{-\alpha t}$ diminishes. In contrast, the circuit's steady-state response, $\frac{\alpha V_{in}}{(\alpha^2 + \omega^2)}(acos\omega t + \omega sin\omega t)$, which is the response to the input signal, $v_{in(t)} = V_{in}\cos\omega t$, lasts as long as $v_{in(t)}$ is applied.

- Using (3S.4), $acos\omega t + bsin\omega t = Acos(\omega t - \theta)$, where $A = \sqrt{(a^2 + b^2)}$ and $\Theta = tan^{-1}\left(\frac{b}{a}\right)$, we can rewrite the answer (8.155) as

$$v_{out}(t) = -\frac{\alpha V_{in}}{(\alpha^2 + \omega^2)}\alpha e^{-\alpha t} + \frac{\alpha V_{in}}{\sqrt{\alpha^2 + \omega^2}}\mathrm{A}\cos(\omega t - \theta), \tag{8.157}$$

where $A = \sqrt{(\alpha^2 + \omega^2)}$ and $\theta = tan^{-1}\left(\frac{\omega}{\alpha}\right)$. Inserting the given values, we attain

$$\begin{aligned} v_{out}(t) &= -\frac{\alpha V_{in}}{(\alpha^2 + \omega^2)}\alpha e^{-\alpha t} + \frac{\alpha V_{in}}{\sqrt{(\alpha^2 + \omega^2)}}\cos(\omega t - \theta) \\ &\approx -0.01975e^{-12.5t} + 0.99224\cos(628t - 88.86^0). \end{aligned} \tag{8.158}$$

- Let's analyze (8.144), $V_{out}(s) = \frac{\alpha V_{in}s}{(s+\alpha)(s^2 + \omega^2)}$. Now, we know that the invert of $\frac{A_1}{(s+\alpha)}$ produces $A_1e^{-\alpha t}$, and a couple of pure imaginary s-domain poles ends as a sinusoidal member in the time domain. Thus, we could *predict* that its inverse Laplace transform will take the form

$$v_{out}(t) = L^{-1}\{V_{out}(s)\} = A_1e^{-\alpha t} + acos\omega t + bsin\omega t.$$

- It's worth concentrating on the imaginary part of this example. Our derivation of the circuit's time-domain response, $v^*_{out}(t) = \frac{\alpha V_{in}}{\sqrt{(\alpha^2 + \omega^2)}}\cos(\omega t - \theta)$, from its s-domain counterpart, $V^*_{out}(s) = \frac{A_2}{(s - j\omega)} + \frac{A_3}{(s + j\omega)}$, given in Equations 8.145 through 8.158 is the proof of the following rule:
If the circuit's s-domain output is given by

$$V^*_{out}(s) = \frac{A_2}{(s - j\omega)} + \frac{A_3}{(s + j\omega)}, \tag{8.159}$$

where A_2 *and* A_3 *are the complex conjugate pair, then the time-domain response of this circuit can be presented as*

$$v_{out}(t) = 2|A_2|\cos(\omega t - \Theta), \tag{8.160}$$

where $|A_2|$ *is the amplitude (modulus) and* Θ *is the phase of* A_2*, whereas* ω *is the magnitude of the imaginary pole* $\pm j\omega$.

Reviewing Sidebar 3S "Basic manipulations with sinusoidal functions and complex numbers" helps in understanding that $A_2 = \frac{\alpha V_{in}(\alpha - j\omega)}{2(\alpha^2 + \omega^2)}$ found in this example can be presented as a complex number in a rectangular form, $A_2 = \frac{\alpha V_{in}(\alpha)}{2(\alpha^2 + \omega^2)} - j\frac{\alpha V_{in}(\omega)}{2(\alpha^2 + \omega^2)}$. Applying (3S.6) gives $|A_2| = \sqrt{\left(\left(\frac{\alpha V_{in}(\alpha)}{2(\alpha^2 + \omega^2)}\right)^2 + \left(\frac{\alpha V_{in}(\omega)}{2(\alpha^2 + \omega^2)}\right)^2\right)} = \frac{\alpha V_{in}}{2\sqrt{(\alpha^2 + \omega^2)}}$ and $\Theta = tan^1\left(-\frac{\omega}{\alpha}\right)$, as was obtained in (8.158).

- Equation 8.157 gives us the other means to verify our result: The steady-state (forced) response of an RC circuit to a sinusoidal signal is a well-known output of an RC filter whose input is a cosine signal. (See, for example, Mynbaev and Scheiner 2020, pp. 336–337.) The latter is given by

$$v_{filter-out}(t) = V_{out}\cos(\omega t + \theta),$$

where $V_{out} = V_{in}\frac{1}{\sqrt{\left(1 + \left(\frac{R}{X_C}\right)^2\right)}}$ and $\theta = -tan^{-1}\left(\frac{R}{X_C}\right)$. Since $X_C = \frac{1}{\omega C}$ and $\alpha = \frac{1}{RC}$, we can rewrite the RC filter's output as

$$v_{filter-out}(t) = \frac{\alpha V_{in}}{\sqrt{(\alpha^2 + \omega^2)}}\cos(\omega t + \theta),$$

which entirely coincides with the RC circuit steady-state response,

$$v_{RCss-out}(t) = \frac{\alpha V_{in}}{\sqrt{(\alpha^2 + \omega^2)}}\cos(\omega t + \theta).$$

- Multisim simulation of this process produces the signals shown in Figure 8.17. The output signal appears after the switch was closed. Analyze the course of the output signals and see how it fits Equation 8.157. Why does the output amplitude be smaller than the input? Can you calculate how much the former is smaller than the latter? Pay attention at the scale of the input signal (left y-axis) and that of the output signal (right y-axis). Does the output phase shift correctly reflect Equation 8.157?

- MATLAB work with the problem of Example 8.7 is presented below:
  For the MATLAB calculation of the inverse Laplace transform, we use Formula 8.144, $V_{out}(s) = \frac{\alpha V_{in} s}{(s + \alpha)(s^2 + \omega^2)}$, where a = 12.5, w = 628, Vin = 5.

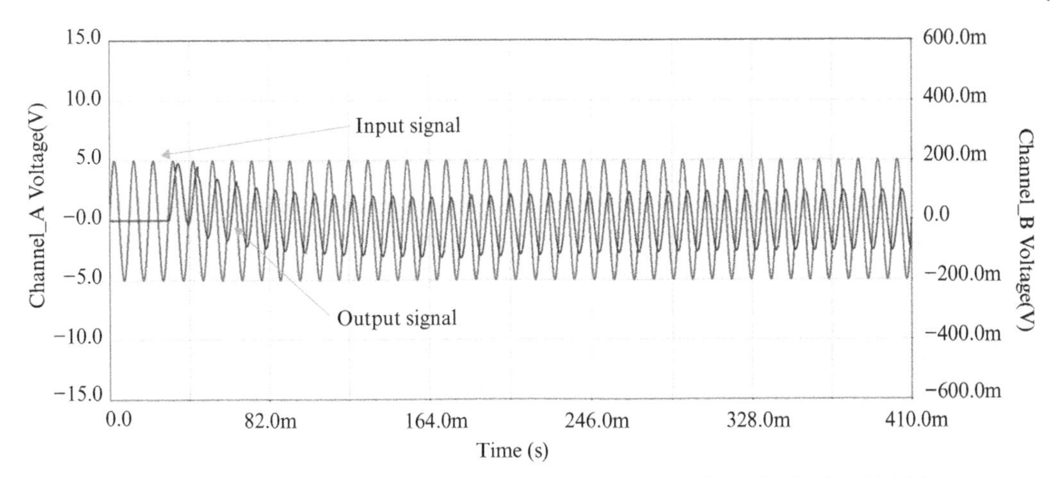

Figure 8.17 Input and output signals of the RC circuit shown in Figure 8.16 obtained by Multisim simulation for Example 8.7.

MATLAB script and answer are as follows:

```
>> syms s t
>> a = 12.5;
>> w = 628;
>> Vin = 5;
>> n = a*Vin*s;
>> d = (s+a)*(s^2+w^2);
>> v = ilaplace(n/d,t);
>> pretty(v);
                            /   25 t \
                         exp| - ---- | 3125
cos(628 t) 3125             \    2   /            sin(628 t) 157000
--------------- - ------------------- + ------------------
    1578161              1578161               1578161
```

Rewriting the MATLAB answer in a common form yields

$$v_{out}(t) = -\frac{3125}{1578161}e^{-\frac{25}{2}t} + \frac{3125}{1578161}\cos(628t) + \frac{157000}{1578161}\sin(628t).$$

Recomputing the coefficients from fractions to the decimals, we obtain

$$v_{out}(t) = -0.00198e^{-\frac{25}{2}t} + 0.00198\cos(628t) + 0.09948\sin(628t),$$

which practically coincides with the manual solution.

Questions

1) Can you find the time-domain response of this RC circuit if $v_{out}(0) = V_0 \neq 0$? Obtain $v_{out}(t)$ when $V_0 = 10V$.

2) What will be $v_{out}(t)$ in Example 8.7 if its input will be $v_{in}(t) = 5sin2\pi100t$? (Consider this question as a mini-project.)

8.6.2 Other Methods of Finding Residues in Case of Pure Imaginary Poles

In (8.48) and (8.49), we discussed the algebraic and direct methods of finding the residues in the case of real poles. Here, we show how both methods can be applied for imaginary poles. (Alexander and Sadiku 2020, pp. 696–697.) To deliberate on this technique, we will consider the sinusoidal response of the RC circuit discussed in Example 8.7.

- **Algebraic method**

The circuit's s-domain response, $V_{out}(s)$, given in (8.144), can be expanded into the partial fractions as

$$V_{out}(s) = \frac{\alpha V_{in} s}{(s+\alpha)(s^2+\omega^2)} = \frac{A_1}{(s+\alpha)} + \frac{Bs}{(s^2+\omega^2)} + \frac{C}{(s^2+\omega^2)}. \quad (8.161)$$

We need three residues because $V_{out}(s)$ has three poles. We choose these forms for the fractions because they enable us to use Table 7.1 to obtain the required inverse Laplace transforms as

$$L^{-1}\left\{\frac{A_1}{(s+\alpha)}\right\} = A_1 e^{-\alpha t}, L^{-1}\left\{B\frac{s}{(s^2+\omega^2)}\right\} = Bcos\omega t, \text{ and } L^{-1}\left\{\frac{C}{\omega}\frac{\omega}{(s^2+\omega^2)}\right\} = \frac{C}{\omega}sin\omega t. \quad (8.162)$$

In other words, this approach allows for avoiding the complex-number manipulations in solving the problem. To find the residues, we need to bring the RHS of (8.161) to the common denominator as

$$\frac{\alpha V_{in} s}{(s+\alpha)(s^2+\omega^2)} = \frac{A_1(s^2+\omega^2)+Bs(s+\alpha)+C(s+\alpha)}{(s+\alpha)(s^2+\omega^2)} = \frac{A_1 s^2 + A_1\omega^2 + Bs^2 + B\alpha s + Cs + C\alpha}{(s+\alpha)(s^2+\omega^2)}, \quad (8.163)$$

and equate the coefficients at the equal power of s to obtain three following algebraic equations:

s^2	$A_1 + B = 0$
s	$\alpha B + C = \alpha V_{in}$
Constant	$A_1\omega^2 + \alpha C = 0$

You will recall that $A_1 = -\frac{\alpha^2 V_{in}}{(\alpha^2+\omega^2)} = -0.00198, \alpha = 12.5, V_{in} = 5$, and $\omega = 628$. Then, we compute B and C as

$$B = -A_1 = \frac{\alpha^2 V_{in}}{(\alpha^2+\omega^2)} = 0.00198 \text{ and } C = -\frac{A_1\omega^2}{\alpha} = \frac{\alpha V_{in}\omega^2}{(\alpha^2+\omega^2)} = 62.475. \quad (8.164)$$

Combining (8.162) and (8.163), we obtain the required result as

$$v_{out}(t) = L^{-1}\{V_{out}(s)\} = L^{-1}\left\{\frac{\alpha V_{in} s}{(s+\alpha)(s^2+\omega^2)}\right\} = L^{-1}\left\{\frac{A_1}{(s+\alpha)} + B\frac{s}{(s^2+\omega^2)} + \frac{C}{\omega}\frac{\omega}{(s^2+\omega^2)}\right\} = \tag{8.165}$$

$$A_1 e^{-\alpha t} + B cos\omega t + \frac{C}{\omega} sin\omega t.$$

Plugging the obtained values into (8.165) yields

$$v_{out}(t) \approx -\ 0.00198e^{-12.5t} + 0.00198cos\omega t + 0.0994 sin\omega t),$$

as in (8.156).

- **Direct method**

Here, we calculate the values with the higher precision, up to fourth digit after the point.

From (8.163), we get

$$\alpha V_{in} s = A_1(s^2+\omega^2) + Bs(s+\alpha) + C(s+\alpha). \tag{8.166}$$

Residue A_1 is found in (8.157) as $A_1 = \frac{-\alpha^2 V_{in}}{(\alpha^2+\omega^2)}$. Then, (8.166) gives

$$\alpha V_{in} s = \frac{-\alpha^2 V_{in}}{(\alpha^2+\omega^2)}(s^2+\omega^2) + Bs(s+\alpha) + C(s+\alpha). \tag{8.167}$$

To find unknown B and C, we can create two equations by plugging consecutively two arbitrary values of s other than poles' values. For computational simplicity, let's use $s=0$ and $s=1$ and calculate:

For $\mathbf{s=0}$, (8.165) becomes

$$0 = \frac{-\alpha^2 V_{in}}{(\alpha^2+\omega^2)}(\omega^2) + C(\alpha),$$

which gives

$$C = \frac{\alpha V_{in}}{(\alpha^2+\omega^2)}(\omega^2),$$

as in (8.164). Hence, C = 62.475.

For $\mathbf{s=1}$, Equation 8.167 turns into

$$\alpha V_{in} = \frac{-\alpha^2 V_{in}}{(\alpha^2+\omega^2)}(1^2+\omega^2) + B(1+\alpha) + C(1+\alpha).$$

After tedious but simple manipulations, we find

$$B = -A_1 = \frac{\alpha^2 V_{in}}{(\alpha^2+\omega^2)} = 0.00198,$$

as in (8.164).

Therefore,

$$V_{out}(s)=\frac{\alpha V_{in}s}{(s+\alpha)(s^2+\omega^2)}=\frac{A_1}{(s+\alpha)}+\frac{Bs}{(s^2+\omega^2)}+\frac{C}{(s^2+\omega^2)}$$
$$=-\frac{0.00198}{(s+12.5)}+\frac{0.00198s}{(s^2+628^2)}+\frac{62.475}{(s^2+628^2)}. \quad (8.168)$$

This approach enables us to verify the correctness of the partial fraction expansion. For this, we take (8.166),

$$\alpha V_{in}s=A_1(s^2+\omega^2)+Bs(s+\alpha)+C(s+\alpha),$$

plug all the given values into this equation,

$$62.5s=-0.00198(s^2+628^2)+0.00198s(s+12.5)+62.475(s+12.5), \quad (8.169)$$

and use the test-value method by plugging, for instance, $s=2$ into (8.169). We compute $125\approx125.06$. The small discrepancy is caused by rounding the computed values.

Therefore, the required inverse Laplace transform is given by

$$v_{out}(t)=L^{-1}\{V_{out}(s)\}=L^{-1}\left\{\frac{\alpha V_{in}s}{(s+\alpha)(s^2+\omega^2)}\right\}$$
$$=L^{-1}\left\{-\frac{0.00198}{(s+12.5)}+\frac{0.00198s}{(s^2+628^2)}+\frac{62.475}{628.32}\frac{628.32}{(s^2+628^2)}\right\} \quad (8.170)$$
$$\approx-0.001978e^{-12.52t}+0.001978\cos(628.32)t+0.099432\sin(628.32)t,$$

as before in (8.156).

- **Modification of the cover-up method**

We can further simplify the residues calculations by using the following approach:

Consider (8.151) in this form

$$\frac{\alpha V_{in}s}{(s+\alpha)(s^2+\omega^2)}=\frac{A_1}{(s+\alpha)}+\frac{A_2}{(s-j\omega)}+\frac{A_3}{(s+j\omega)}.$$

Bringing the RHS to the common denominator results in

$$\frac{\alpha V_{in}s}{(s+\alpha)(s^2+\omega^2)}=\frac{A_1(s-j\omega)(s+j\omega)+A_2(s+\alpha)(s+j\omega)+A_3(s+\alpha)(s-j\omega)}{(s+\alpha)(s-j\omega)(s+j\omega)}. \quad (8.171)$$

or

$$\alpha V_{in}s= A_1(s-j\omega)(s+j\omega)+A_2(s+\alpha)(s+j\omega)+A_3(s+\alpha)(s-j\omega). \quad (8.172)$$

We can find all three residues by letting consecutively $s=-\alpha, s=j\omega$, and $s=-j\omega$. Thus, we attain

for $s=-\alpha$, $A_1=\dfrac{-\alpha^2 V_{in}}{\left(\alpha^2+\omega^2\right)}$, as in (8.145)

for $s=j\omega$, $A_2=\dfrac{j\alpha V_{in}\omega}{(j\omega+\alpha)(j2\omega)}=\dfrac{\alpha V_{in}}{2(\alpha+j\omega)}=\dfrac{\alpha V_{in}(\alpha-j\omega)}{2\left(\alpha^2+\omega^2\right)}$, as in (8.146),

and for $s=-j\omega$, $A_3=\dfrac{-j\alpha V_{in}\omega}{(-j\omega+\alpha)(-j2\omega)}=\dfrac{\alpha V_{in}}{2(\alpha-j\omega)}=\dfrac{\alpha V_{in}(\alpha+j\omega)}{2\left(\alpha^2+\omega^2\right)}$, as in (8.147).

This approach is a modification of the cover-up algorithm. Finding $v_{out}(t)$ must be performed as shown in Example 8.7.

- **Discussion**:

In this subsection, we find *the inverse Laplace transform of the same F(s) by employing three methods*. Which approach is better? The algebraic and direct methods have no rule, algorithm, or established procedure to fulfill the task; it requires imagination, creativeness, and practice. The benefit is straightforward and quick manipulations producing the result. The main one, the cover-up method, relies on a firmly established routine; it requires only the accurate application of the known formulas. Its drawback is a time-consuming procedure with a possibility of errors. It is you, our reader, who has to decide which method to choose for every specific task. Fortunately, choosing an approach is not a crucial decision: If you fail with any alternative methods, you can always resort to partial fraction expansion and the cover-up algorithm.

Example 8.7 and its discussion show that the *pure imaginary poles* of an s-domain function, *F(s)*, result in the *sinusoidal time-domain function, f(t)*, after the inverse Laplace transform is performed.

8.6.3 Circuit Analysis Using the Polar and Exponential Forms of Imaginary Poles

The new approach to finding a transient circuit response will be demonstrated by considering the other example of a real-life problem.

Example 8.8 Using the Laplace transform to find the total (complete) sinusoidal response of an RL circuit.

Problem: Find the total (complete) response of the RL series circuit given in Figure 8.18 by using the Laplace transform with either polar or exponential forms for imaginary poles. Here are the data for this example: $v_{out}(0)=V_0=10V$, $v_{in}(t)=5\cos(4t)$, $R=2\ \Omega$, and $L=1H$. (Though the component values are not realistic, we still take them to stay in line with Example 8.7, review of which is strongly recommended for clear understanding of this example.)

Solution:

- The circuit's differential equation can be derived as follows: The KVL gives

  $v_L(t)+v_R(t)=v_{in}(t)$,

  where $v_L(t)=L\dfrac{di(t)}{dt}$, $v_R(t)=i(t)\cdot R\equiv v_{out}(t)$, and $v_{in}(t)=V_{in}\cos\omega t$. Then,

  $i(t)=\dfrac{v_R(t)}{R}$ and $v_L(t)=\dfrac{L}{R}v_R'(t)$.

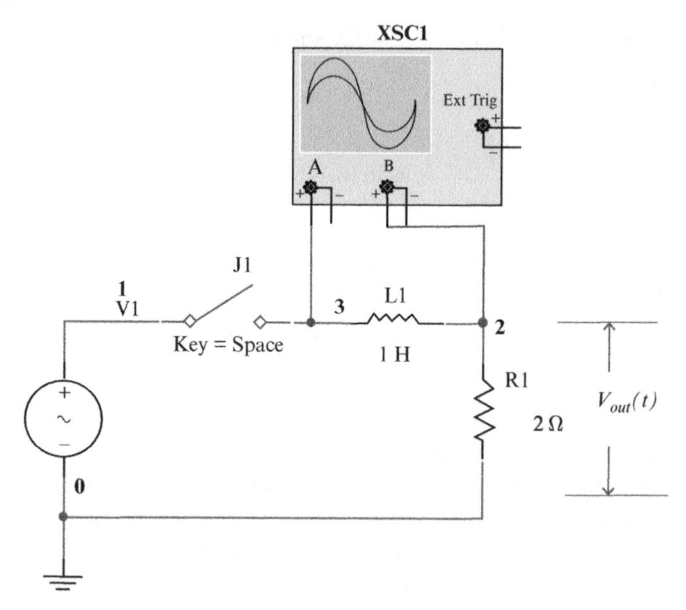

Figure 8.18 A series RL circuit for Example 8.8.

Denoting $\alpha = \frac{R}{L}\left(\frac{1}{s}\right)$ and multiplying through the whole KVL equation by α, we get

$$v'_{out}(t) + \alpha v_{out}(t) = \alpha V_{in} cos\omega t. \tag{8.173}$$

- The Laplace transform of this equation is

$$sV_{out}(s) - V_0 + \alpha V_{out}(s) = \frac{\alpha V_{in} s}{\left(s^2 + \omega^2\right)}. \tag{8.174}$$

- The circuit's response in the s-domain is the solution to (8.174), and it is given by

$$V_{out}(s) = \frac{V_0}{s + \alpha} + \frac{\alpha V_{in} s}{(s + \alpha)\left(s^2 + \omega^2\right)}. \tag{8.175}$$

- Since $p_1 = -\alpha$ and $p_{2,3} = \pm j\omega$, the partial fraction expansion of (8.175) is

$$V_{out}(s) = \frac{V_0}{s + \alpha} + \frac{\alpha V_{in} s}{(s + \alpha)\left(s^2 + \omega^2\right)} = \frac{V_0}{s + \alpha} + \frac{A_1}{s + \alpha} + \frac{A_2}{s - j\omega} + \frac{A_3}{s + j\omega}, \tag{8.176}$$

where A_2 and $A_3 = A_2^*$ are the complex-conjugate numbers.

- To avoid extra step in finding residues V_0 and A_1, we rearrange (8.176) as

$$V_{out}(s) = \frac{V_0 + A_1}{s + \alpha} + \frac{A_2}{s - j\omega} + \frac{A_3}{s + j\omega} \tag{8.177}$$

because $L^{-1}\left\{\frac{V_0+A_1}{s+\alpha}\right\}=(V_0+A_1)e^{-\alpha t}$ is the same result as we would attain in finding V_0 and A_1 individually. Thus,

$$V_0+A_1=(s-p_1)V_{out}(s)\Big|_{s=p_1}=(s+\alpha)\left[\frac{V_0}{s+\alpha}+\frac{\alpha V_{in}s}{(s+\alpha)(s^2+\omega^2)}\right]\Bigg|_{s=-\alpha}=V_0-\frac{\alpha^2 V_{in}}{(\alpha^2+\omega^2)}. \quad (8.178)$$

- For other residues, we need to find only A_2 because A_3 is its complex conjugate.
- Which of the required forms—polar or exponential—should we choose? We know that if an s-domain function has pure imaginary poles, then its time-domain counterpart (that is, its inverse Laplace transform) is a sinusoid. Recall that the direct relationship between an s-domain and sinusoidal functions is given by the Euler formulas,

$$e^{j\varphi}=cos\varphi+jsin\varphi \text{ and } e^{-j\varphi}=cos\varphi-jsin\varphi, \quad (3S.15R)$$

and deduce that we need an exponential form. Follow the procedure developed in Example 8.7, we get A_2 as

$$\begin{aligned}A_2&=(s-p_2)V_{out}(s)\Big|_{s=p_2}=(s-j\omega)\left[\frac{V_0}{s+\alpha}+\frac{\alpha V_{in}s}{(s+\alpha)(s-j\omega)(s+j\omega)}\right]\Bigg|_{s=+j\omega}\\&=\frac{j\alpha V_{in}\omega}{(\alpha+j\omega)(j2\omega)}=\frac{\alpha V_{in}}{2(\alpha+j\omega)}=\frac{\alpha V_{in}(\alpha-j\omega)}{2(\alpha^2+\omega^2)},\end{aligned} \quad (8.179)$$

as in (8.147). Equation 3S.5 shows that $\boldsymbol{Z}=a+jb=Ae^{j\theta}$. Applying this rule to (8.178) yields

$$A_2=\frac{\alpha V_{in}}{2\sqrt{(\alpha^2+\omega^2)}}e^{j\theta}, \quad (8.180)$$

where

$$\theta=-tan^{-1}\left(\frac{\omega}{\alpha}\right). \quad (8.181)$$

Accordingly,

$$A_3=\frac{\alpha V_{in}}{2\sqrt{(\alpha^2+\omega^2)}}e^{-j\theta}. \quad (8.182)$$

- Now, we can find the inverse Laplace transform of the s-domain circuit response by plugging all the residues into (8.177)

$$\begin{aligned}v_{out(t)}&=L^{-1}\{V_{out}(s)\}=L^{-1}\left\{\frac{V_0+A_1}{s+\alpha}+\frac{A_2}{s-j\omega}+\frac{A_3}{s+j\omega}\right\}\\&=\left(V_0-\frac{\alpha^2 V_{in}}{(\alpha^2+\omega^2)}\right)e^{-\alpha t}+\frac{\alpha V_{in}}{2\sqrt{(\alpha^2+\omega^2)}}e^{j\theta}e^{j\omega t}+\frac{\alpha V_{in}}{2\sqrt{(\alpha^2+\omega^2)}}e^{-j\theta}e^{-j\omega t}.\end{aligned} \quad (8.183)$$

According to (3S.16), the sum of the last two RHS members is

$$\frac{\alpha V_{in}}{2\sqrt{\left(\alpha^2+\omega^2\right)}}\left(e^{j(\omega t+\theta)}+e^{-j(\omega t+\theta)}\right)=\frac{\alpha V_{in}}{2\sqrt{\left(\alpha^2+\omega^2\right)}}2\cos\left(\omega t+\theta\right). \tag{8.184}$$

Therefore, the time domain response of the series RL circuit is attained as

$$v_{out}(t)=L^{-1}\left\{V_{out}(s)\right\}=\left(V_0-\frac{\alpha^2 V_{in}}{\left(\alpha^2+\omega^2\right)}\right)e^{-\alpha t}+\frac{\alpha V_{in}}{\sqrt{\left(\alpha^2+\omega^2\right)}}\cos\left(\omega t+\theta\right). \tag{8.185}$$

Plugging the given values into (8.185) yields

$$v_{out}(t)=L^{-1}\left\{V_{out}(s)\right\}=9e^{-2t}+\sqrt{5}\cos\left(4t-63.44^0\right). \tag{8.186}$$

The problem is solved.

Discussion:

- We solved a similar problem for an RC circuit in Example 8.7. Compare the solving procedure and the results obtained in both cases to affirm that the first-order circuits have identical types of the response to a sinusoidal input. Here, the $V_{out}(s)$ of the RC and RL circuits contains the pure imaginary poles, which leads to a sinusoidal output. In particular, liken (8.157) with (8.185) and (8.158) with (8.186).
- Equation 8.158 entirely coincides with solution (5.53) obtained in Section 5.1 by the time-domain approach,

$$v_{out}(t)=\left(V_0-F\cos\Phi\right)e^{-\alpha t}+F\cos\left(\omega t+\Phi\right). \tag{5.53R}$$

Here, $\alpha\left(\frac{1}{s}\right)=\frac{R}{L}$, $F=\frac{V_{in}}{\sqrt{\left(1+\left(\frac{\omega^2}{\alpha}\right)\right)}}=\frac{\alpha V_{in}}{\sqrt{\left(\alpha^2+\omega^2\right)}}$, $\Phi=-tan^{-1}\left(\frac{\omega}{\alpha}\right)=\theta$, and $cos\Phi=\frac{1}{\sqrt{(1+tan^2\Phi)}}=\frac{1}{\sqrt{\left(1+\left(\frac{\omega^2}{\alpha}\right)\right)}}$. Substituting all these expressions into (5.53), we obtain (8.158).

- Can we use a polar form in this example? Of course, we can. For instance, we can present (8.180) as

$$A_2=\frac{\alpha V_{in}}{2\sqrt{\left(\alpha^2+\omega^2\right)}}e^{j\theta}=Ae^{j\theta}=A\angle\Theta. \tag{8.187}$$

However, the polar form of a complex number doesn't provide us with any benefit.

- Build the circuit shown in Figure 8.17 and run the simulation. Quantitatively analyze the obtained input and output signals by comparing their parameters with those given in (8.186).
- The MATLAB solution to the given problem is shown in Figure 8.19. The reader can easily confirm its identity with Equation 8.186. Also, it's advisable to check 5τ criterion for the duration of transient response.

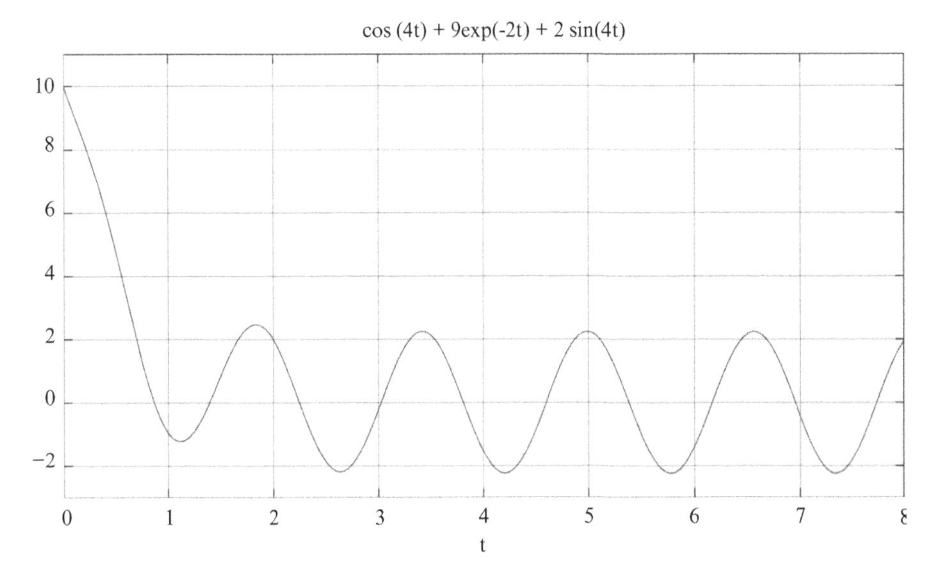

Figure 8.19 MATLAB solution for Example 8.8.

This concludes consideration of the circuits whose s-domain output contains pure imaginary poles.

8.7 Poles Are Complex Numbers

Complex numbers differ from imaginary numbers by the presence of real parts, so that a complex number is written as

$$\mathbf{Z} = a + jb. \tag{3S.5R}$$

Therefore, we can understand the operations with complex numbers by generalizing the manipulations with the imaginary numbers.

It should be realized that the circuits whose s-domain outputs contain the poles with complex numbers must bear at least two reactive components; in other words, they must be the second-order circuits. The full consideration of these circuits will be given in the next chapter; here, we will discuss a numerical example.

Example 8.9 Finding the inverse Laplace transform of an $F(s)$ with complex poles.

Problem: Determine the inverse transform of

$$F(s) = \frac{12(s+1)}{(s+5)(s^2+4s+13)}. \tag{8.188}$$

Solution:

(We urge you to review Examples 8.7 and 8.8, where all the necessary mathematical manipulations are explained in detail.)

- Find the poles: The first pole is

$$p_1 = -5.$$

Two other poles are obtained by solving the characteristic equation of this denominator,

$$\left(s^2 + 4s + 13\right) = 0.$$

The solution gives two complex-conjugate poles

$$p_{2,3} = \frac{-4 \pm \sqrt{\sqrt{(4^2 - 4 \cdot 13)}}}{2} = -2 \pm j3.$$

- Expand the *F(s)* into partial fractions with the appropriate constants as

$$F(s) = \frac{12(s+1)}{(s+5)(s^2+4s+13)} = \frac{A_1}{s-p_1} + \frac{A_2}{s-p_2} + \frac{A_3}{s-p_3} = \frac{12(s+1)}{(s+5)(s+2-j3)(s+2+j3)}$$
$$= \frac{A_1}{s+5} + \frac{A_2}{s+2-j3} + \frac{A_3}{s+2+j3}. \tag{8.189}$$

As always, we follow the general format of a partial fraction expansion given in (8.23).
We remember, of course, that A_3 is a complex conjugate to A_2; that is, $A_3 = A_2^*$. The inverse Laplace transform of (8.189) is

$$f(t) = L^{-1}\left\{\frac{A_1}{s-p_1} + \frac{A_2}{s-p_2} + \frac{A_3}{s-p_3}\right\} = A_1 e^{p_1 t} + A_2 e^{p_2 t} + A_3 e^{p_3 t}. \tag{8.190}$$

- Employing the *cover-up algorithm*, we find

$$A_1 = (s+5)\frac{12(s+1)}{(s+5)(s^2+4s+13)}\Big|_{s=-5} = -\frac{8}{3} \tag{8.191}$$

and

$$A_2 = (s+2-j3)F(s)\Big|_{s=p_2} = (s+2-j3)\frac{12(s+1)}{(s+5)(s+2-j3)(s+2+j3)}\Bigg|_{s=-2+j3} = \frac{2}{3}(2-j). \quad 8.192$$

Of course,

$$A_3 = A_2^* = \frac{2}{3}(2+j). \tag{8.193}$$

Therefore, the inverse Laplace transform of $F(s)$ is given by

$$f(t) = L^{-1}\left\{\frac{A_1}{s-p_1} + \frac{A_2}{s-p_2} + \frac{A_3}{s-p_3}\right\} = A_1 e^{p_1 t} + A_2 e^{p_2 t} + A_3 e^{p_3 t}$$
$$= -\frac{8}{3}e^{-5t} + \frac{2}{3}(2-j)e^{(-2+j3)t} + \frac{2}{3}(2+j)e^{(-2-j3)t}$$
$$= -\frac{8}{3}e^{-5t} + \frac{2}{3}(2-j)e^{-2t}e^{j3t} + \frac{2}{3}(2+j)e^{-2t}e^{-j3t}$$
$$= \frac{8}{3}e^{-5t} + \frac{2}{3}e^{-2t}\left[(2-j)e^{j3t} + (2+j)e^{-j3t}\right]. \tag{8.194}$$

Next, we rearrange the two complex conjugates in (8.194), apply the Euler formula's other form, $\frac{e^{j\varphi} + e^{-j\varphi}}{2} = cos\varphi$ and $\frac{e^{j\varphi} - e^{-j\varphi}}{j2} = sin\varphi$, and find the solution to our problem as

$$f(t) = -\frac{8}{3}e^{-5t} + \frac{2}{3}e^{-2t}\left[2\left(e^{j3t} + e^{-j3t}\right) - j\left(e^{j3t} - e^{-j3t}\right)\right] = -\frac{8}{3}e^{-5t} + \frac{4}{3}e^{-2t}\left(2cos3t + sin3t\right)$$
$$= -\frac{8}{3}e^{-5t} + \frac{4}{3}\sqrt{5}e^{-2t}\cos\left(3t - 26.57^0\right). \tag{8.195}$$

The problem is solved.

Discussion:

- The sinusoidal member of solution (8.195), $\frac{4}{3}\sqrt{5}e^{-2t}\cos\left(3t - 26.57^0\right)$, is an exponentially decaying (damped) cosine caused by the complex poles of $V_{out}(s)$. Compare it to the solutions due to pure imaginary poles obtained in Examples 8.7 and 8.8 as $Acos(\omega t - \theta)$. These results demonstrate that the *circuit output described by the imaginary poles is steady-state sinusoidal oscillations, whereas such an output caused by the complex poles is an exponentially decaying sinusoid diminishing as time tends to infinity.*
- In Example 8.7, we discussed various methods enabling us to avoid the complex numbers algebra to find the residues in case of imaginary poles. Here, we demonstrate the *algebraic* approach for the complex poles (Alexander and Sadiku 2020, pp. 696–697).
  The given s-domain function (8.152) can be expanded into the partial fractions as

$$F(s) = \frac{12(s+1)}{(s+5)\left(s^2+4s+13\right)} = \frac{A_1}{(s+5)} + \frac{Bs}{\left(s^2+4s+13\right)} + \frac{C}{\left(s^2+4s+13\right)}. \tag{8.196}$$

We bring the RHS of (8.196) to the common denominator

$$\frac{12(s+1)}{(s+5)\left(s^2+4s+13\right)} = \frac{A_1s^2 + 4A_1s + 13A_1 + Bs^2 + 5Bs + Cs + 5C}{(s+5)\left(s^2+4s+13\right)}. \tag{8.197}$$

Equating the coefficients at the equal power of s in both sides of (8.197), we obtain three following algebraic equations for finding the constants A, B, and C:

Power of s	Variable	Equation due to equating the coefficients
2	s^2	$A_1 + B = 0$
1	s	$4A_1 + 5B + C = 12$
0	*Constant*	$13A_1 + 5C = 12$

You will recall that $A_1 = -\frac{8}{3}$. Thus,

$$B = -A_1 = \frac{8}{3} \text{ and } C = 12 - 4A_1 - 5B = \frac{28}{3} \tag{8.198}$$

Combining (8.196) and (8.198), we obtain the required result as

$$f(t) = L^{-1}\{F(s)\} = L^{-1}\left\{\frac{12(s+1)}{(s+5)(s^2+4s+13)}\right\}$$

$$= L^{-1}\left\{\frac{-\frac{8}{3}}{(s+5)} + \frac{\frac{8}{3}s + \frac{28}{3}}{(s^2+4s+13)}\right\}$$

$$= L^{-1}\left\{\frac{-\frac{8}{3}}{(s+5)} + \frac{\frac{8}{3}s + \frac{28}{3}}{\left((s+2)^2+3^2\right)}\right\} = L^{-1}\left\{-\frac{8}{3}\frac{1}{(s+5)} + \frac{1}{3}\frac{(8s+16+12)}{\left((s+2)^2+3^2\right)}\right\}$$

$$= L^{-1}\left\{-\frac{8}{3}\frac{1}{(s+5)} + \frac{8}{3}\frac{(s+2)}{\left((s+2)^2+3^2\right)} + \frac{4}{3}\frac{3}{\left((s+2)^2+3^2\right)}\right\}$$

$$= -\frac{8}{3}e^{-5t} + \frac{8}{3}e^{-2t}cos3t + \frac{4}{3}e^{-2t}sin\omega t = -\frac{8}{3}e^{-5t} + \frac{4}{3}\sqrt{}5e^{-2t}(2cos3t + sin\omega t), \tag{8.199}$$

as in (8.195). The sameness of Answers (8.195) and (8.199) assures us that the problem is solved correctly.

Question Would you be able to predict this result by analyzing the given $F(s)$.?

Note that insertion of all the residues into (8.196) turns this equation into the following:

$$\frac{12(s+1)}{(s+5)(s^2+4s+13)} = -\frac{\frac{8}{3}}{(s+5)} + \frac{\frac{8}{3}s + \frac{28}{3}}{(s^2+4s+13)}, \tag{8.200}$$

which enables us to apply the test-value method to confirm the correctness of our partial fraction expansion. Let's plug $s = 2$ into the LHS and RHS of (8.198) and compute $108 = 108$!

You are encouraged to apply direct and other methods considered in Examples 8.7 and 8.8 to find the residues and confirm the obtained solution.

- The MATLAB script for this example is shown below.

```
≫ syms s t
≫ a = 12*(s+1);
≫ b = (s+5)*(s^2+4*s+13);
≫ c = ilaplace(a/b,t);
≫ ezplot(c,[0,3]);
≫ pretty(c);
          / sin(3 t) \
exp(-2 t) | cos(3 t) + ---- | 8
          \    2     / exp(-5 t) 8
--------------------------------------------
           3                  3
```

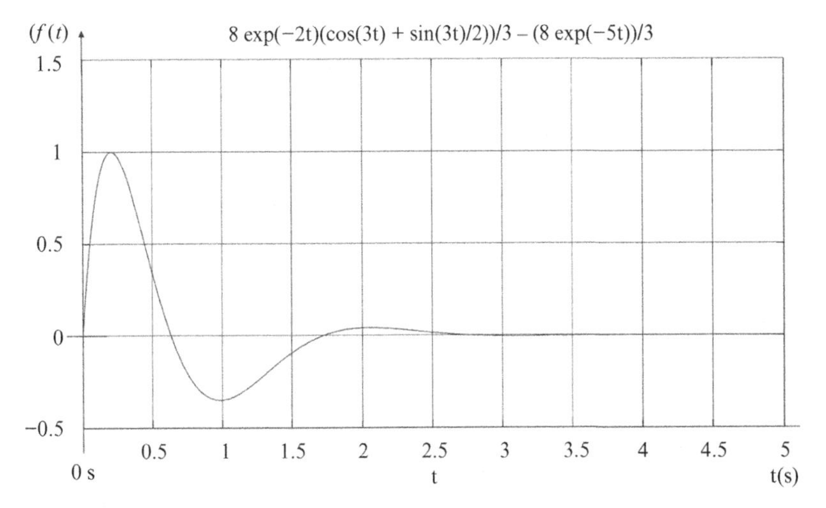

Figure 8.20 MATLAB solution for damping sinusoidal output in Example 8.9.

Rewriting this answer in a common form, we obtain

$$f(t) = \frac{8}{3}\left(cos3t + \frac{1}{2}sin3t\right)e^{-2t} - \frac{8}{3}e^{-5t},$$

which coincides with Equations 8.195 and 8.199.

Compare the formulas given by MATLAB and by manual solution obtained in (8.199).

The graph of $f(t)$ is shown in Figure 8.20. Did you expect to see such a graph for the solution? Explain. Examine the values presenting in this graph: Why is its maximum amplitude equal to 1? The graph comes to the steady-state condition at $t \approx 3s$. Why is it so?

Questions

1) The MATLAB answer for this example shows that f(t) is a difference of two functions. How does Figure 8.20 reflect this fact?
2) Figure 8.20 shows that at $t \approx 0.25s$, *f(t)* reaches its maximum, $f(t) \approx 1$. Can you verify this observation by computations? Show all your calculations.

Conclusion: *When the F(s) poles are the complex numbers, the inverse Laplace transform of the F(s) produces an exponentially decaying sinusoidal signal.*

There is *a useful formula that mitigates the manipulations with the complex poles.* The sinusoidal member of (8.161) can be written in a general form as

$$f(t) = ae^{-\alpha t}cos\omega t + be^{-\alpha t}sin\omega t = Ae^{-\alpha t}\cos(\omega t + \theta). \tag{8.201}$$

Since the exponential member, $e^{-\alpha t}$, and the signal's sinusoid, $cos\omega t$, are known, it's left to determine only two unknown parameters, A and θ, to obtain (8.201) in its entirety. If the denominator of *F(s)* is given by $(s^2 + bs + c)$, then the unknown amplitude A and angle θ in (8.201) are determined as (Stanley, 2003, pp. 285–288)

$$Ae^{j\theta} = A\angle\Theta = \frac{1}{\omega}(s^2 + bs + c)F(s)\Big|_{s=-\alpha+j\omega}. \tag{8.202}$$

A curious reader can derive this formula based on our preceding considerations. Some hints can be found in (Thomas, Rosa, and Toussaint 2019, pp. 471–473). Be aware that the calculations for $s = -\alpha + j\omega$ must be performed at a positive frequency, ω, to obtain the correct phase number. Applying (8.168) to Example 8.9 enables us to find the sinusoidal member of (8.195) as

$$Ae^{j\theta} = A\angle\Theta = \frac{1}{3}\left(s^2+4s+13\right)\left(\frac{12(s+1)}{(s+5)\left(s^2+4s+13\right)}\right)\Bigg|_{s=-2+j3} = \frac{1}{3}\left(\frac{12(s+1)}{(s+5)}\right)\Bigg|_{s=-2+j3}$$
$$= 4\left(\frac{(-1+j3)}{(3+j3)}\right) = 4\frac{\sqrt{10}\angle-71.57^0}{3\sqrt{2}\angle 45^0} = \frac{4}{3}\sqrt{5}\angle(-26.57)^0 \Rightarrow \frac{4}{3}\sqrt{5}e^{-2t}\left(\cos\left(3t-26.57^0\right)\right). \tag{8.203}$$

Application of (8.202) saves time in solving the problem with complex poles. Indeed, instead of performing all sets of the mathematical manipulations from (8.189), $f(t) = L^{-1}\left\{\frac{A_1}{s-p_1} + \frac{A_2}{s-p_2} + \frac{A_3}{s-p_3}\right\} = A_1e^{p_1t} + A_2e^{p_2t} + A_3e^{p_3t}$, through (8.195), $f(t) = -\frac{8}{3}e^{-5t} + \frac{2}{3}(2-j)e^{-2t}e^{j3t} + \frac{2}{3}(2+j)e^{-2t}e^{-j3t} = \ldots = -\frac{8}{3}e^{-5t} + \frac{8}{3}\sqrt{5}e^{-2t}\cos\left(3t-26.57^0\right)$, we can attain the sinusoidal part of the answer by executing two lines of calculations shown in (8.202) and (8.203). This technique becomes especially valuable when the *F(s)* denominator is an irreducible quadratic polynomial and cannot be presented in a factored form.

Question How can we apply (8.202) to pure imaginary poles? Prove your answer by analyzing Examples 8.7 and 8.8.

8.8 Nature of Poles and Nature of Circuit's Response

You will recall the Laplace transform concept shown in Figure 7.3. To summarize the Chapter 8 discussion, we converted Figure 7.3 to Figure 8.21, which shows how this concept is materialized in the case of imaginary poles. In this figure, α is a circuit's parameter; for instance, $\alpha\left(\frac{1}{s}\right) = \frac{1}{RC}$ or $\alpha\left(\frac{1}{s}\right) = \frac{R}{L}$ for series RC or RL circuit, respectively. Also, $V_{in}(V)$ is an amplitude and $\omega\left(\frac{rad}{s}\right)$ is a radian frequency of the input sinusoidal signal. Note that an s-domain function $V_{out}(s)$ contains imaginary poles, and, consequently, the time-domain output, $v_{out}(t)$, includes a sinusoidal member, $\frac{\alpha V_{in}}{\left(\alpha^2+\omega^2\right)}\cos(\omega t-\theta)$. It's worth mentioning that $v_{out}(t)$ is a sum of a transient response, $-\frac{\alpha V_{in}}{\left(\alpha^2+\omega^2\right)}\alpha e^{-\alpha t}$, caused by the input signal's initial push, and a steady-state response, $\frac{\alpha V_{in}}{\left(\alpha^2+\omega^2\right)}\cos(\omega t-\theta)$, supported by the permanently applied input signal.

In conclusion, it's well worthwhile to draw your attention to the relationship between the nature of poles of the s-domain partial fractions and the resulting time-domain signals that would be

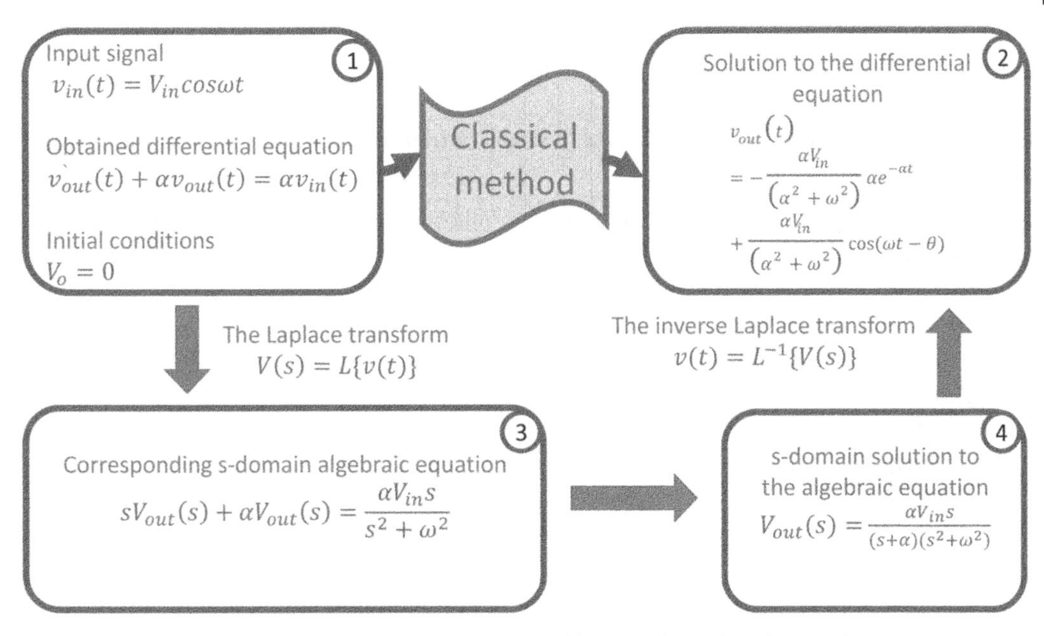

Figure 8.21 The Laplace transform concept implemented in case of pure imaginary poles.

Table 8.2 Poles of partial fractions of the Laplace transform and the nature of corresponding time-domain functions.

Nature of poles of the Laplace transform	**The s-domain function F(s)**	**The inverse Laplace transform $f(t)=L^{-1}\{F(s)\}$**	**Nature of the time-domain functions**
Equal to zero, $p=0$	$F(s)=\frac{A}{s}$	$f(t)=L^{-1}\left\{\frac{A}{s}\right\}=A$	Constant
Pure real, $p=-k$	$F(s)=\frac{A}{s+k}$	$f(t)=L^{-1}\left\{\frac{A}{s+k}\right\}=Ae^{-kt}$	Exponential decay
Pure imaginary, $p=\pm j\omega$	$F(s)=\frac{A_1}{s-j\omega}+\frac{A_2}{s+j\omega}$, $A_2=A_1^*$	$f(t)=L^{-1}\left\{\frac{A_1}{s-j\omega}+\frac{A_1}{s+j\omega}\right\}=A\cos(\omega t\mp\theta)$	Sinusoid
Complex numbers, $p=-\alpha\pm j\omega$	$F(s)=\frac{A_1}{s+\alpha-j\omega}+\frac{A_2}{s+\alpha+j\omega}$, $A_2=A_1^*$	$f(t)=L^{-1}\left\{\frac{A_1}{s+\alpha-j\omega}+\frac{A_1}{s+\alpha+j\omega}\right\}=Ae^{-\alpha t}\cos(\omega t\mp\theta)$	Exponentially decaying sinusoid

obtained by the inverse Laplace transform. This relationship is shown in Table 8.2, which enables us to determine the nature of a circuit's output by merely reviewing its s-domain poles nature.

Example 8.10 Poles of partial fractions of the Laplace transform and the nature of the circuit's time-domain response.

Problem: Applying the inverse Laplace transform, find the time-domain equivalents of the following s-domain functions:

1) $F(s)=\frac{6}{3s}$,

2) $F(s)=\frac{8}{3(s-7)}$,

3) $F(s)=\frac{19}{2s+j8}+\frac{19}{2s-j8}$,

4) $F(s)=\frac{28}{7(s+4-j9)}+\frac{28}{7(s+4+j9)}$.

Solution: By examining Table 7.1, we find

1) $f(t)=L^{-1}\left\{\frac{2}{s}\right\}=2$, as LT-1 shows.

2) $f(t)=\frac{8}{3}L^{-1}\left\{\frac{1}{(s-7)}\right\}=\frac{8}{3}e^{7t}$, as in LT-7.

3) Consider $F(s)=\frac{19}{2}\left(\frac{1}{(s-j4)}+\frac{1}{(s+j4)}\right)$: Since Table 7.1 doesn't explicitly contain complex numbers, we need to perform the additional operation to bring given F(s) to the form of LT-11 and obtain

$$f(t)=\frac{19}{2}\left(L^{-1}\left\{\frac{1}{(s-j4)}+\frac{1}{(s+j4)}\right\}\right)=\frac{19}{2}\left(L^{-1}\left\{\frac{2s}{(s^2+4^2)}\right\}\right)=19\cos(4t).$$

This result can be also attained by applying the rule given in Equations 8.159 and 8.160.

4) In case of $F(s)=\frac{28}{7(s+4-j9)}+\frac{28}{7(s+4+j9)}$, the solution is similar to that of #3: First, bring $F(s)$ to a regular s-domain form as $F(s)=\frac{28}{7(s+4-j9)}+\frac{28}{7(s+4+j9)}=4\frac{2(s+4)}{(s+4)^2+9^2}$.

Secondly, use LT-13 to receive

$$f(t)=4\left(L^{-1}\left\{\frac{2(s+4)}{\left((s+4)^2+9^2\right)}\right\}\right)=8e^{-4t}\cos(9t).$$

The problem is solved.

8.9 A Summary of Chapter 8

- This chapter is devoted to the Laplace transform application for advanced circuit analysis.
- A review of this subject immediately reveals that this application's key and the most demanding procedure is the inverse Laplace transform.
- Chapter 8 carefully considered all possible cases of the inverse Laplace transform that can be encountered in practice. These cases differ by the nature of the roots of the s-domain equivalent equations describing real circuits.
- The discussion of the chapter material is supported by rigorous mathematical consideration and numerous (ten in total) real-life examples.
- MATLAB offers various techniques supporting or replacing the tedious manual mathematical manipulations, and Multisim implements the theoretical considerations at the circuit level and provides independent verification of the calculated results.

Questions and Problems for Chapter 8

1 What are necessary and sufficient conditions for using the Laplace transform in the advanced circuit analysis?

2 List and explain four steps needed to apply the Laplace transform to the advanced circuit analysis.

3 Figure 7.3R shows that a classical method needs only one step to find the circuit's output (that is, the solution to the circuit's differential equation). In contrast, the Laplace transform requires two additional steps to achieve the same goal. Why then is the Laplace transform considering the primary method in the advanced circuit analysis?

4 Figure 7.11 includes a series RC circuit, a switch, a dc source, and an oscilloscope. The schematic in Figure 8.1 includes the same elements and—it seems—in the same order. Then, what is the difference?

5 Compare Equations 7.58, $V_{out}(s) = \frac{V_o}{(s+\alpha)}$, and 8.3, $V_{out}(s) = \frac{\alpha V_{in}}{s(s+\alpha)}$. They differ in only one member, s, in the denominator of (8.3). Why does this member change the approach to the method of finding the inverse Laplace transform, that is, to solving the problem?

6 If the s-domain function can be found in Table 7.1 (the table of the Laplace transform pairs), the inverse Laplace transform is performed by simple examination of the table. However, the majority of the functions encountered in practice can't be found in Table 7.1. How can we find the inverse Laplace transform in these cases?

7 Based on studying Table 8.1, find the inverse Laplace transforms of the following s-domain functions:

a) $F(s) = \frac{3}{s+5} + \frac{24}{(s+2)^5}$

b) $F(s) = \frac{9}{s^2+81} + \frac{6}{s^2}$

c) $F(s) = \frac{s+4}{(s+4)^2+16}$

d) $F(s) = \frac{5}{(s+3)} + \frac{120}{(s+3)^6} + \frac{6}{s^2}$

e) $F(s) = \frac{s+5}{s^2+4s+13}$.

8 What is a partial fraction expansion? Give an example.

9 Explain how a partial fraction expansion enables us to determine the inverse Laplace transform of a given $F(s)$?

10 What are the poles of a rational s-domain function? Give an example.

11 Show the technique for finding the poles of a polynomial. Create an example.

12 Find the poles of the following polynomials:

a) s^2+5s+6
b) $s^2+7s+12$
c) $s^2+9s+20$
d) $s^2+11s+30$
e) $s^2+13s+42$

13 Find the poles, present the following s-domain functions in the factored forms, and find their inverse Laplace transforms in general forms as $\frac{A_1}{(s-p_1)}+\frac{A_2}{(s-p_2)}+\frac{A_3}{(s-p_3)}+\ldots+\frac{A_m}{(s-p_m)}$:

a) $F(s)=\frac{10s+2}{(s^2+6s+8)}=\frac{10s+2}{(s-p_1)(s-p_2)}$

b) $F(s)=\frac{2s+8}{(s^2+8s+15)}=\frac{2s+8}{(s-p_1)(s-p_2)}$

Using the general method (cover-up algorithm), find the residues of the s-domain functions given in Problem 13.

14 Check the results of Problem 13 by using direct and algebraic methods.

15 What techniques for checking the correctness of partial fraction expansion do you know? Give examples.

16 Using the test-value method, prove that the following partial fraction expansion is correct: $\frac{10s+2}{(s^2+6s+8)}=\frac{9}{(s+2)}+\frac{19}{(s+4)}$.

17 Find the partial fraction expansion of $F(s)=\frac{6s+9}{s^2+5s+4}$. Check the result.

18 Find the inverse Laplace transforms of the s-domain function given in Problem 13.

19 Mini-project (lab exercise): Consider the initially relaxed RL circuit shown in Figure 8.P19. The switch is closed at $t=0$ (s), $v_{in}(t)=25(V)$, $R=4$ (Ω), and $L=1$ (H). Find the response of this circuit in the time domain. Analyze the solution's correctness.

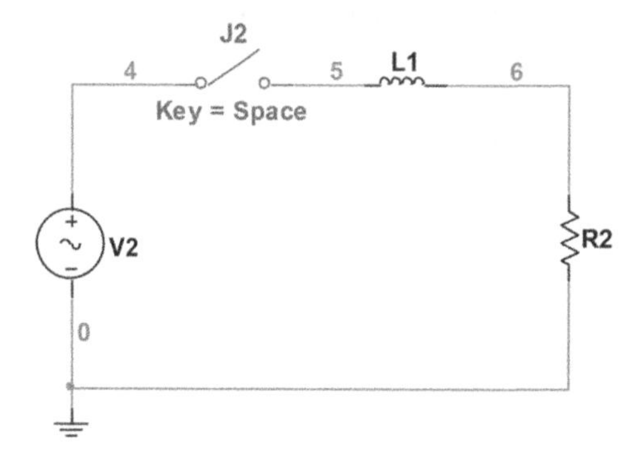

Figure 8.P19 Step response of an RL circuit.

20 Mini-project (lab exercise):

a) Consider the initially relaxed RC circuit shown in Figure 8.P20. The switch is closed at $t=0\,(s)$. Given $v_{in}(t)=15(V), R=4\,(\Omega)$, and $C=\ 1/12\ (F)$. Find the circuit response in the time domain. Analyze the solution's accuracy.

b) Solve Problem 8.20a if the initial voltage is $v_{out}(0)=V_0=5$ V.

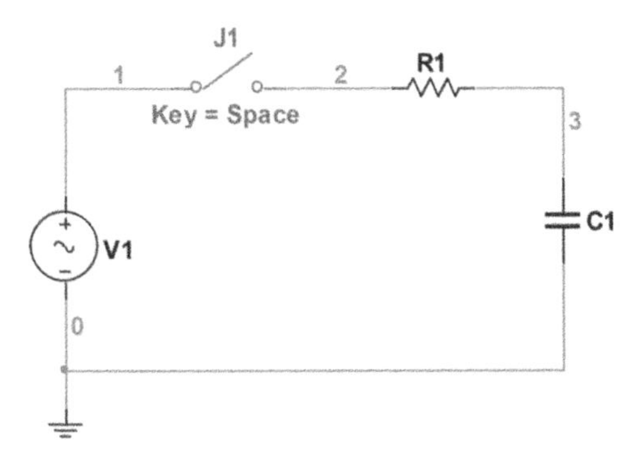

Figure 8.P20 Step response of an RC circuit.

21 What function is called rational? Give examples.

22 Consider the following s-domain function, $\dfrac{10s+2}{\left(s^2+6s+8\right)}$. Use Equation 8.60 to find its inverse Laplace transform.

23 You are assigned to find the inverse Laplace transform of the following s-domain function, $F(s)=(8s+19)/\left(s^2+2s+13\right)$.

a) Can you reduce it to the form shown in Table 7.1 of the Laplace transforms?

b) If the answer to Point 23a is no, find its inverse Laplace transform by using the partial fraction expansion.

24 Find the inverse Laplace transform of $F(s)=\dfrac{10s+2}{\left(s^2+6s+8\right)}$. Show all three steps (finding the poles, determining the residues, and performing the inverse Laplace transform) in solving the problem.

25 Solve Problem 24 by employing MATLAB.

26 What poles of an s-domain function are called simple? Give examples.

27 If an $F(s)$ has real and distinct poles, what is the form of its inverse Laplace transform?

28 Do we always encounter the s-domain responses whose poles are simple? Explain. Give example.

29 Mini-project (laboratory exercise): Consider RL circuit shown in Figure 8.P29. Given $R=10\ \Omega,\ L=0.5$ H and $V_0=0V$, find its response to a step input $V_{in}=8V$ by the following methods:

a) By manual mathematical manipulations.

b) By using MATLAB to determine the formula and plot the graph of the circuit's response.

c) By employing Multisim to obtain the graph of the circuit's response.

Compare all three results and comment their likeness and differences.

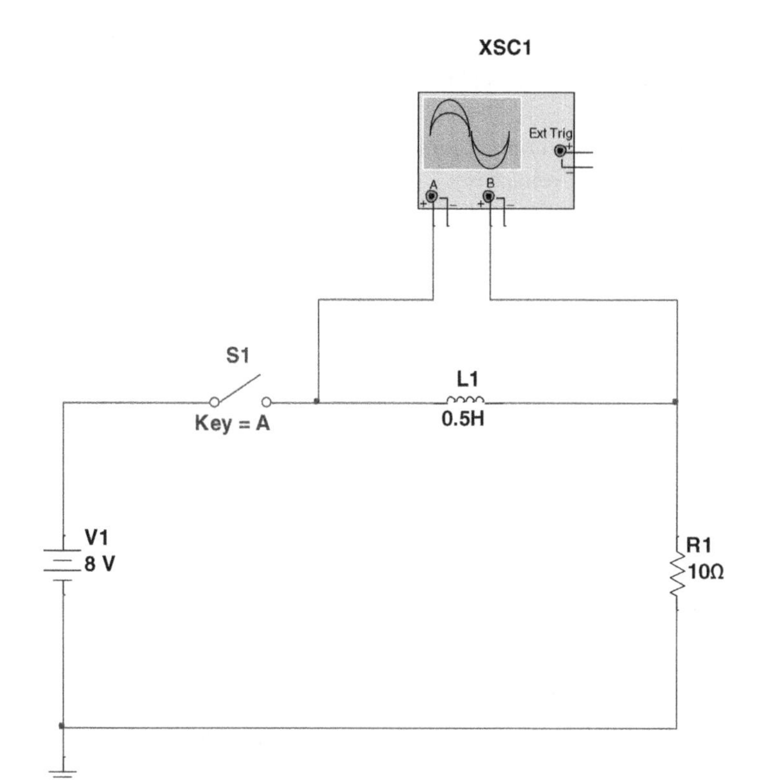

Figure 8.P29 Step response of RL circuit.

30 Contemplate an s-domain function $F(s)$ containing multiple (repeated) poles. Why can't we use the traditional partial-fraction technique for finding the inverse Laplace transform of $F(s)$?

31 Why does Equation 8.80, $F(s)=\frac{N(s)}{(s-p_1)(s-p_2)^2}=\frac{A_{11}}{(s-p_1)}+\frac{A_{21}}{(s-p_2)}+\frac{A_{22}}{(s-p_2)^2}$, require expanding an original s-domain function containing two factors in its denominator into three fractional members?

32 Mini-project (lab exercise):

a) Find the inverse Laplace transform of $F(s)=\frac{2s+8}{(s+3)(s+7)^2}$ by manual calculations.

b) Find the inverse Laplace transform of $F(s)=\frac{2s+8}{(s+3)(s+7)^2}$ using MATLAB.

c) Compare the results and comment discrepancies, if any.

33 Compare two members, $\frac{A}{(s-p)}$ *and* $\frac{B}{(s-p)^2}$: What will be their inverse Laplace transforms? Prove your answer.

34 Mini-project (lab exercise): Consider two cascaded identical RC filters similar to that shown in Figure 8.7. The circuit's parameters are given as $R=1k\Omega$, $C=0.1$ mF, $V_{in}=8V$, and $V_0=0V$.

a) Find the circuit's response to the step input by (1) manual mathematical manipulations, (2) using MATLAB, and (3) Multisim simulation.

b) Compare all three solutions and discuss the discrepancies, if any.

c) Prove the validity of your solutions.

35 *Using "The summary of the discussed technique" in Discussion section of Example 8.4, create your example of F(s) function and find its inverse Laplace transform.

36 Solve Problem 32 using the differentiation method.

37 Find the inverse Laplace transform of the following s-domain function,

$$F(s)=\frac{3(s+2)}{(s+4)(s^2+11s+30)}.$$

38 Find the inverse Laplace transform of $F(s)=\frac{2(s+3)}{(s-4)(s-5)(s-6)^3}$.

39 Apply Equations 8.119 through 8.122 to find the inverse Laplace of

$$F(s)=\frac{7(s+4)}{(s-3)(s-5)(s-7)(s-9)^5}.$$

40 Mini-project (lab exercise): Consider an RC circuit similar to that shown in Figure 8.10 whose R = 2 kΩ and C = 0.25 mF. The excitation to this circuit is a ramp signal similar to that shown in Figure 8.11 and whose parameters are $V_0 = 0$, $t_0 = 1s$, $t_{f1} = 2s$, and $V_f = 5V$. Find the response of this circuit to the given ramp signal.

41 Determine the inverse Laplace transform of $F(s)=\frac{4(s+7)}{(s^2+10s+25)}$ using a common-sense approach.

42 Mini-project (lab exercise): Consider the RC circuit shown in Figure 8.P42. Its parameters are: $R = 10k\Omega$, $C = 2\ \mu F$, $v_{in}(t) = 10\cos(2\pi \cdot 200 \cdot t)$, and $v_{out}(0) = 0V$. Find the circuit's output using the Laplace transform. Verify your solution by finding the answer with MATLAB and performing Multisim simulation.

43 Mini-project (lab exercise): Consider the RC circuit similar to that shown in Figure 8.P42. Its parameters are: $R = 10k\Omega$, $C = 2\ \mu F$, $v_{in}(t) = 10\cos(2\pi \cdot 200 \cdot t)$, and $v_{out}(0) = 0V$. Using the Laplace transform, determine the circuit's output applying algebraic, direct, and modification of the cover-up methods for finding the residues. Verify that all three methods produce identical results.

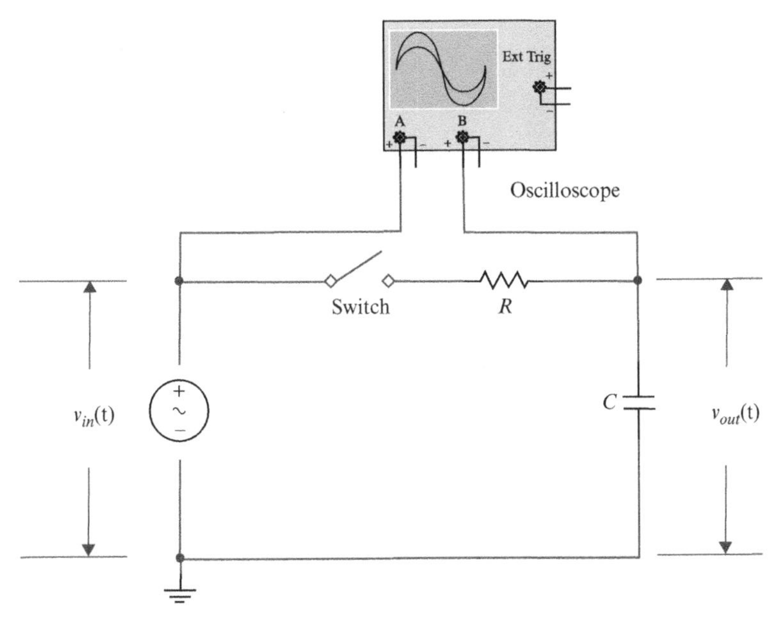

Figure 8.P42 Experimental setup for Problem 8.42.

44 Consider an RC circuit similar to that shown in Figure 8.P42. The circuit's parameters are $R = 1k\Omega, C = 40\,\mu F$, $V_0 = 8V$, and $v_{in}(t) = 10\cos(314t)$. The s-domain response of this circuit clearly contains pure imaginary poles.

a) Find the circuit's response by manual mathematical manipulations using either polar or exponential form of the residues.
b) Find the circuit's response by employing MATLAB. Use Multisim to simulate the circuit's operation and obtain the output's waveform. Compare all results.

45 Repeat Problem 44 if all the parameters remain the same except for $v_{in}(t) = 10\sin(314t)$. (To remind: $R = 1k\Omega, C = 40\,\mu F$, and $V_0 = 8V$).

46 Mini-project (Lab exercise): Consider a series RL circuit where $R = 2.5\Omega, L = 0.1H, v_{in}(t) = 10\cos(314t)$, and $V_0 = 0V$. By manual mathematical manipulations find its response to the input signal after the switch is closed. Verify your answer by using MATLAB and Multisim.

47 Determine the inverse Laplace transform of $F(s) = \dfrac{9(s+2)}{(s+7)(s^2+6s+25)}$. Plot the graph of $v_{out}(t)$.

48 Solve Problem 47 by applying the algebraic approach to find the residues.

49 Solve Problem 47 by finding directly A and θ in $(t) = Ae^{-\alpha t}\cos(\omega t + \theta)$.

50 Answer what will be the formula and sketch the graph of a circuit's time-domain response for the following cases:

a) If $V_{out}(s) = \dfrac{3}{s}$, then $v_{out}(t) = ?$

b) If $V_{out}(s) = \dfrac{3}{(s+5)}$, then $v_{out}(t) = ?$

c) If $V_{out}(s) = \dfrac{3}{(s-j24)}$, then $v_{out}(t) = ?$

d) If $V_{out}(s) = \dfrac{3}{(s+5+j24)}$, then $v_{out}(t) = ?$

9

Advanced Consideration of the Laplace Transform and its Application to Circuit Analysis

9.1 Circuit Analysis in the s-Domain

9.1.1 The Concept

Analysis of the Laplace transform's application to the circuit analysis of the first-order circuits presented in Chapter 8 shows that the first step in the procedure is to derive the circuit's differential equation. Then the Laplace transform converts this differential equation into the s-domain algebraic equation, from which point the circuit analysis with the Laplace transform starts. From the s-domain circuit analysis standpoint, a differential equation's derivation looks like an auxiliary operation. After all, performing the circuit analysis with the Laplace transform requires only the circuit's s-domain equation. *Can we avoid this extra step and obtain the circuit's s-domain equation directly from the circuit's observation?* An additional and essential reason for searching the positive answer to this question is shown in Chapters 5 and 6: They demonstrate that deriving a differential equation, especially for a second-order circuit, might be challenging.

Let's try to find the solution to this problem by asking what information the circuit's s-domain equation delivers. The answer is that it describes the components' properties and their interaction during the circuit's operation. Can we obtain this information in the s-domain equation without referring to the time-domain differential equation? The answer is yes, provided that the circuit's components properties will be given in the s-domain. What properties? The main one is the relationship between the voltage drop over a component and a current flowing through it. This property, as we know from the elemental circuit analysis, is described by a component's impedance. Thus, the problem boils down to presenting the impedances of R, L, and C in the s-domain.

Here is how we can solve this problem:

- You are reminded that the current–voltage ($i - v$) relationship for each component is given by:

$$\begin{aligned} v_R(t) &= Ri_R(t) \\ v_L &= L\frac{di_L(t)}{dt} \\ i_C &= C\frac{dv_C(t)}{dt}. \end{aligned} \tag{9.1}$$

Essentials of Advanced Circuit Analysis: A Systems Approach, First Edition. Djafar K. Mynbaev.

Companion Website: www.wiley.com/go/Mynbaev/AdvancedCircuitAnalysis

(See Table 4.1 and its discussion in Section 4.1 to refresh your memory on this topic.)

- Applying the Laplace transform to (9.1), we can find the equivalent $I(s)-V(s)$ relationship in the s-domain, which immediately gives us the formulas for s-domain impedances as follows:

$$L\{v_R(t)=Ri_R(t)\}\Rightarrow V_R(s)=RI_R(s)\Rightarrow Z_R(s)=\frac{V_R(s)}{I_R(s)}=R$$

$$L\left\{v_L(t)=L\frac{di_L(t)}{dt}\right\}\Rightarrow V_L(s)=LsI_L(s)\Rightarrow Z_L(s)=\frac{V_L(s)}{I_L(s)}=Ls \tag{9.2}$$

$$L\left\{i_C(t)=C\frac{dv_C(t)}{dt}\right\}\Rightarrow V_C(s)=\frac{1}{Cs}I_C(s)\Rightarrow Z_C(s)=\frac{V_C(s)}{I_C(s)}=\frac{1}{Cs}.$$

Table 9.1 displays the relationships between current and voltage for three main circuit components, R, L, and C in the time domain and the s-domain. It also shows the s-domain impedances of these components. Table 9.1 and (9.2) assume zero initial conditions.

Using the s-domain, R, L, and C impedances enable us to build an s-domain equivalent circuit. For example, the schematics of the actual RC circuit shown in Figure 8.1 can be transformed into its s-domain equivalent, as shown in Figure 9.1. Here, the actual dc input, V_{in}, is transformed into $\frac{V_{in}}{s}$, as the operation $L\{V_{in}\}$ requires. Accordingly, the existent resistor R and capacitor C are presented by their s-domain impedances $Z_R(s)=R$ and $Z_C(s)=\frac{1}{Cs}$, respectively.

And here is the main point: *The s-domain circuit's schematic allows for finding the circuit's response in the s-domain using the traditional dc circuit-analysis technique.* Specifically, applying the *voltage-divider rule*, we find

$$V_{out}(s)\equiv V_C(s)=\frac{Z_C(s)}{Z_R(s)+Z_C(s)}V_{in}(s)=\frac{V_{in}\left(\frac{1}{Cs}\right)}{s\left(R+\frac{1}{Cs}\right)}. \tag{9.3a}$$

Simple manipulations enable us to obtain the s-domain response in the form received in (8.3) for this circuit:

Table 9.1 The s-domain impedances of the basic components.

Component	Time-domain voltage-current relationship	s-domain voltage–current relationship	s-Domain impedance
R	$v_R(t)=R\cdot i_R(t)$	$V_R(s)=R\cdot I_R(s)$	$Z_R(s)=\frac{V_R(s)}{I_R(s)}=R$
L	$v_L=L\frac{di_L(t)}{dt}$	$V_L(s)=L_S\cdot I_L(s)$	$Z_L(s)=\frac{V_L(s)}{L(s)}=Ls$
C	$i_c=c\frac{dv_c(t)}{dt}$	$V_C(s)=\frac{1}{Cs}\cdot I_C(s)$	$Z_C(s)=\frac{V_C(s)}{C(s)}=\frac{1}{CS}$

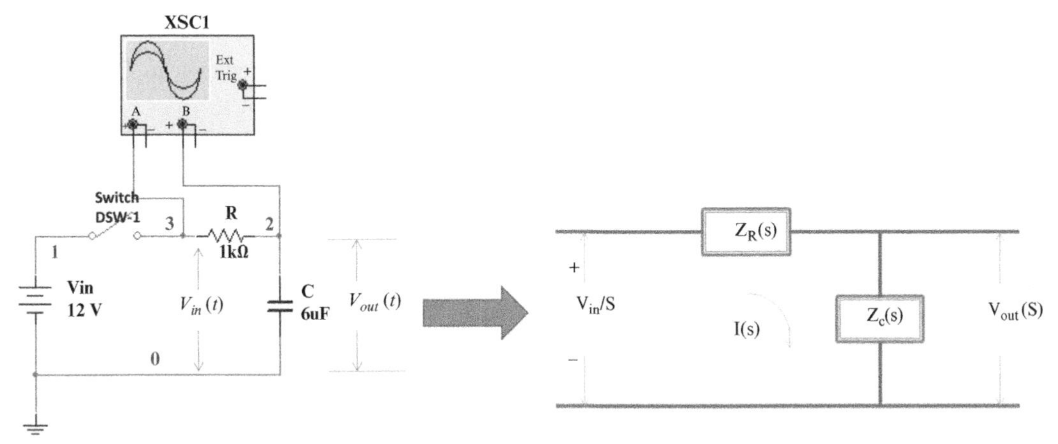

Figure 9.1 Schematics of the actual RC circuit (left) and its s-domain equivalent (right).

$$V_{out}(s) = \frac{V_{in}}{s(RCs+1)} = \frac{V_{in}\left(\frac{1}{RC}\right)}{s\left(s+\frac{1}{RC}\right)} = \frac{\alpha V_{in}}{s(s+\alpha)}. \tag{9.3b}$$

Therefore, using the s-domain schematic of an equivalent circuit and the s-domain impedances of its components, we can find the circuit's s-domain response in one step. For this, we employ the well-known circuit analysis technique developed for analysis of real circuits in the time domain.

Having $V_{out}(s)$ readily obtained, we follow two other steps meticulously discussed in Chapter 8 to complete the task: Resolving $V_{out}(s)$ into partial fractions and finding the time-domain response of the RC circuit, $v_{out}(t)$. For example given in (9.3b), the circuit's s-domain response can be resolved as $V_{out}(s) = \frac{\alpha V_{in}}{s(s+\alpha)} = \frac{A_1}{s} + \frac{A_2}{s+\alpha}$, and its residues can be found as $A_1 = V_{in}$ and $A_2 = -V_{in}$. Then, the time-domain response of the series RC circuit to be determined as $v_{out}(t) = L^{-1}\{V_{out}(s)\} = V_{in}(1-e^{-\alpha t})$. Example 9.1 further clarifies the advantage of performing the circuit analysis in the s-domain.

Since the subjects of our new approach are the same first- and second-order circuits, the results of this chapter's operations will be familiar from the preceding study; nonetheless, the techniques and tools will be very different. Learning these new methods is the main objective of this chapter.

9.1.2 Examples of Circuit Analysis in the s-Domain

Example 9.1 Application of the circuit analysis in the s-domain for finding the step response of a series RC circuit.

Problem: Find the step response of the series RC circuit shown in Figure 9.1 (left) if it was initially relaxed.

Prediction: By inspection of Figure 9.1, we can readily predict that the circuit's response, the voltage across the capacitor, should continuously increase, reaching 12 volts, and then it stays constant.

Solution:

1) Using Table 9.1, draw a schematic of the s-domain equivalent circuit shown in Figure 9.1 (right).
2) Apply voltage-divider rule to find $V_{out}(s)$ as

$$V_{out}(s)=\frac{Z_C(s)}{Z_R(s)+Z_C(s)}V_{in}(s)=\frac{V_{in}}{s(RCs+1)}=\frac{\alpha V_{in}}{s(s+\alpha)},$$

$$\text{where } \alpha\left(\frac{1}{s}\right)=\frac{1}{RC}, \text{ as usual.}$$

3) Use partial fraction expansion to find

$$V_{out}(s)=\frac{A_1}{s}+\frac{A_2}{(s+a)}$$

4) Find the residues, employ the inverse Laplace transform and solve the problem as follows:

$$v_{out}(t)=L^{-1}\{V_{out}(s)\}=V_{in}\left(1-e^{-\alpha t}\right)=12(1-e^{-166.7t}).$$

The problem is solved. (Refer to (8.14) and (8.15) to refresh your memory on transformations in steps 3) and 4)..)

Discussion:

- Compare steps 1 and 2 for obtaining $V_{out}(s)$ in this example with the mathematical manipulations performed in Example 8.1 for the same purpose and see the advantage of using the s-domain circuit analysis.

Let's apply the circuit analysis in the s-domain of this technique for finding $V_{out}(s)$ of a parallel circuit.

Example 9.2 Finding the step response of a parallel RL circuit using the circuit analysis in the s-domain.

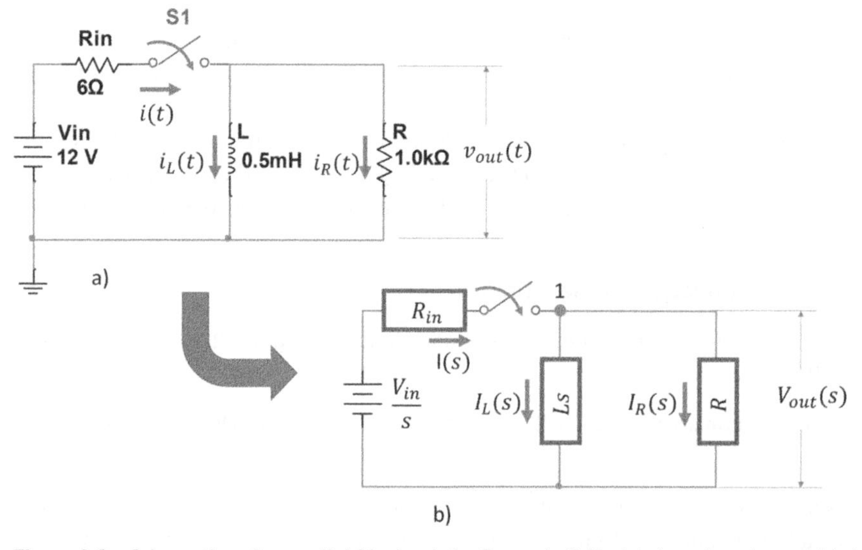

Figure 9.2 Schematics of a parallel RL circuit for Example 9.2in (a) time domain and (b) the s-domain.

Problem: Find the output voltage, $v_{out}(t)$ of the parallel RL circuit shown in Figure 9.2a. Assume $i_L(t)=0A$ before the switch gets closed.

Prediction: Examining the given RL circuit, we observe that the current through the inductor should grow from zero to the maximum value during the change of the input voltage and then retain this value because the applied voltage becomes dc. Consequently, the voltage across the inductor should drop to zero because the pace of the $i_L(t)$ change decreases and eventually becomes zero. (Recall the inductor's governing equation $v_L(t)=\frac{di_L(t)}{dt}$).

Solution: Follow the steps employed in Example 9.1:

1) Draw a schematic of the s-domain equivalent RL circuit shown in Figure 9.2b.
2) Apply the nodal analysis. At Node 1 we have

$$I(s)=I_L(s)+I_R(s),$$

which is

$$\frac{V_{in}-V_{out(s)}}{R_{in}s}=\frac{V_{out}(s)}{Ls}+\frac{V_{out}(s)}{R}.$$

After straightforward manipulations, we obtain

$$\frac{V_{in}}{R_{in}s}=V_{out}(s)\left(\frac{RLs+R_{in}Ls+R_{in}R}{R_{in}Ls}\right).$$

After another set of direct but tedious manipulations, $V_{out}(s)$ is found as

$$V_{out}(s)=V_{in}\left(\frac{R}{R+R_{in}}\right)\left(\frac{1}{s+\frac{R_{in}R}{(R+R_{in})L}}\right).$$

If we accept $\frac{R}{R+R_{in}}\approx 1$ and denote $\frac{R_{in}R}{(R+R_{in})L}\approx\frac{R_{in}}{L}=\alpha$, then the answer to this problem in the s-domain becomes

$$V_{out}(s)\approx V_{in}\left(\frac{1}{s+\alpha}\right).$$

(Note that R_{in}, also called the Thevenin resistor, is typically several ohms in value, which is in the orders of magnitude smaller than the value of a circuit's resistor. In our example, $R_{in}=6\Omega$ and $R=1k\Omega$.)

Finally, the inverse Laplace transform of $V_{out}(s)$ results in

$$v_{out}(t)\approx V_{in}e^{-\alpha t}\approx 12e^{(-12\cdot 10^3)t}.$$

The problem is solved.

Question In the mathematical manipulations for this example, we denoted the s-domain output voltage as $V_{out}(s)$. Why didn't we designate the input voltage as $V_{in}(s)$?

Discussion:

- How can we verify our solution? First, consider its *initial and final values*. (Refer to Section 4.2 for the in-depth discussion of these values.) When the switch is closed at $t=0$, $v_{out}(t)$ should be equal to $V_{in}=12V$ because no change in the applied voltage yet occurs and no energy was stored in the circuit before this instant. After the switch is closed and t tends to infinity, $v_{out}(t)$ should approach zero because an inductor for a dc source becomes a short circuit. We predict this result and the solution confirms it.
- The additional method to check our solution's correctness is to simulate the circuit's operation with Multisim. Figure 9.3a shows the experimental setup, and Figure 9.3b demonstrates the simulation result. Observe the qualitative agreement with our calculations: $v_{out}(t)$ is a damping exponential whose starting amplitude is equal to $V_{in}=12V$ and whose time constant is $\tau=\frac{1}{\alpha}=\frac{L}{R_{in}}\approx 0.08\,ms$. Also, see the quantitative confirmation of our solution: By calculations, $5\tau\approx 0.4\,ms$ and by the experimental graph, $5\tau\approx 51.65ms-51.25ms\approx 0.4ms$.
- Comments on the Figure 9.3b graph: First, note that the detailed view on the graph shows that, in reality, the ascending part of the graph is rather a ramp than a step-like. Refer to Section 7.3 and specifically to the discussion of Figure 7.6, where this point is considered. Secondly, understand that visualization of the ramp-like response's shape depends on the time scale. The fine time scale displays this phenomenon clearly, as Figure 9.3b does, whereas the coarse time scale hides this effect completely. Third, observe that the graph's descending part is a damping exponential function that fully complies with the theoretical result. From the circuit-analysis standpoint, the key is the relationship $v_L(t)=L\frac{di_L(t)}{dt}$. This formula shows that initially, when the inductor current grows rapidly from zero to its maximum value, the voltage increases too, reflecting the positive increase in $i_L(t)$. Very soon, after the transition from the zero state to the dc state will complete, the inductor becomes a short circuit with unchanging $i_L(t)$ and, accordingly, zero $v_L(t)$.
- Still, there is one more way to prove our solution right: The circuit analysis in the s-domain states that $V_{out}(s)=I_R(s)\cdot Z_R(s)$. The current-divider rule for two parallel components gives $I_R(s)=I_T(s)\frac{Z_L(s)}{Z_L(s)+Z_R(s)}$. The total current is given by $I_T(s)=\frac{V_{in}(s)}{Z_T(s)}$. Substituting $Z_R(s)=R$, $Z_L(s)=Ls$, and $Z_T(s)=R_{in}+\frac{R\cdot Ls}{R+ls}$ into $V_{out}(s)$, we find $\mathrm{V_{out}}(s)=\mathrm{V}_{in}\frac{RL}{Ls(R+R_{in})+R\cdot R_{in}}$. Alternatively, after straightforward manipulations, we can bring our initial answer, $V_{out}(s)=V_{in}\left(\frac{R}{R+R_{in}}\right)\left(\frac{1}{s+\frac{R_{in}R}{(R+R_{in})L}}\right)$, to the same form, thus proving the accuracy of our result.

Question Could you guess the nature of the RL circuit step response by simply examining the circuit, i.e., without any calculations?

9.1.3 Initial Conditions

In the preceding discussions of the circuit analysis in the s-domain, we continuously assumed that the initial condition is zero. But what if a capacitor is initially charged and an initial current flow through an inductor? This problem is solved in Examples 7.2 and 8.1 for an RC circuit by a

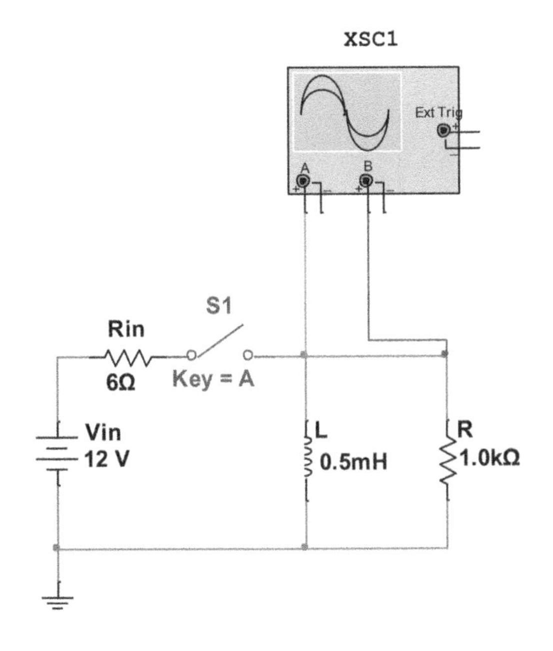

a)

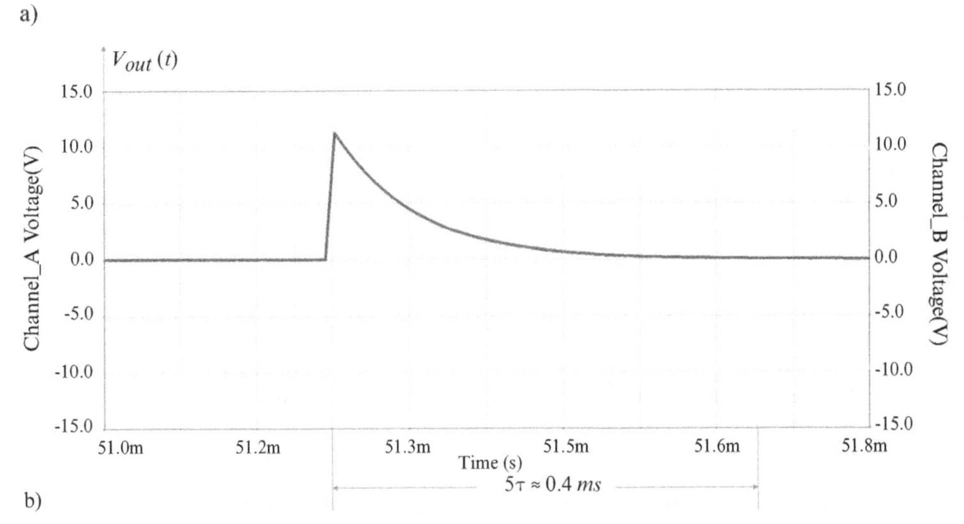

b)

Figure 9.3 Multisim simulation of a step response of the parallel RL circuit: a) Experimental setup; b) graph of the step response.

traditional time-domain approach. Can we consider an initial condition when using the circuit analysis in the s-domain?

Let's review Figures 7.12R and 8.4bR reproduced here for your convenience. These figures show that an initially charged capacitor acts as a source of voltage $V_0 \equiv V_{out}(0)$. Indeed, both graphs start at V_0 instead of $V = 0$. Equations 7.61, $v_{out}(t) = V_0 e^{-\alpha t}$, and 8.20, $V_0 e^{-\alpha t} = V_{in} - (V_{in} - V_0)e^{-\alpha t}$), confirm this statement. Thus, *V_0 can be shown in the circuit schematic as an additional voltage source.*

This conclusion is supported by the formal consideration of the Laplace transform for a capacitor, C:

A capacitor's $i - v$ relationship is given in Table 9.1 as

$$i_C(t) = C\dot{v}_C(t). \tag{9.4}$$

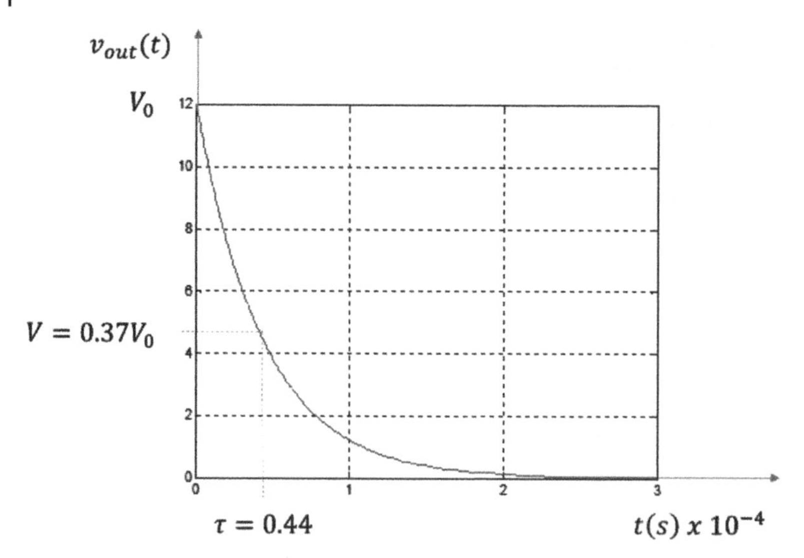

Figure 7.12R Natural (source-free) response of the series RC circuit for Example 7.2 obtained by MATLAB calculations.

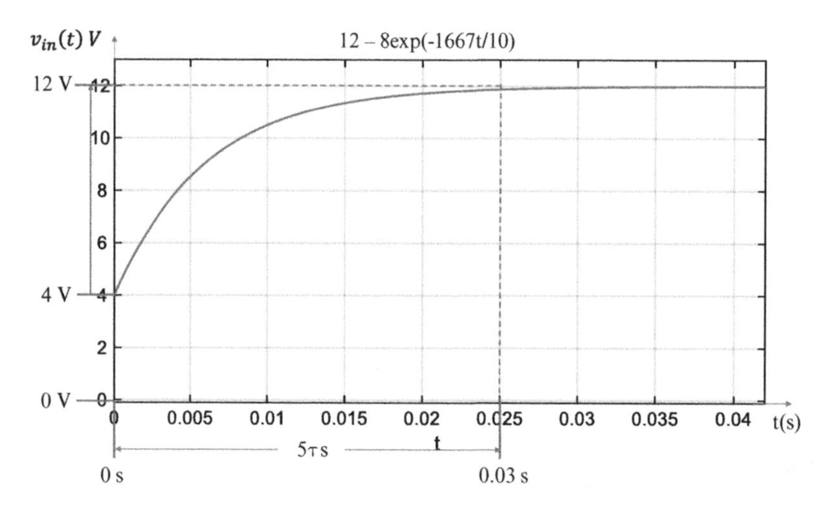

Figure 8.4bR MATLAB solutions for the RC circuit's step responses for ... $(b)\ V_0 = 4V$ in Example 8.1.

Taking its Laplace transform results in

$$L\left\{i_C(t) = C\dot{v_C}(t)\right\}$$

$$\Rightarrow I_c(s) = CsV_C(s) - CV_0 = \frac{V_C(s)}{Z_C(s)} - \frac{\frac{V_0}{s}}{Z_C(s)} \tag{9.5}$$

and

$$V_C(s) = I_C(s) \cdot Z_C(s) + \frac{V_0}{s} = \frac{I_C(s)}{Cs} + \frac{V_0}{s}. \tag{9.6}$$

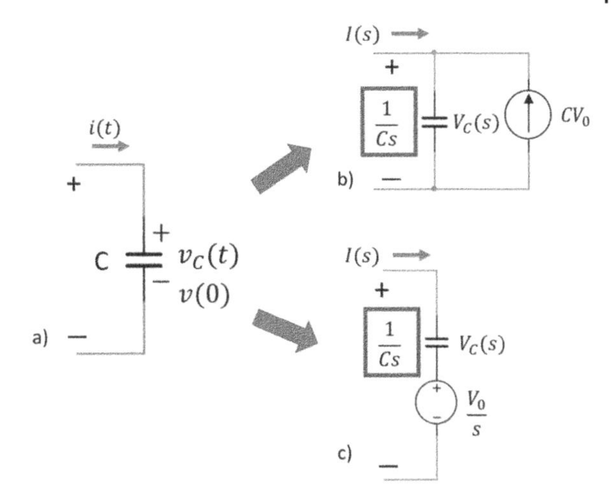

Figure 9.4 An initially charged capacitor a) in the time domain; b) in the s-domain with an initial current source (Norton model); c) in the s-domain with an initial voltage source (Thevenin model).

Graphically, Equations 9.4, 9.5, and 9.6 are presented in Figure 9.4. Specifically, an initially charged capacitor in the time domain described by (9.4) is shown in Figure 9.4a. The s-domain models corresponding to (9.5) and (9.6) are shown in Figures 9.4b and 9.4c.

Equation 9.5 reveals that the total current through a capacitor is a sum of the externally excited current and the current caused by the initial charge of the capacitor. The currents can only be summed when their lines are connected in parallel. The current signs are determined by the member signs in the RHS of (9.5). This is what Figure 9.4b displays. We call it the *Norton model* of an initially charged capacitor in the s-domain.

Equation 9.6 states that the voltage across an initially charged capacitor in the s-domain is the sum of the voltage, $V_C(s)=I_C(s)\cdot Z_C(s)$, caused by an external source, and the initial capacitor's voltage, $\frac{V_0}{s}$. This equation can be visualized as two voltage sources connected in series, which Figure 9.4c shows. This is the *Thevenin model* of a capacitor's initial condition in the s-domain.

It's necessary to remember two fundamental rules regarding the initial conditions: *Voltage across a capacitor and current through an inductor cannot change instantly*; specifically, $v_C(0^-)=v_C(0^+)$ and $i_L(0^-)=i_L(0^+)$ at $t=0$.

It's worth revisiting the discussion of Norton's and Thevenin's theorems in Section 2.2 and initial and final conditions in Section 4.2.

Using a similar approach, we can develop the s-domain analysis of an RL circuit with a non-zero initial condition. Start with an inductor's $i-v$ relationship in the time domain:

$$v_L(t)=Li_L^{'}(t). \tag{9.7}$$

Taking the Laplace transform of this equation gives

$$V_L(s)=LsI_L(s)-LI_0 \tag{9.8}$$

and

$$I_L(s)=\frac{V_L(s)}{Z_L(s)}+\frac{I_0}{s}=\frac{V_L(s)}{Ls}+\frac{I_0}{s}. \tag{9.9}$$

Hence, the voltage across the inductor, $V_L(s)$, is the difference between the voltage drop caused by the induced current, $I_L(s)$, and the voltage drop due to the initial current, $I_0(s)$, as in (9.8). Naturally, these voltage sources are connected in series, as the Thevenin model requires, and

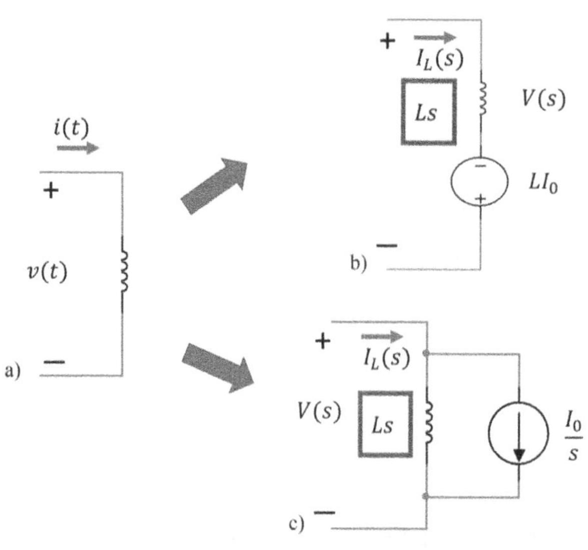

Figure 9.5 An initial condition in an inductor a) in the time domain; b) in the s-domain with an initial voltage source (Thevenin model); c) in the s-domain with an initial current source (Norton model).

Figure 9.5b shows. Equation 9.9 states that the current through the inductor, $I_L(s)$, is a sum of the current due to the applied voltage, $V_L(s)$, and the initial current, $I_0(s)$. This Norton model with parallel lines is shown in Figure 9.5c. These models' voltage polarity and current direction correspond to the member signs in (9.8) and (9.9).

Do Figures 9.4b, 9.4c, 9.5b, and 9.5c look familiar? They should because they resemble the figures demonstrating how a voltage source can be transformed into a current source and vice versa, the topic familiar to us from the basic circuit analysis. See Section 1.4.

To conclude, the Laplace transform enables us to treat the initial conditions as an integral part of the s-domain circuit analysis. Let's consider an example.

Example 9.3 The s-domain analysis of a series RC circuit with non-zero initial conditions.

Problem: Find the sinusoidal response of an RC circuit discussed in Example 8.7, whose $v_{in}(t) = 5cos2\pi100t$, R = 1 kΩ, C = 80 μF, and the capacitor is charged to the initial voltage of $V_0 = 3V$.

Prediction (actually, our educated (literally) guess):

Observe how Figure 9.6a demonstrates the circuit's schematic, Figure 9.6b shows its s-domain model, and Figure 9.6c illustrates the superposition principle for this circuit. Based on these figures, we would be able to predict the solution before starting to solve the problem formally. Recall that we investigated the role of the initial condition in an RC circuit's step response in Example 8.1 and the circuit's sinusoidal response in Example 8.7. So, we expect that the sinusoidal member (steady-state response) should retain its sinusoidal nature though its amplitude and the phase shift will change. The transient response would change compared to Example 8.7 because an initial condition, by its very nature, affects only this part of the complete response. Specifically, the transient response would now include the member depending on the initial capacitor voltage, V_0, in addition to the previous reliance on the initial input value, V_{in}, at $t = 0$. In other words, we would have $V_0e^{-\alpha t}$ and $V_{in}e^{-\alpha t}$.

This guess is also a hint to solving the problem: either (1) use the circuit-analysis technique in the s-domain or (2) apply the superposition principle introduced in Chapter 3.

Solution:

Method 1—Circuit analysis in the s-domain.

The loop equation for the circuit in Figure 9.6b is

$$-\frac{V_{in}s}{\left(s^2+\omega^2\right)}+I(s)R+I(s)\frac{1}{Cs}+\frac{V_0}{s}=0. \tag{9.10}$$

From (9.10),

$$I(s)=\left(\frac{V_{in}s}{\left(s^2+\omega^2\right)}-\frac{V_0}{s}\right)\left(\frac{1}{R+\dfrac{1}{CS}}\right)=\left(\frac{V_{in}s}{\left(s^2+\omega^2\right)}-\frac{V_0}{s}\right)\left(\frac{\alpha Cs}{s+\alpha}\right)$$

$$=\frac{\alpha V_{in}s\cdot Cs}{\left(s+\alpha\right)\left(s^2+\omega^2\right)}-\frac{\alpha CV_0}{s+\alpha} \tag{9.11}$$

where $\alpha=\dfrac{1}{RC}$, as before.

The circuit's response in the s-domain voltage becomes

$$V_{out}(s)=I(s)\left(\frac{1}{Cs}\right)+\frac{V_0}{s}=\frac{\alpha V_{in}s}{\left(s+\alpha\right)\left(s^2+\omega^2\right)}-\frac{\alpha V_0}{s\left(s+\alpha\right)}+\frac{V_0}{s}$$

$$=\frac{\alpha V_{in}s}{\left(s+\alpha\right)\left(s^2+\omega^2\right)}+\frac{V_0}{\left(s+\alpha\right)}. \tag{9.12}$$

Compare (9.12) with (8.144),

$$V_{out}(s)|_{8-144}=\left(\frac{\alpha V_{in}s}{\left(s+\alpha\right)(s^2+\omega^2)}\right), \tag{8.144R}$$

and see that they differ only in the initial voltage, V_0. The expansion of the first member of (9.12) is

$$\frac{\alpha V_{in}s}{\left(s+\alpha\right)\left(s^2+\omega^2\right)}=\frac{A_1}{\left(s+\alpha\right)}+\frac{A_{2,3}}{\left(s^2+\omega^2\right)}.$$

Thus, (9.12) takes the form

$$V_{out}(s)=\frac{A_1}{\left(s+\alpha\right)}+\frac{V_0}{\left(s+\alpha\right)}+\frac{A_{2,3}}{\left(s^2+\omega^2\right)},$$

which becomes

$$v_{out}(t)=L^{-1}\left\{V_{out}(s)\right\}=A_1e^{-\alpha t}+V_0e^{-\alpha t}+NAcos(\omega t-\theta). \tag{9.13}$$

Borrowing Equation 8.157 from Example 8.7,

$$v_{out}(t)=-\frac{\alpha V_{in}}{\left(\alpha^2+\omega^2\right)}\alpha e^{-\alpha t}+\frac{\alpha V_{in}}{\left(\alpha^2+\omega^2\right)}\text{Acos}(\omega t-\theta), \tag{8.157R}$$

we get the answer to our example as

$$v_{out}(t) = \left(V_0 - \frac{\alpha V_{in}}{(\alpha^2 + \omega^2)} \alpha \right) e^{-\alpha t} + \frac{\alpha V_{in}}{(\alpha^2 + \omega^2)} A\cos(\omega t - \theta)$$

$$= V_0 e^{-\alpha t} - \frac{\alpha V_{in}}{(\alpha^2 + \omega^2)} \alpha e^{-\alpha t} + \frac{\alpha V_{in}}{\sqrt{(\alpha^2 + \omega^2)}} \cos(\omega t - \theta)$$

$$\approx 3e^{-12.5t} - 0.01975e^{-12.5t} + 0.99224\cos(628t - 88.86^0). \tag{9.14}$$

Review Example 8.7 to recall how to compute A and θ.

Analyzing the given reasonings, manipulations, and results, we can conclude that our guess was correct, and the initial capacitor's charge affects only transient members. All this means that we could write the answer, Equation 9.14, without all the performed manipulations.

Our solution clearly shows that the simple addition of a new member to the obtained result is evidence of applying the *superposition principle*. The following part will prove this point.

Method 2—Superposition principle.

Based on the superposition principle, the circuit's response in the s-domain is

$$V_{out}(s) = V_{out}^{zsr}(s) + V_{out}^{zir}(s). \tag{9.15}$$

Here $V_{out}^{zsr}(s)$ is a *zero-state response* to an external excitation under zero initial condition, and $V_{out}^{zir}(s)$ is a *zero-input response* caused by the initial condition only. See Figure 9.6c. Term *zero-state response* refers to zero energy stored in the system at $t \le 0$. Term *zero-input response* is another way to denote natural response.

Since the inverse Laplace transform is a linear operation, Equation 9.15 is transformed in the time domain as

$$v_{out}(t) = v_{out}^{zsr}(t) + v_{out}^{zir}(t). \tag{9.16}$$

The *zero-state response* to a cosine input was obtained in Example 8.7 as

$$v_{out}^{zsr}(t) = -\frac{\alpha V_{in}}{(\alpha^2 + \omega^2)} \alpha e^{-\alpha t} + \frac{\alpha V_{in}}{(\alpha^2 + \omega^2)} A\cos(\omega t - \theta). \tag{8.157R}$$

In Example 7.2, we found the RC circuit response to the initial condition V_0 as

$$v_{out}^{zir}(t) = V_0 e^{-\alpha t}. \tag{7.61R}$$

Summing up (8.157R) and (7.61R) yields

$$v_{out}(t) = \left(V_0 - \frac{\alpha^2 V_{in}}{(\alpha^2 + \omega^2)} \right) e^{-\alpha t} + \frac{\alpha V_{in}}{(\alpha^2 + \omega^2)} A\cos(\omega t - \theta), \tag{9.17}$$

as in (9.14).

Thus, the problem is solved.

Discussion:

As an exercise and proof of the results obtained in this example, solve this problem with MATLAB and simulate the RC circuit operation under non-zero initial condition with Multisim.

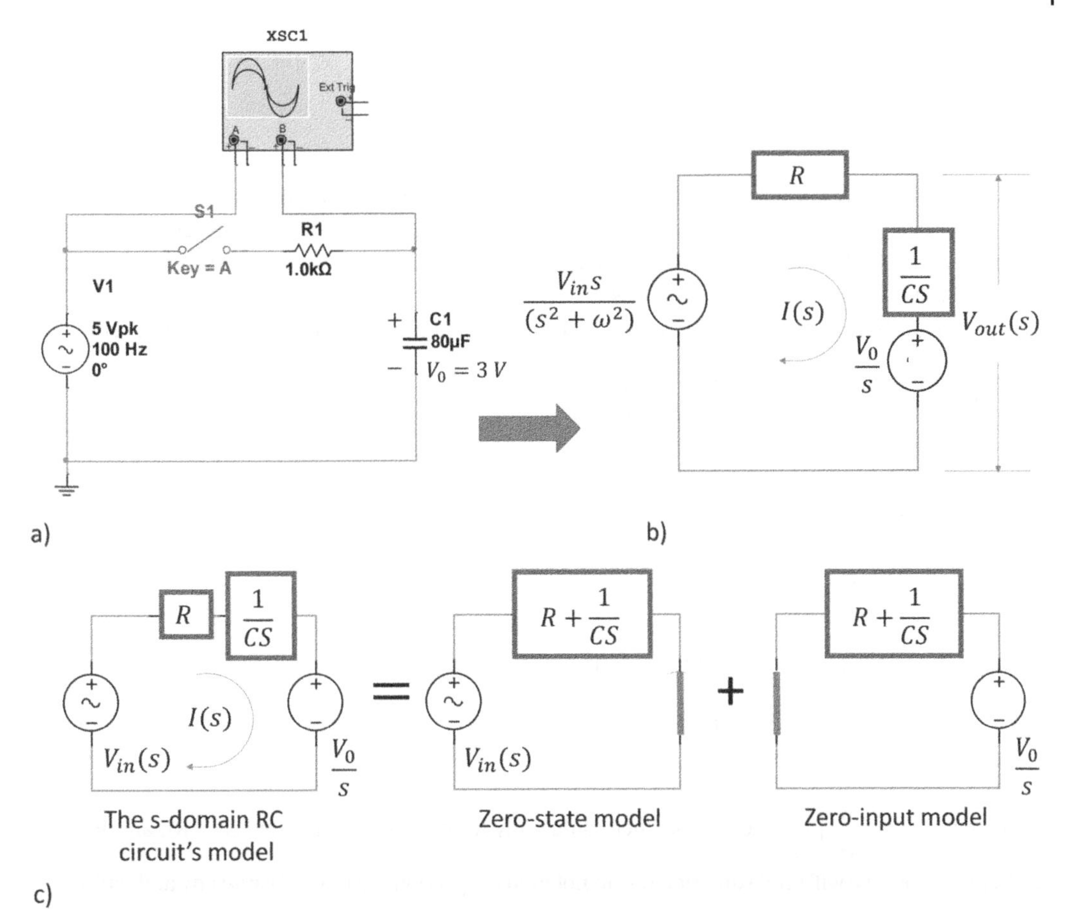

Figure 9.6 Sinusoidal response of an RC circuit with an initial condition in Example 9.3: a) Circuit's schematic; b) the s-domain equivalent model; c) superposition principle.

Exercise 9.1 Use the superposition principle in the s-domain to prove that Equation 8.20 showing an RC circuit's step response with a non-zero initial condition in Example 8.1is correct.

9.2 Analysis of RLC Circuits

9.2.1 The s-Domain Approach

Now we are well equipped to start working with RLC circuits. These circuits contain two reactive components, C and L, and are rightly called the second-order circuits. Their analysis in the time domain is a tedious and time-consuming operation, as Chapter 5 demonstrates; however, their s-domain analysis is much easier, as the following material will show.

Figure 9.7a shows the original, real RLC circuit, and Figure 9.7b presents its s-domain model. Our task is to find the relationship between $v_{in}(t)$ and $v_{out}(t)$, the task we pursue throughout the book. From start, we turn to the s-domain circuit analysis.

The voltage-divider rule being applied to Figure 9.7b gives

$$V_{out}(s) = \frac{Z_C(s)}{Z_R(s) + Z_L(s) + Z_C(s)} V_{in}(s) = \frac{\frac{1}{Cs}}{R + Ls + \frac{1}{Cs}} V_{in}(s). \tag{9.18}$$

(Observe how easily we obtained $V_{out}(s)$ applying the s-domain circuit analysis. Using the traditional approach exploited in Chapter 8, we would have to derive the circuit's differential equation and transform it into the s-domain. Perform all these operations to convince yourself in the advantage of using the s-domain circuit analysis.)

After multiplying (9.18) through by Cs and introducing *damping constant or coefficient* $\alpha_{RLC}\left(\frac{1}{s}\right)$ and *resonant frequency* $\omega_0\left(\frac{rad}{s}\right)$ as

$$\alpha_{RLC} = \frac{R}{2L}\left(\frac{1}{s}\right)$$

and

$$\omega_0^2 = \frac{1}{LC}\left(\frac{rad}{s}\right), \tag{9.19}$$

Equation 9.18 relates input and output of the s-domain model as

$$V_{out}(s) = \frac{\omega_0^2}{\left(s^2 + 2\alpha s + \omega_0^2\right)} V_{in}(s). \tag{9.20}$$

(Note that $\alpha_{RLC} = \frac{R}{2L}\left(\frac{1}{s}\right)$ is determined here differently from the similar parameter of RL and RC circuits. From now, we will omit subscript *RLC* in notation α_{RLC} as long as it won't cause any ambiguity.)

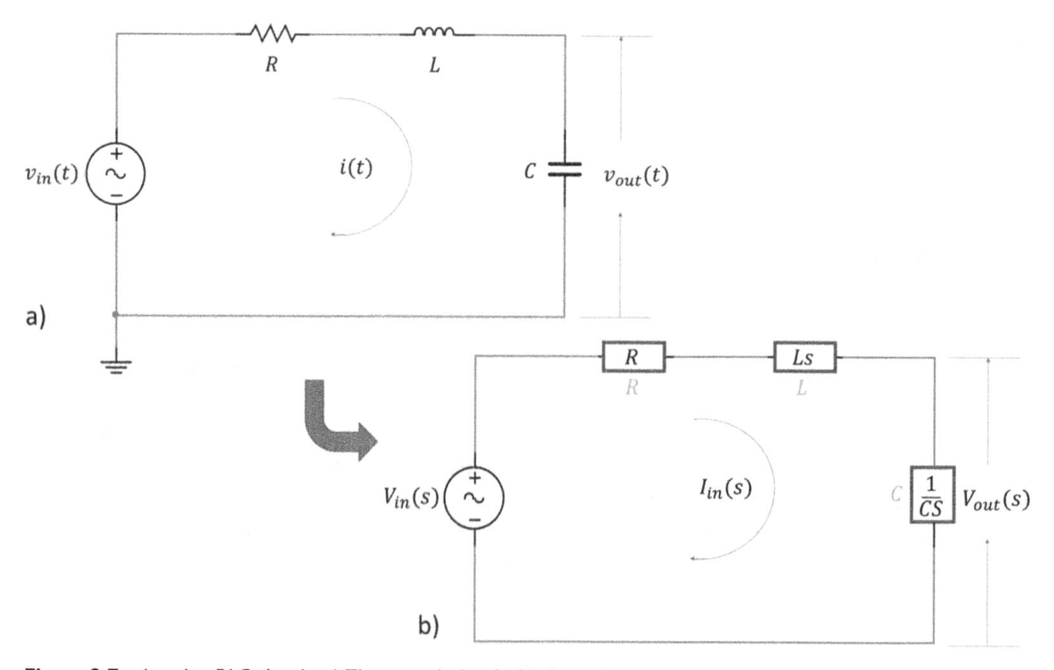

Figure 9.7 A series RLC circuit: a) The actual circuit; b) the s-domain model.

The inverse Laplace transform of (9.18) would produce the desired answer:

$$L^{-1}\{V_{out}(s)\} = L^{-1}\left\{\frac{\omega_0^2}{\left(s^2 + 2\alpha s + \omega_0^2\right)} V_{in}(s)\right\} = v_{out}(t). \tag{9.21}$$

It seems that our task is achieved; however, when we start performing the inverse Laplace transform, we encounter a serious obstacle. Indeed, to expand the kernel of (9.21), $\frac{\omega_0^2}{\left(s^2 + \alpha s + \omega_0^2\right)}$, into the partial fractions, we have to present it in a factored form as

$$\frac{\omega_0^2}{\left(s^2 + 2\alpha s + \omega_0^2\right)} = \frac{\omega_0^2}{(s - p_1)(s - p_2)}. \tag{9.22}$$

To find poles p_1 *and* p_2, we need to solve the *characteristic equation*,

$$s^2 + 2\alpha s + \omega_0^2 = s^2 + \frac{R}{L}s + \frac{1}{LC} = 0. \tag{9.23}$$

Thus, we arrive at

$$p_{1,2} = -\frac{R}{2L} \pm \frac{1}{2}\sqrt{\left(\frac{R}{L}\right)^2 - \frac{4}{LC}} = -\frac{R}{2L} \pm \sqrt{\left(\frac{R}{2L}\right)^2 - \frac{1}{LC}} = -\alpha \pm \sqrt{\alpha^2 - \omega_0^2}. \tag{9.24}$$

Therefore, the nature of poles in (9.24) depends on the relationship between $\left(\frac{R}{2L}\right)^2$ and $\frac{1}{LC}$ or α^2 and ω_0^2. As thoroughly discussed in Chapters 5 and 6, there are three possible cases of this relationship:

Case 1—overdamped:

$$\left(\frac{R}{2L}\right)^2 > \frac{1}{LC} \text{ or } R_{od} > 2\sqrt{\frac{L}{C}} \text{ or } \alpha_{od}^2 > \omega_0^2 \tag{9.25}$$

This case gives *two real distinct poles* p_1 and p_2, which results in a simple solution:

$$v_{out}^{od}(t) = A_1 e^{-p_1 t} + A_2 e^{-p_2 t}. \tag{9.26}$$

Revisit Sections 8.2 and 8.3 for refreshing your memory on a general approach to finding the inverse Laplace transform and Section 8. 4 for specifics on obtaining the inverse Laplace transform in case of real distinct poles.

Case 2—critically damped:

$$\left(\frac{R}{2L}\right)^2 = \frac{1}{LC} \text{ or } R_{cr} = 2\sqrt{\frac{L}{C}} \text{ or } \alpha_{cr}^2 = \omega_0^2. \tag{9.27}$$

In this case, we obtain two *real identical (multiple-order) poles*, $p_1 = p_2 = p$. The solution is

$$v_{out}^{cr}(t) = (A_1 + A_2 t)e^{-pt}. \tag{9.28}$$

Section 8.5 meticulously discusses several techniques of finding the inverse Laplace transform in the case of multiple poles.

Case 3—underdamped:

$$\left(\frac{R}{2L}\right)^2 < \frac{1}{LC} \text{ or } R_{ud} < 2\sqrt{\frac{L}{C}} \text{ or } \alpha_{ud}^2 < \omega_0^2. \tag{9.29}$$

Since poles are determined in (9.24) as $p_{1,2} = -\frac{R}{2L} \pm \sqrt{\left(\frac{R}{2L}\right)^2 - \frac{1}{LC}} = -\alpha \pm \sqrt{\alpha^2 - \omega_0^2}$ and $\left(\left(\frac{R}{2L}\right)^2 - \frac{1}{LC}\right) < 0$ or $\alpha_{ud}^2 < \omega_0^2$, this case results in *complex poles*

$$p_{1,2} = -\frac{R}{2L} \pm \sqrt{\left(\frac{R}{2L}\right)^2 - \frac{1}{LC}} = -\alpha_{ud} \pm \sqrt{\alpha_{ud}^2 - \omega_0^2} = -\frac{\alpha_{ud}}{2} \pm j\omega_d. \tag{9.30}$$

Here, the *damping frequency*, ω_d, is given by

$$\omega_d = \sqrt{\omega_0^2 - \alpha_{ud}^2}. \tag{9.31}$$

As we can see, ω_d is smaller than the resonant frequency ω_0.

Then, the solution is the sinusoidal oscillations

$$v_{out}^{ud}(t) = e^{-\frac{\alpha_{ud}}{2}t}\left[acos(\omega_d t) + bsin(\omega_d t)\right] = e^{-\frac{\alpha_{ud}}{2}t} Acos(\omega_d t - \varphi), \tag{9.32}$$

where $A = \sqrt{(a^2 + b^2)}$ and $\varphi = tan^{-1}\left(\frac{b}{a}\right)$.

Come back to Section 8.7, which considers all the details of finding $v_{out}(t)$ from $V_{out}(s)$ for complex poles.

We summarize our results in Table 9.2. (It's advisable to compare this table with Table 5.1 where these three cases are also displayed.)

Why Table 9.2 summarizes results only of the natural (source-free) responses? Because these responses display the nature of a circuit itself, without involving the circuit's responses to the external excitations.

9.2.2 Three Cases of a Series RLC Circuit Response—Examples

Example 9.4 Natural (source-free) response of an RLC circuit.

Problem: In Figure 9.8a, the switch connects the RLC circuit to the dc source $V_{in} = 4V$ for a long time. At $t = 0$, the switch hooks to the ground, so the circuit becomes source-free. Find three cases of a natural response of this circuit if L = 2 H, C = 5 nF, and R can vary consequently from critical, R_{cr}, to overdamped, R_{od}, to underdamped, R_{od}, values.

Prediction: Since the circuit's only driving force after $t \geq 0$ is the energy previously stored in the circuit, we expect that all the transient processes ignited by this force will diminish to zero as the stored energy will be dissipated by the circuit. Therefore, all the forms of the circuit's responses should be multiplied by a damping exponential function.

Solution: It follows from the condition of this problem that $i_L(0) = 0A$, $v_C(0) = 4V$, and $v_C'(0) = 0V$. (Can you explain why?) Also, we compute $\omega_0^2 = \frac{1}{LC} = 10^8\left(\frac{1}{s}\right)$.

Table 9.2 Three cases of the time-domain zero-input (natural) responses of an RLC circuit.

Poles	Case	Formula of $v_{out}(t)$	Typical graph of $v_{out}(t)$
$R_{od} > 2\sqrt{\frac{L}{C}}$ or $\alpha^2 > \omega_0^2$ real and distinct poles	Overdamped	$v_{out}^{od}(t) = A_1 e^{-p_1 t} + A_2 e^{-p_2 t}$	
$R_{cr} = 2\sqrt{\frac{L}{C}}$ or $\alpha^2 = \omega_0^2 \Rightarrow$ multiple-order poles	Critically damped	$v_{out}^{cr}(t) = (A_1 + A_2 t)e^{-pt}$	
$R_{ud} < 2\sqrt{\frac{L}{C}}$ or $\alpha^2 < \omega_0^2 \Rightarrow$ complex poles	Underdamped	$v_{out}^{ud}(t) = e^{-\frac{\alpha}{2}t} A\cos(\omega_D t + \varphi)S$	

The s-domain model of the RLC circuit is shown in Figure 9.8b. (See Figures 9.4, 9.6b, and 9.7b.) The loop equation for this circuit is

$$I(s)R + I(s)Ls + I(s)\frac{1}{Cs} + \frac{V_0}{s} = 0. \tag{9.33}$$

(See Method 1 in Example 9.2.) The circuit's current is

$$I(s) = \left(-\frac{V_0}{s}\right)\left(\frac{1}{R + LS + \frac{1}{CS}}\right) = \left(-\frac{V_0}{s}\right)\left(\frac{Cs}{LCs^2 + RCs + 1}\right) = -\frac{V_0 C}{LCs^2 + RCs + 1}. \tag{9.34}$$

See (9.18), (9.19), and (9.20). The circuit's output voltage is determined as

$$V_{out}(s) = I(s)Z_C(s) + \frac{V_0}{s} = I(s)\left(\frac{1}{Cs}\right) + \frac{V_0}{s} = -\frac{V_0 C}{LCs^2 + RCs + 1}\left(\frac{1}{Cs}\right) + \frac{V_0}{s} = V_0\left(\frac{s + 2\alpha}{s^2 + 2\alpha s + \omega_0^2}\right) \tag{9.35}$$

where $\alpha = \frac{R}{2L}$ and $\omega_0^2 = \frac{1}{LC}$ as before.

To explore the required three cases, we need to determine the values of R because L and C are given. It's reasonable to start with a critical value, R_{cr}, because two other values will be either greater or smaller than that.

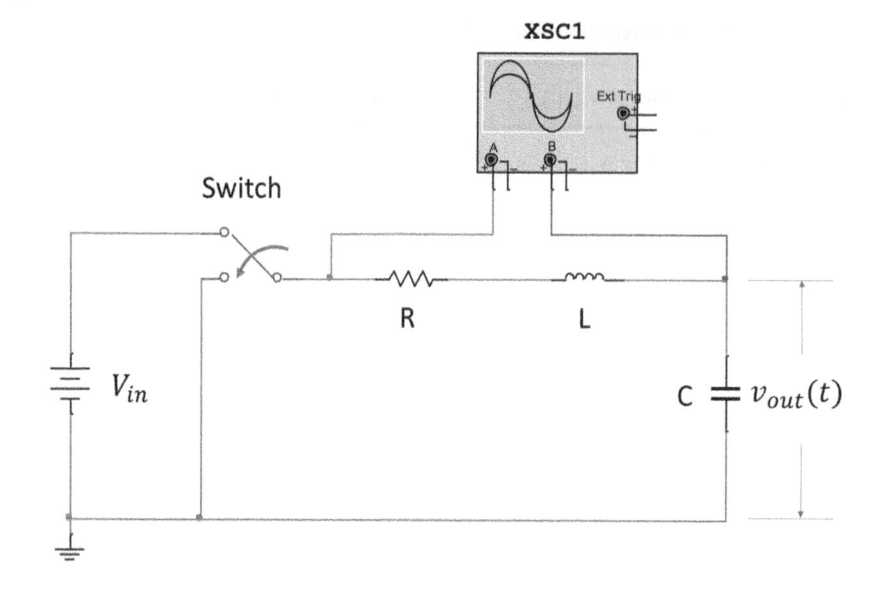

a)

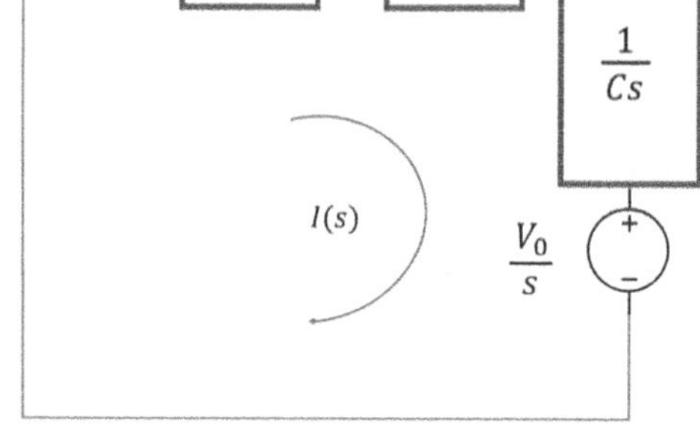

b)

Figure 9.8 Investigating the natural (source-free) response of an RLC circuit in Example 9.4: a) The experimental setup; b) the s-domain model of the actual circuit.

Refer to (9.27) to find

$$R_{cr} = 2\sqrt{\frac{L}{C}} = 40\,k\Omega.$$

For the overdamped case, we assume

$$R_{od} > 2\sqrt{\frac{L}{C}} = 80\,k\Omega,$$

and for the underdamped case, we suppose

$$R_{ud} < 2\sqrt{\frac{L}{C}} = 20\,k\Omega.$$

Note that R_{od} and R_{ud} values are arbitrary. We can choose whatever values we please, provided that $R_{od} > R_{cr}$ and $R_{ud} < R_{cr}$. Practically speaking, we want $R_{od} >> R_{cr}$ and $R_{ud} << R_{cr}$ to see the difference between all cases better.

Therefore, the solutions to our problem are:

Case 1—overdamped: ($R_{od} = 80k\Omega$): In this case, $\alpha_{od} = \frac{R_{od}}{2L} = 2\cdot 10^4\left(\frac{1}{s}\right)$. The characteristic Equation (9.23) takes the form

$$s^2 + 2\alpha_{od}s + \omega_0^2 = s^2 + \frac{R_{od}}{L}s + \frac{1}{LC} = s^2 + 4\cdot 10^4 s + 10^8 = 0.$$

Here, there are two real distinct poles

$$p_{1,2} = -\frac{R}{2L} \pm \sqrt{\left(\frac{R}{2L}\right)^2 - \frac{1}{LC}} = -\alpha_{od} \pm \sqrt{\alpha_{od}^2 - \omega_0^2} = -2\cdot 10^4 \pm \sqrt{\left(2\cdot 10^4\right)^2 - 10^8}.$$

These poles are

$$p_1 = -\alpha_{od} + \sqrt{\alpha_{od}^2 - \omega_0^2} = (-2 + \sqrt{3})10^4 \approx -2.7\cdot 10^3,$$

and

$$p_2 = -\alpha_{od} - \sqrt{\alpha_{od}^2 - \omega_0^2} = (-2 - \sqrt{3})10^4 \approx -37.3\cdot 10^3.$$

Note,

$$(p_1 - p_2) = 2\sqrt{\alpha_{od}^2 - \omega_0^2} = 2\sqrt{3}\cdot 10^4, \text{ and } (p_2 - p_1) = -2\sqrt{\alpha_{od}^2 - \omega_0^2} = -2\sqrt{3}\cdot 10^4.$$

Thus, (9.35) can be presented in the following factored and expanded form:

$$V_{out}^{od}(s) = V_0\left(\frac{s + 2\alpha_{od}}{s^2 + 2\alpha_{od}s + \omega_0^2}\right) = V_0\left(\frac{s + 2\alpha_{od}}{(s - p_1)(s - p_2)}\right) = \frac{A_1}{(s - p_1)} + \frac{A_2}{(s - p_2)}. \quad (9.36)$$

The use of the cover-up algorithm gives the residues A_1 and A_2 as

$$A_1 = \frac{V_0(s + 2\alpha_{od})}{(s - p_1)(s - p_2)}(s - p_1)|_{s=p_1} = \frac{V_0(p_1 + 2\alpha_{od})}{(p_1 - p_2)} = \frac{V_0\left(\alpha_{od} + \sqrt{\alpha_{od}^2 - \omega_0^2}\right)}{2\sqrt{\alpha_{od}^2 - \omega_0^2}} = \frac{4(2 + \sqrt{3})10^4}{2\sqrt{3}\cdot 10^4} \approx 4.31,$$

and

$$A_2 = \frac{V_0(s + 2\alpha_{od})}{(s - p_1)(s - p_2)}(s - p_2)|_{s=p_2} = \frac{V_0(p_2 + 2\alpha_{od})}{(p_2 - p_1)} = \frac{V_0\left(\alpha_{od} - \sqrt{\alpha_{od}^2 - \omega_0^2}\right)}{-2\sqrt{\alpha_{od}^2 - \omega_0^2}} = \frac{4(2 - \sqrt{3})10^4}{-2\sqrt{3}\cdot 10^4} \approx -0.31.$$

Thus, the s-domain natural response of the RLC circuit in the overdamped case is derived as

$$V_{out}^{od}(s) = V_0\left(\frac{s + 2\alpha_{od}}{s^2 + 2\alpha_{od}s + \omega_0^2}\right) = \frac{A_1}{(s - p_1)} + \frac{A_2}{(s - p_2)} \approx \frac{4.31}{\left(s + 2.7\cdot 10^3\right)} - \frac{0.31}{\left(s + 37.3\cdot 10^3\right)}. \quad (9.37)$$

The inverse Laplace transform of (9.37) gives the following time-domain response for Case1:

$$v_{out}^{od}(t) = L^{-1}\left\{V_{out}^{od}(s)\right\} = A_1 e^{p_1 t} + A_2 e^{p_2 t} \approx 4.31e^{-2.7\cdot 10^3 t} - 0.31e^{-37.3\cdot 10^3 t}. \quad (9.38)$$

At $t=0$, $v_{out}^{od}(0)=4V$, and at $t\to\infty$, $v_{out}^{od}(\infty)$ tends to zero. Thus, the initial value is correct, and the final value is as expected.

Case 2—critically damped: ($R_{cr}=40k\Omega$): In this case, $\alpha_{cr}=\frac{R_{cr}}{2L}=10^4\left(\frac{1}{s}\right)$ and $\omega_0^2=10^8\left(\frac{1}{s}\right)$, as before. The characteristic Equation (9.23) takes the form

$$s^2+2\alpha_{cr}s+\omega_0^2=s^2+\frac{R_{cr}}{L}s+\frac{1}{LC}=s^2+2\cdot 10^4 s+10^8=0.$$

Solving it for the poles, we find

$$p_{1,2}=-\frac{R}{2L}\pm\sqrt{\left(\frac{R}{2L}\right)^2-\frac{1}{LC}}=-\alpha_{cr}\pm\sqrt{\alpha_{cr}^2-\omega_0^2}=p=-10\cdot 10^3.$$

Thus, we have two identical (multiple) poles, $p_1=p_2=p=-10\cdot 10^3$.

Note that in a critically damped case,

$$\alpha_{cr}^2=\omega_0^2, \tag{9.39}$$

and the characteristic equation can be written as

$$s^2+2\alpha_{cr}s+\omega_0^2=(s+\alpha_{cr})^2=(s+\omega_0)^2=0. \tag{9.40}$$

(Can you prove (9.39) and (9.40)?) Thus, the other forms of the multiple poles are

$$p=-\alpha_{cr}=-\omega_0=-10\cdot 10^3.$$

Therefore, the s-domain response of the RLC circuit in the critically damped case becomes

$$V_{out}^{cr}(s)=V_0\left(\frac{s+2\alpha_{cr}}{s^2+2\alpha_{cr}s+\omega_0^2}\right)=V_0\left(\frac{s+2\alpha_{cr}}{(s-p)^2}\right). \tag{9.41}$$

The *differentiation method* discussed in Section 8.5 seems to be the best method to find residues A_1 and A_2. (Try the other methods discussed in Section 8.5 to convince yourself of this choice.)

First, we expand this $V_{out}^{cr}(s)$ as

$$V_{out}^{cr}(s)=V_0\left(\frac{s+2\alpha_{cr}}{(s-p)^2}\right)=\frac{A_1}{(s-p)}+\frac{A_2}{(s-p)^2}. \tag{9.42}$$

Next, we find A_2 using the *cover-up method* as

$$A_2=(s-p)^2V_{out}^{cr}(s)\Big|_{s=p}=(s-p)^2V_0\left(\frac{s+2\alpha_{cr}}{(s-p)^2}\right)\Bigg|_{s=p}=V_0\left(\frac{s+2\alpha_{cr}}{1}\right)\Bigg|_{s=-\alpha_{cr}}=\alpha_{cr}V_0.$$

See (8.94) and (8.95).

To find A_1, we need to take the derivative of the product $\left[(s-p_2)^2V_{out}^{cr}(s)\right]$ and evaluate the result at $s=p$, as (8.101) and (8.102) show. Hence,

$$A_1=\frac{d}{ds}\left[(s-p_2)^2V_{out}^{cr}(s)\right]\Big|_{s=p}=\frac{d}{ds}\left[(s-p)^2V_0\left(\frac{s+2\alpha_{cr}}{(s-p)^2}\right)\right]\Bigg|_{s=-\alpha_{cr}}=V_0.$$

Plugging these results in (9.42) yields

$$V_{out}^{cr}(s)=V_0\left(\frac{s+2\alpha_{cr}}{(s-p)^2}\right)=\frac{A_1}{(s-p)}+\frac{A_2}{(s-p)^2}=\frac{V_0}{(s+\alpha_{cr})}+\frac{\alpha_{cr}V_0}{(s+\alpha_{cr})^2}. \tag{9.43}$$

The inverse Laplace transform of (9.43) becomes

$$v_{out}^{cr}(t)=L^{-1}\left\{V_{out}^{cr}(s)\right\}=V_0e^{-\alpha_{cr}t}+\alpha_{cr}V_0te^{-\alpha_{cr}t}. \tag{9.44}$$

Plugging the given values produces the following numerical solution:

$$v_{out}^{cr}(t)=4e^{-10\cdot10^3 t}+4\cdot10^4te^{-10\cdot10^3 t}.$$

A curious reader can prove that the initial and final values of $v_{out}^{cr}(t)$ are appropriate.

Case 3—underdamped: ($R_{ud}=20k\Omega$): In this case, $\alpha_{ud}=\frac{R_{ud}}{2L}=0.5\cdot10^4\left(\frac{1}{s}\right)$. The s-domain response of the RLC circuit now is given by

$$V_{out}^{ud}(s)=V_0\left(\frac{s+2\alpha_{ud}}{s^2+2\alpha_{ud}s+\omega_0^2}\right)=V_0\left(\frac{s+2\alpha_{ud}}{(s-p_1)(s-p_2)}\right)=\frac{A_1}{(s-p_1)}+\frac{A_2}{(s-p_2)}. \tag{9.45}$$

The characteristic equation becomes

$$s^2+2\alpha_{ud}s+\omega_0^2=s^2+\frac{R_{ud}}{L}s+\frac{1}{LC}=s^2+10^4s+10^8=0.$$

Since $\alpha_{ud}^2<\omega_0^2$ ($0.25\cdot10^8<10^8$), we can readily predict that the poles will be the complex numbers. And indeed, the poles are

$$p_{1,2}=-\frac{R}{2L}\pm\sqrt{\left(\frac{R}{2L}\right)^2-\frac{1}{LC}}=-\alpha_{ud}\pm j\sqrt{\omega_0^2-\alpha_{ud}^2}=(-0.5\pm j\sqrt{0.75})10^4.s$$

Then, we can turn to Example 8.9 for guidance in solving this case. It's helpful to calculate $p_1-p_2=j2\sqrt{\omega_0^2-\alpha_{ud}^2}$ and $p_2-p_1=-j2\sqrt{\omega_0^2-\alpha_{ud}^2}$.

Expanding the $V_{out}^{ud}(s)$ given in (9.45) into partial fractions yields

$$V_{out}^{ud}(s)=V_0\left(\frac{s+2\alpha_{ud}}{\left(s+\alpha_{ud}-j\sqrt{\omega_0^2-\alpha_{ud}^2}\right)\left(s+\alpha_{ud}+j\sqrt{\omega_0^2-\alpha_{ud}^2}\right)}\right)$$

$$=\frac{A_1}{\left(s+\alpha_{ud}-j\sqrt{\omega_0^2-\alpha_{ud}^2}\right)}+\frac{A_2}{\left(s+\alpha_{ud}+j\sqrt{\omega_0^2-\alpha_{ud}^2}\right)}. \tag{9.46}$$

Remember that $\boldsymbol{A}_2$ is a complex conjugate to $\boldsymbol{A}_1$, $\boldsymbol{A}_2=\boldsymbol{A}_1^*$. To find the residues, we use the *cover-up algorithm* as follows:

$$\boldsymbol{A}_1=(s-p_1)\left(\frac{V_0(s+2\alpha_{ud})}{(s-p_1)(s-p_2)}\right)\Big|_{s=p_1}=\left(\frac{V_0(p_1+2\alpha_{ud})}{(p_1-p_2)}\right)=2-j\frac{1}{\sqrt{0.75}}.$$

Obviously,

$$A_2 = 2 + j\frac{1}{\sqrt{0.75}}$$

Having the residues A_1 *and* A_2, we can present the inverse Laplace transform of the $V_{out}^{ud}(s)$ as

$$v_{out}^{ud}(t) = L^{-1}\left\{V_{out}^{ud}(s)\right\} = L^{-1}\left\{V_0\left(\frac{s+2\alpha_{ud}}{s^2+2\alpha_{ud}s+\omega_0^2}\right)\right\} = L^{-1}\left\{\frac{A_1}{s-p_1}+\frac{A_2}{s-p_2}\right\} = A_1e^{p_1t}+A_2e^{p_2t}.$$
$$= \left(2-j\frac{1}{\sqrt{0.75}}\right)e^{(-0.5+j\sqrt{0.75})10^4t} + \left(2+j\frac{1}{\sqrt{0.75}}\right)e^{(-0.5-j\sqrt{0.75})10^4t} \qquad (9.47)$$

To present this answer in terms of sine and cosine, we need to use the Euler formulas, $e^{j\varphi} + e^{-j\varphi} = 2cos\varphi$ and $e^{j\varphi} - e^{-j\varphi} = j2sin\varphi$. For this, we regroup (9.47) as

$$v_{out}^{ud}(t) = 2\left(e^{(-0.5+j\sqrt{0.75})10^4t} + e^{(-0.5-j\sqrt{0.75})10^4t}\right) - \left(j\frac{1}{\sqrt{0.75}}\right)\left(e^{(-0.5+j\sqrt{0.75})10^4t} - e^{(-0.5-j\sqrt{0.75})10^4t}\right),$$

which produces

$$v_{out}^{ud}(t) = 4e^{-5000t}\left(cos5000\sqrt{3}t + \frac{1}{\sqrt{3}}sin5000\sqrt{3}t\right) = \frac{8}{\sqrt{3}}e^{-5000t}\cos\left(5000\sqrt{3}t - 30^0\right). \qquad (9.48)$$

The problem is solved.

Discussion:

- MATLAB script, the output formula, and the response graph are as follows:

```
syms s t
V0 = 4;
a = 2*10^4'
w0square = 10^8;
m = V0*(s + 2*a);
n = (s^2+2*a*s+w0square);
c = ilaplace(m/n,t);
ezplot(c,[0, 0.002])
axis([0 0.002 0 5])
grid
```

Note that MATLAB's script for all three cases is very similar; only α's value will change.

Compare the MATLAB's formulas for all three cases with those obtained by manual derivations to verify our solutions.

To obtain the $v_{out}^{od}(t)$ formula without hyperbolic cosine and sine, add the command c1=rewrite(c, 'exp') after `c = ilaplace(m/n,t)`. The result is shown below:

```
c1 = 4*exp(-20000*t)*(exp(-10000*3^(1/2)*t)/2 +
exp(10000*3^(1/2)*t)/2 - (2*3^(1/2)*(exp(-10000*3^(1/2)*t)/2 - exp
(10000*3^(1/2)*t)/2))/3).
```

- Figures 9.9a and 9.9b look very similar; however, the critically damped exponential function diminishes much faster than the overdamped function does. Can you explain why?

- The obtained solutions and the MATLAB's graphs reveal the origin of the results presented in Table 9.2. Pay attention at the fact that the overdamped graph in Table 9.2 crosses zero line, whereas the similar graph in Figure 9.9b does not. Why? It depends on the values of all three parameters, R, L, and C. A curious reader can investigate this point further.

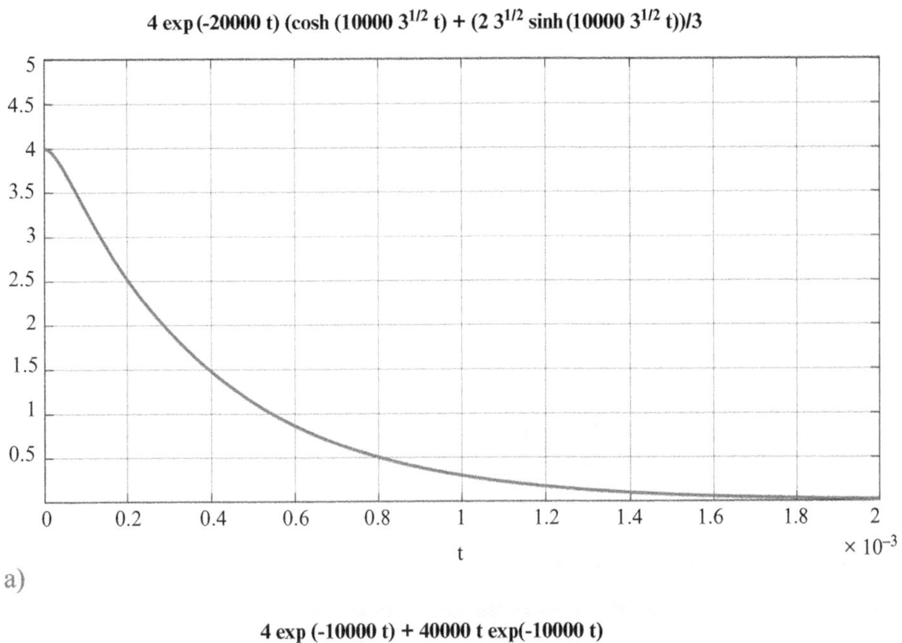

a)

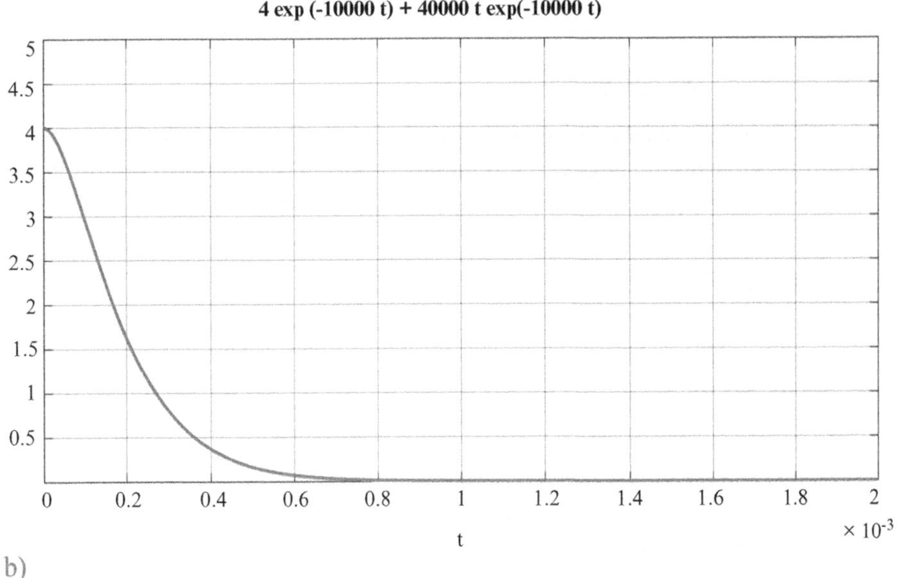

b)

Figure 9.9 MATLAB solution for the RLC circuit's natural responses in Example 9.4: a) The overdamped case; b) critically damped case; c) underdamped case.

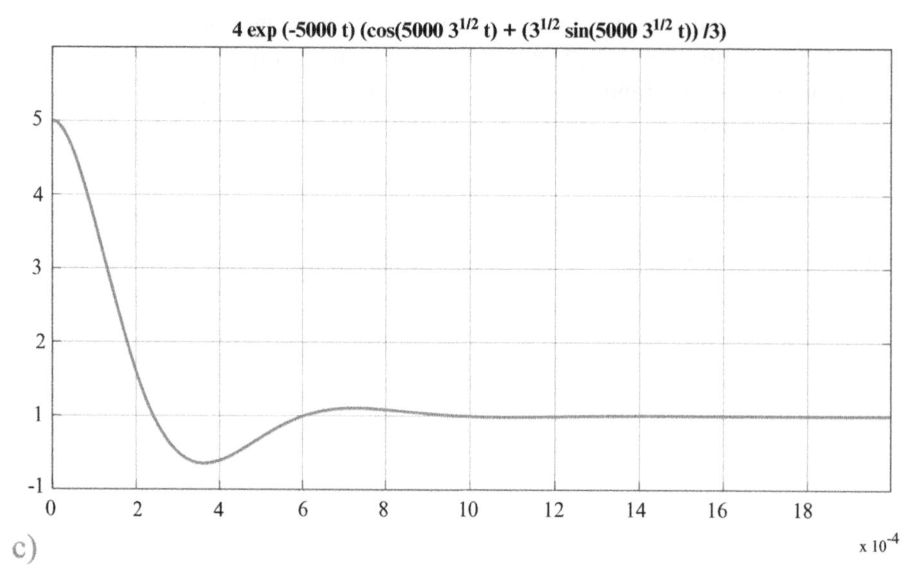

Figure 9.9 (Cont'd)

Question Compare the critically damped graph in Table 9.2 and the underdamped graph in Figure 9.9c. They look similar because they both cross the horizontal (time) line. How can we distinguish them by observation?

>>>

Example 9.5 Step response of a series RLC circuit.

Problem: A series RLC circuit is shown in Figure 9.10a. Values L = 2 H and C = $5 \cdot 10^{-9}$ F are fixed. The resistor's value can be changed. The circuit was initially relaxed. Find the step response of this circuit to $V_{in} = 12V$ for overdamped, critically damped, and underdamped cases.

Solution:

The s-domain model of this circuit enables us to apply the voltage-divider rule to obtain the response $V_{out}(s)$ as

$$V_{out}(s) = \frac{\omega_0^2 V_{in}}{s\left(s^2 + 2\alpha s + \omega_0^2\right)} \tag{9.49}$$

where $\alpha = \frac{R}{2L}\left(\frac{1}{s}\right)$, $\omega_0^2 = \frac{1}{LC}\left(\frac{rad}{s}\right)$, and $V_{in}(s) = \frac{V_{in}}{s} = \frac{12}{s}(V - s)$. (Unhappily, we have to use letter *s* to designate both domain and seconds. So, don't be confused: $V_{in}(s)$ means a voltage function in the s-domain but (V-s) means *volt-second*, the unit of $V_{in}(s)$.)

Expansion of $V_{out}(s)$ into partial fractions yields

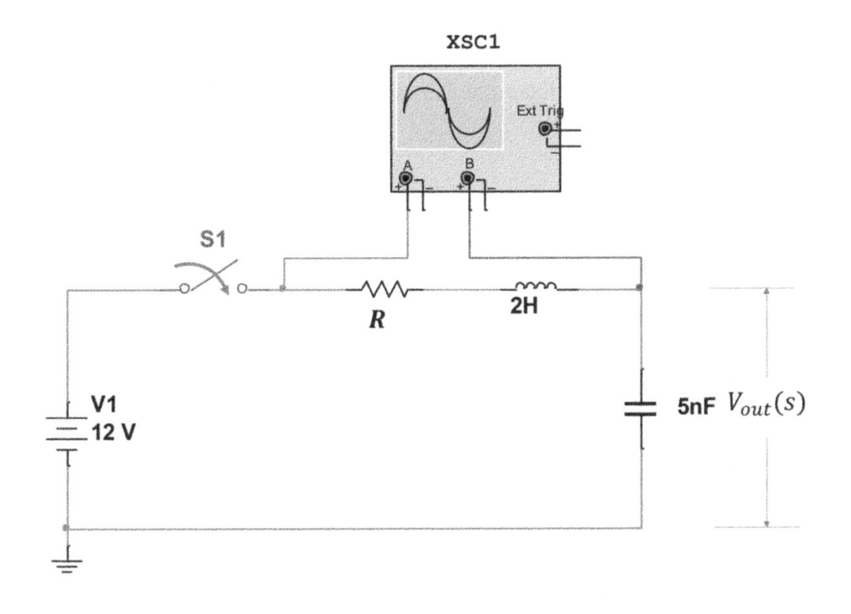

a)

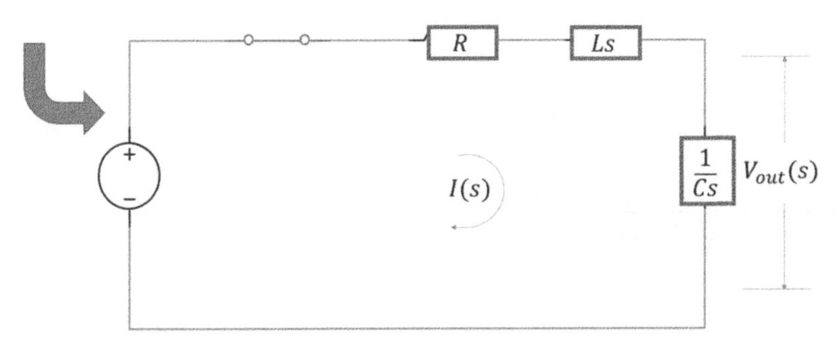

b)

Figure 9.10 Step response of an RLC circuit in Example 9.5: a) Experimental setup; b) the s-domain circuit's model.

$$V_{out}(s) = \frac{\omega_0^2 V_{in}}{s\left(s^2 + 2\alpha s + \omega_0^2\right)} = \frac{\omega_0^2 V_{in}}{(s)(s-p_2)(s-p_3)} = \frac{A_1}{(s)} + \frac{A_2}{(s-p_2)} + \frac{A_3}{(s-p_3)}. \tag{9.50}$$

The first pole is zero, $p_1 = 0$. The poles p_2 and p_3 depend on the values of R, L, and C, as Equations 9.25 through 9.30, Table 9.2, and Example 9.3 describe. The critical resistance, R_{cr}, is calculated in Example 9.3 as

$$R_{cr} = 2\sqrt{\frac{L}{C}} = 40\,k\Omega.$$

We will use the overdamped and underdamped resistances assumed in Example 9.3; i.e. $R_{od} > 2\sqrt{\frac{L}{C}} = 80\,k\Omega$ and $R_{ud} < 2\sqrt{\frac{L}{C}} = 20\,k\Omega$.

Then, the solution to our problem can be found as:

Case 1—overdamped: ($R_{od} = 80\ k\Omega$): From Example 9.3 we transfer $\alpha_{od} = \frac{R_{od}}{2L} = 2 \cdot 10^4 \left(\frac{1}{s}\right)$, $\omega_0^2 = 10^8 \left(\frac{rad}{s}\right)$, the characteristic equation

$$s^2 + 2\alpha_{od} s + \omega_0^2 = s^2 + \frac{R_{od}}{L} s + \frac{1}{LC} = s^2 + 4 \cdot 10^4 s + 10^8 = 0,$$

and two real distinct poles

$$p_{2,3} = -\frac{R}{2L} \pm \sqrt{\left(\frac{R}{2L}\right)^2 - \frac{1}{LC}} = -\alpha_{od} \pm \sqrt{\alpha_{od}^2 - \omega_0^2} = \left(-2 + \sqrt{3}\right)10^4.$$

The pole values are $p_2 \approx -0.27 \cdot 10^4$ and $p_3 \approx -3.73 \cdot 10^4$.
It's useful to calculate $(p_2 - p_3) = 2\sqrt{\alpha_{od}^2 - \omega_0^2}$ and $(p_3 - p_2) = -2\sqrt{\alpha_{od}^2 - \omega_0^2}$.
Using the cover-up algorithm, we find residues A_1, A_2, and A_3 as

$$A_1 = \frac{\omega_0^2 V_{in}}{s(s-p_2)(s-p_3)}(s)\Big|_{s=0} = \frac{\omega_0^2 V_{in}}{(-p_2)(-p_3)} = V_{in} = 12V.$$

Could you be able to predict this result by examining (9.50)? The other two residues are

$$A_2 = \frac{\omega_0^2 V_{in}}{s(s-p_2)(s-p_3)}(s-p_2)\Big|_{s=p_2} = \frac{\omega_0^2 V_{in}}{(p_2)(p_2-p_3)}$$

$$= \frac{\omega_0^2 V_{in}}{\left(-\alpha_{od} + \sqrt{\alpha_{od}^2 - \omega_0^2}\right) 2\sqrt{\alpha_{od}^2 - \omega_0^2}} \approx -12.93,$$

and

$$A_3 = \frac{\omega_0^2 V_{in}}{s(s-p_2)(s-p_3)}(s-p_3)\Big|_{s=p_3} = \frac{\omega_0^2 V_{in}}{(p_3)(p_3-p_2)} = \frac{\omega_0^2 V_{in}}{\left(-\alpha_{od} - \sqrt{\alpha_{od}^2 - \omega_0^2}\right)\left(-2\sqrt{\alpha_{od}^2 - \omega_0^2}\right)} \approx 0.93.$$

Thus, the s-domain response of the RLC circuit in the overdamped case is derived as

$$V_{out}^{od}(s) = \frac{\omega_0^2 V_{in}}{s\left(s^2 + 2\alpha s + \omega_0^2\right)} = \frac{A_1}{(s-p_1)} + \frac{A_2}{(s-p_2)} + \frac{A_3}{(s-p_3)} = \frac{12}{s} - \frac{12.93}{\left(s + 2.7 \cdot 10^3\right)} + \frac{0.93}{\left(s + 37.3 \cdot 10^3\right)}. \quad (9.51)$$

The inverse Laplace transform of (9.34) gives the following time-domain response for Case1:

$$v_{out}^{od}(t) = L^{-1}\{V_{out}(s)\} = A_1 e^{p_1 t} + A_2 e^{p_2 t} + A_3 e^{p_3 t} = 12 - 12.93 e^{-2.7 \cdot 10^3 t} + 0.93 e^{-37.3 \cdot 10^3 t}. \quad (9.52)$$

At $t = 0$, $v_{out}^{od}(0)$ turns to zero, and at $t \to \infty$, $v_{out}^{od}(\infty)$ tends to 12 V. Thus, the initial and final values of our solution are correct.

Case 2—critically damped: ($R_{cr} = 40 k\Omega$): Solving the characteristic equation and borrowing all necessary results from Example 9.3 enable us to obtain poles as $p_1 = 0$ and

$$p_{2,3} = -\frac{R}{2L} \pm \sqrt{\left(\frac{R}{2L}\right)^2 - \frac{1}{LC}} = -\alpha_{cr} \pm \sqrt{\alpha_{cr}^2 - \omega_0^2} = \mathrm{p} = -\alpha_{cr} = -\omega_0 = -10 \cdot 10^3.$$

Thus, the s-domain response of the RLC circuit in this case becomes

$$V_{out}^{cr}(s) = \frac{\omega_0^2 V_{in}}{s\left(s^2 + 2\alpha_{cr} s + \omega_0^2\right)} = \frac{\omega_0^2 V_{in}}{s\left(s + \alpha_{cr}\right)^2} = \frac{\omega_0^2 V_{in}}{s(s-p)^2}. \tag{9.53}$$

We will find residues A_1, A_2, and A_3 using the traditional cover-up method discussed in Section 8.5.
Expansion of $V_{out}^{cr}(s)$ gives

$$V_{out}^{cr}(s) = \frac{\omega_0^2 V_{in}}{s(s-p)^2} = \frac{A_1}{s\left(s-p\right)} + \frac{A_{22}}{\left(s-p\right)^2}. \tag{9.54}$$

Factor the repeated member in (9.54), equate members in the brackets, substitute $p = -\alpha_{cr}$, and obtain

$$\left[\frac{\omega_0^2 V_{in}}{s\left(s+\alpha_{cr}\right)}\right] = \left[\frac{A_1}{s} + \frac{A_{22}}{\left(s+\alpha_{cr}\right)}\right]. \tag{9.55}$$

Refer to (8.83), (8.84), and (8.85).
See (8.86) and (8.87) to find residues A_1 and A_{22} as

$$A_1 = \left.\frac{s\omega_0^2 V_{in}}{s\left(s+\alpha_{cr}\right)}\right|_{s=0} = \left.\frac{\omega_0^2 V_{in}}{\left(s+\alpha_{cr}\right)}\right|_{s=0} = \frac{\omega_0^2 V_{in}}{\left(\alpha_{cr}\right)} = \frac{12 \cdot 10^8}{10 \cdot 10^3} = 12 \cdot 10^4$$

and

$$A_{22} = \left.\frac{\left(s+\alpha_{cr}\right)\omega_0^2 V_{in}}{s\left(s+\alpha_{cr}\right)}\right|_{s=-\alpha_{cr}} = \left.\frac{\omega_0^2 V_{in}}{s}\right|_{s=-\alpha_{cr}} = -\frac{\omega_0^2 V_{in}}{\left(\alpha_{cr}\right)} = -12 \cdot 10^4.$$

Plugging the obtained A_1 into $\frac{A_1}{s\left(s-p\right)}$ and resolving the result into the partial fractions yield

$$\frac{A_1}{s\left(s-p\right)} = \frac{\frac{\omega_0^2 V_{in}}{\left(\alpha_{cr}\right)}}{s(s-p)} = \frac{A_{11}}{s} + \frac{A_{21}}{\left(s-p\right)} = \frac{A_{11}}{s} + \frac{A_{21}}{\left(s+\alpha_{cr}\right)}. \tag{9.56}$$

Repeating the cover-up procedure, we find residues A_{11} and A_{21} from (9.56) as follows:

$$A_{11} = \left.\frac{s\frac{\omega_0^2 V_{in}}{\left(\alpha_{cr}\right)}}{s(s-p)}\right|_{s=0} = \left.\frac{\frac{\omega_0^2 V_{in}}{\left(\alpha_{cr}\right)}}{\left(s+\alpha_{cr}\right)}\right|_{s=0} = \frac{\omega_0^2 V_{in}}{\left(\alpha_{cr}\right)^2} = \frac{12 \cdot 10^8}{10^8} = 12,$$

and

$$A_{21} = \left.\frac{(s-p)\frac{\omega_0^2 V_{in}}{\left(\alpha_{cr}\right)}}{s(s-p)}\right|_{s=p} = \left.\frac{\frac{\omega_0^2 V_{in}}{\left(\alpha_{cr}\right)}}{s}\right|_{s=-\alpha_{cr}} = -\frac{\omega_0^2 V_{in}}{\left(\alpha_{cr}\right)^2} = -12.$$

Therefore, the given $V_{out}(s)$ is expanded as

$$V_{out}^{cr}(s)=\frac{\omega_0^2 V_{in}}{s(s-p)^2}=\frac{A_{11}}{s}+\frac{A_{21}}{(s-p)}+\frac{A_{22}}{(s-p)^2}$$

$$=\frac{\omega_0^2 V_{in}}{(\alpha_{cr})^2 s}-\frac{\omega_0^2 V_{in}}{(\alpha_{cr})^2(s+\alpha_{cr})}-\frac{\omega_0^2 V_{in}}{(\alpha_{cr})(s+\alpha_{cr})^2}=\frac{12}{s}-\frac{12}{(s+10^4)}-\frac{12\cdot 10^4}{(s+10^4)^2}. \quad (9.57)$$

(Compare this result with (8.91).)

Putting it in the other way, $V_{out}^{cr}(s)$, is obtained as:

$$V_{out}^{cr}(s)=\frac{\omega_0^2 V_{in}}{s\left(s^2+2\alpha_{cr}s+\omega_0^2\right)}=\frac{\omega_0^2 V_{in}}{s(s+\alpha_{cr})^2}=\frac{12}{s}-\frac{12}{(s+10^4)}-\frac{12\cdot 10^4}{(s+10^4)^2}. \quad (9.58)$$

Plugging $s=2$ into both sides of (9.58), we can verify our partial fraction expansion by the test-value method. We indeed compute 6≈6, where the approximation rises from neglecting members smaller than the four orders of magnitude.

Finally, the time-domain step response of the RLC circuit in the critically damped case is attained as

$$v_{out}^{cr}(t)=\mathscr{L}^{-1}\{V_{out}(s)\}=\mathscr{L}^{-1}\left\{\frac{\omega_0^2 V_{in}}{(\alpha_{cr})^2 s}-\frac{\omega_0^2 V_{in}}{(\alpha_{cr})^2(s+\alpha_{cr})}-\frac{\omega_0^2 V_{in}}{(\alpha_{cr})(s+\alpha_{cr})^2}\right\}$$

$$=V_{in}-V_{in}e^{-\alpha_{cr}t}-V_{in}\cdot 10^4 te^{-\alpha_{cr}t}$$

$$=12-12e^{-10^4 t}-12\cdot 10^4 te^{-10^4 t}. \quad (9.59)$$

The problem is solved.

Equation 9.59 shows that the initial value of the $v_{out}^{cr}(t)$ is zero, and its final value is $V_{in}=12V$, as it should be.

Case 3—underdamped: ($R_{ud}=20k\Omega$): We again borrow all possible results from Example 9.3. As $\alpha_{ud}=\frac{R_{ud}}{2L}=0.5\cdot 10^4\left(\frac{1}{s}\right)$, the characteristic equation becomes

$$s^2+2\alpha_{ud}s+\omega_0^2=s^2+\frac{R_{ud}}{L}s+\frac{1}{LC}=s^2+10^4 s+10^8=0. \quad (9.60)$$

We expect that the poles will be the complex numbers because $\alpha_{cr}^2<\omega_0^2$ $(0.25\cdot 10^8<10^8)$. And indeed, the poles are

$$p_{2,3}=-\frac{R}{2L}\pm\sqrt{\left(\frac{R}{2L}\right)^2-\frac{1}{LC}}=-\alpha_{ud}\pm\sqrt{\alpha_{ud}^2-\omega_0^2}=\left(-0.5\pm j\sqrt{0.75}\right)10^4.$$

Thus, we have $p_1=0$ and two complex-conjugate poles p_2 and p_3. Refer to Example 8.9 for guidance in solving this case.

It's worth to calculate

$$p_2-p_3=j2\sqrt{(\omega_0^2-\alpha_{ud}^2)}=\left(j2\sqrt{0.75}\right)10^4$$

and $p_3 - p_2 = -j2\sqrt{(\omega_0^2 - \alpha_{ud}^2)} = -\left(j2\sqrt{0.75}\right)10^4$.

Expanding the $V_{out}^{ud}(s)$ into partial fractions yields

$$V_{out}^{ud}(s) = \frac{\omega_0^2 V_{in}}{s\left(s^2 + 2\alpha_{ud}s + \omega_0^2\right)} = \frac{\omega_0^2 V_{in}}{s(s-p_2)(s-p_3)} = \frac{\omega_0^2 V_{in}}{s\left(s + \alpha_{ud} - j\sqrt{\omega_0^2 - \alpha_{ud}^2}\right)\left(s + \alpha_{ud} + j\sqrt{\omega_0^2 - \alpha_{ud}^2}\right)}$$

$$= \frac{A_1}{s} + \frac{\boldsymbol{A}_2}{s - p_2} + \frac{\boldsymbol{A}_3}{s - p_3}. \tag{9.61}$$

Remember that $\boldsymbol{A}_3$ is a complex conjugate to $\boldsymbol{A}_2$, $\boldsymbol{A}_3 = \boldsymbol{A}_2^*$.

The *cover-up algorithm* will be our method for finding all residues. Thus, we calculate

$$A_1 = \frac{s\omega_0^2 V_{in}}{s\left(s^2 + 2\alpha_{ud}s + \omega_0^2\right)}\Big|_{s=0} = V_{in} = 12V.$$

This is a well-expected result because the invert of $\frac{A_1}{s}$ is a constant member caused by the applied excitation V_{in}. Calculating $\boldsymbol{A}_2$ produces

$$\boldsymbol{A}_2 = \frac{(s-p_2)\omega_0^2 V_{in}}{s(s-p_2)(s-p_3)}\Bigg|_{s=p_2} = \frac{\omega_0^2 V_{in}}{p_2(p_2 - p_3)} = -\frac{\omega_0^2 V_{in}}{2(\omega_0^2 - \alpha_{ud}^2) + \left(j2\alpha_{ud}\sqrt{\omega_0^2 - \alpha_{ud}^2}\right)} = -\frac{12}{1.5 + j\sqrt{0.75}}.$$

Obviously,

$$\boldsymbol{A}_3 = -\frac{\omega_0^2 V_{in}}{2(\omega_0^2 - \alpha_{ud}^2) - \left(j2\alpha_{ud}\sqrt{\omega_0^2 - \alpha_{ud}^2}\right)} = -\frac{12}{1.5 - j\sqrt{0.75}}.$$

Thus, the inverse Laplace transform of $V_{out}^{ud}(s)$ turns out to be

$$v_{out}^{ud}(t) = L^{-1}\left\{V_{out}^{ud}(s)\right\} = L^{-1}\left\{\frac{\omega_0^2 V_{in}}{s\left(s^2 + 2\alpha_{ud}s + \omega_0^2\right)}\right\} = L^{-1}\left\{\frac{A_1}{s} + \frac{\boldsymbol{A}_2}{s - p_2} + \frac{\boldsymbol{A}_3}{s - p_3}\right\}$$

$$= A_1 + \boldsymbol{A}_2 e^{p_2 t} + \boldsymbol{A}_3 e^{p_3 t} = V_{in} - \frac{12}{1.5 + j\sqrt{0.75}} e^{(-0.5 + j\sqrt{0.75})10^4 t} - \frac{12}{1.5 - j\sqrt{0.75}} e^{(-0.5 - j\sqrt{0.75})10^4 t}. \tag{9.62}$$

We present the complex residues $\boldsymbol{A}_2$ and $\boldsymbol{A}_3$ in a rectangular form $a+jb$ as

$$\boldsymbol{A}_2 = \frac{12}{1.5 + j2\sqrt{0.75}} = \frac{12(1.5 - j2\sqrt{0.75})}{(1.5 + j\sqrt{0.75})(1.5 - j\sqrt{0.75})} = 6 - j4\sqrt{0.75},$$

and

$$\boldsymbol{A}_3 = \frac{12}{1.5 + j\sqrt{0.75}} = \frac{12(1.5 + j\sqrt{0.75})}{(1.5 - j\sqrt{0.75})(1.5 + j\sqrt{0.75})} = 6 + j4\sqrt{0.75}.$$

Plugging these expressions into (9.62) and presenting the complex numbers in rectangular forms, we attain the $v_{out}(t)$ as

$$v_{out}(t)=12-\frac{12}{1.5+j\sqrt{0.75}}e^{(-0.5+j\sqrt{0.75})10^4t}-\frac{12}{1.5-j\sqrt{0.75}}e^{(-0.5-j\sqrt{0.75})10^4t}=12-\left(6-j4\sqrt{0.75}\right)e^{(-0.5+j\sqrt{0.75})10^4t}$$
$$-\left(6+j4\sqrt{0.75}\right)e^{(-0.5-j\sqrt{0.75})10^4t}.$$

Now, we regroup all the members to present the $v_{out}(t)$ in the form suitable for using the Euler formulas, $e^{j\varphi}+e^{-j\varphi}=2cos\varphi$ and $e^{j\varphi}-e^{-j\varphi}=j2sin\varphi$ as follows:

$$v_{out}^{ud}(t)=V_{in}-\left(6-j4\sqrt{0.75}\right)e^{(-0.5+j\sqrt{0.75})10^4t}-\left(6+j4\sqrt{0.75}\right)e^{(-0.5-j\sqrt{0.75})10^4t}$$
$$=V_{in}-\left[6\left(e^{j\sqrt{0.75}10^4t}+e^{-j\sqrt{0.75}10^4t}\right)+j4\sqrt{0.75}\left(e^{j\sqrt{0.75}10^4t}-e^{-j\sqrt{0.75}10^4t}\right)\right]e^{-0.5\,10^4t}$$

Finally, the answer is

$$v_{out}^{ud}(t)=12-\left[12\cos\left(\sqrt{0.75}\cdot10^4t\right)+8\sqrt{0.75}\sin\left(\sqrt{0.75}\cdot10^4t\right)\right]e^{-0.5\cdot10^4t}$$

$$=12-12e^{-5000t}\left[\cos\left(5000\sqrt{3}t\right)+\frac{\sqrt{3}}{3}\sin\left(5000\sqrt{3}t\right)\right]. \tag{9.63}$$

We can further simplify the answer by presenting it as

$$v_{out}^{ud}(t)=12-8\sqrt{3}e^{-0.5\cdot10^4t}\cos\left(\sqrt{0.75}\cdot10^4t-30^0\right). \tag{9.64}$$

The initial value of the RLC circuit's time-domain response in the underdamped case should be $v_{out}^{ud}(0)=0$ V, and its final value should approach 12 V; Equations 9.63 and 9.64 satisfy these conditions.

The problem is solved in its entirety.

Discussion:

- The s-domain circuit analysis allows for finding $I_{out}(s)$ as soon as $V_{out}(s)$ is determined. In this example,

$$I_{out}(s)=\frac{V_{out}(s)}{Z_C(s)}$$

or

$$I_{out}(s)=\frac{V_{out}(s)}{Z_C(s)}=\frac{\left[\dfrac{\omega_0^2V_{in}}{s\left(s^2+2\alpha s+\omega_0^2\right)}\right]}{\dfrac{1}{Cs}}=\frac{V_{in}}{s}\frac{\dfrac{1}{LC}\cdot Cs}{s^2+\dfrac{R}{L}s+\dfrac{1}{LC}}=\frac{V_{in}}{L}\frac{1}{s^2+2\alpha s+\omega_0^2}. \tag{9.65}$$

This example shows how to deal with such an equation in all three cases.

- MATLAB solutions and Multisim simulations for all three cases are shown in Figures 9.11a, 9.11b, and 9.11c. MATLAB provides a robust check of the obtained results because it relies on mathematical manipulations and delivers formulas and graphs. However, its result is correct if and only if we derived the $V_{out}(s)$ formula correctly. Multisim, on the other hand, simulates the circuit's operation, and the accuracy of its result depends on the exactness of the circuit we build for this simulation. (Also, we should remember that any simulation reflects the actual process as approximately as the mathematical model mirrors the modeled process.) Happily, all Multisim results coincide or are very close to the answers obtained by the mathematical manipulations performed manually.

- For quantitative comparison of the MATLAB, Multisim, and manual results, the 5τ criterion can be applied to the obtained graphs. In MATLAB, to find a 5τ interval, go to Tools window in the figure and click Data Tips. You will see boxes with X (time) and Y (magnitude) values, as shown in all the MATLAB versions of Figure 9.11. Drag the point through the graph to the desired magnitude value: You will obtain the corresponding time (in fact, 5τ) interval. Multisim also enables us to measure the experimentally obtained parameters accurately. For this, one cursor must be positioned at the initial instant t_{in}, where $v_{out}(t)=0V$, and the other must be placed at the final instant t_{fin}, where $v_{out}(t)=V_{in}(1-e^{-5})$, which permits measuring the interval. This is how we determine this time parameter on all Multisim graphs in Figure 9.11. Though MATLAB and Multisim graphs visually look different, their numerical parameters are very close.
- In this example, we deliberately use the traditional, straightforward *cover-up method* for finding the inverse Laplace transform residues. This method allows for demonstration of the relationship between an s-domain response and its residues. Regrettably, the technique requires tedious mathematical manipulations, and it is time-consuming. This statement is especially true for an underdamped case that leads to the complex poles. Fortunately, Chapter 8 discusses many other options to shorten finding the inverse Laplace transform constants. For example, Section 8.7 shows the following *useful formula* for getting A and θ when the answer is in the form $Ae^{-\alpha t}\sin(\omega t+\theta)$ as in the underdamped case:

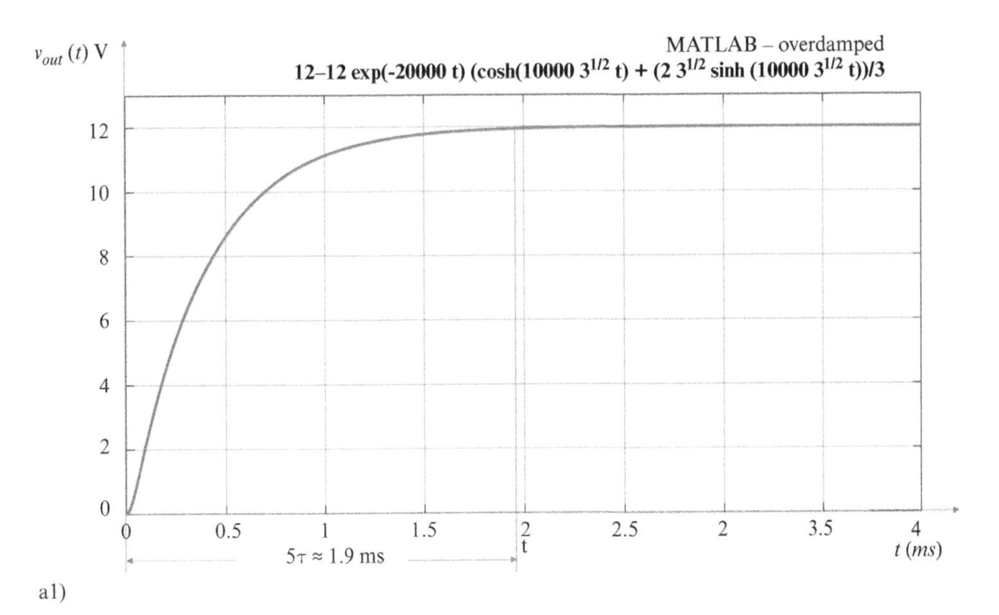

Figure 9.11 Three cases of step response of an RLC circuit in Example 9.5 with MATLAB (top) and Multisim (bottom): Overdamped (a1 and a2); critically damped (b1 and b2); underdamped (c1 and c2).

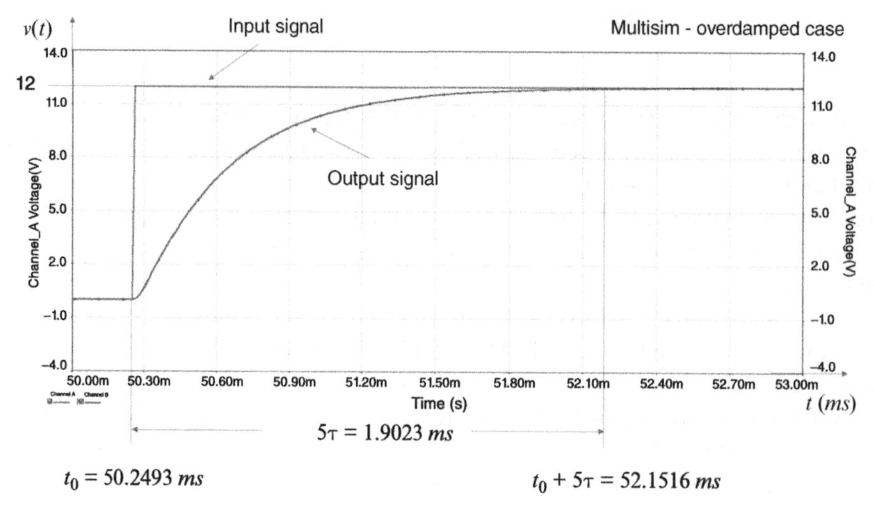

a2)

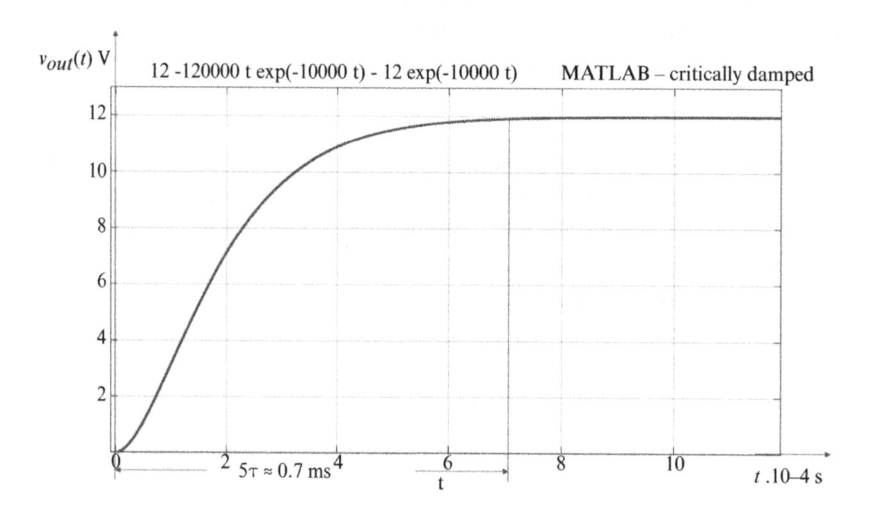

b1)

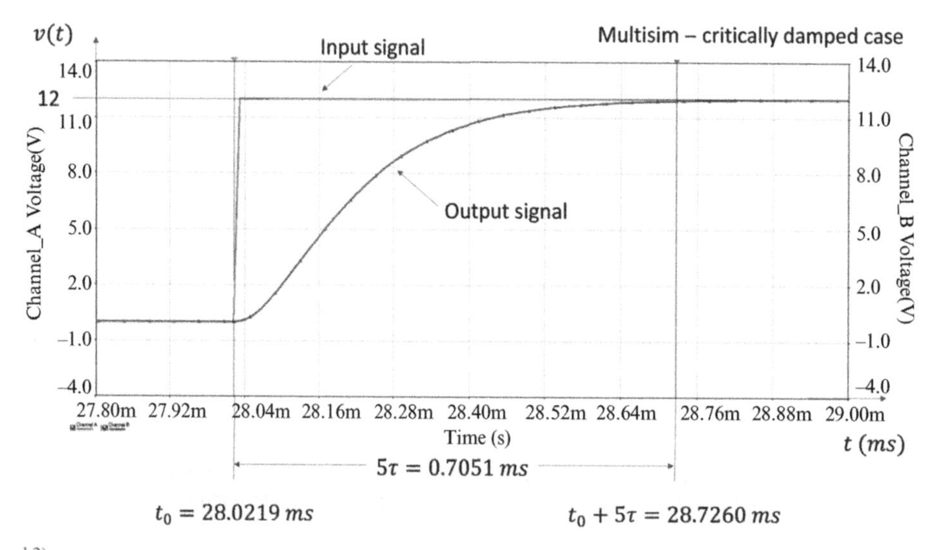

b2)

Figure 9.11 (Cont'd)

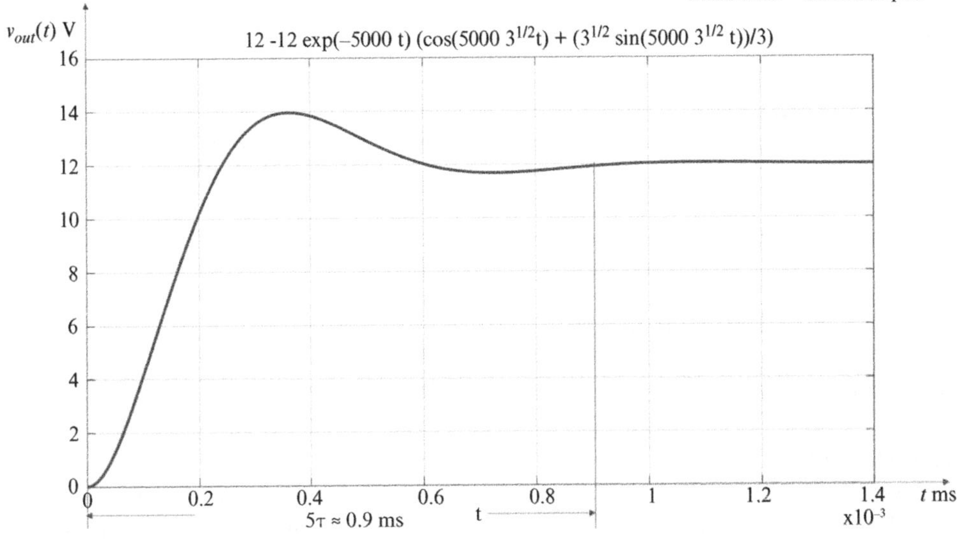

c1)

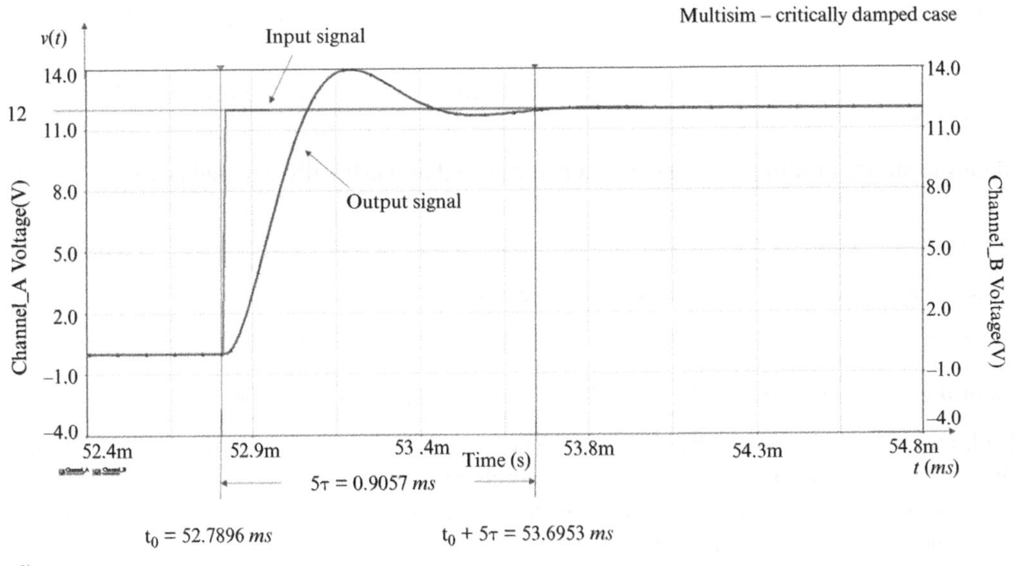

c2)

Figure 9.11 (Cont'd)

$$Ae^{j\theta} = A\angle\Theta = \frac{1}{\omega}\left(s^2 + bs + c\right)F(s)\Big|_{s=-\alpha+j\omega}. \tag{8.202R}$$

Here, $\omega = \omega_d = \sqrt{\omega_0^2 - \alpha_{ud}^2}$ is the radian frequency of the damping oscillations.

Applying (8.202) to Case 3 of Example 9.4 results in

$$Ae^{j\theta} = A\angle\Theta = \frac{1}{\omega_d}\left(s^2 + 2\alpha_{ud}s + \omega_0^2\right)\frac{\omega_0^2 V_{in}}{s\left(s^2 + 2\alpha_{ud}s + \omega_0^2\right)}\Bigg|_{s=\left(-0.5+j\sqrt{0.75}\right)10^4}$$

$$= \frac{1}{\sqrt{0.75}\cdot 10^4}\frac{\omega_0^2 V_{in}}{s}\Bigg|_{s=\left(-0.5+j\sqrt{0.75}\right)10^4} = \frac{1}{\sqrt{0.75}\cdot 10^4}\frac{12\times 10^8}{\left(-0.5\cdot 10^4 + j\sqrt{0.75}\cdot 10^4\right)}$$

$$= \frac{1}{\sqrt{0.75}\cdot 10^4}\frac{12\times 10^8\angle 0^0}{\left(\angle -60^0\right)10^4} = \frac{12\times 10^8\angle 0^0}{0.5\sqrt{3}10^8\angle -60^0} = 8\sqrt{3}\angle 60^0.$$

Thus, the sinusoidal member of the answer is

$$8\sqrt{3}\sin(\sqrt{0.75}\cdot 10^4 t + 60^0) = 8\sqrt{3}\cos\left(\sqrt{0.75}\cdot 10^4 t - 30^0\right),$$

which coincides with (9.64).

We urge you to try other techniques in finding the inverse Laplace transform in the case of complex poles discussed in Section 8.7, particularly the algebraic method. Also, don't forget to use a practical approach where the quadratic polynomial $s^2 + 2\alpha s + \omega$ is presenting as a sum of the full square and free member, $(s+\alpha)^2 + \omega_d$. This approach permits for the use of LT-12 in Table 7.1, where the member of a partial fraction expansion containing the quadratic polynomial is inverted into its sinusoidal time-domain counterpart immediately. Indeed, $L^{-1}\left\{\frac{A}{s^2 + 2\alpha s + \omega} = \frac{B\omega_d}{(s+\alpha)^2 + \omega_d}\right\} = Be^{-\alpha t}\sin(\omega_d t)$. Revisit Section 8.7 for reviving all the techniques to shorten finding the residues in overdamped and critically damped cases.

»»»»»»»»»»»»»»»»»»»»»»»»»»»»»»»»»»»»>

Example 9.6 Step response of a series RLC circuit with an initial condition.

Problem: Find the step response of a series RLC circuit discussed in Examples 9.4 and 9.5 and shown in Figure 9.10a if its capacitor was initially charged up to $V_0 = 4V$.

Prediction: Based on the superposition principle, we expect that the circuit's response will be a sum of natural (source-free) and forced (under an external excitation) responses.

Solution: Example 9.3 demonstrated two methods for solving such a problem. Since both natural and forced responses are found separately in Examples 9.4 and 9.5, applying the superposition method will be the most appropriate approach. The s-domain circuit schematic that includes an initial capacitor's charge is shown in Figure 9.12a. The s-domain schemes illustrating the superposition principle for this circuit are shown in Figure 9.12b.

The superposition principle is given in (9.15) as

$$V_{out}(s) = V_{out}^{zsr}(s) + V_{out}^{zir}(s). \tag{9.15R}$$

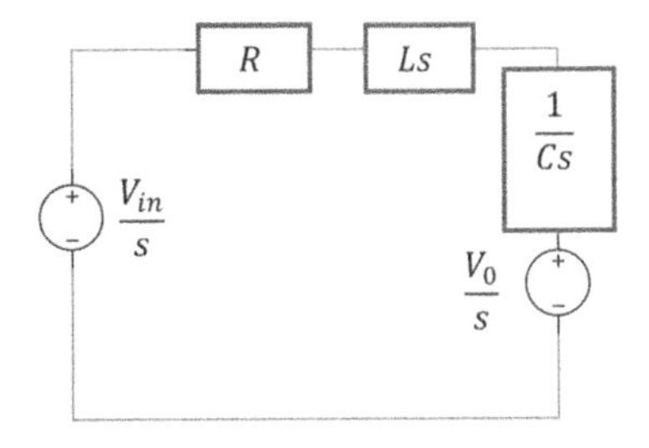

a)

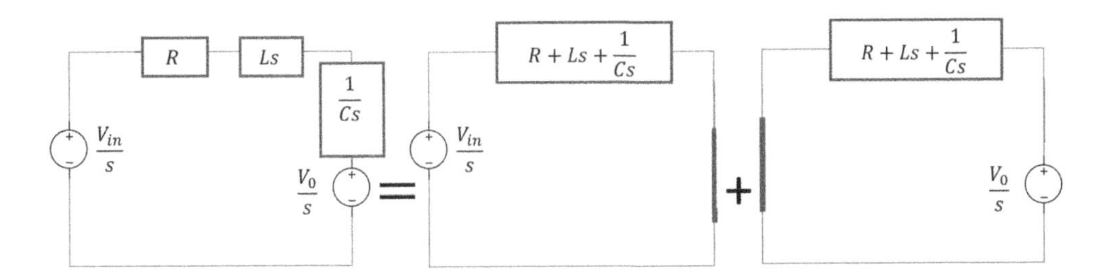

b)

Figure 9.12 The RLC circuit for Example 9.6: a) The s-domain model; b) the application of the superposition principle.

Recall that $V_{out}^{zsr}(s)$ is a *zero-state response* to a step input under zero initial condition, and $V_{out}^{zir}(s)$ is a *zero-input response* to an initial condition only. Figure 9.12b shows how this formula is implemented in the s-domain model.

As (9.15) and (9.16) state, the summation of the s-domain outputs translates to the corresponding sum in the time domain thanks to linearity of the inverse Laplace transform. Thus,

$$v_{out}(t) = v_{out}^{zsr}(t) + v_{out}^{zir}(t). \tag{9.16R}$$

We need to bear in mind that an RLC circuit can have three responses. Three zero-input responses are attained in Example 9.3 as follows:

overdamped case,

$$v_{out}^{od-zir}(t) = L^{-1}\left\{V_{out}^{od}(s)\right\} = A_1 e^{p_1 t} + A_2 e^{p_2 t} \approx 4.31e^{-2.710^3 t} - 0.31e^{-37.310^3 t}, \tag{9.38R}$$

critically damped case,

$$v_{out}^{cr-zir}(t) = L^{-1}\left\{V_{out}^{cr}(s)\right\} = V_0 e^{-\alpha_{cr} t} + \alpha_{cr} V_0 t e^{-\alpha_{cr} t} = 4e^{-10\cdot 10^3 t} + 4\cdot 10^4 t e^{-10\cdot 10^3 t}, \tag{9.44R}$$

and underdamped case,

$$v_{out}^{ud-zir}(t) = 4e^{-5000t}\left(\cos 5000\sqrt{3}t + \frac{1}{\sqrt{3}} \sin 5000\sqrt{3}t\right) = \frac{8}{\sqrt{3}} e^{-5000t} \cos\left(5000\sqrt{3}t - 30^0\right). \tag{9.48R}$$

Example 9.4 delivers all three *zero-state step responses* of this circuit in the following forms:

overdamped case,

$$v_{out}^{od-zsr}(t) = V_{in} + A_2 e^{p_1 t} + A_3 e^{p_2 t} = 12 - 12.93e^{-2.7 \cdot 10^3 t} + 0.93e^{-37.3 \cdot 10^3 t}, \tag{9.52R}$$

critically damped solution,

$$v_{out}^{cr-zsr}(t) = V_{in} - V_{in} e^{-\alpha_{cr} t} - V_{in} \cdot 10^4 te^{-\alpha_{cr} t} = 12 - 12e^{-10^4 t} - 12 \cdot 10^4 te^{-10^4 t}, \tag{9.59R}$$

and the underdamped case,

$$v_{out}^{ud-zsr}(t) = 12 - 12e^{-5000t}\left[\cos\left(5000\sqrt{3}t\right) + \frac{1}{\sqrt{3}}\sin\left(5000\sqrt{3}t\right)\right] \tag{9.60R}$$

or

$$v_{out}^{ud-zsr}(t) = 12 - e^{-5000t}[8\sqrt{3}\cos\left(5000\sqrt{3}t - 30^0\right)]. \tag{9.61R}$$

Now, we need to put together *ZIR* and *ZSR* results.

Applying the routine presented in Example 9.2 and visualized in Figure 9.12b to our case and summing as directed by (9.16) yield:

For the overdamped case,

$$v_{out}^{od}(t) = v_{out}^{od-zir}(t) + v_{out}^{od-zsr}(t) = 12 - 8.62e^{-2.7 \cdot 10^3 t} + 0.62e^{-37.3 \cdot 10^3 t}. \tag{9.66}$$

At $t = 0$, $v_{out}^{od}(0) = 4V$, and when $t \Rightarrow \infty$, $v_{out}^{od}(t) \Rightarrow 12V$, as it must be.

For the critically damped case,

$$v_{out}^{cr}(t) = v_{out}^{cr-zir}(t) + v_{out}^{cr-zsr}(t) = 12 - 8e^{-10 \cdot 10^3 t} - 8te^{-10 \cdot 10^3 t}. \tag{9.67}$$

One can readily verify that at $t = 0$, $v_{out}^{cr}(0) = 4V$, and when $t \Rightarrow \infty$, $v_{out}^{cr}(t) \Rightarrow 12V$. Therefore, both the initial and final conditions are met.

Finally, for the underdamped case,

$$v_{out}^{ud}(t) = v_{out}^{ud-zir}(t) + v_{out}^{ud-zsr}(t)$$

$$v_{out}^{ud}(t) = 12 - \frac{16}{\sqrt{3}}e^{-5000t}\cos\left(5000\sqrt{3}t - 30^0\right). \tag{9.68}$$

At $t = 0$, $v_{out}^{ud}(0) = 12 - \frac{16}{\sqrt{3}}\cos\left(-30^0\right) = 4V$, and when $t \Rightarrow \infty$, $v_{out}^{ud}(t) \Rightarrow 12V$; hence, both conditions are satisfied.

The problem is solved.

Discussion:

- This example highlights time and again that the initial conditions can be naturally integrated into a general solution for an RLC circuit based on the superposition principle. The circuit analysis in the s-domain makes the implementation of this principle transparent and straightforward.

>>

9.2.3 Parallel RLC Circuit

Investigation of a parallel RLC circuit doesn't add, of course, anything new to the concept of the s-domain circuit analysis. Some details, however, are worth the close investigation. In addition, the detailed circuit analysis of both series and parallel RLC circuits opens the gate to the analysis of any series-parallel circuit combinations. Let's start with an example.

Example 9.7 Step response of a parallel RLC circuit.

Problem: Consider the RLC parallel circuit shown in Figure 9.13a. Values R = 1($k\Omega$), L = 1 (mH), and C = 0.5 (mF) are fixed. Use the Laplace transform to find the circuit's response if $I_{in} = 1(A)$, and the circuit was initially relaxed.

Prediction: For this driven parallel RLC circuit, we can readily anticipate that the circuit response, $v_{out}(t) \equiv v_{C1}(t)$, will transit from zero (initial) value to 12 V (final) value and retain the final value as long as the input applies. This prediction, however, is trivial; we made it many times. Now, we can consider a more specific prediction: The switch is turned on at $t = 0^+$; at this instant, the step-like voltage is applied to the circuit. The current $i_L(t)$ starts flowing through an inductor and cause the voltage drop $v_L(t) = L\frac{di_L(t)}{dt}$. When $v_L(t)$ reaches its steady value (12 volts in our example), $i_L(t)$ becomes a dc current, and $v_L(t)$ will go to zero. Such a function is shown in Figure 9.2. During this transient process, the voltage across the capacitor will change too. (Remember, for a parallel circuit $v_L(t) = v_C(t) = v_R(t)$.) Thus, current through a capacitor will flow

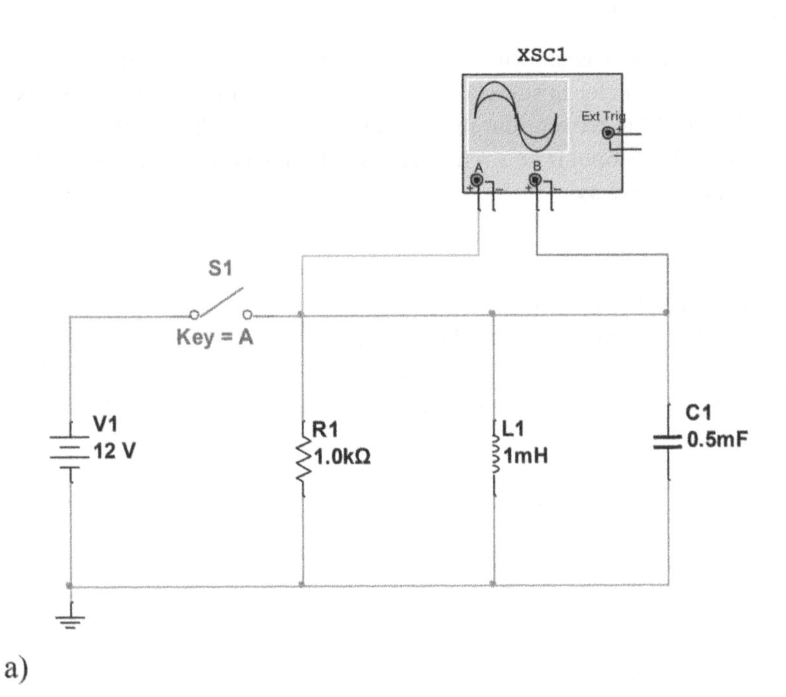

Figure 9.13 Step response of a parallel RLC circuit for Example 9.7: a) Experimental setup; b) the s-domain model.

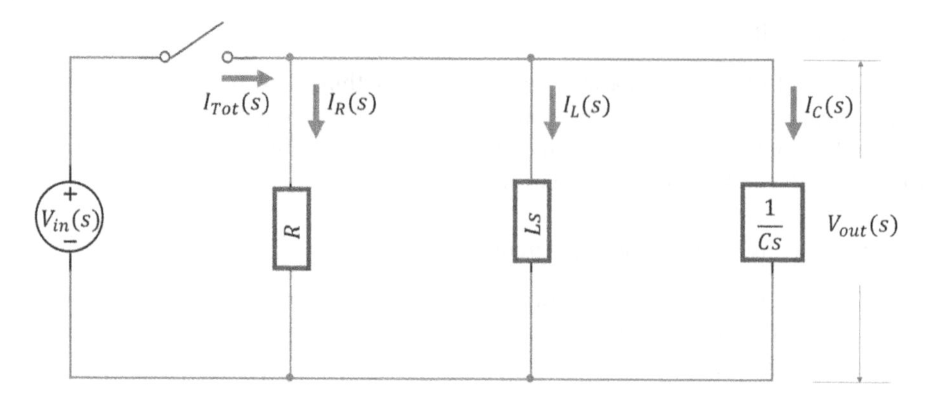

b)

Figure 9.13 (Cont'd)

because $i_C(t) = C\frac{dv_C(t)}{dt}$. However, eventually this current will diminish because $v_C(t)$ reaches its final dc value of 12 V. Therefore, initially, currents and voltages of this parallel RLC circuit will change, and we must calculate how they will do so. Finally, though, an ideal inductor will work as a shorten circuit and draw the maximum current from the source and an ideal capacitor will act as an open circuit and block any current at all.

Solution: The s-domain circuit model is shown in Figure 9.13b. We can derive its s-domain equation using the *duality principle*. Refer to subsection Duality Principle in Section 3.3 and see Section 5.1 and Examples 5.2 and 5.5 to refresh your memory on this subject. The s-domain model of a series RLC circuit with dc voltage input is drawn in Figure 9.10b, and its equation is presented in (9.49), which we slightly modify here as

$$V_{out}(s)\left(s^2 + 2\alpha s + \omega_0^2\right) = \omega_0^2 V_{in}(s). \tag{9.69}$$

According to the duality principle, to transfigure this equation into a parallel RLC circuit equation, we need to replace all electrical entities used for a series circuit with their duals applied in a parallel circuit. See Table 3.6. Following that table, we replace in (9.69) all $V(s)$ by $I(s)$, all $Z(s)$ by $Y(s)$, R by G, L by C and vice versa, division by multiplication, and vice versa. These transfigurations result in

$$I_{out}(s)\left(s^2 + 2\alpha s + \omega_0^2\right) = \omega_0^2 I_{in}(s). \tag{9.70}$$

You recall that we applied the duality principle to derive the differential equation of a parallel RLC circuit in the time domain in Section 5.1. The result was that we can use the same form, Equation 9.69, for both series and parallel RLC circuits. The only difference is that for a *series RLC circuit* $\alpha_{ser} = \frac{R}{2L}$, whereas for a *parallel RLC circuit* $\alpha_{par} = \frac{1}{2RC}$. What's more, Equation 5.6b states we can use (9.69) for a parallel RLC circuit provided that $\alpha \equiv \alpha_{par} = \frac{1}{2RC}$. (Is $\omega_0^2 = \frac{1}{LC}$ the same or different for these circuits? Explain.)

Since (9.69) is the governing equation for a parallel RLC circuit, all the following operations will be the same as the preceding manipulations done for a series circuit. We leave it for our readers to repeat them with the modified $\alpha \equiv \alpha_{par} = \frac{1}{2RC}$. Feel free to change the values of any or all components to achieve the most expressive results. Investigate all three—overdamped, critically damped, and underdamped—cases.

Discussion:

- Remember that if the *initial conditions* are not zero, they can be included in the consideration by the methods displayed in Figures 9.4 and 9.5.
- Below is the other example showing that knowledge of how to deal with a series and parallel RLC enables us to analyze any second-order circuit:

Problem: Derive the s-domain equation of a series-parallel RLC circuit shown in Figure 9.14a.

Solution: To solve this problem, we build the s-domain model of a given circuit shown in Figure 9.14b. This model leads to the following system of the first-order s-domain equations:

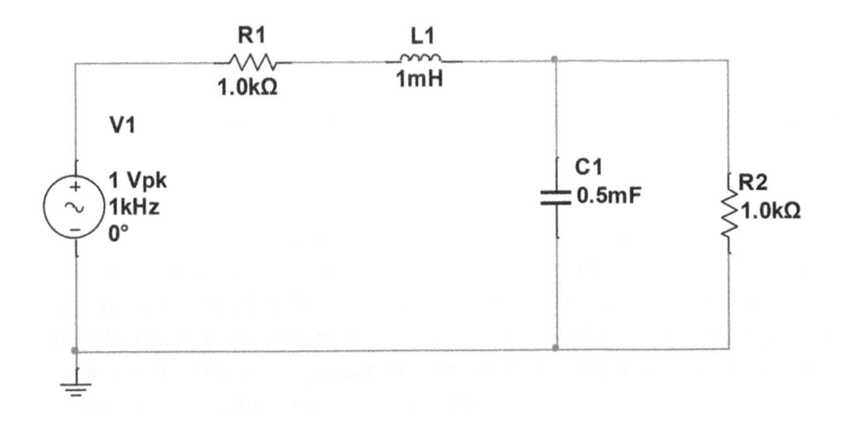

a)

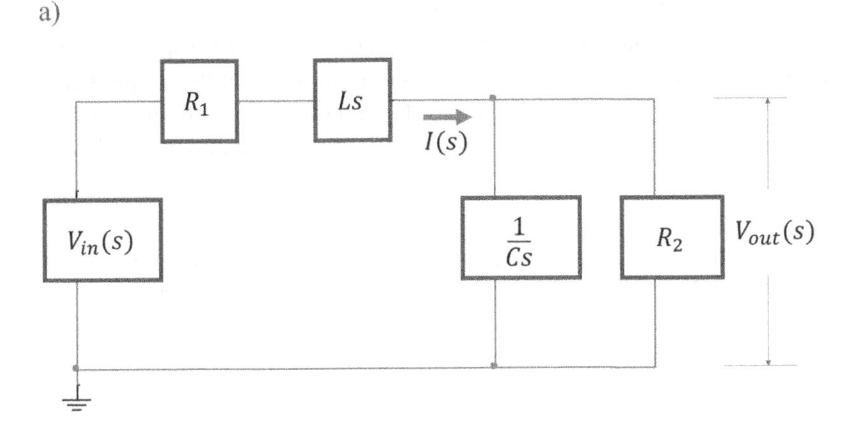

b)

Figure 9.14 Analysis of series-parallel RLC circuit in the s-domain: a) Circuit's schematic; b) the s-domain model of the circuit.

$$\begin{cases} R_1 I(s) + LsI(s) + V_{out}(s) = V_{in}(s) \\ CsV_{out}(s) + \dfrac{1}{R_2} V_{out}(s) = I(s) \end{cases} . \tag{9.71}$$

After straightforward substitutions, (9.71) can be reduced to the following second-order equation:

$$\left(s^2 + 2\alpha s + \left(1 + \frac{R_1}{R_2}\right)\omega_0^2\right) V_{out}(s) = \omega_0^2 V_{in}(s), \tag{9.72}$$

where $\alpha = \frac{1}{2}\left(\frac{R_1}{L} + \frac{1}{R_2 C}\right)$ is new and $\omega_0^2 = \frac{1}{LC}$ is as usual. Note that α contains elements of damping coefficients of series and parallel RLC circuits simultaneously, which is justifiable for a series-parallel circuit.

- Now, Equation 9.72 can be treated as any other second-order equation discussed in this and the preceding chapters, proving this example's point.
- The problem is solved.

9.3 Circuit Analysis in the s-Domain with Transfer Functions

9.3.1 The Circuit's Response—A Review

You will recall that the objective of this book is building the methods and techniques to find a circuit's response to a particular excitation. Obviously, the circuit's topology and element values must be given. This goal was stated numerous times previously and is visualized again in Figure 9.15. An electrical circuit is considered as a system whose input is an excitation, and the output is the circuit's response to this excitation. For example, when an input voltage, $v_{in}(t)$, is applied to a series RLC circuit, it causes output voltage, $v_{out}(t)$, whose waveform and parameters depend on the nature of $v_{in}(t)$, R, L, and C values, and circuit's schematics. Since this book considers a transient response, the circuit's mathematical description requires differential equations.

In this book, we've studied two main methods of finding $v_{out}(t)$: the time-domain approach and the s-domain technique. Studying both methods, we wish to determine and find the simplest and the most effective routine for finding $v_{out}(t) - v_{in}(t)$ relationship. In the time domain, such a strategy seems to be the convolution-integral technique because it gives the relationship in search as

$$v_{out}(t) = h(t) \star v_{in}(t), \tag{6.1R}$$

where $\star$ is the symbol of convolution integral. However, the deliberation in Chapter 6 showed that behind this superficial simplicity lies the complexity of the mathematical manipulations. Is any method in the s-domain looks as straightforward as (6.1) but leads to much simpler mathematical operations? The answer is yes, and this approach is described as follows:

$$V_{out}(s) = H(s) \cdot V_{in}(s). \tag{9.73}$$

Here, $H(s)$ is a *transfer function* that models the circuit's reaction to all possible inputs. This section delves into understanding what the transfer function is, its benefits, and its drawbacks. Note that transfer function is also known as *network function* or *system function*.

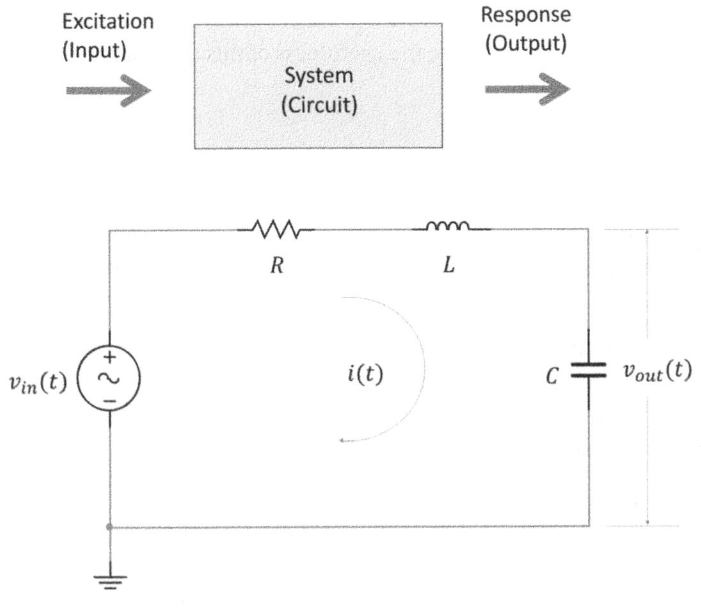

Figure 9.15 The concept of a circuit's response: a) General view (top); b) example with a series RLC circuit (bottom).

9.3.2 Transfer Function—What It Is

Recall the step-response problem of a series RC circuit discussed in Example 9.1. Its s-domain equation is shown in (9.3b), which we repeat here in a slightly modified form:

$$V_{out}(s) = \frac{\alpha}{(s+\alpha)} V_{in}(s). \tag{9.3bR}$$

Denoting

$$\frac{\alpha}{(s+\alpha)} = H(s), \tag{9.74}$$

we receive Equation 9.73. Thus, the *transfer function* of a series RC circuit is given by Equation 9.74. Following this logic, we can derive from (9.69), $V_{out}(s)(s^2 + 2\alpha s + \omega_0^2) = \omega_0^2 V_{in}(s)$,

$$V_{out}(s) = H(s)V_{in}(s),$$

where

$$H(s) = \frac{\omega_0^2}{(s^2 + 2\alpha s + \omega_0^2)}. \tag{9.75}$$

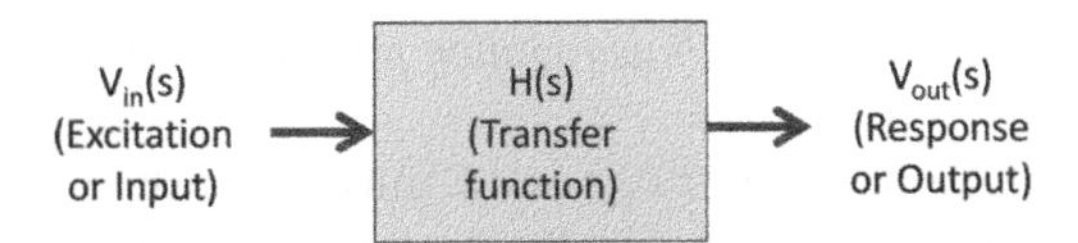

Figure 9.16 Visual presentation of a transfer function.

Is this it? You can exclaim! It seems that this additional notation doesn't bring any new information and knowledge, but the current section should prove the usefulness of this new entity.

Equation 9.73 is often shown in the other form,

$$H(s)=\frac{V_{out}(s)}{V_{in}(s)}. \tag{9.76}$$

Figure 9.16 visualizes both (9.73) and (9.76).

Thinking about Equation 9.73 and Figure 9.16, you comprehend that they carry two important statements:

1) A circuit's transfer function, $H(s)$, contains all information about the circuit; therefore, as soon as a transfer function is known, the circuit's output can be readily found *for any input.*
2) Equation 9.73 and Figure 9.16 demonstrate that a transfer function "transmits" the input signal through a circuit to the output in the s-domain.

The above statements mean that we must have the s-domain equation relating a circuit's input and output, which stems from the circuit's differential equation, a genuinely natural circuit description. Thus, the differential equation is grassroots, the s-domain equation is its transfiguration into the s-domain, and the transfer function is another transformation of the latter in the s-domain. Do we need this third layer of reconversion of the fundamental tool? This section is intended to prove that we do.

Let's consider the above statements closely.

1. The transfer function depends only on the circuit's schematic and its parameters. For example, a series RC circuit has

$$H_{RC}(s)=\frac{V_{out}(s)}{V_{in}(s)}=\frac{\alpha}{s+\alpha}=\frac{1}{RCs+1},$$

and the transfer function of a series RLC circuit is

$$H_{RLC}(s)=\frac{V_{out}(s)}{V_{in}(s)}=\frac{\omega_0^2}{\left(s^2+2\alpha s+\omega_0^2\right)}=\frac{1}{LC\left(s^2+\frac{sR}{L}+\frac{1}{LC}\right)}.$$

Since the values of $R, L,$ and$C$ are given, a transfer function is determined as soon as the s-domain circuit's equation is derived.

2. As the $H(s)$ exists in the s-domain, obtaining the time-domain output, $v_{out}(t)$, requires performing the inverse Laplace transform. This is a price to be paid for the convenience and ease of mathematical operations in the s-domain, as discussed in Chapters 8 and 9.

The concept of transfer function works well when applied to the circuits (systems) under the following *restrictions*:

1) The circuit must satisfy linear, time-invariant (LTI) system requirements:
   a) Linearity property is discussed in Section 3.3 under the title *Superposition principle.* Figure 9.17 reminds this property visually.
   b) Time-invariant practically means that the circuit's parameters $R, L,$ and C must be constant.
2) The circuit must be a lumped system, which means it cannot include distributed parameters.
3) The concept of transfer function works easier at zero initial conditions.

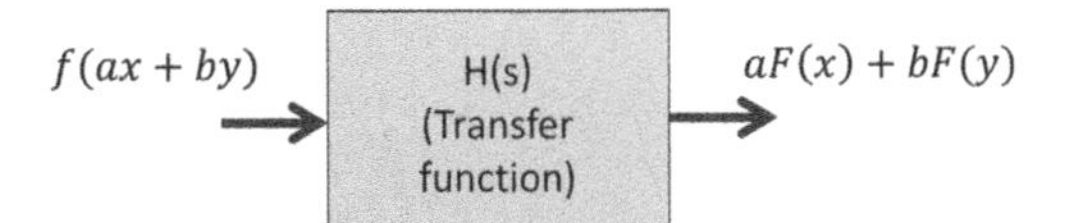

Figure 9.17 Illustration of the linearity concept.

It's time to consider examples.

Example 9.8 Transfer functions of the first-order circuits.

Problem: Find the transfer functions of (a) series RC circuit shown in Figure 9.18a and (b) series RL circuit shown in Figure 9.18b.

Solution:
a) The s-domain equation of a series RC circuit can be found by applying the voltage-divider rule as

$$V_{out}(s) = \frac{Z_C(s)}{Z_R(s) + Z_C(s)} V_{in}(s) = \frac{\alpha}{(s+\alpha)} V_{in}(s)$$

because $\alpha = \frac{1}{RC}$. Following the transfer function definition given in (9.76), we find

$$H_{RC}(s) = \frac{V_{out}(s)}{V_{in}(s)} = \frac{\alpha}{(s+\alpha)} = \frac{2}{s+2}.$$

b) Similarly, voltage-divider rule produces for the RL circuit the following:

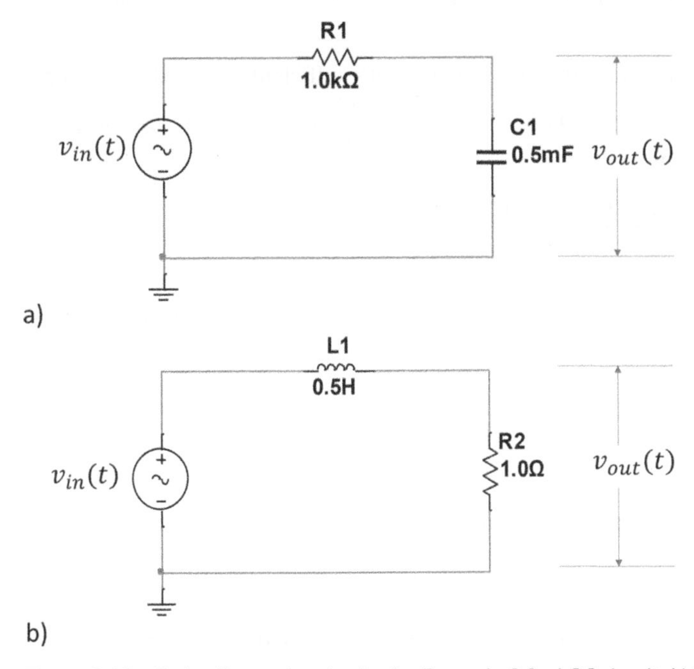

Figure 9.18 Series first-order circuits for Example 9.8: a) RC circuit; b) RL circuit.

$$V_{out}(s) = \frac{Z_R(s)}{Z_R(s) + Z_L(s)} V_{in}(s) = \frac{\alpha}{(\alpha + s)} V_{in}(s)$$

Therefore,

$$H_{RL}(s) = \frac{V_{out}(s)}{V_{in}(s)} = \frac{\alpha}{(\alpha + s)} = \frac{2}{2 + s}.$$

The problem is solved.

Discussion:

- Why do two different circuits have identical transfer functions? Because they perform similar operations. If you present them the same input signal—step, sinusoidal, or anything else—they produce the same result. For example, for a sinusoid, they work as lowpass filters. Thus, this example demonstrates that a transfer function reflects the circuit's nature.
- Do all first-order circuits have identical transfer functions? We suggest you find the answer to this question by comparing, for example, series and parallel RC or RL circuits.

Question What is a dimension (unit) of a transfer function?

How can a transfer function help with finding the circuit's response? Let's consider an example.

Example 9.9 Finding the step and sinusoidal response of RC and RL circuit using their transfer functions.

Problem: Consider RC and RL circuits discussed in Example 9.8 and shown in Figure 9.18. Using their transfer functions, find (a) step response to $V_{in}(t) = 12(V)$ and (b) sinusoidal response to $v_{in}(t)[V] = 12\sin(8t)$.

Solution: Since both circuits have identical transfer functions, it's sufficient to consider just one.

a) A step input is given as $v_{in}(t)[V] = V_{in} \cdot u(t) \Rightarrow V_{in}(s) = \frac{V_{in}}{s} = \frac{12}{s}(volt \cdot second)$. See Table 7.3. Plugging this $V_{in}(s)$ into (9.73) results in

$$V_{out}(s) = H_{RC}(s) \cdot V_{in}(s) = \frac{\alpha}{(s + \alpha)} \frac{V_{in}}{s} = \frac{24}{s(s + 2)}$$

because $\alpha = \frac{1}{RC} = 2\left(\frac{1}{s}\right)$. The time-domain output can be found by finding the inverse Laplace transform of $V_{out}(s)$; i.e.,

$$v_{out}(t) = L^{-1}\{V_{out}(s)\} = L^{-1}\left\{\frac{24}{s(s + 2)}\right\} = L^{-1}\left\{\frac{A_1}{s} + \frac{A_2}{(s + 2)}\right\}.$$

We understand that the time-domain response should take the following form:

$$v_{out}(t) = A_1 e^{p_1 t} + A_2 e^{p_2 t} = A_1 e^{0t} + A_2 e^{-2t},$$

where p_1 and p_2 are the poles. The procedure for finding A_1 and A_2 is well discussed in Chapter 8. Thus,

$$A_1 = \left.\frac{s24}{s(s+2)}\right|_{s=0} = 12$$

and

$$A_2 = \left.\frac{(s+2)24}{s(s+2)}\right|_{s=-2} = -12.$$

Thus,

$$v_{out}(t) = 12\left(1 - e^{-2t}\right),$$

as it should be.

b) A sinusoidal signal must be written in the s-domain as

$$v_{in}(t) = V_{in}\sin(\omega t) \Rightarrow V_{in}(s) = \frac{\omega V_{in}}{(s^2 + \omega^2)}.$$

For this example,

$$v_{in}(t) = 12\sin(8t) \Rightarrow V_{in}(s) = \frac{96}{s^2 + 64}.$$

Hence,

$$V_{out}(s) = H(s) \cdot V_{in}(s) = \frac{\alpha}{(s+\alpha)} \cdot \frac{\omega V_{in}}{\left(s^2 + \omega^2\right)} = \frac{192}{(s+2)\left(s^2 + 64\right)}.$$

To find $v_{out}(t)$, we need to apply the inverse Laplace transform as

$$v_{out}(t) = L^{-1}\left\{V_{out}(s)\right\} = L^{-1}\left\{\frac{192}{(s+2)\left(s^2+64\right)}\right\}.$$

We learn in Chapter 8 that the form of the time-domain response in this case must be

$$v_{out}(t) = Ae^{-\alpha t} + Bsin(\omega t + \theta).$$

(It helps to review Example 8.7.) Hence, for our example,

$$v_{out}(t) = Ae^{-2t} + Bsin(8t + \theta).$$

The residues are found as

$$A = (s+\alpha)(V_{out}(s))\big|_{s=-\alpha} = \left.\frac{\alpha}{(s+\alpha)} \cdot \frac{\omega V_{in}}{\left(s^2+\omega^2\right)}\right|_{s=-\alpha}$$

$$= (s+2)\left.\left(\frac{192}{(s+2)\left(s^2+64\right)}\right)\right|_{s=-2} = 2.82$$

and

$$B\angle\theta = \frac{1}{\omega}\left(\left(s^2+\omega^2\right)(V_{out}(s)\right)\Big|_{s=+j\omega} = \frac{\alpha V_{in}}{(s+\alpha)}\Bigg|_{s=j\omega} = \frac{\alpha V_{in}}{(\alpha+j\omega)} = 0.12\angle-76^0.$$

Therefore, the answer is

$$v_{out}(t) = Ae^{-\alpha t} + Bsin(\omega t+\theta) = 2.82e^{-2t} + 0.12sin\left(8t-76^0\right).$$

The problem is solved.

Discussion:

- As Example 9.8 demonstrates, a transfer function doesn't significantly facilitate finding the time-domain output because we still must carry out all the inverse Laplace transform operations. However, the transfer function's benefits will still be seen soon.
- Examples 9.7 and 9.8 stress that we must be proficient in all the s-domain operations, especially in finding the inverse Laplace transform of the most common signals.

Example 9.10 The use of Transfer Function for Designing a Bandpass Filter.

We borrow the main parts of this example from MATLAB. Follow the link below https://www.mathworks.com/help/control/ug/analyzing-the-response-of-an-rlc-circuit.html.

Problem: Consider an RLC bandpass filter shown in Figure 9.19.
1) Find its transfer function manually and with MATLAB.
2) Use the derived transfer function to improve the filter's characteristic.
Bear in mind that MATLAB denotes transfer function as $G(s)$.

Solution: 1. To derive the RLC filter's transfer function, refer to Figures 9.14a and 9.14b in Example 9.6. Following that pattern, we write the following system of the first-order equations for the s-domain circuit's model shown in Figure 9.19b:

$$\begin{cases} V_{in}(s) = Z_R I(s) + V_{out}(s) \\ I(s) = \dfrac{V_{out}(s)}{Z_L} + \dfrac{V_{out}(s)}{Z_C} \end{cases}. \tag{9.77}$$

It's not difficult to obtain from (9.77) the following second-order s-domain equation describing the circuit in question

$$V_{out}(s) = \frac{\alpha s}{\left(s^2+\alpha s+\omega_0^2\right)} V_{in}(s), \tag{9.78}$$

where $\alpha = \dfrac{1}{RC}$ and $\omega_0^2 = \dfrac{1}{LC}$. The needed transfer function naturally stems from (9.78) as

$$H(s) = \frac{V_{out}(s)}{V_{in}(s)} = \frac{\alpha s}{\left(s^2+\alpha s+\omega_0^2\right)}. \tag{9.79}$$

Equation 9.79 is the transfer function derived manually.

To find the RLC bandpass filter's transfer function with MATLAB, we use the given component values and the following code:

```
>> %|R=L=C=1|:
R = 1; L = 1; C = 1;
G = tf([1/(R*C) 0],[1 1/(R*C) 1/(L*C)])
G =
s
-----
s^2 + s + 1
Continuous-time transfer function.
```

The MATLAB G(s) coincides with the manually-derived H(s).

Exercise Find the transfer function of the RLC circuit in Figure 9.19a whose R = 2 (Ω), L = 3 (H), and C = 4 (F).

2. The main characteristic of a bandpass filter is the width of its passing band, W_{PB}. Improvement of this characteristic means making it as narrow as possible. In our transfer function, *LC*

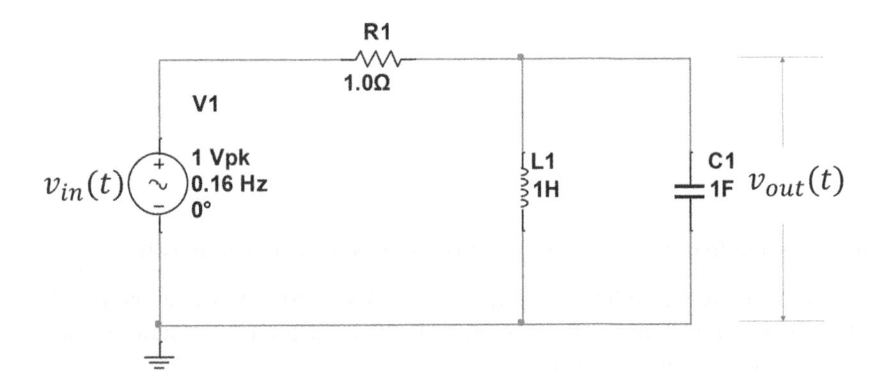

a)

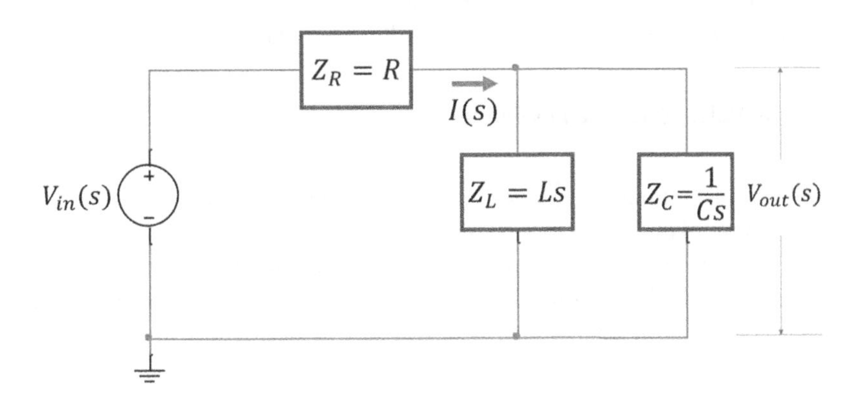

b)

Figure 9.19 The RLC bandpass filter for Example 9.10: a) Circuit's schematic; b) the s-domain model; c) MATLAB built filter's output characteristics.

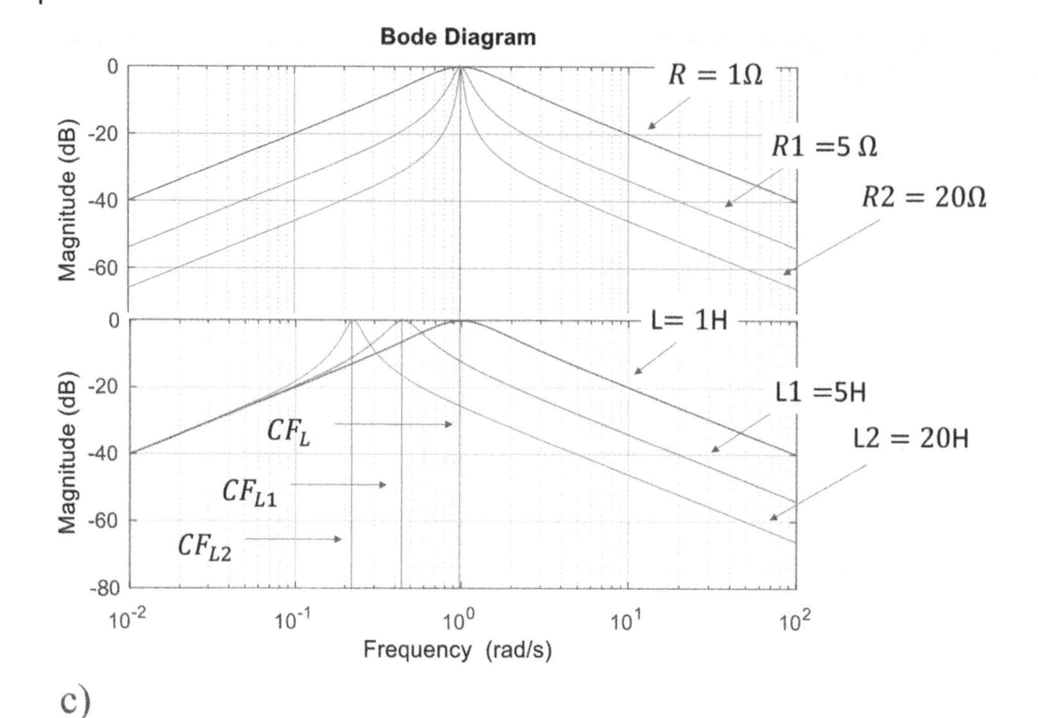

Figure 9.19 (Cont'd)

controls the central passing frequency and RC governs the passing band width as $W_{PB} \sim \frac{1}{RC}$. Therefore, we need to increase RC to improve W_{PB}. Figure 9.19c (top) shows how the filter's characteristics improve with R increasing. Be aware that the Bode diagram is a plot drawn in a logarithmic scale as shown in Figures 9. 19b and 9.19c.

3. We can tune the bandpass filter by changing its central frequency, CF. Figure 9.19c (bottom) shows how CF changes when the inductance varies from 1 to 5 to 20 henry.

The point here is to show how a circuit's transfer function helps to control its output (response) in the desired direction, which is vitally important in designing multistage circuits.

9.3.3 Transfer Function—Poles, Zeroes, and s-Plane

9.3.3.1 Poles and Zeroes

The real benefit of using a transfer function comes from analyzing its poles. But first, let's recall that we already examined the relationship between the partial fractions' poles and the corresponding time-domain function. We found that, depending on the nature of the poles, we can obtain various inverse Laplace transforms (time-domain functions, or waveforms). Specifically, zero pole means a constant in the time domain, negative real pole invokes a decaying exponential function, a pure imaginary pole calls for a sinusoid, and complex pole results in an exponentially decaying sinusoid. All these findings are summarized in Table 8.3; it's worthwhile revisiting this table.

A transfer function, however, enables a deeper look at the relationship between the s-domain poles and the time-domain response of a given circuit. For this, let's formalize the definition of a *transfer function*.

- Transfer function consists of two polynomials: nominator, N(s) and denominator D(s):

$$H(s)=\frac{N(s)}{D(s)}=\frac{a_n s^n+a_{n-1}s^{n-1}+a_{n-2}s^{n-2}+\ldots+a_1 s+a_0}{b_m s^m+b_{m-1}s^{m-1}+b_{m-2}s^{m-2}+\ldots+b_1 s+b_0}, m>n. \quad (9.80)$$

- The solutions to the characteristic equation $N(s)=0$ are the numerator *roots* called *zeros*, $z_1, z_2, z_3,\ldots, z_n$; accordingly, the solutions to $D(s)=0$ are the roots of the denominator called *poles*, $p_1, p_2, p_3,\ldots, p_n$.
- Consequently, a transfer function can be presented in a *factored form* as

$$H(s)=\frac{N(s)}{D(s)}=\frac{A(s-z_1)(s-z_2)(s-z_3)\ldots(s-z_n)}{(s-p_1)(s-p_2)(s-p_3)\ldots(s-p_n)}. \quad (9.81)$$

Here, $A=\frac{a_n}{b_m}$ is called *gain*. For example, consider (9.79), which for $\alpha=5$ and $\omega_0^2=4$ takes the following factored form:

$$H(s)=\frac{\alpha s}{(s^2+\alpha s+\omega_0^2)}\Rightarrow\frac{5s}{(s^2+5s+4)}=\frac{5s}{(s+1)(s+4)},$$

where A = 5, $z_1=0$, $p_1=-1$, and $p_2=-4$.

- It's significant to understand that a transfer function can be traced directly to the circuit's differential equation. For example, the differential equation describing a series RLC circuit is given by

$$v''_{out}(t)+2\alpha v'_{out}(t)+\omega_0^2 v_{out}(t)=\omega_0^2 v_{in}(t). \quad (5.6b)$$

Applying the Laplace transform and taking the output-to-input ratio in the s-domain, we obtain

$$H(s)=\frac{V_{out}(s)}{V_{in}(s)}=\frac{\omega_0^2}{(s^2+\alpha s+\omega_0^2)},$$

as in (9.75).

- Coefficients a_k and b_k in (9.80) are the real numbers, and therefore, z_k and p_k must be either real or complex-conjugate pairs as $p_k=\sigma_k+j\omega_k$ and $p^*_k=\sigma_k-j\omega_k$.

To sum up, the gain, zeroes, and poles fully describe a transfer function, which, in turn, completely describes a circuit because it stems directly from the circuit's differential equation. Therefore, the gain, zeroes, and poles present the circuit in its entirety in the s-domain.

9.3.3.2 The s-Plane

You will recall that *s* is a complex frequency defined as

$$s = \sigma + j\omega. \tag{7.2R}$$

It's customary to present the zeroes and poles graphically using the *s-plane*, whose horizontal axis is σ, the real part of a complex frequency, and the vertical axis is $j\omega$, its imaginary part.

For example, consider the following transfer function,

$$H(s) = \frac{(s-(-3))}{(s-(-2))(s-(-4+j5))(s-(-4-j5))},$$

whose zeroes and poles are shown in Figure 9.20. Specifically, zero -3 is located on the negative half of real axis σ; real pole -2 is nearby; complex pole $-4+j5$ is on the intersection of real negative -4 and imaginary positive 5 lines. Location of the complex conjugate pole $-4-j5$ can be found similarly.

The critical members of the s-plane consideration are *poles* because they are the roots of a characteristic equation and, therefore, determine the output behavior. Indeed, the transfer function of a first-order circuit is

$$H(s) = \frac{\alpha}{s+\alpha},$$

its characteristic equation is

$$s + \alpha = 0,$$

and the root of the characteristic equation is

$$p = -\alpha.$$

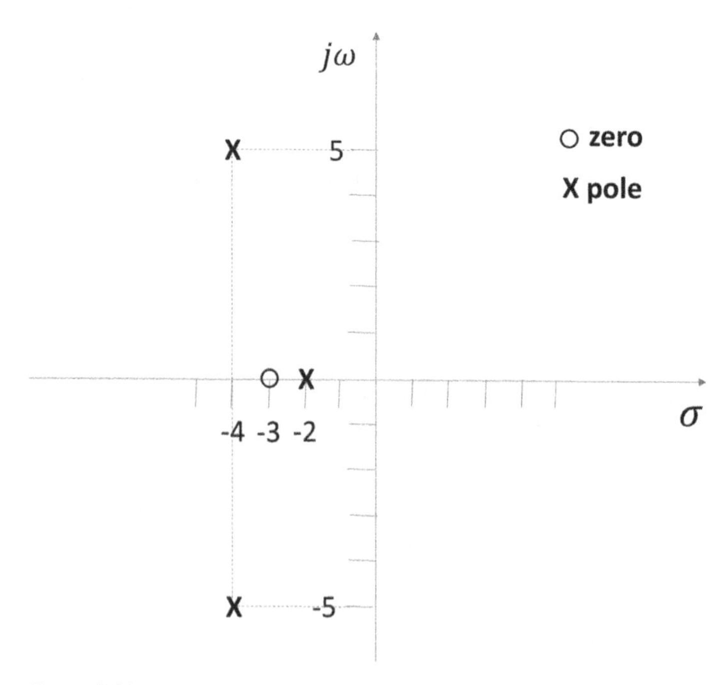

Figure 9.20 The s-plane and examples of locations of zeroes and poles.

Thus, we know that the natural time domain response of such a circuit is

$$v_{out}(t) = Ae^{pt} = Ae^{-\alpha t}.$$

Likewise, the second-order circuit's characteristic equation is

$$s^2 + 2\alpha s + \omega_0^2 = 0,$$

and its roots are

$$p_{1,2} = -\alpha \pm \sqrt{(\alpha^2 - \omega_0^2)}.$$

The time-domain response takes the form

$$v_{out}(t) = A_1 e^{p_1 t} + A_2 e^{p_2 t} = A_1 e^{(-\alpha + \sqrt{(\alpha^2 - \omega_0^2)})t} + A_2 e^{(-\alpha - \sqrt{(\alpha^2 - \omega_0^2)})t},$$

or $v_{out}(t) = A_1 e^{(-4+j5)t} + A_2 e^{(-4+j5)t}$, as Figure 9.20 exemplifies.

Note that a transfer function characterizes the property of a circuit itself. Therefore, the transfer function presents a circuit's natural (source-free) response. To find the circuit's response to a given source, we need to multiply the input and the transfer function, as (9.74) states.

We can reconstruct the transfer function if its *zeroes* and *poles* are known on an s-plane. Consider, for example, Figure 9.21 that shows one zero, $z = 4$, and three poles, $p_1 = 0$, $p_2 = -2 + j3$, and $p_3 = -2 - j3$. From these zero and poles, we can recreate the original transfer function as

$$H(s) = \frac{s - 4}{s\left(s - (-2 + j3)\right)\left(s - (-2 - j3)\right)}.$$

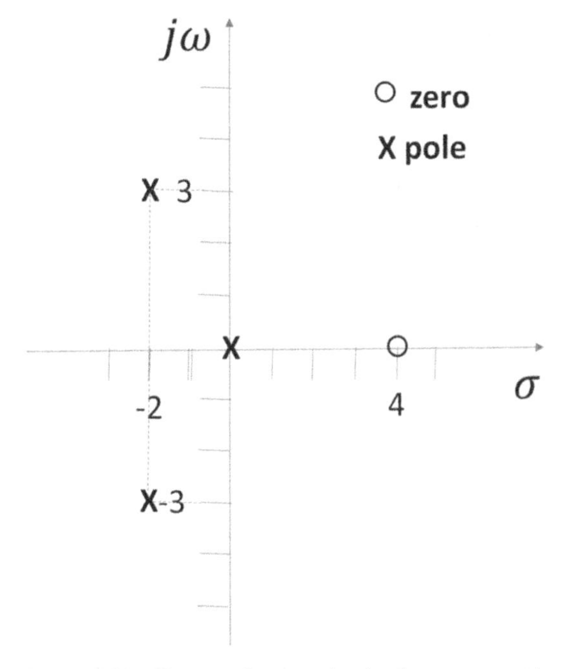

Figure 9.21 The set of zero and poles for reconstruction of the transfer function.

Table 9.3 Poles on the s-plane and corresponding time-domain responses.

Pole nature	Pole location	Poles in s-plane	Natural response $v_{out}(t)$
Zero	Neither	$j\omega$, σ	$v_{out}(t)$, V_{out}, t (s)

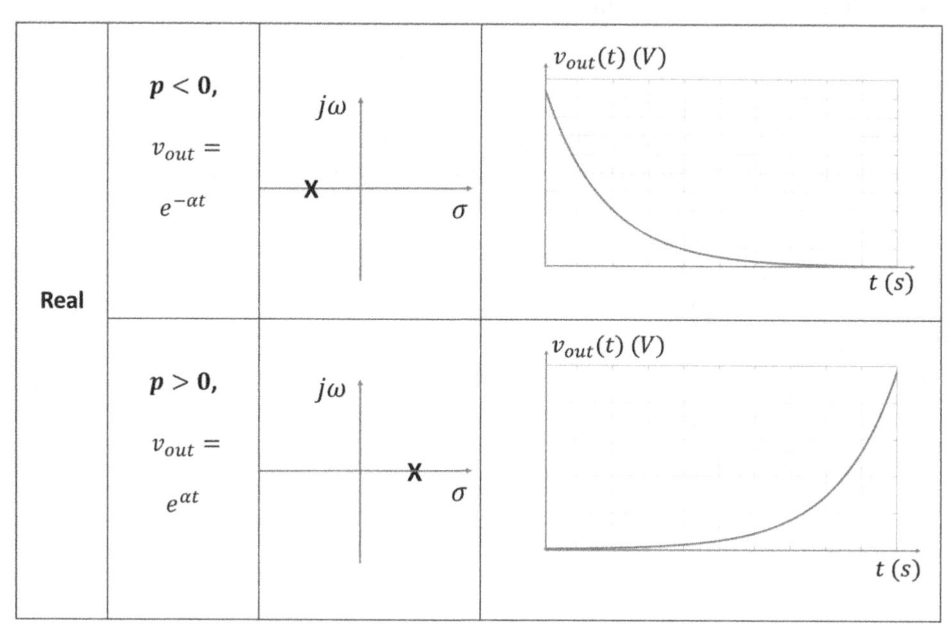

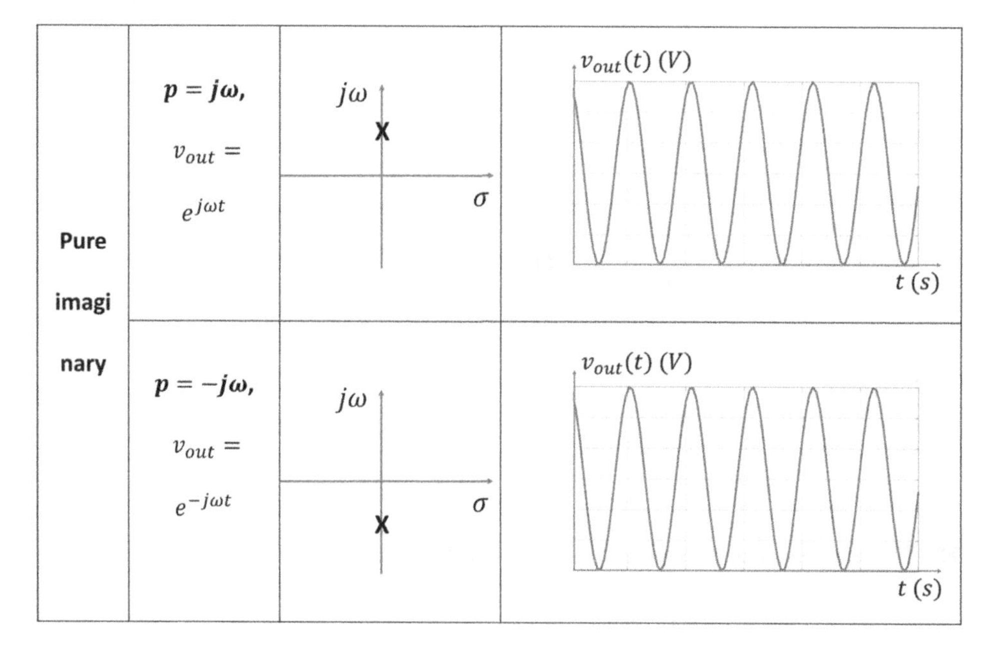

Table 9.3 (Continued)

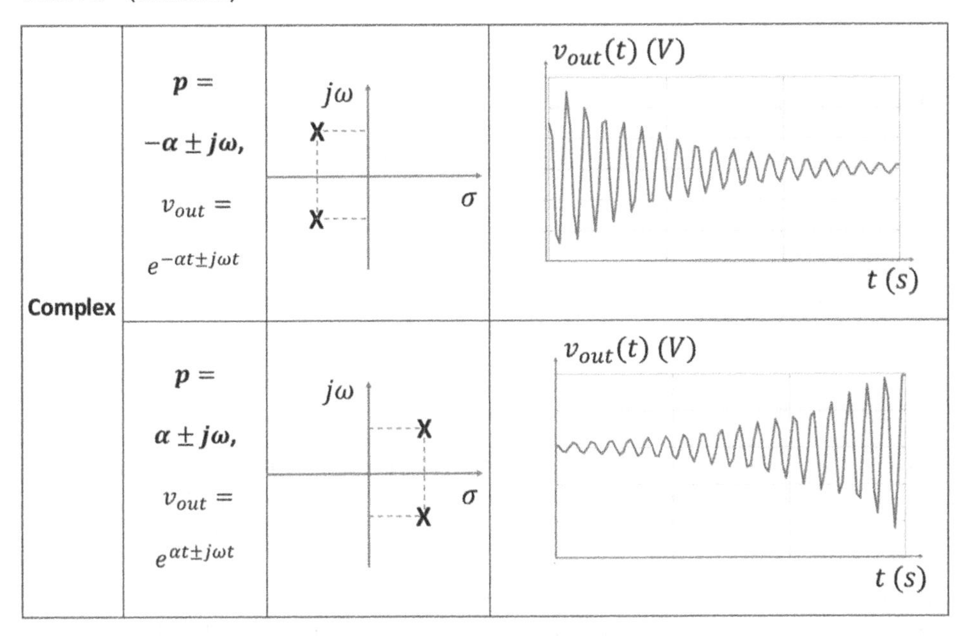

Complex	$p = -\alpha \pm j\omega$, $v_{out} = e^{-\alpha t \pm j\omega t}$	$j\omega$, σ	$v_{out}(t)$ (V), t (s)
	$p = \alpha \pm j\omega$, $v_{out} = e^{\alpha t \pm j\omega t}$	$j\omega$, σ	$v_{out}(t)$ (V), t (s)

Final and probably the most important property of a transfer function and its display on the s-plane: *We can determine the circuit's stability by observation of the location of the poles on the s-plane*:

If all poles are located on the negative half of the s-plane, the circuit is stable, and the output's final value is finite. If all poles are located on the positive half-plane, the circuit is unstable and its response tends to infinity. If poles are equal to zero, then the circuit's response is a constant. Finally, when poles are pure imaginary, the response could be either marginally stable (for a single pole) or unstable (for the multiple poles).

All these situations are shown in Table 9.3.

This table gathers our knowledge about poles and the corresponding time-domain responses in one place. Let's consider the meaning of the table presentations row by row:

- In Chapters 8 and 9, we derived many times that the member $\frac{A}{s}$ in the s-domain corresponds to $B = constant$ in the time domain. This is what the first line of the table, where $p = 0$, shows. Analyze the rest cases in a similar manner.
- Note that members $e^{j\omega t}$ and $e^{-j\omega t}$ produce identical graphs. Can you guess why?
- Table 9.3 shows that a pure imaginary pole's time-domain counterpart is a steady sinusoid. However, this is true only for a single pole. For multiple poles, the time-domain output will be growing unstable sinusoid. MATLAB has difficulty solving such a case and building a proper graph.
- Bear in mind that MATLAB can plot zero-pole diagram on an s-plane.
- Thus, Table 9.3 contains a wealth of useful information; make yourself proficient in working with this table.

9.3.3.3 Transfer Function and Convolution Integral

You recall from Chapter 6 that a convolution integral given in (6.10) enables us to set the circuit's input–output relationship in the time domain as

$$v_{out}(t) = h(t) \star v_{in}(t). \tag{9.82}$$

Compare this equation with (9.73),

$$V_{out}(s) = H(s) \cdot V_{in}(s). \tag{9.73R}$$

We know that an s-domain function is the Laplace-transformed time-domain one. In our case, we have

$$V_{out}(s) = L\{v_{out}(t)\},$$

$$V_{in}(s) = L\{v_{in}(t)\},$$

and

$$H(s).V_{in}(s) = L\{h(t) \star v_{in}(t)\}. \tag{9.83}$$

Therefore, H(s) is the Laplace transform analog of h(t), and convolution in the time domain must be replaced by multiplication in the s-domain. The proof of this statement is an auxiliary exercise for our main course, and we leave it to a curious reader. (Hint: See Alexander and Sadiku, 2020, pp. 677–678.)

The preceding discussion is summarized in Table 9.4.

Let's consider, for example, a step response of a series RC circuit discussed in Example 9.1. In the time domain, $v_{in}(t) = V_{in}u(t)$ and $h(t) = \alpha e^{-\alpha t}$. Thus, $V_{in}(s) = L\{v_{in}(t)\} = \frac{V_{in}}{s}$ and $H(s) = L\{h(t)\} = \frac{\alpha}{s+\alpha}$. According to (9.73), the output signal in the s-domain is

$$V_{out}(s) = H(s) \cdot V_{in}(s) = \frac{\alpha V_{in}}{s(s+\alpha)}.$$

Therefore,

$$v_{out}(t) = L^{-1}\{V_{out}(s)\} = L^{-1}\left\{\frac{V_{in}}{s} - \frac{V_{in}}{s+\alpha}\right\} = V_{in}\left(1 - e^{-\alpha t}\right), \tag{9.84}$$

as expected.

This demonstration clarifies the relationship between a convolution integral and a transfer function.

Table 9.4 Transfer function and convolution integral.

Time domain	s-Domain
$v_{out}(t)$	$V_{out}(s)$
$v_{in}(t)$	$V_{in}(s)$
$h(t)$	$H(s)$
Operation: Convolution	Operation: Multiplication
Equation: $v_{out}(t) = h(t) \star v_{in}(t)$	Equation: $V_{out}(s) = H(s) \cdot V_{in}(s)$

9.3.4 Laconic Summary of Section 9.3

The transfer function describes the properties of a circuit, and therefore, represents the circuit's natural response. To obtain the complete response in the s-domain, a presentation of an input signal must be preliminarily found, and then the total output will be a product of $H(s)$ and $V_{in}(s)$. Of course, to determine $v_{out}(t)$, the inverse Laplace transform must be carried out.

One of the most significant advantages of using a transfer function is that we can make essential conclusions simply by examining $H(s)$ in the s-plane. The circuit's stability is the first and the most important example of that.

A comparison of the results of Chapter 6 and the current chapter shows that a transfer function is an s-domain analog of the convolution integral developed for the time-domain operations.

9.4 Chapter 9—A Concise Conclusion

The Laplace transform application to advanced circuit analysis brings the following benefits:

- It enables us to analyze any circuit by the methods and techniques used for dc circuit analysis, whose simplicity and brevity make such an analysis plain and transparent.
- Analysis of RLC circuits carried out in Chapter 9 demonstrates the effectiveness of using the Laplace transform; however, this advantage is the most pronounced when it relies on the analyst's experience.
- The most rewarding technique of the circuit analysis in the s-domain is a transfer function, which establishes a direct relationship between the input (excitation) and the output (response) of a circuit. This technique enables us to make significant conclusions about the circuit behavior without performing detailed circuit analysis.
- Still, one must bear in mind that obtaining the actual circuit's response in the time domain requires performing the *inverse Laplace transform*, the most challenging operation in the s-domain analysis. Fortunately, accumulating several results of the typical inverse transforms in the engineer's background helps overcome these hurdles easily.

Questions and Problems for Chapter 9

9.1 Circuit analysis in the s-Domain

1. To obtain a circuit's s-domain equation, we must derive the circuit's differential equation and apply the Laplace transform to this equation. Can we obtain the circuit's s-domain equation skipping derivation of its differential equation? Explain.
2. The s-domain looks like an artificial space invented to ease some mathematical operations. Nonetheless, the text states that we can describe the properties of the actual circuits in the s-domain. How is it possible?
3. Why does Table 9.1 present the s-domain impedances, but not s-domain currents and voltages?
4. Derive and sketch the s-domain equivalent to the circuit shown in Figure 9.P4.

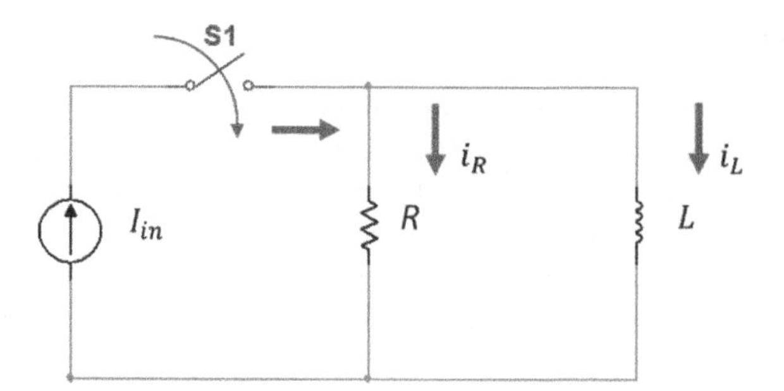

Figure 9.P4 Schematic of the real parallel RL circuit for Problem 9.4.

5 *Consider the circuit in Figure 9.P4 whose $R = 8\Omega$, $L = 2H$, and $I_{in} = 3mA$. Using the s-domain approach, find its output signal by performing the following steps:

a) Sketch the s-domain model of this circuit.
b) Based on the s-domain circuit schematic, compose the s-domain circuit equation.
c) Applying a dc circuit analysis method, derive the formula for the output signal $V_{out}(s)$ in the s-domain.
d) Use the inverse Laplace transform to obtain the formula for $v_{out}(t)$.
e) Compute the signal and plot its graph.
f) Validate your result.

6 *Applying the Laplace transform, determine the natural response of the circuit given in Figure 9.P6 for the underdamped case. Display all steps of your solution similarly to the procedure shown in Problem 9.5.

7 *Consider the circuit shown in Figure 9.P6:

a) What are the initial and final values of its $v_{out}(t)$?
b) For how long does its transient process last?
c) How the nature of its free response change if the value of R will be increased 100 times?
d) How does the circuit's free response change if both initial conditions will be zero?
e) Circuit in Figure 9.P6 show its initial conditions as initially fluxed inductor and initially charged capacitor. Is there any other way to show the circuit initial conditions to reflect their role in the circuit's natural response?

8 For the circuit in Figure 9P6, the natural response displays the circuit's initial conditions in two ways. What are those ways?

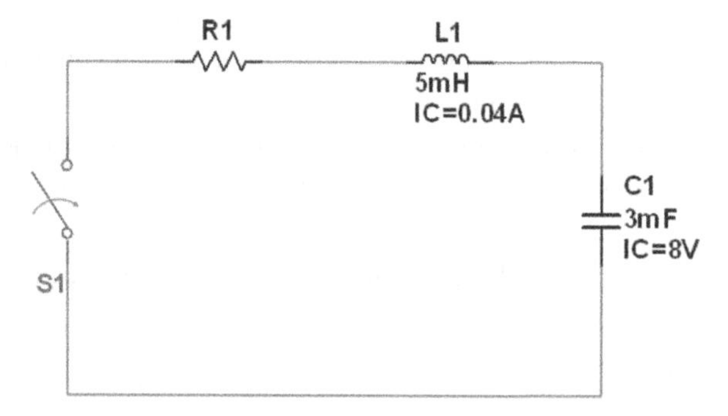

Figure 9.P6 Applying the Laplace transform for finding the natural response of a series RLC circuit.

9 What are the circuit's initial conditions by definition? How can we know the initial state of a resistor, inductor, and capacitor?

10 We study Thevenin's and Norton's theorem in review of basic circuit analysis. However, the text refers to them in subsection *Initial Conditions* of Section 9.1. Why?

11 Apply the superposition principle to find the overdamped natural response of the circuit in Figure 9.P6. Plot and verify your answer. (Hint: See Example 9.3.)

12 Use the superposition principle to find the critically damped step response of the circuit in Figure 9.P6. Draw and validate your answer. (Hint: See Example 9.3.)

13 Use the superposition principle to find the underdamped response of the circuit in Figure 9.P6. Sketch and check your answer. (Hint: See Example 9.3.)

9.2 Analysis of RLC Circuits

14 *Consider the circuit in Figure 9.P14 whose switch was connected to the battery for a long time. At $t=0$, the switch connects the circuit to the ground. Given $V_{in}=12V$, $R=4k\Omega$, and $C=2\mu F$, discover the values of $L(H)$ which creates overdamped, critically damped, and underdamped cases. Assignments:

a) Using the s-domain method, find $v_{out}(t)$ for each case.
b) Plot the response graphs and measure the transient intervals on them.
c) Verify your results by Multisim simulation of these cases and making the measurements of the transient intervals on the experimental graphs.

(Hints: (1) See Example 9.4. (2) As the first step, determine the initial conditions created by the external source.)

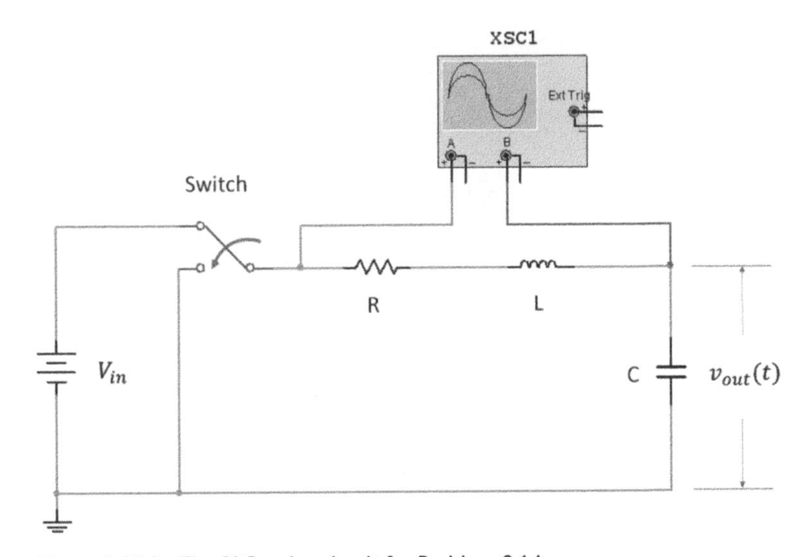

Figure 9.P14 The RLC series circuit for Problem 9.14.

15 Consider circuit in Figure 9.P15. Employing the Laplace transform technique, find its responses for overdamped, critically damped, and underdamped cases, for which calculate the required values of the capacitor C. Validate your results. (Hints: (1) This is the circuit from Problem 9.14 with the initial conditions created for that problem. (2) See Examples 9.5 and 9.6.)

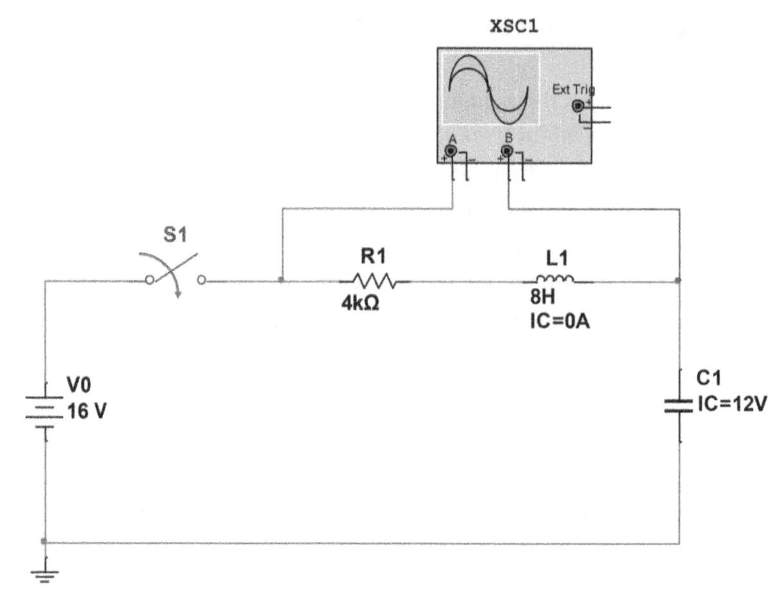

Figure 9.P15 Step response of a series RLC circuit for Problem 9.15.

16 Solve Problem 9.15 for $v_{in}(t) = 5\sin(120\pi t)$.

17 *Figure 9.P17 shows a parallel RLC circuit with a step input. The circuit parameters and the input signal are displayed. The circuit was initially relaxed. Using the s-domain presentation, find the step response of this circuit. (Hint: review Example 9.7.)

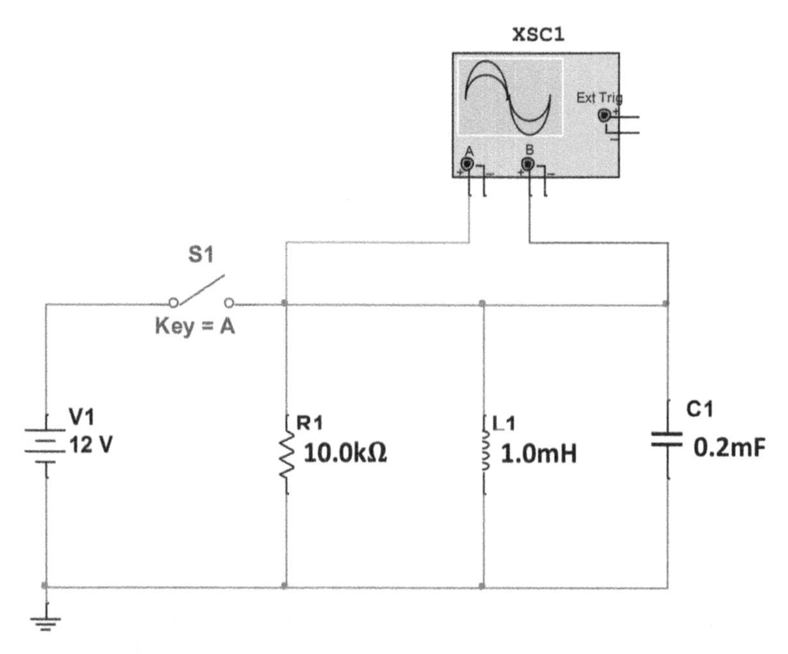

Figure 9.P17 Step response of a parallel RLC circuit for Problem 9.17.

18 Assume that $v_{in}(t) = 5\sin(120\pi t)$ and solve Problem 9.17 keeping all other parameters unchanged.

19 **Determine the initial currents, voltages, and their derivatives for all elements in the circuit shown in Figure 9.P19. (Consider this problem as a mini-project. Hints: (1) Source *I1* is connected forever, but source *I2* is multiplied by the unit step (Heaviside) function $u(t)$ and (2) employ the superposition principle.)

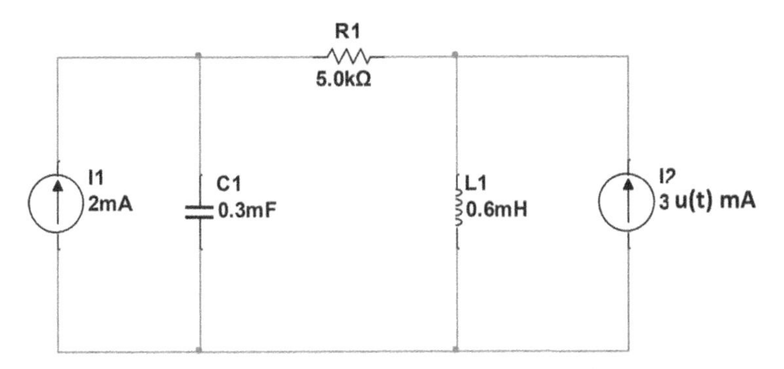

Figure 9.P19 Finding the initial values and their derivatives in a series-parallel RLC circuit. (After Hayt, Kemmerly, and Durbin, 2012, pp. 353–356).

9.3 Circuit Analysis in s-Domain with Transfer Functions

20 What is a transfer function? Give the definition and examples of the transfer function of series RC, parallel RL, series and parallel RLC circuits.

21 Compare Equation 6.1, defining the convolution integral, and 9.73 introducing the transfer function: What do they have in common, and how do they differ?

22 Both a convolution integral and a transfer function provide the circuit's input–output relationship in the most straightforward format. Why do we need two entities to serve an identical purpose?

23 How do convolution integral and a transfer function relate to each other? Explain.

24 A transfer function seems to be a great tool in relating the circuit's input and output. What shortcomings does it have compared with the similar time-domain techniques?

25 Example 9.8 shows that RC and RL circuits have identical transfer functions. How it could be? Is it an advantage or drawback?

26 Examples 8.1 and 8.7 from Chapter 8 find the step and sinusoidal responses of a first-order circuit using the Laplace transform. Example 9.8 solve the same problems employing the transfer function. In your opinion, which approach is preferable and why?

27 Example 9.10 employs a transfer function for designing a bandpass filter. Its conclusion claims that the use of a transfer function is beneficial for the filter design. Show what specific benefits the transfer function application provided.

28 Consider the s-plane shown in Figure 9.P28:

a) What is the order of the circuit which these poles represent?
b) Is this circuit stable? Prove your answer.
c) What response—natural or complete—does the transfer function whose poles and zeroes shown in Figure 9.P28 describe?

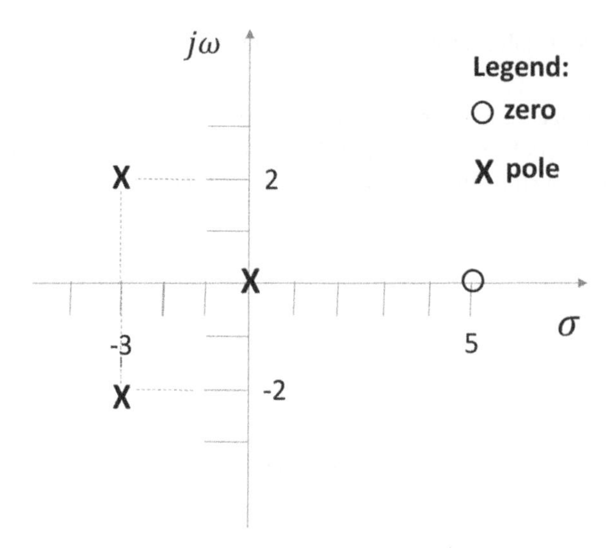

Figure 9.P28 The s-plane for Problem 9.28.

29 Reconstruct the transfer function whose poles and zeroes are shown in Figure 9.P28.

30 Consider Table 9.3, which relates the locations of the poles on the s-plane and the time-domain circuit responses:

a) Why does it show only poles?

b) Why does it pertain to the pole locations only in the natural responses?

31 Consider Table 9.P31: Fill in its two right columns.

Table 9.P31 Pole locations on s-plane and corresponding time-domain presentations.

Pole locations on the s-plane	Formula of the time-domain response	Graph of the time-domain response
$j\omega$ X σ		
$j\omega$ X σ		

Table 9.P31 (Continued)

Pole locations on the s-plane	Formula of the time-domain response	Graph of the time-domain response
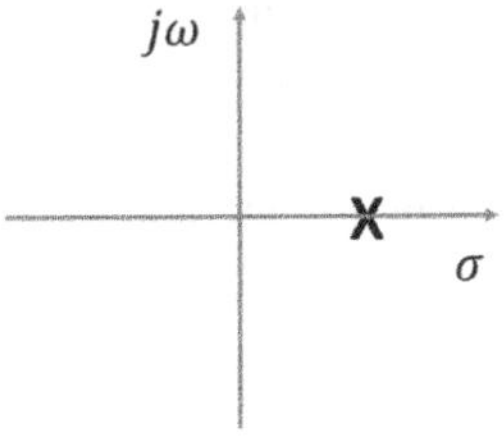 $j\omega$, σ, X		
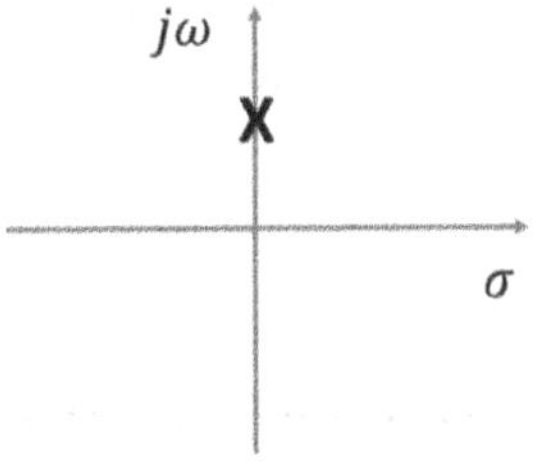 $j\omega$, X, σ		
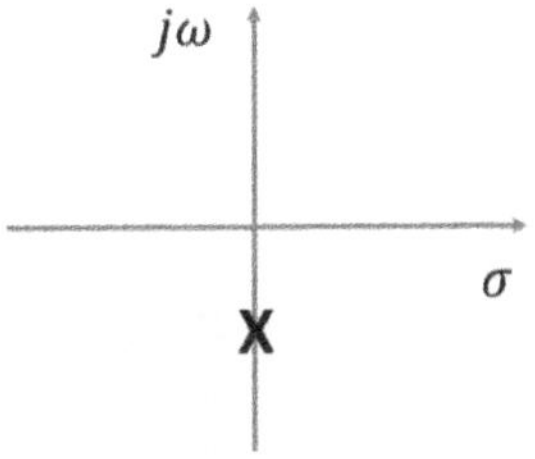 $j\omega$, σ, X		
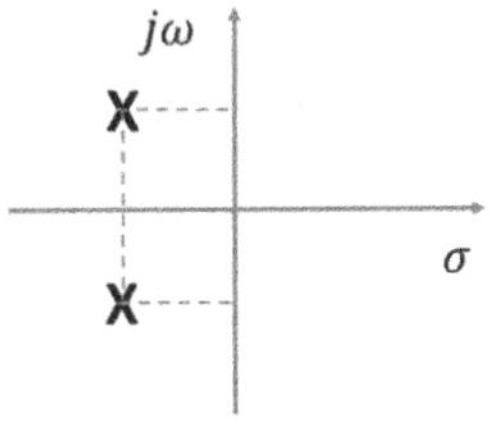 $j\omega$, X, σ, X		
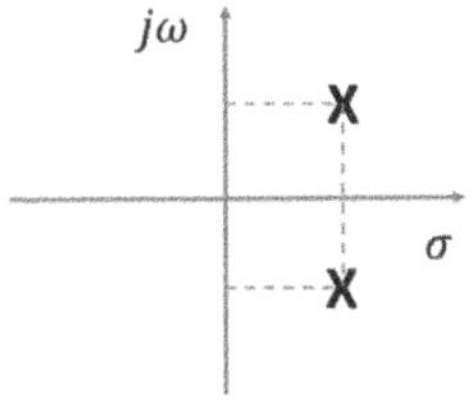 $j\omega$, X, σ, X		

32 The convolution integral operates in the time domain, whereas the transfer function works in the s-domain:

a) Is there any relationship between these two methods of advanced circuit analysis?
b) Which approach is more productive?

33 Example 6.3 demonstrates how to find a pulse response of a series RC circuit employing the convolution integral technique. Utilize this example to compare the convolution-integral approach and the transfer-function method by performing the following steps:

a) Find the pulse response of a series RC circuit by using the transfer-function method.
b) Compare both solutions in a step-by-step format.
c) Determine advantages and shortcomings of each method.
d) Make your conclusion on applicability of each technique.

34 The text repeatedly states that a transfer function describes the properties of a circuit. How is it possible given that a real circuit exists in the time-domain but a transfer function operates in the s-domain?

35 Chapter 9 discusses the application of the Laplace transform to the advanced circuit analysis. In your opinion, is it worth to consider real, time-domain operations in an artificial space such as the s-domain? Explain.

10

Fourier Transform in Advanced Circuit Analysis

In electrical engineering, the name *Fourier*[1] is forever associated with the *spectral analysis*, which analyzes the *distribution of a signal's electromagnetic energy arranged in order of frequencies*. Therefore, this chapter should be entitled "Spectral Analysis in the Advanced Circuit Analysis," which would be a cumbersome header. Hence, we use the displayed title of this chapter.

This subject is extensively discussed in many textbooks listed in the attached bibliography. Naturally, we refer you to Mynbaev and Scheiner, *Essentials of Modern Communications*, Wiley, 2020, Chapters 6 and 7. Reading that book benefits tremendously for a deeper understanding of the current chapter, for which, with the publisher's permission, we borrowed some material and several figures.

10.1 Basics of the Fourier Series with Applications

10.1.1 Periodic and Non-Periodic Signals

All signals in electrical engineering fall into two categories: periodic and non-periodic.

A signal is called periodic if it repeats (copies) itself after a given time interval called a period, T (s). All other signals are called non-periodic (aperiodic).

Examples of periodic signals include sine, cosine, and square-wave signals in Figures 10.1a and 10.1b. Mathematically, periodic signals are described by the following formula:

$$v(t) = v(t + nT), \tag{10.1}$$

1 Jean Baptiste Joseph Fourier (1768–1830) was born in France into a family of tailors and was orphaned at the age of 10. Despite this early hardship, he became a famous scientist and politician. Napoleon himself granted Fourier the title of baron. He studied mathematics in Paris under such famous mathematicians as Lagrange, Laplace, and Monge. In 1797, he became a chair professor at the Ecole Polytechnique, the best technical university in France. He became involved in politics and served as a prefect of Grenoble; however, his interest in mathematics never waned, and he discovered—while studying heat flow—the equation now bearing his name. To solve this equation, he developed an approach based on the function expansion into the series named after him. This achievement came after many corrections he made to meet the criticisms from Lagrange and others; eventually, he published the memoir entitled "The Analytical Theory of Heat," regarded as one of the key contributions to mathematics and engineering. (From Mynbaev and Scheiner, Wiley, 2020, p. 511).

Essentials of Advanced Circuit Analysis: A Systems Approach, First Edition. Djafar K. Mynbaev.

Companion Website: www.wiley.com/go/Mynbaev/AdvancedCircuitAnalysis

where $v(t)$ is the formula describing the signal's waveform, n is the integer (n = 1,2,3...), and T (s) is the period of the signal. For example, if $v(t) = Acos(\omega t)$, then $Acos(\omega t) = Acos(\omega(t+nT))$.

Periodic signals are frequent; we can find them in science and industry. In reality, however, most signals are non-periodic either due to their nature or because they are periodic signals distorted by interference. Consider the non-periodic signal in Figure 10.2a, which is a distorted sinusoidal signal; the distortion typically happens during signal transmission. The signal in Figure 10.2b is noise, which is the most ubiquitous signal in the world.

Know that *all random signals* (and noise is a perfect example of a random signal) are *non-periodic*. Since most practical signals are affected by noise, they are also non-periodic. Nevertheless, the signal generators and other sources produce signals that can be considered periodic to some extent.

10.1.2 Time Domain and Frequency Domain

Part 2 of this textbook discusses the advanced circuit analysis in the *time domain*, where all parameters, functions, and operations depend on time. Hence, this term, time domain that is, doesn't require further explanations. Part 3 is entitled "Frequency-Domain Advanced Circuit Analysis," but so far, it considers only the Laplace transform performed in the domain of the complex

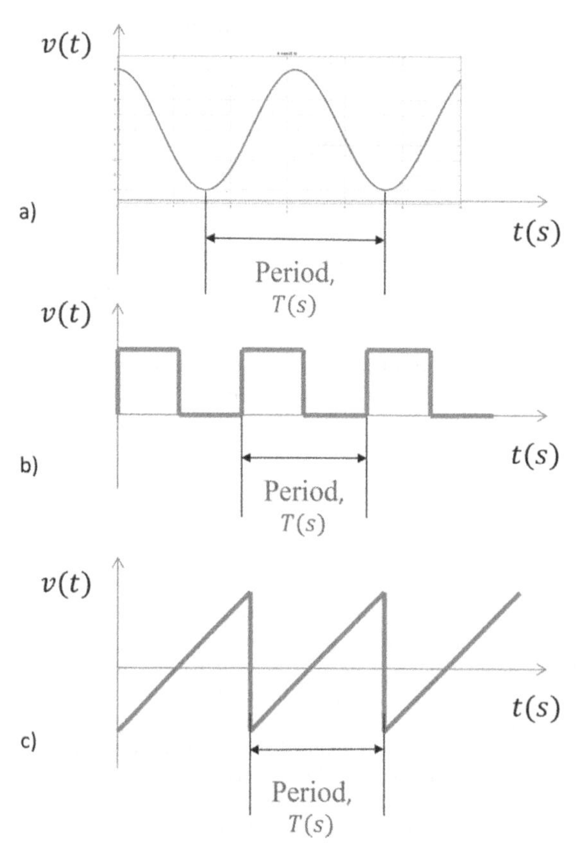

Figure 10.1 Periodic signals: a) Sinusoidal signal; b) square-wave signal; and c) sawtooth signal. (Figure 6.11 from Mynbaev and Scheiner, 2020, Wiley, p. 499.)

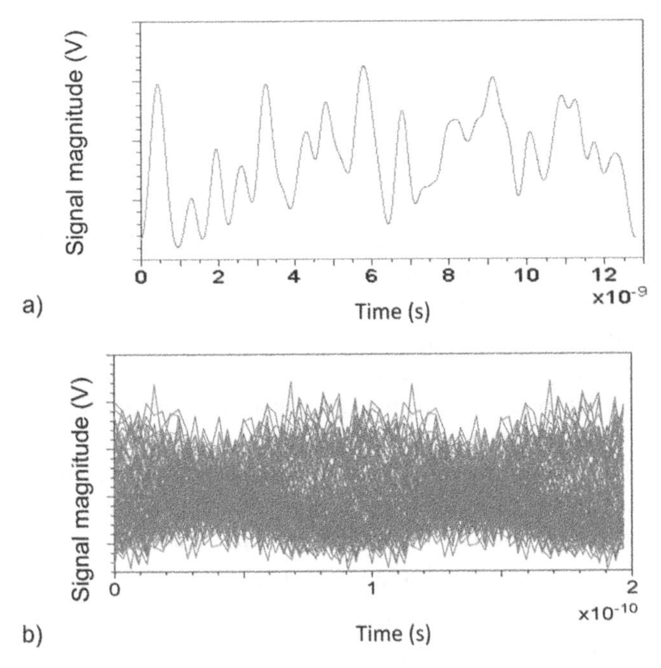

Figure 10.2 Non-periodic signals: a) Sinusoidal signal distorted during transmission; b) noise. (Figure 6.1.2 from Mynbaev and Scheiner 2020, Wiley, p. 501.)

frequency $s\left(\frac{rad}{s}\right)=\sigma+j\omega$. In this chapter, we consider the Fourier transform that performed in a pure frequency realm, where no real part is involved. However, Section 10.1 contemplates only real frequency, $\omega=2\pi f$. Here, $\omega\left(\frac{rad}{s}\right)$ is called *radian or angular frequency*, and $f(Hz)$ is known as *cyclic frequency or frequency*.

To understand the relationship between time domain and frequency domain, let's consider three sinusoidal signals:

$$v_1(t)=10\cos\left(2\pi\cdot 10^3 t\right),$$

$$v_2(t)=3\cos\left(2\pi\cdot 10\cdot 10^3 t\right),$$

$$v_3(t)=0.5\cos\left(2\pi\cdot 100\cdot 10^3 t\right).$$

Note that $v(t)$ is a signal's waveform (cosine in these examples), which shows how the signal changes with respect to time. Thus, $v(t)$ presents a signal in $v-t$ space called the *time domain.* (You will recall that a waveform's formula $v(t)=Acos(\omega t)$ describes how the signal changes with respect to time. If you want to visualize these changes, draw, $v(t)$. Then, a *waveform will be the plot of a signal in the time domain.*)

These three time-depending cosines differ only in their amplitudes and periods, so, in essence, we need only those two parameters to distinguish among these signals. We can achieve such simplification by creating a new $A-f$ (or $A-\omega$) space called the *frequency domain*, where $x\,axis$ displays a signal's frequency, $f\left(Hz\right)$, and $y\,axis$ shows signal's amplitude, $A(V)$.

Unfortunately, the timing parameter, period $T(s)$, cannot be shown in the frequency domain, $A-f$, and frequency $f(Hz)$ cannot be displayed in the time domain. Fortunately, the following *fundamental formula*,

$$T(s)=\frac{1}{f(Hz)}, \tag{10.2}$$

bridges time, $v-t$, and frequency, $A-f$, domains, thus solving the problem. Figure 10.3 visually exemplifies how the same signals can be seen in both domains.

Figure 10.3 displays the concept of the time domain and frequency domain and the relationship between them. Note that the signal's crucial parameters—its amplitude and frequency (period)—are presented in both domains. *The only entity that we can't carry over from time domain to frequency domain is the signal's waveform. But we agree to represent a cosine signal by one line in frequency domain; thus, the signal's waveform is given by default.*

The signal presentation in the time domain and frequency domain are *convention* accepted by engineering community worldwide; these are the convenient tools widely used in practice.

A thoughtful reader should immediately ask, how can we use this concept for a sine? Indeed, how can we distinguish between $v_4(t)=10\sin(2\pi\cdot 10^3 t)$ and $v_1(t)=10\cos(2\pi\cdot 10^3 t)$ depicted next to one another? (Draw them for yourself!) The answer is that a sine differs from a cosine of the same amplitude and frequency by a phase shift equals -90^0, that is,

$$\sin(\omega t)=\cos(\omega t-90^0).$$

Therefore, in addition to $A-f$ axes, we must introduce the other set, $\theta-f$. Figure 10.4 illustrates these explanations. Consequently, the frequency domain must include two sets of axes, $A-f$ (*amplitude spectrum*) and $\theta-f$ (*phase spectrum*).

Why, you can correctly ask, $\cos(\omega t)$ is considered a *phase-reference signal*, so it is depicted in the frequency domain with zero phase shift? The answer is simple: *By agreement.*

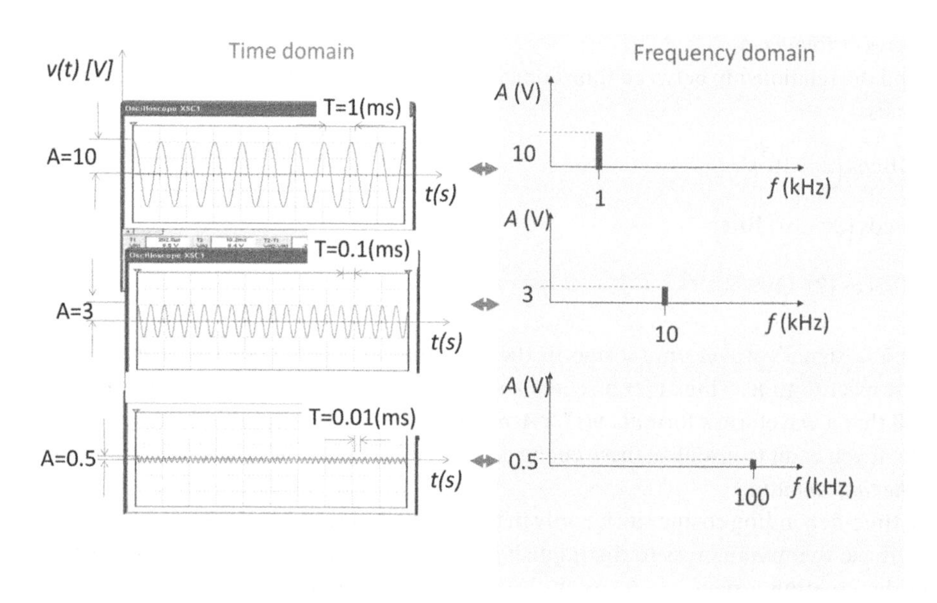

Figure 10.3 Cosine signals in the time domain (left) and frequency domain (right). (Figure 6.1.3 from Mynbaev and Scheiner 2020, Wiley, p. 502.)

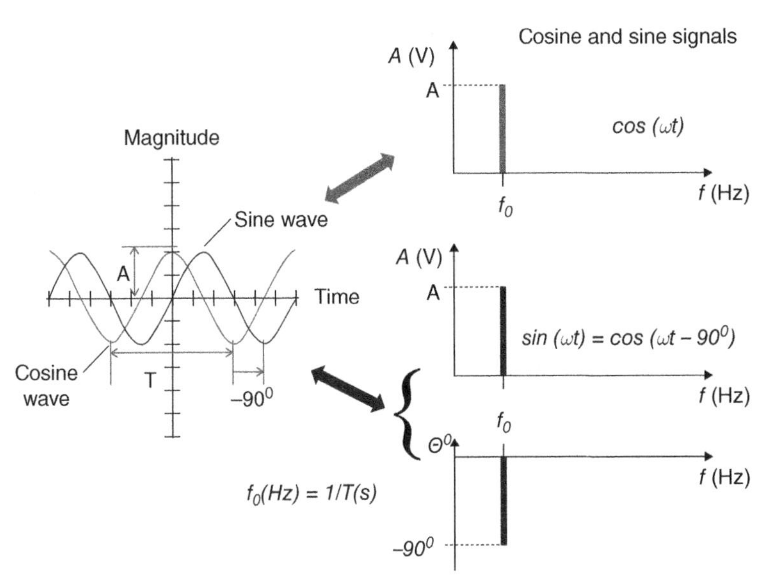

Figure 10.4 Cosine and sine in the time domain (left) and in frequency domain (right). (Figure 6.1.4 from Mynbaev and Scheiner 2020, Wiley, p. 504.)

What is *spectrum*? In our application, it is a range of frequencies that a given signal contains. Hence, the spectrum of a sinusoidal signal, $v(t) = A\cos(\omega t) = A\cos(2\pi ft)$, is one frequency, ω, or f. What is the spectrum of a more complicated signal? To answer, let's consider an example.

Example 10.1 Waveform and spectrum of an analog signal.
(Copy of Example 2.2.4 by Mynbaev and Scheiner 2020, Wiley, p. 154.)

Problem: Sketch the waveform of the following signal and find its line spectrum:

$$v(t) = 8 - 12\cos(60\pi t - 30^0) + 5\sin(180\pi t + 40^0).$$

Solution: The MATLAB-built waveform of this signal is shown in Figure 10.5a. The signal's spectrum is manually drawn in Figure 10.5b.

The problem is solved.

Discussion:

- Examine the waveform of this signal and understand that it is a graphical presentation of the signal's formula.

(Question: Is the signal shown in Figure 10.5a periodic? Explain.)

- Focus on the *spectrum of this signal*, shown in Figure 10.5b. It was plotted by depicting the spectrum of each member of the waveform's formula. The first spectral component is the constant, or dc, member at zero frequency whose amplitude is 8 volts. The second spectral component has $12 - V$ amplitude, $30 - Hz$ frequency, and 150^0 phase shift. Observe that this member is presented as $-12\cos\left(60\pi t - 30^0\right)$, but its spectrum depicts a positive 12-volt amplitude and 150^0 phase shift. These changes reflect the other convention:

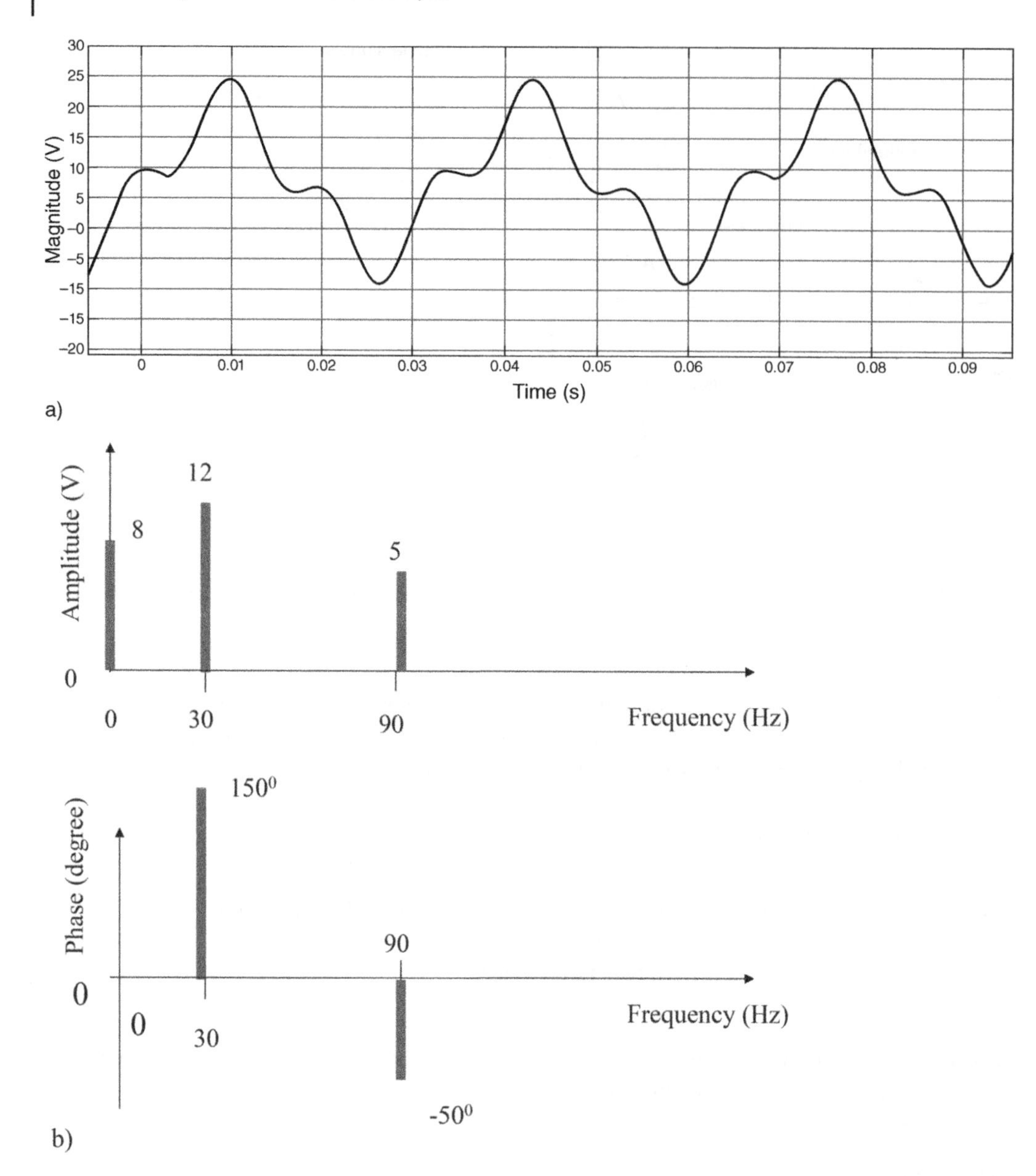

Figure 10.5 Composite signal $v(t) = 8 - 12\cos(60\pi t - 30^0) + 5\sin(180\pi t + 40^0)$: a) The waveform; b) spectrum. (Figure 2.27 from Mynbaev and Scheiner 2020, Wiley, p. 155.)

When depicting a signal in the frequency domain, its amplitude is regarded as a positive quantity and the negative sign is absorbed by the phase as follows:

$$-A\cos(\omega t + \theta) = A\cos(\omega t + \theta \pm 180^0). \tag{10.3}$$

(To visualize (10.3), plot $v(t) = -12\cos(60\pi t - 30^0)$, mark the negative 12-V point, then slide the waveform along the time axis to the left ($+180^0$), or to the right (-180^0), and see where that $-12V$ point will be.)

- The third spectral component in Figure 10.5b has 5 volts amplitude and a 90 hertz frequency. However, the phase shift of its spectrum changes from $+40^0$ in the time domain to -50^0 in the frequency domain to provide the representation of a sine signal through cosine as follows:

$$5\sin((2\pi \cdot 90)t + 40^0) = 5\cos((2\pi \cdot 90)t + 40^0 - 90^0) = 5\cos((2\pi \cdot 90)t - 50^0).$$

  Thus, *in frequency domain we depict the line of a cosine signal, adding* -90^0 *to a sine signal to comply with the agreement.*

- We use the term *line spectrum* when a spectrum is presented as a set of lines, as in Figure 10.5b. This form of a spectrum stems from the fact that each sinusoidal signal is shown in the frequency domain by a line.

Does the frequency-domain signal presentation reflect reality, or is it a mathematical abstraction? Figure 10.6 demonstrates the experimental proof of the existence of the frequency domain.

In Figure 10.6, a sinusoidal signal produced by a signal generator is presented simultaneously to an oscilloscope (the time-domain device) and a spectrum analyzer (the frequency-domain tool). Consequently, the oscilloscope shows the signal's waveform, and the spectrum analyzer displays its spectrum. Thus, Figure 10.6 is a convincing evidence of the frequency-domain reality, strikingly confirming the theoretical view shown in Figure 10.3.

10.1.3 Spectral Analysis

Examine Figures 10.7: A sinusoidal signal's amplitude spectrum is one line in the frequency domain, as in Figure 10.7a, and the amplitude spectrum of the sum of two sinusoidal signals includes two lines shown in Figure 10.7b. But what is the spectrum of a non-sinusoidal periodic

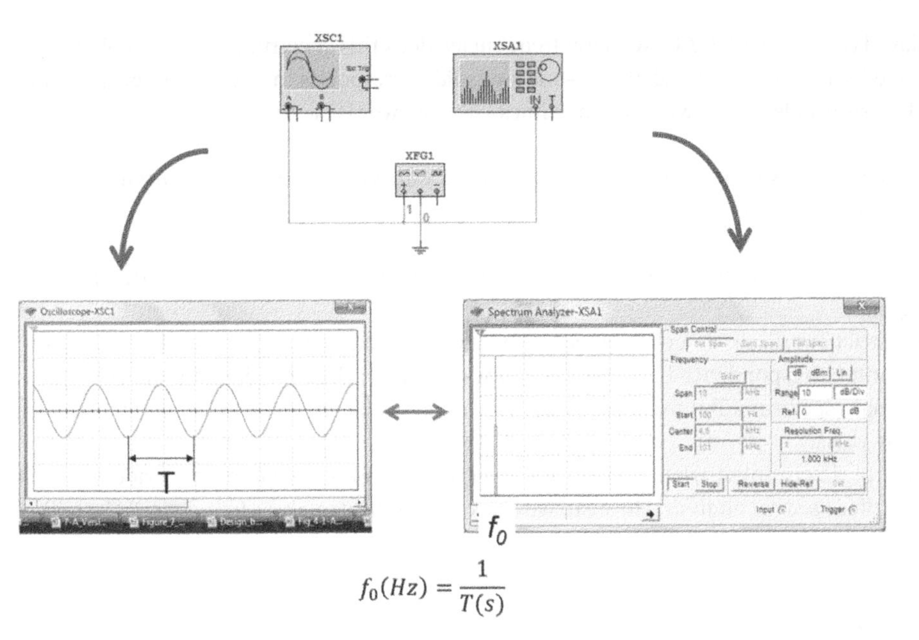

Figure 10.6 Experimental setup (top), the signal's waveform on an oscilloscope's screen (left), and the signal's spectrum shown on the display of a spectrum analyzer (right). (Figure 6.1.6 from Mynbaev and Scheiner 2020, Wiley, p. 505.)

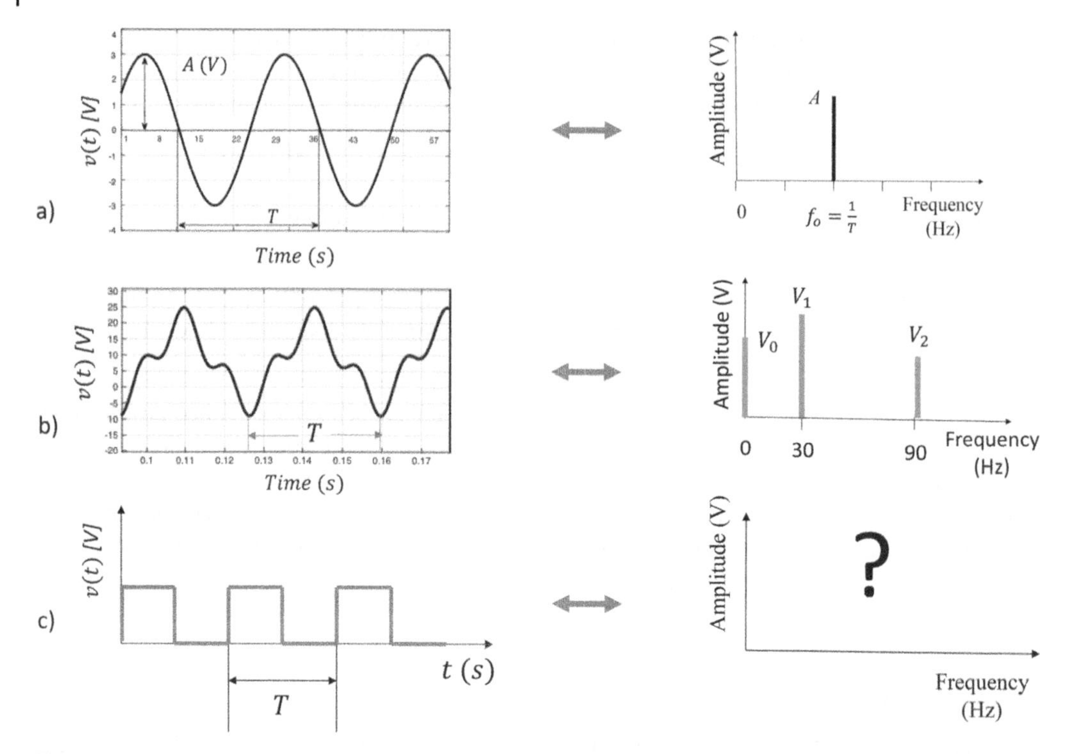

Figure 10.7 The signals in the time domain (left) and their amplitude spectra (right): a) A sinusoidal signal; b) the composition of two sinusoidal signals; c) a square wave with unknown spectrum. (Figure 10.7c is from Figure 6.1.8 from Mynbaev and Scheiner 2020, Wiley, p. 509.)

signal displayed in Figure 10.7c? How many frequencies does this spectrum contain? Follow the logic presented by Figures 10.7a and 10.7b—one sinusoid corresponds to one frequency line, and the sum of two sinusoids needs two frequency lines—the answer is clear:

> *The spectrum of a square wave has as many frequency lines as the number of sinusoids that the square wave consists of!*

This statement is the key to understanding the *Fourier series, the main tool in the spectral analysis of periodic signals.*

Note: We restrict our discussion of Figures 10.7 by the amplitude spectra only to sharp-focus on the crucial point: correspondence between a signal waveform in the time domain and the signal frequency-domain spectrum. But we never forget that a signal's spectrum includes both the amplitude and phase parts.

The process of determining what frequencies the spectrum of a time-domain periodic signal consists of is called *spectral analysis*. How can we carry out spectral analysis in practice? Using the *Fourier series.*

10.1.4 Fourier Series

So, *the problem of determining a signal's spectrum boils down to discovering how many sinusoids this signal consists of.* The theoretical basis that enables us to find it out is the *Fourier theorem*:

Every periodic signal, $v(t)$, can be presented as a sum of cosines and sines.

How? The *Fourier series* shows how:

$$v(t)[V] = A_0 + a_1 cos\omega_0 t + a_2 cos2\omega_0 t + a_3 cos3\omega_0 t + \ldots$$
$$+ b_1 sin\omega_0 t + b_2 sin2\omega_0 t + b_3 sin3\omega_0 t + \ldots \quad (10.4)$$

Here $v(t)$ is the formula of a periodic signal, such as a sinusoid, a square wave, a sawtooth, a half- or full-rectified, or any other waveform. The period of this signal is T (s), and its fundamental frequency is

$$\omega_0\left(\frac{rad}{s}\right) = \frac{2\pi}{T} \, or \, f_0(Hz) = \frac{1}{T}. \quad (10.5)$$

Also, $A_0(V)$ is a dc member, and the coefficients $a_n(V)$ and $b_n(V)$ are the amplitudes of the corresponding cosines and sines, respectively. By definition, Equation 10.4 is a series, and this specific one is called the Fourier series. Each sinusoidal member, $a_n \cos(n\omega_0 t)$ and $b_n \sin(n\omega_0 t)$, is called harmonic; the amplitude of a $n-th$ harmonic is a_n or b_n, and its frequency is $n\omega_0$, where $n = 1,2,3,\ldots.$.

Note that we can write (10.4) in the following concise form:

$$v(t)[V] = A_0 + \sum_{n=1}^{\infty}\left(a_n \cos(n\omega_0 t) + b_n sin(n\omega_0 t)\right). \quad (10.6)$$

The problem seems to be solved because every sinusoid in (10.4) has a unique amplitude and phase lines in the frequency domain; therefore, *the Fourier series delivers the spectrum of the signal $v(t)$*. However, this statement becomes valid only after the amplitudes a_n *or* b_n are determined. Fortunately, the *Fourier theorem* provides us with the following formulas for calculations of these amplitude:

$$A_0(V) = \frac{1}{T}\int_0^T v(t)dt, \quad (10.7)$$

$$a_n(V) = \frac{2}{T}\int_0^T v(t)cos(n\omega_0 t)dt, \quad (10.8)$$

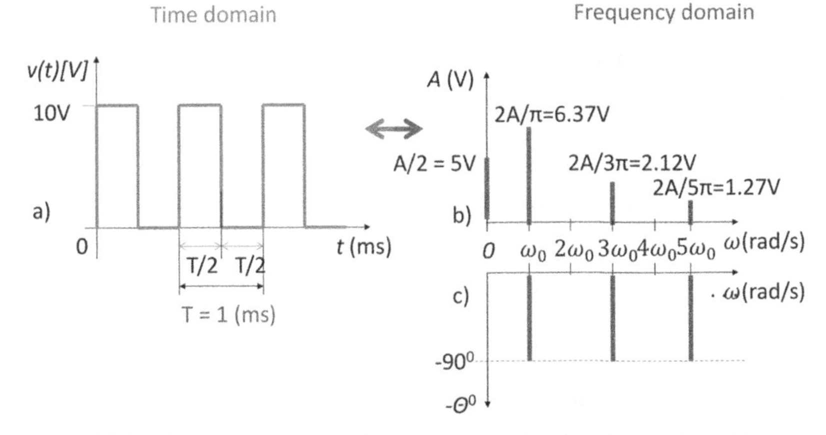

Figure 10.8 The square wave and its spectrum: a) The signal's waveform; b) the amplitude spectrum; c) the phase spectrum. (Figure 6.1.11 from Mynbaev and Scheiner 2020, Wiley, p. 513.)

$$b_n(V) = \frac{2}{T}\int_0^T v(t)\sin(n\omega_0 t)dt. \tag{10.9}$$

Both variables in the RHSs of (10.7), (10.8), and (10.9), $v(t)$ *and* $\omega_0 = \frac{1}{T}$, are known; consequently, (10.4), along with these three formulas, enables us to obtain the entire Fourier series of a given signal and, therefore, its spectrum.

Exercise Verify that the units on the RHS and the LHS of (10.7), (10.8), and (10.9) coincide.

Let's apply the above theory to a practical example.

Example 10.2 The spectrum of a square-wave signal.

Problem: Determine the spectrum of a square-wave signal shown in Figure 10.8a whose $A = 10$ (V) and $T = 1$ (ms).

Solution: To determine the signal's spectrum, we need to expand *v(t)* into the Fourier series. To do so, we have to take the following steps:

1) Define the signal *v*(*t*).
2) Calculate A_0.
3) Calculate a_n and b_n.
4) Obtain the Fourier series.
5) Plot the signal's spectrum.

Let's put this plan into effect:

Step 1: By observing Figure 10.8a, we derive the formula for the signal's waveform as

$$v(t)[V] = \begin{cases} A = 10 \text{ } for \text{ } 0 \le t \le \frac{T}{2} \\ A = 0 \text{ } for \text{ } \frac{T}{2} \le t \le T \end{cases}. \tag{10.10}$$

Step 2: Using (10.7), we find the dc member as

$$A_0 = \frac{1}{T}\int_0^T v(t)dt = \frac{1}{T}\int_0^{T/2} Adt = \left(\frac{1}{T}\right)A\Big|_0^{\frac{T}{2}} = \frac{A}{2} = 5(V). \tag{10.11}$$

Steps 3: Calculations of a_n and b_n is the most laborious step. It is done as follows:

Apply (10.8) to calculate a_n as

$$\begin{aligned} a_n &= \frac{2}{T}\int_0^T v(t)\cos(n\omega_0 t)dt = \frac{2}{T}\int_0^{\frac{T}{2}} A\cos(n\omega_0 t)dt \\ &= \frac{2A}{T}\left(\frac{1}{n\omega_0}\right)\sin(n\omega_0 t)\Big|_0^{\frac{T}{2}} = \frac{2A}{Tn\omega_0}\sin\left(n\omega_0\frac{T}{2}\right) = \frac{A}{n\pi}\sin(n\pi) = 0. \end{aligned} \tag{10.12}$$

This is because $\omega_0 = \frac{2\pi}{T}$, $n = 1, 2, 3, \ldots$, and $\sin(n\pi) = 0$ for every n.

To calculate b_n, we use (10.9) as follows:

$$b_n = \frac{2}{T}\int_0^T v(t)sin(n\omega_0 t)dt = \frac{2}{T}\int_0^{\frac{T}{2}} A\,sin(n\omega_0 t)dt = -\frac{2A}{T}\left(\frac{1}{n\omega_0}\right)cos(n\omega_0 t)\Big|_0^{\frac{T}{2}}$$

$$= -\frac{2A}{Tn\omega_0}\left(\cos\left(n\omega_0\frac{T}{2}\right) - \cos(0)\right) = -\frac{A}{n\pi}(\cos(n\pi) - 1).$$

If n is even, then $\cos(n\pi) = 1$, and $b_n = 0$. For $= 1, 3, 5, \ldots$, $\cos(n\pi) = -1$ and

$$b_n = -\frac{A}{n\pi}(-2) = \frac{2A}{n\pi}. \tag{10.13}$$

Step 4: Inserting calculated A_0, a_n, and b_n into (10.4) produces the *Fourier series* of the given square wave as

$$v(t)[V] = \frac{A}{2} + \frac{2A}{\pi}sin\omega_0 t + \frac{2A}{3\pi}sin3\omega_0 t + \frac{2A}{5\pi}sin5\omega_0 t + \ldots \tag{10.14}$$

Equation (10.14) in concise form is

$$v(t)[V] = \frac{A}{2} + \sum_{n,\ odd\ only}^{\infty} \frac{2A}{n\pi}\sin(n\omega_0 t). \tag{10.15}$$

It's worth comparing the specific square-wave Fourier series to the general formula Fourier series. For this, we write (10.4) on the top and (10.14) beneath, and compare them member by member. The result is shown in (10.16).

$$v_{general}(t) = A_0 + a_1 cos\omega_0 t + a_2 cos2\omega_0 t + \ldots + b_1 sin\omega_0 t + b_2 sin2\omega_0 t + b_3 sin3\omega_0 t + \ldots \tag{10.4R}$$

$$\updownarrow \quad \updownarrow \quad \updownarrow \quad \updownarrow \quad \updownarrow \quad \updownarrow \quad \updownarrow \tag{10.16}$$

$$v_{\underline{square}\atop wave}(t) = \frac{A}{2} \quad + \ 0 \ + \quad 0 + \quad \ldots + \frac{2A}{\pi}sin\omega_0 t \ + \ 0 \ + \ \frac{2A}{3\pi}sin3\omega_0 t + \ldots \tag{10.14R}$$

Thus, the spectrum of this type of square wave includes a constant member and odd sine harmonics only.

Step 5: Remembering that every sine and cosine has a unique line in the frequency domain, we can now depict the *amplitude spectrum* of the given square-wave signal shown in Figure 10.8b. Now we know what the amplitude spectrum of this square wave encompasses. The phase spectrum is shown in Figure 10.8c. (Remember about phase shift for a sine?) Thus, Figures 10.8b *and* 10.8c *display the square-wave spectrum whose waveform is presented in* Figure 10.8a.

Discussion:

- We can conceptually compare the calculated amplitude (Figure 10.8b) and measured square-wave signal's spectra (Figure 10.9). As for the latter, the square wave produced by a signal generator presented simultaneously to an oscilloscope (left) and a spectrum analyzer (right). The oscilloscope shows the signal's waveform (time-domain presentation) and the spectrum

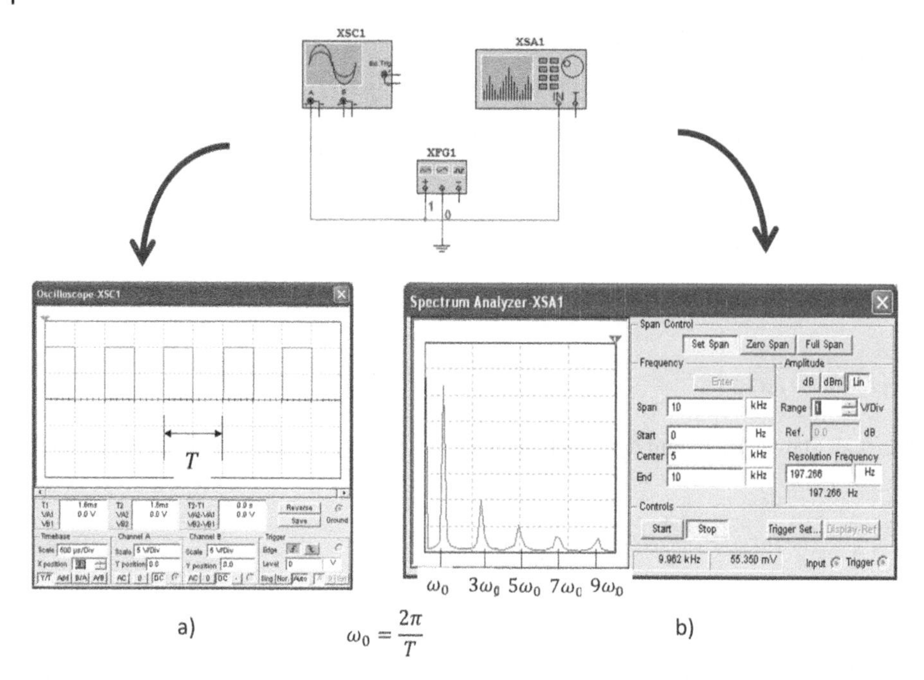

Figure 10.9 Experimental verification of the spectrum of a square wave: a) The time-domain square wave produced by a signal generator; b) amplitude spectrum of the square wave. (Figure 6.1.9 from Mynbaev and Scheiner 2020, Wiley, p. 511.)

analyzer demonstrates the signal's spectrum (frequency-domain display). Now we understand why the measured spectrum includes many frequency lines. (It would be a good *laboratory exercise* to compare calculated and measured spectra quantitatively.)

- An exciting view of the time- and frequency-domain presentations is given in Figure 10.10. It shows how a time-domain waveform is seen as a set of lines from the frequency-domain perspective.
- Note that the *dc member of the Fourier series, A_0, presents the average value of the signal over a period.* In our example, we observe that $A_0 = \frac{A}{2} = 5(V)$, so we could come to this conclusion without calculations.
- Interestingly, the example of a square wave has been used for many years to introduce the concept of a Fourier series; this example can be found in any textbook on communications, electronics, or signals and systems. We probably owe the popularity of this example to the simplicity of calculating the Fourier coefficients. In any event, we follow this academic tradition.
- This example explicitly shows how to find the spectrum of a periodic signal. But what if the signal under investigation is non-periodic? The Fourier series cannot be used in such a case, as the Fourier theorem states; instead, we need to turn to the Fourier transform, the subject of the next section.

10.1.5 Spectral Analysis and Spectral Synthesis

The Fourier theorem and the Fourier series allow us to determine what spectral components (harmonics, that is) a given signal contains. In other words, the Fourier series enables the spectral analysis of a given signal. We call this investigation an *analysis* because we determine the

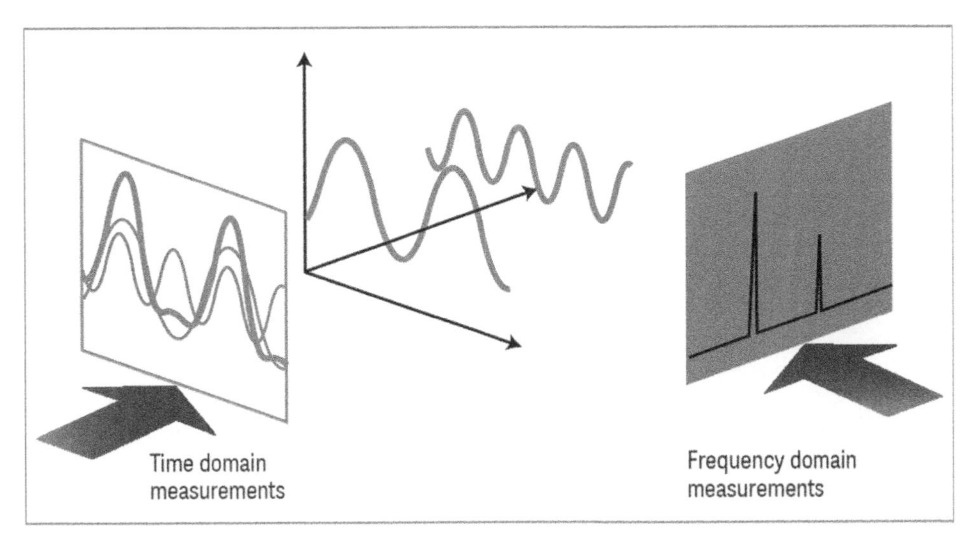

Figure 10.10 Presentations of two harmonics in time domain and frequency domain. (*Source:* © Keysight Technologies, Inc. Reproduced with permission of Keysight Technologies.).

components of a whole entity—the signal; we call it *spectral analysis* because we find the frequency components. An overview of the process of spectral analysis is as follows: In the time domain, we start with a signal whose waveform, amplitude, and period are given, from which we develop the explicit formula of the signal's waveform, $v(t)$. Using this expression, we calculate the coefficients of the Fourier series, which enable us to present $v(t)$ as a sum of specific cosines and sines. Since each cosine and sine has unique lines in the frequency domain, we translate the obtained Fourier series into the lines in the $A-f$ and $\theta-f$ spaces and therefore get the signal's spectrum. Remember that our goal is to find the constituent parts (the harmonics) of the spectrum of a given signal. And again, we accomplish this goal by obtaining these harmonics as individual members of the Fourier series.

The term harmonic is applied to a member of a series whose frequency is a whole-number multiple of a fundamental frequency, as in a Fourier series.

A *spectral analysis* determines the signal's spectrum, that is, the signal's harmonics, from which the given signal's waveform consists of. But we might encounter the opposite problem: Given a signal's spectrum (frequency components), build the signal's waveform. The process of finding the solution to this problem is called *spectral synthesis.*

Figure 10.11 visualizes this process; to simplify the drawings, only amplitude spectra are shown. We remember the key point of spectral analysis: *Every line in frequency domain represents a harmonic signal.* If we take the first two lines from a square-wave spectrum considered in Example 10.2, $\frac{A}{2} = 5(V)\ and\ \frac{2A}{\pi} = 6.37(V)at\ \omega_0 = 2000\left(\frac{rad}{s}\right)$, we can build the waveform $v_a(t)[V] = 5 + 6.37 sin2000\pi t$ shown in Figure 10.11a. The first three spectral lines, $\frac{A}{2}, \frac{2A}{\pi}\ at\ \omega_0,\ and\ \frac{2A}{3\pi}\ at\ 3\omega_0$, enable us to produce the waveform $v_b(t)$ [v] $= 5 + 6.37 sin2000\pi t$ $+ 2.12 sin6000\pi t$ shown in Figure 10.11b. The similar operations are performed in Figures 10.11c and 10.11d. The greater the number of harmonics included in the summation, the closer the

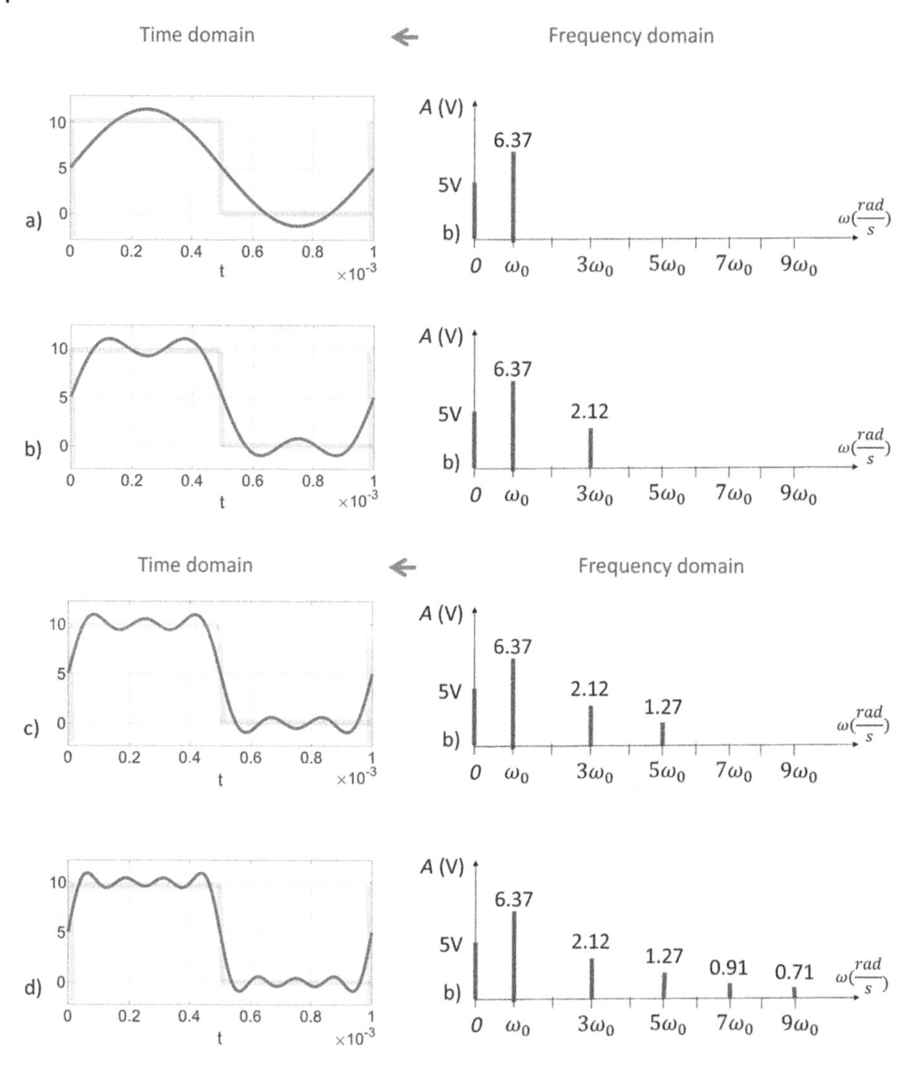

Figure 10.11 Synthesis of a square wave by using its harmonics: a) $v(t) = \frac{A}{2} + \frac{2A}{\pi} sin\omega_0 t$ using the first two spectral lines. b) $v(t) = \frac{A}{2} + \frac{2A}{\pi} sin\omega_0 t + \frac{2A}{3\pi} sin3\omega_0 t$ from three spectral lines; c) $v(t) = \frac{A}{2} + \frac{2A}{\pi} sin\omega_0 t + \frac{2A}{3\pi} sin3\omega_0 t + \frac{2A}{5\pi} sin5\omega_0 t$ from five spectral lines; d) $v(t) = \frac{A}{2} + \frac{2A}{\pi} sin\omega_0 t + \frac{2A}{3\pi} sin3\omega_0 t + \frac{2A}{5\pi} sin5\omega_0 t + \frac{2A}{7\pi} sin7\omega_0 t + \frac{2A}{9\pi} sin9\omega_0 t$ from nine spectral lines. The spectral and time-domain values are taken from Example 10.2. (Replica of Figure 6.1.15 from Mynbaev and Scheiner, Wiley, 2020, pp. 522–523.)

reconstructed waveform will be to the original signal. All Figures 10.11 demonstrate this point convincingly.

For convenience, below is the MATLAB code for synthesizing the waveforms from the first five signal's harmonics.

```
syms t
A = 10;
T = 0.001;
```

```
w0 = (2*pi)/T;
Vout = A/2 +
(2*A/pi).*sin(w0*t)+((2*A)/(3*pi)).*sin(3*w0*t)
+((2*A)/(5*pi)).*sin(5*w0*t);
ezplot(Vout,[0, 0.001])
axis([0 0.001])
grid
```

10.1.6 The Fourier Series Application to Advanced Circuit Analysis

10.1.6.1 The Concept

The main problem we consider in this book is finding the circuit's response to an excitation; it is shown in Figures 4.1a, 9.15, and 9.16 reproduced here. This problem can be solved by the ac circuit analysis in a steady-state case, as Part 1 demonstrates; by using the differential equations and convolution integral enable solving this problem when a transient response is in search, which Part 2 shows; by the Laplace transform that brings a transfer function as an ultimate tool for resolving this issue, clearly demonstrated in Chapters 7, 8, and 9 of Part 3. Now, the Fourier series provides us with a new approach–spectral analysis.

The preceding discussion in this section shows that a square wave can be decomposed into a sum of individual sinusoids (harmonics), as Figures 10.8 and 10.9 visualize; on the other hand, the

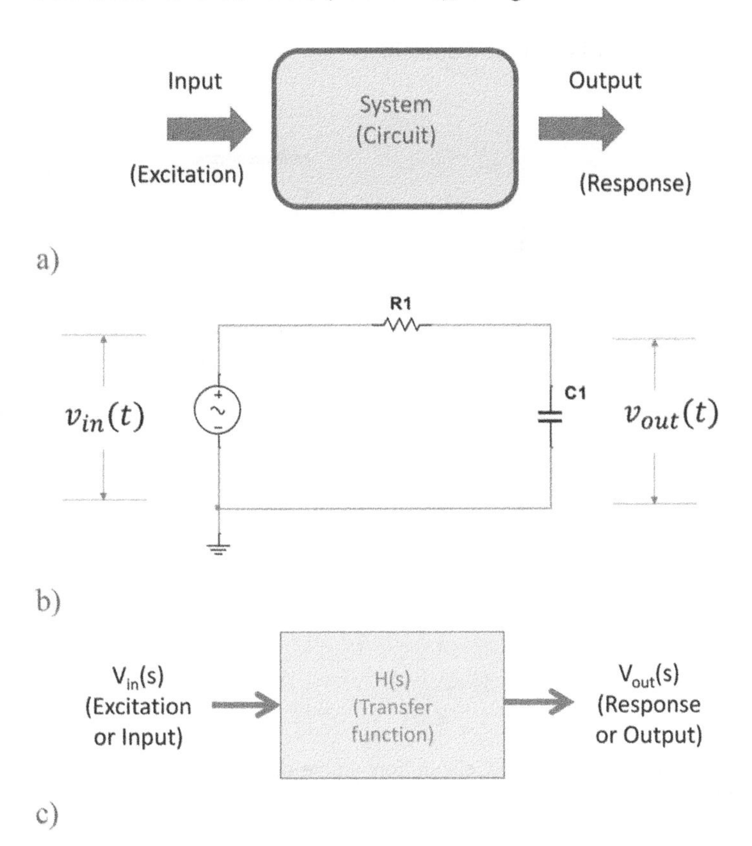

Figures 4.1R, 9.15R, and 9.16R System (circuit) response: a) General view; b) the circuit view; c) the transfer function view in the s-domain.

summation of these harmonics produces a square-wave signal, which is shown in Figures 10.11a through 10.11d. Hence, we can consider the problem of passing a square-wave signal through an RC circuit, shown in Figure 4.1R, from the spectral analysis point of view.

Such an approach boils down to the investigation of passing through the RC circuit an individual harmonic. The problem and its solution are displayed in Figure 10.12.

When a single harmonic (sinusoid, that is) passes through a series RC circuit, its amplitude decreases, and it experiences an additional phase shift, as shown at the top of Figure 10.12. This process and its result are described in the *time domain* as

$$v_n^{in}(t) = A_n^{in} \sin\left(\omega_n t + \theta_n^{in}\right) \Longrightarrow v_n^{out}(t) = A_n^{out} \sin\left(\omega_n t + \theta_n^{out}\right).$$

Here, A_n^{in} and A_n^{out} are the input and output amplitudes of the harmonic, θ_n^{in} and θ_n^{out} are its input and output phase shifts, and n is the harmonic's number.

Why and how much did the output amplitude and phase shift change? The answers are obtained by considering the process in the *frequency domain*, visualized in the shaded bottom part of Figure 10.12.

Mathematically, an RC circuit is described by its transfer function, which we derived in Example 6.2 as

$$H(s) = \frac{\alpha}{s+\alpha}, \tag{10.17}$$

where $\alpha = \dfrac{1}{RC}$. An input sine signal is described in the s-domain as

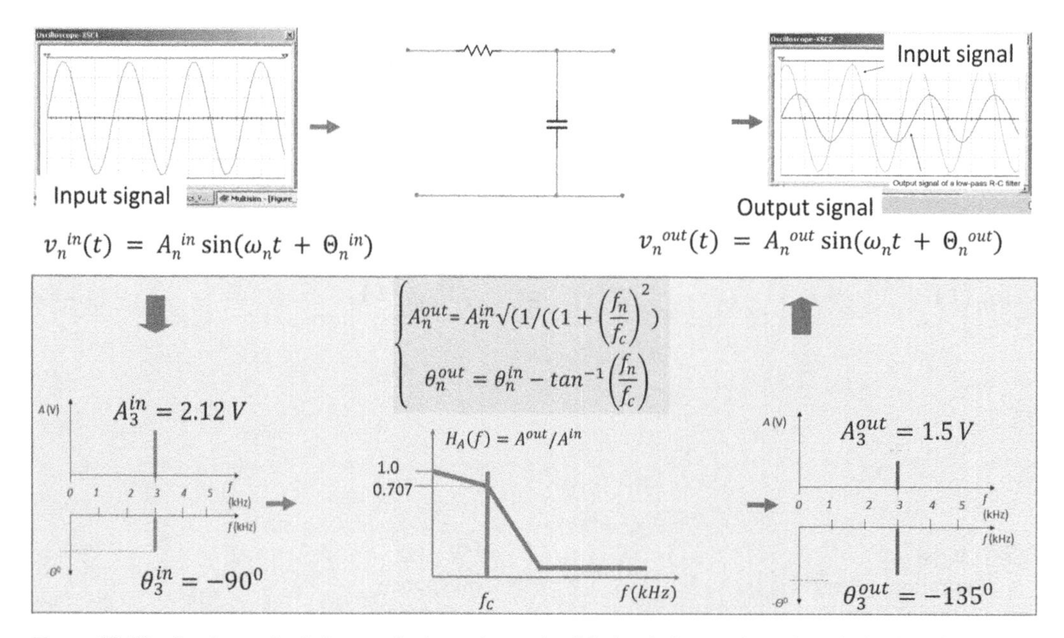

Figure 10.12 Passing a single harmonic through a series RC circuit from a time-domain (top) and frequency-domain points of view (bottom).

$$V_n^{in}(s) = \frac{\omega_n}{s^2 + \omega_n^2},$$

according to Table 7.1. Therefore, the output sine is given by

$$V_n^{out}(s) = H(s) \cdot V_n^{in}(s) = \frac{\alpha}{s+\alpha} \cdot \frac{\omega_n}{s^2 + \omega_n^2}. \tag{10.18}$$

The characteristic equation of the input signal, $s^2 + \omega_n^2 = 0$, gives the solution $s_{1,2} = \pm j\omega$. Substituting $s = j\omega_n$, as we did in Section 9.3, we can translate (10.17) and (10.18) from the s-domain into frequency domain as

$$V_n^{out}(\omega) = H(\omega) \cdot V_n^{in}(\omega) = \frac{\alpha}{\alpha + j\omega} V_n^{in}(\omega). \tag{10.19}$$

Separating the real and imaginary parts of (10.19) results in

$$V_n^{out}(\omega) = \frac{\alpha^2 V_n^{in}(\omega)}{\alpha^2 + \omega_n^2} - j\frac{\alpha \cdot \omega \cdot V_n^{in}(\omega)}{\alpha^2 + \omega_n^2},$$

from which we can obtain the amplitude $A_n^{out} = \left|V_n^{out}(\omega)\right|$ and the additional phase $\Delta\theta_n^{out} = -tan^{-1}\left(\frac{f_n}{f_c}\right)$ of $V_n^{out}(\omega) \equiv V_n^{out}(f)$ as

$$A_n^{out} = A_n^{in}\sqrt{\frac{1}{\left(1 + \left(\frac{f_n}{f_c}\right)^2\right)}} \tag{10.20a}$$

$$\theta_n^{out} = \theta_n^{in} - tan^{-1}\left(\frac{f_n}{f_c}\right). \tag{10.20b}$$

In derivation of (10.20a) and (10.20b) bear in mind that $\alpha = \frac{1}{RC} = f_c$, meaning that $\frac{\omega}{\alpha} = \frac{\omega}{\omega_c} = \frac{f}{f_c}$.

Equations 10.20a and 10.20b are the parts of the RC circuit's transfer function, and therefore, they describe the RC circuit's operation. They enable us to compute the amplitude and phase of a specific $n-th$ harmonic when the input parameters and the circuit's cutoff frequency, f_c, are given. Figure 10.12 visually exemplifies this statement. The left bottom of Figure 10.12 shows the third harmonic of an input signal with $A_3^{in} = 2.12V$, $\theta_3^{in} = -90^0$, and $f_3 = f_c = 3.18kHz$. This harmonic enters the RC circuit whose transfer function graphically shown in the middle bottom of Figure 10.12. Here, A_3^{in} is cut by $\sqrt{\frac{1}{\left(1+\left(\frac{f_3}{f_c}\right)^2\right)}} = 0.707$, which produces $A_3^{out} = 2.12 \cdot 0.707 =$ 1.5 V. This output amplitude is shown at the right bottom of Figure 10.12. Likewise, an additional phase shift is given by $\Delta\theta_3^{out} = -tan^{-1}\left(\frac{f_3}{f_c}\right) = -tan^{-1}(1) = -45°$, from which we calculate $\theta_3^{out} = -90^0 - 45^0 = -135^0$ as shown at the right bottom of Figure 10.12.

This consideration and Equations 10.20a and 10.20b give us the tool for calculating the amplitudes and phase shifts of all harmonics at the output of a series RC circuit, which, in turn, opens the door for finding the waveform of the circuit response when the periodic input signal is given. Indeed, the periodic signal can be presented as a sum of cosines and sines (harmonics, that is), Equations 10.20a and 10.20b enable us to calculate the amplitudes and phase shifts of the output harmonics, and the spectral synthesis gives a mechanism for reconstructing the output waveform. All these operations are based, of course, on the principle of superposition.

The following example shows how to implement this concept in practice.

Example 10.3 Passing the pulse train through a series RC circuit.

Problem: A square-wave signal with $A = 10$ (V) and $T = 1$ (ms) is presented to the series RC circuit with (a) $f_C = 3.18$ (kHz) and (b) $f_C = 0.318$ (kHz). Using the spectral analysis, build the waveforms of the outputs based on the first seven output harmonics.

Solution:

It's necessary to know that the set of identical pulses is called a *pulse train*, and a *square wave* is a particular type of pulse train whose pulse duration is equal to half of a period. Also, it's worth reminding that the spectral analysis of a square wave is considered in the discussion of Figure 10.8, and the spectral synthesis is illustrated in Figure 10.11.

Thus, the solution to this example is obtained as follows:

a) The waveforms of the first seven harmonics of the input square-wave signal presented to the RC circuit are shown in Figure 10.13a. Applying (10.20a) and (10.20b), we compute the amplitudes and phase shifts of the seven output harmonics. The calculated values and the input and output waveforms are shown in Figure 10.13b. We can compute these values using a graphic calculator, MS Excel, or MATLAB. Finally, the response waveform synthesized from the seven output harmonics is displayed in Figure 10.13c.
b) We repeat the above procedure for $f_C = 0.318$ (kHz) and obtain the output signal shown in Figure 10.13d.

The problem is solved.

Discussion:

Equation 10.20a enables us to introduce the amplitude part of an RC circuit's transfer function as

$$H_A(f) = \frac{A_n^{out}}{A_n^{in}} = \frac{1}{\sqrt{\left(1+\left(\frac{f_n}{f_c}\right)^2\right)}}. \tag{10.21}$$

Figure 10.14, where the frequency axis is given in a logarithmic scale, shows the graph of this equation as a thin continuous line. For convenience, the actual incessant graph is often approximated by a piecewise chart shown in Figure 10.14 as shaded straight lines atop the uninterrupted graph. We used this approximation in Figure 10.12 and will continue to employ it in future presentations.

- Using the amplitude transfer function in the piecewise form enables us to demonstrate how the amplitudes of individual harmonics changes after passing through an RC circuit. Also, we can show how the output phase shifts change with respect to the input ones. Figures 10.15a and

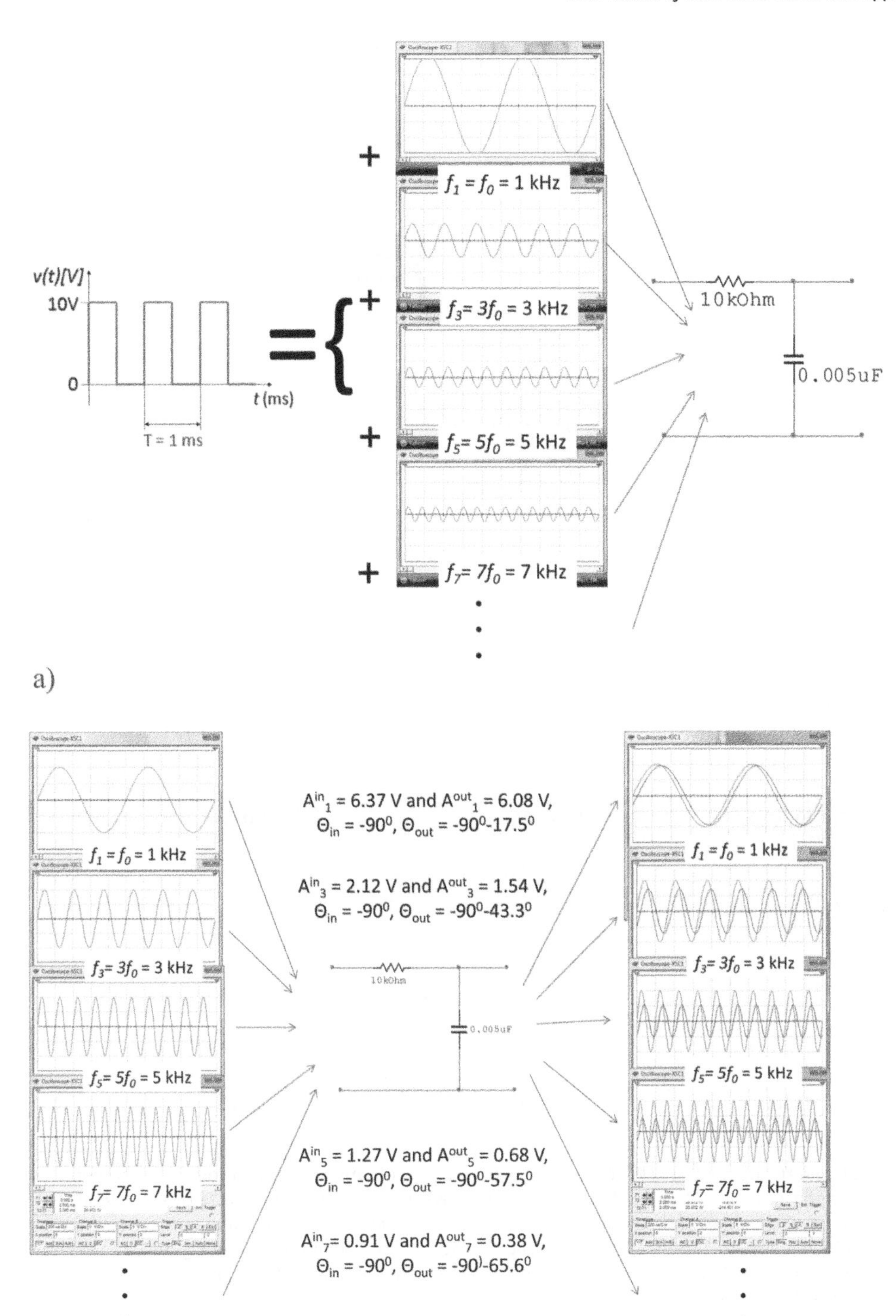

Figure 10.13 Using spectral analysis for evaluation of passing a pulse train through a series RC circuit in Example 10.3: a) Developing a pulse train into a set of seven harmonics by applying the Fourier series;

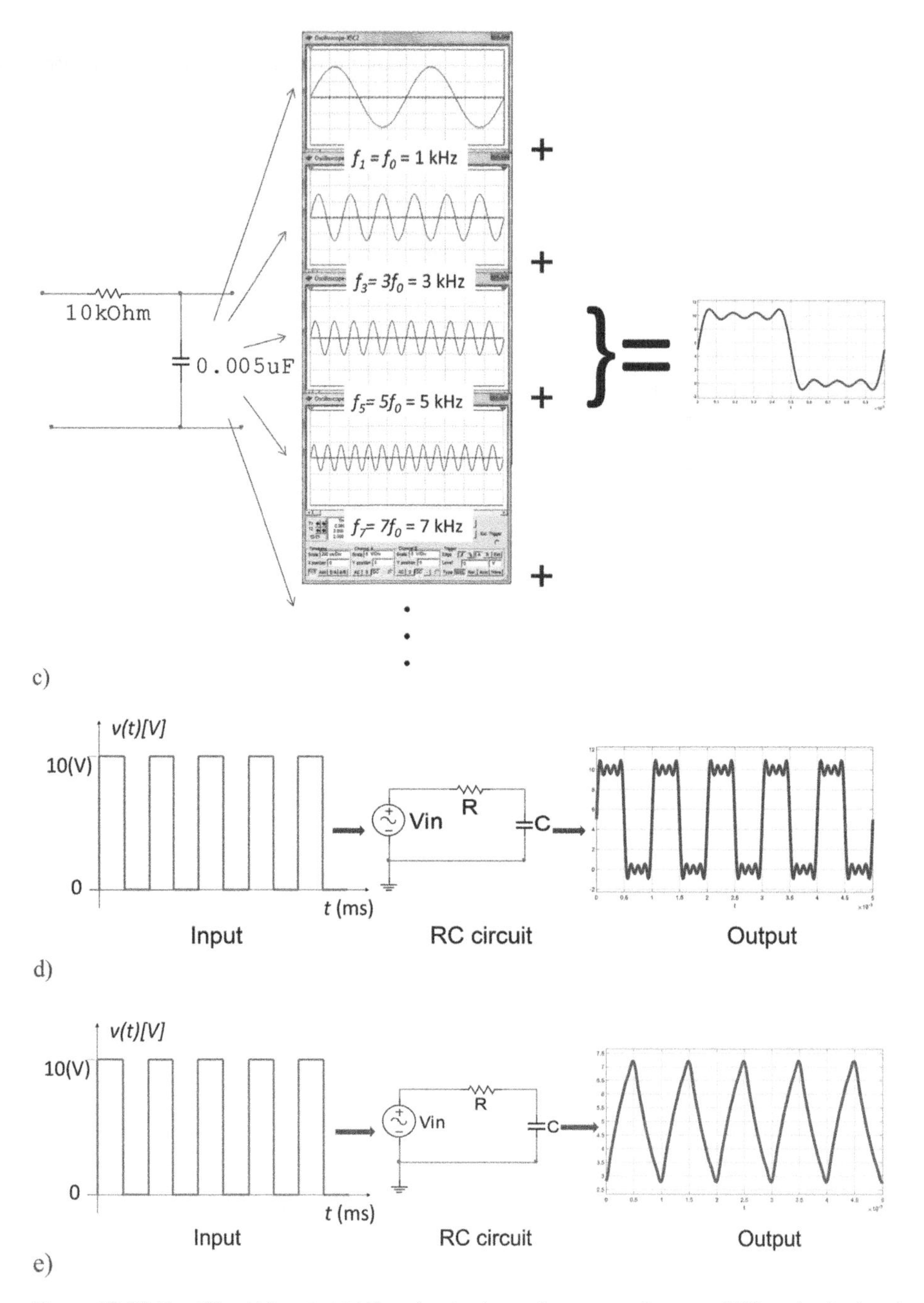

Figure 10.13 (Cond't) b) input at LHS and output over input waveforms at RHS, and calculated output amplitudes and phase shifts (in the middle); c) building the waveform of an output pulse by employing the spectral synthesis; d) the input and output pulse trains at $f_c = 3.18\,kHz$; e) the input and output pulse trains at $f_c = 0.318\,kHz$. (Figures 10.13a, 10.13b, 10.13c, and 10.13d are from Figure 6.2.5, and Figure 10.13e is the part of Figure 6.2.10 from Mynbaev/Scheiner, Essentials of Modern Communications, Wiley, 2020, pp. 555–558 and 564, respectively.)

10.15b demonstrate these phenomena consequently. For example, the 6.27-V amplitude of the first harmonic gets cut to 6.08 V at frequency $f_1 = 1kHz$, whereas the 1.27-V fifth harmonic amplitude drops to 0.68 V at $f_5 = 5kHz$. The amplitude values are computed using (10.20a), and Figure 10.15a visualizes these calculations and displays their meaning. This figure demonstrates

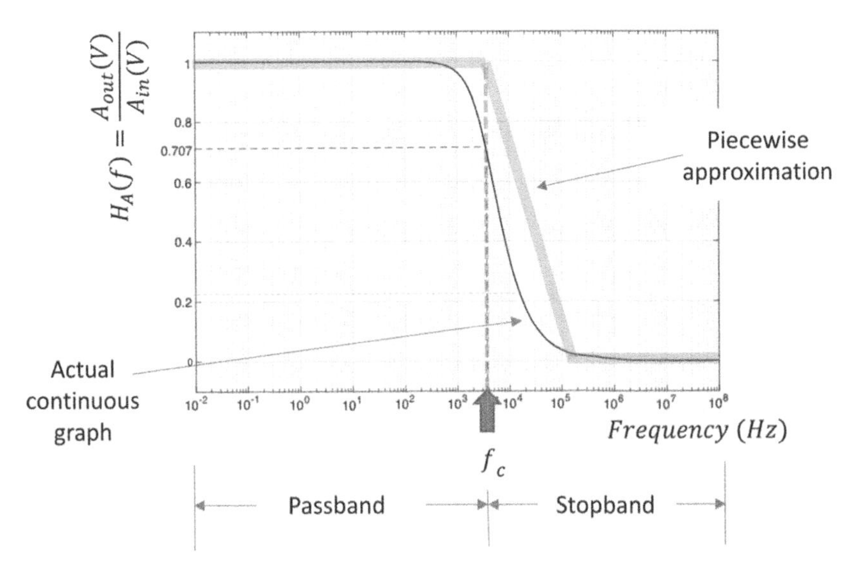

Figure 10.14 The actual continuous graph and its piecewise approximation of the amplitude part of the RC circuit transfer function.

how the amplitude cut depends on the relationship between the harmonic frequency and the circuit cutoff frequency.

- In addition to the amplitudes, we can compute the phase changes applying (10.20b). Graphically, the results are shown in Figure 10.15b. Verify our results by computing, for instance, θ_{out} of the fifth harmonic.
- Compare Figures 10.13b with 10.15a and 10.15b: Figure 10.13b shows the results of the harmonics passing through the RC circuit in the time domain, whereas Figures 10.15a and 10.15b show the same results in the frequency domain. Now, *we understand why the circuit output signal differs from its input and how these changes can be computed!*
- To demonstrate the main point of this discussion—why and how much the periodic signal changes by passing through an RC circuit—let's consider the role of the circuit's cutoff frequency. Figure 10.16a and 10.16b demonstrate how f_C affects the output by comparing the amplitudes and phase changes in the frequency domain. Liken, for example, the output amplitudes of the fifth harmonics: $A_5^{out}(V_{out}) = 0.68\,V\ at\ f_c = 3.18kHz$ and $A_5^{out}(V_{out}) = 0.081V\ at\ f_c = 0.318kHz$. Also, $\theta_5^{out}(V_{out}) = -147.5^0\ at\ f_c = 3.18kHz$ and $\theta_5^{out}(V_{out}) = -176.3^0 at\ f_c = 0.318kHz$. To complete the picture, compare Figures 10.13d and 10.13e with 10.16a and 10.16b: They show the same results but the former does it in the time domain, and the latter displays them in the frequency domain. Analyze these figures closely: *It is the best way to understand the role of spectral analysis in finding the circuit's response!*
- The preceding discussion and Example 10.3 demonstrate how much the spectral analysis enhances our understanding of the input–output problem. Now we know why an input signal changes after passing through an electrical circuit and, most importantly, we can calculate the expected changes. The spectral analysis also serves as a tool for controlling the input–output relationship in the desired way, enabling a meaningful design.

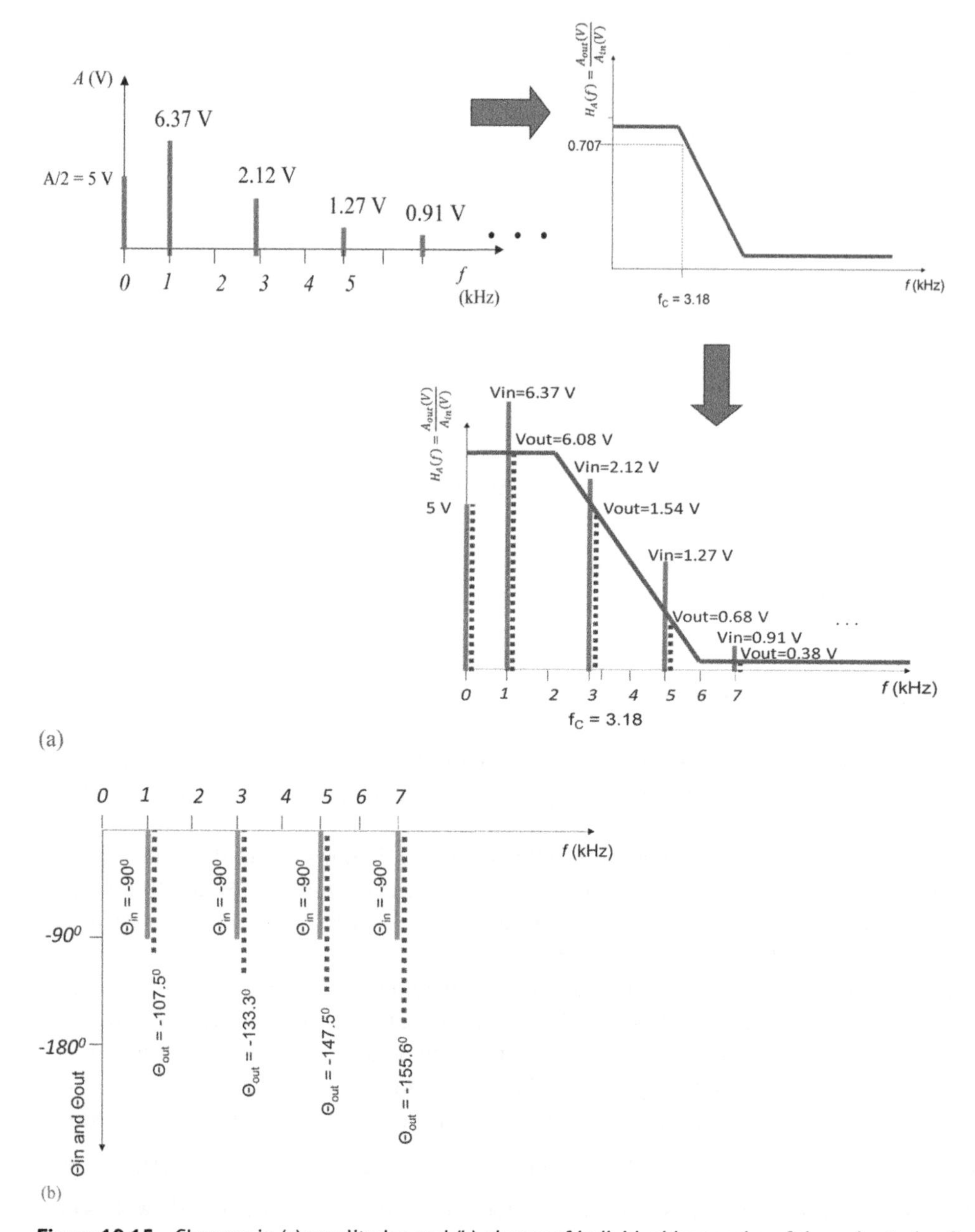

Figure 10.15 Changes in (a) amplitudes and (b) phases of individual harmonics of the pulse train after passing through the RC circuit in Example 10.3. (Figure 6.2.7 from Mynbaev and Scheiner, *Essentials of Modern Communications*, Wiley, 2020, p. 560.)

Questions

1) Review Figures 6.26b, 6.26c, and 6.26d in Section 6.3: They show how the output of an RC circuit degrades with the reduction of the input pulse duration, $\tau(s) \equiv \Delta T$. How will you interpret these results based on the spectral-analysis standpoint? (Here, we use $\tau(s)$ *instead of* ΔT to denote a pulse's duration.)

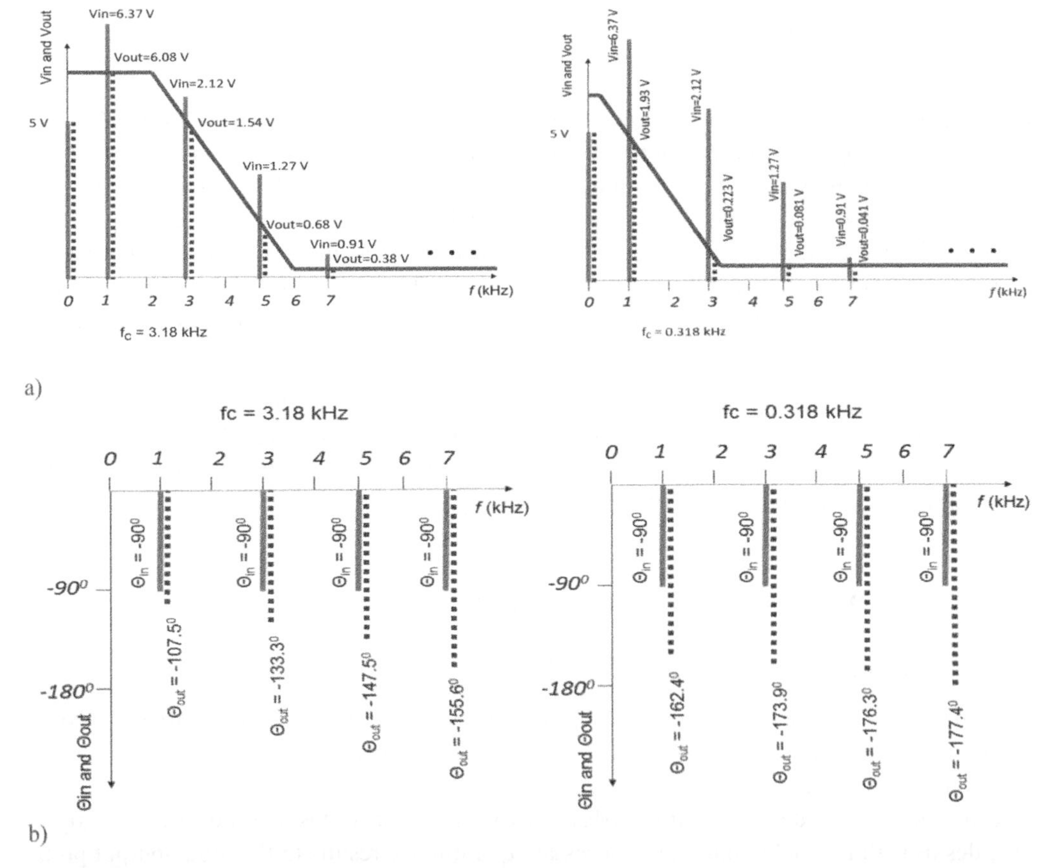

Figure 10.16 The role of cutoff frequency in passing a pulse train through the RC circuit: a) Comparison of the amplitude cuts; b) likening the phase changes. (Figure 6.2.11 from Mynbaev and Scheiner, *Essentials of Modern Communications*, Wiley, 2020, p. 565.)

2) Subsection "*Bit rate and the output waveform*" in Section 6.3 discusses the role of $\frac{\Delta T}{RC} = \frac{\tau}{RC}$ parameter in shaping the output waveform. What is the meaning of this parameter from the spectral-analysis view?

10.1.7 A Brief Conclusion of Section 10.1

To sum up:

- The same signal can be seen in the time domain and frequency domain.
- The basis of such seeing is the agreement of showing an individual sinusoid given in the time domain as a single line with the given amplitude, phase shift, and frequency in the frequency domain. Such frequency-domain presentation implies the signal's sinusoidal waveform. The tools that enable us to materialize the above agreement are the *Fourier theorem* and the *Fourier series.*
- The Fourier theorem states that every periodic signal can be presented as a sum of the sines and cosines, and the Fourier series specifies the amplitudes and frequencies of these sinusoids.
- The Fourier series bridges the presentation of a periodic signal in both domains.

- Finding the frequency domain components (harmonics) of a given time-domain signal is called *spectral analysis*; restoring the signal's waveform from its spectral components is known as *spectral synthesis*.
- Figures 6.18a and 6.18b in Section 6.3 show the changes in the output waveforms of an RC circuit caused by the variations in the duration of an input pulse. The discussion of these figures explains that those changes are due to the finite time needed for charging and discharging the circuit's capacitor. While this discussion reasonably answers such an indicative question, it doesn't provide a constructive technique for predictive calculations of the output signal parameters and the values of R and C parameters needed for the desired pulse response. Only spectral analysis can overcome this shortcoming, as Section 9.3 demonstrates. Nonetheless, based on the Fourier series, this analysis can be applied only to periodic signals.

For a deeper understanding of spectral analysis and synthesis with the Fourier series, we refer our readers to Mynbaev and Scheiner, Wiley, 2020, pp. 497–613, where the material is presented in the same style and approach. Of course, many other books in circuit analysis discuss the Fourier spectral analysis; they are listed in our Bibliography section.

10.2 The Fourier Transform and Its Applications

10.2.1 Why and What of the Fourier Transform

(**Warning:** Be aware that, following the long-standing academic tradition, the Fourier transform and its applications use $\tau(s)$ to designate the pulse duration, whereas this quantity was denoted $\Delta T(s)$ in the preceding chapters.)

Section 10.1, devoted to the Fourier series discussion, demonstrates that the spectral analysis provides us with the tool to obtain the necessary quantitative results in the input–output problem for advanced circuit analysis. These calculations enable us to assess and predict the circuit's behavior when an excitation is a periodic input. But we know that most input signals we meet in practice are non-periodic. How can we apply the spectral analysis in these cases? To find the answer, let's stretch a Fourier series to the case of a non-periodic signal.

10.2.1.1 From a Periodic to Non-Periodic Signal

Consider Figure 10.17a, where a square wave is shown. As mentioned in Section 10.1, this signal is a pulse train whose pulse width (duration), $\tau(s)$, is equal to half of the signal's period, $T(s)$, i.e., $\frac{\tau}{T}=\frac{1}{2}$. Such a ratio, called the *duty cycle*, is the main characteristic of a pulse train,

$$Duty\,cycle=\frac{\tau}{T}. \tag{10.22}$$

Not every pulse train has a duty cycle that equals ½; on the contrary, all others, except the square wave, have duty cycles of different values. For example, the pulse train in Figure 10.17b has a duty cycle equal to $\frac{1}{10}$. In Figure 10.17b, the duty cycle has changed compared to Figure 10.17a by decreasing of τ, whereas the period remains constant.

The signals shown in Figures 10.17a and 10.17b are still periodic and are described by the Fourier series. To find how we can describe a non-periodic signal, let's tend this periodic pulse train to a single pulse (a non-periodic signal, that is) and see how its Fourier series will evolve. Figures 10.18 demonstrate how the amplitude and phase spectra of the given pulse train changes if its τ stays constant but the period increases. A sample of the MATLAB code for building the graphs is shown below.

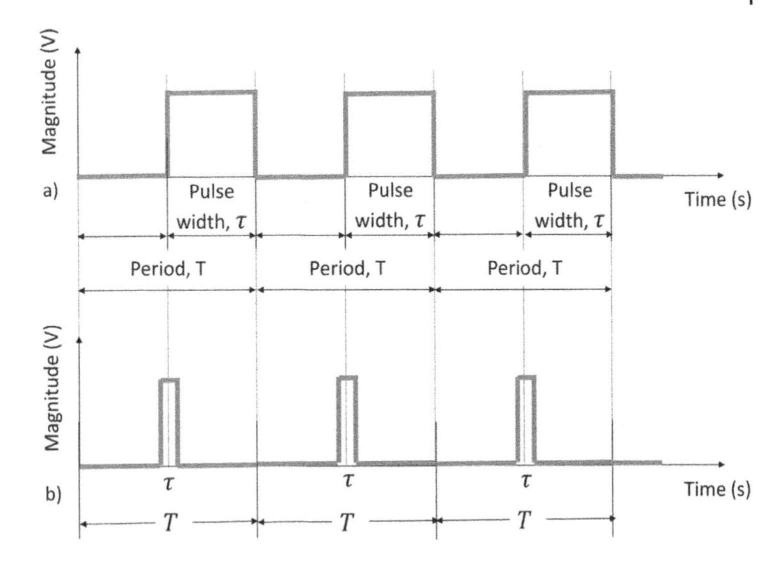

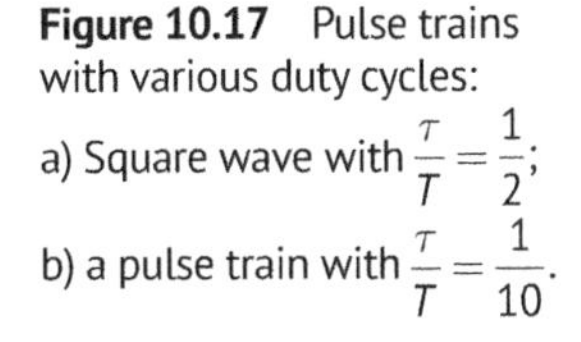

Figure 10.17 Pulse trains with various duty cycles:
a) Square wave with $\frac{\tau}{T} = \frac{1}{2}$;
b) a pulse train with $\frac{\tau}{T} = \frac{1}{10}$.

10.2.1.2 MATLAB Code for Building the Waveforms and Spectra of a Pulse Train for T = 8τ in Figure 10.18b[2]

```
tau=T/8:

V=input(‘Enter the amplitude [V]’);
T=input(‘Enter the pulse period [sec]’);
N=12; tau=T/8; %Pulse period [sec]
fs=1./T; %Sampling frequency [Hz]
D=tau./T; %Duty cycle
N=N/2; t=-0.02:T/512:0.02; %Adjusting time
pulseperiods=[-N:N]*T; %Pulse period
%The function of the pulse train:
v=V.*pulstran(t,pulseperiods,@rectpuls,tau);
subplot(3,1,1), plot(t,v), grid
axis([-0.02 0.02 -1 V+1]);
title(‘The waveform of a pulse train with tau=T/8’);
xlabel(‘Time [sec]’);
ylabel(‘Amplitude [V]’);
%Amplitude spectrum of the pulse train:
a0=V*D; %The DC component a0
n=-12:12; %Adjusting the number of harmonics
an=V*sin(pi*n*D)./(n*pi); %The an coefficients
Phase=atan2(0,an); %Phase in radians
An=sqrt(an.*an); %The bn coefficients are all zero
fn=n./T; %Adjusting frequency
subplot(3,1,2), stem([0 fn],[a0 An]), grid
xlabel(‘Frequency [Hz]’), ylabel(‘Amplitude [V]’)
title(‘Amplitude spectrum of the pulse train with tau=T/8’)
subplot(3,1,3), stem([0 fn],[0 Phase]); grid
title(‘Phase spectrum of the pulse train with tau=T/8’);
xlabel(‘Frequency [Hz]’), ylabel(‘Phase [rads]’)
```

MATLAB code for building the waveforms and spectra of a pulse train for $T \to \infty$ in Figure 10.18d[2]

```
%The waveform of a rectangular pulse:
syms w t
H=heaviside(t+0.5)-heaviside(t-0.5);
ezplot(H, [-10,10])
axis ([-10 10 0 1.1])
grid

% amplitude spectrum continuous
syms w t
H=heaviside(t+0.5)-heaviside(t-0.5);
F=fourier(H);
AF=abs(F);
ezplot(AF,[-7,7])
grid

% phase spectrum continuous
syms w t
H=heaviside(t+1)-heaviside(t-1);
F=fourier(H);
ANF=angle(F);
ezplot(ANF,[-4,4])
grid
```

In analyzing Figures 10.18a through 10.18c, pay close attention to the frequency scale. Also, understand that this figure shows two-sided pictures whose frequencies last from $-\infty$ to $+\infty$, while our preceding discussions were restricted by the one-sided spectra. The pulse width, τ, remains constant and equals 0.00164 (s).

When $T = 2\tau$, there is a classical square wave shown in Figure 10.18a. We can see a regular periodic pulse train whose amplitude spectrum is a set of amplitude-decreasing lines, and the phase spectrum is a set of lines of constant heights. Compare this figure with Figures 10.8a, 10.8b, and 10.8c. Count how many pulses are placed within time interval from $-0.02(s)$ to $+0.02(s)$. Does this number confirm the values of τ and T?

When $T = 8\tau$, the same time interval can accommodate only three pulses, as seen in Figure 10.18b. Observe how frequency intervals change in this figure's amplitude and phase spectra and notice how these spectra forms alter. Note that the amplitude spectrum has the first zero at around $\pm\frac{1}{\tau} = \pm 610(Hz)$, and the phase spectrum is zero within this frequency range.

The main changes can be observed at $T = 64\tau$. Now, only one pulse appears between $-0.02(s)$ and $+0.02(s)$, and the amplitude and phase spectra change drastically within the slightly decreased frequency window, as Figure 10.18c displays. Observe that the amplitude spectrum has a main lobe restricted by $\pm\frac{1}{\tau} = \pm 610\,(Hz)$ window, and the phase spectrum is zero within this frequency scope. But the most significant change is this: *As the period of a pulse train*

2 Developed by Ina Tsikhanava.

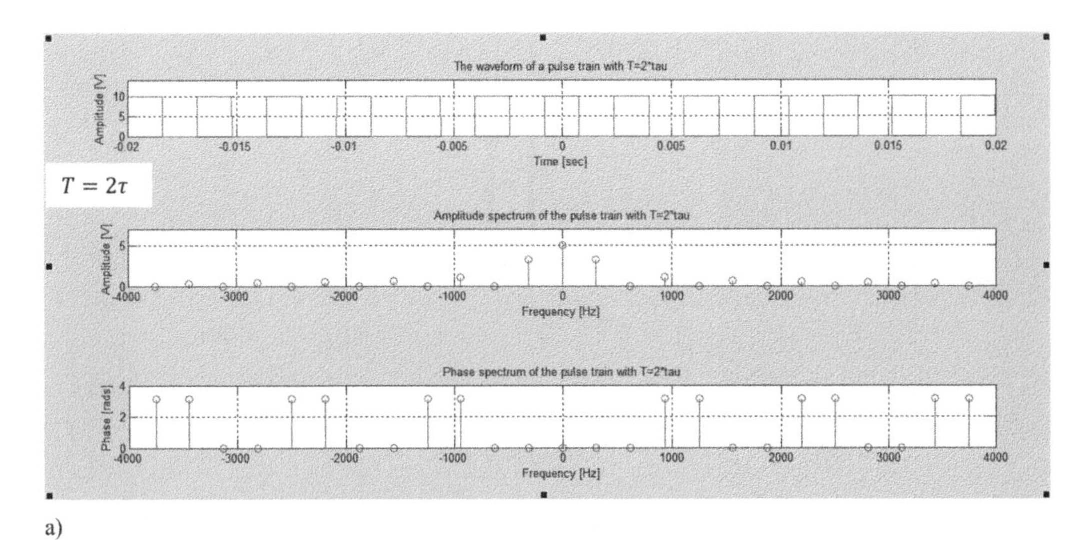

a)

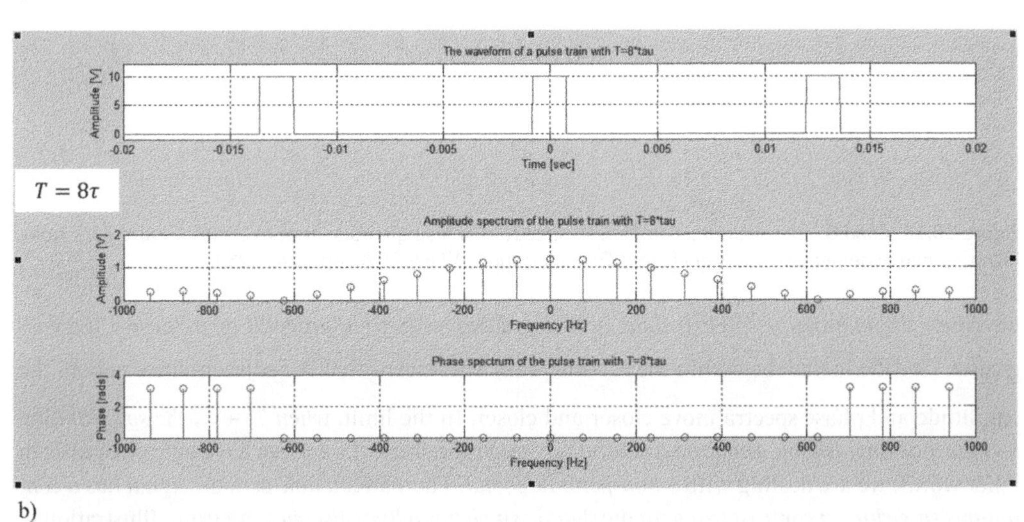

b)

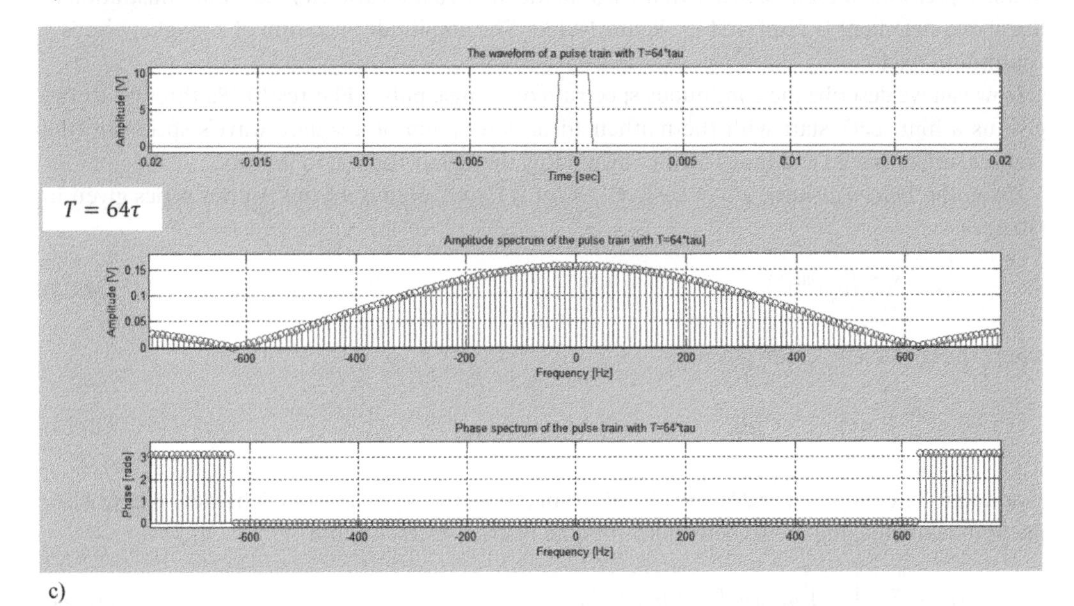

c)

Figure 10.18 Waveforms and spectra of a pulse train when T tends to infinity: a) $T = 2\tau$; b) $T = 8\tau$; c) $T = 64\tau$, but the frequency scale slightly changed to demonstrate the whole picture of the waveform and the spectra of the pulse train;

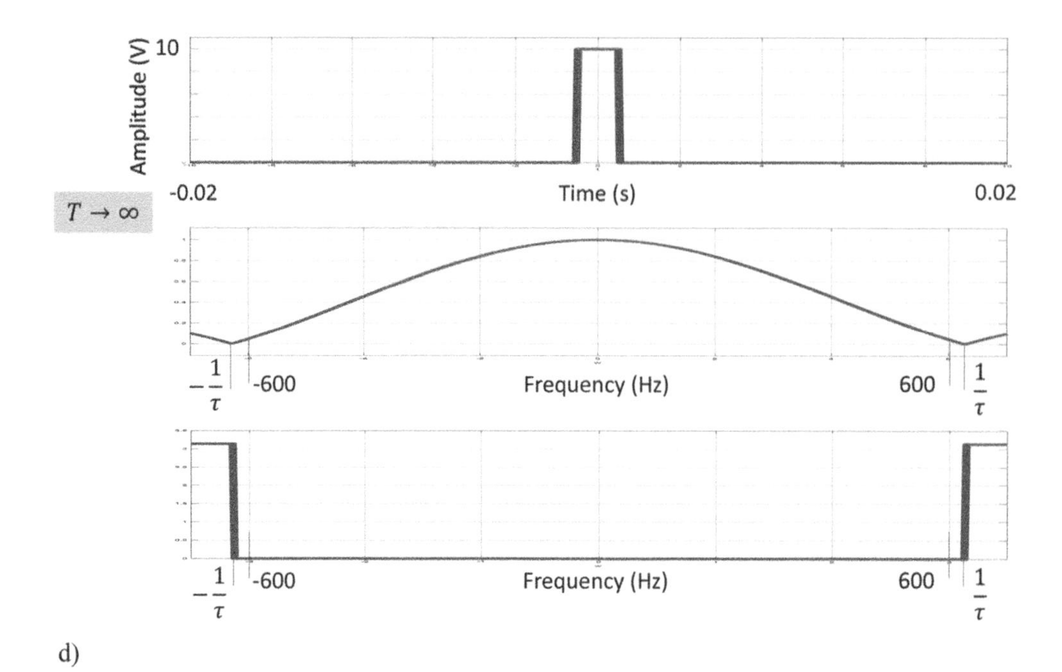

Figure 10.18 (Cond't) d) continuous amplitude spectrum of a single pulse (not to scale). (Figure 7.1.3 from Mynbaev and Scheiner, *Essentials of Modern Communications*, Wiley, 2020, pp. 626–628.)

increases, the number of spectral lines accommodated within one interval of frequency becomes greater. (See the main lobe within the $\pm\frac{1}{\tau}$ interval.) This means that the adjacent lines in both the amplitude and phase spectra move closer and closer. In the limit, *when $T \to \infty$, the spectral lines become indistinguishable and form a continuous spectrum.* But at $T \to \infty$, we have only one pulse; in other words, we are dealing with a *non-periodic signal.* Therefore, *a non-periodic signal has a continuous spectrum, in contrast to a periodic signal, which has a line (discrete) spectrum.* Illustration to the above statement is displayed in Figure 10.18d: The amplitude spectrum of a single pulse is a continuous one!

How can we describe the continuous spectrum of a single pulse? Figures 10.18a through 10.18d give us a hint: Let's start with the mathematical description of a square wave's spectrum (the Fourier series, that is) and move further by tending the signal's period to infinity.

Using the *Euler's identity*, $e^{j\omega t} = \cos(\omega t) + j\sin(\omega t)$, we can present the Fourier series given in (10.6) as

$$v(t) = \sum_{n=-\infty}^{n=\infty} \left(c_n e^{jn\omega_0 t}\right), \tag{10.23}$$

where the coefficients c_n are given by

$$c_n = \frac{1}{T}\int_{t}^{t+T} v(t)\left(e^{-jn\omega_0 t}\right)dt. \tag{10.24}$$

(For details of obtaining (10.23) and (10.24) from (10.6), see Mynbaev and Scheiner, Wiley, 2020, pp. 537–538.) Plugging these coefficients into the Fourier series, we find

$$v(t) = \sum_{n=-\infty}^{n=\infty} \left(\frac{1}{T}\int_{t}^{t+T} v(t)\left(e^{-jn\omega_0 t}\right)dt\right) e^{jn\omega_0 t}). \tag{10.25}$$

The interval between two adjacent spectral lines can be found as

$$\Delta\omega = (n+1)\omega_0 - n\omega_0 = \omega_0 = \frac{2\pi}{T}. \tag{10.26}$$

Now, we can reformat (10.25) with explicit reference to the spacing between the adjacent spectral lines, $\Delta\omega$, as

$$v(t) = \sum_{n=-\infty}^{n=\infty} \left(\frac{1}{2\pi} \int_{t}^{t+T} v(t)(e^{-jn\omega_0 t})dt \right| e^{jn\omega_0 t})\Delta\omega. \tag{10.27}$$

Let's tend the period to infinity, $T \to \infty$, which gets the spectral lines closer and closer, meaning that *in the limit, the incremental spacing, Δω, turns to the differential, dω*. In other words,

$$when\ T \to \infty, \frac{2\pi}{T} \to 0,\ and\ \Delta\omega \to d\omega.$$

This condition also entails changing from discrete harmonics, $n\omega_0$, to continuous frequency, ω, and replacing the summation by integration. Therefore, *when* $T \to \infty$, (10.27) becomes

$$v(t) = \frac{1}{2\pi} \int_{-\infty}^{\infty} \left(\int_{-\infty}^{\infty} v(t)(e^{-j\omega t})dt \right| e^{\omega t} d\omega. \tag{10.28}$$

Let's denote the core of this integral as

$$F(\omega) = \int_{-\infty}^{\infty} v(t)(e^{-j\omega t})dt \tag{10.29}$$

and call it the Fourier transform.

Thus, *the Fourier transform is the integral operation given in (10.29) that takes a time-domain function v(t) and transforms it into the frequency-domain function F(ω).*

Equation 10.30 puts these words in a mathematical format as

$$\mathcal{F}\{v(t)\} = F(\omega) = \int_{-\infty}^{\infty} v(t)(e^{-j\omega t})dt, \tag{10.30}$$

where $\mathcal{F}$ *is the Fourier-transform operator and F(ω) is the result of this operation (the Fourier transform).*

Similarly to the Laplace transform, the *inverse Fourier transform* enables obtaining a time-domain function $v(t)$ from the frequency-domain one F(ω), that is,

$$\mathcal{F}^{-1}\{F(\omega)\} = \frac{1}{2\pi} \int_{-\infty}^{\infty} F(\omega)(e^{j\omega t})d\omega = v(t). \tag{10.31}$$

It's worth revisiting the Laplace transform definition and its discussion in Sections 7.1 and 7.2 of this book to recall such entities as function and transform and the accompanying mathematical operations.

Get back to Figure 10.18d: The first zeros of the amplitude spectrum of a pulse train in this figure are at the frequencies ±1/τ(s). *The frequency range, from –1/τ to 1/τ, can be considered a measure of the pulse bandwidth because most of the signal's power is concentrated here.* Therefore,

$$BW(Hz) \approx \frac{2}{\tau}. \tag{10.32}$$

To understand what the Fourier transform means and does, let's consider an example.

Example 10.4 The Fourier transform of a decaying exponential time-domain function.

Problem: Find the Fourier transform of the decaying exponential function $v(t) = e^{-\alpha t}u(t)$.

Solution: Recall that the multiplication of any function by a unit step (Heaviside) function *u(t)* simply means that the given function exists only on the positive half of axis *t*. Now, refer to the definition of the Fourier transform operation given in (10.30):

$$\mathcal{F}\{v(t)\} = F(\omega) = \int_{-\infty}^{\infty} v(t)\left(e^{-j\omega t}\right)dt. \tag{10.30R}$$

Plug the given time-domain function into (10.30) and find

$$\mathcal{F}\{v(t)\} = \int_{\infty}^{-\infty} e^{-\alpha t}u(t)(e^{-j\omega t})dt = \int_0^{\infty} e^{-(\alpha+j\omega)t}dt = \left[-\frac{e^{-(\alpha+j\omega)t}}{\alpha+j\omega}\right]\Bigg|_0^{\infty} = \frac{1}{\alpha+j\omega}. \tag{10.33}$$

Thus,

$$\mathcal{F}\left\{e^{-\alpha t}u(t)\right\} = F(\omega) = \frac{1}{\alpha+j\omega}. \tag{10.34}$$

The problem is solved.

Discussion:

- These manipulations show that the Fourier transform of a time-domain function $e^{-\alpha t}u(t)$ results in obtaining the frequency-domain function $\frac{1}{\alpha+j\omega}$, as (10.34) demonstrates.
- We can perform these mathematical manipulations by using the Symbolic Math Toolbox. For the time-domain functions considered in this example, we use the following MATLAB code to determine the required Fourier transform:
  a) The Fourier transform of $v(t) = e^{-\alpha t}u(t)$ with $\alpha = 3$:

```
%Fourier transform
syms w t a
a=3;
H=sym('heaviside(t)');
F=fourier(H*exp(-3*t));
pretty(F)
```

  Answer: $F = \frac{1}{wi+3}$

  (Note that a unit step (Heaviside) function *u(t)* is denoted "H" in this MATLAB script.)

  b) As an illustration of how to obtain the *inverse Fourier transform*, $\mathcal{F}^{-1}\{F(\omega)\}$, consider the following MATLAB script for $F(\omega) = 1/(j\omega + 3)$ from this example:

```
%Inverse FT
syms w t a
a=3; %Defining a
v=ifourier(1/(w*i+a),w,t);
v=subs(v,'heaviside(t)','H') %replacing heaviside(t) by H for function v
```

  Answer: v(t) = H*exp(-3*t), as expected.

  Thus, MATLAB makes finding the direct and inverse Fourier transforms of given functions an easy operation.

  (Exercise: Build the graph of v(t) = H*exp(-3*t). (Hint: Use the MATLAB plotting ability.))

10.2.2 The Concept and Table of the Fourier Transforms

Now, we firmly comprehend that

> *the Fourier transform is a mathematical operation that transforms a non-periodic time-domain function into a frequency-domain function.*

Indeed, the function $v(t) = e^{-\alpha t}u(t)$, which is a non-periodic and depends only on t, is transformed by the Fourier integral into the function $F(\omega) = \frac{1}{\alpha + j\omega}$, which depends only on ω. Conceptually, this operation is depicted in Figure 10.19a. The idea of the inverse Fourier transform is shown in Figure 10.19b.

If we continue applying the Fourier integral 10.29 to various time-domain functions, we will obtain their Fourier transforms. The results of such manipulations are typically tabulated; Table 10.1 presents several samples of the Fourier transform pairs.

The first glance at the Table 10.1 reveals that the left column presents the given time-domain functions, and the second right columns exhibits the corresponding frequency-domain functions obtained by applying the Fourier transforms to the given $v(t)$ functions. These transformations from time domain to frequency domain are reciprocal; that is, a time-domain function can be obtained by applying the inverse Fourier transform to the corresponding frequency-domain function. For instance, applying the Fourier transform to $v(t) = e^{-\alpha t}u(t)$, results in $F(\omega) = \mathcal{F}(e^{-\alpha t}u(t)) = \frac{1}{\alpha + j\omega}$. Inversely, $\mathcal{F}^{-1}\left\{\frac{1}{\alpha + j\omega}\right\} = e^{-\alpha t}u(t)$. This result is demonstrated in Example 10.4.

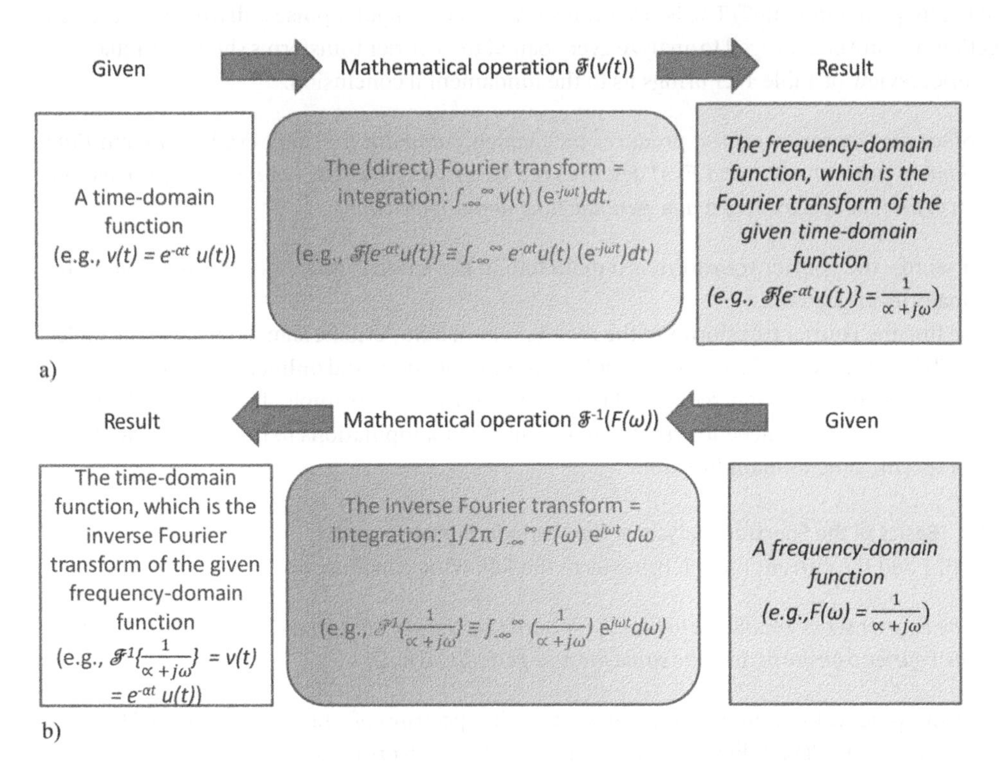

Figure 10.19 Conceptual visualization of (a) the (direct) Fourier transform and (b) the inverse Fourier transform.

Table 10.1 The Fourier Transforms.

Time-domain function, $v(t) = \mathcal{F}^{-1}(F(\omega))$	Fourier → Transform, $\mathcal{F}(v(t))$ ← Inverse Fourier transform, $\mathcal{F}^{-1}(F(\omega))$	Frequency domain function, $F(\omega) = \mathcal{F}(v(t))$	Conditions and comments
$p(t) = \begin{cases} A \text{ for } \lvert t \rvert < \tau \\ 0 \text{ for } \lvert t \rvert > \tau \end{cases}$	←→	$A\tau \cdot sinc\left(\frac{\omega\tau}{2}\right)$	τ-duration of rectangular pulse, p(t)
$e^{-\alpha t}u(t)$	←→	$\frac{1}{\alpha + j\omega}$	$\alpha > 0$, constant
$t^n \cdot e^{-\alpha t}u(t)$	←→	$\frac{n!}{(\alpha + j\omega)^{n+1}}$	$\alpha > 0$, constant
1	←→	$2\pi\delta(\omega)$	$\delta(\omega)$ – delta function
$u(t)$	←→	$\pi\delta(\omega) + \frac{1}{j\omega}$	$u(t)$ – step function
$e^{j\omega_0 t}$	←→	$2\pi\delta(\omega - \omega_0)$	
$\cos(\omega_0 t)$	←→	$\pi[\delta(\omega - \omega_0) + \delta(\omega + \omega_0)]$	
$\sin(\omega_0 t)$	←→	$\pi[\delta(\omega + \omega_0) - \delta(\omega - \omega_0)]$	

The delta function in frequency domain, $\delta(\omega_0)$, is depicted as a vertical line located at ω_0. Thus, the Fourier transform of a sinusoidal signal results in two vertical lines located at $\pm\ \omega_0$, which is a two-sided spectrum of a sinusoidal signal. (Do you remember that the spectrum of a sinusoid is one line in the frequency domain?) The Fourier transform of a rectangular pulse is discussed in the next subsection. You are encouraged to analyze every pair of the Fourier transforms shown in Table 10.1.

This brief review of Table 10.1 brings us to the fundamental conclusion:

> *Since the Fourier transform produces the frequency-domain function, F(ω), of a given time-domain function, v(t), and F(ω) describes the spectrum of v(t), the Fourier transform enables us to find the spectrum of a non-periodic time-domain function.*

In other words, the Fourier transform is a main tool in the spectral analysis of non-periodic (that is, vast majority of) signals.

We can find the Fourier transforms of the most common non-periodic signals by using the tables, similar to Table 10.1, available in any textbook on communications and online. Of course, MATLAB also enables us to obtain the required Fourier transforms, as Example 10.4 shows. Thus, it's extremely seldom that we need to resort to mathematical manipulations to find the Fourier transform of a specific time-domain function.

10.2.2.1 Basics of the Spectral Analysis of Non-Periodic Signals

Section 10.1 and the current section bring us to the following conclusion:

> *The Fourier series enables us to find the spectrum of the periodic time-domain signals, whereas the Fourier transform does the same for non-periodic signals.*

But, from a practical standpoint, how can we find the spectrum of a non-periodic signal by using its Fourier transform? The following three-step algorithm delivers the answer:

1) *Find the Fourier transform, $F(\omega)$, of a non-periodic signal, v(t),*

2) *Determine the signal's amplitude spectrum by finding the absolute value of* $F(\omega)$ *as*

$|F(\omega)|$,

3) *Ascertain the signal's phase spectrum by discovering the argument of* $F(\omega)$, *arg* $[F(\omega)]$.

Let's turn to the examples of finding the spectra of the non-periodic signals.

Example 10.5 The spectrum of the causal exponential decaying signal.

Problem: Find the amplitude and phase spectra of the causal exponential decaying signal, $v(t) = Ae^{-\alpha t}u(t)$, with amplitude $A = 1$(V) and decay constant $\alpha = 3$ (1/s). Plot the waveform, amplitude, and phase spectra of the given function.

Solution:

We found the Fourier transform of this function in Example 10.4 as

$$\mathcal{F}e^{-\alpha t}u(t)\} = \frac{1}{\alpha + j\omega}. \tag{10.33R}$$

Multiplying both sides of (10.33) by constant A results in finding the needed Fourier transform as

$$F(\omega) = \frac{A}{\alpha + j\omega} = \frac{1}{3 + j\omega}.$$

This is a complex function of frequency; its absolute value (modulus) gives us the amplitude spectrum, whereas its argument gives us the *phase spectrum*. To find this modulus and the argument, we present $\frac{A}{\alpha + j\omega}$ in a classical rectangular form, namely

$$F(\omega) = \frac{A}{\alpha + j\omega} = A\left(\frac{\alpha}{(\alpha^2 + \omega^2)} - j\frac{\omega}{(\alpha^2 + \omega^2)}\right). \tag{10.35}$$

The modulus of this complex number is equal to

$$|F(\omega)| = A\sqrt{\left(\frac{\alpha}{(\alpha^2 + \omega^2)}\right)^2 + \left(-j\frac{\omega}{(\alpha^2 + \omega^2)}\right)^2} = \frac{A}{\sqrt{(\alpha^2 + \omega^2)}} = \frac{1}{\sqrt{(3^2 + \omega^2)}} \tag{10.36}$$

and its argument is given by

$$\theta = \arg(F(\omega) = -tan^{-1}\left(\frac{\omega}{\alpha}\right) = -tan^{-1}\left(\frac{\omega}{3}\right). \tag{10.37}$$

The required graphs are presented in Figures 10.20a, 10.20b, and 10.20c. The needed MATLAB code is given below.

```
%Waveform of the causal decaying exponential
%function, v(t) = e-3*t*u(t):
syms w t
v = exp(-3*t);
ezplot(v, [0,3])
axis([0 3 0 1.1])
grid
```

```
xlabel('Time (s)')
ylabel('V(t) [V]')

%Amplitude spectrum of the Fourier transform of the causal
%decaying exponential function, F{v(t) = e-3*t*u(t)}?
syms w t
H=heaviside(t);
F=fourier(H*exp(-3*t));
AF=abs(F);
ezplot(AF,[-40,40])
axis([-40 40 0 0.36])
grid
xlabel('Frequency(rad/s)')
ylabel('Amplitude spectrum of exp(-3*t) [V.s]')

%Phase spectrum of the Fourier transform of the causal
%decaying exponential function, F{v(t) = e-3*t*u(t)}
syms w t
H=heaviside(t);
F=fourier(H*exp(-3*t));
ANF=angle(F);
ezplot(ANF,[-40,40])
axis([-40 40 -1.6 1.6])
grid
xlabel('Frequency(rad/s)')
ylabel('Phase spectrum of exp(-3*t) [rad]')
```

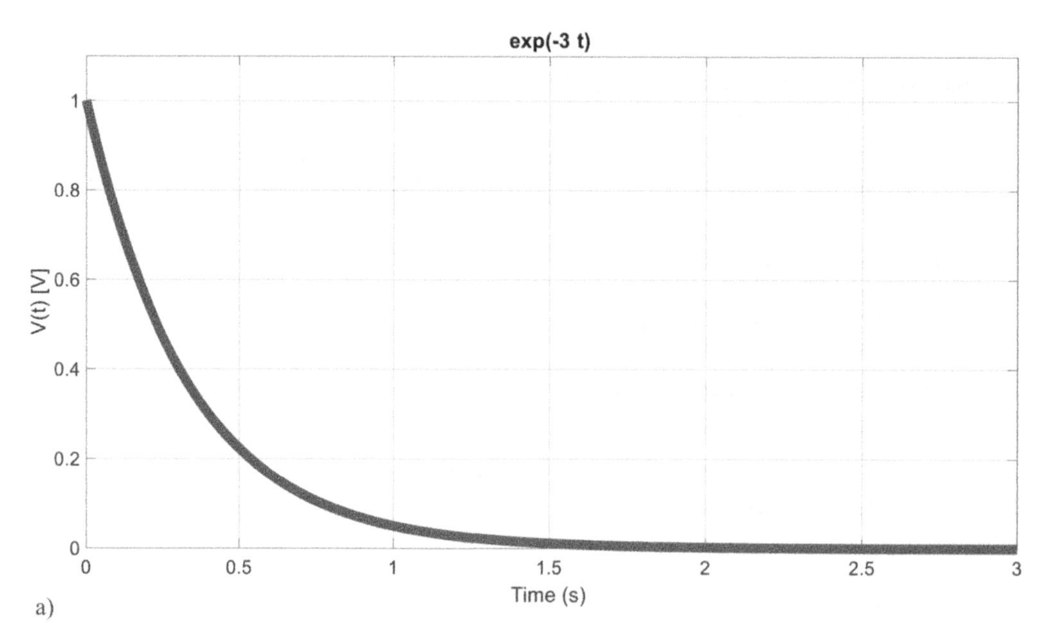

Figure 10.20 Causal decaying exponential signal $Ae^{-\alpha t} \cdot u(t)$ for Example 10.5: a) Waveform; b) amplitude spectrum; c) phase spectrum. (Figure 7.1.6 from Mynbaev and Scheiner, *Essentials of Modern Communications*, Wiley, 2020, pp. 636–637.)

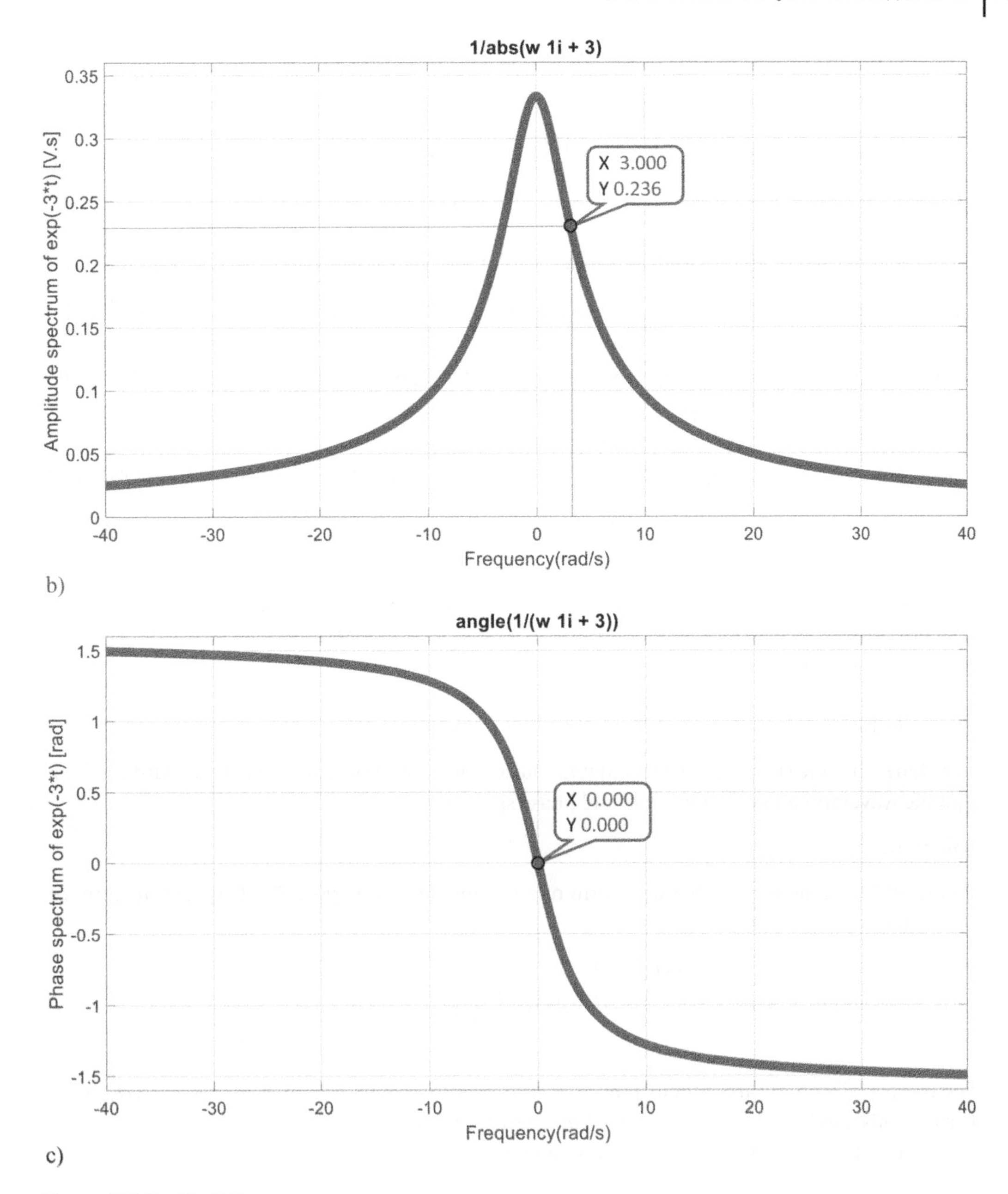

Figure 10.20 (Cont'd)

Discussion:

Let's analyze the graphical presentations of the solution.

The *waveform* of a decaying exponential function given in Figure 10.20a is our old friend and raises no questions. The graph starts at $v(0)=1(V)$, as it must be for the given function, and practically ends at about $t \approx 1.7\,(s)$, which corresponds to the criterion of $5\tau = \frac{5}{\alpha} = \frac{5}{3}\,(s)$.

The amplitude spectrum shown in Figure 10.20b must be equal to $\frac{A}{\alpha}$ at $\omega = 0$. Indeed, (10.36) requires that

$$|F(\omega)| = \frac{A}{\sqrt{(\alpha^2 + \omega^2)}} = \frac{1}{\sqrt{(3^2 + \omega^2)}} \Rightarrow |F(\omega)| = \frac{A}{\alpha} = \frac{1}{3}$$

when $\omega \to 0$.

On the other hand, (10.36) indicates that $|F(\omega)|$ must go to zero when $\omega \to \pm\infty$. Figure 10.20b shows that the graph $|F(\omega)|$ meets these requirements. In addition, when $\omega = \alpha = 3\left(\frac{rad}{s}\right)$, the amplitude must be equal to $\frac{A}{\sqrt{(\alpha^2 + \alpha^2)}} = \frac{1}{3\sqrt{2}} = 0.236$. Figure 10.20b displays this point, and its Data Tips box is placed on the amplitude graph.

The phase spectrum graph shows that when $\omega = 0$, the phase shift is zero, according to (10.37), $\theta = -tan^{-1}\left(\frac{0}{\alpha}\right) = 0$. Calculations show that when $\omega = \pm 40\left(\frac{rad}{s}\right)$, $\theta = \mp 1.55\,(rad)$. All these three points can be seen in Figure 10.20c.

This analysis doesn't verify the correctness of our solution but clarifies its meaning. We can confirm our answer by performing the *Multisim* experiment using a *spectrum analyzer*, but we leave this attempt to our readers as a *laboratory exercise.*

Example 10.6 Spectral analysis of a rectangular pulse.

Problem: Find the spectrum of a rectangular pulse whose amplitude $A = 1$ (V) and width $\tau = 2$ (s). Plot the waveform and the amplitude and phase spectra.

Solution:

Figure 10.21a demonstrates the waveform of a rectangular pulse, *p(t)*. The Fourier transform of this signal is given in Table 10.1 as

$$\mathcal{F}\{p(t)\} = P(\omega) = \frac{A\tau \sin\left(\frac{\omega\tau}{2}\right)}{\frac{\omega\tau}{2}} = A\tau\, sinc\left(\frac{\omega\tau}{2}\right).$$

Focus especially on the amplitude of the Fourier transform, Aτ, whose units are V·s = V/Hz; this is because this amplitude represents a *spectral density* of the signal.

The amplitude spectrum of a single pulse is given by

$$|P(\omega)| = \left|\frac{A\tau \sin\left(\frac{\omega\tau}{2}\right)}{\frac{\omega\tau}{2}}\right| = \left|A\tau \sin c\left(\frac{\omega\tau}{2}\right)\right| \tag{10.38}$$

and its phase spectrum is determined as

$$\arg(P(\omega)) = arg\left(\frac{A\tau \sin\left(\frac{\omega\tau}{2}\right)}{\frac{\omega\tau}{2}}\right) = \arg\left(A\tau\, sinc\left(\frac{\omega\tau}{2}\right)\right). \tag{10.39}$$

Let's insert $A = 1\ (V)$ and $\tau = 2(s)$ and analyze the *amplitude spectrum*, $\left|\frac{A\tau \sin\left(\frac{\omega\tau}{2}\right)}{\frac{\omega\tau}{2}}\right|$. For this example, the amplitude is equal to $A\tau = 2$ (V/Hz). Obviously, the plot of an *amplitude spectrum*, $\left|\frac{A\tau \sin\left(\frac{\omega\tau}{2}\right)}{\frac{\omega\tau}{2}}\right|$, reaches zero every time the argument of sine equals π, 2π, 3π, etc. Thus, the first zero will be reached at $\omega_1 \cdot \frac{\tau}{2} = \pi$; that is, at $\omega_1 = \frac{2\pi}{\tau}$ or $f = \frac{1}{\tau}$. For this example, $\tau = 2\ (s)$ and we find that $\omega_1 = \pi$. Hence, $\omega_2 = 2\pi$, $\omega_3 = 3\pi$, etc. If we want to see these points on a frequency scale, we need to divide ω by 2π to obtain Hz. This is what the graph shows in Figure 10.21b.

The *phase spectrum* of this signal is quite simple: Since arg($A\tau\, sinc\left(\frac{\omega\tau}{2}\right)$ is a real function, the phase shift can take either 0 (rad) or $\pi(rad)$ value. Figure 10.21c shows exactly this picture. Specifically,

θ is equal to 0 (rad) or 0^0 at intervals

- from $\omega = 0$ to $\omega_1 = \frac{\pi}{\tau/2} = \frac{2\pi}{\tau}$
- from $\omega_2 = \frac{4\pi}{\tau}$ to $\omega_3 = \frac{6\pi}{\tau}$, etc.

and $\pi\left(rad\right)$ or 180^0 at intervals

- from $\omega_1 = \frac{2\pi}{\tau}$ to $\omega_2 = \frac{4\pi}{\tau}$
- from $\omega_3 = \frac{6\pi}{\tau}$ to $\omega_4 = \frac{8\pi}{\tau}$, etc.

A similar distribution of the phase angles holds true for negative ω. This spectrum is shown in Figure 10.21c. However, the sign of the phase shift is negative for $\omega\left(\frac{rad}{s}\right) > 0$ and positive for $\omega\left(\frac{rad}{s}\right) < 0$, according to (10.39). This is why the angle at the left half of the phase shifts is shown as $-\pi$, whereas this angle for the RHS is given by π. Unfortunately, MATLAB ignores this difference in the sign.

MATLAB code for plotting the waveforms and spectra of a rectangular pulse for Figure10.21[3]

```
%The waveform of a rectangular pulse:
syms w t
H=heaviside(t+0.5)-heaviside(t-0.5);
ezplot(H, [-10,10])
axis ([-10 10 -0.1 1.1])
grid
xlabel('Time (s)')
ylabel('Rectangular pulse, P(t) [V]')
```

3 Developed by Ina Tsikhanava.

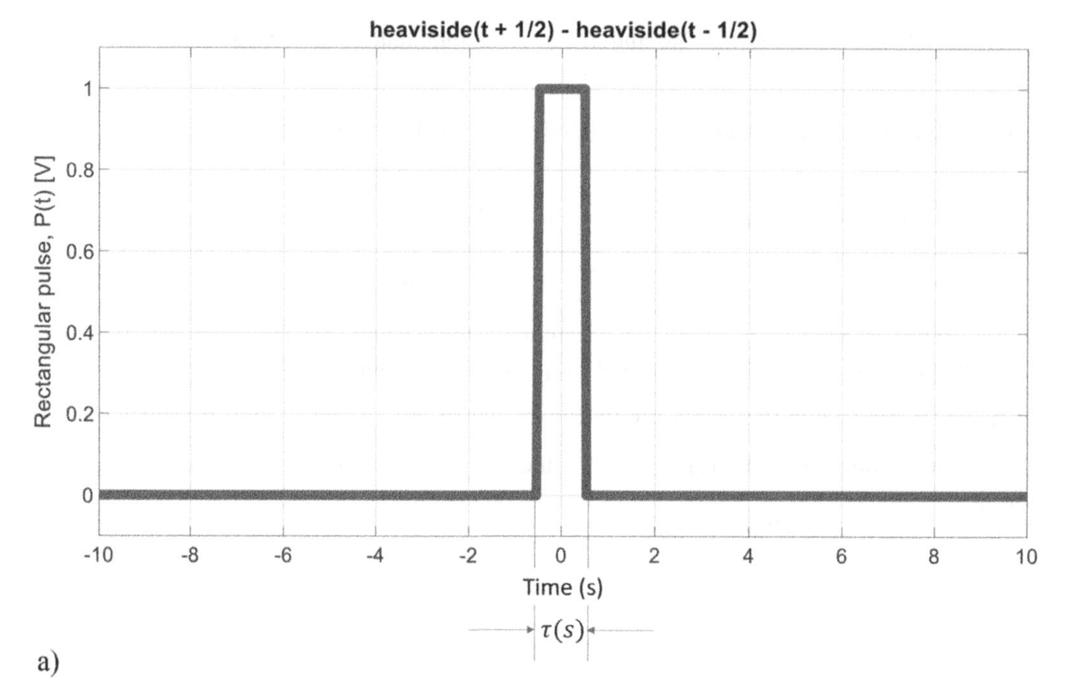

a)

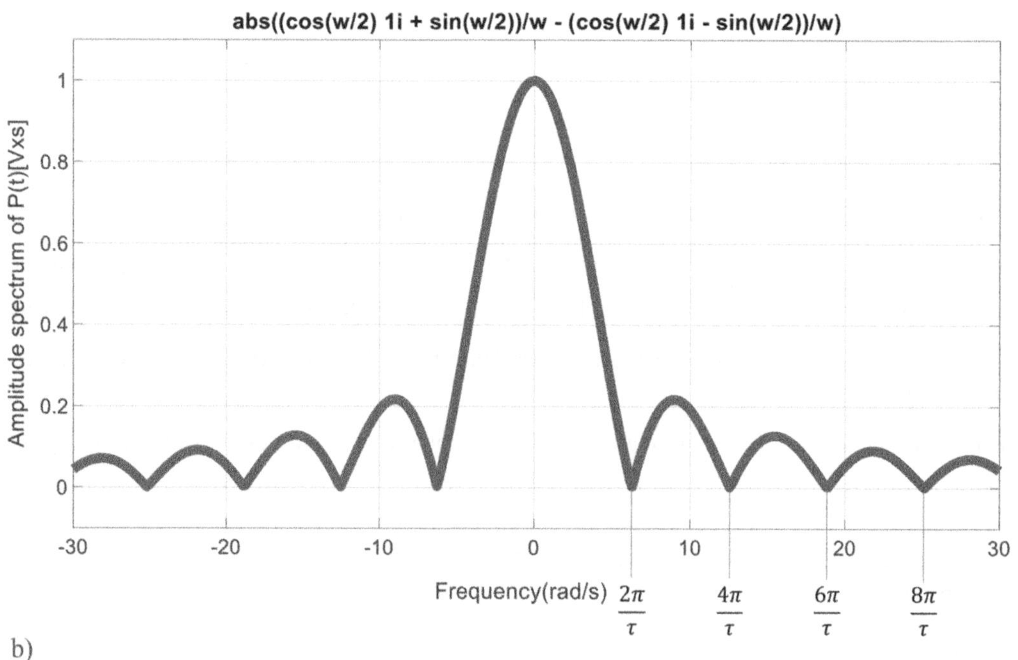

b)

Figure 10.21 Rectangular pulse *p(t) for* Example 10.6: a) Waveform; b) amplitude spectrum; c) phase spectrum. (Figure 7.1.7 from Mynbaev and Scheiner, *Essentials of Modern Communications*, Wiley, 2020, pp. 640–641.)

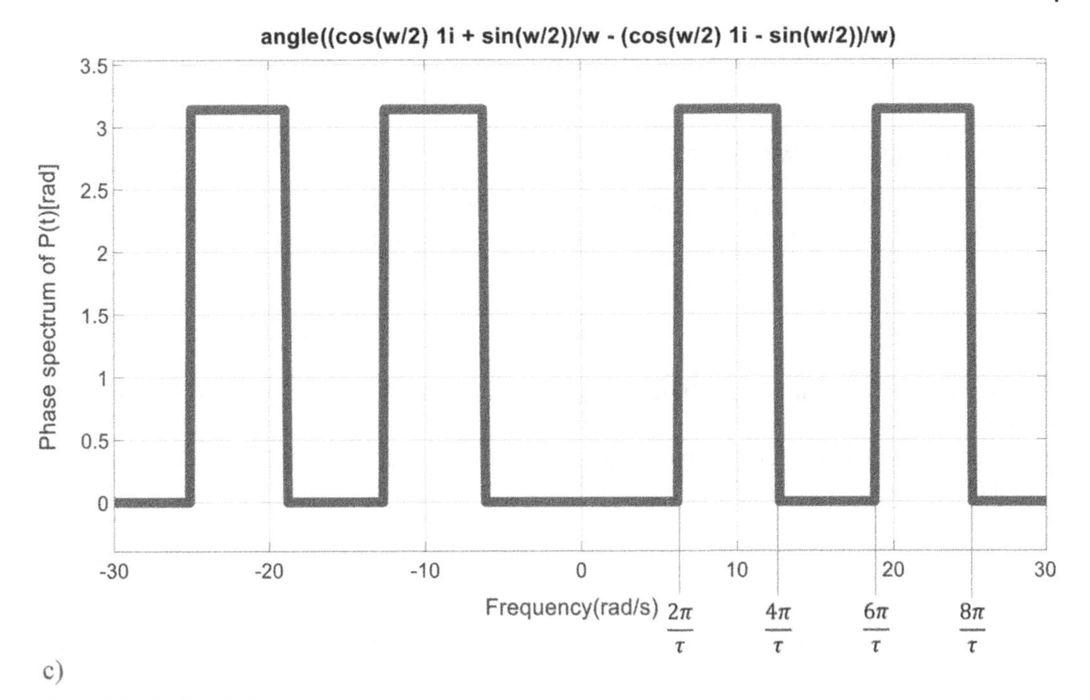

Figure 10.21 **(Cont'd)**

```
% amplitude spectrum of rectangular pulse
syms w t
H=heaviside(t+0.5)-heaviside(t-0.5);
F=fourier(H);
AF=abs(F);
ezplot(AF,[-30,30])
axis ([-30 30 -0.1 1.1])
grid
xlabel('Frequency(rad/s)')
ylabel('Amplitude spectrum of P(t)[Vxs]')

% phase spectrum of rectangular pulse
syms w t
H=heaviside(t+0.5)-heaviside(t-0.5);
F=fourier(H);
ANF=angle(F);
ezplot(ANF,[-30,30])
grid
xlabel('Frequency(rad/s)')
ylabel('Phase spectrum of P(t)[rad]')
```

Discussion:

Refer again to Figures 10.18a and 10.18d and observe how the discrete amplitude and phase spectra of a periodic signal (pulse train) change to continuous ones for a non-periodic signal (a single pulse). A detailed discussion of the single pulse spectra provided by this example clarifies this case's mechanism and numerical peculiarities.

10.2.3 Energy and Power Signals

All electrical signals are divided into two general categories: *energy and power*. Refer to Sections 1.1 and 4.2 for introduction to the topic. The *total energy*, $E(J)$, of a signal, $v(t)$, can be calculated as

$$E(J)=\int_{-\infty}^{\infty} v^2(t)dt, \tag{10.40}$$

where J stands for joule. If (10.40) exist and E is finite; i.e., $0<E<\infty$, then $v(t)$ is an *energy signal*.

What if $v(t)$ doesn't meet the above requirements? Then we must check the other condition given in (10.41):

$$P(W)=\lim_{T\to\infty}\left(\frac{1}{2T}\right)\int_{-T}^{T} v^2(t)dt, \tag{10.41}$$

where $P(W)$ is a signal's total power and W stands for watt. If (10.41) exists and $0<P(W)<\infty$, then $v(t)$ is a *power signal*.

It follows from definitions (10.40) and (10.41) that total power is zero for an energy signal, and total energy tends to infinity for a power signal. Still, some signals don't belong to either category.

Periodic signals are the *power signals*, and the *non-periodic signals* are—you guess that—the energy signals.

The average power per period of a periodic signal is given by

$$P(W)=\left(\frac{1}{T}\right)\int_{t}^{t+T} v^2(t)dt. \tag{10.42}$$

Equation 10.42 can be interpreted as the average power dissipated in an electrical circuit over a 1-Ω resistor with $v(t)$ being either the voltage power $\frac{v^2}{R}$ or the current power $\frac{i^2}{R}$, as we will see shortly. The curious reader can discover that a periodic signal can be expanded into the Fourier series only if this signal has finite average power. Fortunately, all practical periodic signals are power signals; thus, this requirement for the existence of a Fourier series is consistently met.

10.2.3.1 Parseval's Theorem

Remember that the Fourier series shows the signal's voltage or current distribution across its harmonics. Indeed, the amplitudes of the harmonics are in volts and can be directly measured with a spectrum analyzer. If the signal's source produces current, these amplitudes will be measured in amperes and show the entire signal's current distribution across the harmonics. Similarly, we can determine the distribution of the signal's power across the harmonics, thus obtaining the signal's power spectrum. This is how we can do this:

Recall that the periodic signal, $v(t)$, can be presented in the exponential form of the Fourier series as

$$v(t)=\sum_{n=-\infty}^{n=\infty}\left(c_n e^{jn\omega_0 t}\right), \tag{10.23R}$$

where the coefficients c_n are given by

$$c_n = \frac{1}{T}\int_t^{t+T} v(t)\left(e^{-jn\omega_0 t}\right)dt. \tag{10.24R}$$

Plugging this $v(t)$ in (10.42), which describes the average power of a signal, results in

$$P(W) = \left(\frac{1}{T}\right)\int_t^{t+T} v^2(t)dt = \left(\frac{1}{T}\right)\int_t^{t+T}\sum_{n=-\infty}^{n=\infty}\left(c_n e^{jn\omega_0 t}\right)^2 dt$$
$$= \sum_{n=\infty}^{n=-\infty}\left(c_n^* \cdot c_n\right) = \sum_{n=\infty}^{n=-\infty} |c_n|^2. \tag{10.43}$$

Here, c_n^* is complex conjugate to c_n. (Details of this derivation can be found in Mynbaev and Scheiner, Wiley, 2020, p. 595.) Equation 10.43, showing how to calculate the average power of a periodic signal, is called *Parseval's theorem*, after Marc-Antoine Parseval, an 18th-century French mathematician.

Parseval's theorem states that *the power carried by a signal, v(t), is equal to the sum of the powers carried by all the signal's harmonics*. This statement, in essence, is a specific form of the conservation of energy (power) principle.

We can readily express Parseval's theorem through the real coefficients, a_n and b_n, of a trigonometric Fourier series of the signal, *v(t)*. For this, compare two forms of the Fourier series given in (10.6) and (10.23):

$$v(t)[V] = \sum_{-\infty}^{\infty}\left(c_n e^{jn\omega_0 t}\right) = A_0 + \sum_{n=1}^{\infty}\left(a_n \cos(n\omega_0 t) + b_n \sin(n\omega_0 t)\right). \tag{10.44}$$

Alternatively, recalling that

$$a_n \cos(n\omega_0 t) + b_n \sin(n\omega_0 t) = A_n \cos(n\omega_0 t + \theta_n), \tag{10.45}$$

where $A_n = \sqrt{a_n^2 + b_n^2}$ and $\theta_n = -tan^{-1}\left(\frac{b_n}{a_n}\right)$, we can rewrite (10.44) as

$$v(t)[V] = A_0 + \sum_{n=1}^{\infty}\left(a_n \cos(n\omega_0 t) + b_n \sin(n\omega_0 t)\right)$$
$$= A_0 + \sum_{n=1}^{\infty}(A_n \cos(n\omega_0 t + \theta_n). \tag{10.46}$$

Now we can apply these general results to *electrical circuits*. Let $v(t)[V]$ be the voltage across the resistor, *R*(Ω). Then (10.43) presents the power of *root mean square (rms) voltage*, V_{rms}, dissipated over resistor $R(\Omega)$ as

$$P_R(W) = \frac{V_{rms}^2}{R}, \tag{10.47}$$

where V_{rms}^2 value is given by

$$V_{rms}^2 = \left(\frac{1}{T}\right)\int_t^{t+T} v^2(t)dt. \tag{10.48}$$

Using (10.46) enables us to obtain V_{rms}^2 through the harmonics of $v(t)$ signal as

$$V_{rms}^2 = \left(\frac{1}{T}\right)\int_t^{t+T} v^2(t)dt = \left(\frac{1}{T}\right)\int_0^T \left(A_0 + \sum_{n=1}^{\infty}\left(A_n \cos\left(n\omega_0 t + \theta_n\right)\right)\right)^2 dt. \quad (10.49)$$

We change the integration limits in (10.49) to simplify the calculations. Recall the binomial squared rule, $(x+y)^2 = x^2 + 2xy + y^2$, apply it to (10.49), carry out all mathematical manipulations, and get:

$$\begin{aligned} V_{rms}^2 &= \left(\frac{1}{T}\right)\int_t^{t+T} v^2(t)dt = \left(\frac{1}{T}\right)\int_0^T \left(A_0 + \sum_{n=1}^{\infty}\left(A_n \cos\left(n\omega_0 t + \theta_n\right)\right)\right)^2 dt \\ &= A_0^2 + \frac{1}{2}\sum_{n=1}^{\infty} A_n^2 = A_0^2 + \frac{1}{2}\sum_{n=1}^{\infty}\left(a_n^2 + b_n^2\right). \end{aligned} \quad (10.50)$$

Specifics of obtaining of (10.50) are given in the Sidebar "Derivation of Equation 10.50."

Sidebar 10S.1 Derivation of Equation 10.50.

Start with applying binomial squared rule to (10.49) and obtain the following three terms:

$$x^2 \Rightarrow \left(\frac{1}{T}\right)\int_0^T A_0^2 dt = A_0^2.$$

$$2xy \Rightarrow \left(\frac{2}{T}\right)\int_0^T A_0 \sum_{n=1}^{\infty}\left(A_n \cos\left(n\omega_0 t + \theta_n\right)\right) = 2A_0 \sum_{n=1}^{\infty}\left(\frac{A_n}{T}\right)\int_0^T \cos\left(n\omega_0 t + \theta_n\right)dt = 0$$

$$y^2 \Rightarrow \left(\frac{1}{T}\right)\int_0^T \left(\sum_{n=1}^{\infty}(A_n \cos(n\omega_0 t + \theta_n)\right)^2 dt = \left(\frac{1}{T}\right)\int_0^T \left(\sum_{n=1}^{\infty}(A_n \cos(n\omega_0 t + \theta_n)\right) \times$$

$$\sum_{m=1}^{\infty}(A_m \cos(m\omega_0 t + \theta_m)dt = \sum_{n=1}^{\infty} A_n \sum_{m=1}^{\infty} A_m \left(\frac{1}{T}\right)\int_0^T \cos(n\omega_0 t + \theta_n) \times$$

$$\cos(m\omega_0 t + \theta_m)dt = \frac{1}{2}\sum_{n=1}^{\infty} A_n^2.$$

- The integral for the last member is equal to zero for all cosines whose $n \neq m$, and $\left(\frac{1}{T}\right)\int_0^T \left(\cos\left(n\omega_0 t + \theta_n\right)\right)^2 = \frac{1}{2}$.
- Thus, $V_{rms}^2 = \left(\frac{1}{T}\right)\int_0^T v^2(t)dt = A_0^2 + \frac{1}{2}\sum_{n=1}^{\infty} A_n^2$, as in (10.50).

Combining (10.47) and (10.50), we can state that *the power dissipated by the signal $v(t)[V]$ over a resistor R (Ω) is distributed across the signal's harmonics* as

$$P_R(W) = \frac{V_{rms}^2}{R} = \frac{1}{R}\left(A_0^2 + \frac{1}{2}\sum_{n=1}^{\infty} A_n^2\right) = \frac{1}{R}\left(A_0^2 + \frac{1}{2}\sum_{n=1}^{\infty}\left(a_n^2 + b_n^2\right)\right). \quad (10.51)$$

It is common practice to consider power dissipated on the 1-Ω resistor; this power is called *normalized average power, $P_N(W)$*. Thus,

$$P_N(W) = \frac{V_{rms}^2}{1-\Omega} = \left(A_0^2 + \frac{1}{2}\sum_{n=1}^{\infty} A_n^2\right) = A_0^2 + \frac{1}{2}\sum_{n=1}^{\infty}\left(a_n^2 + b_n^2\right) = c_0^2 + 2\sum_{n=1}^{\infty} c_n^2 \qquad (10.52)$$

because $c_0 = A_0$ and $|c_n| = \frac{1}{2}\sqrt{a_n^2 + b_n^2}$.

Apparently, Equation10.52 is the other form of Parseval's theorem. Since c_0^2 is the dc power of a signal and $4c_n^2 = \left(a_n^2 + b_n^2\right)$ is the ac power carried by the nth harmonic, interpretation of Parseval's theorem for electric circuits is clear: *The normalized average power carried by a signal is the sum of its dc and ac power carried by all its harmonics.*

Equation 10.52 shows that Parseval's theorem in any form actually presents the *power spectrum* of a periodic signal, i.e.,

The power spectrum shows what portion of the signal's total power is carried by each harmonic.

We need to remember that in all our future considerations we refer to normalized average power.

Applicability of the Parseval's theorem to periodic signals calls for an appropriate example discussed below.

Example 10.7 The power spectrum of a pulse train.

Problem: Consider the pulse train shown in Figure 10.22a whose amplitude, $V_{pt} = 10$ (V), period, $T = 8$ (ms), pulse duration, $\tau = 1.6$ (ms), and thus duty cycle, $\tau/T = 0.2$. Find the power spectrum of this signal.

Solution: The pulse train is a periodic signal, which calls for using Parseval's theorem. Since power is a real entity, we need to employ a one-sided form with real coefficients, given in (10.46) as

$$v(t)[V] = A_0 + \sum_{n=1}^{\infty}(A_n \cos(n\omega_0 t + \theta_n)) + \sum_{n=1}^{\infty}(a_n \cos(n\omega_0 t) + b_n \sin(n\omega_0 t)). \qquad (10.46R)$$

The amplitude spectrum of the $v(t)$ is shown in Figure 10.22b.

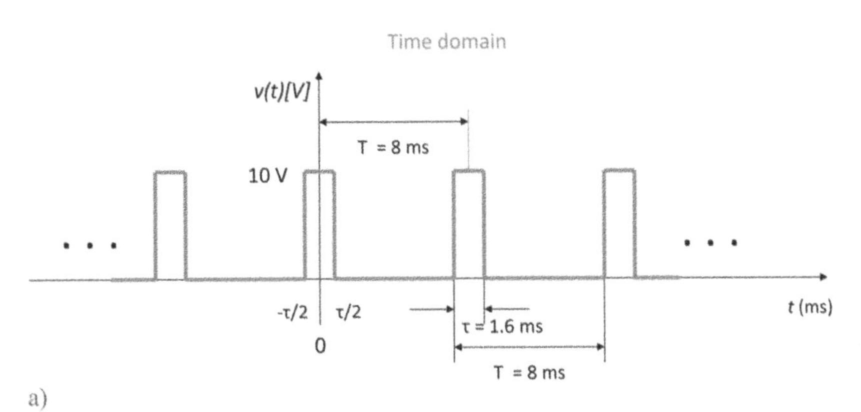

Figure 10.22 The amplitude and power spectra of the periodic pulse train for Example 10.7: a) The pulse train's waveform; b) one-sided amplitude spectrum of the pulse train; c) the pulse train's power spectrum.

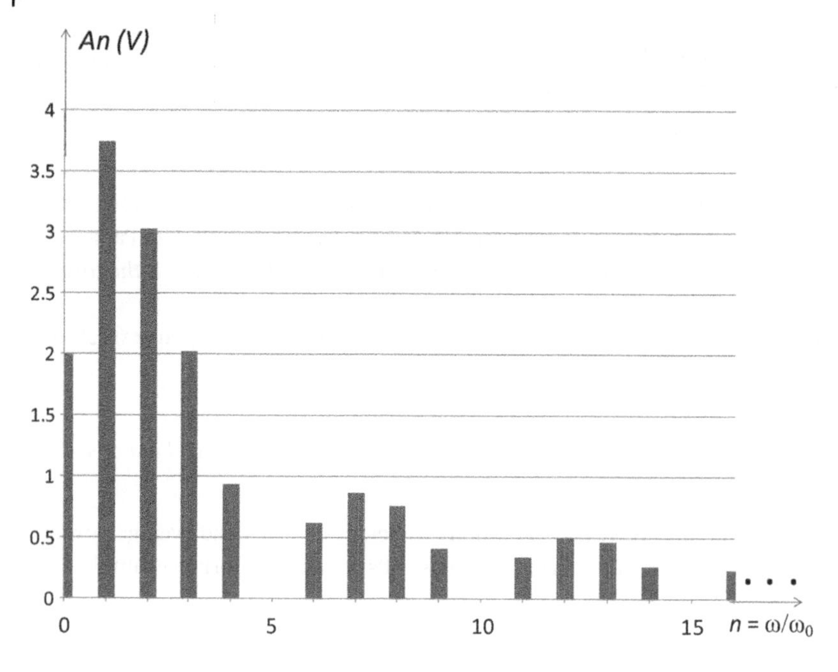

b)

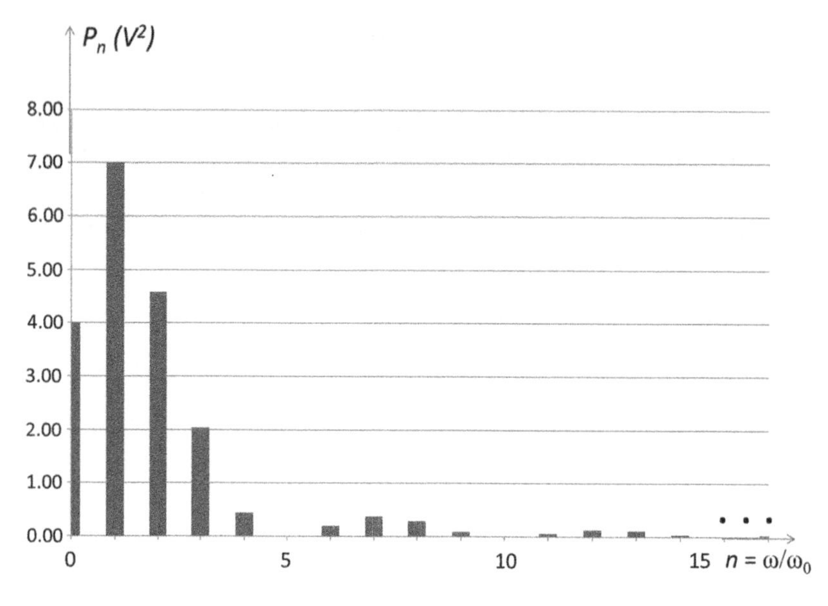

c)

Figure 10.22 (Cont'd)

Then we employ (10.52),

$$P_N(W) = \frac{V_{rms}^2}{1-\Omega} = \left(A_0^2 + \frac{1}{2}\sum_{n=1}^{\infty} A_n^2\right) = A_0^2 + \frac{1}{2}\sum_{n=1}^{\infty}\left(a_n^2 + b_n^2\right), \tag{10.52R}$$

to compute the power of individual harmonics and build the power spectrum of the pulse train shown in Figure 10.22c. You can calculate these coefficients manually, or automate computations using Excel or MATLAB.
The problem is solved.

Discussion:

A pulse train's power spectrum is not the merely subject of academic curiosity; it plays crucial role in transmission of communications signals. To learn more about this important topic, see Mynbaev and Scheiner, Wiley, 2020, pp. 594–608.

10.2.3.2 Rayleigh Energy Theorem

The power of a periodic signal, to be reminded, is determined in Parseval's theorem as

$$P(W)=\left(\frac{1}{T}\right)\int_{t}^{t+T} v^2(t)dt. \tag{10.42R}$$

Can we apply this formula to a non-periodic signal? Of course not because the limits of (10.42) rely on the period. Therefore, the power of a non-periodic signal can't be determined, and, therefore, a *non-periodic signal is an energy signal,* as stated previously.

To remind, the *total energy, E,* of a non-periodic signal, *v(t),* can be calculated as

$$E(J)=\int_{-\infty}^{\infty} v^2(t)dt. \tag{10.40R}$$

Since this signal is related to its Fourier transform, as

$$v(t)=\frac{1}{2\pi}\int_{-\infty}^{\infty} F(\omega)e^{j\omega t}d\omega, \tag{10.53}$$

we can express the signal's energy through its Fourier transform. So, plugging (10.53) into (10.40), we find, after some manipulations (James and Dyke 2018, p. 370),

$$E(J)=\frac{1}{2\pi}\int_{\infty}^{-\infty} \left|F(\omega)\right|^2 d\omega. \tag{10.54}$$

This expression is called the *Rayleigh*[4] *theorem*. Examining (10.54), we immediately recognize that it is analogous to Parseval's theorem, presented in (10.43). In fact, some textbooks refer to (10.54) as Parseval's theorem for non-periodic signals; on the other hand, most books call it the Rayleigh theorem.

The energy distribution of a non-periodic signal along its spectrum is described by the function $|F(\omega)|^2$*(J/Hz), called the energy spectral density; the portion of this energy concentrated in a differential frequency band,* $d\omega$ *(1/s), is given by*

4 Lord Rayleigh (1842–1919), a British scientist and Nobel prize winner (1904), belonged to the higher nobility who did not need to work for a career or money. The science was his passion, and he took this passion seriously. Lord Rayleigh's research covers practically the whole field of physics. His publications are too numerous to be listed, and his influence on classical physics cannot be overestimated. In short, he was one of the most famous scientists of his time, and his scientific legacy lasted until the modern days. The Rayleigh power theorem discussed in this textbook is one of the countless proofs of this statement.

$$dE(J) = |F(\omega)|^2 d\omega. \tag{10.55}$$

The Rayleigh theorem enables us to find the frequency distribution of the signal's energy.

As a brief example, consider the amplitude spectrum of the rectangular pulse shown in Figure 10.21b. The energy concentrated in the main lob of this signal is (Carlson and Grilly, 2009, pp. 50–52.)

$$E_{main} = \frac{1}{2\pi} \int_{\frac{2\pi}{\tau}}^{-\frac{2\pi}{\tau}} |F(\omega)|^2 d\omega = \frac{1}{2\pi} \int_{\frac{2\pi}{\tau}}^{-\frac{2\pi}{\tau}} (A\tau)^2 \, sinc^2(\omega\tau) d\omega \tag{10.56}$$

whereas the total energy of the pulse is $A^2\tau$. Thus, more than 90% of the pulse's energy is concentrated in the main lobe of its spectrum, that is, in the frequency band between $-\frac{2\pi}{\tau}$ and $\frac{2\pi}{\tau}$.

Analogously to Parseval's theorem, the Rayleigh energy theorem calculates the *energy spectrum* of a non-periodic signal.

10.2.4 The Fourier Transform and Circuit's Response

The main subject of this book—to be repeated—is finding the circuit's response (output) to a given input. How can the Fourier transform help in this search? What new features can it bring to advanced circuit analysis? The current subsection answers these questions. However, proceeding with this discussion requires introducing several Fourier transform properties, and they are presented in Table 10.2. Thorough consideration of these properties can be found in Mynbaev and Scheiner, Wiley, 2020, pp. 656–659.

Pay particular attention to two properties—*time differentiation* and *convolution in the time domain*. Observe that the differentiation of a function in the time domain is equivalent to multiplying the Fourier transform of this function by jω. That is, $\boldsymbol{F}\{v'(t)\} = j\omega F(\omega)$. This property helps transmute differential equations of communications systems into the frequency domain. Recall the similar property of the Laplace transform and see the similarity and difference between these two.

Convolution in the time domain requires special attention because it gives a new look at the circuit input–output relationship. Recall that in the time domain, the system's output can be found by using the *convolution integral* defined as

$$v_{out}(t) = \int_{-\infty}^{\infty} v_{in}(t) h(t-\tau) d\tau \equiv v_{in}(t) \star h(t). \tag{6.10R}$$

Here $h(t)$ is the system's response to an impulse $\delta(t)$) and $\star$ is the symbol of the convolution operation. In practice, applying the convolution integral for finding the system's output might be a difficult mathematical operation (though recently, it has become practically achievable thanks to advances in mathematics and computer science). But in the frequency domain, the convolution reduces to simple multiplication, as Table 10.2 for the entry $v_{in}(t) \star h(t)$ shows,

$$\mathcal{F}\{v_{in}(t) \star h(t)\} = V_{in}(\omega) \cdot H(\omega), \tag{10.57}$$

where we denote

Table 10.2 Several properties of the Fourier transforms.

Property	Time domain $v(t)$	Frequency domain $F(\omega) = \mathcal{F}\{v(t)\}$
Signal representation	$v(t)$	$F(\omega)$
Units	Voltage, $v(t)$ [V] Current, $i(t)$ [A]	$\mathcal{F}\{v(t)\} = V(\omega)\ [V{\cdot}s]$ $\mathcal{F}\{i(t)\} = I(\omega)\ [A{\cdot}s]$
Linearity	$a_1v_1(t) + a_2v_2(t)$, a_1 and a_2—constants	$\mathcal{F}\{a_1v_1(t) + a_2v_2(t)\}$ $= a_1F_1(\omega) + a_2F_2(\omega)$
Duality	$v(t)$ $F(t)$	$\mathcal{F}\{v(t)\} = F(\omega)$ $\mathcal{F}\{F(t)\} = 2\pi v(-\omega)$
Time scaling	$v(kt)$, k—constant	$\mathcal{F}\{v(kt)\} = \frac{1}{\lvert k\rvert}F\left(\frac{j\omega}{k}\right)$
Frequency scaling	$(v)t) = \mathcal{F}^{-1}\{F(k\omega)\}$ $= \frac{1}{\lvert k\rvert}v\left(\frac{t}{k}\right)$	$F(k\omega)$
Time shift	$v(t-\tau)$	$\mathcal{F}\{v(t-\tau)\} = F(\omega){\cdot}e^{-j\omega\tau}$
Frequency shift	$\mathcal{F}^{-1}\{F(\omega-\omega_0)\} = v(t)e^{j\omega_0 t}$	$F(\omega-\omega_0)$
Modulation (frequency translation)	$v(t)\cos(\omega_0 t)$	$\mathcal{F}\{v(t)\cos(\omega_0 t)\}$ $= \frac{1}{2}[F(\omega+\omega_0) + F(\omega-\omega_0)]$
Time differentiation	$v'(t)$ .... $v^n(t)$	$\mathcal{F}\{v'(t)\} = j\omega\, F(\omega)$ .... $\mathcal{F}\{v^n(t)\} = (j\omega)^n\, F(\omega)$
Frequency differentiation	$\mathcal{F}^{-1}\left\{(j)^n\frac{d^nF(\omega)}{d\omega^n}\right\} = t^n v(t)$	$(j)^n\frac{d^nF(\omega)}{d\omega^n}$
Integration	$\int_{-\infty}^{t} v(t)dt$	$\mathcal{F}\left\{\int_{-\infty}^{t} v(t)dt\right\}$ $= \frac{F(\omega)}{j\omega} + \pi F(0)\delta(\omega)$
Convolution in time domain	$v_1(t) \ast v_2(t)$	$F_1(\omega){\cdot}\ F_2(\omega)$
Convolution in frequency domain	$v_1(t){\cdot}v_2(t)$	$F_1(\omega) \ast F_2(\omega)$

$$V_{in}(\omega) = \mathcal{F}\{v_{in}(t)\} \tag{10.58}$$

and

$$H(\omega) = \mathcal{F}\{h(t)\}. \tag{10.59}$$

Adding another notation,

$$V_{out}(\omega) = \mathcal{F}\{v_{out}(t)\}, \tag{10.60}$$

enables us to write

$$V_{out}(\omega) = H(\omega)\cdot V_{in}(\omega). \tag{10.61}$$

Thus, the Fourier transform of the whole convolution integral (6.10) results in (10.61).

You will recall that similar formula,

$$V_{out}(s) = H(s)V_{in}(s), \tag{9.73R}$$

is derived in Chapter 9 for the Laplace transform application, and we call $H(s)$ a *transfer function*.

Therefore, $H(\omega)$ *is the circuit's transfer function in the frequency domain. What's more,* Equation 10.61 shows *that this transfer function is the Fourier transform of a system's impulse response. Since the spectrum of delta function includes all possible frequencies, as Section 6.1 demonstrates, the impulse response and, consequently, H(ω) are comprehensively described the system's response to any input frequency.*

Therefore, the convolution property of the Fourier transform enables us to obtain the system's output in frequency domain. The output in time domain, $v_{out}(t)$, can be determined by applying the inverse Fourier transform as

$$\mathcal{F}^{-1}\{V_{out}(s)\} = v_{out}(t). \tag{10.62}$$

The following example supports all the above considerations.

Example 10.8 Amplitude and phase responses of a series RC circuit

Problem: Using the Fourier transform, find the amplitude and phase frequency responses of the series RC circuit with R = 2.2 (kΩ) and C= 0.002 (μF). Its decaying constant and cutoff frequency are $\alpha = \omega_C\,(rad/s) = \frac{1}{RC} \approx 227.27$ (krad/s).)

Solution: We need to find the circuit's transfer function, *H(ω)*, from which the amplitude *response* will be determined as its absolute value, $|H(\omega)|$, and the *phase response* will be obtained as an argument of this function, $\arg H(\omega) = \Theta(\omega)$.

A system's transfer function is given as the Fourier transform of the impulse response, *h(t)*,

$$H(\omega) = \mathcal{F}\{h(t)\}. \tag{10.59R}$$

The series RC circuit's impulse response has been found in Example 6.1 as

$$h(t) = \alpha\, e^{-\alpha t}u(t). \tag{6.28R}$$

Table 10.1 shows that

$$\mathcal{F}\{\alpha e^{-\alpha t}u(t)\} = \frac{\alpha}{\alpha + j\omega}. \tag{10.63}$$

Therefore,

$$H(\omega) = \mathcal{F}\{h(t)\} = \mathcal{F}\{\alpha e^{-\alpha t}u(t)\} = \frac{\alpha}{\alpha + j\omega} = \frac{\alpha^2}{\alpha^2 + \omega^2} - j\frac{\alpha\omega}{\alpha^2 + \omega^2} \tag{10.64a}$$

is the *transfer function of a series RC circuit in the frequency domain.*

The amplitude response is found as

$$|H(\omega)| = |\mathcal{F}\alpha e^{-\alpha t}u(t)\}| = \left|\frac{\alpha^2}{\alpha^2 + \omega^2} - j\frac{\alpha\omega}{\alpha^2 + \omega^2}\right| = \frac{\alpha}{\sqrt{\alpha^2 + \omega^2}}, \tag{10.65}$$

and the phase response is obtained as

$$\theta(\omega) = \arg H(\omega) = \arg\left(\frac{\alpha^2}{\alpha^2+\omega^2} - j\frac{\alpha\omega}{\alpha^2+\omega^2}\right) = -tan^{-1}\left(\frac{\omega}{\alpha}\right). \tag{10.66}$$

The amplitude and phase spectra of a series RC circuit determined by (10.65) and (10.66) coincide with these characteristics found several times in Parts 2 and 3 of this book. Therefore, the results obtained with the transfer function are correct. Figures 10.23a and 10.23b display the graphs of both spectra.

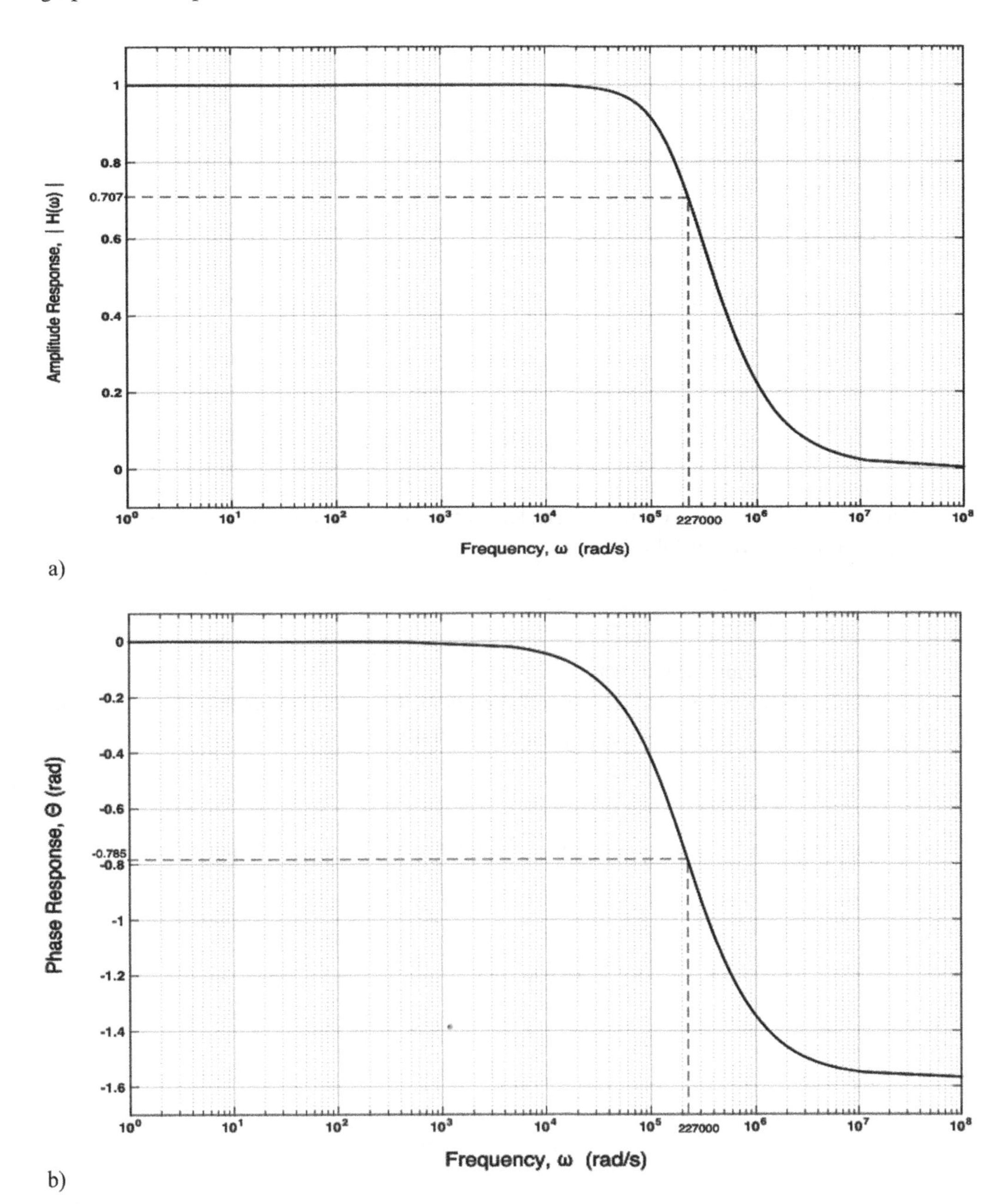

Figure 10.23 Amplitude (a) and phase (b) spectra of the impulse response of a series RC circuit in Example 10.8. (Figure 7.2.5 of Mynbaev and Scheiner, Wiley, 2020, p. 660.)

Discussion:

- We can now better interpret the meaning of the circuit's parameters. We know that the frequency at which the response amplitude value becomes equal to $1/\sqrt{2} = 0.707$ is called the *cutoff frequency,* ω_C. Looking at (10.65), we find that

$$|H(\omega_c)| = H(\omega_c) \frac{\alpha}{\sqrt{\alpha^2 + \omega^2}} = \frac{1}{\sqrt{2}} \tag{10.67}$$

only if $\alpha = \omega$.Thus, the *cutoff angular frequency,* $\omega_c\left(\frac{rad}{s}\right)$, is determined as

$$\omega_c\left(\frac{rad}{s}\right) = \alpha = \frac{1}{RC}, \tag{10.68a}$$

and the cutoff frequency is

$$f_c(Hz) = \frac{1}{2\pi RC}. \tag{10.68b}$$

It follows from (10.66) and (10.68a) that the phase shift at ω_c is equal to

$$\theta(\omega_c) = \arg H(\omega_c) = -tan^{-1}\left(\frac{\omega_c}{\alpha}\right) = -tan^{-1}(1) = -\frac{\pi}{4} = -45^0. \tag{10.69}$$

Both $|H(\omega_c)|$ and $\theta(\omega_c)$ are shown in Figures 10.23a and 10.23b, respectively.

- Using the transfer function enables us to find the circuit's response to any input, as we stress several times in this section. As an example, let's consider the response of a series RC circuit to a rectangular pulse given in Table 10.1 as,

$$p(t) = \begin{cases} A \; for \; |t| < \tau \\ 0 \; for \; |t| > \tau \end{cases},$$

with τ being the pulse width (duration).

To solve this problem, we need to transfer all operations into frequency domain. Thus, we find the Fourier transform of the input signal by referring to Table 10.1 as 7.2.1

$$\mathcal{F}\{p(t)\} = A\tau \cdot sinc\left(\frac{\omega\tau}{2}\right).$$

The circuit's transfer function is the Fourier transform of its impulse response, as shown in (10.59)

$$H(\omega) = \mathcal{F}\{h(t)\} = \mathcal{F}\{\alpha e^{-\alpha t}u(t)\} = \frac{\alpha}{\alpha + j\omega} = \frac{\alpha^2}{\alpha^2 + \omega^2} - j\frac{\alpha\omega}{\alpha^2 + \omega^2}.$$

Therefore, the circuit's response, according to (10.61), is as follows:

$$V_{out}(\omega) = H(\omega) \cdot V_{in}(\omega) = \frac{\alpha}{\alpha + j\omega} \cdot \left(A\tau \cdot sinc\left(\frac{\omega\tau}{2}\right)\right). \tag{10.70}$$

Finding the response in the time domain calls for the inverse Fourier transform,

$$v_{out}(t) = F^{-1}\{V_{out}(\omega)\} = F^{-1}\left\{\frac{\alpha}{\alpha + j\omega} \cdot A\tau \cdot sinc\left(\frac{\omega\tau}{2}\right)\right\}. \tag{10.71}$$

It's worth noting that (10.70) is a straightforward though tedious mathematical manipulation, but (10.71) is far from being straightforward. To grasp the idea what this operation might take, refer to Chapter 8, where the inverse Laplace transform is thoroughly discussed and remember about analogy between the Laplace and Fourier transforms.

From the definition, discussion, and examples of the Fourier transform, we can highlight two critical ***conclusions***:

a) The Fourier transform,

$$\mathcal{F}\{v(t)\} = F(\omega) = \int_{-\infty}^{\infty} v(t)\left(e^{-j\omega t}\right)dt \tag{10.30R}$$

of a time-domain function, $v(t)$, gives the spectrum of this function, which can be presented in a polar form as

$$F(\omega) = |F(\omega)|\angle\theta(\omega) \tag{10.72}$$

where $|F(\omega)|$ is the amplitude spectrum and $\theta(\omega)$ is the phase spectrum. *If the time-domain function, $v_{out}(t)$, is the response of a given device to an impulse input, then its Fourier transform produces the device's amplitude and phase responses in the frequency domain, as shown above.*

b) The fundamental relationship of our consideration,

$$V_{out}(\omega) = H(\omega)\cdot V_{in}(\omega), \tag{10.61R}$$

explains that the circuit's spectral response is determined by its frequency-domain output signal. However, using (10.72), we can write

$$|V_{out}(\omega)|V\angle\theta_{out}(\omega) = H(\omega)\cdot|V_{in}(\omega)|\angle\theta_{in}(\omega)V_{out}(\omega). \tag{10.73}$$

If we can hold $|V_{in}(\omega)| = 1$ *and* $\theta_{in}(\omega) = 0$, then $|V_{out}(\omega)|\angle\theta_{out}(\omega) = H(\omega)$, meaning that *in this case, the transfer function comprehensively describes the circuit's amplitude and phase response.* Multiplying $H(\omega)$ by $|V_{in}(\omega)|$ and $\angle\theta_{in}(\omega)$ will result in scaling the output amplitude and phase.

This critical observation justifies the common expressions that a transfer function describes the circuit's spectra, whereas, formally speaking, it must be said that the circuit's spectra are presented by $V_{out}(\omega)$.

Sidebar 10S.2 Alternative Methods of Finding a Transfer Function.

We've found the circuit's transfer function by applying the Fourier transform to the impulse response. We can achieve this goal, however, by using the circuit's differential equations or the Fourier integral. The first method will be exemplified by considering the transfer function of a series RC LPF.

- Finding the RC circuit's transfer function by using the circuit's differential equation
  This equation is obtained in (4.10) as

$$v'_{out}(t) + \alpha v_{out}(t) = \alpha v_{in}(t), \tag{4.10R}$$

where $\alpha = \frac{1}{RC}$. We can find the RC circuit's transfer function by using the circuit's differential equation as follows:

$$j\omega V_{out}(\omega) + \alpha V_{out}(\omega) = \alpha V_{in}(\omega)$$

or

$$\left(j\omega+\alpha\right)V_{out}\left(\omega\right)=\alpha V_{in}\left(\omega\right). \tag{10S.2.1}$$

Hence, the RC circuit's transfer function in frequency domain is given by

$$H\left(\omega\right)=\frac{V_{out}\left(\omega\right)}{V_{in}\left(\omega\right)}=\frac{\alpha}{\alpha+j\omega}=\frac{\alpha^2}{\alpha^2+\omega^2}-\frac{j\alpha\omega}{\alpha^2+\omega^2}, \tag{10S.2.2}$$

as expected. Observe that in frequency domain the transfer function is obtain by simple mathematical (algebraic, in fact) manipulations.

- We can also find the RC circuit's transfer function by directly applying the Fourier transform integral. A curious reader can find the details of this finding in Mynbaev and Scheiner, Wiley, 2020, pp. 667–668.

10.2.5 Comparison of Laplace and Fourier Approaches to Finding the Circuit's Response

Application of the Laplace transform given in Chapters 8 and 9, and the use of the Fourier transform demonstrated in Chapter 10 present two approaches to finding the circuit's response in the frequency domain. The formal difference between these two transforms is that the Laplace transform operates in the *s-domain*, where *s* is a *complex frequency* given by

$$s\left(\frac{rad}{s}\right)=\sigma+j\omega, \tag{7.3R}$$

whereas the Fourier transform works in the domain of a *radian frequency*, $\omega\left(\frac{rad}{s}\right)$. Recall that if $f(t)$ is a time-domain function, then its Laplace transform, $F(s)$, is

$$L\left\{f\left(t\right)\right\}=F\left(s\right)=\int_{-\infty}^{\infty}e^{-st}f\left(t\right)dt, \tag{7.2R}$$

where L is the symbol of the Laplace transform operation. Likewise, the Fourier transform, $F\left(\omega\right)$, of a time-domain function $f\left(t\right)$ is defined as

$$\mathcal{F}\left\{f\left(t\right)\right\}=F\left(\omega\right)=\int_{-\infty}^{\infty}v\left(t\right)\left(e^{-j\omega t}\right)dt \tag{10.30R}$$

and $\mathcal{F}$ is the symbol of the Fourier-transform operation. As (7.2R) and (10.30R) show, both transforms are integral operations. Consider, for instance, a decaying exponential time-domain function, $f\left(t\right)=e^{-\alpha t}$ whose α is a constant. Application of both transforms results in

$$L\left\{e^{-\alpha t}\right\}=F\left(s\right)=\frac{1}{\alpha+s} \tag{10.74}$$

and

$$\mathcal{F}\left\{e^{-\alpha t}\right\}=F\left(\omega\right)=\frac{1}{\alpha+j\omega}. \tag{10.75}$$

Table 10.3 Comparison of the Laplace and Fourier transforms of several time-domain functions.

Time-domain function	Laplace transform	Fourier transform
Unit step (Heaviside), $u(t)$	$\frac{1}{s}$	$\pi\delta(\omega) + \frac{1}{j\omega}$
Sine, $\sin(\omega t)$	$\frac{\omega}{(s^2 + \omega^2)}$	$\pi[\delta(\omega + \omega_0) - \delta(\omega - \omega_0)]$
Cosine, $\cos(\omega t)$	$\frac{s}{(s^2 + \omega^2)}$	$\pi[\delta(\omega + \omega_0) + \delta(\omega - \omega_0)]$

Comparing the s-domain function, $F(s) \equiv F(\sigma + j\omega)$ with the frequency-domain one, $F(\omega)$, makes an impression that the latter is a streamlined variant of the former. Indeed, if we drop σ from the Laplace transform of $e^{-\alpha t}$, we obtain the Fourier transform of this function denoted as $F(j\omega)$. Equations 10.74 and 10.75 confirm this observation, and it seems that we could conclude that the Fourier transform is the simplified version of the Laplace transform. However, such a conclusion will be superficial. For one, compare the Laplace and the Fourier transforms of the same time-domain functions shown in Table 10.3 and observe the striking difference between them. The examples in Table 10.3 convince us that the Laplace and the Fourier transforms are distinct operations. Nonetheless, they have many common features, which help in the applications of both procedures.

Discussion of the application of the Laplace and Fourier transforms in the advanced circuit analysis provided in Chapters 8, 9, and 10 enables us to summarize the similarity and difference between the Laplace and Fourier transforms approaches to finding the circuit's output as follows:

- The Laplace transform works well in facilitating solving a circuit's differential equation by transmuting the time-domain integro-differential operations into the s-domain algebraic manipulations. In this capacity, the Laplace transform is the best tool for finding the circuit's input–output relationship in the time domain.
- The Fourier transform is indispensable in finding the output's spectral characteristics because this operation shows how the circuit transforms the spectrum of the input signal.
- Despite the difference, both transforms resort to the transfer-function technique for finding the circuit's input–output relationship and facilitating the search process.
- Chapters 8 and 9 show that the Laplace transform handles the initial conditions seamlessly by making them a part of the procedure. Chapter 10 doesn't mention how the Fourier transform treats this issue, which implies the need for a deeper study of the topic.
- Finally, the Laplace transform is the more general method whose advantage becomes increasingly apparent with the growth of the system (circuit) complexity. However, if the system (circuit) spectral analysis is the subject of investigation, the Fourier transform will be the method of choice.

10.2.6 Summary of Section 10.2

Spectral analysis with the Fourier transform

The vast majority of real signals are non-periodic and their spectra can be found only with the Fourier transform. This section discusses how to transit from a periodic to a non-periodic signal, from a Fourier series to a Fourier transform, and how to use the Fourier transform and its inverse version. The important highlights of these considerations are as follows:

- A non-periodic signal has a continuous spectrum, in contrast to a periodic signal, which has a line (discrete) spectrum.
- To find the spectrum of a signal, we need to use a Fourier series for a periodic signal and a Fourier transform for a non-periodic signal.
- The Fourier transform is an integral operation that transforms a time-domain function into a frequency-domain function.
- If $F(\omega)$ is the Fourier transform of a *non-periodic signal*, $v(t)$, then the amplitude spectrum of $v(t)$ is given by $|F(\omega)|$ and the phase spectrum of $v(t)$ is provided by $\arg(F(\omega))$.
- Finding the Fourier transform of a time-domain function by integration is not an easy task. Fortunately, most practical Fourier transforms are tabulated. If the required Fourier transform can't be found in a table, we can obtain it by using MATLAB.
- *Parseval's power theorem* states state that *the normalized average power dissipated by the signal* $v(t)[V]$ on the 1-Ω resistor *is distributed across the signal's harmonics* as

$$P_N(W) = \frac{V_{rms}^2}{1-\Omega} = \left(A_0^2 + \frac{1}{2}\sum_{n=1}^{\infty} A_n^2\right). \tag{10.52R}$$

- Parseval's theorem delivers the *power spectrum of a periodic signal.*
- *Rayleigh energy theorem* gives the energy distribution of a non-periodic signal across the signal's spectra as

$$E(J) = \frac{1}{2\pi}\int_{-\infty}^{\infty} |F(\omega)|^2 \, d\omega. \tag{10.54R}$$

where $|F(\omega)|^2$ (J/Hz) is the *energy spectral density*. The portion of this energy that is concentrated in a differential frequency band, $d\omega$, is given by

$$dE = |F(\omega)|^2 \, d\omega. \tag{10.55R}$$

Thus, the Rayleigh energy theorem delivers the *energy spectrum of a non-periodic signal.*

- The Fourier transform of the convolution integral produces the input signal $V_{in}(\omega)$ and

the transfer function $H(\omega)$ in the frequency domain as

$$\mathcal{F}\{v_{in}(t) \star h(t)\} = V_{in}(\omega) \cdot H(\omega), \tag{10.57R}$$

where $H(\omega)$ is the Fourier transform of the circuit's impulse response,

$$H(\omega) = \mathcal{F}\{h(t)\}. \tag{10.59R}$$

Therefore, the circuit's response can be found in the frequency domain as

$$V_{out}(\omega) = H(\omega) \cdot V_{in}(\omega). \tag{10.61R}$$

Comparison (10.61) with the analogous formula given for the s-domain,

$$V_{out}(s) = H(s) \cdot V_{in}(s), \tag{9.73R}$$

highlights the similarity between the Laplace and Fourier transforms.

Questions and Problems for Chapter 10

Questions marked with an asterisk require a systematic approach to finding the solution.
Many questions and problems, including those marked with an asterisk, imply that you, in addition to reading the textbook, will do research to find the answers. Consider such questions as mini-projects.

10.1 Basics of the Fourier Series with Applications

1 Give definitions and examples of periodic and non-periodic signals.

2 Consider the two signals given in Figure 10.P2. Show which signal is periodic and which is non-periodic and explain your reasoning.

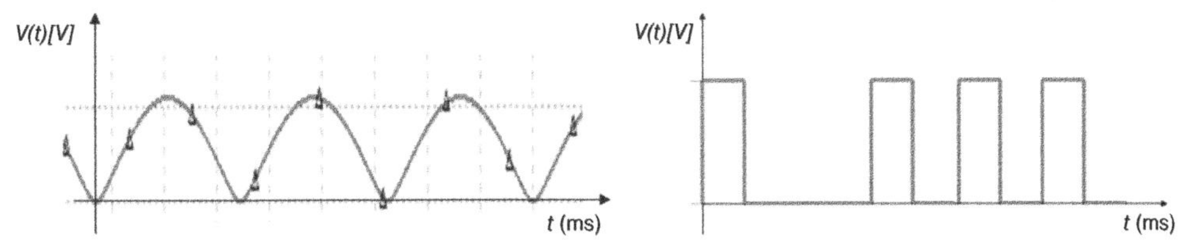

Figure 10.P2 Periodicity of the signals.

3 A majority of what type of signal—periodic or non-periodic—exists in practice? Explain.

4 Define time domain and frequency domain.

5 If the waveform of a signal is given, to what domain—time domain or frequency domain—does this graph belong? Explain.

6 Consider the following signals: $v_1(t) = 12 \cos(2\pi\, 60t)$, $v_2(t) = 6 \cos(2\pi\, 120t)$, and $v_3(t) = 3 \cos(2\pi\, 180\, t)$. Show these signals in time domain and frequency domain.

7 Since the same sinusoidal signal can be presented in time domain and frequency domain, there is a fundamental formula that relates one domain to the other. What is this formula? Give the equation and explain its meaning.

8 *Consider sinusoidal signal $v(t) = A \cos(2\pi f_0\, t)$ with given amplitude A(V) and frequency f_0(Hz). Can you show both of these parameters on a signal waveform? Explain.

9 *What parameter of a sinusoidal signal is shown in both time domain and frequency domain without alteration? Explain by using the figures in both domains.

10 *Depict the following signals in frequency domain:
a) $v(t) = 8 \cos(2\pi\, 60\, t)$.
b) $v(t) = 8 \cos(378\, t)$.
c) $v(t) = 8 \cos(378\, t + 60^0)$.
d) $v(t) = 8 \sin(378\, t)$.

11 *Does the presentation of the same cosine signal in time domain and frequency domain represent a real phenomenon or is it an imaginary mental process developed for the convenience of discussion? Prove your answer.

12 *Can we measure signal parameters in time domain and frequency domain? If the answer is yes, what measuring instruments do we need to use in each domain? Explain.

13 Explain what a spectral analyzer does.

14 Define the spectrum of a signal.

15 Figure 10.P15 shows three periodic signals. Based on the knowledge you obtained so far in this book, can you sketch the spectrum of each of these signals? Comment on your answers.

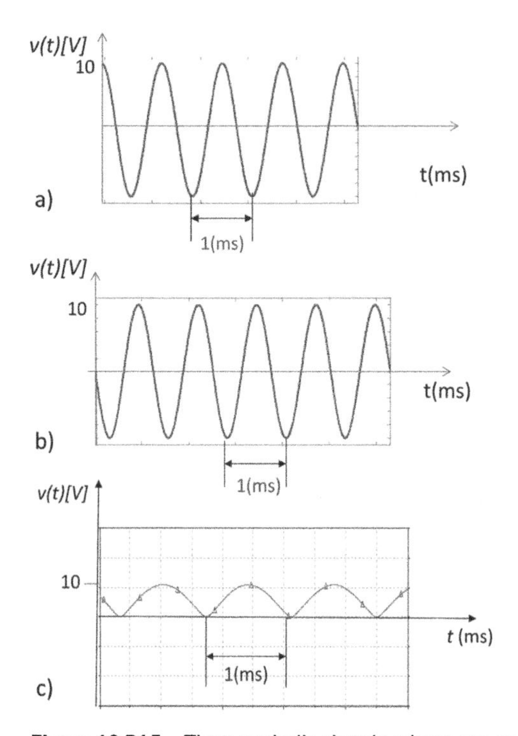

Figure 10.P15 Three periodic signals whose spectra are to be found. (Hint: See Problem 40.)

16 We distinguish between the amplitude spectrum and the phase spectrum of the same signal. Why? Explain and give an example.

17 The spectrum of a periodic non-sinusoidal signal consists of many frequencies, but only one of them is called *fundamental*. What is this frequency?

18 Refer to the signals shown in Figure 10.P15. What are their fundamental frequencies?

19 The Fourier theorem states that "*every periodic signal, v(t), can be represented as a sum of cosines and sines.*" Why do we need to represent a signal as a sum of cosines and sines? What can we achieve with this representation?

20 The Fourier series presents a given non-sinusoidal periodic signal as the following sum of cosines and sines: $v(t) = A_0 + a_1\cos 2\pi f_0 t + a_2\cos 2\pi\, 2f_0 t + a_3\cos 2\pi\, 3f_0 t + \ldots + b_1\sin 2\pi f_0 t + b_2\sin 2\pi\, 2f_0 t + b_3\sin 2\pi\, 3f_0 t + \ldots = A_0 + \sum_{n=1}^{\infty}(a_n\cos 2\pi\, nf_0 t + b_n\sin 2\pi\, nf_0 t)$.

a) What do we know about the given periodic signal?

b) What is its f_0 and how can we compute it?

c) What is the meaning of the number "n" and how does this number change in the Fourier series?

d) What do we know about A_0, a_n, and b_n?

21 Consider the square-wave signal shown in Figure 10.P21. Calculate Ao, the three first cosine and three first sine members of its Fourier series.

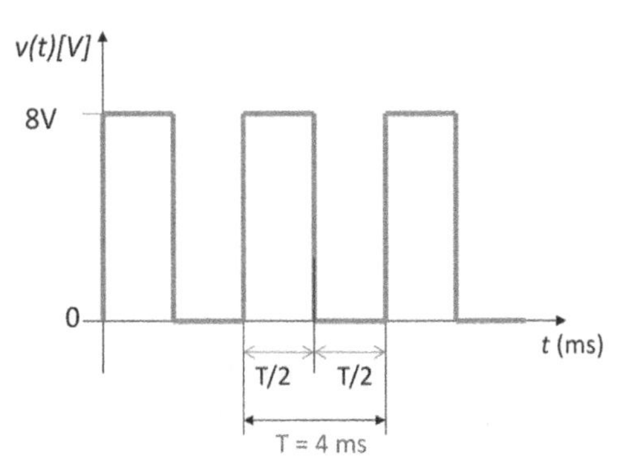

Figure 10.P21 A square-wave signal.

22 The waveform and the Fourier series of the bipolar shifted square wave signal are given in Figure 10.P22:

a) In frequency domain, sketch the first three members and show all their parameters if $A = 2$ (V) and $T = 10$ (ms).
b) Sketch the waveform of the first harmonic of the signal and show its amplitude and period.

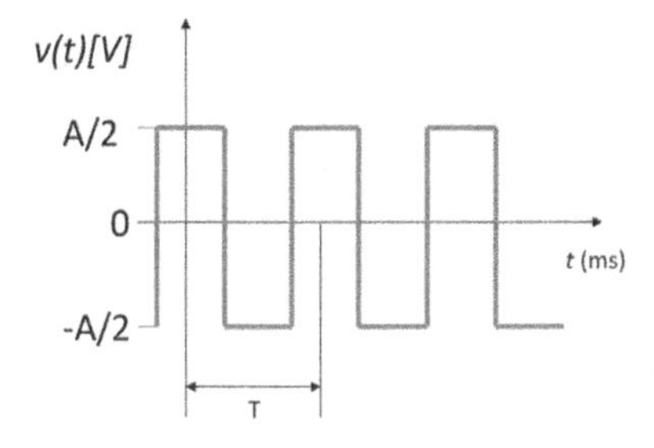

Fourier series

$v(t) = (2A/\pi) \cos 2\pi f_0 t - (2A/3\pi) \cos 2\pi\, (3f_0)t + (2A/5\pi) \cos 2\pi\, (5f_0)t + \ldots$

$= \sum_{n=1}^{\infty} ((A \sin n\pi/2)/(n\pi/2) \cos 2\pi\, (nf_0)t$

Figure 10.P22 A bipolar shifted square-wave signal and its Fourier series.

23 The waveform and the Fourier series of the sawtooth signal are given in Figure 10.P23:

a) In frequency domain, sketch the first three members and show all their parameters if $A = 2$ (V) and $T = 10$ (ms).
b) Sketch the waveform of the first harmonic of the signal and show its amplitude and period.

24 The waveform and the Fourier series of the half-wave signal are given in Figure 10.P24

a) In frequency domain, sketch the first three members and show all their parameters if $A = 2$ (V) and $T = 10$ (ms).
b) Sketch the waveform of the second harmonic of this signal and show its amplitude and period.

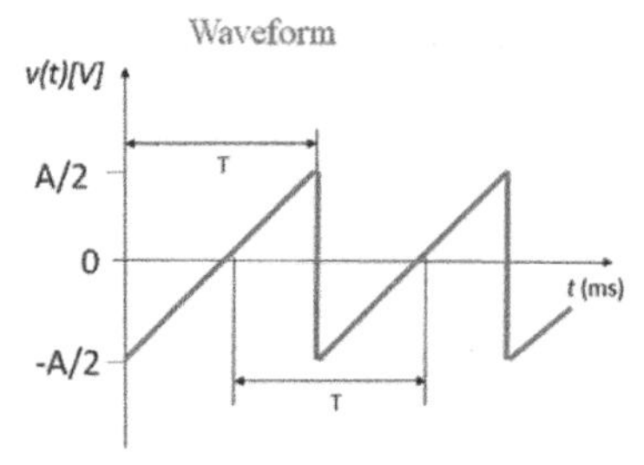

$v(t) = (A/\pi)\sin 2\pi f_0 t + (A/2\pi)\sin 2\pi\,(2f_0)t + (A/3\pi)\sin 2\pi\,(3f_0)t + (A/4\pi)\sin 2\pi\,(4f_0)t \ldots$

Figure 10.P23 A sawtooth signal and its Fourier series.

25 The waveform and the Fourier series of the digital signal are given in Figure 10.P25:

a) In frequency domain, sketch the first three members and show all their parameters if $A = 2$ (V) and $T = 10$ (ms).

b) Sketch the waveform of the third harmonic of this signal and show its amplitude and period.

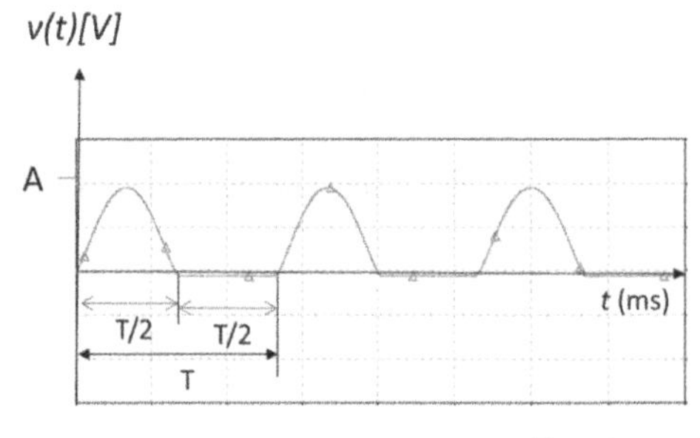

Fourier series

$v(t) = A/\pi + (A/2)\sin 2\pi f_0 t - (2A/3\pi)\cos 2\pi\,(2f_0)t - (2A/15\pi)\cos 2\pi\,(4f_0)t$
$- (2A/35\pi)\cos 2\pi\,(4f_0)t - \ldots$
$= A/\pi + (A/2)\sin 2\pi f_0 t - (2A/\pi)\sum_{n=2,4,6,}^{\infty} (\cos 2\pi\,(nf_0)t/(n^2 - 1))$

Figure 10.P24 A half-wave rectified signal and its Fourier series.

26 *Find the spectrum of the bipolar square-wave signal whose waveform is shown in Figure 10.P26. Display all the calculations. Sketch the first four members of the spectrum.

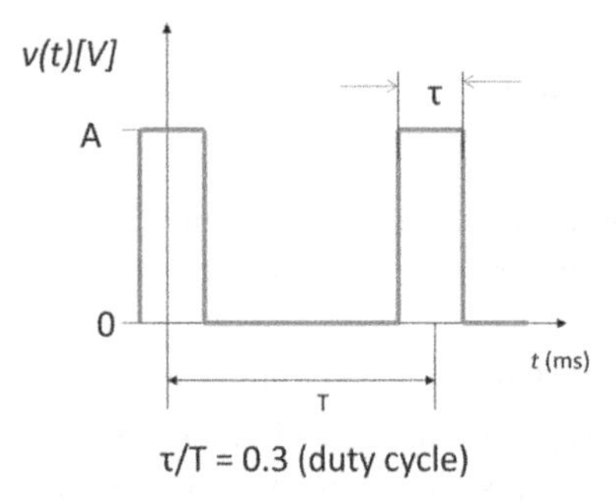

Fourier series

$v(t) = A\,\tau/T + \sum_{n=1}^{\infty} (2A/\,n\pi\,(\sin n\pi\,\tau/T))\cos 2\pi\,(nf_0)t$

Figure 10.P25 A digital signal and its Fourier series.

27 *Find the spectrum of the triangular signal shown in Figure 10.P27. Show all the calculations. Sketch the first four members of the spectrum.

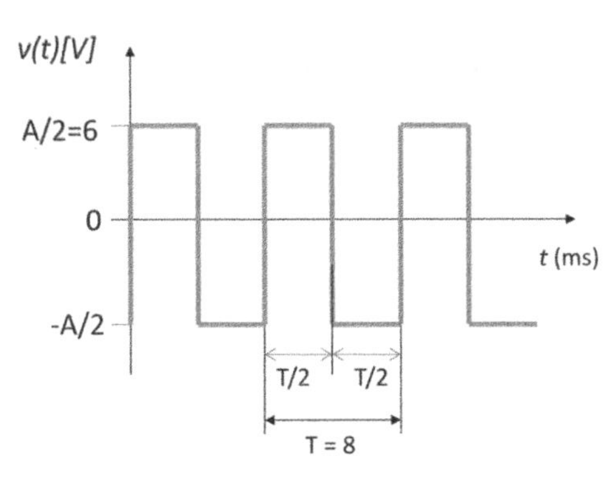

Figure 10.P26 A bipolar square-wave signal.

28 Find the dc member of the Fourier series for the signals shown in Figures 10.P21, 10.P26, and 10.P27.

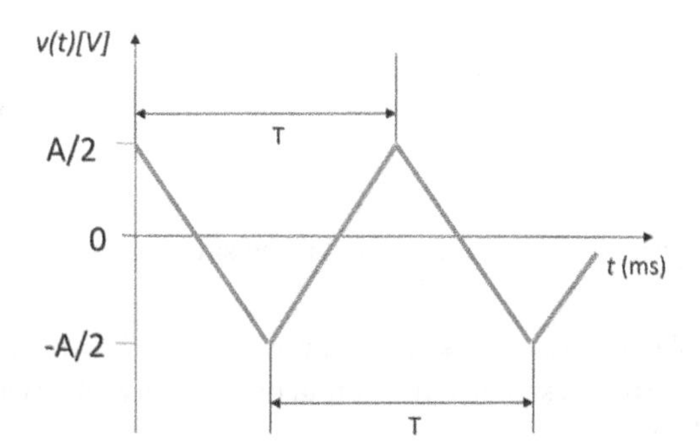

Figure 10.P27 The waveform of a triangular signal.

29 It's said that the Fourier series enables the *spectral analysis* of a given signal: What do we mean by "spectral analysis"?

30 *Describe the process of spectral analysis in conjunction with the meaning of the Fourier series. Refer to Figure 10.9.

31 *Explain the difference between spectral analysis and spectral synthesis.

32 * The Fourier series of a signal is given by $v(t) = (A/\pi) \sin 2\pi f_0 t + (A/2\pi) \sin 2\pi (2f_0)t + (A/3\pi) \sin 2\pi (3f_0)t + (A/4\pi) \sin 2\pi (4f_0)t \ldots = \sum_{n=1}^{\infty} (A/n\pi) \sin 2\pi (nf_0)t$:

a) Sketch the first four spectral components of this Fourier series if A = 2 (V) and T = 1(ms).
b) Applying the spectral-synthesis technique, build the waveform of this signal based on the first four harmonics. (Use any computer tool you are comfortable with such as MS Excel or MATLLAB.)

33 *The spectrum of a signal is shown in Figure 10.P33:

a) Write down the expressions for the first three members of the Fourier series of this signal if $A = 2$ (V) and $f_0 = 1$ (kHz).

b) Applying the spectral-synthesis technique, build the waveform of this signal based on the first three harmonics. (Use any computer tool you are comfortable with such as MS Excel or MATLAB.)

34 *Describe the process of spectral synthesis with reference to Figure 10.11.

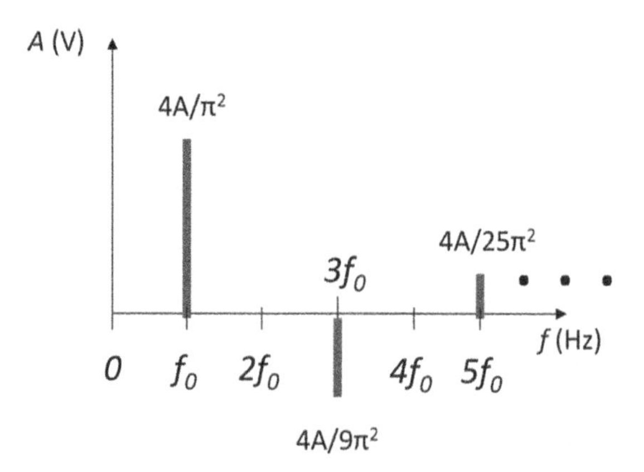

Figure 10.P33 The spectrum of a signal.

35 *Describe the role of individual harmonics and their summation in the processes of spectral analysis and synthesis for a sawtooth signal. Refer to Problem 10.23 and Figure 10.11. Restrict the processes by three spectral components.

36 *Consider Figure 10.10, showing the presentation of two harmonics in time domain and frequency domain. Sketch a similar figure for three harmonics of the sawtooth signal discussed in Problem 10.33.

37 *Consider Figure 10.P37, where a bipolar square-wave signal is presented to a low-pass RC filter: What can you tell about the waveform of a signal at the output of a low-pass filter without any additional investigation? Explain.

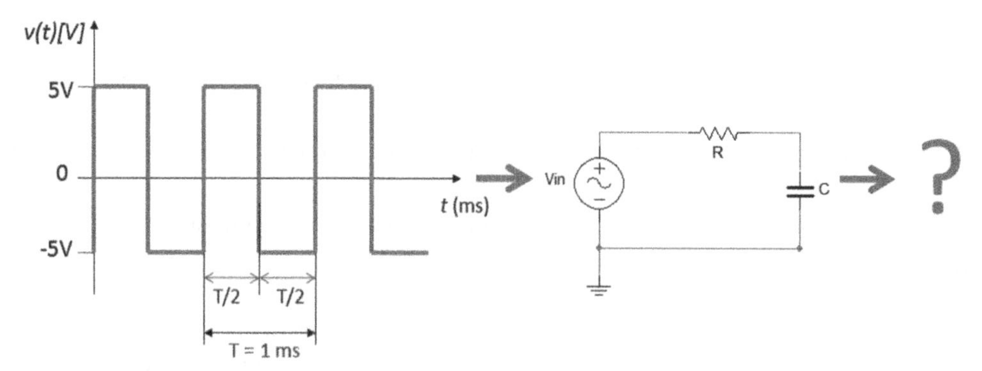

Figure 10.P37 Presenting a bipolar square-wave signal to a low-pass filter.

38 *Consider a sawtooth signal and its Fourier series shown in Figure 10.P38:

a) Show the second spectral component of this signal in frequency domain if $A = 2$ (V) and $T = 10$ (ms).

b) This signal is presented to an RC LPF with critical frequency of 0.2 kHz. Calculate the amplitude and phase of the second spectral component of the output signal and show it alongside the input second component.

c) Qualitatively sketch the waveform of the second output harmonic, show its parameters, and sketch the waveforms of both input and output second harmonic.

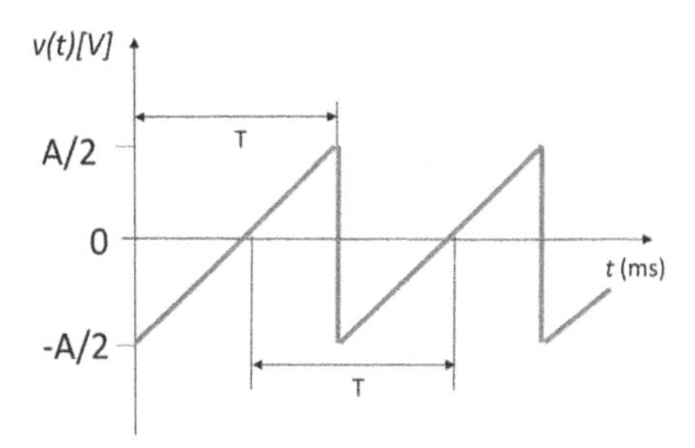

Fourier series:

$v(t) = (A/\pi)\sin 2\pi f_0 t + (A/2\pi)\sin 2\pi(2f_0)t + (A/3\pi)\sin 2\pi(3f_0)t + (A/4\pi)\sin 2\pi(4f_0)t \ldots$

$= \sum_{n=1}^{\infty} (A/n\pi)\sin 2\pi(nf_0)t$

Figure 10.P38 A sawtooth signal to be presented to an RC LPF.

39 *The sawtooth signal shown in Figure 10.P38 is presented to another LPF with $f_C = 0.4$ (kHz). Show the second harmonics of the input and output (filtered) signals in frequency domain and in time domain.

40 Assign A = 3 (V) and T = 0.1 (ms) to a full-wave rectified signal whose waveform and Fourier series are shown in Figure 10.P40 and perform the following tasks:

a) Sketch the first four harmonics of the signal spectrum. Show all of the values.
b) Sketch the waveform of the second harmonic of this signal and show its amplitude and period.
c) This signal is presented to a low-pass filter with a critical frequency of 30 (kHz). Calculate the amplitude and the phase shift of the output second harmonic and show both the input and output of the second harmonic on the same set of axes.
d) On one set of axes, qualitatively sketch the waveforms of the second input (presented to the filter) and output (obtained after filtering) harmonics and show their parameters.

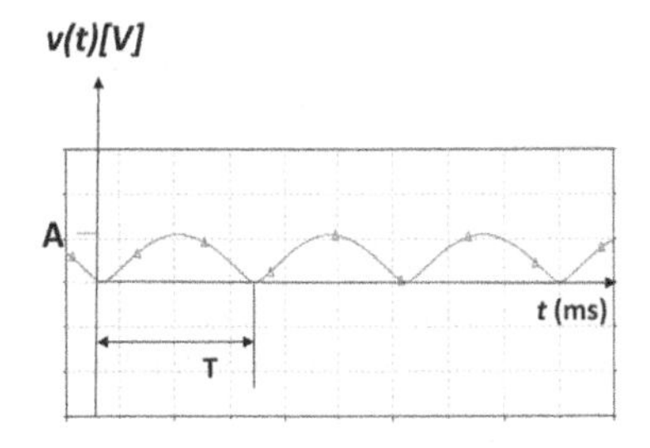

The Fourier series:

$v(t) = 2A/\pi + (4A/3\pi)\cos 2\pi(f_0)t - (4A/15\pi)\cos 2\pi(2f_0)t + (4A/35\pi)\cos 2\pi(3f_0)t - \ldots$

$= 2A/\pi - (4A/\pi)\sum_{n=1}^{\infty} (\cos 2\pi(nf_0)t/(4n^2 - 1))$

Figure 10.P40 Full-wave rectified signal presented to a low-pass filter.

41 *Refer to Questions 39 and 40 and Figure 10.P38, which consider a sawtooth signal with its Fourier series and a low-pass filter with $f_C = 0.2$ kHz: Taking into account only the first three harmonics, sketch the spectra of the input and filtered signals. Show all the numbers.

42 Consider the sawtooth signal shown in Figure 10.P38 and assume $A = 6$(V) and $T = 0.2$(ms). Present this signal to a real LPF with $f_C = 15$(kHz). Calculate the amplitudes of the first six harmonics and sketch the spectra of both the input and output signals.

43 *How many harmonics do we need to sum up to obtain the waveform of a signal? Explain.

44 Use any computerized tool (MS Excel, MALAB, etc.) to do the following:

a) Build the waveform of the filtered sawtooth signal discussed in Question 16 by using the first three harmonics.
b) Repeat this process by using 6, 9, 12, and 15 harmonics.
c) Build this waveform by using 33 and 66 harmonics.
d) What conclusion regarding the quality of the waveform can you make based on your results obtained in all the operations of this problem?

45 *Return to Question 10.38: The answer is given in Figure 10.13, where the process of filtering a square wave is shown. Consider each step of this process and answer the questions in Table 10.P45

Table 10.P45 Questions regarding the process of filtering the signals.

Designation in the caption of Figure 10.13	Step of filtering a signal shown in Figure 10.13	Question
10.13a	A square wave is expanded in the Fourier series (only four harmonics are shown) and each harmonic is presented to an LPF.	Q1: Is this expansion a real process or just a mathematical manipulation? Q2: If this is a real process, what device can perform this operation? Explain your answers.
10.13b	Each harmonic is filtered by the LPF and the output harmonics are obtained.	Q1: Explain how we can find the output harmonics. (Hint: Refer to Table 6.2.2.) Q2: How does an LPF “know” what harmonic to choose and how to filter it?
10.13c	All output harmonics are summarized, creating the entire output signal—the square wave.	Q1: Explain why we need to summarize the output harmonics. Q2: Is this summation a real process or a mathematical manipulation? Explain.

46 *Consider Figure 10.P46 showing the experiment, in which the same square-wave signal passes through two low-pass RC filters. Why are the output signals so different?

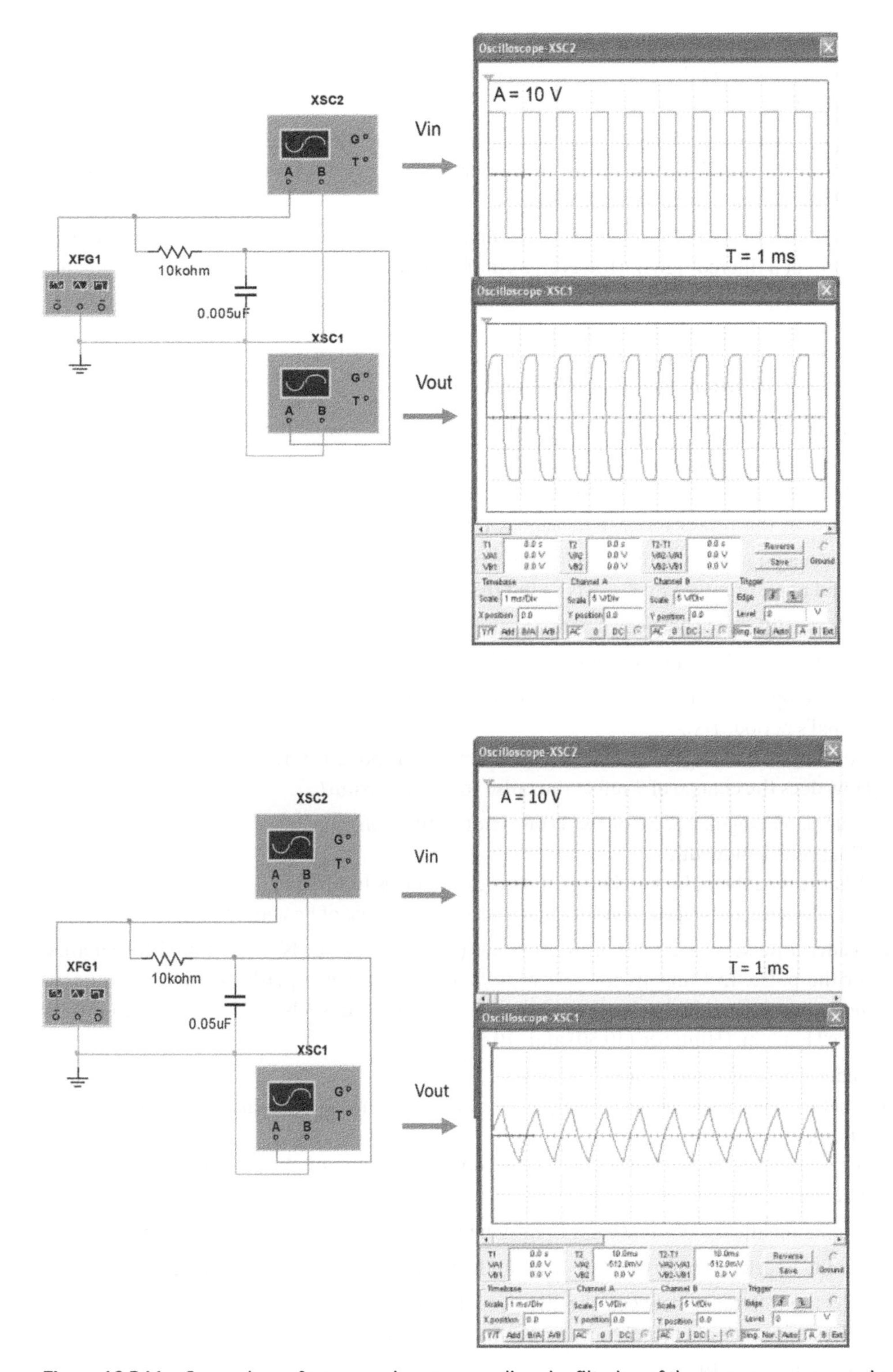

Figure 10.P46 Comparison of two experiments regarding the filtering of the same square-wave signal.

47 *If you aren't satisfied with the waveform of a signal that passed through a low-pass filter, what characteristic of the filter will you change first? In what way will you change this characteristic? Explain.

48 *We know that a waveform is a signal plot in time domain. Nevertheless, this subsection explains that we need to analyze frequency-domain processes to understand and control the filtered waveform. Explain why this is so.

49 *The text says that a cutoff frequency of a filter is the main parameter that determines the waveform of a filtered signal. However, a cutoff frequency is a frequency-domain parameter whereas a waveform is a time-domain characteristic. How can a cutoff frequency determine the waveform?

10.2 The Fourier Transform and its Applications

50 Explain the difference between power and energy signals. Give examples of both types.

51 Consider a periodic electrical signal, $v(t)$ $[V]$:
a) What is its normalized average power?
b) How does this power relate to the signal's *rms* value, V_{rms}?

52 * Consider a pulse train with the following parameters: amplitude $V_{pt} = 2.8\ (V)$, period $T = 4\ (ms)$, and pulse duration $\tau = 0.8\ (ms)$. Find the power spectrum of this signal.

53 Explain the difference between a periodic and a non-periodic signal. Give examples.

54 What is a duty cycle of a non-periodic signal? Explain.

55 We can change the duty cycle of a pulse train by changing either the pulse duration, τ(s), or the signal's period, $T(s)$.
a) How does the change of τ affect the spectrum of this pulse train?
b) How does the change of T affect the pulse train's spectrum?

56 Consider a rectangular pulse of $\tau = 3$ (ms) width (duration):
a) What is its bandwidth?
b) What will be the pulse bandwidth if its width reduces to 0.3 (ms)?
c) Why does the pulse's bandwidth change with the change of its width?

57 *Consider a pulse as a carrier of information (a bit). Its power is proportional to the area under the pulse. In transmission, we want to minimize power-per-bit parameter; thus, it would seem that an ideal pulse should be of zero width. Can such an ideal pulse be built in practice? Explain. (Hint: See (10.50) and (10.51).)

58 Consider a periodic pulse train:
a) What parameters of the periodic signal should we change to turn this periodic signal into a non-periodic one?
b) What kind of non-periodic signal can we obtain by changing the parameters of the periodic pulse train?

59 What mathematical tools do we use to find the spectra of periodic and non-periodic signals?

60 Define the Fourier transform, $F(\omega)$. Explain the meaning of each member of $F(\omega)$.

61 What is the difference between the Fourier series and the Fourier transform? [Hint: Start with the definitions of each of these mathematical tools.]

62 Examine Figure 10.19:
a) Why do we need the inverse Fourier transform?

b) Applying the Fourier transform to a time-domain function, *v(t)*, we obtain *F(ω)*. If we then apply the inverse Fourier transform to frequency-domain function, *F(ω)*, we receive the original *v(t)*. Why do we need to do these seems to be mutually exclusive operations?

63 Using the definition of the Fourier transform, find the Fourier transform of $v(t) = 3e^{-5t}$. Show all your manipulations.

64 Using MATLAB, find the Fourier transform of $v(t) = 3e^{-5t}$. To verify your answer, find the inverse Fourier transform of $F^{-1}(\omega) = F(e^{-5t})$. Show your code and the MATLAB answers.

65 Using the table, find the Fourier transform of
a) $v(t) = t^3 \cdot e^{-2t}u(t)$
b) $p(t) = 6[u(t + 5/2) - u(t - 5/2)]$

66 Mathematically, a rectangular pulse is described as $p(t) = A[u(t + \tau/2) - u(t - \tau/2)]$, where τ is a pulse duration (width), A is its amplitude (height), and *u(t)* is a unit-step function. What is the meaning of this description? Sketch a figure to support your explanations.

67 *Consider a rectangular pulse $p(t) \equiv rect(t) = 5[u(t + 3) - u(t - 3)]$:
a) Sketch this pulse.
b) Find its Fourier transform.
c) Show all your manipulations.

68 *Find the amplitude and phase spectra of $v(t) = 3e^{-5t}$.

69 *Find the amplitude and phase spectra of $p(t) = 15sinc(1.5\omega)$.

70 Can we apply Parseval's theorem to finding the power spectrum of a non-periodic signal? Explain your reasoning.

71 *We know that a periodic signal is a power signal and a non-periodic signal is an energy signal. We can apply Parseval's theorem to find the power spectrum of a periodic signal by calculating the power carried by each individual harmonic. A non-periodic signal, however, has a continuous spectrum, where no individual harmonic exists. How, then, can we find the energy spectrum of a non-periodic signal? Explain.

72 *Consider the rectangular pulse given by $p(t) = 5[u(t + 3) - u(t - 3)]$:
a) Find the energy of this pulse concentrate in the main lob, that is, in the frequency band between $-1/\tau$ and $1/\tau$.
b) What percentage of the pulse's total energy does this main-lob energy constitute? Show your calculations.

73 Using Table 10.1, find the Fourier transform of
a) $v(t) = 3cos(12t)$;
b) $v(t) = \delta(t - 4)$;
c) $v(t) = sin(\omega_0 t)$.

74 Table 10.1 shows that $\mathfrak{F}(cos(\omega_0 t))$ and $\mathfrak{F}(cos(\omega_0 t)u(t))$ are different? Why is this difference?

75 What and why is the difference between $e^{-\alpha t}u(t)$ and $e^{-\alpha|t|}$?

76 *Consider the properties of the Fourier transform:
a) If *v(t)* is an energy signal, what will be the unit of its Fourier transform?
b) What is the Fourier transform of $v(t) = 5cos(3t) + 2sin(7t)$?
c) What is the Fourier transform of $v(t) = e^{j6t}$?
d) What is the Fourier transform of $v(t) = (5 + 4cos(3t)) \cdot cos(33t)$?
e) What is the Fourier transform of $v(t) = \frac{d(2e^{6t})}{dt}$?

77 Determine the impulse response of a series RC LPF whose R = 3.0 (kΩ) and C= 0.006 (μF).

78 *Consider an impulse response of a linear time-invariant communication system:

a) How can we find it? What is the difference between impulse response and free response of an RC LPF? What do these responses have in common?

b) Is an impulse response a real signal or a mathematical model? Explain.

79 Consider a series RC LPF whose R = 3.0 (kΩ) and C= 0.006 (μF). Using the Fourier transform, find the amplitude and phase frequency responses of this filter.

80 *Find the amplitude and phase responses of an RC LPF whose R = 2.2 (kΩ) and C= 0.002 (μF) to a step signal $v(t) = 3u(t)$.

81 What methods of finding the transfer function of an electrical circuit (system) do you know? Give an example.

Conclusion

This book concentrates on the following problem: Finding a circuit output (response) to a given input (excitation) when the circuit's parameters, initial conditions, and all necessary values are provided. The book discusses several approaches to solving this problem, and Table 10.C1 summarizes the result of this discussion.

Table 10.C1 Summary of approaches to finding the response of an electrical circuit.

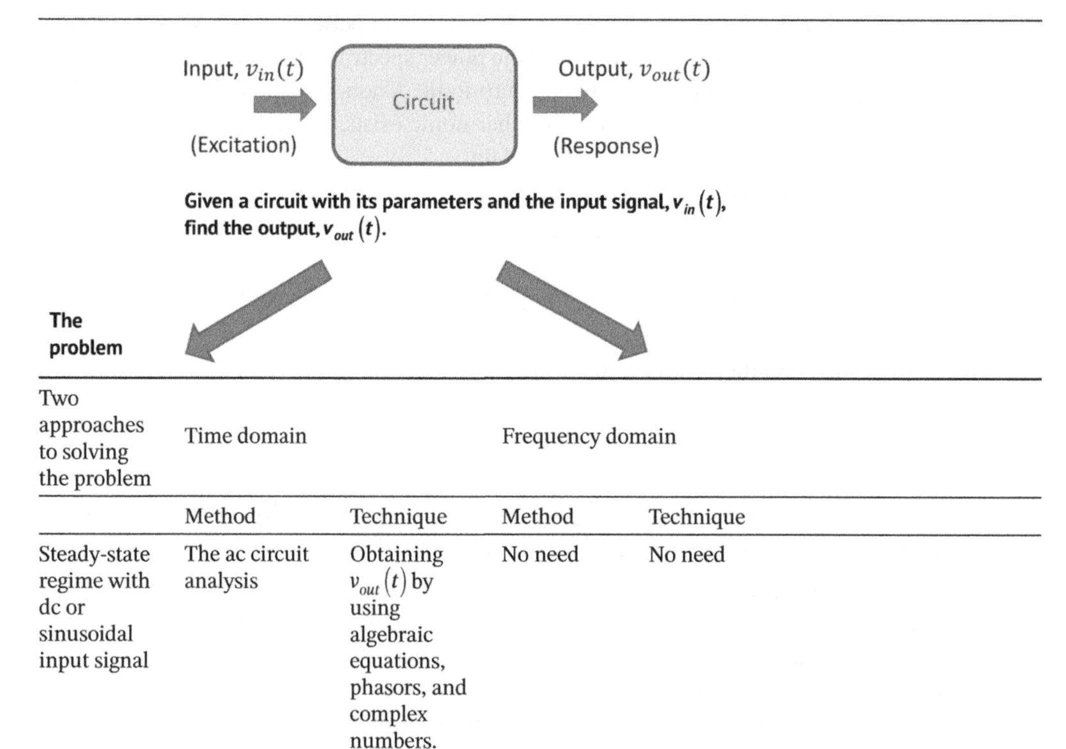

The problem				
Two approaches to solving the problem	Time domain		Frequency domain	
	Method	Technique	Method	Technique
Steady-state regime with dc or sinusoidal input signal	The ac circuit analysis	Obtaining $v_{out}(t)$ by using algebraic equations, phasors, and complex numbers.	No need	No need

Transient regime with an arbitrary input signal	**Differential equation**	Finding $v_{out}(t)$ by integration of a circuit's differential equation.	**Laplace transform** in the s-domain	Attaining $v_{out}(t)$ by transforming circuit's differential equation into its s-domain algebraic equivalent, solving the latter, and using inverse Laplace transform of $V_{out}(s)$ to get the answer. Alternatively, finding a circuit's transfer function, $H(s)$, to facilitate designing the circuit and finding $V_{out}(s)$.
	Convolution integral	Receiving $v_{out}(t)$ by evaluation of the convolution integral based on preliminary determined a circuit's impulse response.	**Fourier transform** in the frequency domain	Getting $v_{out}(t)$ by (a) transforming the convolution integral into the frequency domain, (b) determining $V_{out}(\omega)=H(\omega)V_{in}(\omega)$, where transfer function $H(\omega)$ being Fourier transform of the circuit's impulse response, and (c) obtaining $v_{out} = \mathfrak{F}^{-1}\{V_{out}(\omega)\}$. Since $H(\omega)$ describes the circuit's response to any input frequency, it modulus and argument determine the amplitude and phase circuit's spectra, respectively.

Examining Table 10.C1 helps our thoughtful reader bear in mind the general picture of this treatise, whereas the necessary details will be found inside this volume.

We wish you, our reader, the best of success in your study and your future professional career.

Bibliography and References

This bibliography contains sources concerning general topics of advanced circuit analysis. The references include the entries cited in the text.

Agarwal, A. and Lang, J.H. (2005). *Foundations of Analog and Digital Electronic Circuits*. San Francisco: Morgan Kaufmann Publishers.

Agbo, S.O. and Sadiku, M.N. (2017). *Principles of Modern Communication Systems*. New York, NY: Cambridge University Press.

Alexander, C.K. and Sadiku, M.N. (2020). *Fundamentals of Electric Circuits*, 7e. New York: McGraw-Hill.

Attaway, S. (2017). *MATLAB*. 4e. Oxford, UK: BH/Elsevier.

Bayin, S.S. (2018). *Mathematical Methods in Science and Engineering*, 2e. Hoboken, NJ: Wiley.

Benslama, M., Benslama, A. and Aris, S. (2017). *Quantum Communications in New Telecommunications Systems*. Hoboken, NJ: Wiley.

Boylestad, R.L. (2016). *Introductory Circuit Analysis*, 13e. Upper Saddle River, N.J.: Prentice Hall.

Brillouin, L. (2013). *Science and Information Theory*, 2e. Mineola, NY: Dover.

Carlson, A.B. and Grilly, P. (2009). *Communication Systems*, 5e. New York, NY: McGraw Hill.

Chaparro, L.F. (2015). *Signals and Systems Using MATLAB*. Amsterdam: Elsevier.

Chen, W-K. and Choma, J. (2007). *Feedback Networks: Theory and Circuit Applications*. Hackensack, NJ: World Scientific Publishing Company.

Choma, J. (1985). *Electrical Networks: Theory and Analysis*. New York: Wiley-Interscience.

Chua, L. O., Desoer, C. A., and Kuh, E. S. (1987). *Linear and Nonlinear Circuits*. New York: McGraw-Hill Book Company.

Cooley, J. and Tukey J. (1965). An algorithm for the machine calculation of complex Fourier series. *Mathematics of Computation* 19: 297–301.

Crystal, D. (1994). *Biographical Encyclopedia*. Cambridge, U.K.: Cambridge University Press.

Das, A. (2012). *Signal Conditioning*. Berlin, Germany: Springer.

DeRusso, P.M., Roy, R.J., Close, C.M., and Drarochers, A.A. (1997). *State Variables for Engineers*. Hoboken, N.J.: Wiley.

Dineen, S. (2012). *Analysis – A Gateway to Understanding Mathematics*. Hackensack, N. J.: World Scientific.

Finney, R.L. et al. (1994). *Calculus*. Reading, MA: Addison Wesley Publishing Company.

Floyd, T.L. (2018). *Electronic Devices*, 10e. Upper Saddle River, N.J.: Prentice Hall.

Essentials of Advanced Circuit Analysis: A Systems Approach, First Edition. Djafar K. Mynbaev.

Companion Website: www.wiley.com\go\Mynbaev\AdvancedCircuitAnalysis

Floyd, T.L. and Buchla, D.M. (2020). *Principles of Electric Circuits*, 10e. Upper Saddle River, N.J.: Prentice Hall.

Franco, S. (1999). *Electric Circuits Fundamentals*. New York, NY: Oxford University Press.

Frenzel, L.E., Jr. (2014). *Contemporary Electronics*. New York, NY: McGraw-Hill.

Hambley, A. (2009). *Electronics*, 4e. Upper Saddle River, N.J: Pearson Education.

Hayt, W.H. Jr., Kemmerly, J.E., Phillips, J.D., and Durbin, S.M. (2019). *Engineering Circuit Analysis*, 9e. New York: McGraw Hill.

Horowitz, P. and Hill, W. (1995). *The Art of Electronics*, 2e. Cambridge, UK: Cambridge University Press.

Irwin, J.D. and Nelms, R.M. (2018). *Basic Engineering Circuit Analysis*, 11e. Hoboken, N.J.: John Wiley & Sons.

Israelohn, J. (2004). Noise. 101 (January 8) and Noise 102 (March 18). www.edn.com.

James, G. (2015). *Modern Engineering Mathematics*, 5e. Harlow, England: Pearson (Intl).

James, G. and Dyke, P. (2018). *Advanced Modern Engineering Mathematics*, 5e, Harlow, England: Pearson (Intl).

Kou, J. and Wang, X. (2010). Some improvements of Ostrowski's method. *Applied Mathematics Letters* 23: 92–96.

Lyons, R. G. and Fugal, D. L. (2014). *The Essential Guide to Digital Signal Processing*. Upper Saddle River, NJ: Pearson Prentice Hall.

Maas, S.A. (2005). *Noise in Linear and Nonlinear Circuits*. Boston, MA: Artech House.

Manton, N. and Mee, N. (2017). *The Physical World – An Inspirational Tour of Fundamental Physics*. New York, NY: Oxford University Press.

Morrison, J. C. (2015). *Modern Physics for Scientists and Engineers*, 2e. Amsterdam, The Netherlands: Academic Press/Elsevier.

Mynbaev, D.K. and Scheiner, L.L. (2020). *Essentials of Modern Communications*. Hoboken, NJ: Wiley.

Mynbaev, D.K. (2016). Fundamental and technological limitations of optical communications. *International Journal of High-Speed Electronics and Systems* 25 (1 & 2): 1640010-1–1640010-21.

Mynbaev, D.K. and Scheiner, L.L. (2001). *Fiber-Optic Communications Technology*. Upper Saddle River, NJ: Prentice Hall.

Neamen, D. (2010). *Microelectronics: Circuit Analysis and Design*, 4e. New York, NY: McGraw-Hill.

Newman, M. E. J. (2010). *Networks – An Introduction*. New York, NY: Oxford University Press.

Nilsson, J. W. and Ridel, S. A. (2019). *Electric Circuits*, 11e. New York, NY: Pearson.

Oppenheim, A.V. and Willsky, A.S. (2015). *Systems and Signals*, 2e. New York, NY: Pearson.

Palm III, W. J. (2011). *Introduction to MATLAB for Engineers*, 3e. New York, NY: McGraw Hill.

Raymer, M. G. (2009). *The Silicon Web – Physics for the Internet Age*. Boston, MA: Taylor & Francis.

Razavi, B. (2008). *Fundamentals of Microelectronics*. Hoboken, NJ: Wiley.

Rizzoni, G. (2009). *Fundamentals of Electrical Engineering*. Boston: McGraw-Hill Higher Education.

Roberts, M.J. (2008). *Fundamentals of Signals & Systems*. Boston: McGraw-Hill Higher Education.

Rojo, A. and Bloch, A. (2018). *The Principle of Least Action*. New York, NY: Cambridge University Press.

Schaumann, R., Xiao, H., and Van Valkenburg, M.E. (2010). *Design of Analog Filters*. New York, NY: Oxford University Press.

Sedra, A.S. and Smith, K.C. (2015). *Microelectronic Circuits*, 7e. New York, NY: Oxford University Press.

Sherrick, J.D. (2001). *Concepts in Systems and Signals*. Upper Saddle River, NJ: Prentice Hall.

Siauw, T. and Alexandre M. B. (2015). *An Introduction to MATLAB Programming and Numerical Methods for Engineers*. London, UK: AP/Elsevier.

Smith, S.W. (2003). *Digital Signal Processing (A Practical Guide for Engineers and Scientists).* Newnes/Elsevier.

Stanley, W.D. (2002). *Transform Analysis for Engineering and Technology*, 5e. Upper Saddle River, NJ: Pearson Prentice Hall.

Stanley, W.D. (2003). *Network Analysis with Applications*, 4e. Upper Saddle River, NJ: Pearson Prentice Hall.

Sundararajan, D. (2008). *A Practical Approach to Signals and Systems.* Singapore: John Wiley & Sons (Asia) Pte Ltd.

Taylor, F.J. (1994). *Principles of Signals and Systems.* New York, NY: McGraw-Hill.

Thomas, R.E, Rosa, A.J, and Toussaint, G.J., (2019). *The Analysis and Design of Linear Circuits.* 9e. Hoboken, NJ: Wiley.

Thompson, R.A. et al. (2006). *The Physical Layer of Communications Systems.* Norwood, MA: Artech House.

Trumper, D. (2007). *2.14 Analysis and Design of Feedback Control Systems, Spring 2007.* Massachusetts Institute of Technology: MIT OpenCourseWare. http://ocw.mit.edu. License: Creative Commons BY-NC-SA.

Vorperian, V. (2002). *Fast Analytical Techniques for Electrical and Electronic Circuits.* New York: Cambridge University Press.

Weinan, E., Jiequn, H., and Linfeng, Z. (2021). Machine-learning-assisted modeling. *Physics Today* 74 (7): 36–41.

Wilson, P. (2017). *The Circuit Designer's Companion.* 4e. Oxford, UK: Newness/Elsevier.

Yevick, D. and Yevick, H. (2014). *Fundamental Math and Physics for Scientists and Engineers.* Hoboken, NJ: Wiley.

Ziemer, R.E., Tranfer W.H. and Fannin, D.R. (1993). *Signals and Systems: Continuous and Discrete.* New York: Macmillan Publishing Company.

State variables

1) Course notes
https://eng.libretexts.org/Bookshelves/Industrial_and_Systems_Engineering/Book%3A_Introduction_to_Control_Systems_(Iqbal)/01%3A_Mathematical_Models_of_Physical_Systems/1.06%3A_State_Variable_Models
2) MATLAB tutorial
https://ctools.ece.utah.edu/StateSpace/Circuits/Matlab/StateSpaceCircTutor.pdf
3) DeRusso, P.M., Roy, R.J., Close, C.M. et al. (1997). *State Variables for Engineers*, 2nd ed., Hoboken, NJ:Wiley-Interscience.
4) "State variables," en.wikipedia.org (accessed 13 July 2019.)
5) Glyn J., *Advanced Modern Engineering Mathematics*, 3rd ed., (2004). Pearson/Prentice Hall, Harlow, England.
6) Ziemer, R.E., Tranter, W.H., and D. Fannin, R. (1993). *Signals and Systems: Continuous and Discrete*, 3rd ed., New York: Macmillan.
7) Agarwal, A. and Lang, J.H. (2005). *Foundations of Analog and Digital Electronic Circuits.* San Francisco: Morgan Kaufmann.

Index

Essentials of Advanced Circuit Analysis: A Systems Approach, First Edition. Djafar K. Mynbaev.

Companion Website: www.wiley.com\go\Mynbaev\AdvancedCircuitAnalysis

Printed and bound by CPI Group (UK) Ltd, Croydon, CR0 4YY

16/04/2025

14658421-0003